T0264052

Ring Theory

Student Edition

Ring Theory

Student Edition

Louis H. Rowen
Department of Mathematics and Computer Science
Bar–Ilan University
Ramat–Gan, Israel

ACADEMIC PRESS, INC.
Harcourt Brace Jovanovich, Publishers
Boston San Diego New York
London Sydney Tokyo Toronto

This book is printed on acid-free paper.

Copyright © 1991 by Academic Press, Inc.
All rights reserved.
No part of this publication may be reproduced or
transmitted in any form or by any means, electronic
or mechanical, including photocopy, recording, or
any information storage and retrieval system, without
permission in writing from the publisher.

Cover designed by Elizabeth E. Tustian

ACADEMIC PRESS, INC.
1250 Sixth Avenue, San Diego, CA 92101

United Kingdom Edition published by
ACADEMIC PRESS, LIMITED
24-28 Oval Road, London NW1 7DX

Library of Congress Cataloging-in-Publication Data

Rowen, Louis Halle.
   Ring theory / Louis Halle Rowen. — Student ed.
      p.   cm.
   Includes bibliographic references and index.
   ISBN 0-12-599840-6
   1. Rings (Algebra)   I. Title.
QA247.R68   1991   90-42570
512'.4—dc20   CIP

91 92 93   9 8 7 6 5 4 3 2 1

Transferred to digital printing 2005

Contents

Chapter 2   Basic Structure Theory

Foreword to the
Student Edition

This volume is an abridged edition of Ring Theory, volumes 1 and 2, designed to be more palatable (and affordable) to students. The original two volumes aimed to serve the dual purpose of research guides and texts, and so a layered approach was adopted, consisting of text, supplements, appendices, exercises, and digressions. Since my original manuscript had already been pared down as far as I could bear before publication, I did not cherish the task of deleting any other topics. However, a fair amount of material in the two volumes is aimed at those intending to specialize in ring theory, and might not be missed in a general ring theory course.

Another side benefit of this project is the opportunity to correct a few gross mistakes in the two-volume edition. (See "Errata," following this foreword.) I thank those people who took the time to send in lists of misprints and more serious mistakes.

In line with the relatively modest goals of a graduate text, the more technical exercises and digressions pointing to further research have been deleted; proofs of introductory results, which may be fresh in the mind of a graduate student (and available in standard texts such as Jacobson's Basic Algebra I), have also been removed. In the later chapters, material often has been cut if it was not essential for proving the principal theorems. Nevertheless, the format and numbering of results matches the original two volumes, enabling the reader to consult them for further edification.

Heavy deletions were made from Chapter 7 (division algebras). Nevertheless, virtually all the basic structural results in it can be obtained through elementary means; these techniques sometimes were lost in the original version of the chapter as the theory became increasingly complicated. The necessities of space gave me an excuse to build the theory of simple algebras directly from cyclic algebras; in retrospect, and in view of the Merkurjev–Suslin theorem, one might argue that this is the "right" way to develop the subject. Similarly, I took the opportunity to replace the proof of Shirshov's theorem in §6.3 with Belov's new proof, which is easier and more direct.

Other sections that were deleted:

1. Tangential material not usually considered ring theory, although it pertains to rings, such as the theory of the free group (Appendix A), Banach algebras (Appendix B), invariant theory (§6.4), much of the theory of finite dimensional division algebras, especially applications of the corestriction map and other tools from cohomology (§7.2), and technical results about Lie algebras and superalgebras (§8.4).

2. Much abstract categorical material (§1.5, 3.4, 4.2). Although it is important and formed the basis for much research of the 1970s, this material is often considered too abstruse to be handled in a general course on ring theory.

3. Several interesting examples that require lengthy development. Levy's example (§2.9) of Noetherian rings having f.g. modules failing Krull–Schmidt and related, weaker properties; an indecomposable and non-crossed product division algebra of exponent p; many interesting examples of Bergman; the Goldman–Michler result that Ore extension of Jacobson rings are Jacobson (Exercises 8.4.3ff)

4. Self-contained theories that, although they belong to ring theory, might not normally appear in a general course: rings with involution (§2.13), the Warfield–Stafford–Coutinho version of the Forster–Swan theorem, Jategaonkar's theory of localization without localizing, and the AR–property (all §3.5), Azumaya algebras (§5.3), and beautiful results concerning which projective modules are free (§5.1, §8.4).

5. Representation theory of finite dimensional Artinian algebras (end of §2.9). Although this theory has been one of the more beautiful and active areas during the last twenty years, a treatment appropriate to a graduate course would require much more room, such as in Ringel [84B].

6. The bibliography of articles (except for three recent articles that are especially pertinent).

As one can see, the abridged version lacks what I strived for in the original two volumes, a unified account of the various streams of ring theory. On the other hand, it is hoped that this volume for students will be easier to digest and may whet the appetite of the student or general reader for further reading and research.

Errata from the Original Two-Volume Edition

Volume I

page	line	text
23	7	**Corollary 0.2.20:** If $M = N_1 + \ldots + N_t$ with each N_i Noetherian (resp. Artinian) then M is Noetherian (resp. Artinian).
79	4	Rather than define a self-dual category, we should say a *class* S of categories is self-dual if $C^{op} \in S$ whenever $C \in S$.
99	-8	$\mathscr{A}_1(F)$ is *not* a PLID since the left ideal spanned by λ, μ is not principal.
198	14	*Replace statement in lemma 2.5.50 by the following text:* If $t(R_0)$ is quasi-invertible then $|G|R_0$ is quasi-invertible. *(Proof given in this student edition.)*
397	-10	The uniform dimension is *not* additive (and indeed this was not claimed in 3.2.19).
415	-9	The proof of Proposition 3.5.41 requires: **Remark 3.5.40":** If $M_2 \leq M_1 \leq M$ and $N \leq M$, then there is an exact sequence. $$0 \to M_1 \cap N/M_2 \cap N \to M_1/M_2 \to (M_1 + N)/(M_2 + N) \to 0.$$ *(Proof given in this student edition.)*

Volume II

18	-15	free abelian group
18	-11	in $K_0(R)$, we see
97	2,3	*replace with the following text:* multilinear polynomial f is a weak identity of $M_n(C)$ then $T_k f$ is scalar-valued for all substitutions of matrices of trace 0.
126	10	*replace* (GK dim R_1)(GK dim R_2) *by* GK dim (R_1) + GK dim (R_2)
126	-15	*replace* $\alpha_1 \alpha_2$ *by* $\alpha_1 + \alpha_2$
126	-12	*replace* α, *by* α_1 *(twice)*
129	-7	prime affine In the non-prime case, they show any affine algebra of GK dim 1 is PI, but need not be Noetherian. Still, by (Braun's) theorem 6.3.39 below, the nilradical is nilpotent.
211	5	*replace* GK dim *by* GK tr deg
245, 246		*paragraphs misplaced—take lemma 7.3.6 through bottom of p. 246 and insert them between 7.3.3. and 7.3.4. This entails renumbering these three results and their references, which are used on pages 245, l. -3 and 247, lines 1, 4, 5, 6, 14, -12.*
248	2	*The argument relies on Corollary 7.2.11' of the student edition, which was not proved in the original volumes.*
255	-11	τ does not send H_q to itself. However this can be remedied by using Lagrange resolvents, as noted by Tignol, cf. a forthcoming paper by Rowen and Tignol, which also includes a result for $p = 2$.
257	9	There is a gap here, which could be filled for (odd) $n = p^4$.
353	-7	*replace by the following text:* $\sum r_i a^i$ of $L \cap T_i$}. Note that the leading coefficient is not well-defined on T_i, but may vary according to the way that an element is written. However, it is well-defined modulo lead $_i(0)$, and thus modulo lead$_i(L)$ for any left ideal L.

Introduction: An Overview of Ring Theory

In the solar system of ring theory the Sun is certainly the semisimple Artinian ring, which can be defined most quickly as a finite direct product of matrix rings over division rings. Much of ring theory is involved in measuring how far a ring is from being semisimple Artinian, and we shall describe the principal techniques. The numbers in parentheses refer to locations in these two volumes.

The Structure Theory of Rings

Any simple ring (with 1) having minimal nonzero left ideal is of the form $M_n(D)$, by the celebrated result known as the Wedderburn–Artin theorem (2.1.25′). The general structure theory begins with Jacobson's density theorem (2.1.6), which generalizes the essence of the Wedderburn–Artin theorem to the class of *primitive* rings; using subdirect products (§2.2) one can then pass to semiprimitive rings.

The passage from semiprimitive rings to arbitrary rings leads one to the study of various *radicals*, which are intrinsically-defined ideals which when 0 make the ring much easier to analyze. Thus the radical is the "obstruction" in the structure theory.

In the general structure theory there are several radicals, which we consider in order of decreasing size. The biggest, the *Jacobson* radical (§2.5) denoted Jac(R), is 0 iff the ring is semiprimitive. Since Jac(R) is often nonzero, one considers various *nilradicals* (§2.6), so called because their elements are nilpotent. The largest nilradical is called the *upper nilradical* Nil(R) and the smallest nilradical is called the *prime radical* or *lower nilradical* and is the intersection of the prime ideals; a ring is *semiprime* iff its lower nilradical is 0.

Even in the classical case of finite dimensional algebras over fields (for which semiprimitive rings become semisimple Artinian (§2.3) and for which the radicals all coincide), the structure of the radical itself is extremely complicated; often the radical is considered an encumbrance to be removed as soon as possible. Thus the various structure theorems which enable us to "shrink" the radical take on special significance; most prominent is Amitsur's theorem (2.5.23), which says that if Nil(R) = 0 then Jac($R[\lambda]$) = 0, where λ is a commuting indeterminate over R. Other such results are considered in §2.6.

Certain classes of rings are particularly amenable to the structure theory. If a ring R is Artinian then Jac(R) is nilpotent and R/Jac(R) is semisimple Artinian; furthermore, one can "lift" the idempotents from R/Jac(R) to R to obtain more explicit information about R (§2.7). There are other instances when Jac(R) is nil (§2.5).

One of the principal classes of noncommutative rings is Noetherian rings (§3.5), for which Nil(R) is nilpotent. Fortunately for semiprime Noetherian rings one has Goldie's theorem (§3.2), which shows that the classical ring of fractions exists and is semisimple Artinian; this result makes available the Wedderburn–Artin theorem and thereby rounds out the Wedderburn–Artin–Noether–Jacobson–Levitzki–Amitsur–Goldie structure theory. Goldie's theorem also applies to prime rings with polynomial identities (Chapter 6). Rings of fractions have been generalized to rather broad classes of rings, as described in Chapter 3.

Since the structure theory revolves around primitive and prime rings, one is interested in the set of primitive ideals (of a ring) and the set of prime ideals. These sets are called the *primitive* and *prime* spectra and have a geometry of considerable interest (§2.12). Recent research has focused on the primitive and prime spectra of certain classes of rings.

Since any ring is a homomorphic image of a free ring, the Cohn–Bergman school has conducted research in studying free rings in their own right. Such a project is of immense difficulty, for the very reason that the theory of free rings necessarily includes all of ring theory.

Occasionally one wants to consider an extra bit of structure for a ring, the *involution* (§2.13), which generalizes the transpose of matrices. The structure theory carries over fairly straightforwardly to rings with involution, with involutory analogues of the various structural notions.

The Structure Theory of Modules

Semisimple Artinian rings are also characterized by the property that every module is a direct sum of simple modules (2.4.9), and, furthermore, there are only a finite number of isomorphism classes of simple modules (2.3.13). This raises the hope of studying a ring in terms of its modules, and also of studying a module in terms of simple modules. The "best" modules to study in this sense are Artinian Noetherian modules, for they have *composition series* (§2.3), and these are essentially "unique." Unfortunately there is no obvious way of building a module from its simple submodules and homomorphic images, so one turns to building modules as direct sums of indecomposable modules, thereby leading us to the "Krull-Schmidt" theorem (2.9.17), which says that every module with composition series can be written "uniquely" as a direct sum of indecomposables. Unfortunately this leaves us with determining the indecomposables (§2.9), which is a tremendous project even for Artinian rings and which largely falls outside the scope of this book. Thus one is led to try more devious methods of studying modules in terms of simple modules, such as the noncommutative Krull and Gabriel dimensions (§3.5). The Krull dimension has become an indispensible tool in the study of Noetherian rings.

Category Theory and Homology

When studying rings in terms of their modules, one soon is led to categories of modules and must face the question of when two rings have *equivalent* categories of modules. Fortunately there is a completely satisfactory answer of Morita (§4.1).

Every module is a homomorphic image of a free module, and perhaps one could learn more about modules by studying free modules. It turns out that a more natural notion from the categorical point of view is *projective module* (§2.8), and indeed a ring R is semisimple Artinian iff every module is projective (2.11.7). The interplay between projective and free is very important, leading to the *rank* of a projective module (2.12.19) and the K_0 theory (§5.1).

The path of projective modules can take us to projective resolutions and homological dimension of modules (§§5.1, 5.2). Homological dimension is most naturally described in terms of category theory, which lends itself to dualization; thus one also gets *injective modules* (§2.10), which also play an important role in the more general theories of fractions (§§3.3, 3.4). Homology (and cohomology) lay emphasis on two important functors, $\mathscr{T}or$ and $\mathscr{E}xt$, which are derived respectively from the tensor functor and from the functor $\mathscr{H}om$. These latter two functors are an example of an *adjoint pair* (4.2).

Special Classes of Rings (Mostly Volume II)

Certain classes of rings are of special interest and merit intensive research. Classically the most important are finite dimensional algebras over a field; these fall inside the theory of Artinian rings, but much more can be said. The role of the radical is made explicit by Wedderburn's principal theorem (2.5.37) when the base field is perfect; Wedderburn's result has been recast into the cohomology of algebras (§5.3).

Much of the theory of finite dimensional algebras carries over to the more general realm of rings with polynomial identities (PI-rings, chapter 6). Deep results of representation theory can be obtained using the PI-theory by means of elementary arguments. Moreover, PI-theory is not tied to a base field, and so various generic techniques are available to enrich the PI-theory; relatively free PI-algebras have commanded considerable attention. It turns out that there is just enough commutatively in the PI-theory to enable one to obtain a satisfactory version of much of the theory of commutative algebras and, indeed, to build a noncommutative algebraic geometry (which however lies largely outside the realm of this book).

In a different direction, commutative algebraic geometry leads us to consider *affine* algebras over a field, by definition finitely generated as algebras. Although not much can be said about affine algebras in general, affine algebras become more manageable when they have finite *Gelfand–Kirillov dimension* (§6.2); in particular, affine PI-algebras have finite Gelfand–Kirillov dimension and are a very successful arena for generalizing results from the commutative theory.

Returning to the Wedderburn–Artin theorem, a very natural question is. "What can be said about the division ring D in $M_n(D)$?" This remains one of the more troublesome questions of research today, because tantalizingly little is known about arbitrary division rings. When D is finite dimensional over a field however, a whole world opens up, comprising the theory of (finite dimensional) central simple algebras and the Brauer group (Chapter 7). Much recent research on division algebras is from the standpoint of noncommutative arithmetic and K-theory, but we deal mostly with the general structure theory of division algebras. Most of the exposition is given to understanding the algebraic content of several special cases of the amazing Merkurjev–Suslin theorem.

Our final chapter (8) is about those rings which arise most in representation theory, namely group rings and enveloping algebras of Lie algebras. These rings in fact contain all the information of the respective representation theories, and therefore provide a key link from "pure" ring theory to the outside world. Both areas naturally lie on the border of algebra, drawing also on analysis and geometry. Nevertheless, the ring theory provides a guide to directions of inquiry; the question of central interest in enveloping algebra theory has been to determine the primitive ideals, for these correspond to the irreducible representations. Pure ring theory also yields a surprising amount of information, and much of the theory can be cast in the general framework of Noetherian rings (§8.4). (The structural foundation is laid in §2.12.)

Another topic in modern research is the Galois theory of rings. Jacobson developed a Galois theory to study extensions of division rings, in a similar manner to the Galois theory of extensions of fields. This has led to the study of fixed subrings under groups of automorphisms (end of §2.5), and more recently to Hopf algebras, which are treated in §8.4 as a simultaneous generalization of group rings, enveloping algebras, and algebraic groups.

A word about current research—whereas the 1960s and early 1970s was the era of abstraction and beautiful general theories, the late 1970s and 1980s have displayed a decided return to specific examples. Thus considerable recent attention has turned to Weyl algebras and, more generally, rings of differential polynomials, and the theory of the enveloping algebra of $sl(2, n)$ has produced many interesting examples and a few surprises.

Table of Principal Notation

Note: ! after page reference means the symbol is used differently in another part of the text.

0   General Fundamentals

§0 Preliminary Foundations

The object of this section is to review basic material, including the fundamental results from ring theory. The proofs are sketched, since they can be found in the standard texts on abstract algebra, such as Jacobson [85B]. Two important conventions: $\mathbb{N}$ denotes the natural numbers *including* 0; also, given a function $f: A \to B$ we shall often use fa instead of the more standard notation $f(a)$, and similarly fA denotes $\{fa: a \in A\}$. Given sets A, I define $A^I = \{\text{functions } f: I \to A\}$, which can be identified with the Cartesian product $\prod\{A: i \in I\}$. A special case is $I = \{1, \dots, n\}$; in this case we denote A^I as $A^{(n)}$, which can be identified with $A \times \cdots \times A$ (taken n times) under the bijection sending f to $(f1, \dots, fn)$.

Monoids and Groups

A *monoid* is a semigroup S which has a *unit element* 1, such that $1s = s1 = s$ for each s in S. If a semigroup S lacks a unit element we can adjoin a formal element 1 to produce a monoid $S' = S \cup \{1\}$ by stipulating $1s = s1 = s$ for all s in S.

In verifying that a monoid S is a group, one need only check that each element is left invertible, i.e., for each s there is s' such that $s's = 1$. (Indeed, take a left inverse s'' of s'. Then $s'' = s''(s's) = (s''s')s = s$ proving $ss' = 1$ and s' is also the right inverse of s, i.e., $s' = s^{-1}$.)

Sym(n) denotes the group of permutations, i.e., of 1:1 functions from $\{1, \ldots, n\}$ to itself; Sym(n) is a submonoid of A^A where $A = \{1, \ldots, n\}$.

Although we have used the multiplicative notation for groups, we shall usually write *abelian* groups additively, with $+$ for the group operation, 0 for the unit element, and $-g$ for the inverse of g.

Fundamental Theorem of Abelian groups: *Every finitely generated abelian group is isomorphic to a finite direct sum of cyclic subgroups.*

An element of a group is *torsion* if it has finite order. The set of torsion elements of an abelian group G is obviously a subgroup, and we say G is *torsion-free* if its torsion subgroup is 0. Since any cyclic group is isomorphic either to $(\mathbb{Z}, +)$ or some $(\mathbb{Z}/n\mathbb{Z}, +)$ for some n in $\mathbb{Z}$, we have

Corollary: *Every finitely generated abelian group is the direct sum of a torsion-free abelian group and a finite abelian group.*

Rings and Modules

A *ring* is a set R together with operations $+, \cdot$ (called *addition* and *multiplication*) and distinguished elements 0 and 1, which satisfy the following properties:

$(R, +, 0)$ is an abelian group.
$(R, \cdot, 1)$ is a monoid.
$a(b + c) = ab + ac$ and $(b + c)a = ba + ca$ for all a, b, c in R.

If $ab = ba$ for all a, b in R, we say R is *commutative*, common examples being $\mathbb{Z}, \mathbb{Q}, \mathbb{R}$, and $\mathbb{C}$, as well as rings of polynomials over these rings; matrix rings (cf., §1.1) provide examples of noncommutative rings.

The singleton $\{0\}$ is a ring called the *trivial ring*. In all other rings $1 \neq 0$ (since if $1 = 0$ then any $r = r \cdot 1 = r \cdot 0 = 0$). A *subring* of a ring is a subset which is itself a ring having the same distinguished elements 0, 1; thus $\{0\}$ is *not* a subring of any nontrivial ring. Unless explicitly stated otherwise, rings will be assumed to be nontrivial.

Suppose R, T are rings. A *ring homomorphism* $f: R \to T$ is an additive group homomorphism satisfying $f(r_1 r_2) = (fr_1)(fr_2)$ and $f1 = 1$.

By *domain* we mean a (nontrivial) ring in which each product of nonzero elements is nonzero; R is a *division ring* (or skew-field) if $(R - \{0\}, \cdot, 1)$ is a group, i.e., each nonzero element is invertible. A commutative domain is usually called an *integral domain* (or *entire*); a commutative division ring is called a *field*.

Given a ring R we define a (*left*) *R-module* to be an abelian group M (written additively), together with composition $R \times M \to M$ called *scalar multiplication*, satisfying the following laws for all r_i in R and x_i in M:

$$r(x_1 + x_2) = rx_1 + rx_2; \qquad (r_1 + r_2)x = r_1 x + r_2 x;$$

$$(r_1 r_2)x = r_1(r_2 x); \qquad 1x = x.$$

In other words, every possible associative law and distributive law involving scalar multiplication holds. A *right R-module* M is an abelian group with scalar multiplication $M \times R \to R$ satisfying the right-handed version of these laws. The motivating examples:

(i) R is a left (or right) R-module, where the addition and scalar multiplication are taken from the given ring operations of R.

(ii) The modules over a field F are precisely the vector spaces over F.

Suppose M, N are R-modules. A *module homomorphism* $f: M \to N$, also called a *map*, is a group homomorphism "preserving" scalar multiplication in the sense $f(rx) = rfx$ for all r in R and x in M. In analogy to rings, a map is an *isomorphism* if it has an inverse which is also a map; clearly, this is the case iff the map is bijective. However, maps have additional properties. Write $\operatorname{Hom}_R(M, N)$ for {maps from M to N}, made into an Abelian group under "pointwise" addition of maps, i.e., $(f + g)x = fx + gx$. The *zero map* sends every element of M to 0. Moreover, composition of functions provides a map from $\operatorname{Hom}(N, K) \times \operatorname{Hom}(M, N)$ to $\operatorname{Hom}(M, K)$, which is *bilinear* in the sense

$$(g_1 + g_2)f_1 = g_1 f_1 + g_2 f_2 \qquad \text{and} \qquad g_1(f_1 + f_2) = g_1 f_1 + g_2 f_2$$

for any maps $f_i: M \to N$ and $g_i: N \to K$.

A *submodule* of an R-module M is an additive subgroup N closed under the given scalar multiplication; in this case N is itself a module and we write $N \le M$. A submodule N is *proper* if $N < M$. Viewing R as R-module as above, we say L is a *left ideal* of R if $L \le R$; in other words, a left ideal is an additive subgroup L satisfying $rx \in L$ for all r in R and x in L. Note that for any x in a left R-module M we have $Rx \le M$. The case $M = R$ has special interest, because a left ideal L is proper iff $1 \notin L$. In particular, for $r \in R$ we have $Rr < R$ iff r has no left inverse. *Right ideals* analogously are defined as right submodules of R. A is a *proper ideal* of R (written $A \lhd R$) if A is a proper left and right ideal of R.

If f is a ring homomorphism *or* a module homomorphism, we define its *kernel* ker f to be the preimage of 0. Then rudimentary group theory shows ker $f = 0$ iff f is 1:1. Consequently, f is an isomorphism iff f is onto with ker $f = 0$. Note for a map $f: M \to N$ that ker f is a submodule of M and is proper iff f is nonzero; on the other hand, if $f: R \to T$ is a ring homomorphism then ker $f \lhd R$.

Let us try now to characterize kernels structurally. First take R-modules $N \le M$. Forming the abelian group M/N in the usual way, as {cosets of N}, we can define scalar multiplication by

$$r(x + N) = rx + N \qquad \text{for } r \in R \text{ and } x \in M,$$

thereby making M/N a left R-module called the *quotient module* (or *residue module* or *factor module*). There is a canonical map $\varphi: M \to M/N$ given by $x \to x + N$, and $N = \ker \varphi$. In this way we see every submodule of M is the kernel of a suitable map. Moreover, we have Noether's isomorphism theorems:

Proposition 0.0.1: *Suppose $f: M \to M'$ is a map of R-modules whose kernel contains a submodule N of M. Then there is a map $\bar{f}: M/N \to M'$ given by $\bar{f}(x + N) = fx$, with ker $\bar{f} = (\ker f)/N$. In particular, if $f: M \to M'$ is onto and ker $f = N$ then $\bar{f}$ is an isomorphism.*

Corollary 0.0.2: *If $M_1, M_2 \le M$ then $M_1 + M_2 \le M$ and $M_1 \cap M_2 \le M$, and $(M_1 + M_2)/M_2 \approx M_1/(M_1 \cap M_2)$.*

Corollary 0.0.3: *If $K \le N \le M$ then $(M/K)/(N/K) \approx M/N$.*

One also obtains similar results for rings. Given $I \lhd R$ we define the *quotient ring* (also called *factor ring* or *residue ring*) R/I to have the usual additive group structure (of cosets) together with multiplication

$$(r_1 + I)(r_2 + I) = r_1 r_2 + I.$$

This can be easily verified to be a ring, and there is a canonical ring homomorphism $\varphi: R \to R/I$ given by $\varphi r = r + I$.

Proposition 0.0.4: *Suppose $f: R \to T$ is a ring homomorphism whose kernel contains an ideal A. There is a ring homomorphism $\bar{f}: R/A \to T$ given by $\bar{f}(r + A) = fr$, and ker $\bar{f} = (\ker f)/A$. If f is onto and ker $f = A$ then $\bar{f}$ is an isomorphism.*

Corollary 0.0.5: *If $B \subseteq A$ are proper ideals of R then $(A/B) \lhd R/B$ and $(R/B)/(A/B) \approx R/A$.*

If $f: M \to N$ is a map of R-modules then fM is a submodule of N; thus f is onto iff $N/fM = 0$, providing a useful test. For rings one sees for any homomorphism $f: R \to T$ that fR is a subring of T. Since fR is not an ideal of T we do not have the parallel test for rings, but anyway we shall find it useful at times to replace T by fR, thereby making f onto.

Algebras

In the text C usually denotes a commutative ring. A C-algebra (or algebra over C) is a ring R which is also a C-module whose scalar multiplication satisfies the extra property

$$c(r_1 r_2) = (cr_1)r_2 = r_1(cr_2) \qquad \text{for all } c \text{ in } C, \text{ and } r_1, r_2 \text{ in } R.$$

Any ring R is also a $\mathbb{Z}$-algebra, by taking nr to be $r + \cdots + r$ (taken n times); the formal correspondence is given more formally in example 0.1.10 below.

In general the theories of algebras and of rings are very similar. Indeed, if R is a C-algebra and $A \lhd R$ is an ideal then $A < R$ as C-module (since $ca = c(1a) = (c1)a \in A$ for all c in C and a in A). Thus the ring R/A also has a natural C-module structure, with respect to which R/A is in fact a C-algebra. Put more succinctly, any ring homomorphic image of R is also naturally a C-algebra.

Define the *center* of a ring, denoted $Z(R)$, to be $(z \in R: rz = zr$ for all r in $R\}$, clearly a subring of R. Under its ring operations R is an algebra over every subring of $Z(R)$. Conversely if R is a C-algebra then there is a canonical ring homomorphism $\varphi: C \to Z(R)$ given by $\varphi c = c1$. (Proof: $c1 \in Z(R)$ since $(c1)r = cr = c(r1) = r(c1)$; φ is a ring homomorphism because $(c_1 c_2)1 = c_1(c_2 1) = c_1(1(c_2 1)) = (c_1 1)(c_2 1)$.) In case C is a field we have ker $\varphi \lhd C$ so ker $\varphi = 0$, and we may identify C with a subring of R. Often it is easier to prove theorems about algebras over a field, which one then tries to generalize to algebras over arbitrary commutative rings.

At times we shall need the following generalization of the center of a ring R. Suppose $A \subset R$. The *centralizer of A in R*, denoted $C_R(A)$, is $\{r \in R: ra = ar$ for all a in $A\}$, a subring of R. We say B *centralizes* A if $B \subset C_R(A)$. For example $C_R(R) = Z(R)$. Note that $A \subseteq C_R(C_R(A))$.

Proposition 0.0.6: *Any maximal commutative subring C of R is its own centralizer.*

Proof: Let T be the centralizer of C in R. Then $C \subseteq T$. But for any a in T we see C and a generate a commutative subring C' of R; by maximality of C we have $C' = C$ so $a \in C$; hence $T = C$, as desired.   Q.E.D.

Preorders and Posets

A *preorder* is a relation which is reflexive ($a \leq a$) and transitive (if $a \leq b$ and $b \leq c$ then $a \leq c$). A preorder $\leq$ on S is called a *partial order* (or PO for short) if $\leq$ is antisymmetric; i.e., $a \leq b$ and $b \leq a$ imply $a = b$. In this case $(S, \leq)$ is called a *poset*. We write $a < b$ when $a \leq b$ with $a \neq b$. The following posets are of particular importance to us.

- (i) Every set S has the *trivial* (or *discrete*) PO defined by declaring any two distinct elements are incomparable (i.e., $a \leq b$ iff $a = b$).
- (ii) The *power set* $\mathscr{P}(A)$ of a set A is the set of subsets of A, ordered by set inclusion.
- (iii) If M is an R-module, define $\mathscr{L}(M) = \{$submodules of $M\}$, partially ordered under $\leq$ (so $N_1 \leq N_2$ if N_1 is submodule of N_2). When R is ambiguous we write $\mathscr{L}(_R M)$ for $\mathscr{L}(M)$. One can do the same for $\{$ideals of a ring$\}$.

Upper and Lower Bounds

Suppose S is a set with preorder $\leq$. An *upper bound* for a subset S' of S is an element s in S for which $s' \leq s$ for all s' in S'. Upper bounds need not exist. For example, no pair of distinct elements has an upper bound if $\leq$ is the trivial PO. On the other hand, we call S *directed* (by $\leq$) if every pair of elements has an upper bound. An upper bound s of S' is called a *supremum* if $s \leq s'$ for every upper bound s' of S'. Dually, we define *lower bound* and say S is *directed from below* if every pair of elements has a lower bound. A lower bound s of S' is called an *infimum* if $s \geq s'$ for every lower bound s' of S'. The supremum (resp. infimum) is denoted as $\vee$ (resp. $\wedge$). The supremum and infimum of S (if they exist) are denoted, respectively, as 1 and 0.

For any poset $(S, \leq)$ we can define the *dual poset* $(S, \geq)$ by reversing the inequality, i.e., now $s_1 \geq s_2$ if previously $s_1 \leq s_2$. The technique of passing to the dual poset is extremely useful and ofen produces extra theorems with no extra work.

Lattices

A *lattice* is a poset in which every pair of elements has both a supremum and an infimum; the lattice is *complete* if every subset has both a supremum and an infimum. Passing to the dual reverses $\vee$ and $\wedge$, so we see the dual of a (complete) lattice is also a (complete) lattice.

Proposition 0.0.8: *If $f: M \to N$ is onto then there is a lattice isomorphism from $\mathscr{L}(N)$ to {submodules of M containing $\ker f$}, given by $N' \to f^{-1}N'$. (The inverse correspondence is given by $M' \to fM'$).*

The corresponding result for rings (proposition 0.0.10) can be obtained directly, but for the sake of variety one could make use of the following observation.

Remark 0.0.9: ("Change of rings") Suppose $f: R \to T$ is a ring homomorphism. Any T-module M can be viewed as R-module via the scalar multiplication defining rx to be $(fr)x$ for r in R and x in M.

Proposition 0.0.10: *Any onto ring homomorphism $f: R \to T$ induces a lattice isomorphism of {left ideals of R containing $\ker f$} with {left ideals of T}; likewise for right ideals. In particular $\ker f \lhd R$, and f induces a lattice isomorphism of {ideals of R containing $\ker f$} and {ideals of T}; the inverse correspondences each are given by f^{-1}.*

Modular Lattices

Since proposition 0.0.8 concerns lattices (of submodules) one is led to look for a more lattice-theoretic approach to modules, which stems from the following observation:

Remark 0.0.11: In any lattice $\mathscr{L}$ we have $a \vee (b_1 \wedge b_2) \leq a \vee b_i$ for $i = 1, 2$ so $a \vee (b_1 \wedge b_2) \leq (a \vee b_1) \wedge (a \vee b_2)$. In particular, if $a \leq b$ we have $a \vee (b \wedge c) \leq b \wedge (a \vee c)$.

In certain cases equality actually holds. The most important case is the lattice of submodules $\mathscr{L}(M)$, for if $M_1 \leq M_2$ and $x_2 \in M_2 \cap (M_1 + M_3)$ then $x_2 = x_1 + x_3$ for x_i in M_i so $x_3 = x_2 - x_1 \in M_2$, proving $x_2 \in M_1 + (M_2 \cap M_3)$. We are led to the following definition:

Definition 0.0.12: A lattice $\mathscr{L}$ is *modular* if $a \vee (b \wedge c) = b \wedge (a \vee c)$ for all $a \leq b$ in $\mathscr{L}$.

$\mathscr{L}(M)$ is thus a modular lattice. A more symmetric condition in verifying the modularity of a lattice is given in exercise 2.

· **Definition 0.0.13:** A lattice $\mathscr{L}$ is *complemented* if for each a in $\mathscr{L}$ there is a complement a' with $a \wedge a' = 0$ and $a \vee a' = 1$.

The lattice $\mathscr{P}(S)$ is complemented. On the other hand, our other major example of a lattice $\mathscr{L}(M)$ is in general not complemented. Complements play an important role in module theory because of exercises 12 and 13, which anticipate key structural results in §2.4.

Definition 0.0.14: A *filter* of a lattice $\mathscr{L}$ is a subset $\mathscr{F}$ satisfying the following three conditions:

>(i) If $a \in F$ and $b \geq a$ then $b \in \mathscr{F}$.
>
>(ii) If $a, b \in F$ then $a \wedge b \in \mathscr{F}$.
>
>(iii) $0 \notin \mathscr{F}$.

Filters have several important applications in rings and can be viewed as the "dual" of ideals, c.f., exercises 23*ff.* It is important at times to have maximal filters; these are called *ultrafilters*, cf., remark 0.0.16 below.

Example 0.0.15:

(i) For any $a \in \mathscr{L}$, $\mathscr{F}_a = \{b \in \mathscr{L} : b \geq a\}$ is called the *principal filter* generated by a.

(ii) Suppose $\mathscr{L} = \mathscr{P}(S)$ for some infinite set S.
Then {complements of finite subsets of S} is a filter, called the *Frechet* filter, or *cofinite* filter.

A direct way of obtaining a filter $\mathscr{F}$ of a lattice $\mathscr{L}$ is by finding $\mathscr{B} \subseteq \mathscr{L}$ satisfying properties (ii) and (iii) of definition 0.0.14, i.e., $0 \notin \mathscr{B}$ and if $b, b' \in \mathscr{B}$ then $b \wedge b' \geq b''$ for some b'' in $\mathscr{B}$. Then $\{a \in \mathscr{L} : a \geq b$ for some b in $\mathscr{B}\}$ is a filter $\mathscr{F}$, and $\mathscr{B}$ is called the *base of the filter* $\mathscr{F}$. We shall define many filters via their bases.

Zorn's Lemma

A poset $(S, \leq)$ is a *chain* if for all s_1, s_2 in S we have $s_1 \leq s_2$ or $s_2 \leq s_1$; in this case $\leq$ is called a *total order*; for example, $(\mathbb{Z}, \leq)$ is a chain under the usual (total) order. Given a lattice $(S, \leq)$, one sometimes finds that $\leq$ induces a total order on a certain subset of S. For example, if $A_1 \subseteq A_2 \subseteq A_3 \subseteq \cdots$ are subsets of A then $\{A_i : i \in \mathbb{N}\}$ is a chain in $(\mathscr{P}(A), \subseteq)$. This situation is of great interest because of the following result, often called *Zorn's lemma*. We say a poset $(S, \leq)$ is *inductive* if every chain S' in S has an upper bound in S. For example, $(\mathbb{N}, \leq)$ is not inductive since there exist chains not bounded from above, but $(\mathbb{N}, \geq)$ is inductive since 0 is an upper bound. The real interval $[0, 1]$ is inductive with respect to $\leq$, but $\mathbb{Q} \cap [0, \sqrt{2}]$ is not inductive. An element s of a poset S is *maximal* if there is no element $s' > s$ in S. Minimal elements are defined analogously. (Note that S may have many distinct maximal elements, the most extreme example being when the PO is trivial, in which case every element is both maximal and minimal.

Maximal principle ("Zorn's lemma"):   *If* $(S, \leq)$ *is an inductive poset then* S *has at least one maximal element.*

The key application is as follows: Suppose A is a set and $S \subseteq \mathcal{P}(A)$ such that for any chain $\{A_i : i \in I\}$ in S we have $\bigcup A_i \in S$. Then some subset of A is maximal in S.

The most basic application of the maximal principle in ring theory is that every proper ideal of a ring R is contained in a maximal proper ideal, called a *maximal ideal*. Indeed, an ideal A is proper iff $1 \notin A$, so $\{$proper ideals of $R\}$ is inductive. Likewise any (proper) left ideal is contained in a maximal left ideal. Incidentally this does not hold in general if we do not stipulate the existence of the element 1, which is the main reason we deal with rings with 1, cf., exercise 14.

However, a ring need not have minimal nonzero left ideals (for example $\mathbb{Z}$), since the above argument has no analogue.

Remark 0.0.16:   Any filter is contained in an ultrafilter by the maximal principle.

Remark 0.0.17:   Using remark 0.0.16 one sees that a filter $\mathscr{F}$ of a complemented lattice $\mathscr{L}$ is an ultrafilter iff for all a in $\mathscr{L}$, either $a \in \mathscr{F}$ or $a' \in \mathscr{F}$. One concludes that if $a \vee b$ is in an ultrafilter $\mathscr{F}$ then $a \in \mathscr{F}$ or $b \in \mathscr{F}$.

Another important application of the maximal principle: $\{$commutative subrings of R containing $Z(R)\}$ is inductive and thus has maximal members. In other words, any ring R has maximal commutative subrings, and these are often useful in the study of R, in view of proposition 0.0.6.

The maximal principle is proved by drawing from set theory and, in fact, is equivalent to the *axiom of choice*, which asserts that for any family $\{S_i : i \in I\}$ of sets there is a suitable "choice" function $f : I \to \bigcup S_i$ with $fi \in S_i$ for each i, i.e., f "chooses" one element from each S_i. At first blush this axiom seems obvious; however, the larger cardinality the index set I, the less credible the axiom becomes. P. J. Cohen proved that the axiom of choice is independent of the Zermelo-Fraenkel axioms of set theory, and today it is used freely by algebraists because the maximal principle is so powerful. To understand the connection we must bring in transfinite induction.

Well-Ordered Sets and Transfinite Induction

Many definitions in general ring theory rely on transfinite induction. To understand this process requires some intimacy with the ordinals, and to this end we bring in some formalism from set theory. It is natural to build sets from the bottom up, starting with the empty set and then building sets whose elements themselves are sets. Thus we formally define the symbols

$$\bar{0} = \varnothing, \qquad \bar{1} = \bar{0} \cup \{\bar{0}\} = \{\varnothing\}, \qquad \bar{2} = \bar{1} \cup \{\bar{1}\} = \{\varnothing, \{\varnothing\}\},$$

and so forth. The *axiom of regularity* states for every set $S \neq \varnothing$ there is $s \in S$ with $s \cap S = \varnothing$. This ensures that given two sets S_1, S_2 we cannot have both $S_1 \in S_2$ and $S_2 \in S_1$. (Indeed take $S = \{S_1, S_2\}$.) Thus we can define an anti-symmetric relation $\leq$ on sets by

$$S_1 \leq S_2 \qquad \text{whenever } S_1 \in S_2 \text{ or } S_1 = S_2.$$

Since the elements of a set S are themselves sets, we can view $\leq$ as an anti-symmetric relation on the elements of S. S is called an *ordinal* if $(S, \leq)$ is a chain.

If α is an ordinal then $\alpha^+ = \alpha \cup \{\alpha\}$ is also clearly an ordinal, called the *successor* of α. In particular $\bar{0} = \varnothing$, $\bar{1} = \bar{0}^+$, $\bar{2} = \bar{1}^+, \ldots$ are all ordinals. On the other hand, there are ordinals which are not successors, the first of which is $\{\bar{n} : n \in \mathbb{N}\}$; these are called *limit ordinals*.

A chain is *well-ordered* if every nonempty subset has a minimal element. Every ordinal α is well-ordered under $\leq$ as defined above; indeed if $\varnothing \neq S \subset \alpha$ then any $s \in S$ with $s \cap S = \varnothing$ is minimal in S. Thus we have the following generalization of mathematical induction:

Principle of transfinite induction: *Suppose α is an ordinal and $S \subseteq \alpha$ has the property for every ordinal $\alpha' < \alpha$ that if $\{\beta : \beta < \alpha'\} \subseteq S$ then $\alpha' \in S$. Then $S = \alpha$.*

The proof is rather easy, but the applications are wide-ranging; here are some set-theoretic implications we shall need (cf., exercise 15–20):

(i) Every set can be put into 1:1 correspondence with a suitable ordinal and thus is well-ordered under the corresponding total ordering. (Thus we shall often describe a set S as $\{s_1, s_2, \ldots\}$ even when S is uncountable);

(ii) Zorn's lemma, as stated above.

Fields

In the structure theory of rings one often considers fields as "trivial" since they have no proper ideals $\neq 0$; in fact, most results from ring theory hardly require any knowledge of fields. Nevertheless, fields do play important roles in several key topics (such as division rings), and ideas from field theory provide guidelines for generalization to arbitrary rings. When appropriate we shall assume familiarity with the Galois theory of finite dimensional field extensions, including normal and separable extensions, and the algebraic closure of a field.

Algebraic and Transcendental Elements

Let C-$\mathscr{Alg}$ denote the category of algebras over a commutative ring C.

Given R in C-$\mathscr{Alg}$ and $r \in R$ we write $C[r]$ for $\left\{ \sum_{i=0}^{t} c_i r^i : c_i \in C, t \in \mathbb{N} \right\}$. $C[r]$ is a commutative subalgebra of R, and there is a surjection φ_r: $C[\lambda] \to C[r]$ given by $\varphi_r \lambda = r$, where λ is a commuting indeterminate over C; the elements of $\ker \varphi_r$ are the polynomials *satisfied* by r. We say r is *transcendental* over C if $\ker \varphi_r = 0$; otherwise r is *algebraic* over C, and we say r is *integral* over C iff r satisfies a monic polynomial. R is *integral* (resp. *algebraic*) *over* C if each element of R is integral (resp. algebraic) over C. When C is a field the notions, "algebraic" and "integral" coincide.

More generally suppose the elements $r_1, \ldots, r_t$ of R commute with each other. Let $C[\lambda_1, \ldots, \lambda_t]$ denote the algebra of polynomials in the commuting indeterminates $\lambda_1, \ldots, \lambda_t$ over C. Writing $C[r_1, \ldots, r_t]$ for the (commutative) C-subalgebra of R generated by $r_1, \ldots, r_t$ we have the canonical surjection $\varphi: C[\lambda_1, \ldots, \lambda_t] \to C[r_i, \ldots, r_t]$ given by $\varphi \lambda_i = r_i$ for $1 \le i \le t$; we say $r_1, \ldots, r_t$ are *algebraically independent* (over C) if $\ker \varphi = 0$.

From time to time we shall appeal to the theory of first order logic, described in pp. 13–14 of the unabridged text.

§0.1 Categories of Rings and Modules

The language of categories is useful, particularly in certain aspects of module theory. We presuppose a nodding acquaintance with this language; Jacobson [80B, Chapter 1] more than suffices for this purpose. In particular, the reader should know the definition of category, subcategory, (covariant) functor, contravariant functor, natural transformation (of functors), and natural isomorphism of functors. Given a category $\mathscr{C}$ we write $\mathrm{Hom}_{\mathscr{C}}(A, B)$, or merely $\mathrm{Hom}(A, B)$, for the set of morphisms; $1_A: A \to A$ denotes the unit morphism.

Our interest in categories will focus on $R\text{-}\mathcal{M}od$, where R is a given ring; its objects are the R-modules, and its morphisms are the maps of R-modules. Another category of note is $\mathcal{R}ing$, whose objects are rings and whose morphisms are ring homomorphisms. $R\text{-}\mathcal{M}od$ and $\mathcal{R}ing$ each are subcategories of $\mathcal{Ab}$, the category of abelian groups (written in additive notation); so is $\mathcal{M}od\text{-}R$, the category of *right* R-modules. Often the class of objects of a category is "too large" to be a set, for one can construct distinct objects for each ordinal. Such is the case with $\mathcal{R}ing$ and $R\text{-}\mathcal{M}od$, cf., exercise 1.4.1. Accordingly a category is called *small* when its class of objects is a set.

Definition 0.1.2:   Given a category $\mathcal{C}$ we define the *dual category* $\mathcal{C}^{op}$ by $Ob\,\mathcal{C}^{op} = Ob\,\mathcal{C}$ and $\operatorname{Hom}_{\mathcal{C}^{op}}(A, B) = \operatorname{Hom}_{\mathcal{C}}(B, A)$ with composition in C^{op} given by $g \cdot f = fg$ (where $f \in \operatorname{Hom}_{\mathcal{C}^{op}}(A, B)$ and $g \in \operatorname{Hom}_{\mathcal{C}^{op}}(B, C)$). In other words, we reverse arrows and write things backwards. (In particular $(\mathcal{C}^{op})^{op} = \mathcal{C}$.)

Any general theorem for all categories *a fortiori* holds for the dual categories; translating back to the original category yields a new theorem, the *dual theorem* obtained by switching all arrows. The main problem with this approach is that few theorems hold for *all* categories, and the dual of a well-known category may be quite bizarre. Nevertheless, there are certain important examples for which the dual is well-known and useful.

Monics and Epics

Recall a morphism $f: A \to B$ is *monic* if $fg \neq fh$ for any $g \neq h$ in $\operatorname{Hom}(C, A)$, for all objects C; dually f is *epic* if $gf \neq hf$ for any $g \neq h$ in $\operatorname{Hom}(B, C)$. Clearly the composition of monics (resp. epics) is monic (resp. epic). In any subcategory of $\mathcal{S}et$ each 1:1 morphism is monic, and each onto morphism is epic; we would like to test the converse for $R\text{-}\mathcal{M}od$ and $\mathcal{R}ing$.

Proposition 0.1.3:   *In $R\text{-}\mathcal{M}od$, monics are 1:1, and epics are onto.*

Proof: Suppose we are given $f: M \to N$. If f is monic define g, $h: \ker f \to M$ by taking g to be the identity and $h = 0$; then $fg = fh = 0$ implying $g = h$, so $\ker f = 0$ and f is 1:1. If f is epic then define g, $h: N \to N/fM$ by $g = 0$ and $hy = y + fM$; then $gf = hf = 0$, implying $g = h$ so $N/fM = 0$, i.e., $fM = N$.   Q.E.D.

The story ends differently for $\mathcal{R}ing$.

Example 0.1.4: A 1:1 ring homomorphism which is epic and monic, but *not* onto. Consider the 1:1 ring homomorphism $f: \mathbb{Z} \to \mathbb{Q}$ given by $fn = n$. For any morphism $g: \mathbb{Q} \to R$ in $\mathscr{R}ing$ we have $g(mn^{-1}) = (gm)(gn)^{-1}$ for all $m, n \neq 0$ in $\mathbb{Z}$, implying g is determined by its restriction to $\mathbb{Z}$; it follows at once that if $g \neq h$ then $gf \neq hf$, so f is epic.

In Exercise 1 we see that all monics in $\mathscr{R}ing$ are 1:1. Nevertheless, example 0.1.4 could be called the tragedy of $\mathscr{R}ing$.

Perhaps rings categorically should be viewed in terms of the following example.

Example 0.1.5: Suppose R is a ring. We form a category with only one object, denoted A, and formally define $\mathrm{Hom}(A, A) = R$ where the composition of morphisms is merely the ring multiplication.

To differentiate the approach to rings and to modules, we designate ring homomorphisms as "homomorphisms;" a 1:1 (ring) homomorphism is called an *injection*, and an onto homomorphism is a *surjection*. For modules we adopt, respectively, the more categorical terminology of *map*, *monic*, and *epic*. Nevertheless, since an ideal of R is merely a left and right submodule, we would like to introduce another category.

Definition 0.1.6: Suppose R, R' are rings. An *R-R' bimodule* is a left R-module M which is also a right R'-module satisfying the associativity condition $(rx)r' = r(xr')$ for all r in R, x in M, and r' in R'. R-$\mathscr{M}od$-R' is the category whose objects are R-R' bimodules and whose morphisms $f: M \to M'$ are maps both in R-$\mathscr{M}od$ and $\mathscr{M}od$-R'.

Remark 0.1.7: The R-R sub-bimodules of R are precisely the ideals of R.

Functors

Certain functors arise continually in the study of R-$\mathscr{M}od$. Perhaps the most important are the "Hom" functors:

Example 0.1.8:

(i) $\mathrm{Hom}(A_0, \underline{})$: $\mathscr{C} \to \mathscr{S}et$ is the functor sending an object A to $\mathrm{Hom}(A_0, A)$ and sending a morphism $f: A \to B$ to the function $f_\#$: $\mathrm{Hom}(A_0, A) \to \mathrm{Hom}(A_0, B)$, defined by $f_\# h = fh$ for each h in $\mathrm{Hom}(A_0, A)$. Note $\mathrm{Hom}(A_0, \underline{})$ is covariant because for all h in $\mathrm{Hom}(A_0, A)$ and each $f: A \to B$ and $g: B \to C$ we have

$$(gf)_\# h = gfh = g_\# f_\# h, \qquad \text{so } (gf)_\# = g_\# f_\#$$

(ii) $\mathrm{Hom}(\underline{\quad}, A_0): \mathscr{C} \to \mathscr{S}et$ is the *contravariant* functor sending A to $\mathrm{Hom}(A, A_0)$ and sending f to the function $f^*: \mathrm{Hom}(B, A_0) \to \mathrm{Hom}(A, A_0)$ defined by $f^\# h = hf$ for each h in $\mathrm{Hom}(B, A_0)$.

Other functors we need are the *identity functor* $1_\mathscr{C}$, and the "forgetful functors."

Two categories $\mathscr{C}$, $\mathscr{D}$ are *isomorphic* if there exist functors $F: \mathscr{C} \to \mathscr{D}$ and $G: \mathscr{D} \to \mathscr{C}$ with $GF = 1_\mathscr{C}$ and $FG = 1_\mathscr{D}$. This definition is very stringent, but is useful in identifying a pair of theories.

Example 0.1.10: $\mathscr{A}b$ and $\mathbb{Z}\text{-}\mathscr{M}od$ are isomorphic categories. (Indeed, let $F: \mathbb{Z}\text{-}\mathscr{M}od \to \mathscr{A}b$ be the forgetful functor, and define $G: \mathscr{A}b \to \mathbb{Z}\text{-}\mathscr{M}od$ as follows: Given $M \in \mathscr{A}b$ we view M as $\mathbb{Z}$-module by introducing scalar multiplication $nx = x + \cdots + x$, the sum taken n times for $n \in \mathbb{N}$, and $(-n)x = -(nx)$. Every group homomorphism then becomes a morphism in $\mathbb{Z}\text{-}\mathscr{M}od$, so we have the inverse morphism to F.)

To identify categories of left and right modules, we need the notion of the *opposite ring* R^{op}.

Definition 0.1.11: If R is a ring, R^{op} is the ring obtained by keeping the same additive structure but reversing the order of multiplication, (i.e., the product of r_1 and r_2 in R^{op} is $r_2 r_1$).

R^{op} yields the dual category of the category obtained from R in example 0.1.5, thereby justifying the notation. Note $(R^{op})^{op} = R$.

Proposition 0.1.12: $R\text{-}\mathscr{M}od$ and $\mathscr{M}od\text{-}R^{op}$ are isomorphic categories.

Thus any general theorem about modules is equivalent to a corresponding theorem about right modules. Usually we want to weaken the notion of isomorphism of categories to *equivalent* categories, c.f., Jacobson [80B, p. 27]; we shall see in Chapter 4 that categorical equivalence is a fundamental tool of module theory.

§0.2 Finitely Generated Modules, Simple Modules, and Noetherian and Artinian Modules

Returning to modules, we approach one of the nerve centers of the subject and look at the generation of modules by elements. Finitely generated (f.g.) modules, also frequently called "finite" and "of finite type," turn out to be much more tractable than arbitrary modules. In particular, we shall examine cyclic modules, leading us to simple modules. At the end we introduce the important classes of Artinian modules and Noetherian modules.

Finitely Generated Modules

Given M in R-$\mathcal{Mod}$ and $A \subseteq R$, $S \subseteq M$, define $AS = \left\{ \sum_{i=1}^{t} a_i s_i : t \in \mathbb{N}, a_i \in A, s_i \in S \right\}$; if $M \in \mathcal{Mod}$-R we define SA analogously. For $S = \{s_i : i \in I\}$ we often write $\sum As_i$ instead of AS. Usually A will be an additive subgroup of R, in which case AS is a subgroup of M for any set S; in fact $As = \{as : a \in A\}$.

Proposition 0.2.1:   *Suppose $S \subseteq M$ and $M \in R$-$\mathcal{Mod}$. If $L \leq R$ then $LS \leq M$. RS is the intersection of all submodules of M which contain S. If $M \in R$-Mod-R' then RSR' is the intersection of all sub-bimodules of M containing S.*

Proof: LS is an additive subgroup of M, and for any r in R, $r(LS) = (rL)S \subseteq LS$ proving $LS \leq M$. In particular $RS \leq M$. Now for any $N \leq M$ with $S \subseteq N$ we have $RS \subseteq RN \subseteq N$; since RS itself is a submodule containing each element $s = 1s$ of S we get the second assertion. The last assertion is proved similarly.   Q.E.D.

We say a module M is *spanned by* the subset S if $M = RS$. A module spanned by a finite set is called *finitely generated*, abbreviated as "f.g." throughout the text.

Definition 0.2.2: R-$\mathcal{Fimod}$ is the subcategory of R-$\mathcal{Mod}$ whose objects are the f.g. R-modules; the morphisms are, of course, the maps.

Remark 0.2.3:   If $f: M \to N$ is a map of modules and M is spanned by S then fM is spanned by fS (for if $x = \sum r_i s_i$ then $fx = \sum r_i(fs_i)$). In particular, if M is f.g. then fM is f.g. On the other hand, a submodule of an f.g. module need not be f.g. as seen in the next example.

Example 0.2.4:   Let $\mathbb{Q}[\lambda]$ be the ring of polynomials in one commuting indeterminate over $\mathbb{Q}$ and let $I = \lambda \mathbb{Q}[\lambda]$ and $R = \mathbb{Z} + I \subset \mathbb{Q}[\lambda]$. Then $I \triangleleft R$ but $I \notin R$-$\mathcal{Fimod}$. (Indeed, given any $x_1, \ldots, x_t$ in I, let the coefficient of λ in x_i be m_i/n_i and observe $(2n_1 \cdots n_t)^{-1}\lambda \in I - \sum_{i=1}^{t} Rx_i$.)

Cyclic Modules

Of particular interest are modules spanned by a single element.

Definition 0.2.5: M is a *cyclic* R-module if $M = Rx$ for some x in M.

$R = R1$ is a cyclic R-module, so remark 0.2.3 shows R/L is cyclic for every $L < R$. Conversely, every cyclic module has this form, as we shall see shortly.

Definition 0.2.6: If $M \in R\text{-}\mathcal{M}od$ and $S \subseteq M$, define $\text{Ann}_R S$ (the *left annihilator of* S *in* R) to be $\{r \in R : rs = 0\}$, a proper left ideal of R. If R is understood, we write $\text{Ann}\, S$ for $\text{Ann}_R S$; we also write $\text{Ann}\, x$ for $\text{Ann}\{x\}$.

Lemma 0.2.7: *If* $M \in R\text{-}\mathcal{M}od$, *then for every* x *in* M *there is a map* $f_x: R \to M$ *given by* $f_x r = rx$; $\ker f_x = \text{Ann}\, x$, *implying* $R/\text{Ann}\, x \approx Rx \le M$.

Proof: Clearly f_x is a map, and $\ker f_x = \{r \in R : rx = 0\} = \text{Ann}\, x$, so $Rx = f_x R \approx R/\ker f_x = R/\text{Ann}\, x$.   Q.E.D.

Proposition 0.2.8: $M \in R\text{-}\mathcal{M}od$ *is cyclic iff* $M \approx R/L$ *for some left ideal* L *of* R; *in fact, for* $M = Rx$ *then we can take* $L = \text{Ann}\, x$.

Proof: As noted above, R/L is cyclic; the converse is lemma 0.2.7.   Q.E.D.

Simple Rings and Modules

We shall turn now to a basic philosophy concerning arbitrary categories. One should like to examine objects by taking morphisms to objects whose structure we already know. Then the simplest objects would be those objects from which all morphisms are monic, motivating the following definition.

Definition 0.2.9: A nonzero module M is *simple* if M has no proper nonzero submodules; a ring R is *simple* if R has no proper nonzero ideals. (Simple modules are called *irreducible* in the older literature.)

Remark 0.2.10: In $\mathcal{R}ing$ or in $R\text{-}\mathcal{M}od$, an object A is simple iff every nonzero morphism $f: A \to B$ is 1:1. (*Proof:* ($\Rightarrow$) $\ker f \ne A$ so $\ker f = 0$. ($\Leftarrow$) Any morphism $A \to A/I$ is 1:1 so $I = 0$.)

In particular, we have the following categorical criterion for a module M to be simple: Every nonzero map from M is monic. An analogous criterion holds for $\mathcal{R}ing$ in view of exercise 0.1.1, but the dual criterion only works for modules.

Remark 0.2.11:   An R-module N is simple iff every nonzero map $f: M \to N$ is onto. (*Proof*: ($\Rightarrow$) $0 \neq fM \leq N$ implies $fM = N$; ($\Leftarrow$) if $0 \neq N' < N$ then the injection $N' \to N$ is not onto.)

There are two immediate difficulties in trying to build a structure theory based on simple rings and/or modules:

(i) There must be enough simples to yield general information about rings and modules.

(ii) One needs some technique to study the simples.

The first difficulty can be dealt with by means of maximal left ideals.

Remark 0.2.12:

(i) If L is a maximal left ideal of R then R/L is a simple R-module. (Immediate from the lattice correspondences pertaining to R as R-module.)

(ii) If I is a maximal ideal of R then R/I is a simple ring.

Surprisingly this remark provides all the simple modules.

Lemma 0.2.13:   *If M is a simple R-module then M is cyclic. In fact $Rx = M$ for every $x \neq 0$ in M.*

Proof: $0 \neq Rx \leq M$ so $Rx = M$.   Q.E.D.

Proposition 0.2.14:   *$M \in R\text{-}\mathcal{Mod}$ is simple iff $M \approx R/L$ for a suitable maximal left ideal of R.*

Proof: ($\Rightarrow$) Write $M = Rx$ and define $\varphi: R \to M$ by $\varphi r = rx$. Then φ is onto so $M \approx R/\ker \varphi$, and $\ker \varphi$ is a maximal submodule by proposition 0.0.8. ($\Leftarrow$) Reverse the argument.   Q.E.D.

Before putting this idea aside, we note an important generalization.

Proposition 0.2.15:   *If $M \in R\text{-}\mathcal{Fimod}$ and $N < M$ then N is contained in some maximal (proper) submodule M'. Consequently M/M' is a simple module in which the image of N is 0.*

Proof: Write $M = \sum_{i=1}^{t} Rx_i$, where all $x_i \in M$. For any chain $M_1 \leq M_2 \leq \cdots$ of proper submodules, some $x_i \notin M_j$ for all j (since otherwise *all* $x_i \in M_j$ for large enough j, implying $M = \sum Rx_i \subseteq M_j$ contrary to M_j proper). Thus

$x_i \notin \bigcup M_j$, which hence must be proper (and is clearly a submodule). Taking $M_1 = N$, we have proved {proper submodules of M which contain N} is inductive and thus by Zorn's lemma contains maximal members, which must be maximal submodules of M. The rest is clear.   Q.E.D.

It is difficult to study simple rings without imposing further restrictions because of the intrinsic complexity of ideals. Indeed, the smallest ideal of R containing a given element r is $RrR = \{\sum_{i=1}^{t} r_{i1} r r_{i2} : t \in \mathbb{N}, r_{i1}, r_{i2} \in R\}$ which cannot be described in the first-order theory of rings. On the other hand, the smallest *left* ideal containing r is $Rr = \{r'r : r' \in R\}$, which is much more amenable. When r is in the center then in fact $RrR = Rr$, and the situation is much easier to handle. In general, the classification of simple rings is an immense project, far from completion, but there is the following easy result concerning rings without proper left ideals.

Proposition 0.2.16:

   (i) $Rr = R$ iff r has a left inverse in R;
  (ii) R is a division ring iff R has no proper nonzero left ideals;
 (iii) if R is simple then $Z(R)$ is a field.

Proof: Easy exercise.

Chain Conditions

A poset $(S, \leq)$ satisfies the *maximum* (resp. *minimum*) condition if every non-empty subset has a maximal (resp. minimal) element. $(S, \leq)$ satisfies the *ascending chain condition* (abbreviated ACC) if there is no infinite chain $s_1 < s_2 < s_3 < \cdots$, i.e., if every ascending chain is finite; dually $(S, \leq)$ satisfies DCC if every descending chain is finite.

Proposition 0.2.17:   *A poset $(S, \leq)$ satisfies the maximum condition iff it satisfies ACC. $(S, \leq)$ satisfies the minimum condition iff it satisfies DCC.*

Proof: We prove the first assertion; the second is its dual and follows by passing to the dual poset. First note that an ascending chain is finite iff it contains a maximal element. So if $(S, \leq)$ satisfies the maximum condition then every ascending chain is finite, proving $(S, \leq)$ satisfies ACC. Conversely, if $(S, \leq)$ satisfies ACC then every subset S' is inductive (because every chain of S' is finite) and thus has a maximal element.   Q.E.D.

Thus to verify a chain is well-ordered, we need only check that there is no infinite descending subchain. Also we see that if a lattice satisfies DCC then every subset is well-ordered.

Noetherian and Artinian Modules

The point of studying chain conditions on lattices is in utilizing the lattice $\mathcal{L}(M)$ of submodules.

Definition 0.2.18:   A module M is *Noetherian* if $\mathcal{L}(M)$ satisfies ACC or, equivalently, if $\mathcal{L}(M)$ satisfies the maximum condition. M is *Artinian* if $\mathcal{L}(M)$ satisfies DCC or equivalently the minimum condition.

Proposition 0.2.19:   *Suppose* $M \in R\text{-}\mathcal{M}od$ *and* $N \leq M$. M *is Noetherian iff* N *and* M/N *are Noetherian.* M *is Artinian iff* N *and* M/N *are Artinian.*
   The proof boils down to showing that if $M_i < M_{i+1} < M$ with $M_i \cap N = M_{i+1} \cap N$ and $(M_i + N)/N = (M_{i+1} + N)/N$, then $M_i = M_{i+1}$. This is quite easy; see Remark 3.5.40".

Corollary 0.2.20:   If $M = N_1 + \ldots + N_t$ with each N_t Noetherian (resp. Artinian) then M is Noetherian (resp. Artinian).

Proof: We prove Noetherian; Artinian is analogous. Let $N = \sum_{i=1}^{t-1} N_i$, which is Noetherian by induction on t. But $M/N = (N + N_t)/N \approx N_t/(N \cap N_t)$ is Noetherian so M is Noetherian by the proposition.   Q.E.D.

We say R is *left Noetherian* (resp. *left Artinian*) if R is Noetherian (resp. Artinian) as R-module.

Corollary 0.2.21:   *Every f.g. module over a left Noetherian ring* R *is Noetherian. Every f.g. module over a left Artinian ring* R *is Artinian.*

Proof: We prove Noetherian. Write $M = \sum_{i=1}^{t} Rx_i$. Each $Rx_i \approx R/\text{Ann } x_i$ is Noetherian, so M is Noetherian.   Q.E.D.

Exercises

§0.0

2. A lattice $\mathcal{L}$ is modular iff $\mathcal{L}$ has the property: If $a \leq b$ and $a \wedge c = b \wedge c$ and $a \vee c = b \vee c$ then $a = b$. (Hint: ($\Leftarrow$) Let $a_1 = a \vee (b \wedge c)$ and $a_2 = b \wedge (a \vee c)$. Then $a_1 \leq a_2$ by remark 0.0.11. To show $a_2 \leq a_1$ one needs only show $a_1 \wedge c \geq a_2 \wedge c$ and $a_1 \vee c \geq a_2 \vee c$.)

Modular Lattices In exercises 6 through 13 assume $\mathscr{L}$ is a modular lattice.

6. The dual of $\mathscr{L}$ is modular. Every interval of $\mathscr{L}$ is a modular lattice, which is complemented if $\mathscr{L}$ is complemented.
7. Complements need not be unique in a complemented modular lattice. (Take $\mathscr{L}(\mathbb{R}^{(2)})$.) However, if $a < b$ in $\mathscr{L}$ then no element can be a complement for both a and b.
8. If $(a \vee b) \wedge c = 0$ then $a \wedge (b \vee c) = a \wedge b$.
   (Hint: $a \wedge (b \vee c) \leq (a \vee b) \wedge (b \vee c) = b \vee ((a \vee b) \wedge c) = b$.)
9. An element a of $\mathscr{L}$ is *large* (also called *essential*) if $a \wedge b \neq 0$ for all $b \neq 0$. If b is maximal such that $a \wedge b = 0$ then $a \vee b$ is large. (Hint: use exercise 8.)
13. Suppose $\mathscr{L}$ is modular and complemented. $\mathscr{L}$ satisfies ACC iff $\mathscr{L}$ satisfies DCC. (Hint: Given an infinite chain $a_1 > a_2 > \cdots$ build an infinite ascending chain using complements and exercises 6 thru 9. The reverse direction follows from duality.)

Zorn's Lemma

14. Let R be the set of polynomials (in one indeterminate λ) over $\mathbb{Q}$ having constant term 0. R satisfies all the ring axioms *except* the existence of the unit element 1. If H is an additive subgroup of $\mathbb{Q}$ then $(H\lambda + R\lambda) \lhd R$; since $\mathbb{Q}$ has no maximal additive subgroups, conclude R has no maximal ideals.
15. Prove the validity of transfinite induction. (Hint: If $S \neq \alpha$ take α' minimal in $\alpha - S$ and prove the absurdity $\alpha' \in S$ since $\beta \in S$ for all $\beta < \alpha'$.)
16. If $\alpha \neq \alpha'$ are ordinals then either $\alpha < \alpha'$ or $\alpha' < \alpha$. (Hint: Take a minimal counter-example α'.)
17. Let $\mathcal{O} = \{$class of ordinals$\}$. $\mathcal{O}$ is not a set (for otherwise $\mathcal{O} \in \mathcal{O}$, which is impossible.) Thus to show a class $\mathscr{C}$ is not a set it suffices to find a 1:1 function from $\mathcal{O}$ to $\mathscr{C}$.

§0.1

1. Monics in $\mathscr{Ring}$ are all 1:1. (Hint: If $f: R \to R'$ is monic define a ring structure on the cartesian product $R \times R$ by componentwise operations and let $T = \{(r_1, r_2) \in R \times R : fr_1 = fr_2\}$. Then $f\pi_1 = f\pi_2$ where $\pi_i: T \to R$ is the projection on the i component.) The proof of this exercise introduces two important constructions—the direct product and the pullback.

§0.2

1. By comparing annihilators, show $\mathbb{Z}$ has nonisomorphic simple modules.
2. If $f: M \to N$ is epic and $\ker f$ and N are f.g. R-modules then M is also f.g.
3. If M is a f.g. R-module and $A \lhd R$ is f.g. as left ideal then AM is f.g. (Hint: If $M = \sum Rx_i$ then $AM = \sum Ax_i$.)

1 Constructions of Rings

There are several general constructions of rings and modules which play a central role in the theory of rings and which also provide many interesting examples and applications. In this chapter we consider these basic constructions, along with a little theory needed to shed light on them. The different constructions used are enumerated in the section headings.

§1.1 Matrix Rings and Idempotents

It is fitting to start the main text with a discussion of matrices, since they are undoubtedly the most widely studied class of noncommutative rings. Although matrices are fundamental in the structure theory, we shall postpone most structural considerations until the next chapter, contenting ourselves with examining the elementary properties of matrices. Define the *Kronecker delta* δ_{ij} to be 0, unless $i = j$, in which case $\delta_{ii} = 1$.

Matrices and Matrix Units

Definition 1.1.1: We define $M_n(R)$, the *ring of $n \times n$ matrices* with entries in a given ring R, as follows: each matrix is written as (r_{ij}) where r_{ij} denotes the i-j entry for $1 \leq i, j \leq n$ (with n presumed to be understood); addition and multiplication are given according to the rules:

$$(r_{ij}^{(1)}) + (r_{ij}^{(2)}) = (r_{ij}^{(1)} + r_{ij}^{(2)}) \qquad \text{and}$$

$$(r_{ij}^{(1)})(r_{ij}^{(2)}) = (r_{ij}^{(3)}) \qquad \text{where } r_{ij}^{(3)} = \sum_{k=1}^{n} r_{ik}^{(1)} r_{kj}^{(2)}.$$

We delete the verification that $M_n(R)$ is indeed a ring, analogous to the familiar special case when R is a field; moreover, towards the end of §1.2 we shall lay down general principles which at once imply $M_n(R)$ is a ring.

We obtain a more explicit notation by defining the $n \times n$ *matric unit* e_{ij} to be the matrix whose i-j entry is 1, with all other entries 0. Thus $(r_{ij}) = \sum_{i,j=1}^{n} r_{ij} e_{ij}$; addition is componentwise and multiplication is given according to the rule

$$(r_1 e_{ij})(r_2 e_{uv}) = \delta_{ju}(r_1 r_2) e_{iv}$$

The set of $n \times n$ matric units is a base of $M_n(R)$ as R-module (with scalar multiplication given by $r(r_{ij}) = (rr_{ij})$). Note that if $a = (r_{ij})$ we then see $r_{ij} e_{uv} = e_{ui} a e_{jv}$, a very useful computation which pinpoints the entries of a.

A *scalar matrix* is a matrix of the form $\sum_{i=1}^{n} re_{ii}$. The set of scalar matrices is a subring of $M_n(R)$ which shall be identified with R under the isomorphism $r \to \sum re_{ii}$, and it is the centralizer of $\{e_{ij}: 1 \le 1, j \le n\}$. (Indeed if $a = (r_{ij})$ commutes with each e_{ij} then $\sum_{u=1}^{n} r_{ui} e_{uj} = ae_{ij} = e_{ij}a = \sum_{v=1}^{n} r_{jv} e_{iv}$ so matching entries we get $r_{ui} = 0 = r_{jv}$ unless $u = i$ and $v = j$. For $u = i$ and $v = j$ we get $r_{ii} = r_{jj}$ for all i, j, implying $a = \sum_{i=1}^{n} r_{11} e_{ii}$.)

In particular, $Z(M_n(R)) = \{\sum ze_{ii}: z \in Z(R)\}$ and $\sum e_{ii} = 1$ in $M_n(R)$, leading us to an important internal characterization of matrix rings.

Definition 1.1.2: A set of $n \times n$ *matric units of* a ring T is a set

$$\{e_{ij}: 1 \le i, j \le n\} \subset T$$

such that $\sum_{i=1}^{n} e_{ii} = 1$ and $e_{ij} e_{ku} = \delta_{jk} e_{iu}$ for all i, j, k, u.

Surprisingly, the existence of a set of $n \times n$ matric units makes any ring T an $n \times n$ matrix ring over a suitable ring R; the trick in the proof is to find some intrinsic description of R. Since the idea recurs, let us describe it briefly. If we already know that $T = M_n(R)$ and if $a = (r_{ij}) \in T$ then clearly the scalar matrix corresponding to r_{ij} is $\sum_{u=1}^{n} e_{ui} a e_{ju}$; also it is useful to note $r_{ij} e_{ij} = e_{ii} a e_{jj}$.

Proposition 1.1.3: T has a set of $n \times n$ matric units iff $T \approx M_n(R)$ *for a* suitable ring R.

Proof: ($\Leftarrow$) by definition of the e_{ij}. ($\Rightarrow$) Conversely suppose $\{e_{ij}: 1 \le i, j \le n\}$ is a given set of matrix units of T. Let $R = \{\sum_{u=1}^{n} e_{u1} a e_{1u}: a \in T\}$. R is a subring because it is closed under subtraction, and $1 = \sum_{u=1}^{n} e_{uu} = \sum e_{u1} 1 e_{1u} \in R$ and $\left(\sum e_{u1} a e_{1u}\right)\left(\sum e_{u1} b e_{1u}\right) = \sum e_{u1} (a e_{11} b) e_{1u} \in R$. It remains to define the isomorphism $\varphi: T \to M_n(R)$. Given a in T put $r_{ij} = \sum_{u=1}^{n} e_{ui} a e_{ju} = \sum_{u=1}^{n} e_{u1}(e_{1i} a e_{j1}) e_{1u} \in R$ and define $\varphi a = (r_{ij})$. Clearly φ is an additive group homomorphism. Moreover,

$$r_{ij} e_{ij} = \sum_{u=1}^{n} e_{ui} a e_{ju} e_{ij} = \sum_{u=1}^{n} \delta_{ui} e_{ui} a e_{jj} = e_{ii} a e_{jj},$$

so $\sum_{i,j} r_{ij} e_{ij} = \sum_{i,j=1}^{n} e_{ii} a e_{jj} = (\sum e_{ii}) a (\sum e_{jj}) = a$, implying φ is 1:1; likewise, $\varphi(\sum r_{ij} e_{ij}) = (r_{ij})$ so φ is onto. Finally the i-j term of $\varphi(ab)$ is

$$e_{ii} a b e_{jj} = e_{ii} a \sum_{k=1}^{n} e_{kk} b e_{jj} = \sum_{k=1}^{n} (e_{ii} a e_{kk})(e_{kk} b e_{jj})$$

implying $\varphi(ab) = \varphi a \varphi b$, so φ is a ring isomorphism.   Q.E.D.

There are many close ties between the structure of a ring R and $M_n(R)$, some of which we give now (also, cf., exercise 1).

Proposition 1.1.4:   *Let $\mathcal{M}at_n$ denote the full subcategory of $\mathcal{R}ing$ whose objects are rings of $n \times n$ matrices. Then there is an isomorphism of categories $F: \mathcal{R}ing \to \mathcal{M}at_n$ given by $FR = M_n(R)$ where for any morphism $f: R \to R'$ we define $Ff: M_n(R) \to M_n(R')$ by $(Ff)(r_{ij}) = (fr_{ij})$.*

Proof: Ff is indeed a ring homomorphism, since

$$(fr_{ij}^{(1)})(fr_{ij}^{(2)}) = \left(\sum fr_{ik}^{(1)} fr_{kj}^{(2)}\right) = \left(\sum f(r_{ik}^{(1)} r_{kj}^{(2)})\right) = f((r_{ij}^{(1)})(r_{ij}^{(2)})).$$

It follows at once that F is a functor, and its inverse G is given by $GM_n(R) = R$ where for any morphism $g: M_n(R) \to M_n(R')$, Gg is given by the restriction to the scalar matrices.   Q.E.D.

For any subset S of R we let $M_n(S)$ denote the matrices whose entries lie in S.

Proposition 1.1.5:   *There is a lattice isomorphism $f: \{\text{ideals of } R\} \to \{\text{ideals of } M_n(R)\}$, given by $fA = M_n(A)$.*

Proof: Suppose $A \triangleleft R$. Then $M_n(A)$ is an additive subgroup of R, and the multiplication formula of matrices shows $M_n(A) \triangleleft M_n(R)$. If $B \subset A$

then $M_n(B) \subset M_n(A)$, so f is order-preserving, and it remains to find an order-preserving inverse. If $I \lhd M_n(R)$ define $gI = I \cap R$ (identifying R with the scalar matrices). Clearly, $gf = 1$, and to prove $fg = 1$ we must show that if $a = (r_{ij}) \in I$ then each scalar matrix $r_{ij} \sum_{u=1}^{n} e_{uu} \in I$ for every i, j. But $r_{ij} \sum_{u=1}^{n} e_{uu} = \sum_{u=1}^{n} e_{ui} a e_{ju} \in I$, as desired.   Q.E.D.

There is a useful way to reduce the size of matrices, called *partitioning*, described as follows:

Remark 1.1.6:   Suppose $n = mt$. If $R = M_t(T)$ there is an isomorphism $f: M_n(T) \to M_m(R)$ sending $a = (a_{ij})$ to the $t \times t$ matrix (r_{uv}) where $r_{uv} = \sum_{i=t(u-1)+1}^{tu} \sum_{j=t(v-1)+1}^{tv} a_{ij} e_{ij}$. (Indeed f is clearly an isomorphism of additive groups; the straightforward but messy verification that f preserves multiplication is left to the reader. Note that f merely subdivides the matrix a into m^2 matrices, whence the name "partitioning." The procedure can be done more generally, cf., exercise 3.

$M_n(R)$ has extra structure obtained from the *transpose* $t: M_n(R) \to M_n(R)$ defined by $(r_{ij})^t = (r_{ji})$. Rather than go into the structural implications now, we merely note that the transpose is an additive group homomorphism satisfying the properties $(ab)^t = b^t a^t$ and $(a^t)^t = a$ for all matrices a and b.

The theory of matrices over fields should be familiar to the reader. There are ready extensions of the basic theorems to matrices over arbitrary commutative rings.

Subrings of Matrix Rings

As we shall see later, there are large classes of rings which can be viewed as subrings of a matrix ring over a commutative ring. Here we consider examples of such rings.

Example 1.1.7:   Hamilton's algebra $\mathbb{H}$ of real quaternions over $\mathbb{R}$ is defined as the 4-dimensional vector space (over $\mathbb{R}$) having base $1, i, j, k$, subject to the conditions $i^2 = j^2 = k^2 = -1$ and $ij = -ji = k$. One could, in fact, prove directly that multiplication is associative and distributive over addition, but it is quicker to display $\mathbb{H}$ as an $\mathbb{R}$-subalgebra of $M_2(\mathbb{C})$. Indeed, taking $A = \left\{ \begin{pmatrix} x & y \\ -\bar{y} & \bar{x} \end{pmatrix} : x, y \in \mathbb{C} \right\} \subseteq M_2(\mathbb{C})$, where $^-$ denotes complex conjugation, one sees A is a subring of $M_2(\mathbb{C})$ because it is closed under multiplication, addition and subtraction. Hence A is an $\mathbb{R}$-subalgebra and has base 1, $i' = \begin{pmatrix} i & 0 \\ 0 & -i \end{pmatrix}$, $j' = \begin{pmatrix} 0 & i \\ i & 0 \end{pmatrix}$, and $k' = i'j' = \begin{pmatrix} 0 & -1 \\ 1 & 0 \end{pmatrix} = -j'i'$,

so A is Hamilton's quaternion algebra. This way of viewing $\mathbb{H}$ also shows quickly that $\mathbb{H}$ is a division ring. Indeed, $\begin{pmatrix} x & y \\ -\bar{y} & \bar{x} \end{pmatrix}\begin{pmatrix} \bar{x} & -y \\ \bar{y} & x \end{pmatrix} =$

$\begin{pmatrix} x\bar{x} + y\bar{y} & 0 \\ 0 & x\bar{x} + y\bar{y} \end{pmatrix} \neq 0$ so we have constructed the inverse of each non-zero element.

Example 1.1.8: We give some general classes of subrings of $M_n(R)$

   (i) The ring of scalar matrices (isomorphic to R)

   (ii) The ring of diagonal matrices (i.e., all i-j entries are 0 except for $i = j$)

   (iii) The ring of (upper) triangular matrices (i.e., all i-j entries are 0 except for $i \leq j$)

   (iv) Those matrices (r_{ij}) where $r_{ij} = 0$ unless $i = j$ or $i = 1$

   (v) Those triangular (r_{ij}) such that $r_{11} = r_{22} = \cdots = r_{nn} \in Z(R)$

   (vi) More generally, if T is a subring of the ring of diagonal matrices and $A \lhd R$, take the subring of triangular matrices consisting of each (r_{ij}) whose diagonal lies in T and whose off-diagonal entries each lie in A.

   (vii) Those matrices (r_{ij}) with $r_{ij} = 0$ unless $n_{u-1} < i, j \leq n_u$ for suitable u (where $1 \leq u \leq t$ and we are given $0 = n_0 < n_1 < \cdots < n_t = n$. These matrices decompose into blocks of lengths $n_u - n_{u-1}$.)

Matrices Whose Entries Are Not Necessarily in Rings

One should observe that the ring structure of matrices did not require all the entries to lie in a ring. In particular, if we formally consider $\begin{pmatrix} A & B \\ C & D \end{pmatrix}$, the set of 2×2 matrices of the form $\begin{pmatrix} a & b \\ c & d \end{pmatrix}$ where $a \in A$, $b \in B$, $c \in C$, $d \in D$, we see that addition of matrices require A, B, C, D, to be additive groups; multiplication merely requires various actions on the sets permitting us to define the product, with associativity and distributivity holding wherever possible. More explicitly, we have the following situation.

Example 1.1.9: Suppose R, T are rings and $M \in R\text{-}\mathcal{M}od\text{-}T$. Then the module operations make $\begin{pmatrix} R & M \\ 0 & T \end{pmatrix}$ into a ring. (Indeed, one can define multiplication since if $r_i \in R$, $x_i \in M$, and $s_i \in T$ we have

$$\begin{pmatrix} r_1 & x_1 \\ 0 & s_1 \end{pmatrix}\begin{pmatrix} r_2 & x_2 \\ 0 & s_2 \end{pmatrix} = \begin{pmatrix} r_1 r_2 & r_1 x_2 + x_1 s_2 \\ 0 & s_1 s_2 \end{pmatrix} \in \begin{pmatrix} R & M \\ 0 & T \end{pmatrix}$$

and associativity and distributivity come from the given ring and module properties (exactly as usual for matrices).

This example provides a wealth of interesting examples, several due to Small and Herstein, so we shall analyze its left ideal structure. Suppose Q is a left ideal of $\begin{pmatrix} R & M \\ 0 & T \end{pmatrix}$. If $\begin{pmatrix} x & y \\ 0 & z \end{pmatrix} \in Q$ then

$$\begin{pmatrix} rx & ry \\ 0 & 0 \end{pmatrix} = \begin{pmatrix} r & 0 \\ 0 & 0 \end{pmatrix}\begin{pmatrix} x & y \\ 0 & z \end{pmatrix} \subset Q$$

$$\begin{pmatrix} 0 & Mz \\ 0 & 0 \end{pmatrix} = \begin{pmatrix} 0 & M \\ 0 & 0 \end{pmatrix}\begin{pmatrix} x & y \\ 0 & z \end{pmatrix} \subset Q$$

$$\begin{pmatrix} 0 & 0 \\ 0 & Tz \end{pmatrix} = \begin{pmatrix} 0 & 0 \\ 0 & T \end{pmatrix}\begin{pmatrix} x & y \\ 0 & z \end{pmatrix} \subset Q.$$

Thus $Q = A + Le_{22}$, where A is an R-submodule of $\begin{pmatrix} R & M \\ 0 & 0 \end{pmatrix}$ (viewed as R-module by scalar multiplication, as usual) and L is a left ideal of T such that $MLe_{12} \subseteq A$. Conversely, any such set $A + Le_{22}$ is a left ideal of $\begin{pmatrix} R & M \\ 0 & T \end{pmatrix}$.

Analogously, the right ideals of $\begin{pmatrix} R & M \\ 0 & T \end{pmatrix}$ have the form $B + Pe_{11}$ where B is a right T-submodule of $\begin{pmatrix} 0 & M \\ 0 & T \end{pmatrix}$, P is a right ideal of R, and $PMe_{12} \subseteq B$.

Consequently, the ideals of $\begin{pmatrix} R & M \\ 0 & T \end{pmatrix}$ have the form $\begin{pmatrix} P & N \\ 0 & L \end{pmatrix}$ where $P \lhd R$, $L \lhd T$, and N is an R-T sub-bimodule of M satisfying $ML \subseteq N$ and $PM \subseteq N$.

The reason we are interested in this example is that when R and T are suitably restricted, the left-right ideal structure is reflected in the bimodule structure of M, which can easily be arranged to be very asymmetric; in this manner one can often find a phenomenon occurring in the left ideal structure of $\begin{pmatrix} R & M \\ 0 & T \end{pmatrix}$ but not in the right ideal structure. This theme will be pursued at times in the text; for the time being, the reader is invited to experiment with this example in the following two cases:

(i) $R \subseteq T$ and $M = T$; in particular when T is a field.
(ii) T is a homomorphic image of R and $M = T$ (viewed as R-module as in remark 0.0.9).

Example 1.1.10: (Preview of the Morita Ring) Suppose R, T are rings with $M \in R\text{-Mod-}T$ and $M' \in T\text{-}\mathcal{M}\!od\text{-}R$. In order to define multiplication on $\begin{pmatrix} R & M \\ M' & T \end{pmatrix}$ one needs compositions $M \times M' \to R$ and $M' \times M \to T$ which we denote here respectively as (,) and [,]; i.e., for $x \in M$ and $x' \in M'$ we have $(x, x') \in R$ and $[x', x] \in T$. We define addition componentwise and multiplication in the following natural manner:

$$\begin{pmatrix} r_1 & x_1 \\ x_1' & s_1 \end{pmatrix}\begin{pmatrix} r_2 & x_2 \\ x_2' & s_2 \end{pmatrix} = \begin{pmatrix} r_1 r_2 + (x_1, x_2') & r_1 x_2 + x_1 s_2 \\ x_1' r_2 + s_1 x_2' & [x_1', x_2] + s_1 s_2 \end{pmatrix}.$$

Distributivity of multiplication over addition requires the forms (,) and [,] to be additive in each component, e.g., $(x_1 + x_2, x') = (x_1, x') + (x_2, x')$, whereas associativity requires the following extra conditions:

$$r(x, x') = (rx, x'),\ (xs, x') = (x, sx'),\ (x, x')r = (x, x'r)$$

$$s[x', x] = [sx', x],\ [x'r, x] = [x', rx],\ [x', x]s = [x', xs].$$

The Morita ring plays a key role in the structure of $R\text{-}\mathcal{M}\!od$, in addition to providing a huge class of interesting examples (and in fact yielding example 1.1.9 when we take $M' = 0$).

Idempotents and the Peirce Decomposition

An element e of R is *idempotent* if $e^2 = e$. Idempotents play an important role in ring theory and are closely linked to matrices since the matric units e_{ii} are idempotent. In the next few pages we shall study some general properties of idempotents and their impact on matrices and matric units. First note 0 and 1 are always idempotents and so are called *trivial* idempotents. Idempotents e and e' are *orthogonal* if $ee' = e'e = 0$. If e is idempotent then $1 - e$ is an idempotent orthogonal to e.

Lemma 1.1.11: *Suppose $e \in R$ is idempotent.*

(i) *If L is a left ideal of R then $L \cap eR = eL$.*
(ii) *If $A \lhd R$ then $A \cap eRe = eAe$.*

Proof:

(i) If $er \in L \cap eR$ then $er = e(er) \in eL$, proving ($\subseteq$); ($\supseteq$) is obvious.

(ii) As in (i).   Q.E.D.

Proposition 1.1.12: *Suppose e is an idempotent of a ring R. Then eRe is a ring with multiplicative unit e. Moreover, there is an onto lattice map* ψ: *{ideals of R}* $\rightarrow$ *{ideals of eRe} given by* $A \rightarrow eAe$.

Proof: Clearly eRe is a ring. To see ψ is a lattice map note $\psi(A + B) = eAe + eBe = \psi A + \psi B$ and $\psi(A \cap B) = e(A \cap B)e = A \cap B \cap eRe = (A \cap eRe) \cap (B \cap eRe) = \psi A \cap \psi B$.   If $I \lhd eRe$ then $I = (eRe)I(eRe) = eRIRe = \psi(RIR)$, proving ψ is onto.   Q.E.D.

Thus we see that eRe and likewise $(1 - e)R(1 - e)$ are rings whose ideal structures are at least as nice as that of R, and we are led to study R in terms of these rings. One useful link is the following straightforward result.

Proposition 1.1.13:   ("*Peirce decomposition*") *If e is an idempotent of R then* $R = eRe \oplus (1 - e)Re \oplus eR(1 - e) \oplus (1 - e)R(1 - e)$ *as Abelian groups.*

Proof: For r in R take $r_1 = ere$, $r_2 = (1 - e)re$, $r_3 = er(1 - e)$, and $r_4 = (1 - e)r(1 - e)$; then $(r_1 + r_2) + (r_3 + r_4) = re + r(1 - e) = r$. If $r = r'_1 + r'_2 + r'_3 + r'_4$ is another such decomposition then $er'_2e \in e((1 - e)Re)e = 0$ and likewise $er'_3e = er'_4e = 0$ so $r'_1 = er'_1e = e(r'_1 + r'_2 + r'_3 + r'_4)e = ere = r_1$; likewise $r'_2 = (1 - e)r'_2e = (1 - e)re = r_2$ and so forth.   Q.E.D.

Example 1.1.14:

(i) If $R = M_n(T)$ and $e = e_{11}$ then $eRe \approx T$ and $(1 - e)R(1 - e) \approx M_{n-1}(T)$, and the Peirce decomposition is the usual partitioning of the matrix.

(ii) The Morita ring of example 1.1.10 has an idempotent $e = \begin{pmatrix} 1 & 0 \\ 0 & 0 \end{pmatrix}$, and the Peirce decomposition partitions $\begin{pmatrix} R & M \\ M' & T \end{pmatrix}$ into its components (although of course this ring does *not* have a matric unit e_{12}).

Thus the Peirce decomposition enables us to analyze rings "resembling" matrix rings.

(iii) If $e \in Z(R)$ then $Re \approx eRe$ is a ring. In this situation we can refine the Peirce decomposition.

Proposition 1.1.14: $R \approx R_1 \times \cdots \times R_t$ *as rings iff there are pairwise ortho-*
gonal idempotents e_i *in* $Z(R)$ *such that* $\sum_{i=1}^{t} e_i = 1$ *and* $R_i \approx Re_i$ *for each i.*

Proof: ($\Rightarrow$) Identifying R with $R_1 \times \cdots \times R_t$ take $e_i = (0, \ldots, 0, 1, 0, \ldots, 0)$
where "1" appears in the i position. Clearly $e_i \in Z(R)$.
   ($\Leftarrow$) Let $R_i = Re_i$, and define $\varphi: R \to \prod_{i=1}^{t} R_i$ by $r \mapsto (re_1, \ldots, re_t)$. φ is a
surjection since $(r_1 e_1, \ldots, r_t e_t) = \varphi(\sum r_i e_i); \psi$ is an injection since if $\psi r = 0$ then
each $re_i = 0$ so $r = r\Sigma e_i = 0$.   Q.E.D.

Idempotents and Simple Modules

Proposition 1.1.15: *Suppose e is an idempotent of R. There is a functor*
$F: R\text{-}\mathcal{M}od \to eRe\text{-}\mathcal{M}od$ *given by* $FM = eM$ *where for any map* $f: M \to N$,
we take Ff to be the restriction of f to eM. Moreover, if M is simple in
$R\text{-}\mathcal{M}od$ *with* $eM \neq 0$ *then eM is simple in* $eRe\text{-}\mathcal{M}od$ *(so F "preserves" simple*
modules not annihilated by e).

Proof: The first assertion is immediate since $(eRe)eM \subseteq e(ReM) \leq eM$.
To prove the second assertion note for any x in M with $ex \neq 0$ that $(eRe)ex =$
$e(Rex) = eM$.   Q.E.D.

   There is a converse for idempotents of matrices.

Proposition 1.1.16: *If* $M \in R\text{-}\mathcal{M}od$ *we can view* $M^{(n)}$ *in* $M_n(R)\text{-}\mathcal{M}od$ *by*
defining addition componentwise (i.e.,

$$(x_1, \ldots, x_n) + (x'_1, \ldots, x'_n) = (x_1 + x'_1, \ldots, x_n + x'_n))$$

and multiplication by

$$\left(\sum r_{ij} e_{ij} \right)(x_1, \ldots, x_n) = \left(\sum_{j=1}^{n} r_{1j} x_j, \sum_{j=1}^{n} r_{2j} x_j, \ldots, \sum_{j=1}^{n} r_{nj} x_j \right).$$

Thus we get a functor $G: R\text{-}\mathcal{M}od \to M_n(R)\text{-}\mathcal{M}od$ *by* $GM = M^{(n)}$, *where for any*
map $f: M \to N$ *we define* $Gf: M^{(n)} \to N^{(n)}$ *componentwise.*

Proof: First note that the scalar multiplication we have defined is merely
matrix multiplication if we view the elements of $M^{(n)}$ as $n \times 1$ matrices, so
the straightforward verification that $M^{(n)} \in M_n(R)\text{-}\mathcal{M}od$ is a special case of
a well-known matrix argument. (This idea is generalized in exercise 3.) To
see that G is a functor we note that Gf is a map because putting $h = Gf$

we have

$$h\left(\left(\sum r_{ij}e_{ij}\right)(x_1,\ldots,x_n)\right) = h\left(\sum r_{1j}x_j,\ldots,\sum r_{nj}x_j\right) = \left(\sum r_{1j}fx_j,\ldots,\sum r_{nj}fx_j\right)$$

$$= \left(\sum r_{ij}e_{ij}\right)(fx_1,\ldots,fx_n) = \left(\sum r_{ij}e_{ij}\right)h(x_1,\ldots,x_n).$$

Q.E.D.

Theorem 1.1.17: *R-Mod and $M_n(R)$-Mod are equivalent categories.*

Proof: Identifying R with $e_{11}M_n(R)e_{11}$ let $\mathscr{C} = R$-Mod and $\mathscr{D} = M_n(R)$-Mod. By proposition 1.1.15 and 1.1.16 we have functors $F: \mathscr{D} \to \mathscr{C}$ (taking $e = e_{11}$) and $G: \mathscr{C} \to \mathscr{D}$, so we must prove FG and GF are naturally isomorphic, respectively, to $1_{\mathscr{C}}$ and $1_{\mathscr{D}}$.

If $M \in R$-Mod then $FGM = \{(x,0,\ldots,0): x \in M\}$ so there is a natural transformation $\eta: 1 \to FG$ given by $\eta_M x = (x,0,\ldots,0)$; each η_M is clearly an isomorphism. Conversely, if $M \in M_n(R)$-Mod then $GFM = \{((x_1,0,\ldots,0),\ldots,(x_n,0,\ldots,0)): (x_1,\ldots,x_n) \in M\}$ so there is a natural isomorphism $\eta': 1 \to GF$ given by

$$\eta'_M(x_1,\ldots,x_n) = ((x_1,0,\ldots,0),\ldots,(x_n,0,\ldots,0)). \quad \text{Q.E.D.}$$

This is our first example of equivalent categories which are not isomorphic and will be fundamental to the study of equivalent categories (and the Morita theory). Once we know two categories are equivalent we can use the equivalence to transfer various categorical properties from one category to another, as illustrated in the next result. (Many more examples will be given in the treatment of Morita theory.)

Corollary 1.1.18: *M is simple in R-Mod iff $M^{(n)}$ is simple in $M_n(R)$-Mod.*

Proof: M is simple iff each nonzero map $f: M \to N$ is monic. Since FG and GF are naturally isomorphic to the respective identities (notation as before), one sees easily that G preserves this categorical property, so M is simple in R-Mod iff $M^{(n)} = GM$ is simple in $M_n(R)$-Mod.   Q.E.D.

A more concrete proof of this result is given in exercise 4. The passage from M to $M^{(n)}$ can be applied to reduce proofs about f.g. modules to the cyclic case, along the following lines:

Remark 1.1.19: If M is spanned over R by the elements $x_1,\ldots,x_n$ then $M^{(n)}$ is cyclic in $M_n(R)$-Mod, spanned by the element $\bar{x} = (x_1,\ldots,x_n)$. (Indeed $e_{ij}\bar{x}$ has x_j in the i-th position.)

Primitive Idempotents

An idempotent e is *primitive* if e cannot be written as the sum of two non-trivial orthogonal idempotents.

Remark 1.1.20: If e_1, e_2 are orthogonal idempotents then $Re_1 + Re_2 = R(e_1 + e_2)$ since $e_1 = e_1(e_1 + e_2)$ and $e_2 = e_2(e_1 + e_2)$.

In particular, we see e_{11} is a primitive idempotent of $M_n(F)$ for any field F (because e_{11} has rank 1). To understand this notion better, we look more closely at orthogonal idempotents.

If e_1, e_2 are idempotents with $e_2 e_1 = 0$ then $(1 - e_1)e_2$ is an idempotent orthogonal to e_1, by an easy computation.

Proposition 1.1.21: *An idempotent e of R is primitive iff there is no idempotent $e_1 \neq 0$ such that $Re_1 \subset Re$. In fact, if $Re_1 \subset Re$ then there is an idempotent e_1' such that $Re_1 = Re_1'$ and $e_1', e - e_1'$ are orthogonal.*

Proof: We prove the second assertion, since the first is then an immediate consequence. Take $e_1' = ee_1 \in Re_1 \subset Re$. Then $e_1'e = e_1'$, so $(e_1')^2 = e_1'(ee_1) = (e_1'e)e_1 = e_1'e_1 = ee_1^2 = ee_1 = e_1'$ is idempotent. Also $e_1 = e_1^2 = (e_1 e)e_1 = e_1 e_1' \in Re_1'$ so $Re_1 = Re_1'$. Finally, $e_1' = ee_1'e \in eRe$ implying $e - e_1'$ is an idempotent orthogonal to e_1'.   Q.E.D.

Lifting Matrix Units and Idempotents

Remark 1.1.22: If $\varphi: M_n(R) \to T$ is a ring homomorphism then $T \approx M_n(R')$ for a suitable ring R'. (Indeed $\{\varphi e_{ij}: 1 \leq i, j \leq n\}$ is a set of $n \times n$ matric units of T, so we are done by proposition 1.1.3.)

Next we consider the converse: If $R = T/A$ is an $n \times n$ matrix ring (i.e., has a set of $n \times n$ matric units) then is T an $n \times n$ matrix ring? This question arises often in structure theory and can be handled in three stages. First note for any given set of matric units $\{e_{ij}: 1 \leq i, j \leq n\}$ that the e_{ii} are orthogonal idempotents. Thus we are led to ask (1) if every idempotent e of R has a pre-image in T which also is idempotent, i.e., if e can be *lifted* to an idempotent of T; (2) if a finite set of (pairwise) orthogonal idempotents can be lifted to orthogonal idempotents; and (3) if we finally can fill in a set of matric units. Surprisingly, the first question is hardest to handle, so we start by assuming it is solved.

Definition 1.1.23: $A \lhd T$ is *idempotent-lifting* if the following two conditions are satisfied:

(i) $1 - a$ is invertible for all a in A;
(ii) every idempotent of T/A has the form $x + A$ for suitable x idempotent in T.

Remark 1.1.24:   If A satisfies (i) above then 0 is the only idempotent of A (for if $a^2 = a \in A$ then $a = a(1 - a)(1 - a)^{-1} = 0$).

Proposition 1.1.25:   *Suppose $A \lhd T$ is idempotent-lifting and $R = T/A$.*

(i) *If $x_1 \in T$ is idempotent and $e \in R$ is an idempotent orthogonal to $x_1 + A$ then e can be lifted to an idempotent of T orthogonal to x_1.*
(ii) *Every countable set of (pairwise) orthogonal idempotents of R can be lifted to a set of orthogonal idempotents of T.*
(iii) *Every set of $n \times n$ matric units of R can be lifted to a set of $n \times n$ matric units of T.*

Proof:

(i) Take $x \in T$ idempotent with $e = x + A$. Then $xx_1 + A = e(x_1 + A) = 0$ so $xx_1 \in A$, and we can form $x' = (1 - xx_1)^{-1} x(1 - xx_1)$ which is also idempotent and lies over e. Moreover, since $x(1 - xx_1) = x - xx_1 = x(1 - x_1)$, we see $x'x_1 = 0$ so $x_2 = (1 - x_1)x'$ is idempotent and lies over e, and $x_2 x_1 = 0 = x_1 x_2$.

(ii) Inductively, suppose we have lifted $e_1, \ldots, e_{u-1}$ to orthogonal idempotents $x_1, \ldots, x_{u-1}$ and want to lift e_u to an idempotent x_u orthogonal to $x_1, \ldots, x_{u-1}$. Let $x'_1 = \sum_{i=1}^{u-1} x_i$ and $e = e_u$; by (i) we can lift e_u to an idempotent x_u orthogonal to x'_1. But for all $i \le u - 1$ we have $x'_1 x_i = x_i x'_1 = x_i$, so $x_u x_i = x_u(x'_1 x_i) = 0$ and $x_i x_u = (x_i x'_1)x_u = 0$.

(iii) Having lifted $e_{11}, \ldots, e_{nn}$ to orthogonal idempotents $x_{11}, \ldots, x_{nn}$, we must also lift the e_{ij} to x_{ij} for $i \ne j$. Take b_{ij} in T with $b_{ij} + A = e_{ij}$. Define $x_{i1} = x_{ii}b_{i1}x_{11}$ and $a_i = x_{11} - b_{1i}x_{i1}$. Then $a_i + A = e_{11} - e_{1i}e_{i1} = 0$ so $a_i \in A$; let $d_i = (1 - a_i)^{-1}$ and put $x_{1i} = d_i b_{1i}x_{ii}$. Then x_{1i} lifts e_{1i} and $x_{1i}x_{i1} = d_i b_{1i}x_{ii}x_{i1} = d_i b_{1i}x_{i1}x_{11} = d_i(x_{11} - a_i)x_{11} = d_i(1 - a_i)x_{11} = x_{11}$. Hence $(x_{ii} - x_{i1}x_{1i})^2 = x_{ii}^2 - x_{ii}x_{i1}x_{1i} - x_{i1}x_{1i}x_{ii} + (x_{i1}x_{1i})^2 = x_{ii} - x_{i1}x_{1i}$ is an idempotent in A so is 0, implying $x_{i1}x_{1i} = x_{ii}$. Then put $x_{ij} = x_{i1}x_{1j}$, a set of matric units lifting e_{ij}.   Q.E.D.

The main tool for lifting idempotents involves an algebraicity condition.

Proposition 1.1.26: *Suppose* $r \in R$ *and there exists* $a = \sum_{i=0}^{t} z_i r^i$ *for* z_i *in* $Z(R)$ *such that* $r^n = ar^{n+1}$. *Then* $e = (ar)^n$ *is idempotent and, moreover, in any homomorphic image* $\bar{R}$ *in which* $\bar{r}$ *is idempotent we have* $\bar{e} = \bar{r}$.

Proof: By induction we have $r^n = a^m r^{m+n}$ for all $m \geq 1$. Hence

$$e^2 = (ar)^{2n} = a^n(a^n r^{n+n}) = a^n r^n = e.$$

If $\bar{r}$ is idempotent then $\bar{e} = \bar{a}^n \bar{r}^{2n} = \bar{r}^n = \bar{r}$.   Q.E.D.

To continue this idea to its conclusion we call an element r of R *nilpotent* if some power $r^t = 0$; $S \subset R$ is *nil* if every element is nilpotent (not necessarily with respect to the same t).

Remark 1.1.27: If $a^t = 0$ then $(1 - a)(1 + a + \cdots + a^{t-1}) = 1$ so $1 - a$ is invertible.

Corollary 1.1.28: *Every nil ideal* A *of a ring* R *is idempotent-lifting. In fact, if* $\bar{r} = r + A$ *is idempotent in* R/A *then there is an idempotent* e *in the subring of* R *generated by* r, *satisfying* $\bar{e} = \bar{r}$.

Proof: Condition (i) follows from remark 1.1.27; to see (ii) we note that if $x + A$ is idempotent then $x^2 - x \in A$ so $0 = (x - x^2)^n = x^n - nx^{n+1} + \cdots$ for suitable n; hence proposition 1.1.26 produces an idempotent lifting $x + A$.

Surprisingly, one may not be able to lift an uncountable set of orthogonal idempotents modulo an idempotent-lifting ideal.

Example 1.1.29: (Zelinsky [54]) A ring R with $N \lhd R$ such that $N^2 = 0$ (so that N is idempotent-lifting), where R/N has an uncountable set of idempotents which *cannot* be lifted to an uncountable set of idempotents of R. Let F be any field and let R be a vector space over F with basis $\{b_{ij}, e_i : i, j \in I\}$ given multiplication by the rules $e_i^2 = e_i$, $e_i e_j = b_{ij}$ for $i \neq j$, $e_i b_{jk} = \delta_{ij} b_{ik}$, $b_{ij} e_k = \delta_{jk} b_{ik}$, and $b_{ij} b_{uv} = 0$. One checks readily $R \in F\text{-}\mathscr{A}lg$ when multiplication is extended by distributivity to all of R. (To verify this the reader could use the criteria given towards the end of the next section.) Moreover, the subspace spanned by the b_{ij} is clearly an ideal N with $N^2 = 0$, and $\bar{R} = R/N$ has a set $\bar{S} = \{\bar{e}_i : i \in I\}$ of orthogonal idempotents. We shall show $\bar{S}$ lifts to an orthogonal set of idempotents $\{d_i : i \in I\}$ only if I is countable. Indeed, suppose we have the d_i, and put $x_i = d_i - e_i \in N$ since $\bar{x}_i = 0$. Let $I(i) = \{j \in I : e_i x_j = 0\}$. Since any x_i is a linear combination of a finite number of b_{uv}

we see by the multiplication table that for any j the set $\{i \in I : e_i x_j \neq 0\}$ is finite; hence for any infinite subset K of I we must have $\bigcup_{i \in K} I(i) = I$. Also define $I'(i) = \{ j \in I : x_i e_j \neq 0 \}$ which likewise is finite. But for $i \neq j$

$$0 = d_i d_j = (e_i + x_i)(e_j + x_j) = b_{ij} + e_i x_j + x_i e_j.$$

If $e_i x_j = 0$ then $x_i e_j = -b_{ij} \neq 0$ so each $I(i) \subseteq I'(i)$ is finite. Therefore $I = \bigcup_{i \in K} I(i)$ is countable (seen by taking K infinite and countable), as desired.

Eckstein [69] has a different method of obtaining the above results (and more) by means of semigroups.

§1.2 Polynomial Rings

This section features a very important construction, the monoid ring, which leads to huge classes of important examples. Of particular significance is the polynomial ring, a special case. Later on we shall see how to view the monoid ring in a more general setting. We start with a module construction over a ring R.

Definition 1.2.1:   Given a set S we form the *R-module freely generated by S*, denoted as RS, to be the set of formal expressions $\{ \sum_{s \in S} r_s s : r_s \in R$, almost all $r_s = 0\}$, endowed with addition and scalar multiplication defined "componentwise," i.e.,

$$\sum r_s s + \sum r_s' s = \sum (r_s + r_s')s \qquad r \sum r_s s = \sum (r r_s) s,$$

where $r \in R$. $\{ s \in S : r_s \neq 0 \}$ is called the *support* of $\sum r_s s$. For notational convenience, we identify each element s' in S with the expression $\sum_{s \in S} r_s s$ where $r_{s'} = 1$ and all other r_s are 0. Thus we view S as a subset of RS.

Remark 1.2.2: RS is indeed a module. (Verifications are componentwise.)

Later on we shall identify RS with the "free" module, but let us note that any element $f = \sum r_s s$ of RS can now be viewed as the finite sum of those $r_s s$ for which $r_s \neq 0$; these $r_s s$ are called *the terms of f*, and r_s is called the *coefficient of s* (in R). Thus any element of RS is written *uniquely* as a linear combination of elements of S.

Monoid Rings

We are interested in defining a ring structure on RS. To this end we first define multiplication on terms and extend it to arbitrary elements by distributivity. This is done easily when S is a monoid. Then we define the product of terms

$$(r_s s)(r'_t t) = (r_s r'_t) st$$

and more generally

$$\left(\sum_{s \in S} r_s s \right) \left(\sum_{t \in S} r'_t t \right) = \sum_{u \in S} \left(\sum_{st=u} r_s r'_t \right) u,$$

where the inner sum is evaluated over all s, t in the respective supports such that $st = u$. We denote this ring structure as $R[S]$, called the *monoid ring*. When S is a group $R[S]$ is called a *group ring*.

Proposition 1.2.3: *The monoid ring $R[S]$ is indeed a ring, whose unit element is 1_S, the unit element of S. Furthermore if $R \in C\text{-}\mathcal{A}lg$ then $R[S]$ is an algebra with scalar multiplication $c \sum r_s s = \sum (cr_s) s$. There is an injection $R \to R[S]$ given by $r \mapsto r1_S$. Hence S centralizes R in $R[S]$.*

We leave the straightforward but messy proof to the reader, noting that it is essentially the same proof the reader has had to plough through to show that the polynomials in one indeterminate over a field form a ring.

There is one monoid of special interest. Assume throughout I is a well-ordered set.

Example 1.2.4: The *word monoid* on a well-ordered set I. A *word* is a formal string $w = (i_1 \cdots i_m)$; m is called the *length*, and $\max\{i_u : 1 \le u \le m\}$ is called the *height*. Multiplication of words is formed by juxtaposition, i.e., $(i_1 \cdots i_m)(i'_1 \cdots i'_n) = (i_1 \cdots i_m i'_1 \cdots i'_n)$, and is obviously associative. We get the word monoid by formally adjoining a unit element, written as ( ) and called the *blank word*. One may think of the blank word as having nothing in it and thus being of length 0, thereby yielding

Remark 1.2.5: For any two words w_1, w_2 we have length $(w_1 w_2) =$ length w_1 + length w_2 and height $(w_1 w_2) = \max(\text{height } w_1, \text{height } w_2)$.

In case i repeats in a word, we write i^t for $i \cdots i$, taken t times.

Polynomial Rings

Example 1.2.6: Suppose I is a singleton $\{\lambda\}$. Then every word has the form λ^i for some $i \ge 0$ (writing λ^0 for the blank word), and putting $S = \{\lambda^i : i \in \mathbb{N}\}$ we write $R[S]$ as $R[\lambda] = \{\sum_{i=0}^u r_i \lambda^i : u \in \mathbb{N}, r_i \in R\}$, the *polynomial ring over R* (in one commuting indeterminate), denoted $R[\lambda]$. Note that $\lambda \in Z(R[\lambda])$.

When trying to generalize example 1.2.6 to form polynomials in several indeterminates, one has to decide whether or not one wants the indeterminates to commute with each other. We consider the cases separately.

Example 1.2.7:   Define $X = \{X_i : i \in I\}$, where the X_i are formal symbols, called (noncommuting) *indeterminates*. Clearly X and I are in 1:1 correspondence, so we can well-order X by means of the well-ordering on I and form a word monoid S on X. The words (in the X_i) are called *pure monomials*; if C is a commutative ring then $C[S]$ is written instead as $C\{X\} = \{\sum_{\text{finite}} c_{(i)} X_{i_1} \cdots X_{i_{m(i)}}$, where (i) denotes the word $i_1 \cdots i_{m(i)}$ of length $m(i)$ (allowing repetitions), and $c_{(i)} \in C$ for each $(i)\}$. The elements of $C\{X\}$ are called *noncommutative polynomials* and $C\{X\}$ is called the *free C-algebra*. In case I is finite, i.e., $I = \{i_1, \ldots, i_t\}$, we shall denote $C\{X\}$ as $C\{X_1, \ldots, X_t\}$.

In the current literature $C\{X\}$ is usually denoted as $C\langle X\rangle$, cf. example 1.9.20 below.

To obtain commutative polynomials we could define inductively $R[\lambda_1, \ldots, \lambda_t] = (R[\lambda_1, \ldots, \lambda_{t-1}])[\lambda_t]$, iterating example 1.2.6. However, from our point of view we should like to display this ring as a monoid ring preferably using words. To do this we examine words a little closer.

Definition 1.2.8:   A partial order $\leq$ on words is defined inductively on the length, as follows: For arbitrary words $w = (i_1 \cdots i_m)$ and $w' = (i'_1 \cdots i'_n)$ which are *not* blank, $w < w'$ if either $i_1 < i'_1$ (in I) or $i_1 = i'_1$ and $(i_2 \cdots i_m) < (i'_2 \cdots i'_n)$ (defined inductively).

For example, if $I = \{A, B, \ldots, Z\}$ with the alphabetic ordering then $ACT < AT$, but ACT and ACTION are not comparable. The reason we do not use the customary lexicographic (total) order is to make the following remark true:

Remark 1.2.9:   If $w < w'$ and w_1, w_2 are arbitrary words then $w_1 w < w_1 w'$ and $w w_2 < w' w_2$. Nevertheless, $\leq$ induces a total order on the set of words of a given length.

A word w is *basic* if for each w' obtained by interchanging two letters, we have $w \leq w'$. For example, BEEN is basic since BEEN $\leq$ EBEN, BEEN, EEBN, NEEB, BNEE, BENE.

Remark 1.2.10: $(i_1 \cdots i_m)$ is basic iff $i_1 \leq i_2 \leq \cdots \leq i_m$; hence for every word w there is a basic word which we call Bw, obtained by rearranging its letters in increasing order. If w' is obtained by rearranging the letters of w then $Bw' = Bw$.

Definition 1.2.11: The *symmetric word monoid* on I is $\{Bw: w$ is a word on $I\}$ (including the blank word) under multiplication defined by $w_1 \cdot w_2 = B(w_1 w_2)$.

The symmetric word monoid on I is a commutative monoid, since associativity and commutativity follow from remark 1.2.10: $B(B(w_1 w_2)w_3) = B(w_1 w_2 w_3) = B(w_1 B(w_2 w_3))$ and $B(w_1 w_2) = B(w_2 w_1)$.

Definition 1.2.12: Put $\lambda = \{\lambda_i : i \in I\}$ and let S be the symmetric word monoid on λ. Then $R[S]$ is denoted as $R[\lambda]$ and is called the *polynomial ring over R on the commuting indeterminates* λ_i, $i \in I$. The words of S are called *pure monomials*. If $\lambda = \{\lambda_1, \ldots, \lambda_t\}$ we write $R[\lambda_1, \ldots, \lambda_t]$ instead of $R[\lambda]$.

To obtain further results we examine special kinds of monoid rings, introducing the notions of ordered monoid and filtration.

Definition 1.2.13: A monoid S is *ordered* if S has a total ordering $\leq$ such that whenever $s < s'$ we have $s_1 s < s_1 s'$ and $s s_1 < s' s_1$ for all s_1 in S. A C-algebra A has a *filtration* by an ordered monoid S if A is the union of C-submodules $\{A(s): s \in S\}$ such that

(i) $A(s) \supset A(s')$ for all $s < s'$ in S,

(ii) $A(s_1)A(s_2) \subseteq A(s_1 s_2)$ for all s_1, s_2 in S,

(iii) $\bigcap_{s \in S} A(s) = 0$.

Note that if $s_1 < s_1'$ and $s_2 < s_2'$ then $s_1 s_2 < s_1' s_2 < s_1' s_2'$.

Remark 1.2.13': The most important case in the literature is for $S = (\mathbb{Z}, +)$, i.e., $\cdots A(-1) \supseteq A(0) \supseteq A(1) \supseteq \cdots$. Often $A(0) = A(-1) = A(-2) = \cdots = A$, so we actually have a filtration over $\mathbb{N}$. However, there are situations in which $A(1) = A(2) = \cdots = 0$, so we have instead a filtration by the negative integers. In the latter case we could write A_n for $A(-n)$, and so (i) becomes $A_m \subseteq A_n$ for $m \leq n$. At times this is used as the definition for filtration in the literature.

Definition 1.2.14: The filtration of A by S is *valuated* if for each $a \neq 0$ the set $\{s \in S : a \in A(s)\}$ has a maximal element, which we denote va, thereby defining a function $v: A - \{0\} \to S$.

In order to include 0 in the domain of v, we adjoin a formal element ∞ to S with $\infty\infty = \infty$ and $s\infty = \infty s = \infty > s$ for all s in S, and write $S_\infty = S \cup \{\infty\}$. Putting $v0 = \infty$ we now have a function $f: A \to S_\infty$, where $va = \infty$ iff $a = 0$. One also calls v a *pseudo-valuation*. Its worth is seen in the following observation.

Remark 1.2.15: For any a_i in A, $v(a_1 a_2) \geq (va_1)(va_2)$. If v is a monoid homomorphism then for all nonzero a_1, a_2 in A, $v(a_1 a_2) = (va_1)(va_2) \in S$ implying $a_1 a_2 \neq 0$; thus A is a domain.

We shall use this remark to obtain many examples of domains and start by applying it to monoid rings.

Example 1.2.16: If $R \in C\text{-}\mathcal{A}lg$ and S is an ordered monoid then $A = R[S]$ has the filtration given by $A(s) = \sum(Rs': s' \geq s)$, which is valuated since for $a = \sum r_s s$ we see va is the minimal s in the support of a; the corresponding $r_s s$ will also be called the *lowest order term* of a.

Proposition 1.2.17: *If R is a domain and S is an ordered monoid then $R[S]$ is a domain.*

Proof: Let $r_i s_i$ denote the lowest order term of a_i in $A = R[s]$ for $i = 1, 2$. Then $(r_1 s_1)(r_2 s_2) = (r_1 r_2)(s_1 s_2) \neq 0$ and is the lowest order term of $a_1 a_2$ since S is ordered. Thus v of remark 1.2.15 is a monoid homomorphism implying $R[S]$ is a domain.   Q.E.D.

As we see, the idea of using the lowest order term to examine monoid rings (over ordered monoids) is very useful, and we shall also have occasion to use the *highest order term*.

Example 1.2.18: (Making the word monoid an ordered monoid). Define the following order on words: $w < w'$ if length $(w) <$ length (w'), or if length w $=$ length w' and w is less than w' according to definition 1.2.8. In other words, we sort words *first* according to length and *afterwards* we order words of the same length. In view of remark 1.2.9 one sees easily that this ordering is total and satisfies the requirement of definition 1.2.13.

Moreover, ordering the basic words in this manner, we also see the commutative word monoid also becomes an ordered monoid. Thus the non-commutative polynomial ring and commutative polynomial ring are monoid rings over ordered monoids, proving

Proposition 1.2.19: *If R is a domain then $R\{X\}$ and $R[\lambda]$ are domains* (*Instant application of proposition 1.2.17.*)

Some results relating R and $R[S]$ do not require any additional assumptions on S. For any $A \subseteq R$ we write $A[S]$ in place of $AS = \{\sum_{\text{finite}} a_i s_i : a_i \in A, s_i \in S\}$.

Proposition 1.2.20:

(i) *If $I \lhd R$ then $I[S] \lhd R[S]$ and $R[S]/I[S] \approx (R/I)[S]$;*

(ii) *$Z(R[S]) \subseteq Z(R)[S]$, equality holding iff S is commutative.*

Proof:

(i) Define $\varphi: R[S] \to (R/I)[S]$ by $\varphi(\sum r_i s_i) = \sum \bar{r}_i s_i$ where $\bar{r}_i = r_i + I$. By inspection φ is a ring surjection and $I[S] = \ker \varphi \lhd R[S]$.

(ii) Write $Z = Z(R)$. If $\sum r_i s_i \in Z(R[S])$ then for all r in R, $\sum (rr_i)s_i = (r1)\sum r_i s_i = (\sum r_i s_i)r1 = \sum (r_i r)s_i$ implying $rr_i = r_i r$ for each i, so $r_i \in Z$ and $\sum r_i s_i \in Z[S]$. Conversely, for S commutative clearly $S \subseteq Z(R[S])$ so $Z[S] \subseteq Z(R[S])$.   Q.E.D.

Rings of Formal Power Series and of Laurent Series

Often we are interested in embedding the monoid ring $R[S]$ into a division ring. To this end we want to start with a larger module than RS. The obvious candidate is R^S, the Cartesian product of copies of R indexed by S. Denoting a typical element of R^S as an "S-tuple" (r_s), we can define R-module operations "componentwise," i.e.,

$$(r_s) + (r'_s) = (r_s + r'_s) \qquad \text{and} \qquad r(r_s) = (rr_s).$$

(Note $RS \leq R^S$, equality holding iff S is finite.) If $f = (r_s)$ we define $\text{supp}(f) = \{s \in S : r_s \neq 0\}$. We should like to define a ring multiplication as before, i.e., $(r_s)(r'_s) = (r''_u)$ where $r''_u = \sum_{st = u} r_s r'_t$. Unfortunately, this sum need no longer be finite, and so is meaningless in R. However, the sum is defined in one important special case. Suppose S has a given order and define $R((S)) = \{f \in R^S : \text{supp}(f)$ is well-ordered (under the order of S)$\}$. Clearly $R((S)) \leq R^S$ since it is closed under the module operations. Reverting to the previous notation we have

Proposition 1.2.22: $R((S))$ *is a ring, where multiplication is defined by* $(\sum r_v v)(\sum r'_w w) = \sum_{s \in S}(\sum_{vw=s} r_v r'_w)s$, *and becomes a* C-algebra if $R \in C\text{-}\mathcal{Alg}$ *(via scalar multiplication* $c \sum r_s s = \sum (cr_s)s$); *moreover,* $A = R((S))$ *has a filtration by* S *given by* $A(s) = \{f \in A : s \leq \text{each element of } \text{supp}(f)\}$. (*In particular,* $R[S]$ *is a subring of* $R((S))$ *and inherits its filtration.*)

Proof: We claim first that if S_1 and S_2 are two well-ordered subsets of S then for any s in S there are only a finite number of pairs (s_1, s_2) in $S_1 \times S_2$ with $s_1 s_2 = s$. Indeed, let $S'_1 = \{s_1 \in S_1 : s_1 s_2 = s$ for some s_2 in $S_2\}$, take $s_{10} \in S'_1$ minimal; then $s_{10}s_{20} = s$ for some s_{20} in S_2. Inductively, given $s_{10}, \ldots, s_{1,m-1}$ take s_{1m} minimal in $S'_1 - \{s_{10}, \ldots, s_{1,m-1}\}$ and take s_{2m} such that $s_{1m}s_{2m} = s$. Then $s_{10} < s_{11} < s_{12} < \cdots$, implying $s_{20} > s_{21} > s_{22} > \cdots$; since S_2 is well-ordered this process must terminate after a finite number of steps, proving S'_1 is finite, yielding the claim.

Thus we see that the proposed definition of the product does yield an element of R^S, whose support we claim is well-ordered. Indeed, otherwise there is an infinite chain $s_{10}s_{20} > s_{11}s_{21} > s_{12}s_{22} > \cdots$ with $s_{1i} \in S_1$, $s_{2i} \in S_2$. Take $u_{-1} = 0$ and, inductively, given u_{i-1} choose u_i such that s_{1, u_i} is minimal among $\{s_{1u} : u > u_{i-1}\}$. Then $s_{1u_0} \leq s_{1u_1} \leq s_{1u_2} \leq \cdots$ implying $s_{2u_0} > s_{2u_1} > s_{2u_2} > \cdots$ is an infinite descending chain, contrary to S_2 being well-ordered.

We now see the product yields a member of $R((S))$. To check associativity and distributivity we note that any term in the product involves only a finite number of terms in the multiplicands and hence multiplication works just as in $R[S]$. Associativity and distributivity follow from the corresponding properties in $R[S]$. The rest of the proposition is straightforward.   Q.E.D.

Corollary 1.2.23: $R((S))$ *is a domain for any domain* R *and ordered monoid* S.

Proof: Exactly as in the proof of proposition 1.2.17.   Q.E.D.

The ring $R((S))$ is often useful because it has enough room to invert elements of $R[S]$. For example, the only invertible polynomials in $F[\lambda]$ (for a field F) are the nonzero constants; as we shall see presently, our new construction provides a method for inverting elements of $F[\lambda]$. We shall discuss one case separately, since it is considerably easier but encompasses many examples. Let us say an ordered monoid S is *strongly archimedian* if given v in S we can find k in $\mathbb{N}$ with $s^k > v$ for all $s > 1$ in S.

Proposition 1.2.24:   *Suppose* R *is a domain,* S *is an ordered monoid, and* $f \in R((S))$ *has lowest-order term* rs. *Then* f *is invertible in* $R((S))$ *iff* $r^{-1} \in R$ *and* $s^{-1} \in S$.

Proof: Suppose f^{-1} exists in $R((S))$ and let $r's'$ denote its lowest order term. Then the lowest order term of ff^{-1} is $rr'ss'$. But $ff^{-1} = 1$ so $rr' = 1$ and $ss' = 1$. Likewise $r'r = 1$ and $s's = 1$ proving $r' = r^{-1}$ and $s' = s^{-1}$.

Conversely suppose $r^{-1} \in R$ and $s^{-1} \in S$ and let $g = 1 - r^{-1}s^{-1}f$. We shall conclude the proof by showing $r^{-1}s^{-1}f = (1-g)$ is invertible, for then $f^{-1} = (1-g)^{-1}r^{-1}s^{-1}$. For any v in S let $r_v^{(i)}$ denote the coefficient of v in g^i (where $g^0 = 1$); putting $h_j = \sum_{i=0}^{j} g^i = \sum_{v \in S}(\sum_{i=0}^{j} r_v^{(i)})v$ we have $(1-g)h_j = h_j(1-g) = 1 - g^{j+1} = 1 - \sum_{v \in S} r_v^{(j+1)}v$. As $j \to \infty$ we shall see that the h_j "approach" $(1-g)^{-1}$, so the natural candidate for $(1-g)^{-1}$ is $h = \sum_{v \in S}(\sum_{i=0}^{\infty} r_v^{(i)})v$. We claim

(1) for each v in S there exists some $k(v)$ in $\mathbb{N}$ such that $r_v^{(i)} = 0$ for all $i > k(v)$
(2) supp(h) satisfies DCC.

Indeed (1) permits us to define h, and (2) then shows $h \in R((S))$, and a term-by-term check using (1) implies $(1-g)h = h(1-g) = 1$. As we shall see, the proofs of (1) and (2) hinge on the fact that supp(g) = $\{s^{-1}s' : s' \in \text{supp}(f)\} - \{1\}$, i.e., every element of supp(g) is greater than 1. At this point we could give the general proof, but present first an important special case which covers most applications.

Special Case. S is strongly archimedian. Take $k = k(v)$ such that $w^k > v$ for all $w > 1$ in S. Then $w_1 \cdots w_i > (\min\{w_1, \dots, w_i\})^i > v$ for every $i > k$, so (1) is immediate; likewise if $v = v_1 > v_2 > \cdots$ in supp(h) then each $v_j \in \text{supp}(h_k)$ which is well-ordered (since $h_k \in R((S))$), so the chain must be finite, proving (2).

General Case. Let Q denote the set of finite sequences of elements of supp(g) and, given a sequence $q = (s_1, s_2, \dots, s_u)$, write $\text{val}(q) = s_1 s_2 \cdots s_u \in \text{supp}(q^u)$, and length($q$) = u. Then Q has a partial order given by $q_1 < q_2$ if either $\text{val}(q_1) < \text{val}(q_2)$ or $\text{val}(q_1) = \text{val}(q_2)$ with length(q_1) > length (q_2).

Claim. Q has no infinite descending chain $q_1 > q_2 > \cdots$. Otherwise pick a descending chain such that length(q_1) is minimal; of all such possible descending chains we pick one with length(q_2) minimal, and so on. Thus we have a chain $q_1 > q_2 > \cdots$ where (given $q_1, \dots, q_{u-1}$) q_u has minimal possible length. Write s_u for the first element of q_u and q_u' for the remainder of the sequence. Take $u_0 = 0$ and, inductively, given u_{j-1} take $u_j > u_{j-1}$ with s_u minimal possible. We can do this since supp(g) is well-ordered, and have $s_{u_1} \le s_{u_2} \le \cdots$, implying $q_1 > \cdots > q_{u_1 - 1} > q_{u_1}' > q_{u_2}' > \cdots$ is an infinite descending chain. Since q_{u_1}' has smaller length than q_{u_1} we have a contradiction which yields the claim.

If (1) were false we would have an infinite sequence $i_1 < i_2 < \cdots$ in $\mathbb{N}$ with v in $\mathrm{supp}(g^{i_j})$ for each j; taking q_j to be a sequence with $\mathrm{length}(q_j) = i_j$ and $\mathrm{val}(q_j) = v$, we get $q_1 > q_2 > q_3 > \cdots$ contrary to the claim.

If (2) were false we would have a chain $\mathrm{val}(q_1) > \mathrm{val}(q_2) > \mathrm{val}(q_3) > \cdots$ for suitable q_i in Q, so $q_1 > q_2 > q_3 > \cdots$ contrary to the claim.   Q.E.D.

We now have the following immediate corollary:

Theorem 1.2.24′:   *If G is an ordered group and D is a division ring then $D[G]$ can be embedded in the division ring $D((G))$.*

In the unabridged edition (1.3.34–1.3.39 and Appendix A), we define the *free group* on the alphabet $\{X_i : i \in I\}$ and prove it is ordered. Thus, the free algebra $F\{X\}$ over a field is contained in a division ring $F((X))$; this construction is also used in example 2.1.36.

Remark 1.2.25:   If w_1 is a word of length k then for every nonblank word w clearly the length of w^{k+1} exceeds k. This trivial observation shows that the monoids of the next three examples are strongly archimedian.

Example 1.2.26:   Define the monoid $S = \{\lambda^i : i \in \mathbb{Z}\}$ with multiplication given by $\lambda^i \lambda^j = \lambda^{i+j}$, ordered by $\lambda^i < \lambda^j$ if $i < j$. S is strongly archimedian (as noted above). $R((S))$ is denoted $R((\lambda))$ and called the *ring of Laurent series over R* because every element has the form $f = \sum_{i=t}^{\infty} r_i \lambda^i$ for suitable t in $\mathbb{Z}$ corresponding to the minimal element λ^t of $\mathrm{supp}(f)$. For a field F, we see $F((\lambda))$ is a field containing $F[\lambda]$, so that one could form the field of fractions K of $F[\lambda]$ inside $F((\lambda))$, and at times we shall use this technique. Note that $K \subset F((\lambda))$ since there are Taylor series expansions of functions which are not quotients of polynomials. (For example, take $f = \sum_{n=0}^{\infty} \lambda^n / n!$ The formal derivative of f is f itself, implying f cannot be the quotient of two polynomials.)

Example 1.2.27:   Take $S = \{\lambda^i : i \in \mathbb{N}\}$, a submonoid of the monoid of example 1.2.26. Now $R((S))$ is denoted $R[[\lambda]]$ and called the *ring of formal power series over R*. Note that $R \subset R[\lambda] \subset R[[\lambda]] \subset R((\lambda))$.

Example 1.2.28: S is the (noncommutative) word monoid, with the strongly archimedian order of example 1.2.18. For any commutative ring C we denote $C((S))$ as $C\{\{X\}\}$.

§1.3 Free Modules and Rings

In this section, we shall study and construct free modules and rings. There is an extensive theory in Cohn [85B], which largely lies outside the scope of this book; in this section we cover only the most basic properties, leaving free products for §1.4 and §1.9.

Intuitively, "free" objects in a category are constructed in the most general possible way, without any extra conditions, and thus "lie" above all other objects. It turns out that we have already constructed the free R-modules in §1.2, and we shall also use their properties to understand the nature of "free." Then we construct the free monoid and the free ring; at the end we construct the free group, which is needed in several ring-theoretic constructions.

Every free module has a base, and technical problems arise concerning the uniqueness of the cardinality of a base. This matter is addressed in the concept of IBN (invariant base number) in definition 1.3.24ff.

In order to verify the theories we give a category-theoretic definition of free, which is put into context towards the end of the section when we introduce "universals," which are applicable in a wide variety of situations. However, the reader should already be apprised that the definition is ill-suited to category theory, and in §2.8 we shall discuss a related notion, "projective," which fits much better into a categoric approach.

Free Modules

Since the module freely generated by a set has no superfluous conditions in its construction, we examine it to discover the nature of "free."

Definition 1.3.1: A *base* for an R-module M is a subset S of M such that every element of M has the form $\sum_{s \in S} r_s s$ where the $r_s \in R$ are almost all 0 and are uniquely determined. A module with a base is called *free*.

In order to prove S is a base it is enough to show S spans M and is independent, for then if $\sum r_s s = \sum r'_s s$ then $\sum (r_s - r'_s)s = 0$ implying each $r_s - r'_s = 0$ so $r_s = r'_s$.

Remark 1.3.2: S is a base of the module RS freely generated by S, as noted in definition 1.2.1; thus RS is free.

Proposition 1.3.3: *Suppose a module F has a base $S = \{s_i : i \in I\}$. For any module M and any $\{x_i : i \in I\} \subset M$ there is a unique map $f: F \to M$ such that $fs_i = x_i$ for all i in I.*

Proof: Any such f must satisfy $f(\sum r_i s_i) = \sum r_i f s_i = \sum r_i x_i$, proving uniqueness, and we can in fact use this formula to define f since S is a base. Clearly f is a map. Q.E.D.

Example 1.3.4:

(i) $M_n(R)$ is free in R-$\mathcal{M}od$ with base $\{e_{ij}: 1 \leq i, j \leq n\}$ (which has n^2 elements).

(ii) The ring of triangular matrices over R is free in R-$\mathcal{M}od$ with base $\{e_{ij}: 1 \leq i \leq j \leq n\}$ which has $n(n + 1)/2$ elements.

(iii) $R[\lambda]$ is a free R-module, having base $\{\lambda^i: i \in \mathbb{N}\}$.

(iv) $C\{X\}$ is a free C-module having base the words in the alphabet X.

(v) $R^{(n)}$ is the free R-module.

The concept of dependence of elements can be generalized to modules, enabling us to understand free modules in terms of "direct sums."

Independent Modules and Direct Sums

Definition 1.3.5:   A set $\{M_i: i \in I\}$ of submodules of a module M is *independent* if $M_i \cap \sum_{j \neq i} M_j = 0$ for each i in I; *dependent* means not independent. If $M = \sum_{i \in I} M_i$ and $\{M_i: i \in I\}$ is independent we say M is an (*internal*) *direct sum* of the M_i and write $M = \bigoplus M_i$. If each $M_i \approx M$ we write $M^{(I)}$ for $\bigoplus_{i \in I} M_i$.

Proposition 1.3.6:   *Suppose $M \in R$-$\mathcal{M}od$ and $S \subset M$. S is independent iff $\{Rs: s \in S\}$ is an independent set of submodules with $\text{Ann}_R s = 0$ for each s in S.*

Proof: ($\Rightarrow$) If $0 \neq rs \in Rs \cap \sum_{s' \neq s} Rs'$ then we can write rs in two different ways in terms of the independent set S, which is impossible, so $Rs \cap \sum_{s' \neq s} Rs' = 0$. Likewise if $r \in \text{Ann}_R s$ then $rs = 0 = 0s$ so $r = 0$.

($\Leftarrow$) If $\sum_{\text{finite}} r_s s = 0$ then for each s we have $r_s s = -\sum_{s' \neq s} r_{s'}$ so $r_s s \in Rs \cap \sum_{s' \neq s} Rs' = 0$; hence $r_s \in \text{Ann } s = 0$, as desired.   Q.E.D.

Corollary 1.3.7:   *If S is a base of M then $M = \bigoplus_{s \in S} Rs$ and $R \approx Rs$ for each s via the map $r \rightarrow rs$. In other words, every free module is a direct sum of copies of R.*

Remark 1.3.8:

(i) Every dependent set of submodules can be shrunk to a *finite* dependent set. (Indeed if $M_i \cap \sum_{j \neq i} M_j \neq 0$ then $0 \neq x_i = \sum_{u=1}^{t} x_{j_u}$ for suitable t in $\mathbb{N}$, x_i in M_i, and x_{j_u} in M_{j_u}; then $\{M_i, M_{j_1}, \ldots, M_{j_t}\}$ are dependent.) It follows at once that {independent sets of submodules of M} is inductive, and thus each independent set of submodules can be expanded to a *maximal* independent set. In the same way, each independent set of elements of M is contained in a maximal independent set.

(ii) If $\mathscr{S} = \{M_i : i \in I\}$ is independent and $N < M$ with $N \cap \sum_{i \in I} M_i = 0$ then $\mathscr{S} \cup \{N\}$ is independent. (We must show $M_i \cap (N + \sum_{j \neq i} M_j) = 0$ for each i. Taking x_j in M_j and y in N suppose $x_i = y + \sum_{j \neq i} x_j$. Then $y = x_i - \sum_{j \neq i} x_j \in N \cap \sum_{i \in I} M_i = 0$, implying $x_i = \sum_{j \neq i} x_j \in M_i \cap \sum_{j \neq i} M_j = 0$, as desired.)

Modules Over Division Rings Are Free

Modules over division rings are often called *vector spaces* since many techniques from linear algebra are applicable, largely because of the following easy results.

Remark 1.3.9: Suppose D is a division ring and $M \in D\text{-}\mathscr{M}od$. Then $\text{Ann}_D x = 0$ for all x in M. Moreover, if $S \subset M$ and $dx \in DS \subseteq M$ with $0 \neq d \in D$ then $x = d^{-1} dx \in DS$. Put another way, if $x \notin DS$ then $Dx \cap DS = 0$.

Proposition 1.3.10: *Every module M over a division ring D is free. In fact, the following statements are equivalent for any subset S of M:*

(i) *S is a minimal generating set (of M).*
(ii) *S is a maximal independent set.*
(iii) *S is a base.*

Proof: Remark 1.3.8 (i) shows maximal independent sets exist, so if we can establish the equivalence we shall see they are all bases (so M is free).

(i) $\Rightarrow$ (ii) If $\sum d_i s_i = 0$ with some $d_i \neq 0$ then $s_i \in \sum_{j \neq i} D_j s_j$, contrary to minimality.

(ii) $\Rightarrow$ (i) For each x in M we have $Dx \cap DS \neq 0$, so $x \in DS$, proving $DS = M$. Hence S spans M and is minimal since if $S' \subset S$ then $S \not\subseteq DS'$ (by independence).

(i), (ii) $\Rightarrow$ (iii) is true by definition.

(iii) $\Rightarrow$ (ii) S is independent, by definition. For any $x \in S$ we still have $x \in DS$, so $S \cup \{x\}$ is dependent.   Q.E.D.

Free Objects

We draw on proposition 1.3.3 to obtain a general definition of free, for any category whose objects are sets (perhaps with extra structure), i.e. there is a forgetful functor from $\mathscr{C}$ to $\mathscr{Set}$. We shall call such a category *concrete*.

Definition 1.3.11: A *free object* $(F; X)$ of $\mathscr{C}$ is an object F with a subset $X = \{X_i : i \in I\}$ whereby for every A in $Ob\,\mathscr{C}$ and every $\{a_i : i \in I\} \subseteq A$ there is a *unique* morphism $f: F \to A$ such that $fX_i = a_i$ for all i in I.

Proposition 1.3.12: *Suppose $\mathscr{C}$ and I are given.*

(i) *Any two free objects are canonically isomorphic.*

(ii) *If a free object $(F; X)$ exists then every object of cardinality $\leq |I|$ is an image of F under a suitable morphism.*

Proof: Left for reader, cf. example 1.3.22.

We shall now show free objects exist in those other categories of interest to us.

Free Rings and Algebras

If C is a commutative ring then $C\{X\}$ is the free C-algebra. This is intuitively clear since for any C-algebra R and $\{r_i : i \in I\} \subseteq R$ we can "specialize" $X_i \mapsto r_i$. We shall prove this fact from the previous module results, aided by the following tool:

Definition 1.3.13: Suppose $\varphi: R \to T$ is a ring homomorphism, and $M \in R\text{-}\mathscr{Mod}$ and $N \in T\text{-}\mathscr{Mod}$. An additive group homomorphism $f: M \to N$ is called a φ-*map* if $f(rx) = (\varphi r)fx$ for all r in R, x in M. φ-maps are also called *semilinear maps*.

Remark 1.3.14: Any φ-map $f: M \to N$ becomes a module map in the usual sense when N is viewed as R-module by remark 0.0.9 (defining ry as $(\varphi r)y$ for y in N). Thus for M free with base S, any function $f: S \to N$ can be extended uniquely to a φ-map $\bar{f}: M \to N$ by putting $\bar{f}(\sum r_s s) = \sum(\varphi r_s)fs$.

This remark enables us to apply our module-theoretic results to rings, taking $M = R[S]$.

Proposition 1.3.15: *Suppose $\varphi: R \to T$ is a homomorphism of C-algebras. Also suppose S is a monoid and $f: S \to T$ is a monoid homomorphism such that φR*

centralizes fS. Then there is a unique algebra homomorphism $\bar{f}: R[S] \to T$ whose restriction to R is φ and whose restriction to S is f.

Proof: By remark 1.3.14 we need check merely that $\bar{f}$ is a monoid homomorphism, which is true since $\bar{f}((r_1 s_1)(r_2 s_2)) = \bar{f}(r_1 r_2 s_1 s_2) = \varphi(r_1 r_2) f(s_1 s_2) = \varphi r_1 \varphi r_2 f s_1 f s_2 = (\varphi r_1 f s_1)(\varphi r_2 f s_2) = \bar{f}(r_1 s_1) \bar{f}(r_2 s_2)$.   Q.E.D.

If $\varphi R \subseteq Z(T)$ then the condition of the proposition is satisfied; in particular, this is the case if $R = C$ is commutative, $T \in C\text{-}\mathcal{A}lg$, and $\varphi: C \to T$ is the canonical homomorphism $c \to c1$. This special case is worth restating explicitly:

Proposition 1.3.16: *Suppose S is a monoid and T is a C-algebra. Any monoid homomorphism $f: S \to T$ can be extended uniquely to an algebra homomorphism $\bar{f}: C[S] \to T$.*

In view of proposition 1.3.16, we expect to find a free ring if we can locate the free monoid.

Proposition 1.3.17: *Suppose X is an arbitrary set, and S is the word monoid built from X (i.e., X is the alphabet). Then $(S; X)$ is free in $\mathcal{M}on$. Moreover, $(C\{X\}; X)$ is free in $C\text{-}\mathcal{A}lg$ for any commutative ring C, and $(\mathbb{Z}\{X\}; X)$ is free in $\mathcal{R}ing$.*

Proof: For any monoid V and $\{x_i : i \in I\} \subseteq V$, define $f: S \to V$ by $f(X_{i_1} \cdots X_{i_m}) = x_{i_1} \cdots x_{i_m}$. Clearly, f is a morphism of monoids and is unique such that $fX_i = x_i$, proving $(S; X)$ is free. Thus $(C\{X\}; X)$ is free by proposition 1.3.16. Hence $(\mathbb{Z}\{X\}; X)$ is free in $\mathbb{Z}\text{-}\mathcal{A}lg$ which is isomorphic to $\mathcal{R}ing$.   Q.E.D.

Corollary 1.3.18: Any C-algebra is isomorphic to $C\{X\}/A$ for suitable $A \vartriangleleft C\{X\}$, where X is a suitably large set of noncommuting indeterminates.

One can use corollary 1.3.18 as one cornerstone of the study of C-algebras. Given any subset S of R, we write $\langle S \rangle$ for the ideal RSR, called the ideal *generated* by S; one views $R/\langle S \rangle$ as the ring formed by sending the elements of S to 0. Let us examine the situation more closely for $R = C\{X\}$. Writing x_i for the image of X_i in $C\{X\}/\langle S \rangle$ (where $X = \{X_i : i \in I\}$), we can write $C\{X\}/\langle S \rangle$ as $C\{x_i : i \in I\}$, or $C\{x\}$ for short; and we say $(x_i : i \in I)$ *generate* $C\{x\}$ as an algebra (since no proper subalgebra of $C\{x\}$ contains all the x_i).

On the other hand, any polynomial f in $C\{X\}$ involves only a finite number of indeterminates X_i, say $X_{i_1}, \ldots, X_{i_t}$; we denote the image of f in $C\{X\}/\langle S \rangle$ as $f(x_{i_1}, \ldots, x_{i_t})$, which intuitively should be thought of as "substituting" x_i for X_i. Then $f(X_{i_1}, \ldots, X_{i_t}) \in \langle S \rangle$ iff $f(x_{i_1}, \ldots, x_{i_t}) = 0$, so the elements of S can be thought of as *relations* among the x_i. Thus defining a C-algebra as $C\{X\}/\langle S \rangle$ is sometimes called "defining an algebra by generators and relations." Such a construction is often very useful but involves the difficulty that perhaps $\langle S \rangle$ is much larger than one might expect and perhaps even $\langle S \rangle = C\{X\}$. Sometimes this difficulty can be overcome by building a concrete example to go along with the abstract definition.

Free Commutative Ring

When working with commutative rings one would like to have an analogue of corollary 1.3.18 using the free commutative ring, which we shall now identify.

Proposition 1.3.19:

(i) *The commutative word monoid is free in the category of commutative monoids.*

(ii) *For any ring homomorphism $f: R \to R'$ and any set $\{z_i : i \in I\} \subseteq Z(R')$, there is a unique homomorphism $\bar{f}: R[\lambda] \to R'$ whose restriction to R is f, such that $\bar{f}\lambda_i = z_i$ for all i. (Here $\lambda = \{\lambda_i : i \in I\}$ is a set of commuting indeterminates over R.)*

Proof:

(i) Let M be the commutative word monoid on the set I. Given a map $f: I \to S$ where S is a commutative monoid, we can extend f to a monoid homomorphism $f: M \to S$ given by $f(i_1 \cdots i_t) = (fi_1) \cdots f(i_t)$, where $i_1 \leq \cdots \leq i_t$, and such an extension is clearly unique. Thus M is free.

(ii) follows from (i) and proposition 1.3.16.   Q.E.D.

Corollary 1.3.20:   *For any set λ of commuting indeterminates, $(C[\lambda], \lambda)$ is free in the category of commutative C-algebras, and $(\mathbb{Z}[\lambda], \lambda)$ is free in the category of commutative rings. Any commutative C-algebra has the form $C[\lambda]/A$ where $A \lhd C[\lambda]$.*

Let us exemplify these results. Using well-known properties of polynomials in one indeterminate over $\mathbb{Z}$, one can easily verify $\mathbb{Z}[i] \approx \mathbb{Z}[\lambda]/\langle \lambda^2 + 1 \rangle$,

$\mathbb{Z}[\omega] \approx \mathbb{Z}[\lambda]/\langle \lambda^2 + \lambda + 1 \rangle$ for ω a primitive cube root of 1, and $\mathbb{Z}[\sqrt{n}] \approx \mathbb{Z}[\lambda]/\langle \lambda^2 - n \rangle$ when n is not a square. The ease in handling commutative examples such as these is largely lacking in the noncommutative case.

Diagrams

Before continuing our study of modules, let us develop some more abstract theory to enable us to see the broader picture. One important new technique is the use of commutative *diagrams*, cf., Jacobson [80B, p. 10]. A dotted line is drawn to show there exists a morphism such that the ensuing diagram obtained by drawing a solid line is commutative. Thus

means there is $h: A \to C$ such that $h = gf$; we then say h *completes* this diagram. This notion enables us to put free objects in a general categorical context.

Universals

Definition 1.3.21: Suppose we are given $G: \mathscr{C} \to \mathscr{D}$ a functor of categories and $X \in Ob(\mathscr{D})$. A *universal from X to G* is an object U of $\mathscr{C}$ together with a morphism $u: X \to GU$ such that for every A in $Ob(\mathscr{C})$ and every morphism $f: X \to GA$ there is a unique morphism $f': U \to A$ such that $f = (Gf')u$, i.e., Gf' completes the diagram

In case this definition appears dry, let us whet the reader's appetite by observing that intuitively U is the "free" object of $\mathscr{C}$ with respect to being built up from X. In particular we have

Example 1.3.22: Let $\mathscr{C}$ be a concrete category and $\mathscr{D} = \mathscr{Set}$. For any set X the free object $(F; X)$ of $\mathscr{C}$ is the universal from X to the forgetful functor (where u is the identification of X as a subset of F). Thus the next result generalizes proposition 1.3.12(i).

Proposition 1.3.23: ("*Abstract nonsense*") *Any two universals from X to G are isomorphic.*

Proof: Let $\{U; u: X \to GU\}$ and $\{U'; u': X \to GU'\}$ be two universals. Applying the definition with $A = U'$ and $f = u'$ we get $f': U \to U'$ with $u' = (Gf')u$; interchanging the roles of U and U' we get $f'': U' \to U$ with $u = (Gf'')u'$. Hence $u = (Gf'')(Gf')u = G(f''f')u$. On the other hand, taking the universal U and $A = U$ and $f = u$, we get a *unique* $f''': U \to U$ with $u = (Gf''')u$; 1_U and $f''f'$ each fill this role so $1_U = f''f'$. Analogously, $1_{U'} = f'f''$, proving $f': U \to U'$ is an isomorphism.   Q.E.D.

The reader should realize that there may not exist a universal from X to G. However, universals exist in many settings and help organize seemingly disparate constructions.

Note: It is often easier to recognize that an object is a universal than to pinpoint the corresponding functor and category. (This is certainly the case with direct products, to be defined shortly.) Hence it is customary to write a diagram resembling that of definition 1.3.21 and to say a given construction "solves a universal mapping problem" and is therefore unique in this respect (up to isomorphism). It is a good exercise to write down each universal in precise terms.

Invariant Base Number

We shall find many occasions to deal with free modules and will want a workable "dimension," usually called *rank*. The obvious definition of rank is the cardinality of a base, which leads us to the fundamental question of whether different bases can have different cardinalities.

Definition 1.3.24: R has *invariant base number* (abbreviated IBN) if $R^{(m)} \approx R^{(n)}$ implies $m = n$.

Lemma 1.3.25: $R^{(m)} \approx R^{(n)}$ *iff there is an $m \times n$ matrix A and an $n \times m$ matrix B, each with entries in R, such that $AB = 1_m$ (the identity $m \times m$ matrix) and $BA = 1_n$.*

Proof: Let $x_1, \ldots, x_m$ be a base of $R^{(m)}$ and let $y_1, \ldots, y_n$ be a base of $R^{(n)}$. ($\Rightarrow$) Let $\psi: R^{(m)} \to R^{(n)}$ be an isomorphism. Put $\psi x_i = \sum_{j=1}^{n} a_{ij} y_j$ and $\psi^{-1} y_j =$

$\sum_{k=1}^{m} b_{jk}x_k$, so $x_i = \psi^{-1}\psi x_i = \psi^{-1}\sum_{j=1}^{n} a_{ij}y_j = \sum a_{ij}\psi^{-1}y_j = \sum_{j,k} a_{ij}b_{jk}x_k$ for each i, from which one readily concludes $AB = 1_m$; likewise $BA = 1_n$.

($\Leftarrow$) There are maps $\psi: R^{(m)} \to R^{(n)}$ given by $\psi x_i = \sum_{j=1}^{n} a_{ij}y_j$, and $\psi': R^{(n)} \to R^{(m)}$ given by $\psi'y_j = \sum_{k=1}^{m} b_{jk}x_k$. Then $\psi'\psi x_i = x_i$ for each i, so $\psi'\psi = 1$ (since the two maps agree on a base). Likewise $\psi\psi' = 1$ so ψ and ψ' are inverses, proving ψ is an isomorphism.   Q.E.D.

Proposition 1.3.26: *R lacks IBN iff there are $m \neq n$ with an $m \times n$ matrix A and an $n \times m$ matrix B having entries in R such that $AB = 1_m$ and $BA = 1_n$.*

Corollary 1.3.27:

(i) *If R has IBN then $M_n(R)$ has IBN.*

(ii) *If R lacks IBN and $\psi: R \to T$ is a ring homomorphism then T lacks IBN.*

(iii) *Every homomorphic image of a ring lacking IBN also lacks IBN.*

Proof:

(i) Follows at once.

(ii) If A, B are matrices over R with $AB = 1_m$ and $BA = 1_n$ for $m \neq n$ then applying ψ gives the corresponding equations for matrices over T.

(iii) Is a special case of (ii).   Q.E.D.

This last result is very powerful. For example, any field has IBN, by proposition 1.3.26, so any nontrivial commutative ring C has IBN (since C/M is a field for any maximal ideal M). We can refine the method still further.

Definition 1.3.28: A *trace map* on a ring R is an additive group homomorphism $t: R \to H$ for an abelian group H, satisfying $t(r_1r_2) = t(r_2r_1)$ for all r_i in R.

Proposition 1.3.29: *If R has a trace map t such that $t1$ has no $\mathbb{Z}$-torsion then R has IBN.*

Proof: Suppose $A = (a_{ij})$ is an $m \times n$ matrix and $B = (b_{ij})$ is an $n \times m$ matrix with $AB = 1_m$ and $BA = 1_n$, and suppose $e = t1$. Then

$$me = \sum_{i=1}^{m} t1 = \sum_{i=1}^{m}\sum_{k=1}^{n} t(a_{ik}b_{ki}) = \sum_{k=1}^{n}\sum_{i=1}^{m} t(b_{ki}a_{ik}) = \sum_{k=1}^{n} t1 = ne$$

implying $m = n$ as desired.   Q.E.D.

We shall see examples of trace maps later, particularly in the discussion of group rings.

There are, indeed, rings lacking IBN (cf., example 1.3.33 below), and thus their simple homomorphic images lack IBN. On the other hand, theorem 0.3.2 implies division rings do have IBN, and virtually every class of rings we shall study (Artinian rings, Noetherian rings, PI-rings, etc.) will be seen to have IBN.

Weakly Finite Rings

There is a slightly stronger property than IBN which often is easier to verify.

Definition 1.3.30: A ring R is *weakly n-finite* if for all matrices A, B in $M_n(R)$ one has $AB = 1$ implies $BA = 1$. R is *weakly finite* (also called *Dedekind finite* or *von Neumann finite*) if R is weakly n-finite for all n.

Remark 1.3.31: If R is weakly finite then R has IBN, by proposition 1.3.26.

In the course of the text we shall see that many classes of rings are weakly finite, and thus satisfy IBN. At the moment let us note that IBN has the following connection to idempotents; further results will be collected in Chapter 3.

Proposition 1.3.32: (*Jacobson*). *If R is not weakly 1-finite then R has an infinite set of orthogonal idempotents. More generally, if $ab = 1$ and $ba \neq 1$ in R then putting $e_{ij} = b^i(1 - ba)a^j$ we have $e_{ij}e_{uv} = \delta_{ju}e_{iv}$ for all i, j, u, v.*

Proof: Note $a^j b^j = a^{j-1}(ab)b^{j-1} = a^{j-1}b^{j-1} = \cdots = 1$ for all j. Also $a(1 - ba) = 0 = (1 - ba)b$. Thus $(1 - ba)a^j b^u(1 - ba) = 0$ unless $j = u$, in which case

$$(1 - ba)a^j b^j(1 - ba) = (1 - ba)^2 = 1 - ba, \text{ so } e_{ij}e_{uv} = \delta_{ju}b^i(1 - ba)a^v = \delta_{ju}e_{iv}.$$

$$\text{Q.E.D.}$$

Example 1.3.33: Let us search for an example of a ring R lacking IBN. Of course it cannot be weakly finite, and proposition 1.3.32 leads us to look for some infinite analog of matrices. Let V be a countably infinite dimensional vector space over a field F, and let $R = \text{Hom}_F(V, V)$. Then R is a monoid under composition of functions and becomes a ring when we define $f + g$ by $(f + g)v = fv + gv$ for all v in V. (This construction will be studied shortly in great detail.) Let $B = \{v_i : i \in \mathbb{N}\}$ be a base of V over F.

We claim R is not weakly 1-finite. Indeed, since every map is determined by its action on B we need only find a 1:1 function $f: B \to B$ which is not onto (for then f will have a left inverse which is not a right inverse). Define f to be the *right shift*, i.e., $fv_i = v_{i+1}$ for each i.

Actually R lacks IBN; in fact $R^{(i)} \approx R^{(j)}$ in $R\text{-}\mathcal{M}od$, for all i, j in $\mathbb{N}\text{-}\{0\}$. To prove this we need only show $R \approx R^{(2)}$. Define $\Phi: R \to R^{(2)}$ by defining $\Phi f = (f_1, f_2)$ for each f in R, where the f_i are given according to the following action on B:

$$f_1 v_i = f v_{2i} \quad \text{and} \quad f_2 v_i = f v_{2i-1} \quad \text{for each } i \text{ in } \mathbb{N}.$$

If $(f_1, f_2) = (0, 0)$ then $fv_i = 0$ for all i, implying $f = 0$, so Φ is 1:1; conversely Φ is onto since given (f_1, f_2) we recover f by putting $fv_{2i} = f_1 v_i$ and $fv_{2i-1} = f_1 v_i$ for all i. Clearly Φ is a module map and is thus the desired isomorphism.

This example is also important in other aspects of ring theory.

§1.4 Products and Sums

Direct Products and Direct Sums

The direct product construction is very straightforward and probably familiar to the reader. Given a set of objects $\{A_i : i \in I\}$ in a category $\mathcal{C}$ we can often view the cartesian product $\prod_{i \in I} A_i$ as an object of $\mathcal{C}$, where the relevant properties are defined componentwise. Rather than formalize this idea now, we present now the two examples of importance to us and defer a formal discussion until the treatment of reduced products.

Definition 1.4.1:

(i) The *direct product* of a set $\{R_i : i \in I\}$ of rings, denoted $\prod_{i \in I} R_i$, is the cartesian product which is a ring endowed with componentwise operations $(r_i) + (r_i') = (r_i + r_i')$ and $(r_i)(r_i') = (r_i r_i')$ where $r_i, r_i' \in R_i$ for all i.

(ii) The *direct product* of a set $\{M_i : i \in I\}$ of R-modules, denoted $\prod_{i \in I} M_i$, is the cartesian product which is in $R\text{-}\mathcal{M}od$ when endowed with componentwise operations $(x_i) + (x_i') = (x_i + x_i')$ and $r(x_i) = (rx_i)$ where $r \in R$ and $x_i, x_i' \in M_i$ for all i.

The same sort of definition "works" in $\mathcal{A}b$, $\mathcal{G}rp$, $\mathcal{M}on$, and $\mathcal{P}os$ as is easily seen. The following observation is crucial for understanding how direct products work.

Remark 1.4.2: Suppose $\{A_i : i \in I\}$ are objects in a category for which direct products have been defined (e.g., $\mathcal{R}ing$, $R\text{-}\mathcal{M}od$, etc.). For each j in I there is an onto *projection morphism* $\pi_j: \prod_{i \in I} A_i \to A_j$ given by reading the j-th component i.e., $\pi_j(a_i) = a_j$ for each (a_i) in $\prod A_i$. Moreover, given any object A and morphisms $f_i: A \rightarrow A_i$ for each i in I, one has a unique morphism $f: A \rightarrow \prod A_i$ such that $\pi_i f = f_i$ for all i in I. (Indeed, such f can be defined by the rule $fa = (f_i a)$, the element whose i-th component is $f_i a$, and one sees easily that any such f must satisfy this rule and thus is uniquely determined.)

We turn to the (external) direct sum of modules. The construction involves the direct product, as follows: Define $\mu_j: M_j \rightarrow \prod_{i \in I} M_i$ in $R\text{-}\mathcal{M}od$ by taking $\mu_j x$ to have all components 0 except the j-th component, which is x. (Here $x \in M_j$.) Then $\pi_j \mu_j = 1_{M_j}$ and $\pi_j \mu_i = 0$ for all $i \neq j$.

Definition 1.4.3: The *direct sum* $\bigoplus M_i$ of a set $\{M_i : i \in I\}$ of R-modules is the sum of the $\mu_i M_i$ in $\prod M_i$. In other words, $\bigoplus M_i = \{(x_i) \in \prod M_i : \text{almost all } x_i \text{ are } 0\}$. (Thus the direct sum and direct product of $\{M_i : i \in I\}$ are the same iff I is finite.)

Remark 1.4.4: The notation $\bigoplus M_i$ is justified since $\bigoplus M_i$ is the internal direct sum of the $\mu_i M_i$. (Indeed to see the $\mu_i M_i$ are independent, note that $\mu_i x = \mu_i \pi_i (\mu_i x)$ for any x in M_i, so

$$\mu_i M_i \cap \sum_{j \neq i} \mu_j M_j = \mu_i \pi_i \left(\mu_i M_i \cap \sum_{j \neq i} \mu_j M_j \right) = 0$$

since the $\pi_i \mu_j = 0$.)

Exact Sequences

To understand direct sums better we shall introduce sequences, one of the basic tools of category theory. A sequence of maps $\cdots M' \xrightarrow{f} M \xrightarrow{g} M'' \rightarrow \cdots$ is *exact at M* if $fM' = \ker g$; an *exact sequence* is a sequence exact at each of its modules. Thus $0 \rightarrow M \xrightarrow{f} N$ is exact iff f is monic, and $M \xrightarrow{f} N \rightarrow 0$ is exact iff f is epic.

It turns out that exact sequences contain precisely the information one needs for proofs. The *length* of an exact sequence is the number of arrows.

Remark 1.4.5: The following information characterizes exact sequences of length m which start and end with 0.

  (i) $m = 2$. $0 \rightarrow M \rightarrow 0$ is exact iff $M \approx 0$.
  (ii) $m = 3$. $0 \rightarrow M \xrightarrow{f} N \rightarrow 0$ is exact iff $f: M \rightarrow N$ is an isomorphism.
  (iii) $m = 4$. $0 \rightarrow K \xrightarrow{g} M \xrightarrow{f} N \rightarrow 0$ is exact iff g is monic, $N = fM$, and

$N \approx M/K$ (using g to view K as a submodule of M, and using f to induce the isomorphism). This is called a *short exact sequence*.

(iv) $m = 5$. $0 \rightarrow K \rightarrow L \xrightarrow{f} M \rightarrow N \rightarrow 0$ is exact iff $K \approx \ker f$ and $N \approx M/fL$ (using the given maps for the identifications).

We should observe that the information becomes weaker as the length increases. On the other hand, any exact sequence starting and ending with zeroes can be made longer by adding extra zeroes and arrows denoting the zero map, enabling us at times to apply results about exact sequences of longer length.

Split Exact Sequences

Viewing a commutative diagram as a traffic flow, where the arrows denote one-way streets, one might sometimes want to go the "wrong way" on a one-way street to arrive at a given desired object. More formally, given $g: M \rightarrow N$ one might want to find $h: N \rightarrow M$ such that $gh = 1_N$ and/or $hg = 1_M$. These notions can be formulated most fruitfully for short exact sequences.

Definition 1.4.6: An exact sequence $0 \rightarrow K \xrightarrow{f} M \xrightarrow{g} N \rightarrow 0$ is *split* if there is a map $i: N \rightarrow M$ such that $gi = 1_N$.

Definition 1.4.7: If $M = \bigoplus_{i \in I} M_i$ we say each M_i is a *summand* of M and $\sum_{j \neq i} M_j$ is the *complement* of M_i. In the literature, summands often are called *direct summands*.

Clearly if M_i is a summand of M then M is a direct sum of M_i and its complement. Thus the case of direct sums of two submodules is of special significance, and we shall now consider it closely. Note $M = M_1 \oplus M_2$ iff M_1 and M_2 are summands of M and complements of each other.

Remark 1.4.8: If $M = M_1 \oplus M_2$ then any map $f: M_1 \rightarrow N$ can be extended to a map $f: M \rightarrow N$ by putting $f(x_1 + x_2) = fx_1$ for all x_i in M_i (i.e., $fM_2 = 0$).

Lemma 1.4.9: *Suppose* $g: M \rightarrow N$ *and* $h: N \rightarrow M$ *are maps with* $gh = 1_N$. *Then* g *is epic; likewise* h *is monic, and* $M = \ker g \oplus hN$.

Proof: If $fg = f'g$ then $f = f1_N = fgh = f'gh = f'$, proving g is epic; likewise h is monic. For any x in M, $g(x - hgx) = gx - ghgx = gx - gx = 0$, implying $x - hgx \in \ker g$, so $M = \ker g + hN$. On the other hand, if $x \in (\ker g) \cap hN$ then writing $x = hy$ we have $0 = gx = ghy = y$, implying $x = 0$.

Q.E.D.

Proposition 1.4.10: *The following assertions are equivalent for an exact sequence* $0 \rightarrow K \xrightarrow{f} M \xrightarrow{g} N \rightarrow 0$:

   (i) *The exact sequence is split.*
   (ii) *There is a map* $p: M \rightarrow K$ *with* $pf = 1_K$.
   (iii) *fK is a summand of M.*
   (iv) *$M \approx K \oplus N$; in fact $M = fK \oplus iN$ for suitable $i: N \rightarrow M$ with $gi = 1_N$.*

Proof: (i) $\Rightarrow$ (iv). By lemma 1.4.9, taking $i: N \rightarrow M$ with $gi = 1_N$, we have i monic and $M = \ker g \oplus iN$. But $\ker g = fK \approx K$ and $iN \approx N$. (ii) $\Rightarrow$ (iv) is analogous, and (iv) $\Rightarrow$ (iii) is *a fortiori*. So it suffices to assume (iii), and prove (i) and (ii). Write $M = fK \oplus M'$. Since $\ker g = fK$, the restriction of g to M' is a map $M' \rightarrow N$ which is monic as well as epic and thus has an inverse $i: N \rightarrow M'$, yielding (i). Likewise, f induces an isomorphism $K \rightarrow fK$ with some inverse $p: fK \rightarrow K$ which extends by remark 1.4.8 to a map $p: M \rightarrow K$ with $pf = 1$, yielding (ii).   Q.E.D.

In view of this result, we introduce the following terminology: A monic $f: M \rightarrow N$ is *split* if there exists $h: N \rightarrow M$ with $hf = 1_M$; an epic $g: M \rightarrow N$ is *split* if there exists $h: N \rightarrow M$ with $gh = 1_N$.

Reduced Products

We can generalize direct products by using filters. Since this method generates interesting examples and also provides a technique for shortening proofs, we shall examine it in some detail.

Definition 1.4.11: Suppose $\mathscr{F}$ is a given filter of the power set $\mathscr{P}(I)$, and we are given sets $\{A_i : i \in I\}$. The *filtered product* or *reduced product* of the A_i with respect to $\mathscr{F}$, written $\prod A_i / \mathscr{F}$, is the set of equivalence classes of the cartesian product $\prod A_i$ under the equivalence given by $(a_i) \sim (a_i')$ iff $\{i \in I : a_i = a_i'\} \in \mathscr{F}$. (This is seen to be an equivalence as a consequence of the definition of filter.) To simplify notation, we still write (a_i) when technically we mean the equivalence class of (a_i).

Note that the reduced product becomes the direct product when $\mathscr{F}$ is the trivial filter $\{I\}$. Of course, we want the reduced product to inherit the algebraic structure of the A_i, as illustrated in the following straightforward observation.

Remark 1.4.12:

(i) Suppose the A_i each are rings. Then $A = \prod A_i/\mathscr{F}$ is a ring under componentwise operations, i.e., $0 = (0)$, $1 = (1)$, $(a_i)(a_i') = (a_i a_i')$, and $(a_i) - (a_i') = (a_i - a_i')$; these are well-defined since filters are closed under finite intersections.

(ii) If, moreover, $M_i \in A_i\text{-}\mathcal{M}od$ then $(\prod M_i)/\mathscr{F} \in A\text{-}\mathcal{M}od$ by defining scalar multiplication $(a_i)(x_i) = (a_i x_i)$.

The verifications are easy and are examples of the following general considerations:

Suppose in general each A_i lies in a concrete category $\mathscr{C}$. We want to see when the reduced product $A = \prod A_i/\mathscr{F}$ remains in $\mathscr{C}$. It is convenient to assume $\mathscr{C}$ is axiomatizable in a first-order language $\mathscr{L}$, for then at least we can define a structure in $\mathscr{L}$ having A as the underlying set, as follows:

Any constant symbol is assigned to (a_i) in A, where a_i is the assignment in A_i:

Any n-ary relative symbol R is assigned to the relation on A defined by saying $R((x_i), (y_i), (z_i), \ldots)$ holds iff $\{i: R(x_i, y_i, z_i, \ldots)$ holds in $A_i\} \in \mathscr{F}$;

Any n-ary function symbol f is assigned to $f: A^{(n)} \to A$ given by $f((x_i), (y_i), (z_i), \ldots) = (f(x_i, y_i, z_i, \ldots)))$. Again these definitions are well-defined because filters are closed under finite intersections.

Proposition 1.4.13: *Suppose φ is an elementary sentence whose only connective is $\wedge$. φ holds in $(\prod A_i)/\mathscr{F}$ iff $\{i: \varphi$ holds in $A_i\} \in \mathscr{F}$.*

Proof: By definition the proposition holds for atomic formulas, so we argue by induction on the rank of φ (given in the definition of elementary sentences):

Case I. $\varphi = \varphi_1 \wedge \varphi_2$. φ holds in A iff φ_1 and φ_2 each holds in A, iff (by induction) $\{i: \varphi_1$ holds in $A_i\} \in \mathscr{F}$ and $\{i: \varphi_2$ holds in $A_i\} \in \mathscr{F}$, iff $\{i: \varphi_1$ and φ_2 holds in $A_i\} \in \mathscr{F}$.

Case II. $\varphi = (\exists x)\varphi_1(x)$. φ holds in A iff $\varphi_1(s)$ holds for some $s = (s_i)$ iff (inductively) $\{i: \varphi_1(s_i)$ holds for some $s_i\} \in \mathscr{F}$, iff $\{i: (\exists x)\varphi_1(x)$ holds in $A_i\} \in \mathscr{F}$.

Case III. $\varphi = (\forall x)\varphi_1(x)$. If $\{i: \varphi$ holds in $A_i\} \in \mathscr{F}$ then φ holds in A, proved as in Case II. Conversely, suppose φ holds in A and let $J = \{i: \varphi$ holds in $A_i\}$

and $J' = I - J$. For each i in J' there is s_i in A_i such that $\varphi_1(s_i)$ does not hold. Taking s_i arbitrarily in J and $s = (s_i)$, nevertheless, $\varphi_1(s)$ holds; so letting $J'' = \{i: \varphi_1(s_i) \text{ holds in } A_i\}$ we have $J'' \in \mathscr{F}$ by induction. Since $J'' \cap J' = \varnothing$ by choice of s we conclude $J \supseteq J'' \in \mathscr{F}$.   Q.E.D.

Since the axioms for rings and modules are atomic sentences (suitably quantified) we see that any reduced product of rings (resp. R-modules) is a ring (resp. R-module). Nevertheless, certain sentences do not carry over. For example, the direct product of domains is not a domain. The simplest sentence which does not carry over is the following:

Example 1.4.14:   Let $R_1 = \mathbb{Z}/2\mathbb{Z}$ and $R_2 = \mathbb{Z}/3\mathbb{Z}$, Then R_1 and R_2 each satisfies the sentence $(\forall x)((2x = 0) \vee (3x = 0))$ which nevertheless fails in $R_1 \times R_2$.

A general class of sentences preserved under reduced products is given in exercise 6. Filters and reduced products can be described in purely ring-theoretic terms (cf., exercises 9 through 12) and this interplay enchances both ring theory and the study of filters.

Ultraproducts

This technique becomes powerful when we restrict the particular filter under consideration. Recall that any filter can be embedded in a maximal filter, called an ultrafilter. A reduced product $\prod A_i/\mathscr{F}$ is called an *ultraproduct* when $\mathscr{F}$ is an ultrafilter. Thus we know that ultraproducts are also plentiful, although it is usually near impossible to describe a given ultraproduct precisely. Nevertheless, they are important because of the following result.

Theorem 1.4.15:   (*Łoś's theorem*) *Suppose* $A = \prod A_i/\mathscr{F}$, *where each* A_i *is a structure in the same first-order language, and* $\mathscr{F}$ *is an ultrafilter. A first-order sentence* φ *holds in* A *iff* $\{i \in I: \varphi \text{ holds in } A_i\} \in \mathscr{F}$.

Proof: By induction on the rank of φ. Proposition 1.4.13 has disposed of every case except $\varphi = \neg \varphi_1$, which holds in A iff φ_1 does *not* hold, iff $\{i \in I: \varphi_1 \text{ holds in } A_i\} \notin \mathscr{F}$ (by induction), iff $\{i \in I: \neg \varphi_1 \text{ holds in } A_i\} \in \mathscr{F}$ (by remark 0.0.17).   Q.E.D.

Remark 1.4.16:   The principal ultrafilters are rather dull. If $\mathscr{F}$ is a principal ultrafilter on $\mathscr{P}(I)$, generated by $I_0 \subseteq I$, then remark 0.0.17 implies I_0

is a suitable singleton $\{i_0\}$; thus $\prod A_i/\mathcal{F} \approx A_{i_0}$ and we have nothing new. However, every nonprincipal ultrafilter is interesting, as we see now.

Lemma 1.4.17: *An ultrafilter $\mathcal{F}$ is nonprincipal iff $\mathcal{F}$ contains the cofinite filter.*

Proof: ($\Leftarrow$) $\mathcal{F}$ does not have any finite set, so is nonprincipal.

($\Rightarrow$) We prove the contrapositive. If $\mathcal{F}$ does not contain the cofinite filter then some cofinite set is not in $\mathcal{F}$, and hence its complement is a finite set in $\mathcal{F}$. Take a finite set I_0 minimal in $\mathcal{F}$. If $J \in \mathcal{F}$ then $\varnothing \neq I_0 \cap J \in \mathcal{F}$, so $I_0 \subseteq J$ by hypothesis, proving I_0 generates $\mathcal{F}$.   Q.E.D.

Proposition 1.4.18: *Suppose $\mathcal{F}$ is a nonprincipal ultrafilter on I, and φ is an elementary sentence holding in almost all A_i. Then φ holds in the ultraproduct $\prod A_i/\mathcal{F}$.*

Proof: $\{i : \varphi$ holds in $A_i\} \in \mathcal{F}$ by the lemma, so φ holds in A by Łoś' theorem.   Q.E.D.

Example 1.4.19: Suppose R_i is a ring of characteristic m_i for each $i \in \mathbb{N}$ where $m_1 < m_2 < m_3 < \cdots$. If $\mathcal{F}$ is a nonprincipal ultrafilter on $\mathbb{N}$ then $\prod R_i/\mathcal{F}$ has characteristic 0. (Indeed for each $m > 0$ the statement $m \cdot 1 \neq 0$ holds in almost all R_i.

Proposition 1.4.20: *Suppose φ is a first-order sentence which is true for every ring of characteristic 0 in a given axiomatizable category. Then φ is true for all rings in this category having suitably large characteristic.*

Proof: Otherwise, there is a sequence $m_1 < m_2 < \cdots$ in $\mathbb{N}$ and rings R_i in this theory having characteristic m_i in which φ fails. Then φ fails for their ultraproduct, which has characteristic 0.   Q.E.D.

This is usually called the "compactness argument" in logic and can be applied to almost any "finiteness" condition which is not bounded; it shall be refined later on (cf., remark 2.12.37' and its applications in §8.1 and §8.3).

Also of importance is the following tool which often enables us to assume an algebra is finitely generated.

Products and Coproducts

We return now to direct products, using remark 1.4.2 to provide the following categorical characterization:

Remark 1.4.22: Suppose $A_i : i \in I$ are rings. Given any ring A and homomorphisms $f_i : A \to A_i$ for each i, there is a *unique* homomorphism $f : A \to \prod A_i$ which simultaneously completes the following diagrams for each i:

One should bear in mind that the projection morphisms π_i go from the direct product to the components. More generally we have

Definition 1.4.23: A *product* of *objects* $\{A_i : i \in I\}$ in a category $\mathscr{C}$ is an object denoted $\prod A_i$, together with morphisms $\pi_i : \prod A_i \to A_i$ such that for any object A and morphisms $f_i : A \to A_i$ for each i there is a unique morphism $f : A \to \prod A_i$ with $\pi_i f = f_i$ for each i in I.

Thus the direct product (in categories for which it is defined) is an example of a product in a category. To get back to direct sums we dualize the notion of product by reversing the arrows.

Definition 1.4.24: A *coproduct* of objects $\{A_i : i \in I\}$ in a category $\mathscr{C}$ is an object denoted $\coprod A_i$, together with morphisms $\mu_i : A_i \to \coprod A_i$, such that for any object A and morphisms $f_i : A_i \to A$ for each i there is a unique morphism $f : \coprod A_i \to A$ with $f\mu_i = f_i$ for each i. In other words, f simultaneously completes the diagrams

Example 1.4.25: (The direct sum as a coproduct) Notation as in definition 1.4.24, we see that $\bigoplus M_i$ together with the $\mu_j : M_j \to \bigoplus M_i$ is the coproduct of the M_i. (Indeed for any maps $f_i : M_i \to M$ we define $f : \bigoplus M_i \to M$ by $fx = \sum f_i \pi_i x$, which makes sense since almost all $\pi_i x = 0$; clearly, f is uniquely determined such that $f_i = f\mu_i$ for all i.)

Surprisingly, coproducts of rings may be trivial. $R_1 = \mathbb{Z}/2\mathbb{Z}$ and $R_2 = \mathbb{Z}/3\mathbb{Z}$ have coproduct 0 since any ring R having homomorphisms $\mu_i : R_i \to R$ would satisfy $1 + 1 = \mu_1(1 + 1) = 0$ and $1 + 1 + 1 = \mu_2(1 + 1 + 1) = 0$, implying $1 = 0$, so $R = 0$. In fact, the element 1 is precisely what ruins coproducts for $\mathscr{Ring}$; this is one of the few instances in which we might rather have defined rings *not* to have 1, c.f., exercise 1.5.7. Nevertheless, we can construct a ring "like" a coproduct which is very useful in providing examples.

Example 1.4.27: (Direct sums of rings) Suppose $\{R_i : i \in I\}$ is an infinite set of rings and define the function $\mu_j : R_j \to \prod R_i$ by taking $\mu_j r$ to have all components 0 except the j-th component, which is r. The μ_j satisfy all requirements of ring homomorphism *except* preserving 1; in fact, μ_j identifies R_j with an ideal $\tilde{R}_j = \mu_j R_j \lhd \prod R_i$. Define the *direct sum* $\bigoplus R_i$ of the R_i to be the subring of $\prod R_i$ generated by all the $\tilde{R}_i$. In other words,

$$\bigoplus R_i = \{(m) + \textstyle\sum_{\text{finite}} r_i : r_i \in R_i \text{ and } m \in \mathbb{Z}\}.$$

Since $\pi_j \mu_j = 1$ and $\pi_j \mu_i = 0$ for all $i \neq j$ the direct sum satisfies the following universal property: If $f_i : R_i \to T$ are ring homomorphisms, then there exists a unique ring homomorphism $f : \bigoplus R_i \to T$ such that $f_i = f\mu_i$ for all i.

Indeed, define f by $f((m) + \sum_{\text{finite}} r_i) = m + \sum f_i \pi_i r_i$; to check f is a homomorphism we note

$$f((m) + \textstyle\sum r_i)((m') + \sum r_i') = f((mm') + \sum(mr_i' + m'r_i + r_i r_i'))$$
$$= mm' + \textstyle\sum(mf_i \pi_i r_i' + m'f_i \pi_i r_i + f_i \pi_i(r_i r_i'))$$
$$= (m + \textstyle\sum f_i \pi_i r_i)(m' + \sum f_i \pi_i r_i')$$
$$= f((m) + \textstyle\sum r_i)f((m') + \sum r_i').$$

(The other verifications are instant.) This construction foreshadows the adjunction of 1, to be described in §1.5. See exercise 5 for an application.

Despite our initial disappointment, coproducts of algebras are quite useful in several instances. We need an example for guidance.

Example 1.4.28: Let X be a set of indeterminates and partition $X = \cup X^{(i)}$ where the i merely are superscripts. For any commutative ring C, the free C-algebra $C\{X\}$ is the coproduct of the $C\{X^{(i)}\}$ in C-$\mathcal{Alg}$, where the $\mu_i : C\{X^{(i)}\} \to C\{X\}$ are the canonical inclusion maps. (Since $C\{X^{(i)}\}$ are free C-algebras, the identity action on $X^{(i)}$ produces μ_i.) To see this, suppose we are given $g_i : C\{X^{(i)}\} \to R$ for a C-algebra R; we want to show there is a unique $g : C\{X\} \to R$ satisfying $f_i = g\mu_i$ for each i. Indeed, g and g_i must agree on $X^{(i)}$, so for each indeterminate $X_j^{(i)}$ in $X^{(i)}$, we define the action of g on $X_j^{(i)}$ by $gX_j^{(i)} = g_i X_j^{(i)}$. These combine to give a set-theoretic map $g : X \to R$ which uniquely yields the desired homomorphism $g : C\{X\} \to R$.

Example 1.4.29:   For one indeterminate X_i we see $C\{X_i\}$ is a commutative polynomial algebra in one indeterminate. Thus $C\{X\} = \coprod_{i \in I} C\{X_i\}$ is a coproduct of free *commutative* C-algebras. This example merits further study, but first we note an important application.

Theorem 1.4.30:   *Coproducts exist in C-$\mathcal{Alg}$.*

Proof: Given C-algebras $\{R_i : i \in I\}$, write $R_i = C\{X^{(i)}\}/A_i$ for suitable disjoint sets of indeterminates $X^{(i)}$ and ideals A_i of $C\{X^{(i)}\}$. Let $X = \bigcup X^{(i)}$, so that $C\{X\} = \coprod C\{X^{(i)}\}$ and let $\mu_i : C\{X^{(i)}\} \to C\{X\}$ be the canonical inclusion map (c.f., example 1.4.28). Letting A be the ideal of $C\{X\}$ generated by the $\mu_i A_i$, we put $R = C\{X\}/A$. The composite homomorphism $C\{X^{(i)}\} \to C\{X\} \to C\{X\}/A$ has kernel which contains A_i, so we have an induced homomorphism $\bar{\mu}_i : R_i \to R$. We claim that R together with the $\bar{\mu}_i$ is the coproduct of the R_i.

Suppose we have another C-algebra R' with homomorphisms $\mu_i' : R_i \to R'$. Then we have the homomorphism $v_i : C\{X^{(i)}\} \to R'$ given by the composite $C\{X^{(i)}\} \to R_i \to R'$, which yields a unique homomorphism $v : C\{X\} \to R'$ such that $v_i = v\mu_i$ for each i. Thus $0 = v_i A_i = v(\mu_i A_i)$ for each i, implying $A \subseteq \ker v$. Hence v induces a homomorphism $\bar{v} : R \to R'$ (since $R = C\{X\}/A$), and one has $\mu_i' = \bar{v}\bar{\mu}_i$. The uniqueness of $\bar{v}$ follows from the uniqueness of v, so we have proved the claim.   Q.E.D.

The coproduct of algebras over a field F is called the *free product*.

§1.5 Endomorphism Rings and the Regular Representation

Although we have a fair number of rings at our disposal by now, we lack a unified framework in which to view them. The goal for this section is to introduce endomorphism rings and show that every ring is a subring of a suitable endomorphism ring, thereby providing a theorem parallel to Cayley's theorem in group theory. The benefits are similar, enabling us to construct rings as subrings of known rings without having to verify associativity and distributivity. We shall also see how matrix rings are a special class of endomorphism rings. The underlying idea has its roots in abelian groups. In the foregoing discussion we always assume $M \neq 0$. Any module M can be viewed as an abelian group and thus as $\mathbb{Z}$-module.

Endomorphism Rings

Remark 1.5.1: $\text{Hom}_{\mathbb{Z}}(M, M)$ is a ring whose addition is sum of maps and whose multiplication is composition of maps. Moreover $M \in \text{Hom}_{\mathbb{Z}}(M, M)$-

$\mathcal{M}od$ where the scalar multiplication fx is taken to be the action of f as a map.

When $M \in R\text{-}\mathcal{M}od$ we can transfer the elements of R to $\text{Hom}_{\mathbb{Z}}(M, M)$ as follows:

Definition 1.5.2: For M in $R\text{-}\mathcal{M}od$ and $r \in R$, define the *left multiplication* (also called *left homothety*) map $\rho_r \colon M \to M$ by $\rho_r x = rx$. Likewise, for M in $\mathcal{M}od\text{-}R$ define the *right multiplication map* $\rho_r' \colon M \to M$ by $\rho_r' x = xr$.

Remark 1.5.3: For every r in R, the left multiplication map ρ_r is in $\text{Hom}_{\mathbb{Z}}(M, M)$, and there is a ring homomorphism $\rho \colon R \to \text{Hom}_{\mathbb{Z}}(M, M)$ given by $\rho r = \rho_r$; moreover $\ker \rho = \text{Ann}_R M$. (Verifications are straightforward.)

We should like ρ to be an injection, which is the case iff $\text{Ann}_R M = 0$. Accordingly, we say M is *faithful* if $\text{Ann}_R M = 0$. When $M \in R\text{-}\mathcal{M}od$ is faithful, we call ρ the *(left) regular representation* of R in $\text{Hom}_{\mathbb{Z}}(M, M)$. This can all be done from the right as well. Namely, if $M \in \mathcal{M}od\text{-}R$ define $\rho' \colon R \to \text{Hom}_{\mathbb{Z}}(M, M)$ by $\rho' r = \rho_r'$. However ρ' reverses the order of multiplication $(\rho'(r_1 r_2) = (\rho' r_2)(\rho' r_1))$. Such a map is called an *anti-homomorphism*. Note that in general a map $f \colon R_1 \to R_2$ is a ring anti-homomorphism iff the induced map $f \colon R_1 \to R_2^{\text{op}}$ is a homomorphism, so in order to avoid anti-homomorphisms one must introduce opposite rings from time to time. For example we shall view the right (regular) representation as a homomorphism $\rho' \colon R \to \text{Hom}_{\mathbb{Z}}(M, M)^{\text{op}}$.

The regular representation has many interesting applications.

Example 1.5.4: Viewing R in $R\text{-}\mathcal{M}od$ we have $\text{Ann}_R R = 0$ since $\text{Ann}_R\{1\} = 0$. Thus the regular representation identifies R with a subring of $\text{Hom}_{\mathbb{Z}}(R, R)$.

The regular representation is one of the most important tools in ring theory, and we shall try now to hone it further to obtain many interesting applications.

Proposition 1.5.5: *If $M \in R\text{-}\mathcal{M}od$ then $\text{Hom}_R(M, M)$ is the centralizer of ρR in $\text{Hom}_{\mathbb{Z}}(M, M)$, where $\rho \colon R \to \text{Hom}_{\mathbb{Z}}(M, M)$ is the regular representation.*

Proof: Suppose $f \in \text{Hom}_{\mathbb{Z}}(M, M)$. Then $f \in \text{HOM}_R(M, M)$ iff $f(rx) = rfx$ for all r in R, x in M, iff $f\rho_r = \rho_r f$ for all r in R.   Q.E.D.

In particular, $\mathrm{Hom}_R(M, M)$ is a ring. Furthermore, one sees at once that $M \in R\text{-}\mathscr{Mod}\text{-}\mathrm{Hom}_R(M, M)^{\mathrm{op}}$. Analogously, if $M \in \mathscr{Mod}\text{-}T$ then $\mathrm{Hom}_T(M, M)$ is a ring, the centralizer of $\rho'T$ in $\mathrm{Hom}_{\mathbb{Z}}(M, M)$; consequently, $M \in \mathrm{Hom}_T(M, M)\text{-}\mathscr{Mod}\text{-}T$. To unify these two situations we make the following definitions.

Definition 1.5.6:

(i) If M is an R-module define $\mathrm{End}_R M$ to be $\mathrm{Hom}_R(M, M)^{\mathrm{op}}$, i.e., multiplication in $\mathrm{End}_R M$ is composition of maps in the *reverse* order.

(ii) If M is a right T-module define $\mathrm{End}\, M_T$ to be $\mathrm{Hom}_T(M, M)$.

We hope the asymmetry of this definition will not bother the reader, but it makes later assertions flow more smoothly.

Proposition 1.5.7: *Suppose* $M \in R\text{-}\mathscr{Mod}\text{-}T$ *is faithful as* R-module. *Then* $R \subseteq \mathrm{End}\, M_T$ *by the (left) regular representation.*

Proof: By proposition 1.5.3, we have the injection $\rho: R \to \mathrm{Hom}_{\mathbb{Z}}(M, M)$ which, by hypothesis, centralizes $\rho'T$; so $\rho R \subseteq \mathrm{End}\, M_T$ by proposition 1.5.5.
                                                                                                    Q.E.D.

This method gives us many useful techniques and examples.

Example 1.5.8: If $R \in C\text{-}\mathscr{Alg}$ then taking $M = R$ and $T = C$ in proposition 1.5.7 (identifying $C\text{-}\mathscr{Mod}$ and $\mathscr{Mod}\text{-}C$), we have $R \subseteq \mathrm{End}\, R_C$. However, $\mathrm{End}\, R_C$ is defined using only the C-*module* structure of R. Thus, forgetting the original multiplication of R, we see that every possible multiplication on R which makes R a C-algebra produces a suitable subalgebra of $\mathrm{End}\, R_C$. This result can be used to construct the algebra structure by viewing $R \subseteq \mathrm{End}\, R_C$ in the suitable way, just as Cayley's theorem is used to construct finite groups. Then distributivity and associativity need not be verified since they are inherited from $\mathrm{End}\, R_C$.

Example 1.5.9: $R \approx \mathrm{End}\, R_R$ for every ring R. (Indeed, it suffices to prove the regular representation ρ is onto. Take any f in $\mathrm{End}\, R_R$ and put $r = f1$. For all r' in R we have

$$fr' = f(1r') = (f1)r' = rr' = \rho_r r',$$

proving $f = \rho_r$. Likewise the right regular representation gives an isomorphism from R to $(\mathrm{Hom}_R R)^{\mathrm{op}} = \mathrm{End}_R R$.

Remark 1.5.10: Taking $M = R$ in proposition 1.5.5 we see the centralizer of ρR in $\mathrm{Hom}_Z(R, R)$ is $\mathrm{Hom}_R(R, R) = \rho' R$. By symmetry we see the left and right representations of R are the centralizers of each other in $\mathrm{Hom}_Z(R, R)$.

When R is a C-algebra (cf., example 1.5.8), we conclude *a fortiori* that the left and right representations of R are the centralizers of each other in $\mathrm{Hom}_C(R, R)$. *Thus we see the right regular representation of R gives us a copy* of R in $\mathrm{End}_C R$, whereas the left regular representation gives a copy of R^{op}, and these centralize each other.

Example 1.5.11: If $M \in R\text{-}\mathcal{M}od$ is faithful then taking $T = \mathrm{End}_R M$ we have $R \subseteq \mathrm{End}\, M_T$. An important part of the structure theory is the study of R in terms of $\mathrm{End}\, M_T$, which is called the *biendomorphism ring* of M for obvious reasons.

Example 1.5.12: Suppose $M = R$ and T is a subring of R. Then $R \subseteq \mathrm{End}\, R_T$. This is particularly useful when R is f.g. as T-module, as we shall now see.

Endomorphisms as Matrices

Theorem 1.5.13: *Suppose* $M \in \mathcal{M}od\text{-}T$ *is spanned by* $x_1, \ldots, x_n$. *Then* $\mathrm{End}\, M_T$ *is a homomorphic image of a subring of* $M_n(T)$. *In fact, if* $\{x_1, \ldots, x_n\}$ *is a base of* M *then* $\mathrm{End}\, M_T \approx M_n(T)$.

Proof: Write $R = \{r = (a_{ij}) \in M_n(T)$: there exists β_r in $\mathrm{End}\, M_T$ with $\beta_r x_j = \sum_{i=1}^n x_i a_{ij}$ for all $j\}$. Note that the matrix r determines the action of β_r on the spanning set $\{x_1, \ldots, x_n\}$ and thus on all of M, so we have a function $\varphi: R \to \mathrm{End}\, M_T$ given by $\varphi r = \beta_r$. Moreover, φ is onto since for any β in $\mathrm{End}\, M_T$ we have $\beta x_j = \sum_{i=1}^n x_i a_{ij}$ for suitable a_{ij} in T.

To show R is a ring we note for $r=(a_{ij})$ and $r'=(a'_{ij})$ that $\beta_r - \beta_{r'} = \beta_{r-r'}$ and $\beta_{rr'} = \beta_r \beta_{r'}$ since

$$\beta_r \beta_{r'} x_j = \beta_r\left(\sum_{k=1}^n x_k a'_{kj}\right) = \sum_{k=1}^n (\beta_r x_k)a'_{kj} = \sum_{k=1}^n \sum_{i=1}^n x_i a_{ik} a'_{kj}$$

$$= \sum_{i=1}^n x_i\left(\sum_{k=1}^n a_{ik} a'_{kj}\right) = \beta_{rr'} x_j \qquad \text{for all } x_j.$$

This verification also shows φ is a homomorphism, so we have displayed $\mathrm{End}\, M_T$ as a homomorphic image of $R \subseteq M_n(T)$, as desired. If $\{x_1, \ldots, x_n\}$ is a base then $(M; \{x_1, \ldots, x_n\})$ is free so every matrix $r = (a_{ij})$ defines an endomorphism β_r, implying $R = M_n(T)$; moreover, $\ker \varphi = \{(a_{ij}): \sum_{j=1}^{n} x_i a_{ij} = 0$ for all $i\} = 0$, so φ is an isomorphism from $M_n(T)$ to $\operatorname{End} M_T$. 　　Q.E.D.

Corollary 1.5.14: *Suppose R is a ring with subring T, and as a right T-module R is free with a base of n elements. Then $R \subseteq M_n(T)$.*

Proof: Example 1.5.12 shows $R \subseteq \operatorname{End} R_T \approx M_n(T)$. 　　Q.E.D.

Corollary 1.5.15: *Every finite dimensional algebra over a field is isomorphic to a subalgebra of a suitable matrix algebra.*

Corollary 1.5.16: *If $M \in D\text{-}\mathcal{F}imod$ for a division ring D then $\operatorname{End}_D M \approx M_n(D)$ for some n.*

Proof: Every module over a division ring is free, so apply the theorem.

These important corollaries begin to show us the power of theorem 1.5.13 even in the case when M is free. In fact, the relation between endomorphisms and matrices elevates matrix rings to perhaps the most prominent position in the structure theory of noncommutative rings, leading us to study endomorphisms in terms of matrix properties as in the following remark:

Remark 1.5.17: Suppose M has a base $\{x_1, \ldots, x_n\}$ in $\mathcal{M}od\text{-}T$ and $(a_{ij}) \in M_n(T)$. Then $\{\sum_{i=1}^{n} x_i a_{ij}: 1 \leq j \leq n\}$ spans M iff (a_{ij}) is right invertible (since then we can recover the x_i).

When $M \neq N$ then $\operatorname{Hom}_R(M, N)$ is not a ring, but is nevertheless an abelian group which is also of considerable interest.

Remark 1.5.18: Suppose $M \in R\text{-}\mathcal{M}od$ and $N \in R\text{-}\mathcal{M}od\text{-}T$. Then $\operatorname{Hom}_R(M, N)$ becomes a right T-module with the scalar multiplication fa (for $f: M \to N$ and $a \in T$) given by $(fa)x = (fx)a$ for all x in M. (The verifications are straightforward and left to the reader.) Likewise if $M \in \mathcal{M}od\text{-}T$ then $\operatorname{Hom}_T(M, N) \in R\text{-}\mathcal{M}od$ with scalar multiplication rf given by $(rf)x = r(fx)$. (Incidentally, this latter version "looks like" associativity.)

Remark 1.5.18': (For later use) Suppose $M \in R\text{-}\mathcal{M}od\text{-}S$ and $N \in R\text{-}\mathcal{M}od$.

Then $\operatorname{Hom}_R(M, N) \in S\text{-}\mathscr{M}od$ under the scalar multiplication sf defined by $(sf)x = f(xs)$. (Indeed, $((s_1 s_2)f)x = f(xs_1 s_2) = (s_2 f)(s_1 x) = (s_1(s_2 f))x$.)

Finally, if $M \in R\text{-}\mathscr{M}od\text{-}S$ and $N \in R\text{-}\mathscr{M}od\text{-}T$ then simultaneously we can view $\operatorname{Hom}_R(M, N)$ as S-module (as above) or as right T-module (as in remark 1.5.18), whereby, in fact, $\operatorname{Hom}_R(M, N)$ is an S-T bimodule.

We continue this brief digression to record a useful generalization of example 1.5.9.

Proposition 1.5.19: $\operatorname{Hom}_R(R, M) \approx M$ in $R\text{-}\mathscr{M}od$ where we view $\operatorname{Hom}_R(R, M)$ as R-module via the action rf defined by $(rf)x = f(xr)$ for r in R and $f: M \to R$.

Proof: Define $\psi: M \to \operatorname{Hom}_R(R, M)$ by $\psi y = \rho_y$ where ρ_y is right multiplication by y. For all r, x in R we then have $\psi(ry)x = \rho_{ry}x = xry = \rho_y(xr) = (r\rho_y)x$ so $\psi(ry) = r\psi y$, proving ψ is a map. ψ is monic since $y \neq 0$ implies $0 \neq y = \rho_y 1$; ψ is epic because $f = \rho_{f1}$ for any $f: R \to M$.   Q.E.D.

Proposition 1.5.20: *For any n in $\mathbb{N}$ and M in $R\text{-}\mathscr{M}od\text{-}T$ there is an isomorphism* $\Phi: (\operatorname{End}_R M)^{(n)} \to \operatorname{Hom}_R(M^{(n)}, M)$ *in* $\mathscr{M}od\text{-}T$, *sending* $(f_1, \ldots, f_n)$ *to the map* f *defined by* $f(x_1, \ldots, x_n) = \sum_{i=1}^{n} f_i x_i$.

Proof: Write $\pi_i: M^{(n)} \to M$ for the projection onto the i-th component, and $\mu_i: M \to M^{(n)}$ for the map $\mu_i x = (0, \ldots, 0, x, 0, \ldots, 0)$ where x appears in the i-th component. Then $\Phi(f_1, \ldots, f_n) = \sum f_i \pi_i$. Defining $\Psi: \operatorname{Hom}_R(M^{(n)}, M) \to (\operatorname{End}_R M)^{(n)}$ by $\Psi f = (f\mu_1, \ldots, f\mu_n)$, we see $\Phi \Psi f = \sum_{i=1}^{n} f\mu_i \pi_i = f \sum \mu_i \pi_i = f$ and $\Psi \Phi(f_1, \ldots, f_n) = \Psi \sum f_i \pi_i = (f_1, \ldots, f_n)$. Thus $\Phi = \Psi^{-1}$ and Ψ, Φ are both isomorphisms.   Q.E.D.

The Dual Base

Taking $M = R$ and $T = \mathbb{Z}$ in proposition 1.5.20 and identifying $\operatorname{End}_R R$ with R as in example 1.5.9, we see $R^{(n)} \approx \operatorname{Hom}_R(R^{(n)}, R)$ as abelian groups. This identification can be made more explicit.

Definition 1.5.21: Suppose $M \in R\text{-}\mathscr{M}od$. Taking $N = R \in R\text{-}\mathscr{M}od\text{-}R$ in the natural way, we put $\hat{M} = \operatorname{Hom}_R(M, R)$, viewed in $\mathscr{M}od\text{-}R$ as in remark 1.5.18. If M is free in $R\text{-}\mathscr{F}imod$ with base $\{x_1, \ldots, x_n\}$, define $\hat{x}$ in $\hat{M}$ for each i by $\hat{x}_i \sum_{j=1}^{n} r_j x_j = r_i$ (so that $\hat{x}_i x_j = \delta_{ij}$). We call $\{\hat{x}_1, \ldots, \hat{x}_n\}$ the *dual base* of $x_1, \ldots, x_n$. (An analogous construction exists for M in $\mathscr{M}od\text{-}R$, in which case $\hat{M} \in R\text{-}\mathscr{M}od$.)

Proposition 1.5.22: *If M is free in $R\text{-}\mathcal{F}imod$ with base $\{x_1,\ldots,x_n\}$ then $\{\hat{x}_1,\ldots,\hat{x}_n\}$ is a base of $\hat{M}$ (so, in particular, $\hat{M}$ is free in $\mathcal{M}od\text{-}R$).*

Proof: Any map $f\colon M \to R$ in $\hat{M}$ satisfies $f\left(\sum_{i=1}^n r_i x_i\right) = \sum r_i f x_i$, so letting $a_j = f x_j \in R$ we have $f = \sum_{j=1}^n \hat{x}_j a_j$ (because $\left(\sum \hat{x}_j a_j\right)\left(\sum r_i x_i\right) = \sum \hat{x}_j \left(\sum r_i x_i\right) a_j = \sum r_i a_i = f\left(\sum r_i x_i\right)$). This proves that $\hat{x}_1,\ldots,\hat{x}_n$ span $\hat{M}$. Independence is straightforward; if $\sum_{i=1}^n \hat{x}_i r_i = 0$ then for each i we have $0 = \left(\sum \hat{x}_j r_j\right) x_i = \sum_{j=1}^n (\hat{x}_j x_i) r_j = \sum_{j=1}^n \delta_{ij} r_j = r_i$.   Q.E.D.

There is an enlightening connection between M and $\hat{M}$ using category theory, given in exercises 2, 3. Note that for any map $f\colon M \to N$ we have (by example 0.1.8(ii)) a map $\hat{f}\colon \hat{N} \to \hat{M}$ given by $\hat{f}h = hf$ for any $h\colon N \to R$. This map is especially important when R is a field and M is a finite dimensional vector space, for then the full force of linear algebra applies. In the literature, $\hat{M}$ is often designated as M^*, and we shall also feel free to use the latter notation.

Adjunction of 1

Although we consider only rings with 1, there is an extensive literature on rings without 1 (i.e., all axioms are satisfied except for the existence of 1), which forms a category $\mathcal{R}ng$ whose morphisms φ satisfy all requirements of ring homomorphisms except we obviously need not have $\varphi 1 = 1$. Thus the forgetful functor (forgetting that 1 exists) enables us to view $\mathcal{R}ing$ as a subcategory of $\mathcal{R}ng$, which is not full since the function $\mu\colon \mathbb{Z} \to \mathbb{Z} \times \mathbb{Z}$ given by $\mu n = (n, 0)$ is a morphism in $\mathcal{R}ng$ but not in $\mathcal{R}ing$. The first question one may ask is, given a ring R_0 without 1, what is the universal from R_0 to the forgetful functor?

Definition 1.5.23: The ring R_0' obtained by *adjoining* 1 *formally* to R_0 is the additive group $\mathbb{Z} \oplus R_0$ together with multiplication defined by

$$(m_1, r_1)(m_2, r_2) = (m_1 m_2, m_1 r_2 + m_2 r_1 + r_1 r_2).$$

(The trick here is to think of (m, r) as $m + r$ where $m \in \mathbb{Z}$ and $r \in R$.)

Proposition 1.5.24: *The ring R_0' is indeed a ring and, together with the canonical injection $u\colon R_0 \to R_0'$ given by $ur = (0, r)$, is a universal from R_0 to the forgetful functor.*

Proof: We demonstrate R_0' as a ring by means of the regular representation. Namely, for every (m, r) in $\mathbb{Z} \oplus R_0$ define $\overline{(m, r)}$ in $\text{End}_{\mathbb{Z}}(\mathbb{Z} \oplus R_0)$ by

$(m, r)(m', r') = (mm', mr' + m'r + rr')$ for all (m', r') in $\mathbb{Z} \oplus R_0$. We need to show $(m_1, r_1)(m_2, r_2) = \overline{(m_1, r_1)}\overline{(m_2, r_2)}$ so that R_0' will be identified with a subring of $\mathrm{End}_{\mathbb{Z}}(\mathbb{Z} \oplus R_0)$. Well

$$\overline{(m_1, r_1)}\overline{(m_2, r_2)}(m', r') = \overline{(m_1, r_1)}(m_2 m', m_2 r' + m'r_2 + r_2 r')$$

$$= (m_1 m_2 m', m_1(m_2 r' + m'r_2 + r_2 r') + m_2 m'r_1$$

$$+ r_1(m_2 r' + m'r_2 + r_2 r'))$$

$$= (m_1 m_2, m_1 r_2 + m_2 r_1 + r_1 r_2)(m', r')$$

$$= \overline{(m_1, r_1)(m_2, r_2)}(m', r')$$

as desired. Note that $(1, 0)$ is the multiplicative unit.

To show universality, suppose R is a ring and $f: R_0 \to R$ is a morphism in $\mathscr{Rng}$. We want to extend f to a ring homomorphism $f': R_0' \to R$. Clearly, then, $f'(1, 0) = 1$, so $f'(m, r) = m + fr$ for all r in R_0, and we see this in fact defines a homomorphism since

$$f'((m_1, r_1)(m_2, r_2)) = f'(m_1 m_2, m_1 r_2 + m_2 r_1 + r_1 r_2)$$

$$= m_1 m_2 + f(m_1 r_2 + m_2 r_1 + r_1 r_2)$$

$$= m_1 m_2 + m_1 fr_2 + (fr_1)m_2 + fr_1 fr_2$$

$$= (m_1 + fr_1)(m_2 + fr_2) = f'(m_1, r_1)f'(m_2, r_2) \qquad \text{Q.E.D.}$$

The importance of the injection u is that it permits us to identify R_0 as an ideal of R_0'. Thus rings without 1 should be viewed instead as ideals of rings. One problem with this construction is that if R_0 happens to be a ring (with 1) then R_0' has a different 1. To correct this problem we call R_0 *nondegenerate* if $\mathrm{Ann}_{R_0} R_0 = 0$. In this case the regular representation $R_0 \to \mathrm{End}_{R_0} R_0$ is an injection.

Remark 1.5.25: Let $T = \mathrm{End}_{R_0} R_0$. By proposition 1.5.24 the regular representation $\rho: R_0 \to T$ extends to a homomorphism $\rho': R_0' \to T$ given by $\rho'(m, r) = m + \rho r$, and $\ker \rho' = \{(m, r): mr_0 + rr_0 = 0 \text{ for all } r_0 \text{ in } R_0\} = \mathrm{Ann}_{R_0'} R_0$. Taking $R = R_0'/\ker \rho'$ we have a ring injection $\overline{\rho'}: R \to T$. Let $\bar{u}: R_0 \to R$ be the composite of the canonical homomorphisms $u: R_0 \to R_0'$ and $R_0' \to R_0'/\ker \rho' = R$.

When R_0 is nondegenerate we have $0 = \mathrm{Ann}_{R_0} R_0 = R_0 \cap \ker \rho' = \ker \bar{u}$, so $\bar{u}$ is an injection; we call R the *reduced ring with 1 adjoined to R_0*. This construction has the advantage that if R_0 happens already to have an element 1 then $\bar{u}1 = 1$ because $(1, -1)R_0 = 0$, implying $\bar{u}$ is a ring isomorphism.

§1.6 Automorphisms, Derivations, and Skew Polynomial Rings

In this section we construct a noncommutative analogue of $R[\lambda]$ which still has many of its nice properties. In order to view the construction in context, we consider automorphisms and derivations of rings (leading to Lie algebras).

Automorphisms

Definition 1.6.1:   An *automorphism* is an isomorphism $\sigma: R \to R$. Write R^σ for $\{r \in R: \sigma r = r\}$, a subring of R called the *fixed subring* (under σ).

Remark 1.6.2:   Any automorphism σ of R restricts to an automorphism of $Z(R)$. (Indeed, for any z in $Z(R)$ and r in R, $(\sigma z)r = \sigma(z\sigma^{-1}r) = \sigma((\sigma^{-1}r)z) = r(\sigma z)$, proving $\sigma z \in Z(R)$, so σ restricts to a homomorphism of $Z(R)$, whose inverse, likewise, is the restriction of σ^{-1}.)

   Automorphisms of fields give rise to classical Galois theory. On the other hand, there are many automorphisms of R which in fact fix $Z(R)$. For example, if $a \in R$ is invertible there is an automorphism φ_a of R defined by $\varphi_a r = ara^{-1}$ for all r in R; if $z \in Z(R)$ then $\varphi_a z = aza^{-1} = zaa^{-1} = z$. Automorphisms of the form φ_a for invertible a are called *inner automorphisms*.

Definition 1.6.3:   A *unit* of a ring R is an invertible element. Unit$(R) = \{$units of $R\}$, a multiplicative group; Inn Aut$(R) = \{$inner automorphisms of $R\}$.

Proposition 1.6.4:   *There is a group homomorphism* $\Phi: \text{Unit}(R) \to \text{Aut}(R)$ *given by* $\Phi a = \varphi_a$ *(as defined above). Then* $\Phi(\text{Unit}(R)) = \text{Inn Aut}(R)$ *and* $\ker \Phi = \text{Unit}(R) \cap Z(R)$, *implying* Inn Aut$(R) \approx \text{Unit}(R)/(\text{Unit}(R) \cap Z(R))$ *as groups.*

Proof: For any a, b in Unit(R) and all $r \in R$, $\varphi_a \varphi_b r = \psi_a(brb^{-1}) = abrb^{-1}a^{-1} = \varphi_{ab}r$ proving Φ is a homomorphism; $\ker \Phi = \{a \in R: ara^{-1} = r$ for all r in $R\} = \text{Unit}(R) \cap Z(R)$. The last assertion is now clear.   Q.E.D.

Derivations, Commutators, and Lie Algebras

Definition 1.6.5:   A *derivation* of a ring R is an additive group homomorphism $d: R \to R$ satisfying $d(r_1 r_2) = (dr_1)r_2 + r_1 dr_2$. It is important to note that this definition does *not* depend on the associativity of multiplication, and, in fact, we shall have occasion to deal with derivations of nonassociative algebras, cf., definition 1.6.6.

One motivating example is the usual derivative d on the polynomial ring $R = W[\lambda]$ given by $d(\sum_{i=0}^t w_i \lambda^i) = \sum_{i=1}^t i w_i \lambda^{i-1}$. However, there is another fundamental class of derivations, for which we now aim. Given a ring R and $r_i \in R$, define the *commutator* $[r_1, r_2] = r_1 r_2 - r_2 r_1$. We shall rely heavily on the following computation:

$$[r_1, r_2 r_3] = r_1 r_2 r_3 - r_2 r_3 r_1 = [r_1, r_2]r_3 + r_2[r_1, r_3].$$

In particular, defining the map $d_a : R \rightarrow R$ by $d_a r = [a, r]$, we see d_a is a derivation, called the *inner derivation given by a*. Write $[a, R]$ for $\{[a, r] : r \in R\}$. Then $d_a = 0$ iff $[a, R] = 0$ iff $a \in Z(R)$. Thus we have a map $R \rightarrow \text{Deriv}(R)$ given by $a \rightarrow d_a$, whose kernel is $Z(R)$. To give this map algebraic significance we introduce a new structure.

Definition 1.6.6: A module M over a given commutative ring C is called a *nonassociative algebra* if M has binary multiplication satisfying the following properties for all c in C, x_i in M:

$$c(x_1 x_2) = (cx_1)x_2 = x_1(cx_2).$$

In other words, nonassociative algebras share all properties of associative algebras *except* associativity of multiplication and the existence of the multiplicative unit 1.

Definition 1.6.7: A *Lie algebra* is a nonassociative algebra A where multiplication, written $[a_1 a_2]$ instead of $a_1 a_2$, satisfies the following laws for all a_i in A:

$$[a_1 a_1] = 0$$

$$[[a_1 a_2]a_3] + [[a_2 a_3]a_1] + [[a_3 a_1]a_2] = 0 \text{ (called the Jacobi identity).}$$

To avoid confusion, sometimes we write $[a_1, a_2]$ in place of $[a_1 a_2]$.

Remark 1.6.8: In any Lie algebra, $0 = [a_1 + a_2, a_1 + a_2] = [a_1 a_1] + [a_2 a_1] + [a_1 a_2] + [a_2 a_2] = [a_2 a_1] + [a_1 a_2]$, implying $[a_2 a_1] = -[a_1 a_2]$ for all a_i. This is called the *anticommutative law*.

The reason [ ] is used for multiplication in a Lie algebra arises from the first of the following important examples.

Example 1.6.9: Replacing the usual multiplication of an associative algebra A by the commutator $[a_1, a_2]$ yields a nonassociative algebra, denoted A^-. Clearly $[a, a] = 0$ for all a in A, and, in fact A^- is a Lie algebra; the easy verification of the Jacobi identity is left to the reader. For this reason, the commutator is also called the *Lie product*.

Of course, we want to deal with categories, so we need the morphisms.

Definition 1.6.10: If L_1, L_2 are Lie algebras over C, a *Lie homomorphism* is a C-module map $f: L_1 \to L_2$ satisfying the condition $f[a_1 a_2] = [fa_1, fa_2]$ for all a_1, a_2 in L_1. C-$\mathcal{L}ie$ is the category whose objects are Lie algebras over C and whose morphisms are Lie homomorphisms. If $L \in C$-$\mathcal{L}ie$, a *Lie subalgebra* of L is a C-submodule which is closed under the Lie multiplication of L (and thus is also in C-$\mathcal{L}ie$, because the conditions of definition 1.6.7 follow *a fortiori*).

When mentioning an associative algebra A in the context of C-$\mathcal{L}ie$, we usually mean A^-. For example, a *Lie homomorphism* $f: R \to T$ of (associative) C-algebras is a module map satisfying $f[r_1, r_2] = [fr_1, fr_2]$.

Example 1.6.11: If $R = M_n(C)$ for a commutative ring C, then $M_n(C)^-$ has the following Lie subalgebras:

   (i) the set of matrices of trace 0 (for $\text{tr}[a, b] = \text{tr}(ab) - \text{tr}(ba) = 0$).
   (ii) the set of skew-symmetric matrices.

Proposition 1.6.12: Deriv(R) *is a Lie subalgebra of* $(\text{End}_C(R)^-$, *for every algebra* R.

Proof: (Note that we shall not use associativity of R.) If $d_1, d_2 \in \text{Deriv}(R)$ then

$$d_1 d_2 (r_1 r_2) = d_1((d_2 r_1) r_2 + r_1 d_2 r_2)$$
$$= (d_1 d_2 r_1) r_2 + (d_2 r_1)(d_1 r_2) + (d_1 r_1)(d_2 r_2) + r_1 d_1 d_2 r_2$$

and, symmetrically,

$$d_2 d_1 (r_1 r_2) = (d_2 d_1 r_1) r_2 + (d_1 r_1)(d_2 r_2) + (d_2 r_1)(d_1 r_2) + r_1 d_2 d_1 r_2,$$

implying

$$[d_1, d_2](r_1 r_2) = ([d_1, d_2] r_1) r_2 + r_1([d_1, d_2] r_2),$$

so $[d_1, d_2]$ is a derivation.   Q.E.D.

If L is a Lie algebra, define a *Lie ideal* I to be a submodule satisfying $[LI] \subseteq I$, i.e., $[ax] \in I$ for all a in L, x in I. It is straightforward then to verify that the quotient module L/I has the natural Lie multiplication $[a_1 + I, a_2 + I] = [a_1 a_2] + I$, under which L/I becomes a Lie algebra; if $\Phi: L \to T$ is an onto Lie homomorphism then $\ker \Phi$ is a Lie ideal and $L/\ker \Phi \approx T$ as Lie algebras.

Proposition 1.6.13: *There is a Lie homomorphism* $\Phi: R^- \to \mathrm{Deriv}(R)$ *given by* $a \to d_a$ *(the inner derivation corresponding to a).Writing* $\mathrm{Inn}\ \mathrm{Der}(R)$ *for the set of inner derivations, we have* $\Phi(R) = \mathrm{Inn}\ \mathrm{Der}(R)$ *and* $\ker \Phi = Z(R)$, *implying* $\mathrm{Inn}\ \mathrm{Der}(R) \approx R^-/Z(R)$.

Skew Polynomial Rings and Ore Extensions

We have noted the polynomial ring $R[\lambda]$ over a domain R satisfies

$$\deg(fg) = \deg f + \deg g \tag{1}$$

for all $0 \neq f, g \in R[\lambda]$, and deg 0 is undefined.

In fact, (1) formally implies $R[\lambda]$ is a domain; hence we are led to ask under what general conditions $R[\lambda]$, viewed *only* as an R-module, can be endowed with multiplication making it a ring T graded by $\mathbb{N}$ which satisfies (1). Then T would automatically be a domain and might provide interesting new examples. We start by noting some necessary conditions obtained from polynomials of degree 0 and 1; where $r_i \in R\text{-}\{0\}$ are arbitrary:

(i) $\deg(r_1 r_2) = \deg(r_1) + \deg(r_2) = 0 + 0 = 0,$ so $R \subseteq T$ as domains.

(ii) $\deg(\lambda r) = \deg(\lambda) + \deg(r) = 1 + 0 = 1,$ so $\lambda r = (\sigma r)\lambda + \delta r$
   for suitable maps $\sigma, \delta: R \to R$ with $\ker \sigma = 0$.

(iii) $\sigma(r_1 r_2)\lambda + \delta(r_1 r_2) = \lambda r_1 r_2 = (\lambda r_1)r_2 = ((\sigma r_1)\lambda + \delta r_1)r_2$
$$= (\sigma r_1)(\sigma r_2)\lambda + (\sigma r_1)\delta r_2 + (\delta r_1)r_2;$$

matching components we get

$$\sigma(r_1 r_2) = (\sigma r_1)(\sigma r_2) \quad \text{and} \quad \delta(r_1 r_2) = (\sigma r_1)\delta r_2 + (\delta r_1)r_2.$$

In line with definition 1.6.5, we are led to the following definition.

Definition 1.6.14: If $\sigma: R \to R$ is a homomorphism, a *σ-derivation* is a ($\mathbb{Z}$-module) map $\delta: R \to R$ satisfying $\delta(r_1 r_2) = (\sigma r_1)\delta r_2 + (\delta r_1)r_2$.

Thus for $\sigma = 1$, a σ-derivation is merely a derivation. The properties described above in fact characterize T, as we shall now see:

Proposition 1.6.15:   *If the new ring structure on $R[\lambda]$ satisfies property (1) above, then for all r in R, $\lambda r = (\sigma r)\lambda + \delta r$ for a suitable injection $\sigma: R \to R$ and suitable σ-derivation δ on R. Conversely, if $\sigma: R \to R$ is an injection and $\delta: R \to R$ is a σ-derivation, then there is a unique new multiplication on $R[\lambda]$ (extending the R-module action) such that $\lambda r = (\sigma r)\lambda + \delta r$ for all r in R; moreover, this defines a ring T satisfying (1) and is thus a domain.*

Proof: The first assertion was prove above. For the converse we shall use the regular representation to obtain T as a subring of $\text{End}_Z R[\lambda]$. Define $\bar\lambda$ in $\text{End}_Z R[\lambda]$ by the rule

$$\bar\lambda\left(\sum_{i=0}^{t} r_i\lambda^i\right) = \sum_{i=0}^{t}\left((\sigma r_i)\lambda^{i+1} + (\delta r_i)\lambda^i\right).$$

Writing $\bar r$ for left multiplication by r define $T = \left(\sum_{i=0}^{t} \bar r_i\bar\lambda^i : t \in \mathbb{N}, \text{ all } r_i \in R\right)$. If $\sum \bar r_i\bar\lambda^i = 0$ then $0 = \left(\sum \bar r_i\bar\lambda^i\right)\lambda = \sum r_i\lambda^{i+1}$ so each $r_i = 0$; hence $\{\bar\lambda^i : i \in \mathbb{N}\}$ is a base of T as module over $\bar R = \{\bar r : r \in R\}$.

$$\overline{\lambda r}\left(\sum_{i=0}^{t} r_i\lambda^i\right) = \bar\lambda\left(\sum_{i=0}^{t} rr_i\lambda^i\right) = \sum\left(\sigma(rr_i)\lambda^{i+1} + \delta(rr_i)\lambda^i\right)$$

$$= \sum\left((\sigma r)(\sigma r_i)\lambda^{i+1} + ((\sigma r)\delta r_i + (\delta r)r_i)\lambda^i\right)$$

$$= \overline{\sigma r}\sum(\sigma r_i\lambda^{i+1} + \delta r_i\lambda^i) + \overline{\delta r}\sum r_i\lambda^i = (\overline{\sigma r}\bar\lambda + \overline{\delta r})\sum r_i\lambda^i,$$

implying $\overline{\lambda r} = \overline{\sigma r}\bar\lambda + \overline{\delta r}$. Therefore, T is closed under multiplication and is thus a subring of $\text{End}_Z R[\lambda]$; identifying R with $\bar R$ we see at once that T satisfies (1) (and thus is a domain). Uniqueness of the multiplication follows easily from (1).   Q.E.D.

Definition 1.6.16:   Suppose $\sigma: R \to R$ is an injection and δ is a σ-derivation. The *Ore extension* $R[\lambda; \sigma, \delta]$ is the ring obtained by giving $R[\lambda]$ the new multiplication $\lambda r = (\sigma r)\lambda + \delta r$, as in proposition 1.6.15. If $\delta = 0$ we write $R[\lambda; \sigma]$ for $R[\lambda; \sigma, \delta]$; this is called a *skew polynomial* ring. $R[\lambda; 1, \delta]$ is called a *differential polynomial ring* in the literature.

Note that an Ore extension of a domain is a domain, by proposition 1.6.15.

Remark 1.6.17:   A necessary and sufficient condition that each element of $R[\lambda; \sigma, \delta]$ can be written in the form $\sum_{i=0}^{t} \lambda^i r_i$ is that σ is onto (i.e., σ is an automorphism). (Indeed, if any $r\lambda$ can be written as $\sum_{i=0}^{t} \lambda^i r_i$, then clearly $t = 1$, so $r\lambda = (\sigma r_1)\lambda + (\delta r_1) + r_0$, proving $\sigma r_1 = r$ and thus σ is onto. Conversely, if σ is an automorphism then $r\lambda = \lambda\sigma^{-1}r - \delta\sigma^{-1}r$, and then one sees easily by induction on j that each term $r\lambda^j$ can be written in the form $\sum_{i=0}^{j} \lambda^i r_i$, and thus each element of $R[\lambda; \sigma, \delta]$ can be written in this form.)

Principal Left Ideal Domains (PLID's)

Let us now study the structure of skew polynomial rings more closely. Recall that $F[\lambda]$ is a PID (principal ideal domain) whenever F is a field, and this property is used repeatedly in the commutative theory. The noncommutative analogue is a straightforward generalization, but we shall see later that the asymmetry of the construction of skew polynomial rings gives them some fascinating properties.

Definition 1.6.18: A *principal left ideal domain* (PLID) is a domain in which every left ideal is principal (i.e., of the form Rr).

Remark 1.6.19: Every PLID satisfies the *left Ore* property that $Rr_1 \cap Rr_2 \neq 0$ for any nonzero elements r_1, r_2 of R. (Indeed, we are done unless $r_2 \notin Rr_1$, in which case $Rr_1 + Rr_2 = Ra$ with $a \notin Rr_1$. Then $a = x_1r_1 + x_2r_2$ for x_i in R with $x_2 \neq 0$. But for some $r \neq 0$, we have

$$r_1 = ra = r(x_1r_1 + x_2r_2) = rx_1r_1 + rx_2r_2$$

implying $0 \neq rx_2r_2 = (1 - rx_1)r_1 \in Rr_1 \cap Rr_2$.) The left Ore property is of utmost importance in constructing rings of fractions, c.f., §3.1.

The link between skew polynomial rings and PLIDs lies in the *Euclidean algorithm*.

Remark 1.6.20: Suppose $0 \neq f, g \in R[\lambda; \sigma, \delta]$ where the leading coefficient of g is invertible in R. Then $f = qg + p$ for suitable q, p in $R[\lambda; \sigma, \delta]$ such that either $p = 0$ or $\deg p < \deg g$. (Indeed, let m, n be the respective degrees of f and g. We proceed inductively on m. If $n > m$ then we are done with $q = 0$ and $p = f$. Otherwise, letting $a\lambda^m, b\lambda^n$ be the respective leading terms of f, g, we note $a\lambda^m = (a\lambda^{m-n}b^{-1})b\lambda^n$, so putting $h = f - a\lambda^{m-n}b^{-1}g$, we see $\deg h < m$ and, inductively, $h = q'g + p$ for suitable q', p with $p = 0$ or $\deg p < n$. Hence $f = (a\lambda^{m-n}b^{-1} + q')g + p$.)

Of course, the hypothesis of remark 1.6.20 holds when R is a division ring, leading to the next result.

Proposition 1.6.21:   *Suppose D is a division ring. Then $T = D[\lambda; \sigma, \delta]$ is a PLID. Moreover, if $Tf = Tg$ then $f = dg$ for some d in D; in particular, $\deg f = \deg g$ and f, g have the same number of (nonzero) terms.*

Proof: A left ideal L is generated by any element g in L of lowest degree (since if $f \in L$ we write $f = qg + p$ implying $p \in L$ so $p = 0$). Now if $Tf = Tg$ we see $f = qg$ implies $\deg f \geq \deg g$; likewise, $\deg g \geq \deg f$; hence $\deg q = 0$ so $q \in D$ as desired.   Q.E.D.

It is of interest to check when T is a principal *right* ideal domain (PRID).

Proposition 1.6.22:   *The following assertions are equivalent for $T = D[\lambda; \sigma, \delta]$:*

(i) *σ is onto.*
(ii) *T is a PRID.*
(iii) *T is right Ore.*

Proof:

(i) $\Rightarrow$ (ii) follows from remark 1.6.17 since the right-handed analogue of proposition 1.6.21 is applicable.

(ii) $\Rightarrow$ (iii) is remark 1.6.19.

(iii) $\Rightarrow$ (i). For any d in D we have some nonzero $h \in \lambda T \cap d\lambda T$, so $h = \lambda f = d\lambda g$ for suitable f, g in T. Let f, g have leading terms $a\lambda^n$ and $b\lambda^n$, respectively; then the leading term of h is $\sigma a \lambda^{n+1} = d\sigma b \lambda^{n+1}$, so $\sigma a = d\sigma b$, implying $d = \sigma(ab^{-1}) \in \sigma D$, proving σ is onto.   Q.E.D.

A minor modification of the Euclidean algorithm permits us to generalize proposition 1.6.21.

Remark 1.6.23:   Suppose $0 \neq f, g \in R[\lambda; \sigma, \delta]$ with $\deg f > \deg g$, and σr is invertible in R where r is the leading coefficient of g. Then we can write $f = qg + p$ where either $p = 0$ or $\deg p \leq \deg g$. (Indeed, write $g = r\lambda^n + \sum_{i=0}^{n-1} r_i \lambda^i$. The leading term of λg is $(\sigma r)\lambda^{n+1}$, so apply remark 1.6.20 to f and λg.)

Proposition 1.6.24:   *If R is a PLID with $\sigma(R - \{0\}) \subseteq \text{Unit}(R)$ then $T = R[\lambda; \sigma, \delta]$ is a PLID.*

Proof: Given a left ideal L of T let $L_0 = \{$leading coefficients of polynomials of L having minimal degree, $n\}$. Then $L_0 < R$ so $L_0 = Tr_0$ for some r_0 in R. Take $g \in L$ of degree n having leading coefficient r_0. We claim $L = Tg$. Indeed take any $f \in L\text{-}Tg$. Using remark 1.6.23 we may replace f by p and assume $\deg f = n$ so the leading coefficient of f is rr_0 for some r in R; then $f - rg$ is an element of L of degree $< n$ so $f - rg = 0$, i.e., $f = rg \in Tg$.   Q.E.D.

Of course this result only adds interest when σ is *not* onto. However, it is important for two reasons: (i) the converse is true (cf., exercise 3) and the condition $\sigma(R - \{0\}) \subseteq \text{Unit}(R)$ permits an iterative construction which provides many interesting examples of Jategaonkar [69]; consequently, the condition is called *Jategaonkar's condition*. Presently we shall restrict our attention to special cases in order to introduce several important classes of rings.

Skew Polynomial Rings (Without Derivation) Over Fields

First we consider the case $\delta = 0$, i.e., $T = K[\lambda; \sigma]$. Clearly, $T\lambda^i \lhd T$ for i, and, more generally, $Tz\lambda^i \lhd T$ for every z in $Z(T)$ and all i. Surprisingly enough, every ideal has this form. Let K^σ denote $\{a \in K : \sigma a = a\}$

Proposition 1.6.25: *Suppose K is a field having an endomorphism σ. Let $T = K[\lambda; \sigma]$ and $F = K^\sigma$.*

(i) *$Z(T) = F$ unless σ is an automorphism of order n, in which case $Z(T) = F[\lambda^n]$.*

(ii) *Every ideal of T has the form $Tp\lambda^m$ for p in $Z(T)$.*

Proof:

(i) Obviously $F[\lambda^n] \subseteq Z(T)$. Conversely, suppose $f = \sum a_i \lambda^i \in Z(T)$ where $a_i \in K$. Then $0 = [\lambda, f] = \sum (\sigma a_i - a_i)\lambda^{i+1}$ proving $\sigma a_i = a_i$ so $a_i \in F$ for each i. Moreover, for every a in K we have $0 = [a, f] = \sum a_i(a - \sigma^i a)\lambda^i$, implying $a = \sigma^i a$ whenever $a_i \neq 0$, i.e., $\sigma^i = 1$ so n divides i. Hence $f \in F[\lambda^n]$ as desired.

(ii) Suppose $A \lhd T$ and choose $f \neq 0$ in A of minimal degree t. By proposition 1.6.21 we have $A = Tf$ and f has the same number of (nonzero) terms as any other polynomial in A of degree t. Replacing f by $a_t^{-1}f$ we may assume $a_t = 1$. Now $[\lambda, f] = \sum_{i=0}^{t-1}(\sigma a_i - a_i)\lambda^{i+1}$ has fewer terms than f and thus must be 0 by the above discussion. Hence each $a_i \in F$. Moreover, for every a in K we see $(\sigma^t a)f - fa$ has degree $\leq t - 1$ and is in A, so must be 0, implying

$(\sigma^t a)a_i = a_i \sigma^i a$ for each i. Hence $\sigma^t a = \sigma^i a$ whenever $a_i \neq 0$, implying $\sigma^{t-i} = 1$ so $n \mid (t - i)$. Taking m to be the degree of the lowest order term of f we conclude $f \in F[\lambda^n]\lambda^m = Z(T)\lambda^m$.   Q.E.D.

(The proof given above is redundant, for the proof of (ii) also implies (i). Indeed if $f \in Z(T)$ then $Tf \lhd R$ and thus $f = ap\lambda^m$ for some a in K and p in $F[\lambda^n]$, from which it readily follows $a \in F$ and $m = 0$. However, the argument of (ii) is also subtler than that of (i).)

Example 1.6.26: Let K be a field with an endomorphism σ which is *not* onto, e.g., $K = \mathbb{Q}(\lambda)$ and σ is given by $\sigma\lambda = \lambda^2$. (For $\sigma(f(\lambda)g(\lambda)^{-1}) = f(\lambda^2)g(\lambda^2)^{-1}$; one sees at once $\lambda \notin \sigma K$, for if $\lambda = \sigma(fg^{-1})$ then $g(\lambda^2)\lambda = f(\lambda^2)$ contradicting the parities of degree.) $K[\lambda; \sigma]$ is a PLID which is *not* a PRID by proposition 1.6.22; using proposition 1.6.25, we see the ideals only have the form $\langle \lambda^i \rangle$, so, in particular, the ideals form a chain.

Corollary 1.6.27:   *Suppose σ is an automorphism of K of order n, and $F = K^\sigma$. If $p(\lambda) \neq \lambda$ is an irreducible polynomial of $F[\lambda]$, then $\langle p(\lambda^n) \rangle$ is a maximal ideal of $T = K[\lambda; \sigma]$ and $T/\langle p(\lambda^n) \rangle$ is a simple ring.*

Proof: Any ideal containing $\langle p(\lambda^n) \rangle$ has the form $Tq(\lambda^n)\lambda^m$ where $q \in F[\lambda]$, so $p(\lambda^n) \in Tq(\lambda^n)\lambda^m$. If f is an irreducible factor of $q(\lambda^n)$ in $F[\lambda]$ then $f \mid p(\lambda^n)$. If there were g, h in $F[\lambda]$ such that $gp + hq = 1$, then $g(\lambda^n)p(\lambda^n) + h(\lambda^n)q(\lambda^n) = 1$ contrary to the existence of f. Hence, p, q are *not* relatively prime in $F[\lambda]$ so $p \mid q$ implying $\langle p(\lambda^n) \rangle$ indeed is maximal.   Q.E.D.

Example 1.6.28: (Cyclic algebras) Let $R = K[\lambda; \sigma]/\langle \lambda^n - \alpha \rangle$ where σ is an automorphism of order n and $\alpha \in F = K^\sigma$. Since $\lambda - \alpha$ is obviously irreducible in $F[\lambda]$, we see that R is a simple ring which we call the *cyclic algebra* (K, σ, α).

Let z be the canonical image of λ in R and identify K with its image in R. Then $R = \sum_{i=0}^{n-1} Kz^i$, and $\{z^i : 0 \leq i < n\}$ is a base for R as a vector space over K (for if $\sum_{i=0}^{n-1} a_i z^i = 0$ with some $a_i \neq 0$ then $\sum_{i=0}^{n-1} a_i \lambda^i \in \langle \lambda^n - \alpha \rangle$ which is impossible by a degree comparison). Multiplication is given by the straightforward rule

$$a_i z^i a_j z^j = \begin{cases} (a_i \sigma^i a_j)z^{i+j} & \text{if } i + j < n \\ (\alpha a_i \sigma^i a_j)z^{i+j-n} & \text{if } i + j \geq n. \end{cases}$$

In particular, $z^n = \alpha \in F$. Moreover, $F = Z(R)$ (for if $\sum_{i=0}^{n-1} a_i z^i \in Z(R)$, then $0 = [z, \sum a_i z^i] = \sum(\sigma a_i - a_i)z^{i+1}$, showing each $a_i \in F$; then $0 = [a, \sum a_i z^i] = \sum a_i(a - \sigma^i a)z^i$ for every a in K, so taking a with $\sigma^i a \neq a$ we see $a_i = 0$ for each $i \neq 0$.) Now $n = [K:F] = [R:K]$ so $[R:F] = n^2$.

Thus each cyclic algebra is a simple algebra of dimension n^2 over its center for suitable n. This construction generalizes $\mathbb{H} = (\mathbb{C}, \sigma, -1)$ where σ is complex conjugation, and is the key to much of the theory of finite dimensional division algebras, which will be discussed in Chapter 7 in much greater detail.

Differential Polynomial Rings Over Fields

Now we consider the opposite extreme, where $\sigma = 1$. Some of the methods are similar, but first we need to see how a derivation δ "works" on a ring.

Remark 1.6.29: ("Leibniz's formula") The following rule is seen by induction on n:

$$\delta^n(ab) = \sum_{i=0}^{n} \binom{n}{i} \delta^i a \delta^{n-i} b$$

Leibniz's formula is used extensively in studying derivations. Our next result relies on the obvious consequence that if $\delta^i a = 0$ and $\delta^j b = 0$ then $\delta^{i+j-1}(ab) = 0$.

Remark 1.6.30: Suppose δ is a derivation of a ring R, and put $R_i = \{r \in R : \delta^{i+1} r = 0\}$ and $R_\infty = \bigcup_{i \in \mathbb{N}} R_i \subseteq R$. Then R_∞ is a ring and satisfies $R_i R_j \subseteq R_{i+j}$. In particular, R_0 is a subring; if $r \in R_0$ and $r^{-1} \in R$ then $r^{-1} \in R_0$. (Indeed, first note $\delta 1 = \delta(1 \cdot 1) = \delta 1 + \delta 1$ implying $\delta 1 = 0$ and R_0 is a subring. Then $0 = \delta(rr^{-1}) = (\delta r)r^{-1} + r\delta r^{-1} = r\delta r^{-1}$ for any r in R_0 which is invertible, implying $\delta r^{-1} = 0$.)

Lemma 1.6.31: *Notation as in remark* 1.6.30. *Suppose* $R = R_\infty$ *and* $T = R[\lambda; 1, \delta]$. *if* A *is any subset of* T *such that* $[\lambda, A] \subseteq A$ *then*

$$A \cap R_0[\lambda] \neq 0.$$

(*Note* $R_0[\lambda]$ *is the usual polynomial ring.*)

Proof: Of all the nonzero elements of A of minimal degree t, take $f = \sum_{i=0}^{t} r_i \lambda^i$ such that $\delta^n r_t = 0$ for n minimal possible. (By hypothesis some $\delta^n r_t = 0$.) Then $\sum_{i=0}^{t}(\delta r_i)\lambda^i = [\lambda, f] \in A$ and $\delta^{n-1}(\delta r_t) = 0$, so by assumption we see $[\lambda, f] = 0$, i.e., all $\delta r_i = 0$.   Q.E.D.

The Weyl Algebra

We are now going to define a ring which is very important both in the structure of enveloping algebras (§8.3 below) and also in its own right.

Example 1.6.32: Suppose C is a commutative ring. The *Weyl algebra* $\mathscr{A}_1(C) = C\{\mu, \lambda\}/\langle \lambda\mu - \mu\lambda - 1\rangle$ where μ, λ are indeterminates over C. In other words, we have $\overline{\lambda\mu} = \overline{\mu\lambda} + 1$ in $\mathscr{A}_1(C)$. More generally, taking $\mu_1, \ldots, \mu_n, \lambda_1, \ldots, \lambda_n$ to be indeterminates over C, we define $\mathscr{A}_n(C) = C\{\mu_1, \ldots, \mu_n, \lambda_1, \ldots, \lambda_n\}/B$ where $B = \langle [\lambda_i, \mu_i] - 1, [\lambda_i, \mu_j], [\lambda_i, \lambda_j], [\mu_i, \mu_j]$ for all $i \neq j\rangle$. When C is understood, we merely write $\mathscr{A}_n$ for $\mathscr{A}_n(C)$.

There is an alternate, more explicit description. Let $W_1 = R[\lambda; 1, \delta]$ where $R = C[\mu]$ and δ is differentiation with respect to μ. Note that $\lambda f - f\lambda = \delta f$, so, in particular, $\lambda\mu = \mu\lambda + 1$, and we have a canonical isomorphism $\mathscr{A}_1 \approx W_1$. Proceeding inductively we put $R_{(n-1)} = \mathscr{A}_{n-1}[\mu_n]$ and have $\mathscr{A}_n \approx R_{(n-1)}[\lambda_n; 1, \delta_n]$ where δ_n is partial differentiation with respect to μ_n.

Because of this description, the notation $C\left[x_1, \ldots, x_n, \dfrac{\partial}{\partial x_1}, \ldots, \dfrac{\partial}{\partial x_n}\right]$ for $\mathscr{A}_n(C)$ has become widespread in the literature. (Thus x_i replaces μ_i, and $\partial/\partial x_i$ replaces λ_i.) More generally, there has been increasing algebraic interest in *differential operator* rings, cf., recent work by K. Brown, J. McConnell, S. P. Smith, and T. Stafford.

On the other hand, $-\mu\lambda = \lambda(-\mu) + 1$ so $\mathscr{A}_n \approx (\mathscr{A}_{n-1}[\lambda_n])[-\mu_n; 1, \delta'_n]$ where δ'_n is partial differentiation with respect to λ_n. (By convention take $\mathscr{A}_0 = C$). This is used to prove the following key result.

Proposition 1.6.33: *If C is a commutative $\mathbb{Q}$-algebra then every Lie ideal L of $\mathscr{A}_n$ intersects C nontrivially.*

Proof: In the notation of example 1.6.32 we have $L \cap R_{(n-1)} \neq 0$ by lemma 1.6.31. But viewing instead $\mathscr{A}_n$ as $(\mathscr{A}_{n-1}[\lambda_n])[-\mu_n; 1, \delta'_n]$ we apply lemma 1.6.31 to $L \cap R_{(n-1)}$ to get $0 \neq (L \cap R_{(n-1)}) \cap \mathscr{A}_{n-1}[\lambda_n] = L \cap \mathscr{A}_{n-1}$, a Lie ideal of $\mathscr{A}_{n-1}$. Continuing inductively we conclude $0 \neq L \cap \mathscr{A}_0 = L \cap C$.   Q.E.D.

Corollary 1.6.34: *$\mathscr{A}_n(F)$ is a simple domain, for any field F of characteristic 0. More generally, if C is a commutative $\mathbb{Q}$-algebra then every nonzero ideal B of $\mathscr{A}_n(C)$ is generated by central elements, and $\mathscr{A}_n(C)/B \approx \mathscr{A}_n(C/(B \cap C))$.*

Proof: Let $B' = B \cap C \neq 0$, and write R for $\mathscr{A}_n(C)$. The canonical map $R \to \mathscr{A}_n(C/B')$, obtained by taking coefficients modulo B', clearly has kernel RB', so $R/RB' \approx \mathscr{A}_n(C/B')$. If $RB' \subset B$ then B/RB' corresponds to a nonzero ideal of $\mathscr{A}_n(C/B')$, which contains a nonzero element of $C/B' =$

$C/(RB' \cap C) \approx (C + RB')/RB'$, so B contains an element of $C - RB'$, contradiction. Hence $RB' = B$ and $R/B \approx \mathscr{A}_n(C/B')$ as desired.   Q.E.D.

Let us return to the general setting.

Lemma 1.6.35: *Suppose* $R = D[\lambda; 1, \delta]$ *where* D *is a division ring, and* $0 \neq A \lhd R$. *Then* $A = Ra$ *for suitable* a *which centralizes* D; *if* $\mathrm{char}(D) = 0$ *then* δ *is inner.*

Proof: By proposition 1.6.21 we have $A = Ra$ where $a = \lambda^t + \sum_{i=0}^{t-1} d_i \lambda^i$. Then for each d in D we have

$$ad = d\lambda^t + (t\delta d + d_{t-1}d)\lambda^{t-1} + \cdots$$

Now $ad \in A = Ra$ and $\deg(ad) = t = \deg(a)$ so $ad \in Da$; comparing coefficients of λ^t proves $ad = da$.

Now comparing coefficients of λ^{t-1} shows $t\delta d + d_{t-1}d = dd_{t-1}$; so $\delta d = [d, t^{-1}d_{t-1}]$ if $t \nmid \mathrm{char}(D)$.   Q.E.D.

Proposition 1.6.36: *Suppose* δ *is a derivation on a division algebra* D.

(i) *If* δ *is inner given by* $[d, \underline{\hphantom{x}}]$ *then there is an isomorphism from the polynomial ring* $D[\lambda]$ *to* $D[\lambda; 1, \delta]$ *given by* $\lambda \to \lambda + d$, *and every ideal is generated by a central element.*

(ii) *If* δ *is not inner and* $\mathrm{char}(D) = 0$ *then* $D[\lambda; 1, \delta]$ *is simple.*

Proof: (i) $[\lambda - d, \sum d_i \lambda^i] = \sum([\lambda, d_i]\lambda^i - [d, d_i]\lambda^i) = \sum(\delta d_i \lambda^i - \delta d_i \lambda^i) = 0$, so $\lambda - d$ acts like a commuting indeterminate, and we have the first assertion. Thus the second assertion can be checked in $D[\lambda]$. Let $A \lhd D[\lambda]$. By the lemma we have $A = D[\lambda]a$ where a centralizes D and thus is in $Z(D[\lambda])$.

(ii) The contrapositive of the lemma.   Q.E.D.

Skew Power Series and Skew Laurent Series

When $\delta = 0$ we can define a filtration on $R[\lambda; \sigma]$ according to the lowest order monomial, as in example 1.2.26, which leads us to look for a generalization of power series rings and Laurent series in this case.

Definition 1.6.37: The *skew power series ring* $R[[\lambda; \sigma]]$ has the same additive structure as $R[[\lambda]]$ but with multiplication given by

$$\left(\sum_{i=0}^{\infty} r_i \lambda^i \right)\left(\sum_{i=0}^{\infty} r_i' \lambda^i \right) = \sum_{i=0}^{\infty} \left(\sum_{u=0}^{i} r_u \sigma^u r_{i-u}' \right)\lambda^i.$$

(We assume $\sigma: R \to R$ is an injection.)

Proposition 1.6.38: $A = R[[\lambda; \sigma]]$ *is indeed a ring containing* $R[\lambda; \sigma]$ *canonically and has a filtration defined by putting* $A(t) = \{\sum_{i=0}^{\infty} r_i \lambda^i : r_i = 0 \text{ for all } i < t\}$. *Consequently* A *is a domain if* R *is a domain.*

Proof: We prove A is a ring by using the regular representation as in the proof of proposition 1.6.15. The remainder of the proposition follows the discussion of filtrations in §1.2.   Q.E.D.

Definition 1.6.39: The *skew Laurent series ring* $R((\lambda; \sigma))$, is the ring whose additive structure is that of $R((\lambda)) = \{\sum_{i=m}^{\infty} r_i \lambda^i : m \in \mathbb{Z}, r_i \in R\}$ but with multiplication given by $(\sum_{i=m}^{\infty} r_i \lambda^i)(\sum_{i=n}^{\infty} r_i' \lambda^i) = \sum_{i=m+n}^{\infty} (\sum_{u=m}^{i-n} r_u \sigma^u r_{i-u}') \lambda^i$.

There is an important subring denoted $R[\lambda, \lambda^{-1}; \sigma]$ consisting of those elements having finite support, i.e., $\sum_{i=m}^{n} r_i \lambda^i$ where $m \leq n$ are in $\mathbb{Z}$. Unfortunately $R[\lambda, \lambda^{-1}; \sigma]$ is called a *Laurent extension of* R in the literature, but at least the notation is unambiguous. We write $R[\lambda; \lambda^{-1}]$ for $R[\lambda; \lambda^{-1}; 1]$.

Skew Group Rings

It is easy to check that $R[\lambda, \lambda^{-1}]$ is isomorphic to the group ring $R[\mathbb{Z}]$, so we are led to try to "skew" the group ring construction.

Definition 1.6.40: Suppose G is a group of automorphisms of R. Define the *skew group ring* $R * G = \{\sum_{g \in G} r_g g : r_g \in R\}$ with multiplication given by the rule $(rg)(sh) = (rs^g)gh$, where gh is the product in G, and s^g denotes the element $g(s)$ in R.

We leave it to the reader to check that $R * G$ is indeed a ring and that $R * \langle \sigma \rangle \approx R[\lambda, \lambda^{-1}; \sigma]$ when σ is an automorphism of R having infinite order. We shall refer to skew group rings only occasionally in the sequel.

§1.7 Tensor Products

The next construction to be considered in this chapter is the tensor product. This is an extremely important tool, some of whose uses will be outlined following the definition. We start with M in $\mathscr{M}od$-R and N in R-$\mathscr{M}od$. Given an abelian group A we say $\psi: M \times N \to A$ is a *balanced map* if ψ satisfies the three conditions

$$\psi(x_1 + x_2, y) = \psi(x_1, y_1) + \psi(x_2, y_1)$$

$$\psi(x_1, y_1 + y_2) = \psi(x_1, y_1) + \psi(x_1, y_2)$$

$$\psi(x_1 r, y_1) = \psi(x_1, r y_1)$$

for all x_i in M, y_i in N, and r in R. This definition closely resembles that of bilinear form (cf., exercise 3). The tensor product turns out to be a universal for a suitable category involving balanced maps.

Definition 1.7.1: The *tensor product* $M \otimes N$ is the abelian group G/H, where G is freely generated (as $\mathbb{Z}$-module) by the Cartesian product $M \times N$, and H is the subgroup of G generated by all elements of the form

$$(x_1 + x_2, y_1) - (x_1, y_1) - (x_2, y_1)$$

$$(x_1, y_1 + y_2) - (x_1, y_1) - (x_1, y_2)$$

$$(x_1 r, y_1) - (x_1, r y_1)$$

for x_i in M, y_i in N, and r in R. Write $x \otimes y$ for the image of (x, y) in $M \otimes N$. If R is in doubt, we write $M \otimes_R N$ in place of $M \otimes N$.

Digression 1.7.2: Historically, the tensor product arose from a need to "multiply" two finite dimensional algebras A and B over a field F. Explicitly, given respective bases $a_1, \ldots, a_m$ and $b_1, \ldots, b_n$ of A and B over F, we could define $A \otimes B$ to be the vector space of dimension mn over F, whose base is written formally as $\{a_i \otimes b_t : 1 \le i \le m, 1 \le t \le n\}$. To define multiplication, one notes that multiplication in A and B are determined, respectively, by $a_i a_j = \sum_{k=1}^{m} \alpha_{ijk} a_k$ and $b_t b_u = \sum_{v=1}^{n} \beta_{tuv} b_v$ for suitable $\alpha_{ijk}, \beta_{tuv}$ in F; then one could define

$$(a_i \otimes b_t)(a_j \otimes b_u) = \sum_{k,v} \alpha_{ijk} \beta_{tuv} a_k \otimes b_v,$$

and extend multiplication distributively to $A \otimes B$.

The standard present-day treatment is much more general and elegant and still enables us to recover the above construction (cf., corollary 1.7.23 below). The approach is completely different. Whereas the $x \otimes y$ span $M \otimes N$ in $\mathbb{Z}$-*Mod* they are not independent, so we cannot use them to define operations on $M \otimes N$. Instead, we rely almost exclusively on the following universal property.

Proposition 1.7.3: *There is a balanced map* $\psi : M \times N \to M \otimes N$ *given by* $\psi(x, y) = x \otimes y$. *Moreover, if* $\varphi : M \times N \to A$ *is any balanced map there is a unique group homomorphism* $\bar{\varphi} : M \otimes N \to A$ *such that*

$$\bar{\varphi}(x \otimes y) = \varphi(x, y).$$

Proof: The balanced map conditions are automatic from definition 1.7.1. To prove the other assertion, note φ extends naturally to a group homomorphism $\varphi: G \to A$ given by $\varphi \sum(x_i, y_i) = \sum \varphi(x_i, y_i)$ (notation as in definition 1.7.1); then $H \subseteq \ker \varphi$ yielding the desired $\bar{\varphi}: G/H \to A$ by Noether's structure theorem.   Q.E.D.

Corollary 1.7.4: *Suppose $f: M \to M'$ and $g: N \to N'$ are maps in $\mathcal{M}od$-R and R-$\mathcal{M}od$, respectively. Then there is a group homomorphism denoted $f \otimes g$: $M \otimes N \to M' \otimes N'$ such that $(f \otimes g)(x \otimes y) = fx \otimes gy$.*

Proof: Define $\varphi: M \times N \to M' \otimes N'$ by $\psi(x, y) = fx \otimes gy$. Then φ is balanced; take $f \otimes g$ to be $\bar{\varphi}$.   Q.E.D.

Most proofs about tensor products rely on this result, and are omitted in this Student Edition; the reader can check the unabridged edition or Jacobson's *Basic Algebra II*.

Remark 1.7.5:   Tensor products serve the following purposes: (i) keeping track of balanced maps; (ii) changing rings of scalars (for modules and algebras); (iii) providing a "multiplication" of algebras (over a common base ring); and (iv) providing universal constructions, such as the "tensor algebra".

These uses of tensor product will occupy us for much of the book and could easily fill several volumes. For example, (iii) is the setting for the Brauer group and the ensuing theory of central simple algebras (Chapter 7) and Azumaya algebras (§5.3). For the remainder of this section we shall introduce some of the main themes involving tensor products, largely postponing their very important categorical role until Chapter 4.

Remark 1.7.6:   If $f: M \to M'$ and $f': M' \to M''$ and $g: N \to N'$ and g': $N' \to N''$ are maps then $f'f \otimes g'g = (f' \otimes g')(f \otimes g)$. (Just check this on $x \otimes y$, which span $M \otimes N$.) Consequently, there is a tensor functor $\bigotimes_R$ from $(\mathcal{M}od$-$R) \times (R$-$\mathcal{M}od)$ to $\mathcal{A}b$ given by $(M, N) \to M \otimes N$ and $(f, g) \to f \otimes g$. (Note $1_M \otimes 1_N = 1_{M \otimes N}$.)

Tensor Products of Bimodules and of Algebras

Proposition 1.7.7:   *Suppose $M \in T$-$\mathcal{M}od$-R and $N \in R$-$\mathcal{M}od$. Then $M \otimes N \in$ T-$\mathcal{M}od$ by the scalar multiplication $a(x \otimes y) = ax \otimes y$; notation as in corollary 1.7.4, if f is also a map in T-$\mathcal{M}od$-R then $f \otimes g$ is a map in T-$\mathcal{M}od$.*

Corollary 1.7.8: *If $M \in T$-$\mathcal{M}od$-R then there is a functor $M \otimes \_\_$ from R-$\mathcal{M}od$ to T-$\mathcal{M}od$ given by $N \to M \otimes N$ and $g \to 1 \otimes g$. Analogously, if $N \in R$-$\mathcal{M}od$-T there is a functor $\_\_ \otimes N$ from $\mathcal{M}od$-R to $\mathcal{M}od$-T given by $M \to M \otimes N$ and $f \to f \otimes 1$.*

Of course, we can always return to the original case by taking $T = \mathbb{Z}$. For greater generality we may assume M and N are both bimodules.

Proposition 1.7.9: *If $M \in R$-$\mathcal{M}od$-S and $N \in S$-$\mathcal{M}od$-T then $M \otimes N \in R$-$\mathcal{M}od$-T (as in proposition 1.7.7).*

A particular instance is when a ring T is viewed as a T-R-bimodule (for instance if $R \subseteq T$). Then $T \otimes_R \_\_$ is an important functor from R-$\mathcal{M}od$ to T-$\mathcal{M}od$, which we shall study closely, and is called *changing the ring of scalars from R to T*. This technique is also applicable to algebras.

Proposition 1.7.10: *Suppose R and S are C-algebras. Then $R \otimes S$ is a C-algebra with multiplication $(r_1 \otimes s_1)(r_2 \otimes s_2) = r_1 r_2 \otimes s_1 s_2$.*

Remark 1.7.11: *If $f: R \to R'$ and $g: T \to T'$ are C-algebra homomorphisms then $f \otimes g$ is also a C-algebra homomorphism. (This follows at once from proposition 1.7.7.)*

Properties of the Tensor Operation

Having obtained the tensor product for modules and algebras, we are ready to ascertain some of its most fundamental properties.

Proposition 1.7.12: *Suppose R_i are rings for $i = 0, 1, 2, 3$ and $M_i \in R_{i-1}$-$\mathcal{M}od$-R_i for $1 \leq i \leq 3$. Then*

$$(M_1 \otimes M_2) \otimes M_3 \approx M_1 \otimes (M_2 \otimes M_3) \text{ in } R_0\text{-}\mathcal{M}od\text{-}R_3$$

under an isomorphism sending $(x_1 \otimes x_2) \otimes x_3$ to $x_1 \otimes (x_2 \otimes x_3)$.

Corollary 1.7.13: *If R_i are C-algebras then $(R_1 \otimes R_2) \otimes R_3 \approx R_1 \otimes (R_2 \otimes R_3)$ canonically as C-algebras.*

Proposition 1.7.14: *If $M, N \in C$-$\mathcal{M}od$ for C commutative then there is a C-module isomorphism $M \otimes N \to N \otimes M$ given by $x \otimes y \to y \otimes x$; this is an algebra isomorphism if M, N are C-algebras.*

One also has distributivity of the tensor product over direct sums (What else?). This is a special case of results in Chapter 4, but a short proof is available along the same lines as before.

Proposition 1.7.15: *There is an isomorphism $M \otimes (\oplus_{i \in I} N_i) \to \oplus (M \otimes N_i)$ sending $x \otimes (y_i)$ to $(x \otimes y_i)$.*

Proof: Define the balanced map $\psi: M \times (\oplus N_i) \to \oplus (M \otimes N_i)$ by $\psi(x,(y_i)) = (x \otimes y_i)$. Then $\bar{\psi}$ is the desired map. To construct $\bar{\psi}^{-1}$, let $\mu_i: N_i \to \oplus N_i$ be the canonical injection and let $f_i = 1 \otimes \mu_i: M \otimes N_i \to M \otimes (\oplus_{i \in I} N_i)$. Piecing together the f_i yields a map $f: \oplus (M \otimes N_i) \to M \otimes (\oplus N_i)$ as in example 1.4.25, and $f(x \otimes y_i) = x \otimes (y_i)$, proving f and ψ are inverses.   Q.E.D.

Corollary 1.7.16: *If F is a free W-module with base $\{x_i : i \in I\}$ and R is a ring containing W, then $R \otimes_W F$ is a free R-module with base $\{1 \otimes x_i : i \in I\}$.*

Proof: View F as $\oplus Wx_i$ and R in R-$\mathcal{M}od$-W in the natural way.

Q.E.D.

Tensors and Centralizing Extensions

One of the best ways of finding useful examples of tensor products is by means of the following notion.

Definition 1.7.17: We say a ring T is a *centralizing extension* of R if $T = C_T(R)R$, i.e., if T is generated by R and its centralizer (in T). T is a *central extension* if $T = Z(T)R$.

Clearly central extensions are centralizing; centralizing extensions are also called *extensions* in the literature and are called *liberal extensions* in Robson-Small [31].

Example 1.7.18:

(i) For any monoid S the monoid ring $R[S]$ is a centralizing extension of R. Likewise $R[\lambda]$, $R[[\lambda]]$, and $R((\lambda))$ are all central extensions of R.

(ii) $M_n(R)$ is a centralizing extension of R (identifying R with the scalar matrices, whose centralizer contains the matric units).

Tensor products are also centralizing extensions, in a very special way. If R, R' are C-algebras there is a homomorphism $f: R \to R \otimes R'$ given by

$fr = r \otimes 1$, and we write $R \otimes 1$ for fR. (We shall see below that f need not be injective in general.) We shall carry this notation for the next few pages.

Proposition 1.7.19: $R \otimes R'$ *is a centralizing extension of* $R \otimes 1$, *for any* C-*algebras* R *and* R'. *Moreover, if* T *is a centralizing extension of* R *and* $R' = C_T(R)$ *then there is an algebra surjection* $h: R \otimes_C R' \to T$ *satisfying* $h(r \otimes r') = rr'$.

Proof: Note $(1 \otimes r')(r \otimes 1) = r \otimes r' = (r \otimes 1)(1 \otimes r')$. Hence $1 \otimes R'$ centralizes $R \otimes 1$, and $R \otimes R' = (1 \otimes R')(R \otimes 1)$. To prove the second assertion, we define the (onto) balanced map $\psi: R \times R' \to T$ by $\psi(r, r') = rr'$ and take $h = \bar{\psi}: R \otimes R' \to T$. To check h is a ring homomorphism we note

$$h((r_1 \otimes r_1')(r_2 \otimes r_2')) = h(r_1 r_2 \otimes r_1' r_2') = r_1 r_2 r_1' r_2'$$

$$= r_1 r_1' r_2 r_2' = h(r_1 \otimes r_1')h(r_2 \otimes r_2'). \qquad \text{Q.E.D.}$$

Corollary 1.7.20: *Suppose* T *is a centralizing extension of a* C-*algebra* R *and* $R' = C_T(R)$. *If* R' *contains a base* B *of* T *as* R-*module, then* $T \approx R \otimes R'$.

Proof: Let $h: R \otimes R' \to T$ be the surjection of proposition 1.7.19. If $\sum_{i=1}^t r_i \otimes x_i \in \ker h$ for $x_i \in B$ then $\sum r_i x_i = 0$, implying each $r_i = 0$ by hypothesis; hence $\ker h = 0$, proving h is an isomorphism. Q.E.D.

This somewhat technical result has several immediate applications.

Example 1.7.21: Assume R is a C-algebra, where C is an arbitrary commutative ring. Tensors below are taken over C.

(i) The monoid ring $R[S]$ has S as a base (as R-module), implying $R[S] \approx R \otimes C[S]$.

(ii) $M_n(R)$ has the e_{ij} as a base, implying $M_n(R) \approx R \otimes M_n(C)$.

(iii) Let us partition $M_{mn}(R)$ into blocks of size $n \times n$. Thus there are m^2 blocks, and to the u-v block we define $E_{uv} = \sum_{i=1}^n e_{(u-1)n+i, (v-1)n+i}$ for $1 \le u, v \le m$. There is an injection $M_n(R) \to M_{mn}(R)$ given by $a \to \begin{pmatrix} a & & \\ & \ddots & \\ & & a \end{pmatrix}$, i.e., the block a is repeated m times along the diagonal and we write 0's elsewhere. Identifying $M_n(R)$ with its image, we see easily that its centralizer contains E_{uv}, and these E_{uv} are a base for $M_{mn}(R)$ over $M_n(R)$. Since the E_{uv} are a set of $m \times m$ matric units we conclude that $M_{mn}(R) \approx M_n(R) \otimes M_m(C)$. In particular, $M_{mn}(C) \approx M_m(C) \otimes M_n(C)$.

(iv) The Weyl algebra $\mathcal{A}_n(C) \approx \mathcal{A}_1(C) \otimes \cdots \otimes \mathcal{A}_1(C)$, cf., exercise 1.

Example 1.7.21':   Another useful example of tensor product is as follows: Suppose $A \lhd R$. Then $M/AM \approx (R/A) \otimes_R M$ for any R-module M. (Indeed, the map $M \to (R/A) \otimes_R M$ given by $x \to 1 \otimes x$ has kernel containing AM, so we get a map $M/AM \to (R/A) \otimes_R M$. Its inverse is constructed from the balanced map $R/A \times M \to M/AM$ given by $(r + A, x) \to rx + AM$.) Connections of this kind between tensor products and changes of ring are very important because they give us the opportunity of obtaining general results about rings from properties of the tensor functor (corollary 1.7.8). See proposition 2.11.9ff and §4.2 to see how this idea unfolds.

Tensor Products Over Fields

The classical theory of tensor products is for C a field, in which case the subject becomes quite straightforward.

Corollary 1.7.23:   *Suppose C is a field, and $T = R \otimes R'$. The canonical maps $R \to R \otimes 1$ and $R' \to 1 \otimes R'$ are isomorphisms; thus we may view $R \subset T$ and $R' \subset T$, and T is a centralizing extension of R. Moreover, if B, B' are respective bases of R, R' over C then $\{b \otimes b' : b \in B, b' \in B'\}$ is a base of T over C.*

Proof: Left for the reader.

Corollary 1.7.24:   *Viewed in $R \otimes R'$, we have $Z(R) \otimes Z(R') \subseteq Z(R \otimes R')$, equality holding if C is a field.*

Proof: The first assertion is obvious. To obtain the reverse inclusion, expand a base B_0 of $Z(R)$ over C to a base B of R (over C) and expand a base B_0' of $Z(R')$ to a base B' of R' (over C). If $\sum c_{ij} b_i \otimes b_j' \in Z(R \otimes R')$ then $0 = [\sum c_{ij} b_i \otimes b_j', r \otimes 1] = \sum [c_{ij} b_i, r] \otimes b_j'$ for each r in R, implying (for each j) $0 = [\sum_i c_{ij} b_i, r]$ for each r in R, so each $\sum_i c_{ij} b_i \in Z$. Thus $c_{ij} = 0$ whenever $b_i \notin B_0$; likewise $c_{ij} = 0$ whenever $b_j \notin B_0'$, implying $\sum c_{ij} b_i \otimes b_j' \in Z(R) \otimes Z(R')$.   Q.E.D.

We would like to match the ideal structure of $R \otimes R'$ with those of R and R'. Of course one should not be too optimistic. Indeed if K is a finite field extension of F then the canonical surjection $K \otimes_F K \to K$ of proposition 1.7.19 sending $a_1 \otimes a_2 \mapsto a_1 a_2$ is *not* 1:1, by a dimension count. Thus $K \otimes_F K$ is *not* a field and thus has zero-divisors. Our next example should be compared to theorem 2.5.36 below.

Example 1.7.25: Suppose K is a purely inseparable field extension of a field F; then $K \otimes_F K$ has nonzero nilpotent elements. (Indeed, take $a \in K$ such that $a^p \in F$, where $p = \text{char}(F)$. Then $(a \otimes 1 - 1 \otimes a)^p = a^p \otimes 1 - 1 \otimes a^p = 0$.)

Nevertheless, there are positive results to be had when $C = Z(R)$.

Lemma 1.7.26: *Suppose R is simple and $a, b \in R$ are linearly independent over $Z(R)$. Then there are w_{i1}, w_{i2} in R, $1 \le i \le k$ (for suitable k) such that $\sum_{i=1}^{k} w_{i1} a w_{i2} \ne 0$ and $\sum_{i=1}^{k} w_{i1} b w_{i2} = 0$.*

Proof: Otherwise there is a well-defined map $f: RbR \to R$ given by

$$f\left(\sum r_{i1} b r_{i2}\right) = \sum r_{i1} a r_{i2}$$

for all r_{i1}, r_{i2} in R, and f is clearly an $R - R$ bimodule map. But $RbR = R$. Writing $z = f1$ we have $zr = (f1)r = f(1r) = f(r1) = rz$, implying $z \in Z(R)$; then $a = f(1b) = zb$, contrary to hypothesis.   Q.E.D.

Theorem 1.7.27: *Suppose $C = Z(R)$ and R is a simple C-algebra. Then every nonzero ideal A of $R \otimes_C R'$ contains an element of the form $1 \otimes r' \ne 0$; in particular if R' is also simple then $R \otimes_C R'$ is simple.*

Proof: Choose $0 \ne a = \sum_{j=1}^{u} r_j \otimes r'_j \in A$ with u minimal possible. If $u \ge 2$ then there are elements w_{i1}, w_{i2} in R for $1 \le i \le k$ such that letting $s_j = \sum_{i=1}^{k} w_{i1} r_j w_{i2}$ we have $s_{u-1} \ne 0$ and $s_u = 0$. Then

$$\sum_{j=1}^{u-1} s_j \otimes r'_j = \sum_{j=1}^{u} \sum_{i=1}^{k} w_{i1} r_j w_{i2} \otimes r_{j'} = \sum(w_{i1} \otimes 1) a (w_{i2} \otimes 1) \in A,$$

contrary to minimality of u. Thus $u = 1$, and $a = r_1 \otimes r'_1$. Then $1 \otimes r'_1 \in (R \otimes 1) a (R \otimes 1) \subseteq A$, proving the first assertion.

If R' too is simple then $1 \otimes 1 \in (1 \otimes R')(1 \otimes r'_1)(1 \otimes R') \subseteq A$ proving $R \otimes R'$ is simple.   Q.E.D.

Corollary 1.7.28: *If $T = RW$ where R is simple and W is a $Z(R)$-algebra centralizing R then $T \approx R \otimes_{Z(R)} W$.*

Proof: Otherwise the canonical epic $\varphi: R \otimes W \to RW$ given by $\varphi(r \otimes w) = rw$ is not monic. But then there is some $0 \ne 1 \otimes w \in \ker \varphi$ which is absurd since $0 = \varphi(1 \otimes w) = w$.   Q.E.D.

Corollary 1.7.29: *Suppose T is a finite dimensional subalgebra over a field F, and $T = R \otimes_F R'$ where R is a simple subalgebra. Then $R' = C_T(R)$.*

Proof: Let $R'' = C_T(R)$. Obviously $R' \subseteq R''$. But then $T = RR''$ so $T \approx R \otimes_F R''$ by corollary 1.7.28, yielding $[R:F][R'':F] = [T:F] = [R:F][R':F]$. Hence $[R'':F] = [R':F]$ and thus $R' = R''$.   Q.E.D.

In describing subsets of $M \otimes N$ it is convenient to write $A \otimes B$ for $\{\sum_i a_i \otimes b_i : a_i \in A, b_i \in B\}$, where $A \subset M$ and $B \subset N$. One must be careful *not* to view this as the tensor product of A and B even though the notation is the same. For example, the tensor product $2\mathbb{Z} \otimes_{\mathbb{Z}} (\mathbb{Z}/2\mathbb{Z}) \neq 0$; however, viewed inside $\mathbb{Z} \otimes_{\mathbb{Z}} (\mathbb{Z}/2\mathbb{Z})$ we have $2\mathbb{Z} \otimes (\mathbb{Z}/2\mathbb{Z}) = 0$ since any $2m \otimes \bar{n} = m \otimes 2\bar{n} = m \otimes 0 = 0$.

Proposition 1.7.30: *If $f: M \to N$ is an epic R-module map and $M' \in \mathcal{M}od\text{-}R$ then $\ker(1_{M'} \otimes f) = M' \otimes \ker f$ (viewed in $M' \otimes_R M$).*

Proof: Write $K = M' \otimes \ker f$. Then $K \leq \ker(1 \otimes f)$ so $1 \otimes f$ induces a map $\bar{f}: (M' \otimes M)/K \to M' \otimes N$ satisfying $\bar{f}(x' \otimes x + K) = x' \otimes fx$. We shall show $K = \ker(1 \otimes f)$ by proving $\bar{f}$ is an isomorphism, i.e., by constructing $\bar{f}^{-1}$. Define the balanced map $\psi: M' \times N \to (M' \otimes N)/K$ by $\psi(x', fx) = x' \otimes x + K$. Then $\bar{\psi}: M' \otimes N \to (M' \otimes N)/K$ satisfies $\bar{\psi}(x' \otimes fx) = x' \otimes x + K$ as desired.   Q.E.D.

Tensor Products and Bimodules

We have seen already that when transferring properties between rings $R \subseteq R'$ one often relies only on the $R - R$ bimodule structure of R'. However, bimodules are very complicated; for example, try describing the free $R - R$ bimodule in terms of a base! At times the tensor product is a useful tool in studying bimodules, in view of the following fact:

Proposition 1.7.31: *Any $R \otimes_{\mathbb{Z}} R^{op}$-module M can be viewed as $R - R$ bimodule under the scalar multiplications $rx = (r \otimes 1)x$ and $xr' = (1 \otimes r')x$, leading to an isomorphism of the categories $R \otimes_{\mathbb{Z}} R^{op}\text{-}\mathcal{M}od$ and $R\text{-}\mathcal{M}od\text{-}R$, for any ring R.*

Proof: The two verifications of interest in proving M is an $R - R$ bimodule are

$$(rx)r' = ((r \otimes 1)x)r' = (1 \otimes r')(r \otimes 1)x = (r \otimes 1)(1 \otimes r')x = r(xr').$$

$$x(r'_1 r'_2) = (1 \otimes r'_1 r'_2)x = ((1 \otimes r'_2)(1 \otimes r'_1))x = (1 \otimes r'_2)(xr'_1) = (xr'_1)r'_2.$$

Any $R \otimes R^{op}$-module map $f: M \to N$ becomes an $R - R$ bimodule map, so we have a functor $F: R \otimes R^{op}\text{-}\mathcal{M}od \to R\text{-}\mathcal{M}od\text{-}R$.

To obtain the inverse functor one takes an $R - R$ bimodule M and defines a scalar multiplication as follows: The usual balanced map argument yields maps $\psi_x: R \otimes R^{op} \to M$ (for fixed x in M) given by $\psi_x(r \otimes r') = rxr'$; now allowing x to vary, define $(r \otimes r')x = \psi_x(r \otimes r')$. To check M is a module over $R \otimes R^{op}$ we observe

$$(r \otimes r')(x_1 + x_2) = r(x_1 + x_2)r' = rx_1r' + rx_2r' = (r \otimes r')x_1 + (r \otimes r')x_2.$$

$$((r_1 \otimes r'_1)(r_2 \otimes r'_2))x = (r_1r_2 \otimes r'_2r'_1)x = r_1r_2xr'_2r'_1$$

$$= r_1(r_2xr'_2)r'_1 = (r_1 \otimes r'_1)((r_2 \otimes r'_2)x).$$

The other verifications are clear, and any $R - R$ bimodule map becomes an $R \otimes R^{op}$-module map under this action, so we have the inverse functor to F, as desired.   Q.E.D.

Corollary 1.7.32:   Let $F = (R \otimes R^{op})^{(n)}$, with base $e_1, \ldots, e_n$. Then F can be viewed as free $R - R$ bimodule under the action $\sum_{i=1}^{n} r_i e_i r'_i = \sum (r_i \otimes r'_i)e_i$. Likewise, free $R - R$ bimodules of infinite rank can be constructed.

§1.8 Direct Limits and Inverse Limits

In §1.10 we shall explore localization over a central monoid. The proof that localization exists is cumbersome, and it actually can be obtained more easily via "direct limits," a general construction which is useful also in many other settings. We shall also consider the dual construction, inverse limits, as well as completions as a special case.

Direct Limits

As motivation, recall how the coproduct of $\{A_i : i \in I\}$ in a category $\mathscr{C}$ is the "smallest" object A containing each A_i "independently" (e.g., when $\mathscr{C} = R\text{-}\mathcal{M}od$). We would like to perform a similar construction when the A_i are "glued together" in some way. An extreme case is when $I = \mathbb{N}$ and $A_1 \subseteq A_2 \subseteq A_3 \subseteq \cdots$; then one would take $A = \bigcup A_i$. Of course, $A_i \subseteq A_{i+1}$ really means there is a canonical injection $\varphi: A_i \to A_{i+1}$, and there is also the problem of finding the suitable set in which to form the union A. To handle these two difficulties in a manner which also embraces the coproduct, we assume the index set I is given an arbitrary preorder $\leq$ (i.e., $\leq$ is reflexive and transitive).

Definition 1.8.1: A *system* $(A_i; \varphi_i^j)$ is a set of objects $\{A_i : i \in I\}$ in a category $\mathscr{C}$, together with φ_i^j in $\mathrm{Hom}(A_i, A_j)$ whenever $i \leq j$, subject to the following two rules:

$$(1) \quad \varphi_j^k \varphi_i^j = \varphi_i^k \qquad \text{whenever } i \leq j \leq k.$$

$$(2) \quad \varphi_i^i = 1_{A_i} \qquad \text{for all } i.$$

Examples 1.8.2: (i) if I has the trivial preorder then the φ_i^j only exist for $i = j$ since any $i \neq j$ are incomparable. (ii) Any chain is a system, where I is given the usual order of ordinals, and φ_i^j is the inclusion map.

Definition 1.8.3: Suppose I is a preordered set, as before. The *direct limit* of the system $(A_i; \varphi_i^j)$ is an object denoted $\varinjlim A_i$ (or, more precisely, $\varinjlim(A_i; \varphi_i^j)$ if necessary) together with morphisms $\mu_i : A_i \to \varinjlim A_i$ satisfying $\mu_i = \mu_j \varphi_i^j$ for all $i \leq j$; for which given any family of morphisms $g_i : A_i \to A$ satisfying $g_i = g_j \varphi_i^j$ for all $i \leq j$ we have a unique $g : \varinjlim A_i \to A$ such that $g\mu_i = g_i$ for each i, i.e., g completes the diagram

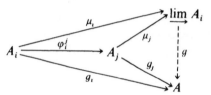

for each $i \leq j$.

Direct limits are defined by a universal property, and so are unique *if they exist.*

Example 1.8.4: We interpret the examples of example 1.8.2. (when they exist).

(i) If I has the trivial preorder then $\varinjlim A_i$ is the coproduct $\coprod A_i$.
(ii) If I is a chain then $\varinjlim A_i$ can be interpreted as $\bigcup A_i$.

Thus we see the direct limit unifies several examples, but to obtain full benefit we want to prove that in these cases the limit must exist.

Proposition 1.8.5: *Direct limits always exist in R-$\mathcal{M}od$.*

Proof: Let $(M_i; \varphi_i^j)$ be a system of R-modules. Letting $\mu_j : M_j \to \bigoplus M_i$ be the injection into the j-th component, let $N = \sum_{i \leq j} (\mu_j \varphi_i^j - \mu_i) M_i \leq \bigoplus M_i$. Each μ_j induces a map $\bar{\mu}_j : M_j \to (\bigoplus M_i)/N$, and the constructions show

$\bar{\mu}_j \varphi_i^j = \bar{\mu}_i$ for all $i \le j$. We shall show $\varinjlim(M_i; \varphi_i^j)$ is $(\bigoplus M_i)/N$ together with the $\bar{\mu}_i$. Indeed given $g_i: M_i \to M$ satisfying $g_j \varphi_i^j = g_i$ we first define the unique map $g': \bigoplus M_i \to M$ satisfying $g'\mu_i = g_i$ for all i. Then $0 = g_j \varphi_i^j - g_i = g'\mu_j \varphi_i^j - g'\mu_i = g'(\mu_j \varphi_i^j - \mu_i)$ so $N \le \ker g'$. Hence g' induces a map $g: \bigoplus M_i/N \to M$ satisfying $g\bar{\mu}_i = g_i$, as desired; g is unique since the $\bar{\mu}_i M$ span $\bigoplus M_i/N$.   Q.E.D.

Example 1.8.9: (A module is the direct limit of its f.g. submodules). Suppose $M \in R\text{-}\mathcal{M}od$ and $\{S_i : i \in I\}$ are the finite subsets of M, viewed as a directed poset under set containment. Writing M_i for the submodule spanned by S_i, one sees at once that $M = \lim M_i$ since any maps $g_i: M_i \to N$ can be "pasted" together to a map $g: M \to N$.

Proposition 1.8.10: *If $(N_i; \bar{\mu})$ is a system in $R\text{-}\mathcal{M}od$ with direct limit N and $K \in S\text{-}\mathcal{M}od\text{-}R$, then $(K \otimes_R N_i; 1 \otimes \varphi_i^j)$ is a system in $S\text{-}\mathcal{M}od$ and $K \otimes_R N = \varinjlim(K \otimes_R N_i)$.*

Proof: $(K \otimes_R N_i; 1 \otimes \varphi_i^j)$ is a system in $S\text{-}\mathcal{M}od$. If $\mu_i: N_i \to N$ are the canonical morphisms then we have morphisms $1 \otimes \mu_i: K \otimes N_i \to K \otimes N$ satisfying $1 \otimes \mu_i = 1 \otimes \mu_j \varphi_i^j = (1 \otimes \mu_j)(1 \otimes \varphi_i^j)$ for $i \le j$, so it remains to verify the universality condition. Given $g_i: K \otimes N_i \to M$ satisfying $g_i = g_j(1 \otimes \varphi_i^j)$ we need $g: K \otimes N \to M$ satisfying $g(1 \otimes \mu_i) = g_i$ for each i. Fixing a in K define $h_{ia}: N_i \to M$ by $h_{ia}x = g_i(a \otimes x)$ for x in N_i. Clearly $h_{ia} = h_{ja}\varphi_i^j$ for $i \le j$, so by hypothesis there is $h_a: N \to M$ satisfying $h_a \mu_i = h_{ia}$ for each i. Note for any r in R that $h_{i,ar}x = h_{ia}(rx)$, from which it follows $h_{ar}x = h_a(rx)$. Thus we have a balance map $h: K \times N \to M$ given by $h(a, x) = h_a x$, yielding a map $g: K \otimes N \to M$ given by $g(a \otimes x) = h(a, x)$.

Now $g(1 \otimes \mu_i)(a \otimes x) = g(a \otimes \mu_i x) = h_a(\mu_i x) = h_{ia}x = g_i(a \otimes x)$ for all a in K and x in N, proving $g(1 \otimes \mu_i) = g_i$; uniqueness is seen by fixing a and noting the h_a are uniquely given.   Q.E.D.

D Supplement: Projective Limits

The dual construction to direct limits is called "projective" or "inverse" limits. Whereas the definition is obtained formally by reversing arrows, the flavor is different. Direct limits involve "pasting together" objects, whereas inverse limits often involve a topological completion.

Let us start with a category $\mathcal{C}$ and a preordered set I. An *inverse system* is a set $\{A_i : i \in I\}$ of objects in $\mathcal{C}$ together with morphisms $\varphi_j^i: A_j \to A_i$ for each $i \le j$ satisfying

$$\varphi_k^i = \varphi_j^i \varphi_k^j \qquad \text{whenever } i \leq j \leq k.$$

$$\varphi_i^i = 1_{A_i} \qquad \text{for all } i.$$

Definition 1.8.11: The *inverse limit* (denoted $\varprojlim F$ and also called the *projective limit*) of the inverse system $(A_i; \varphi_j^i)$ is an object $\varprojlim A_i$ of $\mathscr{C}$, together with morphisms $v_i: \varprojlim A_i \to A_i$ with $\varphi_j^i v_j = v_i$ for all $i \leq j$, which satisfy the following property:

For every A in $Ob\,\mathscr{C}$ and every set of morphisms $g_i: A \to A_i$ satisfying $\varphi_j^i g_j = g_i$ for all $i \leq j$, there is a unique $g: A \to \varprojlim A_i$ satisfying $v_i g = g_i$ for each i in I.

If I has the trivial preorder we have redefined the direct product. The usual argument shows that if the inverse limit exists it is unique up to isomorphism. We want to find situations in which it is guaranteed to exist; to obtain a good general condition we assume the inverse limit exists and see where it leads us.

Suppose $\mathscr{C}$ is a concrete category with direct products, and suppose $(A_i; \varphi_j^i)$ is an inverse system having an inverse limit. From the characterization of product, letting $\pi_j: \prod A_i \to A_j$ denote the projection, we have a unique morphism $f: \varprojlim A_i \to \prod A_i$ such that $\pi_i f = v_i$ for each i. For each $i \leq j$ we then have $\pi_i f = v_i = \varphi_j^i v_j = \varphi_j^i \pi_j f$; thus each (a_i) in $f(\varprojlim A_i) \subseteq \prod A_i$ satisfies $a_i = \varphi_j^i a_j$ for all $i \leq j$. In particular, we have the following necessary nondegeneracy condition:

$$\{(a_i) \in \prod A_i : a_i = \varphi_j^i a_j \text{ for } i \leq j\} \neq \varnothing$$

We shall call this the *nondegeneracy condition* (for inverse limits), and now show it often is sufficient as well as necessary in the important case that $\mathscr{C}$ is a category defined in a first-order language. Suppose $(A_i; \varphi_j^i)$ is an inverse system satisfying the nondegeneracy condition, so that $L = \{(a_i) \in \prod A_i : a_i = \varphi_j^i a_j \text{ for } i \leq j\} \neq \varnothing$. Also suppose $L \in \mathscr{C}$, and let $v_i: L \to A_i$ be the restriction of π_i to L. If $A \in Ob\,\mathscr{C}$ and $g_i: A \to A_i$ satisfy $\varphi_j^i g_j = g_i$ then define $g: A \to L$ by $ga = (g_i a)$. By definition we have $v_i ga = g_i a$ for all a in A, proving $L = \lim A_i$. This defines inverse limits in $R\text{-}\mathscr{M}\!od$, $\mathscr{R}ing$, and $\mathscr{A}\!b$, and more generally for "universal algebra," c.f., Jacobson [80B, Chapter 2].

The Completion

We turn now to the most important special case of inverse limit. We shall carry it out first for abelian groups and then consider rings and modules as special cases.

Example 1.8.12: Suppose we are given a set of subgroups $\{B_i : i \in I\}$ of an abelian group G. Preordering I by putting $i \leq j$ iff $B_j \subseteq B_i$, we get an inverse system $(G_i; \varphi_j^i)$ where $G_i = G/B_i$ and $\varphi_j^i : G_j \rightarrow G_i$ is the canonical map (viewing G_i as $G_j/(B_i/B_j)$). The nondegeneracy condition is satisfied by $(g + B_i)$ for each g in G, so we can form $\hat{G} = \varprojlim G_i$, called the *completion* of G with respect to the B_i. We have a canonical group homomorphism $\psi : G \rightarrow \hat{G}$ given by $\psi g = (g + B_i)$; ψ is an injection iff $\bigcap B_i = 0$.

Explicitly $\hat{G} = \{(g_i + B_i) \in \prod G_i : g_j - g_i \in B_i \text{ for all } j \geq i\}$. Thus $\psi : G \rightarrow \hat{G}$ is onto iff for every such sequence $(G_i + B_i)$ there is g in G with $g - g_i \in B_i$ for each i; in this case we say G is *complete*.

Another description of the completion comes from topology. We say (g_i) in G^I is a *Cauchy net* if for each i there is some i' for which $g_j - g_{i'} \in B_i$ for all $j \geq i'$. (Then for all $j, j' \geq i'$ we have $g_j - g_{j'} = (g_j - g_{i'}) - (g_{j'} - g_{i'}) \in B_i$.) {Cauchy nets} form an abelian group Cauchy(G) under component-wise addition, and are the set of convergent nets under the topology obtained by viewing $\{B_i : i \in I\}$ as a neighborhood system of 0. There is a group homomorphism Cauchy$(G) \rightarrow \hat{G}$ obtained by "weeding out" the terms of a Cauchy net so that i' becomes i for each i. The kernel N is the set of Cauchy nets which converge to 0; i.e., for each i there is some i' for which $g_j \in B_i$ for all $j \geq i'$. Thus $\hat{G} \approx$ Cauchy$(G)/N$, which is the topological completion of G.

Digression 1.8.13: The topology on G (defined above) is Hausdorff iff $\bigcap_{i \in I} B_i = 0$. (Proof: ($\Rightarrow$) If $g \in \bigcap B_i$ then 0 lies in the closure of $\{g\}$. ($\Leftarrow$) If $g \neq h \in G$ then $g - h \notin B_i$ for some i, so $g + B_i$ and $h + B_i$ are disjoint open sets, proving the space is Hausdorff.)

Example 1.8.14: Suppose R is a ring with filtration over $(\mathbb{Z}, +)$. Taking $B_i = R(i)$ we have $B_j \subseteq B_i$ for all $i \leq j$ so we can form $\hat{R}$ as above. In fact $\hat{R}$ is a ring with multiplication by $(r_i + B_i)(r_i' + B_i) = (r_i r_i' + B_i)$, in the case that each $B_i \lhd R$. There are two cases of interest:

(i) $R(i) = 0$ for all $i > 0$; then any Cauchy net $(r_i + B_i)$ is equivalent to $(r_0 + B_i)$, so $\hat{R} \approx R$ canonically, and R is complete.

(ii) R has a filtration over $\mathbb{N}$ (i.e., $R(i) = R$ for all $i \leq 0$); then $\hat{R}$ has a filtration over $\mathbb{N}$, given by $\hat{R}(n) = \{(r_i + B_i) : \text{each } r_i \in B_n\}$. Note this implies $r_i + B_i = 0$ for $i \leq n$ (since $B_i \supseteq B_n$). Consequently, $R \cap \hat{R}(n) = B_n = R(n)$. Also if $(r_i + B_i) \in \hat{R}(n)$ then $(r_i + B_i) - (r_{n+1} + B_i) \in \hat{R}(n+1)$; since $r_{n+1} \in B_n$ we see $\hat{R}(n) = R(n) + \hat{R}(n+1)$.

Suppose $M = M_0 > M_1 > M_2 > \cdots$ in $R\text{-}\mathcal{M}od$. We can form the completion $\hat{M}$ of M with respect to the M_i. If $M_i = B_i M$ then $\hat{M}$ is in fact an $\hat{R}$-module under scalar multiplication $(r_i + B_i)(x_i + M_i) = (r_i x_i + M_i)$.

In particular, we have the following special case of example 1.8.14(ii).

Definition 1.8.15: Suppose $A \lhd R$. The *A-adic completion* of R is the completion of R with respect to the *A-adic filtration* (over $\mathbb{N}$) given by $R(0) = R$ and $R(i) = A^i$ for $i > 0$. Likewise the *A-adic completion* $\hat{M}$ of an R-module M is its completion with respect to the submodules $M_i = A^i M$.

This definition was motivated by the p-adic integers; here $R = \mathbb{Z}$ and $A = p\mathbb{Z}$.

The A-adic completion $\hat{R}$ of R has the ideal $\hat{A} = \{(a_i + A^i) \in \hat{R} : a_1 \in A\}$ which contains ψA (recalling $\psi : R \to \hat{R}$ is the canonical ring homomorphism given by $\psi r = (r + A^i)$) and plays the following key role:

Theorem 1.8.16: *Notation as above, $R/A \approx \hat{R}/\hat{A}$ canonically and $\hat{A}$ is an idempotent-lifting ideal of $\hat{R}$.*

Proof: Composing ψ with the canonical map $\hat{R} \to \hat{R}/\hat{A}$ yields a homomorphism $R \to \hat{R}/\hat{A}$ whose kernel is A, yielding an injection $\bar{\psi} : R/A \to \hat{R}/\hat{A}$. But $\bar{\psi}$ is onto, for if $(r_i + A^i) \in \hat{R}$ then $(r_i + A^i) - (r_1 + A^i) \in \hat{A}$, so $(r_i + A^i) + \hat{A} = \bar{\psi}(r_1 + A)$.

Thus $R/A \approx \hat{R}/\hat{A}$. Next we show $1 - \hat{a}$ is invertible for any element $a = (a_i + A^i)$ in $\hat{A}$. Note if $(a_i + A^i) \in \hat{A}$ then $a_1 \in A$ so $a_i \in a_1 + A^1 \subseteq A$ for each i, implying $a_i^i \in A^i$ and $(1 - a_i)(\sum_{u=0}^{i-1} a_i^u) = 1 - a_i^i \in 1 + A^i$. Thus putting $b_i = \sum_{u=0}^{i-1} a_i^u$ we see $(b_i + A^i)$ is the desired inverse of $1 - \hat{a}$.

It remains to show every idempotent $e = (r_i + A^i) + \hat{A}$ of $\hat{R}/\hat{A}$ can be lifted to $\hat{R}$. We want to write $e = (e_i + A^i) + \hat{A}$ where each $e_i + A^i$ is idempotent in R/A^i. Since $r_1^2 - r_1 \in A$ we can take $e_1 = r_1$, and we work inductively on i. Suppose we have found $e_1, \ldots, e_i$ for $i \geq 1$. A^i/A^{i+1} is an idempotent-lifting ideal of R/A^{i+1} since $(A^i/A^{i+1})^2 = 0$. Viewing $R/A^i \approx (R/A^{i+1})/(A^i/A^{i+1})$ we can lift $e_i + A^i$ to an idempotent $e_{i+1} + A^{i+1}$ of R/A^{i+1}, thereby completing the induction step.   Q.E.D.

Proposition 1.8.17: *Let $\hat{M}$ be the A-adic completion of M, and $f : M \to \hat{M}$ be the canonical map given by $fx = (x + A^i M)$. If $M = \sum_{j=1}^t Rx_j$ then $\hat{M} = \sum_{j=1}^t \hat{R} f x_j$. In particular, if M is f.g. then $\hat{M}$ is an f.g. $\hat{R}$-module.*

Proof: Given $\hat{y} = (y_i + A^i M)$ in $\hat{M}$ we write $y_{i+1} - y_i = \sum_{u=1}^k a_{ui} z_{ui}$ where $a_{ui} \in A^i$ and $z_{ui} \in M$. Then $z_{ui} = \sum_{j=1}^t r_{uij} x_j$ for suitable r_{uij} in R, so

$$y_{i+1} - y_i = \sum_{j=1}^{t}\left(\sum_{u=1}^{k} a_{ui}r_{uij}\right)x_j.$$

Putting $\hat{b}_j = (b_{ij} + A^i)$ where $b_{0j} = 0$ and, inductively, $b_{i+1,j} = b_{ij} + \sum_{u=1}^{k} a_{ui}r_{uij}$, we see $\hat{b}_j \in \hat{R}$ and $y_i = \sum_{j=1}^{t} b_{ij}x_j$, implying $\hat{y} = \sum_{j=1}^{t} \hat{b}_j\hat{x}_j$.   Q.E.D.

§1.9 Graded Rings and Modules

In the last 20 years considerable interest has been attracted to the theory of graded rings and modules, primarily because any ring with filtration over $\mathbb{Z}$ has an "associated" graded ring whose structure bears heavily on the structure of the original ring. This construction is deferred until definition 3.5.30, since it is most relevant to Noetherian rings; here we lay out the basic framework of the theory of graded rings and modules, including the category and its "free" objects. Then we construct the "tensor ring," an important source of examples of graded rings, and apply it to the free product construction.

The reader interested in graded rings should pursue this subject in theorem 2.5.40 (the Jacobson radical), proposition 3.5.31 (the associated graded ring), definition 5.1.36 thru theorem 5.1.41 (the K_0-theory), theorem 6.2.9 (the Golod-Shafarevich counterexample), and §8.3, 8.4 (enveloping algebras). Also, cf., exercises 2.6.11ff.

Definition 1.9.1:   Suppose S is a monoid. A C-algebra R is S-*graded*, or *graded over* S, if $R = \bigoplus\{R_s : s \in S\}$ for C-submodules R_s for which $R_s R_{s'} \subseteq R_{ss'}$. *Graded ring* means graded $\mathbb{Z}$-algebra (i.e., the R_s are abelian groups). Suppose R is S-graded. An R-module M is *graded* if $M = \bigoplus_{s \in S} M_s$ and $r_s x_{s'} \in M_{ss'}$ for all r_s in R_s and $x_{s'}$ in $M_{s'}$. R-$\mathcal{Gr}$-$\mathcal{Mod}$ is the category of graded R-modules, whose morphisms are the maps $f : M \to N$ satisfying $fM_s \subseteq N_s$ for each s. (These are called *graded maps*.) R-$\mathcal{Gr}$-$\mathcal{Fimod}$ is the subcategory consisting of graded f.g. submodules. Likewise $\{S$-graded C-algebras$\}$ is a subcategory of C-$\mathcal{Alg}$ whose morphisms are the algebra homomorphisms $f : A \to B$ satisfying $fA_s \subseteq B_s$ for each s.

If M is an S-graded module we call each M_s a *homogeneous component* and call its elements *homogeneous*; writing $x = \sum_{\text{finite}} x_s$ for x_s in M_s, we define the *support* $\text{supp}(x) = \{s \in S : x_s \neq 0\}$, a finite set which is nonempty iff $x \neq 0$.

Example 1.9.2:

(i) $A = R[S]$ is graded by S, where $A_s = Rs$.

(ii) $A = M_n(R)$ is $\mathbb{Z}$-graded, where $A_u = \sum_{j=i+u} Re_{ij}$; thus $A_u = 0$ for $|u| \geq n$.

(iii) Suppose ζ is a primitive n-th root of 1. $A = \mathbb{Z}[\zeta]$ is graded by $\mathbb{Z}/n\mathbb{Z}$, where $A_u = \zeta^u \mathbb{Z}$.

$(\mathbb{N}, +)$ and $(\mathbb{Z}, +)$ are the grading monoids of greatest interest, largely in view of the following tie to filtrations.

Remark 1.9.3: If S is an ordered monoid then any S-graded algebra A has the filtration $A(s) = \sum_{s' \geq s} A_{s'}$, which is valuated by $va = \bigwedge\{\text{supp}(a)\}$ for $a \neq 0$. Similarly, one can define the degree function $\deg a = \bigvee\{\text{supp}(a)\}$.

Remark 1.9.4: In view of remark 1.2.15 the following conditions are equivalent when the grading monoid S of A is ordered: (i) A is a domain; (ii) $v: A \to S_\infty$ is a monoid homomorphism; (iii) the product of nonzero homogeneous elements must be nonzero; (iv) $\deg(ab) = (\deg a)(\deg b)$ for all nonzero a, b in A (i.e., if one side is defined then both sides are defined and equal).

It is easy to characterize the kernels in graded categories. If M is an S-graded module and $N < M$, we write N_s for $N \cap M_s$. Clearly, $\sum N_s \leq N$; we say N is *graded* if $\sum N_s = N$.

Remark 1.9.5:

(i) The kernel of a graded map is a graded submodule. Conversely, if M is an S-graded module and N is a graded submodule then M/N is graded (by putting $(M/N)_s = M_s/N_s$) and the canonical map $M \to M/N$ is graded. (The easy proofs are left for the reader.)

(ii) The kernel of a graded homomorphism is a graded ideal; conversely, if R is an S-graded algebra and $A \lhd R$ is graded then R/A is a graded algebra (as in (i)) and the canonical homomorphism $R \to R/A$ is graded.

Remark 1.9.6: Suppose $g: M \to N$ is a graded epic which splits (but not necessarily by a graded map). Then g does split by a graded map, i.e., there is graded $f: N \to M$ for which $gf = 1_N$. (Indeed, we take ungraded $f': N \to M$ for which $gf' = 1_N$, and define $f_s: N_s \to M_s$ to be the composition $N_s \to N \xrightarrow{f'} M \to M_s$ where the outside arrows are the canonical maps. Then $f = \bigoplus f_s: N \to M$ is graded and $gf x_s = (gf'x)_s = x_s$ for all homogeneous x_s, implying $gf = 1_N$.)

The category $R\text{-}\mathcal{G}\imath\text{-}\mathcal{M}od$ turns out to be quite useful, especially for $S = (\mathbb{N}, +)$. If R is S-graded then $R^{(n)} \in R\text{-}\mathcal{G}\imath\text{-}\mathcal{F}imod$, viewing $R_s^{(n)}$ as $(R_s)^{(n)}$. However, $R^{(n)}$ is not necessarily free in $R\text{-}\mathcal{G}\imath\text{-}\mathcal{F}imod$ since its generators are in $R_1^{(n)}$, whereas a graded module could be generated by homogeneous elements from other components. To rectify this situation we need a larger module. Given any S-graded module M we write $M(s)$ for M with the new grade given by $M(s)_{s'} = M_{ss'}$. Then $\bigoplus_{s \in S} R(s)$ is a graded module which has the following "freeness" property.

Remark 1.9.7: Suppose $M = \sum_{i=1}^{t} Rx_i$ is S-graded. If $x_i = \sum x_{is}$ then each $x_{is} \in M$ so M is generated by homogeneous elements. Moreover, if $f_s : R^{(t)} \to \sum_{i=1}^{t} Rx_{is}$ are epics then $\bigoplus f_s : \bigoplus R^{(t)}(s) \to M$ is a graded epic. (Note $\bigoplus R^{(t)}(s) \approx (R^{(t)})^{(S)}$ is a free R-module.)

Finite grading sets are important largely because of the following link to $\mathbb{Z}$ and $\mathbb{N}$.

Remark 1.9.8: Suppose $\psi : S \to S'$ is a monoid surjection. Any S-graded algebra R can be S'-graded by defining $R_u = \sum_{\psi s = u} R_s$ for u in S'. This yields a functor from $\{S$-graded C-algebras$\}$ to $\{S'$-graded C-algebras$\}$.

Example 1.9.9: Applying this technique to example 1.9.2 (ii) yields a $\mathbb{Z}/2\mathbb{Z}$-grading of $M_n(R)$ called the *checkerboard grade*: e_{ij} is in the 0 (resp. 1) component iff $i + j$ is even (resp. odd).

As an immediate consequence of proposition 1.7.15 we have the following tensor product connection:

Proposition 1.9.10: *Suppose R is an S-graded ring. Then R_1 is a ring, and for any subring W of R_1 there is a functor $R \otimes_W —$ from W-$\mathcal{M}od$ to R-$\mathcal{G}r$-$\mathcal{M}od$. In case $W = R_1$, this functor identifies R_1-$\mathcal{M}od$ with the full subcategory of R-$\mathcal{G}r$-$\mathcal{M}od$ consisting of those graded modules M for which $M = RM_1$.*

Note that we used the multiplicative notation for S; in many applications S is written additively, with 0 as the neutral element, so we would write R_0 and M_0 instead of R_1 and M_1.

Proof: R_1 is closed under multiplication since $1 \cdot 1 = 1$. Given N in W-$\mathcal{M}od$ we grade $R \otimes_W N$ by $(R \otimes N)_s = R_s \otimes N$, proving the first assertion. If $W = R_1$ then $(R \otimes N)_1 = R_1 \otimes N \approx N$, so the functor $M \to M_1$ (going the other direction) shows R_1-$\mathcal{M}od$ is categorically equivalent to the specified subcategory of R-$\mathcal{G}r$-$\mathcal{M}od$.   Q.E.D.

Remark 1.9.11: Let us refine proposition 1.9.10. If R is S-graded then we can grade R_1 trivially (i.e., $(R_1)_s = 0$ for all $s \neq 1$) and thereby have R_1-$\mathcal{G}r$-$\mathcal{M}od$.

These are merely the R_1-modules, graded as a direct sum of submodules indexed over S. There is a functor $F = R \otimes_{R_1} - : R_1\text{-}\mathcal{G}\imath\text{-}\mathcal{M}od \to R\text{-}\mathcal{G}\imath\text{-}\mathcal{M}od$ whereby we grade $R \otimes_{R_1} M$ by putting $(r \otimes x)_s = \sum_{uv=s} r_u \otimes x_v$. In the other direction, there is a functor $G = R_1 \otimes_R - : R\text{-}\mathcal{G}\imath\text{-}\mathcal{M}od \to R_1\text{-}\mathcal{G}\imath\text{-}\mathcal{M}od$ whereby $R_1 \otimes_R N$ is graded via $(r \otimes x)_s = r \otimes x_s$. GF is naturally isomorphic to the identity under the usual tensor product isomorphism. However, FG is *not* naturally isomorphic to the identify since the grade may be changed. We shall investigate this further in the proof of theorem 5.1.34.

Tensor Rings

One useful type of graded ring arises from tensor products.

Definition 1.9.12: Suppose M is an $R-R$ bimodule. Define the *tensor power* $M^{\otimes n}$ by $M^{\otimes 0} = R$, $M^{\otimes 1} = M$ and, inductively, $M^{\otimes(n+1)} = M \otimes_R M^{\otimes n}$, viewed as R-bimodule in the natural way. Define the tensor ring $T(M) = \bigoplus_{n \in \mathbb{N}} M^{\otimes n}$. When R is ambiguous we write $T_R(M)$ for $T(M)$.

Clearly $T(M)$ is an $R - R$ bimodule; to show $T(M)$ is a ring we need a multiplication. First we note $M^{\otimes n}$ is spanned by elements $x_1 \otimes (x_2 \otimes (\cdots \otimes x_n))$, which we write more simply as $x_1 \otimes \cdots \otimes x_n$. By proposition 1.7.12 we have an isomorphism $\varphi_{mn}: M^{\otimes m} \otimes M^{\otimes n} \to M^{\otimes(m+n)}$ given by $(x_1 \otimes \cdots \otimes x_m) \otimes (y_1 \otimes \cdots \otimes y_n) \to x_1 \otimes \cdots \otimes x_m \otimes y_1 \otimes \cdots \otimes y_n$. Applying proposition 1.7.15 twice we have a map

$$\varphi: T(M) \otimes T(M) \to T(M)$$

given by $a_m \otimes b_n \to \varphi_{mn}(a_m \otimes b_n)$ for a_m in $M^{\otimes m}$ and b_n in $M^{\otimes n}$.

Now we have a multiplication in $T(M)$ by defining the product of a and b to be $\varphi(a \otimes b)$, and it is easy to verify the ring axioms. Summarizing, we have

Proposition 1.9.13: $T(M)$ is an $\mathbb{N}$-graded ring, with $T(M)_0 = R$, $T(M)_1 = M$, and, in general, $T(M)_n = M^{\otimes n}$, multiplication given by

$$(x_1 \otimes \cdots \otimes x_m)(y_1 \otimes \cdots \otimes y_n) = x_1 \otimes \cdots \otimes x_m \otimes y_1 \otimes \cdots \otimes y_n.$$

If R is a C-algebra then $T(M)$ is also a C-algebra under multiplication $c(x_1 \otimes \cdots \otimes x_n) = (c1x_1) \otimes \cdots \otimes x_n$.

The customary special case is when C is a commutative ring and $M \in C\text{-}\mathcal{M}od$ is viewed symmetrically as a bimodule. Then $T_C(M)$ is called the *tensor algebra* of M.

Example 1.9.15: If $M = C^{(n)}$ then the tensor algebra $T_C(M) \approx C\{X_1, \ldots, X_n\}$. Indeed, taking a base $x_1, \ldots, x_n$ for M we define $f: M \to C\{X_1, \ldots, X_n\}$ by $fx_i = X_i$ and use the proposition to obtain a homomorphism $f': T_C(M) \to C\{X_1, \ldots, X_n\}$ which is an isomorphism since the $x_{i_1} \otimes \cdots \otimes x_{i_u}$ form a base of T_u over C.

This example is the motivating example in the study of tensor algebras, enabling us to obtain other constructions.

Example 1.9.16: Let I be the ideal of $T_C(M)$ generated by all $x \otimes x$ for x in M. Then I is graded, so $T_C(M)/I$ is an N-graded algebra called the *exterior algebra* of M, written as $\Lambda(M)$. Note for every x in M that $x^2 = 0$ in $\Lambda(M)$, and thus $xx' = -x'x$ for all x, x' in M. (Indeed, $0 = (x + x')^2 = x^2 + xx' + x'x + (x')^2$ so $xx' + x'x = 0$.) This property can be used to characterize the exterior algebra as a suitable universal (cf., exercise 4) and leads to important applications in the theory of determinants (cf., Jacobson [85B, §7.2]).

Example 1.9.17: Let I be the (graded) ideal of $T_C(M)$ generated by all $x \otimes y - y \otimes x$ for x, y in M. $T_C(M)/I$ is called the *symmetric algebra* of M and is commutative; commutative polynomial rings are obtained as a special case when M is free in $C\text{-}\mathcal{M}od$.

Example 1.9.18: (Upper triangular matrices as a tensor ring). Let R be a direct product of n copies of W, and let $M = W^{(n-1)}$ as W-bimodule, made into an R-bimodule as follows: Take a base of M over W and label it $\{e_{12}, e_{23}, \ldots, e_{n-1,n}\}$; take a base of R over W and label it $\{e_{11}, \ldots, e_{nn}\}$, and define

$$(we_{ii})(w'e_{j,j+1}) = \delta_{ij} ww' e_{j,j+1} \quad \text{and} \quad (w'e_{j,j+1})(we_{ii}) = \delta_{i,j+1} w' w e_{j,j+1}.$$

In other words, the action from each side is matric multiplication. Then $M \otimes M$ has base $\{e_{i,i+1} \otimes e_{i+1,i+2}\}$ since for $j \neq i$ we have

$$e_{i-1,i} \otimes e_{j,j+1} = e_{i-1,i} e_{ii} \otimes e_{j,j+1} = e_{i-1,i} \otimes e_{ii} e_{j,j+1} = e_{i-1,i} \otimes 0 = 0.$$

In this way we identify the u component of $T_R(M)$ as the u diagonal in the ring of upper triangular matrices, and $T_R(M)$ is easily seen to be isomorphic to the ring of upper triangular matrices over W.

§1.10 Central Localization (also, cf., §2.12.9ff.)

One of the most fundamental problems in ring theory is to "invert" a subset S of R in the sense of finding a ring homomorphism $\varphi: R \to T$ such that s is invertible in T for all s in S. One would ideally want φ to be an injection, in which case a necessary condition is certainly that each s in S is *regular* in the sense that $rs \neq 0$ and $sr \neq 0$ for all $r \neq 0$ in R. Two familiar instances from commutative algebra, when R is an integral domain (so that every nonzero element is regular):

(i) One can form the field of fractions F of R.
(ii) Given a prime ideal P of R let $S = R - P$ and let R_P denote $\{s^{-1}r: s \in S, r \in R\} \subset F$.

R_P is *local*, i.e., its unique maximal ideal is $\{s^{-1}r: s \in S, r \in P\}$. Consequently, the process we are about to describe is called *localization*. Before proceeding, let us describe the situation more formally in terms of universals.

Definition 1.10.1:   Let $\mathscr{D}$ denote the category whose objects are pairs (R, S) where R is a ring and $S \subset R$; *morphisms* $f: (R, S) \to (R', S')$ are ring homomorphisms $f: R \to R'$ for which $fS \subseteq S'$. Then $\mathscr{D}$ has the full category $\mathscr{C}$ consisting of all (R, S) for which each element of S is invertible in R, and we can define the forgetful functor $F: \mathscr{C} \to \mathscr{D}$ which "forgets" this condition on S. A universal from (R, S) to F is called a *localization of R at S* and is denoted $S^{-1}R$ if it exists.

Explicitly there is a given homomorphism $v: R \to S^{-1}R$ such that vs is invertible for each s in S, and, moreover, for any ring homomorphism $\psi: R \to T$ such that s is invertible in T for all s in S we have a unique homomorphism completing the diagram

By abstract nonsense $S^{-1}R$ is unique up to isomorphism. There is a general construction of $S^{-1}R$, but it is rather unwieldy and resists computation. (For example, it is difficult to determine even when v is 1:1.) However, when $S \subseteq Z(R)$ we can mimic the construction from the commutative theory and bypass many of the difficulties. We shall see in corollary 1.10.18' that the construction actually can be achieved directly from the commutative theory, merely by using the tensor product.

Consequently we shall perform the construction here for $S \subset Z(R)$, called *central localization*, and defer noncommutative generalizations until Chapter 3. Throughout, S denotes a submonoid of $Z(R)$.

In the literature R_S or $R[S^{-1}]$ are also written in place of $S^{-1}R$. We shall take advantage of this notational ambiguity by introducing different constructions by different notations and then showing they are the same. The first construction is by brute force.

Construction 1.10.2: Define an equivalence $\sim$ on $S \times R$ by $(s_1, r_1) \sim (s_2, r_2)$ if $s(s_1 r_2 - s_2 r_1) = 0$ for suitable s in S; letting $s^{-1}r$ denote the equivalence class of (s, r) define $S^{-1}R$ as $S \times R/\sim$ made into a ring under the operations

$$s_1^{-1}r_1 + s_2^{-1}r_2 = (s_1 s_2)^{-1}(s_2 r_1 + s_1 r_2)$$

$$-(s^{-1}r) = s^{-1}(-r)$$

$$(s_1^{-1}r_1)(s_2^{-1}r_2) = (s_1 s_2)^{-1}(r_1 r_2).$$

This construction is very intuitive, but the verifications that $S^{-1}R$ is a ring (done more generally in §3.1) can best be described as "tedious," especially checking $\sim$ is indeed an equivalence. (Soon we shall bypass these verifications.) Let us see $S^{-1}R$ has the desired universal property. Indeed define $v: R \to S^{-1}R$ by $vr = 1^{-1}r$, obviously a homomorphism, and note $(vs)^{-1} = s^{-1}1$. Given $\psi: R \to T$ with ψs invertible for each s in S, we have

$$\psi r \psi s = \psi(rs) = \psi(sr) = \psi s \psi r, \qquad \text{implying } (\psi s)^{-1}\psi r = (\psi r)(\psi s)^{-1}$$

for all r in R, s in S. If there exists ψ_S satisfying $\psi = \psi_S v$ then $\psi_S(s^{-1}r)\psi s = \psi_S(s^{-1}r)\psi_S(1^{-1}s) = \psi r$, implying

$$\psi_S(s^{-1}r) = (\psi s)^{-1}\psi r, \tag{1}$$

proving ψ_S is unique. Conversely, we show ψ_S exists by defining it via this equation; it is easy to show ψ_S is a homomorphism with $\psi = \psi_S v$, once we have shown ψ_S is well-defined. But if $s_1^{-1}r_1 = s_2^{-1}r_2$ then $s(s_1 r_2 - s_2 r_1) = 0$ for some s in S and

$$0 = \psi(s(s_1 r_2 - s_2 r_1)) = \psi(ss_1 s_2)((\psi s_2)^{-1}\psi r_2 - (\psi s_1)^{-1}\psi r_1),$$

implying $(\psi s_2)^{-1}\psi r_2 = (\psi s_1)^{-1}\psi r_1$ since $\psi(ss_1 s_2)$ is invertible. Thus we have constructed $S^{-1}R$ explicitly.

A quicker construction can be obtained using exercise 1, but it is instructive to consider a third construction using direct limits.

Construction 1.10.3: For each s in S take $R_s = \text{Hom}_R(Rs, R)$. Preordering S as usual, i.e., $s \leq t$ iff s divides t, we define $\varphi_s^t: R_s \to R_t$ (for $s \leq t$) by sending a map $f: Rs \to R$ to its restriction to Rt. Now $(R_s; \varphi_s^t)$ is a directed system of abelian groups, whose direct limit we denote R_S, together with the canonical maps $\mu_s: R_s \to R_S$. If $f: Rs \to R$ we denote $\mu_s f$ as the element $s^{-1}r$ where $r = fs$; note s and r determine f since $f(as) = ar$ for any a in R.

We want to verify in general that R_S serves as $S^{-1}R$ and to prove this directly so as to have a self-contained proof of the existence of $S^{-1}R$. Of course, we must define multiplication on R, but first we want some preliminary observations.

(i) $s^{-1}r = (s's)^{-1}(s'r)$ for all s' in S. (Indeed, writing $s^{-1}r = \mu_s f$ we have $fs = r$ so $f(s's) = s'(fs) = s'r$ and taking $t = s's$ we have $t^{-1}s'r = \mu_t \varphi_s^t f = \mu_s f = s^{-1}r$.)

(ii) $s_1^{-1}r_1 - s_2^{-1}r_2 = (s_1s_2)^{-1}(s_2r_1 - s_1r_2)$, seen by passing to s_1s_2 using (i).

(iii) $\ker \mu_s = \{ f \in \mathrm{Hom}(Rs, R): s'(fs) = 0 \text{ for some } s' \text{ in } S\}$. (Proof ($\supseteq$) Take $t = s's$ and note $0 = (s's)^{-1}(s'fs) = \mu_t \varphi_s^t f = \mu_s f$. ($\subseteq$) If $\mu_s f = 0$, then by the proof of proposition 1.8.5 we can view everything in $\bigoplus R_s$ and write f as a sum of terms $\varphi_i^j f_i - f_i$, where for notational convenience we wrote i instead of s_i, and $s_j \geq s_i$. Let t be the product of s and all the s_i and s_j appearing in these expressions. Then $\varphi_i^j f_i(t) = f_i(t)$ for all these i, j, so $0 = f(t)$. Writing $t = s's$ we get $0 = s'fs$ as desired.

(iv) $s_1^{-1}r_1 = s_2^{-1}r_2$ iff $s(s_1r_2 - s_2r_1) = 0$ for some s. (Indeed $s_1^{-1}r_1 - s_2^{-1}r_2 = (s_1s_2)^{-1}(s_2r_1 - s_1r_2)$ so apply (iii).)

Now define multiplication by $(s_1^{-1}r_1)(s_2^{-1}r_2) = (s_1s_2)^{-1}(r_1r_2)$. This is well-defined in view of (iv) and is distributive over addition by a straightforward verification; $1^{-1}1$ is the multiplicative unit. At this stage we have verified all the conditions necessary to carry over the verification of construction 1.10.2, to define the homomorphism $v: R \to R_S$ by $vr = 1^{-1}r$, and to show (1) holds, so R_S is identified naturally with $S^{-1}R$.

This construction emphasizes the R_s, which could be thought of as modules $s^{-1}R$, and this vantage point will be used to good effect later.

Remark 1.10.4: Notation as in (1), $\ker \psi_s = S^{-1}\ker \psi$. (Indeed, $s^{-1}r \in \ker \psi_s$ iff $(\psi s)^{-1}\psi r = 0$, iff $\psi r = 0$.)

Proposition 1.10.5: If $S \subseteq S'$ there is a canonical homomorphism $\varphi: S^{-1}R \to (S')^{-1}R$ given by $s^{-1}r \to s^{-1}r$. If, moreover, $v_S s$ is invertible for all s in S' then φ is an isomorphism.

Proof: Take $\psi: R \to (S')^{-1}R$ to be the canonical homomorphism and $\varphi = \psi_S$. Under the additional hypothesis we can go in the opposite direction, starting with the canonical homomorphism $\psi: R \to S^{-1}R$; then $\varphi^{-1} = \psi_{S'}$.

                                                                          Q.E.D.

In order to utilize central localization, we need to see first how structure passes from R to $S^{-1}R$, then to note how $S^{-1}R$ may be in better form, and finally to show how one does not lose information when localizing wisely. Throughout S is a central submonoid of $Z(R)$.

Structure Passing From R to $S^{-1}R$

We start by passing structure from R to $S^{-1}R$, with special attention paid to ideals and the center. Given $A \subseteq R$ write $S^{-1}A$ for $\{s^{-1}a: s \in S$ and $a \in A\}$, and $s^{-1}A$ for $\{s^{-1}a: a \in A\}$.

Remark 1.10.6: If $L < R$ then $S^{-1}L \leq S^{-1}R$, by direct computation.

Remark 1.10.7: For any finite set of elements $x_1, \ldots, x_n$ of $S^{-1}L$ there is some s such that each $x_i \in s^{-1}L$. (Indeed, if $x_i = s_i^{-1}a_i$ take $s = s_1 \ldots s_n$: then $x_i = s^{-1}(s_1 \ldots s_{i-1}s_{i+1} \ldots s_n a_i)$.)

Note in remark 1.10.7 that we could multiply each x_i by s to produce elements with denominator 1; this common but important technique is called *clearing denominators*, and often is used to pass from $S^{-1}R$ back to R.

Write $\mathscr{L}(R)$ for the lattice of proper left ideals of R, viewed as R-modules. If $L < R$ and $s \in L \cap S$ then $1 = s^{-1}s \in S^{-1}L$, so $S^{-1}L = R$. Thus every left ideal intersecting S "blows up", and we focus on $\mathscr{L}_S(R) = \{L < R: L \cap S = \varnothing\}$.

Proposition 1.10.8: *There is an onto lattice homomorphism* $\Phi: \mathscr{L}_S(R) \to \mathscr{L}(S^{-1}R)$ *given by* $L \to S^{-1}L$.

Proof: Suppose $L_1, L_2 < R$. Then $\Phi(L_1 + L_2) = \Phi L_1 + \Phi L_2$ and $\Phi(L_1 \cap L_2) \subseteq \Phi L_1 \cap \Phi L_2$, by inspection. On the other hand, if $x = s_1^{-1}x_1 = s_2^{-1}x_2 \in \Phi L_1 \cap \Phi L_2$ then there is s in S such that $ss_2 x_1 = ss_1 x_2 \in L_1 \cap L_2$, so $x = (ss_1 s_2)^{-1}(ss_2 x_1)$. This proves Φ is a lattice homomorphism, which is onto since if $L' < S^{-1}R$ we take $L = \{r \in R: 1^{-1}r \in L'\}$; then $L < R$ and $L' = S^{-1}L$.   Q.E.D.

In general Φ is not 1:1, but it does act 1:1 on certain left ideals. We say a left ideal L is *S-prime* if $sr \in L$ implies $r \in L$, whenever $s \in S$ and $r \in R$. The importance of this concept will not be seen until we consider the prime spectrum in §2.12, but let us record some results here.

Lemma 1.10.9: *If $L < R$ is S-prime and $s^{-1}r \in S^{-1}L$ then $r \in L$.*

Proof: Write $s^{-1}r = s_1^{-1}x_1$ where $s_1 \in S$ and $x_1 \in L$. Then $s'(sx_1 - s_1r) = 0$ for some s' in S so $s's_1r = s'sx_1 \in L$ implying $r \in L$.   Q.E.D.

Proposition 1.10.10:   *Suppose* $L_1, L < R$ *and* L *is* S-*prime. If* $S^{-1}L_1 \leq S^{-1}L$ *then* $L_1 \leq L$.

Proof: If $r \in L_1$ then $1^{-1}r \in S^{-1}L_1 \leq S^{-1}L$ so $r \in L$.   Q.E.D.

Note that $\{S$-prime left ideals of $R\}$ is not necessarily a sublattice of $\mathcal{L}_S(R)$, since it need not be closed under sums. However, it is a lattice in its own right, isomorphic to $\mathcal{L}(S^{-1}R)$, as we see in the next result.

Theorem 1.10.11:   *The lattice* $\mathcal{L}_S(R)$ *has a closure operation* $^-$ *given by* $\bar{L} = \{r \in R : Sr \cap L \neq \varnothing\}$, *under which the "closed" left ideals are precisely the* S-*prime left ideals. Thus* $\{S$-*prime left ideals of* $R\}$ *is a lattice (where* $L_1 \vee L_2 = \overline{L_1 + L_2}$) *which is isomorphic to* $\mathcal{L}_S(R)$ *under* Φ (*of 1.10.8*).

Proof: If $s_1r_1 \in L$ and $s_2r_2 \in L$ then $s_1s_2(r_1 + r_2) \in L$, from which one sees readily $\bar{L}$ is a left ideal. Moreover, $\bar{\bar{L}} = \bar{L}$ and $\overline{L_1 \cap L_2} = \bar{L}_1 \cap \bar{L}_2$ (for if $r \in \bar{L}_1 \cap \bar{L}_2$ then $s_i r \in L_i$ for suitable s_i in S so $s_1s_2r \in L_1 \cap L_2$ and $r \in \overline{L_1 \cap L_2}$). Clearly $L = \bar{L}$ iff L is S-prime. By proposition 1.10.10 the restriction Φ' of Φ to S-prime left ideals of R is 1:1. Moreover, $\Phi L = \Phi \bar{L}$ since if $sr \in L$ then $1^{-1}r \in S^{-1}L$. Thus Φ' is onto.

In summary, we see $\{S$-prime left ideals of $R\}$ is a poset under $\subseteq$ and is naturally a lattice since any L_1, L_2 have a sup (namely $\overline{L_1 + L_2}$) and an inf (namely $L_1 \cap L_2$). Φ' is a lattice homomorphism since $\Phi'(\overline{L_1 + L_2}) = \Phi(L_1 + L_2) = \Phi L_1 + \Phi L_2 = \Phi' \bar{L}_1 + \Phi' \bar{L}_2$.   Q.E.D.

Corollary 1.10.12: Φ *restricts to an onto lattice homomorphism from* $\{Ideals$ *of* R *disjoint from* $S\}$ *to* $\{Ideals$ *of* $S^{-1}R\}$. Φ' *restricts to a lattice iso-morphism from* $\{S$-*prime ideals of* $R\}$ *to* $\{Ideals$ *of* $S^{-1}R\}$.

Proposition 1.10.13: $S^{-1}Z(R) \subseteq Z(S^{-1}R)$, *equality holding if* S *is regular.*

Proof: Suppose $s_1 \in S$ and $z_1 \in Z(R)$. For any $s^{-1}r$ in $S^{-1}R$ we have

$$[s_1^{-1}z_1, s^{-1}r] = (s_1s)^{-1}[z_1, r] = 0 \quad \text{so} \quad s_1^{-1}z_1 \in Z(S^{-1}R).$$

Conversely, if S is regular and $s_1^{-1}r_1 \in Z(S^{-1}R)$, then for all r in R we have $0 = [s_1^{-1}r_1, 1^{-1}r] = s_1^{-1}[r_1, r]$ so $[r_1, r] = 0$ proving $r_1 \in Z(R)$.   Q.E.D.

Examples of Central Localization

Example 1.10.14: If every ideal (resp. every left ideal) of R contains an element of S then $S^{-1}R$ is simple (resp. a division ring). This fact is used most often when S is all regular elements of R, in which case $S^{-1}R$ is the *ring of central quotients of R*. This coincides with the commutative ring of quotients when R is commutative.

Example 1.10.15: Suppose $S = \{s^i : i \in \mathbb{N}\}$ for some fixed s in $Z(R)$. Then we write $R[s^{-1}]$ for $S^{-1}R$, and $A[s^{-1}]$ for $S^{-1}A$ where $A \subset R$. For $L < R$ note $L[s^{-1}] \neq R[s^{-1}]$ iff L contains no power of s.

Our final observations concern S-prime left ideals.

Remark 1.10.16: By Zorn's lemma any left ideal disjoint from S is contained in a left ideal L maximal with respect to $L \cap S = \varnothing$. L is then S-prime since $\bar{L} = \{r \in R : sr \in L$ for suitable s in $R\}$ is a left ideal containing L, disjoint from S, so $\bar{L} = L$. Thus we see S-prime left ideals exist in abundance, and for L as defined here $S^{-1}L$ is a maximal left ideal of $S^{-1}R$. The same remark holds if we replace "left ideal" throughout by "ideal."

These three examples are the basic tools for central localization and shall be examined closer in §2.1.2.

Central Localization of Modules

Much of the previous discussion holds more generally for modules and when viewed in this way actually yields a tensor product connection between central localization and commutative localization, as we shall see.

Definition 1.10.17: Suppose S is a submonoid of $Z(R)$ and $M \in R\text{-}\mathcal{M}od$. For each s in S take $M_s = \text{Hom}_R(Rs, M)$ and under the preorder $s \leq t$ iff s divides t, we define $\varphi_s^t : M_s \to M_t$ (for $s \leq t$) by sending a map $f : Rs \to M$ to its restriction $f : Rt \to M$. Define $S^{-1}M = \varinjlim(M_s; \varphi_s^t)$, together with the canonical maps $\mu_s : M_s \to S^{-1}M$; given $f : Rs \to M$ we denote $\mu_s f$ as the element $s^{-1}x$ where $x = fs$.

We now make a series of observations analogous to those of construction 1.10.3 and omit the proofs when they are identical. First note if s is regular then $Rs \approx R$ so $M_s \approx M$ as an abelian group. In general the following also hold for s_i in S and x_i in M:

(i) $s^{-1}x = (s's)^{-1}(s'x)$ for all s' in S.

(ii) $s_1^{-1}x_1 - s_2^{-1}x_2 = (s_1 s_2)^{-1}(s_2 x_1 - s_1 x_2)$.

(iii) $\ker \mu_s = \{f \in \operatorname{Hom}(Rs, M): s'(fx) = 0 \text{ for some } s' \text{ in } S\}$.

(iv) $s_1^{-1}x_1 = s_2^{-1}x_2$ iff $s(s_1 x_2 - s_2 x_1) = 0$ for some s in S.

(v) There is a canonical map $v_S: M \to M_S$ given by $v_S x = 1^{-1}x$, and $\ker v_S = \{x: S \cap \operatorname{Ann}_R x \neq \varnothing\}$ (immediate from (ii) and (iv)).

Thus the theory of central localization of modules parallels that of rings, with the simplification that we need not worry about ring multiplication. Indeed the following result holds:

Proposition 1.10.18: $S^{-1}R \otimes_R M \approx S^{-1}M$, by an isomorphism $s^{-1}r \otimes x \to s^{-1}(rx)$.

Proof: There is a balanced bilinear map $S^{-1}R \times M \to S^{-1}M$ given by $(s^{-1}r, x) \to s^{-1}(rx)$, which thus induces a map $\varphi: S^{-1}R \otimes M \to S^{-1}M$ given by $s^{-1}r \otimes x \to s^{-1}(rx)$. On the other hand, φ^{-1} is defined by the map $s^{-1}x \to s^{-1} \otimes x$, which is well-defined, since if $s_1^{-1}x_1 = s_2^{-1}x_2$ then some $s(s_1 x_2 - s_2 x_1) = 0$, so

$$s_1^{-1} \otimes x_1 = (ss_1 s_2)^{-1}ss_2 \otimes x_1 = (ss_1 s_2)^{-1} \otimes ss_2 x_1$$

$$= (ss_1 s_2)^{-1} \otimes ss_1 x_2 = s_2^{-1} \otimes x_2.$$

Hence φ is the desired isomorphism.   Q.E.D.

Corollary 1.10.18': If S is a submonoid of $Z \approx Z(R)$ then $S^{-1}Z \otimes_Z R \approx S^{-1}R$ as Z-algebras.

Proof: The map of proposition 1.10.18 respects multiplication.   Q.E.D.

Proposition 1.10.19: There is an onto lattice homomorphism $\Phi: \mathscr{L}(M) \to \mathscr{L}(S^{-1}M)$ given $N \to S^{-1}N$ for $N < M$.

Proof: Duplicate proof of proposition 1.10.8, writing N in place of L.

Instead of copying out all the previous results for M replacing R, we shall now formulate them in a categorical setting.

Theorem 1.10.20: There is a functor $F: R\text{-}\mathscr{M}od \to S^{-1}R\text{-}\mathscr{M}od$ given by

$FM = S^{-1}M$ where for any map $f: M \to N$ we define $Ff: S^{-1}M \to S^{-1}N$ by $Ff(s^{-1}x) = s^{-1}fx$. This functor satisfies the following important properties:

(i) $\ker(Ff) = S^{-1}\ker f$ for every map f.

(ii) If $K \xrightarrow{f} M \xrightarrow{g} N$ is exact then $S^{-1}K \xrightarrow{Ff} S^{-1}M \xrightarrow{Fg} S^{-1}N$ is exact.

(iii) F induces an onto lattice homomorphism $\mathscr{L}(M) \to \mathscr{L}(S^{-1}M)$.

(iv) Define N to be an S-prime submodule of M if $sx \in N$ implies $x \in N$ for each s in S. Then $L(M)$ has a closure operation given by $\bar{N} = \{x \in M : sx \in N$ for some s in $S\}$; {S-prime submodules of M} is a lattice isomorphic to $\mathscr{L}(S^{-1}M)$ under F.

Proof: Ff is the composition $S^{-1}M \approx S^{-1}R \otimes M \xrightarrow{1 \otimes f} S^{-1}R \otimes N \approx S^{-1}N$. It follows at once F is a functor. Identify $S^{-1}M$ with $S^{-1}R \otimes M$ in what follows.

(i) Clearly $S^{-1}(\ker f) = S^{-1}R \otimes \ker f \subseteq \ker(Ff)$. Conversely if $s^{-1}x \in \ker(Ff)$ then $s^{-1}fx = 0$; hence there is s' in S with $0 = s'fx = f(s'x)$, so $s'x \in \ker f$ implying $s^{-1}x \in S^{-1}(\ker f)$.

(ii) $\ker(Fg) = S^{-1}\ker g = S^{-1}(fK) = (Ff)(S^{-1}K)$.

(iii) Duplicate proof of proposition 1.10.8, where here $L_i < M$.

(iv) Duplicate proof of theorem 1.10.11.   Q.E.D.

One nice feature of the central localization functor is that even noncategorical properties are preserved.

Remark 1.10.21: If F is free in R-$\mathscr{M}od$ with base $\{x_1, \ldots, x_n\}$ then $S^{-1}F$ is free in $S^{-1}R$-$\mathscr{M}od$ with base $\{1^{-1}x_1, \ldots, 1^{-1}x_n\}$. (Either verify this directly or recall we have proved the more general result that tensor products preserve freeness.)

Remark 1.10.22: If $S \subseteq S'$ are submonoids of $Z(R)$ and $M \in R$-$\mathscr{M}od$ then there is a canonical $S^{-1}R$-module map $S^{-1}M \to (S')^{-1}M$ given by $s^{-1}x \to s^{-1}x$; this is an isomorphism if $1^{-1}S'$ is invertible in $S^{-1}R$. (Clear, by applying proposition 1.10.5 to tensor products.)

Proposition 1.10.23: If $z, s \in Z(R)$, then $(R[z^{-1}])[s^{-1}] \approx R[(zs)^{-1}]$ as rings and $M[z^{-1}][s^{-1}] \approx M[(zs)^{-1}]$ for any R-module M.

Proof: Take $S = \{(zs)^i : i \in \mathbb{N}\}$ and $S' = \{z^i s^j : i, j \in \mathbb{N}\}$. Then $S \subseteq S'$ and for $z^i s^j \in S'$ we have $(z^i s^j)^{-1} = (zs)^{-(i+j)}z^j s^i$ in $S^{-1}R$. Hence $R[z^{-1}][s^{-1}] \approx R[(zs)^{-1}]$ by proposition 1.10.5, and remark 1.10.22 gives the isomorphism $M[z^{-1}][s^{-1}] \approx M[(zs)^{-1}]$.

Exercises

§1.1

1. Suppose $f: M_n(R) \to T'$ is a ring homomorphism. Then $T' = M_n(R')$ for some ring R', and f is given componentwise by a suitable homomorphism $\bar{f}: R \to R'$. (Hint: Define the matric units $e'_{ij} = fe_{ij}$. Now writing $T' = M_n(R')$ let $\bar{f}r$ be that r' such that $f(\sum re_{ii}) = \sum r'e'_{ii}$.)

2. If e_1, e_2 are idempotents of R with $e_2 \in Re_1$ and if $e_2 + A = e_1 + A$ for some idempotent-lifting ideal A then $e_1 = e_1 e_2$. (Hint: $e_1(1 - e_2)$ is idempotent.)

3. Write $M_{m,n}(A)$ for the set of $m \times n$ matrices with entries in A, made componentwise into an abelian group if A is an abelian group. $M_{m,n}(R) \in M_m(R)$-$\mathcal{M}od$-$M_n(R)$ where the scalar multiplication is taken to be multiplication of matrices; more generally, if $N \in R$-$\mathcal{M}od$-T then $M_{m,n}(N) \in M_m(R)$-$\mathcal{M}od$-$M_n(T)$ via matrix multiplication. In particular, any matrix of $M_{m+n}(R)$ can be partitioned in the form $\begin{pmatrix} A & B \\ C & D \end{pmatrix}$ where $A \in M_m(R)$, $B \in M_{m,n}(R)$, $C \in M_{n,m}(R)$ and $D \in M_n(R)$ and this provides a special case of example 1.1.10.

4. If $M \in M_n(R)$-$\mathcal{M}od$ and $e_{ii}M = 0$ for some i then $M = 0$. Use this fact to prove corollary 1.1.18 directly. (Hint: Suppose M is simple in R-$\mathcal{M}od$ and $0 \neq N \leq M^{(n)}$. Then $0 \neq e_{ii}N \leq e_{ii}M^{(n)} \approx M$ so $e_{ii}N \approx M$; hence $N = M^{(n)}$. The converse is even easier.)

Infinite Matrices

5. $M_\infty(R)$ is defined to be the set of infinite matrices which are row-finite, i.e., every matrix has a countably infinite number of rows, but almost all entries of each row are 0. Show $M_\infty(R)$ is a ring whose opposite is isomorphic to the ring of column-finite matrices.

6. Defining $M_{\infty,1}(N)$ to be the set of columns with entries in N, show there is a functor $F: R$-$\mathcal{M}od \to M_\infty(R)$-$\mathcal{M}od$ given by $N \to M_{\infty,1}(N)$.

7. Suppose $W \subseteq R$ are rings and $T = \{w \cdot 1 + \sum_{\text{finite}} r_{ij}e_{ij}: w \in W, r_{ij} \in R\}$, the subring of $M_\infty(R)$ generated by W and the e_{ij}. Show $Z(T) \approx Z(W)$.

§1.2

1. (Incidence algebras) Suppose I is a finite poset, viewed as a small category $\mathscr{C}$ as in example 0.1.1, and F is a field; then $F[\mathscr{C}]$ is called an *incidence algebra*. Interpret the incidence algebras (in terms of matrices) arising from the following partial orders on $\{1, \ldots, n\}$: (1) the usual total order; (2) $i \leq j$ iff $i = 1$ or $i = j$; (3) $i \leq j$ iff $i \leq j$ and $i < i_0$ for some fixed i_0. Show an element is invertible iff each "diagonal" entry is invertible.

§1.3

1. If a ring R is not a division ring then each simple module is not free. Conclude that a ring is a division ring iff each module is free.

2. The free algebra $C\{X_1, X_2\}$ in two indeterminates contains the free algebra $C\{X\}$ in a countable number of indeterminates. (Hint: Define an injection $\psi: C\{X\} \to C\{X_1, X_2\}$ by $\psi X_i = X_1^i X_2$.) The free group in two indeterminates contains the free group in a countable number of indeterminates. (Hint: $X_i \to X_1^i X_2 X_1^i$.)

3. $f: R^{(n)} \to R^{(n)}$ is epic iff gf is the identity for some map $g: R^{(n)} \to R^{(n)}$. What is the situation for monics?

4. View $M = R^{(n)}$ as a *right* R-module with base $\{e_1, \ldots, e_n\}$, where $e_i = (0, \ldots, 0, 1, 0, \ldots, 0)$. Every $n \times n$ matrix $A = (a_{ij})$ over R corresponds to a unique map $f: M \to M$ given by $fe_j = \sum_{i=1}^n e_i a_{ij}$. Likewise, if the matrix B corresponds to g, show AB corresponds to fg. Conclude f is epic iff $BA = 1$ in $M_n(R)$ for a suitable matrix B.

5. Using exercise 4 show R is weakly n-finite iff every epic $f: R^{(n)} \to R^{(n)}$ is an isomorphism.

6. R is weakly n-finite iff $R^{(n)} \approx R^{(n)} \oplus M$ implies $M = 0$. (*Hint:* ($\Rightarrow$) by exercise 4. ($\Leftarrow$) If $f: R^{(n)} \to R^{(n)}$ is epic then apply exercise 3 to the exact sequence $0 \to \ker f \to R^{(n)} \xrightarrow{f} R^{(n)} \to 0$ to show $R^{(n)} \approx R^{(n)} \oplus \ker f$. (If necessary refer to proposition 1.4.10.)

7. Define $t: R \to R/[R, R]$ by $t1 = 1 + [R, R]$. This is a trace map, called the *Stallings trace map*, and has the universal property that any trace map of R factors through it.

§1.4

1. There is a 1:1 function from {ordinals} to {rings}, proving {rings} is not a set. (Hint: Define $f: \{\text{ordinals}\} \to \{\text{rings}\}$ by $f\alpha = \mathbb{Z}^\beta$ for suitable β defined by transfinite induction such that if $\alpha_1 = \alpha_2^+$ then $|\mathbb{Z}^{\beta_1}| > |\mathbb{Z}^{\beta_2}|$.)

2. The following examples can be constructed easily using the direct product:
   (i) A ring which is not an integral domain but which has no nilpotent elements $\neq 0$
   (ii) An infinite ring having prime characteristic $p > 0$.
   (iii) A ring in which for each k in $\mathbb{N}$ there exists r nilpotent with $r^k \neq 0$.

3. If $e = (r_i) \in \prod R_i$ and each $r_i = 0$ or 1 then e is a central idempotent; e is minimal under the *PO* of exercise 0.0.24 iff all but one r_i are 0.

4. There exists an infinite chain of nonisomorphic fields $F_1 \subset F_2 \subset \cdots$ (Hint: Let F_i be a finite algebraic extension of F_{i-1} and count dimensions.)

5. (An example of nonisomorphic commutative rings R_1, R_2 with injections $R_1 \to R_2$ and $R_2 \to R_1$) Take fields $F_1 \subset F_2 \subset \cdots$ as in exercise 4 and let $R_1 = \bigoplus F_{2i-1}$ and $R_2 = \bigoplus F_{2i}$. The inclusion maps $F_{2i-1} \to F_{2i}$ give an injection $R_1 \to R_2$ and, likewise, the inclusions $F_{2i} \to F_{2i+1}$ gives an injection $R_2 \to R_1$. But there is no isomorphism $\psi: R_1 \to R_2$. (Hint: let e be one of the minimal idempotents of exercise 3. Then ψe is minimal since ψ is an isomorphism, but this implies $\psi: F_i \to F_j$ for some $j \geq i + 1$. Applying the same argument to ψ^{-1} shows $F_i \approx F_j$ which is impossible.)

Ultraproducts

7. The ultraproduct of division rings is a division ring.

7′. Suppose $R_i \approx M_n(W_i)$ for each i in I, and $\mathscr{F}$ is an ultrafilter on I. Then $(\prod R_i)/\mathscr{F} \approx M_n(\prod W_i/\mathscr{F})$. In particular, if each W_i is a division ring then $(\prod R_i)/\mathscr{F}$ is $n \times n$ matrices over a division ring.

Reduced Products As Rings In exercises 9–12 suppose $\{R_i : i \in I\}$ are rings and let $R = \prod_{i \in I} R_i$. Write $Z(r) = \{i \in I : r_i = 0\}$ and for any $A \subset R$, define $Z(A) = \{Z(r) : r \in A\} \subseteq \mathscr{P}(I)$. Conversely, for any $\mathscr{F} \subseteq \mathscr{P}(I)$, define $\mathscr{I}(\mathscr{F}) = \{r \in R : Z(r) \in \mathscr{F}\}$.

9. For any set $J \subset I$ there is a central idempotent $e(J)$ with $Z(e(J)) = J$. (As in exercise 3). If $\mathscr{F}$ is a filter of I then $\mathscr{I}(\mathscr{F}) = \sum_{J \in \mathscr{F}} Re(J)$.

10. If $\mathscr{F}$ is a filter of $\mathscr{P}(I)$ then $\mathscr{I}(\mathscr{F}) \lhd R$ and $Z(\mathscr{I}(\mathscr{F})) = \mathscr{F}$, and $R/\mathscr{I}(\mathscr{F})$ is isomorphic to the reduced product $\prod R_i/\mathscr{F}$. (Hint: $Z(r + r') \supseteq Z(r) \cap Z(r')$ and $Z(rr') \supseteq Z(r) \cup Z(r')$ for r, r' in R, so $\mathscr{I}(\mathscr{F}) \lhd R$ with $Z(\mathscr{I}(\mathscr{F})) = \mathscr{F}$ by exercise 9. Define the obvious surjection $\psi : R \to \prod R_i/\mathscr{F}$ and show ker ψ is the set of (r_i) such that $\{i \in I : r_i = 0\} \in I$, which is $\mathscr{I}(\mathscr{F})$.)

11. $\mathscr{I}(\ )$ gives a lattice injection from $\{$filters of $\mathscr{P}(I)\}$ to $\{$ideals of $R\}$. This is a lattice isomorphism if each R_i is a field. However, if $I = \mathbb{N}$ and $R_i \approx M_i(\mathbb{Q})$ for each i then $\{(r_i):$ the rank of the r_i is bounded$\}$ is an ideal of R not in the image of $\mathscr{I}(\ )$.

12. The ultrafilters of a set I correspond to the maximal ideals of F^I where F is an arbitrary field.

Rings of Continuous Functions The following exercises were chosen to give a taste of Gillman and Jerison [60B], which is a study of the ring R of continuous real-valued functions on a topological space X. (Gillman-Jerison denote this as $C(X)$.) Typical elements of R are continuous $f : X \to \mathbb{R}$, and we write $Z(f)$ for $\{x \in X : fx = 0\}$, called the *zero set* of f. A Z-set is a zero set of suitable f.

13. $\{Z$-sets$\}$ is a sublattice of $\mathscr{P}(X)$.

14. Each Z-set is a countable intersection of open sets of X. (Hint: $Z(f) = \bigcap_{n \in \mathbb{N}} \{x \in X : |fx| < n^{-1}\}$.)

15. If X is compact then $\{Z$-sets$\}$ is closed under countable intersections but not countable unions. (Hint: One may assume $0 \leq f_i x \leq 1$ for each f_i and each x in X, and note $\bigcap Z(f_i) = Z(\sum f_i/2^i)$.)

16. A Z-filter is a sublattice of $\{Z$-sets$\}$ such that if $Z_1 \in \mathscr{F}$ and $Z_1 \subset Z_2$ then $Z_2 \in \mathscr{F}$. If $A \lhd R$ define $Z(A) = \{Z(f) : f \in A\}$ and show it is a Z-filter. Also given a Z-filter $\mathscr{F}$ define $\mathscr{I}(\mathscr{F}) = \{f \in R : Z(f) \in \mathscr{F}\}$. Show $\mathscr{I}(\mathscr{F}) \lhd R$, and $R/\mathscr{I}(\mathscr{F})$ has no nilpotent elements $\neq 0$.

17. If A is a maximal ideal of R then $Z(A)$ is an ultra-filter; if $\mathscr{F}$ is an ultra-Z-filter then $\mathscr{I}(\mathscr{F})$ is a maximal ideal. In the same way, prime ideals of R correspond to prime filters (in the sense that $I_1 \cup I_2 \in \mathscr{F}$ implies $I_1 \in \mathscr{F}$ or $I_2 \in \mathscr{F}$).

18. There is a 1:1 correspondence from $\{$ultra-Z-filters$\}$ to the points of the Stone-Cech compactification of X. In this way, one can use geometry to study R, and, in fact, this example embodies many features of algebraic geometry.

§1.5

0. If N is an R-module then End $N^{(n)} \approx M_n(\text{End } N)$. (Hint: Modify the proof of theorem 1.5.13, writing $N = Rx$ and sending $(f_{ij}) \in M_n(\text{End } N)$ to the map $\beta : N^{(n)} \to N^{(n)}$ given by $\beta(x_1, \ldots, x_n) = (\sum_{i=1}^n f_{i1} x_i, \ldots, \sum_{i=1}^n f_{in} x_i)$.)

1. Define the *idealizer* $\mathscr{I}(L)$ of a left ideal L of R to be $\{r \in R : Lr \subseteq L\}$. Then $\mathscr{I}(L)$ is

the largest subring of R in which L is an ideal; also $\mathscr{I}(L)/L \approx \text{End}(R/L)$, by sending an element r to its right multiplication map.

2. (The *dual functor*) Given f in $\text{Hom}_R(M, N)$ define $\hat{f}: \hat{N} \to \hat{M}$ by taking $\hat{f}h$ to be the composition hf for any $h: N \to R$. This gives a contravariant functor $F: R\text{-}\mathscr{M}od \to \mathscr{M}od\text{-}R$ given by $FM = \hat{M}$ and $Ff = \hat{f}$, called the *dual* functor. Likewise, there is a contravariant dual functor $G: \mathscr{M}od\text{-}R \to R\text{-}\mathscr{M}od$, and the covariant functor $GF: R\text{-}\mathscr{M}od \to R\text{-}\mathscr{M}od$ is called the *double dual*.

3. (Ellenberg-MacLane) There is a natural transformation η from the identity functor of $R\text{-}\mathscr{M}od$ to the double dual functor. (Hint: Define $\eta_M: M \to \hat{\hat{M}}$ by taking $\eta_M x$ to be the map $(\eta_M x)f = fx \in R$ for f in $\hat{M}$. Show $(\hat{\hat{g}}(\eta_M x))f = (\eta_{\hat{M}}(gx))f$ for each f in $\hat{M}$ and x in M, proving $\hat{\hat{g}}\eta_M = \eta_{\hat{M}}g$.)

4. If R is a division ring in exercise 3 then η restricts to a natural isomorphism in $R\text{-}\mathscr{F}imod$. (Hint: Each f.g. R-module M is free; to show η_M is 1:1 show that if $x \neq 0$ there is f in $\hat{M}$ with $fx \neq 0$. A dimension count shows each η_M is an isomorphism.)

5. Suppose K, M, N are finite dimensional vector spaces over a field F, and $K \xrightarrow{f} M \xrightarrow{g} N$ is exact. Then $\hat{N} \xrightarrow{\hat{g}} \hat{M} \xrightarrow{\hat{f}} \hat{K}$ is exact. (Hint: Certainly $\ker \hat{f}$ contains the dual of $\text{coker } f$ so equality holds by a dimension count; likewise $\hat{g}\hat{N}$ is dual of $M/\ker g$, implying $\hat{g}\hat{N}$ and $\ker \hat{f}$ have the same dimension.) In the terminology of §2.11 below, this says the dual functor is exact for $F\text{-}\mathscr{F}imod$; a more general result is given in exercise 2.11.0.

§1.6

1. In any domain if $ab = 1$ then $ba = 1$. (Hint: $ba(ba - 1) = 0$.)
2. If $Ra = Rb$ in a domain R then $b = ua$ for some unit u.
3. If $T = R[\lambda; \sigma, \delta]$ is a PLID then R is a PLID satisfying Jategaonkar's condition. (Hint: specialize $\lambda \to 0$ to show R is a PLID. To verify Jategaonkar's condition suppose $0 \neq r \in R$. Then $Tr + T\lambda = Tb$ for some b in T; $r \in Tb$ implies $b \in R$, and $\lambda \in Tb$ implies σb is invertible. Specialize $\lambda \to 0$ to get $b \in Rr$.)
4. If F is a field of characteristic p then $Z(\mathscr{A}_1(F)) \approx F[\lambda^p, \mu^p]$, so $\mathscr{A}_1(F)$ is not simple.

Derivations

5. If δ is a derivation on R then $\delta(Z(R)) \subseteq Z(R)$. (Hint: $0 = \delta[z, r] = [z, \delta r] + [\delta z, r]$.)
6. Using Leibniz' rule show that if $\text{char}(R) = p$ and $\delta \in \text{Deriv}(R)$ then $\delta^p \in \text{Deriv}(R)$.
7. Suppose $\delta^2 a = 0$. Then $\delta^n(a^n) = n!(\delta a)^n$.
8. (Hirano-Yamakawa [84]) For any derivation δ of R and for each positive integer k there are integers n_{ki} such that $(\delta^k a)b = \sum_{i=0}^{k} n_{ki}\delta^{k-i}(a\delta^i b)$ for all a, b in R. (Proof by induction).
9. (Hirano-Yamakawa [84]) Let $R = M_\infty(F[\lambda])$ and δ be defined by taking differentiation at each component. Then δ is nil on $\text{soc}(R)$ but not on R. However, using exercise 8, one can easily prove the following result: Suppose the derivation δ of R is nil on an ideal A. If $\delta^k S = 0$ for some subset S of A for which $\text{Ann}_R S = 0$, then δ is nil on R. In particular, this holds if $\text{Ann}_R S = 0$ for some finite subset S of A. Moreover, if $\delta^k A = 0$ and $\text{Ann } A = 0$ then $\delta^{3k-2}R = 0$.

Suppose σ is an automorphism of R. A σ-*ideal* of R is an ideal A of R such that $\sigma A = A$; we denote this by $A \lhd (R, \sigma)$.

10. (Goldie-Michler [74]) The σ-ideals are the kernels in the category of rings with automorphism σ, where the morphisms preserve σ. Explicitly, if $A \lhd (R, \sigma)$ then σ acts naturally on R/A; conversely, if $f: R \to T$ is a ring homomorphism and R, T have automorphisms both denoted σ and $f(\sigma r) = \sigma(fr)$ for all r in R then $\ker f \lhd (R, \sigma)$.

   A ring is *prime* if the product of any two nonzero ideals is nonzero.

11. (Pearson-Stephenson [77]) Suppose R has an automorphism σ. A σ-ideal P is called σ-*prime* if whenever $A \lhd R$ and $B \lhd (R, \sigma)$ with $AB \subseteq P$, one must have $A \subseteq P$ or $B \subseteq P$. R is called σ-*prime* iff 0 is a σ-prime ideal. Show $T = R[\lambda; \sigma]$ is a prime ring iff R is σ-prime. (Hint: ($\Leftarrow$) If $AB = 0$ and $B \lhd (R, \sigma)$ then $(TAT)(TBT) \subseteq TABT = 0$. ($\Rightarrow$) If $A', B' \lhd T$ with $A'B' = 0$ then take $A = \{$leading coefficients of elements of $A'\}$ and likewise for B.)

12. (Cauchon [79]) Suppose $T = R[\lambda; \sigma, \delta]$ where R is simple and σ is an automorphism. If $A \lhd T$ then A contains a monic polynomial f of minimal degree, and then $A = Tf$. Furthermore, if we take a monic polynomial g of minimal degree > 0 such that $Tg \lhd T$ then $A = Tzg^n$ for suitable n and suitable z in $Z(T)$. (Hint: The first assertion follows easily from the Euclidean algorithm; note if f in A has minimal degree t then RfR contains a monic polynomial of degree t. To prove the second assertion, write $f = hg^n$ where $h \notin Tg$; then $Th \lhd T$. It suffices to prove $Th = Tz$ for some z in $Z(T)$. Write $h = qg + p$, with $\deg(p) < \deg(g)$. Write $n = \deg h$ and $m = \deg(g)$. Note $\deg q = n - m$. For all r in R one has $gr - (\sigma^m r)g \in Tg$ implying $gr = (\sigma^m r)g$; likewise, $hr = (\sigma^n r)h$, and $pr = (\sigma^n r)p$. Similarly, there are r_i in R such that $g\lambda = (\lambda + r_1)g$ and $h\lambda = (\lambda + r_2)h$; if $(\deg p) < m - 1$ then $p\lambda = (\lambda + r_2)p$, so $Tp \lhd T$, and by minimality of $\deg g$ we get $\deg p = 0$, so $p \in R$ is invertible and $p^{-1}h$ is central. If $\deg p = m - 1$ then $g = \lambda' p + p'$ where $\deg(\lambda') = 1$ and $\deg p' < \deg p$. Taking $\sigma' = \sigma^{n-m}$ one has $\lambda' r = (\sigma' r)\lambda'$ for all r in R, so $T \approx R[\lambda'; \sigma']$, and we are reduced to the case $\delta = 0$; in this case, $T\lambda \lhd T$ so $m = 1$, and one can use the proof of proposition 1.6.25(ii).) Cauchon also describes the center.

§1.7

1. The Weyl algebra $\mathscr{A}_n(C) \approx \bigotimes_{i=1}^{n} \mathscr{A}_1(C)$. (Hint: Take the natural homomorphism $\bigotimes_{i=1}^{n} \mathscr{A}_1(C) \to \mathscr{A}_n(C)$ sending λ and μ of the i-th component to λ_i and μ_i. This map sends a base to a base so is an isomorphism.)

2. Suppose $A, B \in C\text{-}\mathscr{A}lg$ and H is a commutative C-algebra. Then $(A \otimes_C H) \otimes_H (B \otimes_C H) \approx (A \otimes_C B) \otimes_C H$.

3. Given $M \in R\text{-}\mathscr{M}od$ write M^{op} for the corresponding right R^{op}-module (i.e., same addition and opposite scalar multiplication). If $M, N \in R\text{-}\mathscr{M}od$ the bilinear forms $B: M \times N \to R$ correspond to the balanced maps $M^{op} \times N \to R$ (over R) and thus induce group homomorphisms $M^{op} \otimes_R N \to R$. This explains the tie between tensor products and bilinear forms.

4. Suppose C is a commutative ring and A is an abelian group. If $A \in C\text{-}\mathscr{M}od$ then

scalar multiplication induces a map $\mu: C \otimes A \to A$ given by $\mu(c \otimes a) = ca$. If, moreover, $A \in C\text{-}\mathcal{Alg}$ then there is a map $\pi: A \otimes A \to A$ given by $\pi(a_1 \otimes a_2) = a_1 a_2$, and $\varepsilon: C \to A$ given by $\varepsilon c = c1$. Show the following diagrams commute:

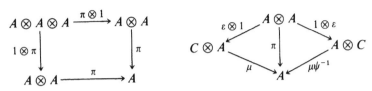

where $\psi: C \otimes A \to A \otimes C$ is given by $\psi(c \otimes a) = a \otimes c$. Conversely, show that any maps π, ε as described above give rise to an algebra structure by defining $a_1 a_2 = \pi(a_1 \otimes a_2)$ and $ca = \pi(\varepsilon c \otimes a)$. Thus a C-algebra is given by the triple (A, π, ε).

7. If $M \in \mathcal{Mod}\text{-}R$ and $N \in R\text{-}\mathcal{Mod}$ such that R is a summand of N then the canonical map $M \to M \otimes 1 \subseteq M \otimes N$ is monic. (Hint: Construct the left inverse map $M \otimes N \to M$.)

8. Suppose $f_i: R_i \to T_i$ are homomorphisms for $i = 1, 2$. Then $\ker(f_1 \otimes f_2) = (\ker f_1) \otimes R_2 + R_1 \otimes \ker f_2$ viewed in $R_1 \otimes R_2$.

§1.8

1. Define the *pushout* of morphisms $f_1: A_0 \to A_1$ and $f_2: A_0 \to A_2$ to be the direct limit of the following system: $i = \{0, 1, 2\}$ with PO defined by $0 < 1$ and $0 < 2$; the A_i are as given with $\varphi_0^i = f_i$ for $i = 1, 2$ and $\varphi_i^i = 1_{A_i}$ for all i. Thus the pushout is an object P together with morphisms $\mu_i: A_i \to P$ for $i = 1, 2$ satisfying $\mu_1 f_1 = \mu_2 f_2$. (Take $\mu_0 = \mu_1 f_1$.) Show by example that the pushout need not exist in $\mathcal{Ring}$. In $R\text{-}\mathcal{Mod}$ show $P = (A_0 \oplus A_1 \oplus A_2)/(\sum R(-a, f_1 a, 0) + R(-a, 0, f_2 a)) \approx (A_1 \oplus A_2)/\sum R(-f_1 a, f_2 a)$, the sums taken over all a in A_0. Associate the pushout to the following diagram and universal:

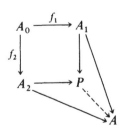

1'. (The pullback) Given morphisms $f_i: A_i \to A_0$ for $i = 1, 2$ define the *pullback* of f_1 and f_2 to be the inverse limit of the system dual to that of exercise 1. Thus it is an object P together with morphisms $\rho_i: P \to A_i$ for $i = 1, 2$ such that $f_1 \rho_1 = f_2 \rho_2$, i.e., according to the diagram of exercise 1 with the arrows reversed. Show

$$P = \{(a_0, a_1, a_2) \in \textstyle\prod A_i : a_0 = f_1 a_1 = f_2 a_2\} \approx \{(a_1, a_2) \in A_1 \times A_2 : f_1 a_1 = f_2 a_2\}$$

in $R\text{-}\mathcal{M}od$, $\mathcal{R}ing$, and $\mathcal{Ab}$.

2. If $f: A \to B$ is a map then cok f is a pushout of the system $A \xrightarrow{f} B$, and ker f is a

$$\begin{array}{cc} 0 & \downarrow \\ \downarrow & 0 \end{array}$$

pullback of the inverse system $A \xrightarrow{f} B$. In particular, cok is a direct limit and ker is an inverse limit.

3. (Reduced products as direct limits) Any filter $\mathcal{F}$ of subsets of a set I is partially ordered via set containment. Letting $\pi_S^T: \prod_{i \in S} A \to \prod_{i \in T} A_i$ denote the natural projection for $T \subset S$ in $\mathcal{F}$, we get a system $(\prod_{i \in S} A_i; \pi_S^T)$ whose direct limit is $\prod A_i / \mathcal{F}$ together with the natural maps $\mu_S: \prod_{i \in S} A_i \to \prod A_i / \mathcal{F}$.

§1.9

1. The Morita ring $A = \begin{pmatrix} R & M \\ M' & T \end{pmatrix}$ is graded by

$$A_0 = \begin{pmatrix} R & 0 \\ 0 & T \end{pmatrix}, \quad A_1 = \begin{pmatrix} 0 & M \\ 0 & 0 \end{pmatrix}, \quad A_{-1} = \begin{pmatrix} 0 & 0 \\ M' & 0 \end{pmatrix},$$

and all other $A_\mu = 0$.

2. Suppose A is an algebra over an algebraically closed field F, and let $\sigma \in \text{End}_F A$. We define the *characteristic subspace* $A_\alpha = \{a \in A : (\sigma - \alpha)^t a = 0$ for suitable $t\}$ for α in F and recall by linear algebra that $A = \bigoplus A_\alpha$. If σ is a *ring* endomorphism as well then $A_\alpha A_\beta \subseteq A_{\alpha\beta}$ (hint: $(\sigma - \alpha\beta)(ab) = (\sigma - \alpha)a\sigma b + \alpha a(\sigma - \beta)b$ for a in A_α and b in A_β), so A is graded by the multiplicative structure of F. If, moreover, σ is an automorphism algebraic over F (viewing $\text{End}_F A$ as an F-algebra) then there are only a finite number of eigenvalues, since each eigenvalue is a root of the minimal polynomial, so there are only a finite number of nonzero components. If, instead, σ is a derivation then A is graded by the *additive* structure of F (hint: show $A_\alpha A_\beta \subseteq A_{\alpha+\beta}$ by means of the equation $(\sigma - \alpha - \beta)(ab) = ((\sigma - \alpha)a)b + a(\sigma - \beta)b)$.

3. Suppose A is graded. There is a lattice injection from {left ideals of A_1} to {graded left ideals of A} given by $L \to AL$ (since $(AL)_1 = L$). This foreshadows a connection between the structures of A_1 and of A which will be of focal interest at times. Often A_1 has a special interpretation. For example, in exercise 2 if σ is an automorphism which satisfies a separable polynomial then $A_1 = A^\sigma$.

Tensor Algebras

4. For every C-algebra A and every map $f: M \to A$ such that $(fx)^2 = 0$ for all x in M there is a unique algebra homomorphism $f: \Lambda(M) \to A$ extending f. Use this property to write $\Lambda(M)$ as a suitable universal.

5. Suppose M is a free C-module with base indexed by I. Then $\Lambda(M) \approx C\{X\}/A$ where A is the ideal generated by $\{X_i^2, X_iX_j + X_jX_i : i, j \in I\}$. (Hint: Verify the universal property.) Furthermore, letting x_i denote the canonical image of X_i, one has a base $B = \{1\} \cup \{X_{i_1} \cdots x_{i_u} : 0 < u < \infty$ and $i_1 < \cdots < i_u\}$ of $\Lambda(M)$ over C. The *degree* of the monomial $x_{i_1} \cdots x_{i_u}$ is defined to be u and is well-defined. Letting $B_0 = \{b \in B : \deg(b)$ is even$\}$ show $\Lambda(M)$ is $\mathbb{Z}/2\mathbb{Z}$-graded, where the 0-component is $CB_0 = Z(\Lambda(M))$. Thus $\Lambda(M)$ is a free module over its center with base $\{1\} \cup \{x_i : i \in I\}$.

6. The symmetric algebra is a universal from a given module M to the forgetful functor (from {commutative C-algebras} to C-$\mathcal{Mod}$.)

6'. (Tensor rings as universals) Define the category $\mathcal{Ring\text{-}Bimod}$ whose objects are pairs $\varphi(R, M)$ with M in R-$\mathcal{Mod}$-R, and whose morphisms are pairs (φ, f) where $\varphi : R \to R'$ is a ring homomorphism and $f : M \to M'$ is a φ-map on each side (i.e., $\varphi f(r_1 x r_2) = (\varphi r_1)(fx)(\varphi r_2)$. There is a functor $F : \{N\text{-graded rings}\} \to \mathcal{Ring\text{-}Bimod}$ satisfying $FT = (T_0, T_1)$. Taking $u : (R, M) \to T_R(M)$ by sending R to the 0-component and M to the 1-component, show $(T_R(M), u)$ is the universal from (R, M) to F.

7. Define the algebra of *dual numbers* $\hat{C}$ over C to be free with base $1, x$ and satisfying $x^2 = 0$. (This is the exterior algebra of a C-module isomorphic to C.) If $R \in C$-$\mathcal{Alg}$ define a correspondence $\mathrm{Deriv}(R) \to \mathrm{Aut}(R \otimes \hat{C})$ by sending δ to the automorphism $\hat{\delta}$ of $R \otimes \hat{C}$ given by $\hat{\delta}(r_1 \otimes 1 + r_2 \otimes x) = r_1 \otimes 1 + (r_2 + \delta r_1) \otimes x$; inner derivations of R are sent to inner automorphisms of $R \otimes \hat{C}$.

§1.10

1. Suppose S is a submonoid of $Z(R)$, and let $\lambda = \{\lambda_s : s \in S\}$ be a commutative set of indeterminates. Then $R_S \approx R[\lambda]/\langle \lambda_s s - 1 : s \in S \rangle$, seen by verifying the universal property.

2 Basic Structure Theory

This chapter contains the general results on rings and their modules which have come to be known as the "structure theory." One of the main themes is Jacobson's structure theory, via primitive rings and the Jacobson radical. This approach has been spectacularly successful in the 40 years since its inception. The approach is threefold: First one studies "primitive" rings, which are the building blocks, then one obtains the "semiprimitive rings," and finally one studies the Jacobson radical J, the smallest ideal of a ring R such that R/J is semiprimitive.

Since the inception of this theory, prime rings have come to the fore, especially with Goldie's theorem, which characterizes left orders in simple Artinian rings. Although Goldie's theorem and its consequences are not studied until Chapter 3, we shall also recast the structure theory in terms of prime rings; the radical associated with this theory is the "lower nilradical." We shall also examine other nilradicals.

Another side to the structure theory is the study of rings in terms of their modules. This can be done through the Jordan-Hölder-Schreier theorem and the Krull-Schmidt theorem. New classes of rings arising in such a discussion are semiprimary, perfect, and seimiperfect rings, which are useful generalizations of Artinian rings. On the other hand, perhaps the best approach to module theory is via projective modules and injective modules, thereby laying the foundation for homological algebra (Chapter 5). Sections 2.8–2.12 deal with this very important aspect of the structure theory.

At the end of the chapter we consider involutions on rings. This extra piece of structure turns out to be very useful in studying those rings that possess it.

§2.1 Primitive Rings

Our object in this section is to become familiar with the class of "primitive" rings which is general enough to embrace a very large assortment of examples but is amenable to study by means of Jacobson's "density theorem." Our approach will be to develop the theory of primitive rings first and then describe some of the examples in detail. In the course of investigation we shall introduce some important tools of structure theory, including prime rings, minimal left ideals, and the socle. The definition of primitive ring itself is rather straightforward.

Definition 2.1.1:   A ring is *primitive* if it has a faithful simple module.

Remark 2.1.2:

(i) Any simple ring R is primitive. Indeed every nonzero module M is faithful since Ann $M \lhd R$ and is thus 0; hence R/L is faithful simple in R-$\mathcal{M}od$ for any maximal left ideal L.

(ii) If D is a division ring and M is a right vector space over D then End M_D is primitive (since M is faithful simple viewed as a module over End M_D).

Our principal objective is to relate primitive rings to End M_D. First we locate D.

Proposition 2.1.3:   *("Schur's lemma") if $M \in R$-$\mathcal{M}od$ is simple then $D = \text{End}_R M$ is a division ring.*

Proof: By remarks 0.2.10, 0.2.11 every nonzero element of D is an isomorphism and is thus invertible.   Q.E.D.

If R has a faithful simple module M then by means of the regular representation (cf., example 1.5.11) we can view $R \subseteq \text{End } M_D$ where $D = \text{End}_R M$. In fact, R takes up a lot of room, as we shall see now.

Jacobson's Density Theorem

Definition 2.1.4: Suppose R is an arbitrary subring of End M_D where M is a vector space over a division ring D, and view $M \in R\text{-}\mathcal{M}od$ by the given action of R on M. We say R is *dense* if for every n in $\mathbb{N}$ and every D-independent set $\{x_1, \ldots, x_n\}$ in M we have $Rx = M^{(n)}$ where $x = (x_1, \ldots, x_n) \in M^{(n)}$. (In other words, given $y_1, \ldots, y_n$ in M one can find r in R such that $rx_i = y_i$ for $1 \le i \le n$.)

Remark 2.1.5: Every dense subring R of End M_D is primitive. (Indeed, M is obviously faithful in $R\text{-}\mathcal{M}od$ since $fM = 0$ implies $f = 0$; M is simple by the case $n = 1$ of definition 2.1.4.)

The term "density" stems from the following topological considerations: Given $x \in M$ and $f \in$ End M_D define $B(x; f) = \{g \in \text{End } M_D : gx = fx\}$. The $B(x; f)$ are a sub-base for a topology called the *finite topology* of End M_D; R is dense in this topology iff R is a dense subring. The converse of remark 2.1.5 is the key to the subject.

Theorem 2.1.6: *(Jacobson's density theorem) Suppose R has a faithful simple module M and $D = \text{End}_R M$. Then R is dense in End M_D.*

Proof: We shall prove that $R(x_1, \ldots, x_n) = M^{(n)}$ for every D-independent set $\{x_1, \ldots, x_n\}$ in M. For $n = 1$ this is true since M is simple, so we proceed by induction on n.

Claim: *There is r in R such that $rx_n \ne 0$ and $rx_i = 0$ for all $i < n$.*

Otherwise $r(x_1, \ldots, x_{n-1}) = 0$ always implies $rx_n = 0$ so we have a well-defined map $f : M^{(n-1)} \to M$ given by $f(r(x_1, \ldots, x_{n-1})) = rx_n$ (since $R(x_1, \ldots, x_{n-1}) = M^{(n-1)}$ by induction hypothesis). Then by proposition 1.5.20 there are $d_1, \ldots, d_{n-1}$ in D such that $\sum_{i=1}^{n-1} x_i d_i = f(x_1, \ldots, x_{n-1}) = x_n$, contrary to the assumed independence of $x_1, \ldots, x_n$.

Having established the claim, we see by symmetry that for each $j \le n$ there exists $r_j \ne 0$ with $r_j x_j \ne 0$ and $r_j x_i = 0$ for all $i \ne j$. Given arbitrary $y_1, \ldots, y_n$ in M we find r'_j in R with $r'_j(r_j x_j) = y_j$ and put $r = \sum_{j=1}^{n} r'_j r_j$. Then $rx_i = \sum r'_j r_j x_i = r'_i r_i x_i = y_i$ for each i, as desired.   Q.E.D.

An elegant proof of a generalization of the density theorem is given in exercise 2.4.1. One can also formulate these results in terms of matrices.

Proposition 2.1.7: *If R is a dense subring of* End M_D *then one of the following situations holds:*

(i) $[M:D] = n$ *for some* $n < \infty$, *in which case* $R \approx M_n(D)$;
(ii) *For every n in* $\mathbb{N}$ *there is a subring of R having* $M_n(D)$ *as a homomorphic image.*

Proof: If $[M:D] = n$ then $R \approx$ End $M_D \approx M_n(D)$. Otherwise, given n pick D-independent elements $x_1, \ldots, x_n$ of M, put $V = \sum_{i=1}^{n} x_i D$ and let $R_1 = \{r \in R : rV \subseteq V\}$. Then $V \in R_1\text{-}\mathcal{M}od$ so the regular representation gives us a homomorphism $R_1 \to$ End $V_D \approx M_n(D)$ which is a surjection by the density theorem.   Q.E.D.

The structure of End M_D is much less amenable when M is infinite dimensional over D; in this case End M_D even lacks *IBN* (cf., example 1.3.33). One immediate consequence of Jacobson's density theory is the following extremely important structure theorem.

Theorem 2.1.8: *(Wedderburn-Artin) If R is primitive and satisfies DCC on left ideals then* $R \approx M_n(D)$ *for a suitable division ring D.*

Proof: Write R as dense subring of End M_D, and let $\{x_i : i \in I\}$ be a base of M over D. Let $L_i = \text{Ann}\{x_j : j \leq i\}$. Then $L_1 > L_2 > \cdots$ by density, contrary to hypothesis unless $|I|$ is finite. But then we are done by proposition 2.1.7(i).
                                                                                          Q.E.D.

The usual formulation of the Wedderburn-Artin theorem is "any simple ring satisfying DCC on left ideals has the form $M_n(D)$." Since we are assuming $1 \in R$ this statement can be improved (corollary 2.1.25′) and, furthermore, there are several short proof (exercises 8, 9). Nevertheless this treatment shows how the density theorem gives us a firm grasp of primitive rings.
There is a useful internal characterization of primitive rings. Define the *core* of a left ideal L (written core(L)) to be the sum of all those (two-sided) ideals of R contained in L. Thus the core is the unique largest ideal of R contained in L.

Proposition 2.1.9: (i) Core$(L) = \text{Ann}_R(R/L)$. (ii) *if* $L < R$ *is maximal then* $R/\text{Core}(L)$ *is primitive ring.* (iii) R *is a primitive ring iff* R *has a maximal left ideal L whose core is 0.*

Proof: Let $A = \text{Core}(L)$.

(i) $\text{Ann}_R(R/L)$ is an ideal of R, and for any r in $\text{Ann}_R(R/L)$ we have $r = r1 \in rR \subseteq L$, implying $\text{Ann}_R(R/L) \subseteq A$. Conversely, $AR = A \subseteq L$ so $A \subseteq \text{Ann}_R(R/L)$.

(ii) R/L is a simple module over R and is thus faithful simple over R/A.

(iii) ($\Leftarrow$) follows from (ii); conversely, if M is a faithful simple R-module then $M \approx R/L$ for some maximal left ideal L, implying $0 = \text{Ann}_R M = \text{Ann}_R(R/L) = \text{core}(L)$ by (i).   Q.E.D.

Corollary 2.1.10:   *Every commutative primitive ring has no nonzero left ideals and is thus a field.*

There is a related criterion for primitivity which is sometimes easier to verify. We say a left ideal L is *comaximal with all ideals* if $L + A = R$ for all $0 \neq A \lhd R$. Clearly, a maximal left ideal L has core 0 iff L is comaximal with all ideals, leading us to

Proposition 2.1.11:   *R is a primitive ring iff R has a left ideal L comaximal with all ideals.*

Proof: ($\Rightarrow$) Take L maximal with core 0.

($\Leftarrow$) Take a maximal left ideal $L' \supset L$. Then L' is comaximal with all ideals, so core $(L') = 0$.   Q.E.D.

Prime Rings and Minimal Left Ideals

The density theorem relies heavily on the choice of faithful simple module, and one may ask whether all faithful simple R-modules need be isomorphic. In general the answer is "no" but an affirmative answer is available in one very important situation. We shall say a (nonzero) left ideal is *minimal* if it is minimal as a nonzero left ideal. $\mathbb{Z}$ is an example of a ring without minimal (left) ideals since if $n \in I \lhd \mathbb{Z}$ then $2n\mathbb{Z} \subset I$. Primitive rings having a minimal left ideal are particularly well-behaved, and to elucidate the situation we introduce a more general class of rings.

Definition 2.1.12: R is a *prime* ring if $AB \neq 0$ for all $0 \neq A, B \lhd R$.

In generalizing definitions from elementary commutative ring theory, one often does best in replacing elements by ideals. We shall see that prime rings form the cornerstone of ring theory, being the "correct" noncommutative generalization of integral domain.

Proposition 2.1.13: *The following conditions are equivalent: (i) R is prime; (ii) $\text{Ann}_R L = 0$ for every nonzero left ideal L of R; (iii) $r_1 R r_2 \neq 0$ for all non-zero $r_1, r_2 \in R$.*

Proof: (i)$\Rightarrow$(ii) $\text{Ann}_R L$ and LR are ideals of R with $(\text{Ann}_R L)(LR) = 0$, and $0 \neq L \subseteq LR$, so $\text{Ann}_R L = 0$.

(ii) $\Rightarrow$ (iii) if $r_2 \neq 0$ then $r_1 \in \text{Ann}_R R r_2$ implying $r_1 = 0$ by (ii).

(iii) $\Rightarrow$ (i) if $0 \neq A, B \lhd R$ then taking nonzero $a \in A$, $b \in B$ yields $0 \neq (aR)b \subseteq AB$.   Q.E.D.

Remark 2.1.14: Every primitive ring is prime. (Indeed, if $M \in R\text{-}\mathcal{M}od$ is faithful and simple and $0 \neq A, B \lhd R$ then $0 \neq BM \leq M$ so $BM = M$; implying $0 \neq AM = A(BM)$, so $AB \neq 0$.)

The converse of remark 2.1.14 is false even for commutative rings, by corollary 2.1.10. The "missing link" is minimal left ideals.

Proposition 2.1.15: *Suppose R is prime and has a minimal left ideal L. Then L is faithful simple in R-$\mathcal{M}od$, implying R is primitive; moreover, every faithful simple R-module M is isomorphic to L, implying $\text{End}_R M \approx \text{End}_R L$.*

Proof: $\text{Ann}_R L = 0$ by proposition 2.1.13, so L is faithful simple, and R is primitive. Suppose $M \in R\text{-}\mathcal{M}od$ is also faithful simple. Then $LM \neq 0$ so $a_0 x_0 \neq 0$ for some a_0 in L, x_0 in M. By remarks 0.2.10,11 the map $L \to M$ given by $a \to a x_0$ is an isomorphism. Thus $L \approx M$ and $\text{End}_R L \approx \text{End}_R M$.
                                                                                                      Q.E.D.

Finite-Ranked Transformations and The Socle

To study minimal left ideals more closely, we view a primitive ring R as a dense subring of $\text{End } M_D$, where M is faithful simple in R-$\mathcal{M}od$ and $D = \text{End}_R M$. For any r in R define $rank(r) = [rM:D]$, viewing rM as a D-subspace of M. We are interested in the elements of finite rank, because they permit great flexibility in the study of R inside $\text{End } M_D$. (We shall see below that "finite rank" is independent of the choice of M.)

Remark 2.1.16: Suppose r_1, r_2 in R have respective finite ranks t_1, t_2. Then $rank(r_1 + r_2) \leq t_1 + t_2$ and $rank(r_1 r_2) \leq \min(t_1, t_2)$ by remark 0.2.3.

The rank 1 elements play a special role.

Lemma 2.1.17: *If* rank$(r) = t \geq 1$ *then there are* rank 1 *elements* $r_1, \ldots, r_t$ *in* Rr *such that* $r = \sum_{i=1}^{t} r_i$.

Proof: Write $rM = \sum_{i=1}^{t} x_i D$, and for each i pick a_i in R such that $a_i x_j = \delta_{ij} x_i$ for $1 \leq j \leq t$. Given x in M there are d_i in D with $rx = \sum_{j=1}^{t} x_j d_j = \sum_{i,j=1}^{t} a_i x_j d_j = \sum_{i=1}^{t} a_i rx$, so $r = \sum_{i=1}^{t} a_i r$. Taking $r_i = a_i r$ we note $r_i M = x_i D$ so each r_i has rank 1. Q.E.D.

Lemma 2.1.18: *Suppose* rank$(r) = t$. Rr *is a minimal left ideal of* R *iff* $t = 1$.

Proof: ($\Rightarrow$) Take $0 \neq r_1 \in Rr$ of rank 1. $0 \neq Rr_1 \leq Rr$, so by hypothesis $Rr_1 = Rr$ and $r \in Rr_1$; r has rank 1 by remark 2.1.16.

($\Leftarrow$) We prove Rr is minimal by showing $r \in Ra$ for every $0 \neq a \in Rr$. Indeed, suppose $a = r'r \neq 0$ and pick x in M with $0 \neq ax = r'rx$. Let $y = rx \neq 0$. Then $rM = yD$. Since M is simple there is r'' in R with $r''(ax) = y$. Take x' arbitrary in M and write $rx' = yd'$ for suitable d' in D; then

$$rx' = yd' = (r''ax)d' = r''r'(rx)d' = r''r'yd' = r''r'rx' = r''ax'$$

proving $r = r''a \in Ra$. Q.E.D.

To make use of these lemmas we must show that the existence of a minimal left ideal L implies the existence of finite-ranked elements. But proposition 2.1.15 shows $M \approx L$ in $R\text{-}\mathcal{Mod}$, so we take $M = L$, and $D = \text{End}_R L$. The crucial step is finding an idempotent in L.

Lemma 2.1.19: *Suppose* L *is a minimal left ideal of* R *and* $a \in L$ *with* $La \neq 0$. *Then* $La = L$ *and* $\text{Ann}_L a = 0$, *and* L *has an idempotent* e *with* $ea = a = ae$.

Proof: $La \neq 0$ and $\text{Ann}_L a$ are left ideals of R contained in L. Since L is minimal we get $La = L$ and $\text{Ann}_L a = 0$. Thus $a \in La$ so $a = ea$ for some $e \in L$. Then $ea = e^2 a$ so $(e^2 - e) \in \text{Ann}_L a = 0$, proving e is idempotent. Hence $\text{Ann}_L e = 0$ (as above) so $a - ae \in \text{Ann}_L e = 0$. Q.E.D.

Corollary 2.1.20: *Suppose* L *is a minimal left ideal. If* $L^2 \neq 0$ *then* L *has a nonzero idempotent; in particular, this is the case if* R *is prime.*

Having produced an idempotent we are ready to utilize it in the structure theory.

Proposition 2.1.21: *Suppose* R *is any ring with idempotent* e, *and put* $L = Re$.

Then there is an isomorphism from $\mathrm{End}_R L$ *to* eRe *preserving the right module action of* L.

If L *is a minimal left ideal then* eRe *is a division ring, and, viewing* L *as module over the primitive ring* $\bar{R} = R/\mathrm{Ann}_R L$, *we see all elements of* L *have rank* 1.

Proof: Recall $\mathrm{End}_R L = \mathrm{Hom}_R(L, L)^{\mathrm{op}}$. For any f in $\mathrm{End}_R L$ we have $fe = f(e^2) = efe \in eL = eRe$, so we can define $\psi: \mathrm{End}_R L \to eRe$ by $\psi f = fe$. Then $\psi(f_1 + f_2) = (f_1 + f_2)e = f_1 e + f_2 e = \psi f_1 + \psi f_2$ and $\psi(f_1 f_2) = (f_2 f_1)e = f_2((f_1 e)e) = (f_1 e)f_2 e = \psi f_1 \psi f_2$, so ψ is a homomorphism, with $\ker \psi = \{f \in \mathrm{End}_R L: fe = 0\} = 0$. Moreover, given ere in eRe we could define f_r in $\mathrm{End}_R L$ as right multiplication by re; then $\psi f_r = f_r e = ere$, proving ψ is onto and thus is an isomorphism.

If L is a minimal left ideal then L is a faithful simple $\bar{R}$-module and so $eRe \approx \mathrm{End}_R L \approx \mathrm{End}_{\bar{R}} L$ is a division ring; $eL = eRe$ so $\mathrm{rank}(e) = 1$. Hence $\mathrm{rank}(re) = 1$ for any $re \neq 0$.   Q.E.D.

We have prepared the ground for the definition of an important ideal, defined for an arbitrary ring R.

Definition 2.1.22: The *socle* of a ring R, denoted $\mathrm{soc}(R)$, is the sum of the minimal left ideals of R if R has minimal left ideals; otherwise $\mathrm{soc}(R) = 0$.

Remark 2.1.23: If L is a minimal left ideal of R then so is Lr for any r in R such that $Lr \neq 0$. (Indeed if $0 \neq ar \in Lr$ for a in L then $L = Ra$ so $Lr = Rar$.) Consequently $\mathrm{soc}(R) \triangleleft R$.

Lemma 2.1.24: *If* R *is prime and* $\mathrm{soc}(R) \neq 0$ *then* $\mathrm{soc}(R) = \bigcap\{nonzero\ ideals$ *of* $R\}$.

Proof: Suppose $0 \neq A \triangleleft R$. For any minimal left ideal L of R we have $0 \neq AL \leq A \cap L \leq L$; hence $A \cap L = L$ and $L \subseteq A$, proving $\mathrm{soc}(R) \subseteq A$. But $\mathrm{soc}(R) \triangleleft R$ by remark 2.1.23.   Q.E.D.

Theorem 2.1.25: *If* R *is a dense subring of* $\mathrm{End}\ M_D$ *then* $\mathrm{soc}(R) = \{elements$ *of* R *having finite rank*\}.

Proof: {elements of R having finite rank} is clearly an ideal A which by lemmas 2.1.17 and 2.1.18 is contained in $\mathrm{soc}(R)$. To prove $\mathrm{soc}(R) \subseteq A$ we may assume $\mathrm{soc}(R) \neq 0$. Then $A \neq 0$ by corollary 2.1.20 and proposition 2.1.21, so $\mathrm{soc}(R) \subseteq A$ by lemma 2.1.24.   Q.E.D.

Corollary 2.1.25':   *Any simple* ring R *with minimal left ideal* L *has the form* $M_n(D)$ *for a suitable division* ring D.

Proof: R is dense in End L_D; the socle, being nonzero, must be the whole ring. In particular rank$(1) = n$ for some n, implying $[L:D] = n$, so $R \approx M_n(D)$.
$$\text{Q.E.D.}$$

Remark 2.1.26:   If R is dense in End M_D and soc$(R) \neq 0$ then soc(R) is also dense in End M_D. (Indeed take any rank 1 idempotent e and take $x \in M$ with $ex \neq 0$. Now given independent $x_1, \ldots, x_t$ in M and arbitrary $y_1, \ldots, y_t$ in M we take r_{i1} in R such that $r_{i1} x_i = x$ and $r_{i1} x_j = 0$ for all $j \neq i$, and take r_{i2} in R such that $r_{i2} ex = y_i$; then letting $r = \sum_{i=1}^{t} r_{i2} e r_{i1} \in \text{soc}(R)$ we clearly have $r x_i = y_i$ for $1 \leq i \leq t$.) Actually, a much more general result (exercise 2) is available if we are willing to move to the category of rings without 1, noting the proof of the density theorem is still valid.

We conclude this line of exploration by showing the socle of a prime ring is left-right symmetric.

Proposition 2.1.27:   *Suppose R is prime. Rr is a minimal left ideal of R iff rR is a minimal right ideal.*

Proof: ($\Rightarrow$) Suppose $0 \neq r' \in rR$. We need to prove $r \in r'R$. Take an idempotent e of Rr; then $r \in Re$ and $r = r_1 e$ for some r_1 in R. Write $r' = ra = r_1 ea$. Then $r_1 e a R r_1 e a \neq 0$ so there is r_2 in R with $0 \neq r_1 e a r_2 r_1 e \in r'R$. Since eRe is a division ring we can multiply $ear_2 r_1 e$ by some element to get e, implying $r_1 e \in r'R$; which is what we were trying to prove.
    ($\Leftarrow$) By left-right symmetry.   Q.E.D.

Corollary 2.1.28:   *A prime ring R has minimal left ideals iff it has minimal right ideals, in which case* soc$(R) = \sum$(*minimal right ideals of R*).

Let us pause to list the classes of rings described above, from general to specific:

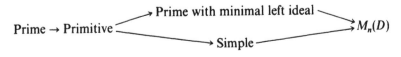

Examples of Primitive Rings

The class of primitive rings is fascinating since they vary greatly despite the unified framework imposed by the density theorem. We shall examine some examples in detail, starting with primitive rings with socle. The socle is very useful for constructing examples because of the following easy observation.

Remark 2.1.29: If R is a subring of End M_D containing soc(End M_D) then R is dense in End M_D. In particular, if R_1 is any subring of End M_D then $R_1 + $ soc(End M_D) is a dense subring of End M_D. (It is a ring since soc(End M_D) is an ideal.)

This remark is very useful in constructing examples of primitive rings resembling given examples. For example, if C is an arbitrary integral domain, then taking an infinite dimensional right vector space M over the field of fractions F of C, we see $C + $ soc(End M_F) is a primitive ring whose center is C. (Here we identify each element c of C with $c \cdot 1$, where 1 denotes the identity transformation of End M_F.) The following important example is also obtained from this remark, cf., exercise 6.2.6 below.

Example 2.1.30: Suppose F is a field and T is an infinite dimensional F-algebra. Viewing $T \subseteq $ End T_F by the regular representation, let $R = T + $ soc(End T_F). Then $R \in F\text{-}\mathcal{A}lg$ and is a dense subalgebra of End T_F.

Having seen how to construct interesting primitive rings *with* socle $\neq 0$ we turn to primitive rings having zero socle. In fact, we already have many examples and need merely to recognize them by means of the following result.

Proposition 2.1.31: *If R is a domain having a minimal left ideal then R is a division ring.*

Proof: There is a short structural proof using proposition 2.1.15 but we may as well use the following easy argument: Suppose $L = Rx$ is a minimal left ideal and $0 \neq a \in R$. Then $0 \neq Rax \subseteq Rx$ so $Rax = Rx$ and $x \in Rax$, implying $x = bax$ for some b in R. Then $1 = ba$ so we have found a left inverse for every nonzero element.   Q.E.D.

Corollary 2.1.32: *If R is a simple domain which is not a division ring then R is primitive with socle 0.*

Proof: Simple implies primitive, so we apply the contrapositive of the proposition.   Q.E.D.

Example 2.1.33: The Weyl algebra over a field of characteristic 0 is a simple domain, so is primitive with socle 0.

Example 2.1.34: For any field F the free algebra $F\{X_1, X_2\}$ is not simple but, nevertheless, is primitive with socle 0. To see this, we take a right vector space M with countable base $\{v_0, v_1, \ldots\}$ over F and define the transformations r_j in End M_F by $r_1 v_0 = 0$, $r_1 v_{i+1} = v_i$ and $r_2 v_i = v_{i^2+1}$. We see easily that $F\{X_1, X_2\}$ is isomorphic to the subalgebra R of End M_F generated by r_1 and r_2, by the map $X_j \mapsto r_j$ (because the "monomials" in r_1 and r_2 are linearly independent). But M is obviously faithful in R-$\mathcal{M}od$ and M is simple (because for any x in M we get some $r_1^m x = v_0 \alpha$ for suitable $\alpha \neq 0$ implying $v_0 \in Rx$, and clearly each $v_i \in Rv_0$ implying $M = Rv_0$). Hence R is primitive. $F\{X_1, X_2\}$ is a domain and thus has socle 0.

Example 2.1.35: Suppose K is a field with an endomorphism σ not of finite order. Then $R = K[\lambda; \sigma]$ is primitive; indeed $R(\lambda - 1)$ is a maximal left ideal with core 0 since by proposition 1.6.25 all two sided ideals have the form $R\lambda^i$. On the other hand, R is a domain and thus has socle 0.

More examples are given in the exercises, as well as in the supplements here.

E Supplement: A Right Primitive Ring Which is Not Primitive

Many basic concepts of ring theory are left-right symmetric, but not primitivity. While an undergraduate, Bergman [64] produced a right primitive ring which is not (left) primitive. His example relies on the asymmetry inherent in the skew polynomial construction. Later Jategaonkar found a general construction which has other applications, and we shall present Jategaonkar's ring. (Also cf. Cohn [85B].)

Example 2.1.36: (Jategaonkar's counterexample) A *Jategaonkar sequence* of rings is a collection $\{R_\beta : \beta \leq \alpha\}$ of rings for a given ordinal α satisfying the following conditions:

(1) R_0 is a division ring D, and $D \subset R_\beta$ for each β.
(2) For every successor ordinal $\gamma = \beta^+$ there is an injection $\sigma_\beta : R_\beta \to D$ such that R_γ is the skew polynomial ring $R_\beta[\lambda_\beta; \sigma_\beta]$.
(3) $R_\gamma = \bigcup_{\beta < \gamma} R_\beta$ for every limit ordinal γ.

To define a Jategaonkar sequence it suffices by transfinite induction to define the σ_β, but first we shall study the properties of Jategaonkar sequences, assuming that they exist. In the above notation we call R_α a *Jategaonkar ring* if α is a limit ordinal. All proofs are straightforward by transfinite induction unless otherwise stated.

(i) $R_\beta \subset R_\gamma$ for every $\beta < \gamma \leq \alpha$.

(ii) Every nonzero element of R_α can be written uniquely in *standard form* $f = \sum_{i=0}^t d_i \lambda_{i_1} \cdots \lambda_{i_t}$ for suitable t, suitable d_i in D, and suitable ordinals $i_1 \leq i_2 \leq \cdots \leq i_t < \alpha$. In particular, we can define $v(f) = \{$largest $\beta : \lambda_\beta$ appears in a monomial of $f\}$.

(iii) If $f \in R_\gamma$ and $\beta > v(f)$ then $\lambda_\beta = (\sigma_\beta f)^{-1}\lambda_\beta f \in R_\gamma f$.

(iv) Each R_γ is a PLID. (Proof by induction: For γ a successor ordinal this follows from proposition 1.6.24. For γ a limit ordinal and $L < R_\gamma$ we take $\beta < \gamma$ with $L \cap R_\beta \neq 0$ and, inductively, take $L \cap R_\beta = R_\beta f$ for suitable f in R_β; for every $\beta' > \beta$ we have $\lambda_{\beta'} \in R_\gamma f$ by (iii), so, in fact, $L = R_\gamma f$.) In particular, R_α is a PLID.

(v) $R_\alpha \lambda_\gamma \lhd R_\alpha$ for each γ (since $\lambda_\gamma \lambda_\beta = (\sigma_\gamma \lambda_\beta)\lambda_\gamma$ for $\beta < \gamma$ and $\lambda_\gamma \lambda_\beta \in \lambda_\gamma R_\alpha \lambda_\gamma \subset R_\alpha \lambda_\gamma$ for $\beta \geq \gamma$ by (iii)).

(vi) For each $\gamma < \alpha$ there is a homomorphism $R_\alpha \to R_\gamma$ obtained by specializing for $\lambda_\beta \to 0$ for each $\beta > \gamma$. (since $\sum_{\beta > \gamma} R_\alpha \lambda_\beta \lhd R_\alpha$ by (v))

(vii) Every nonzero left ideal L of R_α contains a non-zero two-sided ideal of the form $R_\alpha \lambda_\beta$. (Indeed, write $L = R_\alpha f$ and take $\beta > v(f)$. Then $\lambda_\beta \in L$ by (iii) and $R_\alpha \lambda_\beta \lhd R_\alpha$ by (v).)

(viii) R_α is *not* primitive since every maximal left ideal has a nonzero core.

(ix) If $\beta_1 < \beta_2 < \cdots < \beta_t$ and $\sum_{i=1}^t (1 + \lambda_{\beta_i})f_i \in D$ for f_i in R_α then each $f_i = 0$. (Otherwise, take such an equation with γ minimal where $\gamma = \max\{v(f_i): 1 \leq i \leq t\}$. If $\gamma > \beta_t$ then specializing $\gamma \to 0$ gives $\sum(1 + \lambda_{\beta_i})f_i' \in D$ with $f_i = g_i \lambda_\gamma + f_i'$. But each $v(f_i') < \gamma$, contrary to hypothesis unless each $f_i' = 0$; now, $f_i = g_i \lambda_\gamma$, so $\sum(1 + \lambda_{\beta_i})g_i = 0$, and by induction on the degree we conclude each $g_i = 0$, so $f_i = 0$.

This proves $\gamma \leq \beta_t$. If $\gamma < \beta_t$ then λ_{β_t} does not appear in any f_i; matching terms in λ_{β_t} shows $\lambda_{\beta_t} f_t = 0$ implying $f_t = 0$, again impossible. Hence $\gamma = \beta_t$. Let d_i denote the degree of λ_γ in f_i, and put $d = d_t$. Writing $f_i = g_i \lambda_\gamma^{d+1} + f_i'$ where f_i' has degree $\leq d$ in λ_γ (where possibly some $g_i = 0$) and $f_t = g_t \lambda_\gamma^d + f_t'$ we have

$$\lambda_\gamma g_t \lambda_\gamma^d + \sum_{i=1}^{t-1}(1 + \lambda_{\beta_i})g_i \lambda_\gamma^{d+1} = 0, \qquad \text{so } \rho_\gamma g_t + \sum_{i=1}^{t-1}(1 + \lambda_{\beta_i})g_i = 0.$$

Thus $\sum_{i=1}^{t-1}(1 + \lambda_{\beta_i})g_i \in D$, so by induction on t, we see each $g_i = 0$; so, $\rho_\gamma g_t = 0$, implying $g_t = 0$, contrary to definition of d_t.

(x) R_α is *right* primitive. (Indeed, let $A = \sum_{\beta < \alpha}(1 + \lambda_\beta)R_\alpha$. $1 \notin A$ by (ix); A is comaximal with all ideals for if $0 \neq I \triangleleft R_\alpha$ then some $\lambda_\beta \in I$ by (vii), implying $1 = (1 + \lambda_\beta) - \lambda_\beta \in A + I$. Hence R_α is primitive by proposition 2.1.11.)

In particular, R_α is right primitive but not (left) primitive. R_α has other interesting properties which shall be derived later. We turn now to the existence of Jategaonkar rings. This may seem to be a formidable task, but in fact can be handled quite easily by means of the "generic approach," one of the themes of this book. The underlying idea, well illustrated in this example, is to use indeterminates to satisfy a given restriction in the most general possible manner.

In the case at hand, we shall show that any given division ring D_0 is contained in a suitable Jategaonkar ring R_α, for any ordinal α. Let $X = \{X_\beta : 1 \leq \beta < \alpha\}$ be a set of noncommuting indeterminates over D_0. We shall build R_α as a subring of a larger ring, with the X_β taking the place of the λ_β. Since one large ring we know how to build from D_0 and X is the Laurent series ring $D_0((X))$ of theorem 1.2.24', we shall take each R_γ to be a suitable subring. For γ a limit ordinal we clearly want $R_\gamma = \bigcup\{R_\beta : \beta < \gamma\}$, so we shall henceforth assume γ is a successor ordinal, say $\gamma = \mu^+$. Now we want $R_\gamma = R_\mu[X_\gamma; \rho_\gamma]$ for a suitable homomorphism $\rho_\gamma : R_\mu \to D$ where $D = R_0$; hence for all $\beta < \gamma$ we must have $X_\gamma X_\beta X_\gamma^{-1} = \rho_\gamma X_\beta \in D$. The "generic" approach mentioned earlier leads us to use this condition to define D as follows:

Let H be that subgroup of the free group $\mathscr{G}(X)$ generated by all $X_\nu X_\beta X_\nu^{-1}$ for all $\beta < \nu$. Then H inherits the ordering of $\mathscr{G}(X)$ and so $D_0((X))$ has a division subring $D = D_0((H))$ in view of theorem 1.2.24'. Let R_γ be the subring of $D_0((X))$ generated by D and all X_β for $\beta \leq \gamma$. Then there is an injection $\rho_\gamma : R_\mu \to D$ given by $\rho_\gamma f = X_\gamma f X_\gamma^{-1}$, and it remains to show that the X_γ act like indeterminates over R_μ or, more formally, that there is an isomorphism $\psi_\gamma : R_\mu[\lambda; \rho_\gamma] \to R_\gamma$ given by $\psi_\gamma \sum_{i=0}^t f_i \lambda^i = \sum_{i=0}^t f_i X_\gamma^i$ for each t. Clearly ψ_γ is a surjection, so it suffices to prove $\ker \psi_\gamma = 0$, i.e., if $\sum_{i=0}^t f_i X_\gamma^i = 0$ then each $f_i = 0$.

Assume, on the contrary, that $\sum_{i=0}^t f_i X_\gamma^i = 0$, with each f_i in R_μ, and $f_t \neq 0$, and choose t minimal such. Then $0 = X_\gamma(\sum f_i X_\gamma^i)X_\gamma^{-1} = \sum(\rho_\gamma f_i)X_\gamma^i$, so we may assume each $f_i \in D$. Multiplying through by f_t^{-1} we now may assume $f_t = 1$. But now $X_\gamma^t + \sum_{i=0}^{t-1} f_i X_\gamma^t = 0 = X_\gamma^t + \sum_{i=0}^{t-1}(\rho_\gamma f_i)X_\gamma^i$ (again conjugating by X_γ) so $\sum_{i=0}^{t-1}(f_i - \rho_\gamma f_i)X_\gamma^i = 0$; by minimality of t we see $f_i = \rho_\gamma f_i$ for each i.

Claim: *If $f \in D$ and $\rho_\gamma f = f$ then $f \in D_0$. (If we can prove this claim then* *each $f_i \in D_0$, contrary to X_γ being an indeterminate over D_0, so we conclude that* ker $\psi_\gamma = 0$.)

Proof of Claim: Write f as a Laurent series $\sum d_h h$ where the $d_h \in D_0$ and $h \in H$. Now every element of H is a string of $X_\nu X_\beta X_\nu^{-1}$ or $(X_\nu X_\beta X_\nu^{-1})^{-1} = X_\nu X_\beta^{-1} X_\nu$ for $\beta < \nu$; so any $h \neq 1$ starts with some X_{ν_1} and ends with some $X_{\nu_2}^{-1}$, and by inspection cannot commute with X_γ. If $\mathrm{supp}(f) \neq \{1\}$ then we can take $h \neq \{1\}$ minimal in $\mathrm{supp}(f)$ (which is well-ordered). Then $X_\gamma h < h X_\gamma$ or $h X_\gamma < X_\gamma h$. In the former case, we have $X_\gamma h X_\gamma^{-1} < h$ and $X_\gamma h X_\gamma^{-1} \in \mathrm{supp}(X_\gamma f X_\gamma^{-1}) = \mathrm{supp}(f)$; in the latter case, $X_\gamma^{-1} h X_\gamma < h$ and $X_\gamma^{-1} h X_\gamma \in \mathrm{supp}(f)$. At any rate, we have contradicted the minimality of h, so indeed $\mathrm{supp}(f) = \{1\}$, proving $f \in D_0$ as claimed.

Having established the claim, we conclude as noted above that ker $\psi_\gamma = 0$, so that $R_\gamma = R_\mu[X_\gamma; \rho_\gamma]$ as desired.   Q.E.D.

§2.2 The Chinese Remainder Theorem and Subdirect Products

This short section provides a general method for simplifying the study of rings by breaking a ring into more manageable rings. Our motivation comes from a venerable result of number theory, often called the "Chinese Remainder Theorem:" Given $m_1, \ldots, m_t, n_1, \ldots, n_t$ in $\mathbb{N}$ where $m_1, \ldots, m_t$ are pairwise relatively prime, one can find some $n \in \mathbb{N}$ such that $n \equiv n_i$ (modulo m_i) for $1 \leq i \leq t$. We shall find a very easy proof of this fact by restating it in terms of the notions we previously encountered. Namely, given a collection $\{A_i : i \in I\}$ of ideals of R, we define the *canonical homomorphism* $R \to \prod_{i \in I} R/A_i$ by $r \to (r + A_i)$. The Chinese Remainder Theorem says that if $n_1, \ldots, n_t$ are pairwise relatively prime then the canonical homomorphism $\mathbb{Z} \to \prod_{i=1}^{t} \mathbb{Z}/n_i \mathbb{Z}$ is onto. In fact, this result can be stated much more generally. We say ideals $A_1, \ldots, A_t$ of R are *comaximal* if $A_i + A_j = R$ whenever $i \neq j$.

Proposition 2.2.1: ("*Chinese Remainder Theorem*") *Suppose $A_1, \ldots, A_t$ are comaximal ideals of R. Then the canonical homomorphism $\varphi : R \to \prod_{i=1}^{t} R/A_i$ is onto.*

Proof: It suffices to show that $(0, \ldots, 0, r + A_i, 0, \ldots, 0) \in \varphi R$ for each i, which is clearly the case if $(0, \ldots, 0, 1, 0, \ldots, 0) \in \varphi R$. By symmetry we need only show $(1, 0, \ldots, 0) \in \varphi R$, i.e., there exists r in R such that $1 - r \in A_1$ and $r \in A_2 \cap \cdots \cap A_t$. For each $j > 1$ write $1 = a_{1j} + a_j$ for suitable a_{1j} in A_1 and

a_j in A_j. Then

$$1 = 1^{t-1} = (a_{12} + a_2)(a_{13} + a_3) \cdots (a_{1t} + a_t) = a + a_2 \cdots a_t,$$

where a is a sum of terms from A_1 (since each term has some a_{1j} as a multiplicand). Thus, letting $r = a_2 \cdots a_t$ we see $1 - r = a \in A_1$ and $r \in A_2 \cap \cdots \cap A_t$, as desired.   Q.E.D.

Subdirect Products

Since proposition 2.2.1 gives a sufficient condition for φ to be onto, we now ask when φ is 1:1.

Proposition 2.2.2:   If $\{A_i : i \in I\}$ are ideals of R then the kernel of the canonical map $\varphi : R \to \prod_{i \in I} R/A_i$ is $\bigcap_{i \in I} A_i$. In particular, if $A_1, \ldots, A_t$ are comaximal then $R/ \bigcap_{i=1}^{t} A_i \approx \prod_{i=1}^{t} R/A_i$.

Proof: $\varphi r = 0$ iff $r + A_i = 0$ (iff $r \in A_i$) for each i, proving the first assertion. The rest follows from Proposition 2.2.1.   Q.E.D.

Corollary 2.2.3:   If $A_1, \ldots, A_t$ are distinct maximal ideals of R having intersection 0 then $R \approx \prod_{i=1}^{t} R/A_i$.

Definition 2.2.3':   An ideal P of R is *prime* if R/P is a prime ring. The following conditions are easily seen to be equivalent to this definition, cf. proposition 2.1.13:

  (i) If $A, B \triangleleft R$ and $AB \subseteq P$ then $A \subseteq P$ or $B \subseteq P$
  (ii) If $A \supset P$ and $B \supset P$ are ideals then $AB \not\subseteq P$
  (iii) If $aRb \subseteq P$ then $a \in P$ or $b \in P$.

A *semiprime* (resp. *semiprimitive*) ideal is an intersection of prime (resp. primitive) ideals. Prime ideals are one of the keys to the theory of rings. Let us record a useful general result relating maximal ideals to prime ideals.

Remark 2.2.4:   Suppose $A_1, \ldots, A_t$ are distinct maximal ideals of R and $A_1 \cdots A_t \subseteq P$ for some prime ideal P. Then $P = A_i$ for some i. (Indeed, $A_1 \cdots A_{t-1} \subseteq P$ or $A_t \subseteq P$; by induction on t we conclude some $A_i \subseteq P$, and thus $A_i = P$ since A_i is maximal.)

This remark sheds further light on corollary 2.2.3. If $A_1, \ldots, A_t$ are maximal ideals of R with $\bigcap_{i=1}^{t} A_i = 0$, then any prime ideal P contains $0 = A_1 \cdots A_t$

and thus $P = A_i$ for some i. Thus $A_1, \ldots, A_t$ are the only prime ideals of R. For future reference we digress a bit, with a module-theoretic version of corollary 2.2.3.

Proposition 2.2.5: *If M has maximal submodules $N_1, \ldots, N_t$ with $\bigcap_{i=1}^{t} N_i = 0$ where t is minimal, then $M \approx \prod_{i=1}^{t} M/N_i$.*

Proof: Put $M_i = M/N_i$. Define $\varphi: M \to \prod_{i=1}^{t} M_i$ by $\varphi x = (x + N_1, \ldots, x + N_t)$. Then $\ker \varphi = \bigcap_{i=1}^{t} N_i = 0$ so it suffices to prove φ is epic. By symmetry it suffices to show for each element x in M there exists x' in $N_2 \cap \cdots \cap N_t$ with $x + N_1 = x' + N_1$. By hypothesis $N_2 \cap \cdots \cap N_t$ has some nonzero element y. Since M_1 is simple we have $x + N_1 = r(y + N_1)$ for some r in R, so we take $x' = ry$.   Q.E.D.

Definition 2.2.6: Given a collection $\{A_i : i \in I\}$ of ideals of R, we say R is a *subdirect product* of $\{R/A_i : i \in I\}$ if the canonical homomorphism $\varphi: R \to \prod_{i \in I} R/A_i$ is an injection.

Remark 2.2.7: By proposition 2.2.2 we see R is a subdirect product of the R/A_i iff $\bigcap A_i = 0$. Thus we have an internal condition (in terms of ideals) for when R can be studied in terms of direct products of homomorphic images.

Subdirect products are much more general than direct products, as we see by the following examples.

$\mathbb{Z}$ is a subdirect product of the finite fields $\mathbb{Z}/p\mathbb{Z}$ for p prime, since $\bigcap p\mathbb{Z} = 0$. For any product of distinct primes $m = p_1 \cdots p_t$, $\mathbb{Z}/m\mathbb{Z} \approx \prod_{i=1}^{t} \mathbb{Z}/p_i\mathbb{Z}$.

The polynomial ring $F[\lambda]$ is a subdirect product of the fields $F[\lambda]/\langle p \rangle$ where p runs over the irreducible polynomials. For any product of distinct irreducible polynomials $f = p_1 \cdots p_t$ we have $F[\lambda]/\langle f \rangle \approx \prod_{i=1}^{t} F[\lambda]/\langle p_i \rangle$.

Semiprime Rings

Often "semi" means "subdirect product."

Definition 2.2.8: R is *semiprime* if R is a subdirect product of prime rings.

Remark 2.2.9: By remark 2.2.7, R is semiprime iff $\bigcap \{\text{prime ideals of } R\} = 0$.

Semiprime rings have a very useful property which gives them a central role in the structure theory.

Remark 2.2.10: If L is a nonzero (or right) ideal in a semiprime ring R then $L^2 \neq 0$. (Indeed $L \not\subseteq P$ for some prime $P \lhd R$, so the image $\bar{L} \neq 0$ in the prime ring $\bar{R} = R/P$, implying $\bar{L}^2 \neq 0$ by proposition 2.1.13; hence $L^2 \neq 0$.)

In §2.6 we shall see that this property actually characterizes semiprime rings. For the time being we utilize it as follows:

Proposition 2.2.11: *Any minimal left ideal L of a semiprime ring contains a primitive idempotent.*

Proof: By corollary 2.1.20 L has a nonzero idempotent, which is primitive by proposition 1.1.21.   Q.E.D.

§2.3 Modules with Composition Series and Artinian Rings

This section features the Jordan-Hölder-Schreier theory of composition series. As an application, we determine the basic structure of left Artinian rings and also introduce central simple algebras. We shall rely on §0.2.17ff.

The Jordan-Hölder and Schreier Theorems

Let us say a chain $M > M_1 > \cdots > M_t$ has *length* t. When verifying M is Artinian we shall often find it convenient to prove the stronger assertion that every chain in $L(M)$ has length bounded by some t. This can usually be done directly, but we prove now that it is enough to examine one particular chain.

Definition 2.3.1: A *composition series of length t* for $M \in R\text{-}\mathcal{M}od$ is a chain of submodules $M = M_0 > M_1 > \cdots > M_t = 0$ such that each M_{i-1}/M_i is simple in $R\text{-}\mathcal{M}od$; each M_{i-1}/M_i is called a *factor* of the composition series. Two composition series $M = M_0 > M_1 > \cdots > M_t = 0$ and $M = N_0 > N_1 > \cdots > N_u = 0$ are *equivalent* if $u = t$ and there is a permutation π such that $N_{\pi i - 1}/N_{\pi i} \approx M_{i-1}/M_i$ for $1 \leq i \leq t$. (In other words, the two composition series have the same sets of factors, possibly permuted.) A *refinement* of a chain of submodules is a new chain obtained by inserting additional submodules.

Remark 2.3.2: A module M has a composition series iff M is both Noetherian and Artinian. (($\Rightarrow$) follows at once from theorem 2.3.3 below. ($\Leftarrow$) M is Noetherian so has a maximal (proper) submodule M_1; M_1 has a maximal submodule M_2, etc; the chain $M > M_1 > M_2 > \cdots$ terminates since M is Artinian, thereby yielding a composition series.)

Theorem 2.3.3: *Suppose $M \in R\text{-}\mathcal{M}od$ has a composition series $\mathscr{C} = (M = M_0 > M_1 > \cdots > M_t = 0)$. Then any chain of submodules of M has a refinement equivalent to $\mathscr{C}$. (In particular, every chain has length $\leq t$.)*

Proof: Write $\sim$ to denote equivalent chains, and let $\mathscr{D} = (M > N_1 > N_2 > \cdots)$ be an arbitrary chain. We aim to prove $\mathscr{D}$ has a refinement $\mathscr{C}' \sim \mathscr{C}$. We introduce the following general notation to permit manipulation of chains: picking N_u and letting $\mathscr{D}' = (M > N_1 > \cdots > N_u)$ and $\mathscr{D}'' = (N_u > N_{u+1} > \cdots)$ we write $\mathscr{D} = \mathscr{D}' + \mathscr{D}''$ and $\mathscr{D}'' = \mathscr{D} - \mathscr{D}'$. Also define $\mathscr{C}_1 = \mathscr{C} - (M > M_1) = (M_1 > M_2 > \cdots > M_t = 0)$, a composition series of length $t - 1$.

The proof is best visualized by means of the following picture:

The proof is by induction on t, the length of the composition series. If $N_1 \leq M_1$ then by induction we can refine $\mathscr{D}' - (M > N_1) + (M_1 \geq N_1) = (M_1 \geq N_1 > N_2 > \cdots)$ to a composition series $\mathscr{C}'_1 \sim \mathscr{C}_1$, so $\mathscr{D}$ has the refinement $(M > M_1) + \mathscr{C}'_1 \sim (M > M_1) + \mathscr{C}_1 = \mathscr{C}$, and we are done.

Thus we may assume $N_1 \not\leq M_1$. Then $N_1 + M_1 = M$ since $(N_1 + M_1)/M_1$ is a nonzero submodule of the simple module M/M_1. By induction on t, we can refine the chain $M_1 > N_1 \cap M_1 \geq 0$ to a composition series $\mathscr{C}'_1 \sim \mathscr{C}_1$, and we can surely write $\mathscr{C}'_1 = \mathscr{C}'_2 + \mathscr{C}'_3$, where $\mathscr{C}'_3$ is a composition series for $N_1 \cap M_1$ and $\mathscr{C}'_2$ is a chain from M_1 to $N_1 \cap M_1$ whose factors are all simple. But $M_1/(N_1 \cap M_1) \approx (M_1 + N_1)/N_1 = M/N_1$, so we can translate $\mathscr{C}'_2$ to a chain $\mathscr{D}_0$ from M to N_1 having the same (simple) factors.

Denoting the length of $\mathscr{C}'_2$ as m, we see $\mathscr{C}'_3$ has length $(t - 1) - m$. Also $N_1/(N_1 \cap M_1) \approx (N_1 + M_1)/M_1 = M/M_1$ is simple. Hence $(N_1 > N_1 \cap M_1) + \mathscr{C}'_3$ is a composition series for N_1 having length $t - m$, which, by induction, is equivalent to some refinement $\mathscr{D}_1$ of $\mathscr{D} - (M > N_1)$.

$\mathscr{D}_0 + \mathscr{D}_1$ is a composition series which is a refinement of $\mathscr{D}$, so we need merely show $(\mathscr{D}_0 + \mathscr{D}_1) \sim \mathscr{C}$. To see this we examine the sets of factors, letting

$\mathcal{F}(\ )$ denote the set of factors of a chain.

$$\mathcal{F}(\mathcal{D}_0 + \mathcal{D}_1) = \mathcal{F}(\mathcal{D}_0) \cup \mathcal{F}(\mathcal{D}_1) = \mathcal{F}(\mathcal{C}'_2) \cup \mathcal{F}((N_1 > N_1 \cap M_1) + \mathcal{C}'_3)$$
$$= \mathcal{F}(\mathcal{C}'_2) + \{N_1/N_1 \cap M_1\} \cup \mathcal{F}(\mathcal{C}'_1 - \mathcal{C}'_2)$$
$$= \mathcal{F}(\mathcal{C}'_1) \cup \{N_1/N_1 \cap M_1\}$$
$$= \mathcal{F}(\mathcal{C}_1) \cup \{M/M_1\} = \mathcal{F}(\mathcal{C}). \qquad \text{Q.E.D.}$$

This theorem combines the "Jordan-Hölder theorem" (stating the equivalence of any two composition series) with the "Schreier refinement theorem" (stating that any chain can be refined to a composition series).

This very fundamental theorem enables us to define the *length* of M, written $\ell(M)$, to be the length of its composition series. (If M has no composition series we write $\ell(M) = \infty$.)

Corollary 2.3.4: *Suppose* $\ell(M) = n < \infty$.

(i) *Every chain of submodules has length* $\leq n$.
(ii) $\ell(M) = \ell(N) + \ell(M/N)$ *for every* $N < M$. *(In particular,* $\ell(N) < \ell(M)$.)

Proof: (i) is immediate; (ii) is clear when we refine the chain $M > N$.
                                                                    Q.E.D.

A related result to be used later connects composition series to primitive ideals, cf., remark 2.5.0 below.

Remark 2.3.5: If an f.g. module M has a composition series of length m then there are primitive ideals $P_1, \ldots, P_m$ such that $P_m \cdots P_1 M = 0$. (Indeed, if $M = M_0 > M_1 > \cdots > M_m = 0$ is a composition series, take $P_i = \text{Ann}_R(M_{i-1}/M_i)$; then $P_i M_{i-1} \subseteq M_i$ for each i so $P_m \cdots P_1 M \subseteq M_m = 0$.)

Artinian Rings

Definition 2.3.6: A ring R is *left Artinian* (resp. *right Artinian*) if R is Artinian as left (resp. right) R-module. R is *Artinian* if R is left and right Artinian.

It is easy to see that any homomorphic image of a left Artinian ring is left Artinian, because any chain of left ideals can be lifted up from a homomorphic image. We are ready to determine much of the structure of left Artinian rings.

Example 2.3.7: Every finite dimensional algebra R over a field F is Artinian. (Indeed, if $[R:F] = t$ and $R > L_1 > L_2 > \cdots$ then viewing the L_i as subspaces of R we have $t = [R:F] > [L_1:F] > [L_2:F] > \cdots$ implying $[L_j:F] = 0$ for some $j \leq t$, so $L_j = 0$.)

Remark 2.3.8: Every simple left Artinian ring has a minimal left ideal and thus by Wedderburn-Artin is of the form $M_n(D)$ for suitable n and suitable division ring D. Conversely, $R = M_n(D)$ is simple Artinian since we have the composition series of R-modules

$$R > L_{n-1} > L_{n-2} > \cdots > L_0 = 0 \qquad \text{where } L_u = \sum_{j=1}^{u} Re_{jj}$$

(so that $L_u/L_{u-1} \approx Re_{uu}$ is a simple R-module); R is simple by proposition 1.1.5.)

Thus simple Artinian is just a different name for a ring of the form $M_n(D)$. This characterization, resulting from work of Wedderburn, Artin, Hopkins, and Levitzki, is so fundamental that rings of the form $M_n(D)$ are thereby called *simple Artinian rings*. Their module structure is very straightforward.

Remark 2.3.8′: Suppose e is a primitive idempotent of a simple Artinian ring R. Then Re is a simple R-module, and any simple R-module is isomorphic to Re, by proposition 2.1.15.

Theorem 2.3.9: *Suppose R is a left Artinian ring.*

(i) *Every prime ideal of R is maximal; in particular, if R is prime then R is simple Artinian.*

(ii) *There are only a finite number of distinct prime ideals $P_1, \ldots, P_t$, and $R/\bigcap_{i=1}^{t} P_i \approx \prod_{i=1}^{t} R/P_i$ a direct product of simple Artinian rings.*

Proof:

(i) Suppose $P \lhd R$ is prime. Then R/P is prime Artinian and thus primitive by proposition 2.1.15. R/P is simple Artinian by theorem 2.1.8, so, in particular, P is a maximal ideal.

(ii) Suppose $P_1, P_2, \ldots$ are distinct prime (and thus maximal) ideals. Then the chain

$$P_1 \supset P_1 \cap P_2 \supset P_1 \cap P_2 \cap P_3 \supset \cdots$$

terminates at some $P_1 \cap \cdots \cap P_t$, implying for each prime ideal P that $P_1 \cap \cdots \cap P_t \subseteq P$. Hence $P = P_i$ for some $i \leq t$ by remark 2.2.4; the rest is proposition 2.2.2.   Q.E.D.

Armed with this result we shall now characterize a wide class of Artinian rings.

Theorem 2.3.10:   *The following conditions are equivalent for a ring R:*

(i) *R is a finite direct product of simple Artinian rings.*

(ii) *$R = \text{Soc}(R)$.*

(iii) *R is semiprime and left Artinian.*

Proof:

(i) $\Rightarrow$ (ii) $M_n(D)$ is the sum of its minimal left ideals, and this property passes to finite direct products.

(ii) $\Rightarrow$ (iii) Write $R = \sum_{\gamma \in \Gamma} L_\gamma$. Then $1 = \sum_{i=1}^{t} a_{\gamma_i}$ for suitable L_{γ_i} and $a_{\gamma_i} \in L_{\gamma_i}$; writing L_i place of L_{γ_i} yields $1 \in \sum_{i=1}^{t} L_i$ so $R = \sum_{i=1}^{t} L_i$. Putting $L'_j = \sum_{i=1}^{j} L_j$ gives a composition series $R > L_{t-1} > L'_{t-2} > \cdots > L'_1 > 0$ for R in $R\text{-}\mathcal{M}od$, proving R is left Artinian.

To show R is semiprime, we note that $0 = \text{Ann}\{1\} \supset \bigcap_{i=1}^{t} \text{Ann}\, L_i$. Since each L_i is simple in $R\text{-}\mathcal{M}od$, we see each $\text{Ann}\, L_i$ is a primitive ideal; hence R is a subdirect product of the primitive (and thus prime) rings $R/\text{Ann}\, L_i$.

(iii) $\Rightarrow$ (i) By theorem 2.3.9, letting $P_1, \ldots, P_t$ be the prime (and thus maximal) ideals of R, we have $R \approx \prod_{i=1}^{t}(R/P_i)$, and each R/P_i is simple Artinian.
$$\text{Q.E.D.}$$

A ring R satisfying the conditions of theorem 2.3.10 is called *semisimple Artinian*, or *semisimple*. Note that condition (i) is left-right symmetric, implying that the right-handed versions of (ii) and (iii) are also necessary and sufficient for a ring to be semisimple Artinian.

Corollary 2.3.11:   *Suppose R is a semiprime ring in which every nonzero left ideal contains a minimal left ideal. If $\text{soc}(R)$ is a sum of a finite number of minimal left ideals then R is semisimple Artinian.*

Proof: If $\text{soc}(R) = \sum_{i=1}^{t} L_i$ for minimal left ideals L_i, then we have a composition series $\text{soc}(R) = M_{t-1} > \cdots > M_1 = 0$ where each $M_u = \sum_{i=1}^{u} L_i$ so $\ell(\text{soc}(R)) \leq t$. We shall use this fact to show $\text{soc}(R) = Re$ for some idempotent e.

Indeed, put $e_0 = 0$ and use the following inductive procedure: Given e_u idempotent take a minimal left ideal $L < R(1 - e_u)$ and write $L = Ra_u$ for a suitable idempotent a_u of R (by proposition 2.2.11). Writing $a_u = r_u(1 - e_u)$ we see $a_u e_u = 0$ but $(1 - e_u)a_u \neq 0$; hence $a'_u = (1 - e_u)a_u$ is an idempotent of L orthogonal to e_u, so $e_{u+1} = a'_u + e_u$ is idempotent in $\text{soc}(R)$.

The chain $0 < Re_1 < Re_2 < \cdots$ in soc(R) must have length $\leq t$, so $Re_t =$ soc(R). If soc(R) $\neq R$ then by hypothesis $R(1 - e_t)$ contains a minimal left ideal *not* in the socle, which is absurd. Hence soc(R) $= R$.   Q.E.D.

Corollary 2.3.12: *Suppose a semiprime ring R has orthogonal primitive idempotents* $e_1, \ldots, e_t$ *with* $\sum_{i=1}^{t} e_i = 1$. *Then R is semisimple Artinian and each* Re_i *is a minimal left ideal.*

Proof: Put $A = $ soc(R) and $L_i = A \cap Re_i$. Then $A = \sum_{i=1}^{t} L_i$ since any $a = a \sum e_i = \sum ae_i \in \sum L_i$. But L_i, being in the socle, contains a minimal left ideal $L_i' = Re_i'$ where e_i' is idempotent (cf., proposition 2.2.11), and proposition 1.1.21 shows $Re_i = L_i'$ is minimal. Hence $R = $ soc(R) is semisimple Artinian.   Q.E.D.

Semisimple Artinian rings are so important in the structure theory that we shall summarize their basic properties in terms of idempotents.

Theorem 2.3.13: *Write a semisimple Artinian ring R as* $\prod_{i=1}^{t} R_i$ *for suitable* $R_i = M_{n_i}(D_i)$ *called the simple components of R.*

  (i) *Let* $z_i = (0, \ldots, 0, 1, 0, \ldots, 0)$ *where 1 appears in the i position. Then each* z_i *is a central idempotent of R with* $\sum_{i=1}^{t} z_i = 1$; $A_i = Rz_i$ *are the unique minimal ideals of R. Each* A_i *is itself a ring with multiplicative unit* z_i, *and* $R_i \approx A_i$ *as rings.*

  (ii) *Any simple R-module M is isomorphic to Re for some primitive idempotent e of R.*

  (iii) *Every primitive idempotent e of R lies in some* A_i. *Two primitive idempotents e, e' lie in the same* A_i *iff Re* $\approx$ *Re' as modules, iff* $eRe' \neq 0$.

  (iv) *There are precisely t isomorphic classes of simple R-modules, one for each simple component.*

Proof:

  (i) by proposition 1.1.14.

  (ii) $M = \sum_{i=1}^{t} A_i M$ so some $A_i M \neq 0$ implying $A_i M = M$. Let $A = \sum_{j \neq i} A_j$. Then $AA_i = 0$ so $AM = AA_i M = 0$ and thus we view M as module over $R/A \approx R_i$. M is a simple R_i-module, by remark 0.0.9, so by remark 2.3.8' is isomorphic (in R_i-$\mathcal{M}od$) to $R_i e$ for some primitive idempotent e of R_i. Viewing e in A_i we see $M \approx Re$ in R-$\mathcal{M}od$.

  (iii) Write $e = (e_1, \ldots, e_t)$ where $e_i \in A_i$. The e_i are orthogonal idempotents so $e = e_i$ for some i, and all other $e_j = 0$. Clearly e, e' are in the same component iff $eRe' \neq 0$ (since A_i is simple as a ring). If $e, e' \in A_i$ then $Re \approx Re'$ as shown in (ii); conversely, if $f: Re \to Re'$ is an isomorphism and $e \in A_i$, then $0 \neq fe = f(z_i e) = z_i fe \in z_i Re'$ implying $z_i e' \neq 0$, so $e' \in A_i$.

(iv) Immediate from (ii) and (iii).   Q.E.D.

Let us disgress for a moment to show that the left-right symmetry of semi-simple Artinian rings does not hold for left Artinian rings in general.

Example 2.3.14: Suppose F is a field and K is a subfield, and let $R = \begin{pmatrix} F & F \\ 0 & K \end{pmatrix}$. By example 1.1.9, the only proper left ideals of R are $0, Fe_{12} + Ke_{22}$, Fe_{12}, and $\{\alpha' e_{11} + \alpha' \beta e_{12} : \alpha' \in F\}$ for each β in F, so R has length 2 in $R\text{-}\mathcal{M}od$ and is left Artinian. On the other hand, for any K-subspace J of F, we see $\begin{pmatrix} 0 & J \\ 0 & 0 \end{pmatrix}$ is a *right* ideal. Thus R is *not* right Artinian if $[F:K] = \infty$.

Rather than enter here into the consequences of these basic results on Artinian rings, we close the discussion with a typical argument concerning an Artinian-like condition.

Proposition 2.3.15: *Suppose R has no infinite chain $Rr \supset Rr^2 \supset Rr^3 \supset \cdots$ for r in R. If $a \in R$ with $\text{Ann}_R a = 0$ then a is invertible in R.*

Proof: By hypothesis $Ra^i = Ra^{i+1}$ for some i, so $a^i \in Ra^{i+1}$. Write $a^i = ba^{i+1}$. Then $(1 - ba)a^i = 0$, implying $1 - ba = 0$ (seen inductively on i); hence $ba = 1$. But now $(ab - 1)a = a(ba) - a = 0$ so $ab - 1 \in \text{Ann } a = 0$ and $ab = 1$.   Q.E.D.

Split Semisimple Artinian Algebras

Definition 2.3.16: Suppose R is a finite dimensional F-algebra. R is *split* if $R \approx \prod_{u=1}^{t} M_{n_u}(F)$ for suitable t, u in N. A field extension K of F is a *splitting field* for R if $R \otimes_F K$ is split as K-algebra.

One of the prime uses of tensor products is in splitting algebras, largely because of the following few results. Note that if R is simple and split then $R \approx M_n(F)$ for some n.

Proposition 2.3.17: *Any finite dimensional semisimple algebra R over an algebraically closed field F is split.*

Proof: Write $R = \prod_{u=1}^{t} R_u$ where each R_u is simple Artinian. Then $R_u \approx M_{n_u}(D_u)$ where D_u is a division algebra finite dimensional over F. For any d in D_u, we see $F[d_u]$ is a finite extension of F and is an integral domain (in fact a field); since d is algebraic over F we conclude $d \in F$. This proves $D_u = F$ for each u, so $R \approx \prod_{u=1}^{t} M_{n_u}(F)$.   Q.E.D.

Corollary 2.3.18: *The algebraic closure K of F is a splitting field for a finite dimensional F-algebra R iff $R \otimes_F K$ is semiprime.*

Proof: ($\Rightarrow$) is *a fortiori.* ($\Leftarrow$) $[R \otimes_F K : K] = [R : F]$ so $R \otimes_F K$ is Artinian; if semiprime then $R \otimes_F K$ is semisimple and thus split.   Q.E.D.

The condition in corollary 2.3.18 does not always hold (cf., example 1.7.25), but an important positive result is available.

Corollary 2.3.19: *If $R = \prod_{u=1}^{t} R_u$ where each R_u is a simple finite-dimensional F-algebra with center $\approx F$, then the algebraic closure K of F is a splitting field for R.*

Proof: By proposition 1.7.15 we have $R \otimes_F K \approx \prod R_u \otimes_F K$, each component of which is simple by theorem 1.7.27, so $R \otimes K$ is split by corollary 2.3.18.   Q.E.D.

Remark 2.3.20: The general procedure for splitting R (if possible) is to write $R = \prod R_u$ and take a field compositum F' of the $Z(R_u)$ over F. If $R \otimes_F F'$ is still semiprime then we can split it by corollary 2.3.19.

Given a splitting field we can find a splitting field which is a finite extension of F, by the following.

Proposition 2.3.21: *If K is a splitting field for R over F then K contains a splitting field which is finitely generated (as a ring) over F.*

Proof: Write $R \otimes_F K = \prod_{u=1}^{t} M_{n_u}(K)$. Letting $\{e_{ij}^{(u)} : 1 \leq i, j \leq n_u\}$ be a set of matric units for R_u we write $e_{ij}^{(u)} = \sum_k r_{ijk}^{(u)} \otimes a_{ijk}^{(u)}$ for $r_{ijk}^{(u)}$ in R and $a_{ijk}^{(u)}$ in K. Let K_0 be the subfield of K generated over F by all the $a_{ijk}^{(u)}$. Then K_0 is a finitely generated extension of F, and $R \otimes_F K_0$ contains enough matric units to be split (by a dimension check).   Q.E.D.

Central Simple Algebras and Splitting

Definition 2.3.22: R is a *central simple* F-algebra if R is simple, $F = Z(R)$, and $[R:F] < \infty$. (In the literature, the "finite dimensionality" criterion is not always included in the definition.)

Every central simple algebra is clearly Artinian, but the theory of central simple algebras is considerably richer than that of simple Artinian rings. We shall develop some of the basic splitting results here, but leave the remainder of the theory for Chapter 7.

Remark 2.3.23:

(i) If R is a prime algebra which is finite dimensional over its center then R is simple; indeed R is Artinian so apply theorem 2.3.9.

(ii) Every subalgebra of a central simple division algebra is a division algebra. (Clear by (i) since it is a domain.)

Proposition 2.3.24: *If R is a central simple F-algebra then the algebraic closure $\bar{F}$ of F is a splitting field for R.*

Proof: By theorem 1.7.27 $R \otimes_F F'$ is simple and thus split, by proposition 2.3.17.   Q.E.D.

(See proposition 6.1.24 below for a related result.)

Corollary 2.3.25: *If R is a central simple algebra then $[R:F] = n^2$ for suitable n, in particular, for n such that $R \otimes_F \bar{F} \approx M_n(\bar{F})$.*

Remark 2.3.26: Writing a central simple algebra R in the form $M_t(D)$ for D a division ring, we have the following information:

(i) D itself is a central simple F-algebra, with $[D:F] = (n/t)^2$ where $n^2 = [R:F]$.

(ii) $R \approx M_n(F) \otimes_F D$ by example 1.7.21.

(iii) R^{op} is also a central simple algebra, and $R^{op} \approx M_n(D^{op}) \approx M_n(F) \otimes_F D^{op}$

Continuing this notation, we know by propositions 2.3.24 and 2.3.21 that R has a splitting field finite over F, but we shall improve this result in §7.0.

We conclude this short discussion with a theorem so important that (according to Zelinsky) the master of central simple algebras, A. A. Albert, called it the most important theorem in the subject.

Theorem 2.3.27:   *If R is a central simple F-algebra and $[R:F] = n$ then $R \otimes_F R^{op} \approx M_n(F)$.*

Proof: $M_n(F) \approx \operatorname{End}_F(R)$ contains both R and R^{op} via the right and left regular representations of R, respectively, cf., remark 1.5.10, thereby yielding a ring homomorphism $\varphi: R \otimes_F R^{op} \to M_n(F)$ given by $r \otimes r' \to rr'$. But $R \otimes R^{op}$ is simple by theorem 1.7.27 so φ is an injection; counting dimensions shows φ is an isomorphism.   Q.E.D.

Zariski Topology (for Finite Dimensional Algebras)

Definition 2.3.28:   Suppose F is an infinite field, and let $V = F^{(n)}$. Given a subset S of the polynomial ring $F[\lambda_1, \ldots, \lambda_n]$ define $V(S) = \{(\alpha_1, \ldots, \alpha_n) \in V: f(\alpha_1, \ldots, \alpha_n) = 0 \text{ for all } f \text{ in } S\}$. Then the $\{V(S): S \subseteq F[\lambda_1, \ldots, \lambda_n]\}$ constitute the closed sets of a topology on V called the Zariski topology. The *Zariski topology* of a finite dimensional algebra over an infinite field is its Zariski topology as a vector space. Write $V(f)$ for $V(\{f\})$.

The open sets have the form $V - V(S) = V - \bigcap\{V(f): f \in S\} = \bigcup\{V - V(f)\}: f \in S\}$, so the sets $V - V(f)$ form a base of the open sets.

Remark 2.3.29: $V(f) \neq V$ for every $0 \neq f \in F[\lambda_1, \ldots, \lambda_n]$, by the fundamental theorem of algebra, cf., Jacobson [85B, theorem 2.19]. It follows that any two nonempty open sets intersect nontrivially, seen by taking complements of $V(f_1) \cup V(f_2) \subseteq V(f_1 f_2) \neq V$. In particular, *every open set is dense.* This simple fact has far-reaching consequences.

The idea in using the Zariski topology is to show various natural conditions are Zariski open and thus Zariski dense; hence any finite number of these conditions define a Zariski dense set. In this manner one can add on certain conditions at no expense. In what follows, R is a finite dimensional F-algebra, and we fix a base $b_1, \ldots, b_t$.

Definition 2.3.30:   Let $\mu_1, \ldots, \mu_t$ be commuting indeterminates over R, and define $R' = R[\mu_1, \ldots, \mu_t]$ and $F' = F[\mu_1, \ldots, \mu_t]$. The element $x = \sum_{i=1}^{t} b_i \mu_i$ in R' is called the *generic element of R* (with respect to the base $b_1, \ldots, b_t$). Viewing $b_1, \ldots, b_t$ as a base of R' over F' we can embed R' in $M_t(F')$ by mean of the regular representation; the characteristic polynomial of the matrix corresponding to x is called the *generic characteristic polynomial $p_x(\lambda)$*.

Remark 2.3.31: Because F' is a unique factorization domain and p_x is monic, the monic irreducible factors of p_x in $F'[\lambda]$ are actually irreducible in $F(\mu_1,\ldots,\mu_t)[\lambda]$, so in particular the *minimal polynomial* m_x of x over $F(\mu_1,\ldots,\mu_t)$ has coefficients in F'. This is called the *generic reduced characteristic polynomial*.

Definition 2.3.32: Suppose $r = \sum \alpha_i b_i \in R$, and $m_x = \sum_{i=0}^d f_i(\mu_1,\ldots,\mu_t)\lambda^i$ where $f_i \in F'$ (so $f_d = 1$). The *reduced characteristic polynomial* m_r of r is $\sum_{i=0}^d f_i(\alpha_1,\ldots,\alpha_t)\lambda^i$. The *reduced norm of* r is $(-1)^d f_0(\alpha_1,\ldots,\alpha_t)$ and the *reduced trace of* r is $-f_1(\alpha_1,\ldots,\alpha_t)$.

Remark 2.3.33': These definitions are independent of the choice of base. (Indeed a change of base corresponds to conjugation of matrices.)

Remark 2.3.34:

(i) Suppose K is a field extension of F. The reduced norm (resp. trace) of r in R is the same as the reduced norm (resp. trace) of $r \otimes 1$ in $R \otimes_F K$, where we use the base $b_1 \otimes 1,\ldots,b_t \otimes 1$.

(ii) If $R = M_n(F)$ the generic reduced characteristic polynomial is easily seen to be the characteristic polynomial of x. Thus, in this case the reduced trace is the trace, and the reduced norm is the determinant, and these are independent of the choice of base.

Proposition 2.3.35: *In the above notation every r in R has degree $\leq d$ over F, and $\{r \in R$: the minimal polynomial of r has degree $d\}$ is a nonempty Zariski-dense open subset of R.*

Proof: As above let x be the generic element with respect to the base $b_1,\ldots,b_t$. Since x has degree d we have $1, x,\ldots,x^{d-1}$ linearly independent so writing $x^k = \sum_{i=1}^t h_{jk} b_k$ for h_{jk} in F', we see the $t \times d$ matrix (h_{jk}) has a nonzero $d \times d$ minor q in F'. Letting $q_1 = q, q_2,\ldots,q_u$ denote the $d \times d$ minors in (h_{jk}) we see that a suitable specialization of the $\mu_i \mapsto \alpha_i$ in F will give $q(\alpha_1,\ldots,\alpha_t) \neq 0$ and thus the matrix $(h_{jk}(\alpha_1,\ldots,\alpha_t))$ has a nonzero $d \times d$ minor; letting $r = \sum_{i=1}^t \alpha_i b \in R$, we conclude that $1, r,\ldots,r^{d-1}$ are linearly independent, so r has degree d.

An arbitrary element $r = \sum \alpha_i b_i$ will have degree d iff $1, r,\ldots,r^{d-1}$ are independent, iff one of the minors $q_1(\alpha_1,\ldots,\alpha_t),\ldots,q_u(\alpha_1,\ldots,\alpha_{tj})$ is nonzero, i.e., iff $(\alpha_1,\ldots,\alpha_t)$ belongs to the union of the Zariski open sets $V - V(q_1)$, $\ldots, V - V(q_u)$ where $V - R$. Q.E.D.

Proposition 2.3.36:   *An element r of R is invertible iff its reduced norm is non-zero. In particular {invertible elements of R} is Zariski open.*

Proof: Let $p_x(\lambda)$ and $m_x(\lambda)$ be the generic characteristic and generic reduced characteristic polynomials. Then m_x divides p_x, so specializing $x \mapsto r$ we see $m_r(\lambda)$ divides $p_r(\lambda)$. Let $q_r(\lambda)$ be the minimal polynomial of r. Then $q_r \mid m_r$. But p_r is the characteristic polynomial of the matrix corresponding to r under the regular representation, so q_r and p_r have the same irreducible factors; consequently, m_r and p_r have the same irreducible factors.

Now r is invertible iff $\lambda \nmid m_r(\lambda)$, iff $\lambda \nmid p_r(\lambda)$, iff the reduced norm of r is nonzero.   Q.E.D.

§2.4 Completely Reducible Modules and the Socle

In §2.1 we introduced the socle of a ring in order to study primitive rings. The careful reader will have noticed that a much more general definition was given in exercise 0.0.12, and we shall now examine (from scratch) how this definition applies to modules, leading to several important characterizations of semi-simple Artinian rings.

Definition 2.4.1:   The *socle* of an R-module M, written $\mathrm{soc}(M)$, is the sum of the simple submodules of M (so $\mathrm{soc}(M) = 0$ if M has no simple sub-modules). M is *completely reducible* (also called *semisimple*) if $\mathrm{soc}(M) = M$.

In examining $\mathrm{soc}(M)$ we are led to the *complement* of a submodule N in M, which we recall is $N' < M$ such that $M = N \oplus N'$ (where $\oplus$ denotes the internal direct sum throughout this section). Of course, the complement need not be unique, as evidenced by the example $M = \mathbb{R}^{(2)}$ and $N = \{(\alpha, 0): \alpha \in \mathbb{R}\}$. We say M is *complemented* if every nonzero submodule has a complement.

Modules are not necessarily complemented, so we introduce a related concept.

Definition 2.4.2:   A submodule L of M is *large* (also called *essential*) if $L \cap N \neq 0$ for all $0 \neq N < M$. An *essential complement* of $N \leq M$ is some $N' \leq M$ such that $N \cap N' = 0$ and $N + N'$ is large.

Thus every complement is an essential complement. Essential complements are very important in module theory because they exist, as we shall see now.

Lemma 2.4.3:   *Suppose $A, B, C \leq M$.*

(i) *If $(A + B) \cap C \neq 0$ and $A \cap C = 0$ then $B \cap (A + C) \neq 0$.*
(ii) *If $A \leq C \leq A + B$ then $C = A + (B \cap C)$.*

Proof: Each result follows instantly from the observation if $0 \neq c = a + b$ then $b = c - a \in A + C$.   Q.E.D.

Proposition 2.4.4:   *Every submodule of M has an essential complement. In fact, if $N, Q \leq M$ with $N \cap Q = 0$ then N has an essential complement containing Q.*

Proof: We prove the second assertion, since the first then follows for $Q = 0$. By Zorn's lemma there is $N' \leq M$ containing Q, maximal with respect to $N \cap N' = 0$. We claim $N + N'$ is large. We must show for $0 \neq P < M$ that $P \cap (N + N') \neq 0$. This is obvious unless $P \not\leq N'$ in which case $N' + P > N'$ and so $(N' + P) \cap N \neq 0$. Hence $P \cap (N' + N) \neq 0$ by lemma 2.4.3(i).   Q.E.D.

Proposition 2.4.5:   *If $A \leq N \leq M$ and B is a complement (resp. essential complement) of A in M then $B \cap N$ is a complement (resp. essential complement) of A in N.*

Proof: The assertion about complements is a special case of lemma 2.4.3(ii) taking $N = C$ and $M = A + B$. To prove the assertion for essential complements suppose $0 \neq P \leq N$; then $P \leq M$ so $0 \neq (A + B) \cap P$. Taking $0 \neq p = a + b$ we have $b = p - a \in N$ so $p \in A + (B \cap N)$.   Q.E.D.

When verifying a submodule N of M is large it is enough to show $N \cap Rx \neq 0$ for every $0 \neq x \in M$, since every nonzero module contains a nonzero cyclic submodule. This trivial observation will be used frequently, especially when we study large submodules in earnest in §3.3, but let us make several easy observations.

Proposition 2.4.5′:

(i) *Suppose $N \leq N' \leq M$. N is large in M iff N is large in N' and N' is large in M.*

(ii) *If $M = M_1 + M_2$ and N_i are large submodules of M_i then $N_1 + N_2$ is large in M.*

Proof:

(i) Immediate from the definition.

(ii) First consider the case $N_1 \cap N_2 = 0$. If $x = x_1 + x_2$ where $x_i \in M_i$ we need r in R such that $0 \neq rx \in N_1 + N_2$. This is trivial unless $x_1 \neq 0$, in which case $0 \neq r_1 x_1 \in N_1$ for some r_1 in R. We are done unless $r_1 x_2 \neq 0$, in which case $0 \neq r_2 r_1 x_2 \in N_2$ for some r_2, and then we take $r = r_2 r_1$.

In general let M'_2 be an essential complement of $M_1 \cap M_2$ in M_2. Then $M_1 + M'_2$ is large in $M_1 + M_2$ by our special case; likewise $N_1 + (N_2 \cap M'_2)$ is large in $M_1 + M'_2$ and so is large in $M_1 + M_2$ by (i). *A fortiori* $N_1 + N_2$ is large in $M_1 + M_2 = M$.   Q.E.D.

Armed with these basic tools we shall now find the link between completely reducible modules and semisimple

Proposition 2.4.6: *If M is complemented then M is completely reducible.*

Proof: Let M' be a complement of $\operatorname{soc}(M)$ in M. We shall show $M' = 0$ (so that $\operatorname{soc}(M) = M$). Indeed, if there is $0 \neq x \in M'$ we can take $N \leq M'$ maximal with respect to $x \notin N$ (by Zorn's lemma), and N has a complement N' in M' by proposition 2.4.5. We claim N' is simple, contrary to $M' \cap \operatorname{soc} M = 0$.

To prove the claim, suppose to the contrary that $0 < A < N'$. Then A has a complement A' in N', and by choice of N we have $x = y_1 + a = y_2 + a'$ for suitable y_i in N, a in A, and a' in A'. Then $y_1 - y_2 = a' - a \in N \cap N' = 0$, so $a' = a \in A \cap A' = 0$ and thus $x = y_1 \in N$, a contradiction. Thus the claim is proven, and we must conclude $M' = 0$ as desired.   Q.E.D.

Theorem 2.4.7: $\operatorname{Soc}(M) = \bigcap \{$ *large submodules of* $M\}$.

Proof: If $0 \neq S \leq M$ is simple and $L \leq M$ is large then $0 \neq S \cap L \subseteq S$ implying $S \cap L = S$ and so $S \subseteq L$. Hence $\operatorname{soc}(M) \subseteq L$. Letting $P = \bigcap \{$large submodules of $M\}$ we thus have $\operatorname{soc}(M) \subseteq P$. To prove the opposite inclusion we shall show P is complemented (and then we are done by proposition 2.4.6) Indeed if $N \leq P$ then N has an essential complement N'; thus $N + N'$ is large, implying $N \leq P \leq N + N'$. By lemma 2.4.3(ii) we see $P = N \oplus (N' \cap P)$, so N has a complement in P, as desired.   Q.E.D.

Theorem 2.4.8:   *The following assertions are equivalent for M in R-$\mathcal{M}od$:*

(i) *M is completely reducible.*

(ii) *M has no large proper submodules.*

(iii) *M is complemented.*

Proof: (i) ⇒ (ii). $M = \text{soc}(M) = \bigcap\{\text{large submodules of } M\}$, so M is its only large submodule.

(ii) ⇒ (iii) Every submodule has an essential complement, which is thus a complement.

(iii) ⇒ (i) By proposition 2.4.6.   Q.E.D.

Since R is semisimple Artinian iff R is completely reducible in R-*Mod*, we have a new characterization of semisimple Artinian rings, with which we conclude this discussion.

Theorem 2.4.9: *The following assertions are equivalent for a ring R:*

(i) *R is semisimple Artinian.*

(ii) *R is completely reducible in R-Mod.*

(iii) *Every R-module is completely reducible.*

(iv) *Every R-module is complemented.*

(v) *The only large left ideal of R is R itself.*

(vi) *Every short exact sequence of R-modules splits.*

Proof: We already have (i) ⇔ (ii) and (iii) ⇔ (iv), and (iii) ⇒ (ii) is *a fortiori*. It remains to prove (ii) ⇒ (iii). Write $R = \sum L_i$ where L_i are minimal left ideals, and suppose $M \in R$-*Mod*. For any x in M we have $Rx = \sum L_i x$. If $L_i x \neq 0$ then $a \to ax$ gives an epic from L_i to $L_i x$, implying $L_i x$ is a simple module; hence $x \in \text{soc}(M)$ for each x in M, proving M is completely reducible. (ii) ⇔ (v) By theorem 2.4.8. (iv) ⇔ (vi) by proposition 1.4.10.   Q.E.D.

§2.5 The Jacobson Radical

We return to the general structure theory of rings. Having the Jacobson density theory at out disposal, it is natural to focus on primitive rings. This leads us to call an ideal P *primitive* if R/P is a primitive ring. Using subdirect products we can then study *semiprimitive* rings, i.e., rings R which are subdirect products of primitive rings, i.e., $\bigcap\{\text{primitive ideals}\} = 0$. We shall need an easy criterion for an ideal to be primitive.

Remark 2.5.0: $P \lhd R$ is a primitive ideal iff P is the annihilator of a simple R-module M. (Proof: (⇒) R/P has a faithful simple module M, which is then a simple R-module with annihilator P, by remark 0.0.9; ($\Leftarrow$) if $P = \text{Ann}_R M$ then M is a faithful simple R/P-module.)

The subject of this section is the obstruction to a ring being semiprimitive.

Definition 2.5.1:   The *Jacobson radical* $\text{Jac}(R) = \bigcap\{\text{primitive ideals of } R\}$.
   Clearly R is semiprimitive iff $\text{Jac}(R) = 0$, so we shall try to "remove" the Jacobson radical.

Proposition 2.5.1':

   (i) *If* $A \lhd R$ *and* $A \subseteq \text{Jac}(R)$ *then* $\text{Jac}(R/A) = \text{Jac}(R)/A$; *in particular* $R/\text{Jac}(R)$ *is semiprimitive.*
   (ii) *If* $\text{Jac}(R/A) = 0$ *then* $A \supseteq \text{Jac}(R)$.

Proof: Write $\bar{\ }$ for the image in $\bar{R} = R/A$.

   (i) If P is a primitive ideal of R then $A \subseteq \text{Jac}(R) \subseteq P$ so $\bar{R}/\bar{P} \approx R/P$ is primitive, implying $\bar{P}$ is a primitive ideal of $\bar{R}$. Thus

$$\text{Jac}(\bar{R}) = \bigcap\{\bar{P} : P \text{ is a primitive ideal of } R\}$$
$$= \overline{\bigcap\{\text{primitive ideals of } R\}} = \overline{\text{Jac}(R)}$$

   (ii) If $\text{Jac}(\bar{R}) = 0$ then $0 = \overline{\bigcap\{\bar{P} : P \text{ is a primitive ideal containing } A\}} = \overline{\bigcap\{\text{primitive ideals containing } A\}} \supseteq \overline{\text{Jac}(R)}$ implying $\text{Jac}(R) \subseteq A$.   Q.E.D.

Proposition 2.5.2: $\text{Jac}(R) = \bigcap\{\textit{maximal left ideals of } R\}$.

Proof: ($\subseteq$) By proposition 2.1.9 every maximal left ideal contains a primitive ideal and thus contains $\text{Jac}(R)$.
   ($\supseteq$) If $P \lhd R$ is primitive then for some simple M in $R\text{-}\mathcal{M}od$ we have $P = \text{Ann}_R M = \bigcap\{\text{Ann } x : x \in M\} \supseteq \bigcap\{\text{maximal left ideals of } R\}$ since each $\text{Ann } x$ is a maximal left ideal.   Q.E.D.

Quasi-Invertibility

In order to study $\text{Jac}(R)$ as a set, we need a description in terms of elements. To this end let us say $a \in R$ is left *quasi-invertible* if $1 - a$ is left invertible in R, i.e., if $1 \in R(1 - a)$; a is *quasi-invertible* if $1 - a$ is invertible (from both sides). *A subset of R is quasi-invertible* if each element is quasi-invertible. (Compare with definition 1.1.23(i).)

Lemma 2.5.3: *Any left ideal L of left quasi-invertible elements is quasi-invertible. In fact if $a \in L$ and $r(1 - a) = 1$ then $1 - r \in L$.*

Proof: $1 - r = -ra \in L$, so $r = 1 - (1 - r)$ has a left inverse b. Hence the right and left inverses of r are equal, i.e., $1 - a = b$ is invertible.   Q.E.D.

Proposition 2.5.4: $Jac(R)$ *is a quasi-invertible ideal of R which contains every quasi-invertible left ideal.*

Proof: For any $a \in Jac(R) = \bigcap\{\text{maximal left ideals of } R\}$ we cannot have $1 - a$ in a maximal left ideal, so $R(1 - a) = R$, proving a is left quasi-invertible. Hence $Jac(R)$ is quasi-invertible by Lemma 2.5.3.

Now suppose B is a quasi-invertible left ideal. If there were some maximal left ideal $L \not\supseteq B$ we would have $B + L = R$, so $b + a = 1$ for some b in B, a in L, and then $a = 1 - b$ would be invertible (since b is quasi-invertible), contrary to $L \neq R$. Hence B is contained in every maximal left ideal, so $B \subseteq Jac(R)$.
    Q.E.D.

In view of this result, $Jac(R)$ is the same as what we would get from the right-handed analogue of definition 2.5.1. Indeed, calling this "right-handed" Jacobson radical J, we see J is quasi-invertible (by the right-handed version of proposition 2.5.4). Thus $J \subseteq Jac(R)$, and symmetrically $Jac(R) \subseteq J$.

Remark 2.5.4': Recall a left ideal L of R is *nil* if every element of L is nilpotent. Remark 1.1.27 applied to proposition 2.5.4 show $Jac(R)$ contains every nil left (or right) ideal.

The characterization given in 2.5.4 is also useful because invertibility (and thus quasi-invertibility) passes to homomorphic images and sometimes back again, as we shall see now.

Lemma 2.5.5: *If J is a quasi-invertible ideal of R and r is an element of R whose canonical image in R/J is invertible then r is invertible in R.*

Proof: Take r' in R such that $1 - r'r \in J$ and $1 - rr' \in J$. Then these elements are quasi-invertible, so $r'r$ and rr' are invertible in R. It follows at once that r is left and right invertible, so r is invertible.   Q.E.D.

Our final basic result concerning the Jacobson radical involves passing to homomorphic images.

Proposition 2.5.6:

  (i) *If* $\varphi: R \to T$ *is a ring surjection then* $\varphi(\mathrm{Jac}(R)) \subseteq \mathrm{Jac}(T)$.

  (ii) *If* $A \lhd R$ *then* $(\mathrm{Jac}(R) + A)/A \subseteq \mathrm{Jac}(R/A)$, *equality holding if* $A \subseteq \mathrm{Jac}(R)$.

Proof: Write $J = \mathrm{Jac}(R)$.

  (i) If $a \in J$ then $\varphi(1 - a)^{-1}(1 - \varphi a) = 1$, implying φa is left quasi-invertible; thus $\varphi J \subseteq \mathrm{Jac}(T)$.

  (ii) The first assertion follows from (i); the second assertion is proposition 2.5.1'.   Q.E.D.

Examples

To give the reader the flavor of the Jacobson radical, we present some results concerning large classes of examples.

Proposition 2.5.7: *Suppose* $0 \cup \{invertible\ elements\ of\ R\}$ *is a division* ring D. *Then* $\mathrm{Jac}(R) = 0$.

Proof: Let $J = \mathrm{Jac}(R)$. Then $J \cap D \lhd D$ implying $J \cap D = 0$. But if $a \in J$ then $1 - a$ is invertible so $1 - a \in D$ and $a \in D \cap J = 0$.   Q.E.D.

Corollary 2.5.8: *Suppose A is a ring with a filtration by* $\mathbb{N}$ *such that $A(0)$ is a division ring. Then* $\mathrm{Jac}(R) = 0$.

  In particular, every skew polynomial ring over a division ring is semi-primitive. Let us now consider the opposite extreme.

Definition 2.5.9: A ring R is *local* if $\{noninvertible\ elements\ of\ R\} \lhd R$.

  (Some authors call these rings "quasi-local," reserving the use of "local" for when the ring is Noetherian.)

Proposition 2.5.10: *The following statements are equivalent:*

  (i) *R is local.*

  (ii) *$\{noninvertible\ elements\ of\ R\}$ is a maximal left ideal of R.*

  (iii) *R has a unique maximal left ideal.*

  (iv) *$\mathrm{Jac}(R)$ is a maximal left ideal (and thus the unique maximal left ideal).*

  (v) *$R/\mathrm{Jac}(R)$ is a division ring.*

Proof: Let $J = \{$noninvertible elements of $R\}$. If $r \notin J$ then $Rr = R$, so r cannot be contained in any proper left ideal of R.

(i) $\Rightarrow$ (ii) $\Rightarrow$ (iii) is immediate from the above observation.

(iii) $\Rightarrow$ (i) Let J' be the unique maximal left ideal of R. Obviously $J' \subseteq J$. But if $r \in J$ then Rr is a proper left ideal and thus contained in a maximal left ideal, which is J', proving $J \subseteq J'$.

(iii) $\Leftrightarrow$ (iv) by proposition 2.5.2.

(iv) $\Leftrightarrow$ (v) A ring is a division ring iff it has no nonzero left ideals.   Q.E.D.

Example 2.5.11: If D is a division ring then the ring of formal power series $D[[\lambda]]$ is local (since the noninvertible elements are the power series having constant term 0, clearly an ideal).

Remark 2.5.12: The easiest condition to verify that R is local is the following: If $a + b = 1$ then a or b is invertible. (For suppose R is not local. Then $\{$noninvertible elements$\}$ is not closed under addition, so there are non-invertible elements x and y with $u = x + y$ invertible. Hence $xu^{-1} + yu^{-1} = 1$, violating the condition.)

Idempotents and the Jacobson Radical

The Jacobson radical is particularly apt for dealing with idempotent-lifting since we now see that condition (i) of definition 1.1.23 is that every idempotent-lifting ideal is contained in the Jacobson radical. This connection will be examined closely in §2.7; here we collect a few basic facts.

Remark 2.5.13: $\mathrm{Jac}(R)$ contains no nontrivial idempotent $e \neq 0$ by remark 1.1.24. It follows that a local ring R has no nontrivial idempotents.

Proposition 2.5.14: If $e \in R$ is idempotent then $\mathrm{Jac}(eRe) = e\mathrm{Jac}(R)e$.

Proof: Write $J = \mathrm{Jac}(R)$. ($\supseteq$) If $x \in eJe$ and $1 = y(1 - x)$ then $e = ey(1 - x)e = eye - eyxe = eye(e - x)$ since $xe = x = ex$, proving eye is the left quasi-inverse of x in eRe, so eJe is left quasi-invertible in eRe.

($\subseteq$) Proposition 1.1.15 implies $\mathrm{Jac}(eRe)eM = 0$ for all simple M in R-$\mathcal{M}od$, so $\mathrm{Jac}(eRe) = \mathrm{Jac}(eRe)e \subseteq J$.   Q.E.D.

Corollary 2.5.15: $\mathrm{Jac}(M_n(R)) = M_n(\mathrm{Jac}(R))$.

Proof: This follows from corollary 1.1.18, but we now have a more elementary proof. $\mathrm{Jac}(M_n(R)) = M_n(J)$ for some $J \lhd R$; identifying R with $eM_n(R)e$ for $e = e_{11}$ yields $J = e\mathrm{Jac}(M_n(R))e = \mathrm{Jac}(R)$.   Q.E.D.

This result leads us to the question of $\mathrm{Jac}(M_\infty(R))$, cf., exercise 1.1.5. The solution is given in exercises 2.8.29ff.

Weak Nullstellensatz and the Jacobson Radical

One of the traditional aims of ring theory is to extract the essence of some well-known theorem and then find a straightforward generalization to non-commutative rings. One favorite such theorem has been Hilbert's renowned Nullstellensatz, which is the statement,

"Suppose F is an algebraically closed field, and consider polynomials $f, f_1, \ldots, f_t$ in the commutative polynomial ring $R = F[\lambda_1, \ldots, \lambda_n]$. If f vanishes at all the common zeroes of $f_1, \ldots, f_t$, then $f^m \in \sum Rf_i$ for some m."

In other words, for $A = \sum Rf_i$ the conclusion of the Nullstellensatz is that $f + A$ is nilpotent in R/A. As in Kaplansky [70B, theorem 33], the hypotheses show f is in every maximal ideal of R containing A, so $f + A \in \mathrm{Jac}(R/A)$. Thus the key structural fact needed to complete the proof is that $\mathrm{Jac}(R/A)$ is nil. In other words, we want to show the Jacobson radical of every homomorphic image of R is nil. In general, one might say any ring R having this property satisfies the "weak Nullstellensatz." Our next results give broad instances of when the Jacobson radical is nil, thereby providing the "weak Nullstellensatz" for left Artinian rings and for arbitrary algebras over suitably large fields. We shall return to the Nullstellensatz at definition 2.12.26ff.

In case R is left Artinian the radical predates Jacobson, and we have the following result.

Theorem 2.5.16: $\mathrm{Jac}(R)$ *is nilpotent for any left Artinian ring* R.

Proof: Let $J = \mathrm{Jac}(R)$. Then $J \supseteq J^2 \supseteq J^3 \supseteq \cdots$ so $J^t = J^{t+1}$ for some t. Let $N = J^t$; then $N = N^2$. We claim $N = 0$. Otherwise N contains a nonzero left ideal L minimal with respect to $L = NL$. Then $0 \neq Na \subseteq L$ for some a in L, and $Na = N^2a = N(Na)$ implies $L = Na$ (by assumption on L). Hence $a = ra$ for some r in N. But then $(1 - r)a = 0$, contrary to r quasi-invertible. Thus $0 = N = J^t$ as claimed.   Q.E.D.

Note that $\mathrm{Jac}(T) \cap R$ need not be quasi-invertible when R is a subring of

T; for example let T be a localization of $\mathbb{Z}$, and $R = \mathbb{Z}$. Nevertheless, there is a useful generalization of theorem 2.5.16:

Proposition 2.5.17: *Suppose R is a subring of T.*

(i) *If every element of R which is invertible in T is already invertible in R, then $R \cap \mathrm{Jac}(T) \subseteq \mathrm{Jac}(R)$.*

(ii) *If R is left Artinian then $R \cap \mathrm{Jac}(T) \subseteq \mathrm{Jac}(R)$ is nilpotent.*

Proof: (i) If $r \in R \cap \mathrm{Jac}(T)$ then $(1 - r)^{-1} \in T$ so $(1 - r)^{-1} \in R$, proving $R \cap \mathrm{Jac}(T)$ is quasi-invertible in R.

(ii) If $r \in R$ is invertible in T then $\mathrm{Ann}_R r = 0$, implying r is invertible in R (by proposition 2.3.15). Now (i) implies $R \cap \mathrm{Jac}(T) \subset \mathrm{Jac}(R)$ which is nilpotent by the theorem.   Q.E.D.

Corollary 2.5.18: *Suppose R is an algebra over a field F. Then every r in $\mathrm{Jac}(R)$ is nilpotent or transcendental.*

Proof: If r is not transcendental then $F[r]$ is Artinian and $r \in F[r] \cap \mathrm{Jac}(R)$, a nil ideal by proposition 2.5.17(ii).   Q.E.D.

Corollary 2.5.19: *If R is an algebraic algebra over a field then $\mathrm{Jac}(R)$ is nil.*

Our further study of the radical of an algebra R over a field F requires the notion of the *resolvent set* of r in R, defined as $\{\alpha \in F : (r - \alpha)$ is invertible$\}$ (identifying F with a subfield of R). Write $\sigma(r)$ for the complement (in F) of the resolvent set.

Lemma 2.5.20: *If $r \in R$ is algebraic (over a field F) with minimal polynomial p then $\sigma(r) = \{\alpha \in F : p(\alpha) = 0\}$.*

Proof: $F[r] \approx F[\lambda]/\langle p \rangle$ for some polynomial $p = p(\lambda)$. We note a cyclical set of implications which implies all assertions are equivalent for arbitrary $q(\lambda)$ in $F[\lambda]$:

$q(r)$ is invertible in $R \Rightarrow q(r)$ is not a zero divisor in $F[r] \Rightarrow p(\lambda), q(\lambda)$ are relatively prime $\Rightarrow 1 = g(\lambda)p(\lambda) + h(\lambda)q(\lambda)$ for suitable $g(\lambda), h(\lambda)$ in $F[\lambda] \Rightarrow 1 = h(r)q(r) \Rightarrow q(r)$ is invertible in $F[r]$.

Now taking $q(\lambda) = \lambda - \alpha$ (which is irreducible) we see $q(r) \in \sigma(r)$ iff $\lambda - \alpha$ and p are *not* relatively prime, iff $(\lambda - \alpha) \mid p$, iff $p(\alpha) = 0$.   Q.E.D.

Proposition 2.5.21: *Suppose* $[F[r]:F] \geq t$. *Then for any distinct* $\alpha_1,\ldots,\alpha_t$ *in the resolvent set of* r *we have* $\{(r - \alpha_1)^{-1},\ldots,(r - \alpha_t)^{-1}\}$ *is F-independent.*

Proof: Suppose $\sum_{i=1}^{t} \beta_i(r - \alpha_i)^{-1} = 0$ for β_i in F. Defining

$$q_i = \prod_{j \neq i} (\lambda - \alpha_j)$$

and multiplying through by $\prod_{j=1}^{t}(r - \alpha_j)$ yields $\sum_{i=1}^{t} \beta_i q_i(r) = 0$. On the other hand, $\deg(\sum \beta_i q_i) \leq t - 1$, so, by hypothesis, $\sum \beta_i q_i = 0$. Thus for each u we get $0 = \sum \beta_i q_i(\alpha_u) = \beta_u q_u(\alpha_u) = \beta_u \prod_{j \neq u}(\alpha_u - \alpha_j)$, implying each $\beta_u = 0$ for $1 \leq u \leq t$.   Q.E.D.

Theorem 2.5.22: *Suppose F is a field,* $R \in F\text{-}\mathcal{A}\ell g$, *and* $|F| - 1 > [R:F]$ (*as cardinal numbers). Then* $\mathrm{Jac}(R)$ *is nil.*

Proof: If $r \in \mathrm{Jac}(R)$ then $(1 - \alpha r)^{-1}$ exists for all α in F, implying $r - \alpha = -\alpha(1 - \alpha^{-1}r)$ is invertible for all $0 \neq \alpha \in F$. If r were not algebraic then every finite subset of $S = \{(r - \alpha)^{-1} : \alpha \in F\}$ would be F-independent by proposition 2.5.21, implying S is F-independent, contrary to hypothesis. Hence r is algebraic and thus nilpotent by Corollary 2.5.18.   Q.E.D.

Remark 2.5.22: It is very easy to construct a commutative algebra R of countable dimension over an arbitrary field F such that $\mathrm{Jac}(R)$ is not nilpotent; let $C = F[\lambda_1, \lambda_2,\ldots]$ and consider C/A where $A = \sum_{n \in \mathbb{N}} C\lambda_n^n$.

The Structure Theoretical Approach to Rings

Let us outline a sequence of steps which one often follows in proving a theorem about a class of rings. The philosophy is to prove the theorem for a ring R in increasing generality, as in the following cases:

1. R is a division ring.
2. R is simple Artinian (matrices over a division ring).
3. R is primitive (use the density theorem to reduce to 2).
4. R is semiprimitive (reduce to 3 using subdirect products).
5. R has no nil ideals $\neq 0$.
6. R is semiprime.
7. General case.

We shall encounter several instances of this method. Perhaps the most straightforward is the class of "commutativity theorems," which enable one to conclude that rings are commutative under rather weak hypotheses.

Although the list of people working in this area in the 1950s and 1960s reads like a "Who's Who in Algebra," interest has waned considerably since 1970, and, aside from a few technical questions, this area is largely of historical interest. Nevertheless, it is instructive in the use of structure theory and is included in a series of exercises. One of the most successful applications of the structure theory has been in the theory of polynomial identity rings, to be described in Chapter 6.

When performing the reductions listed above, one can proceed immediately from 4 to 5 (or 6) when the Jacobson radical is nil (or nilpotent); several instances of this were just seen. Even when the Jacobson radical is not nil, there is a method of passing from 4 to 5, due to the following important theorem of Amitsur:

Theorem 2.5.23 (Amitsur): *If R has no nonzero nil ideals then $R[\lambda]$ is semiprimitive.*

Often we can transfer the conclusion of an assertion from $R[\lambda]$ to R, and in this way pass from 4 to 5. We now give a short direct proof of Amitsur's theorem; a second proof is given below in the context of graded rings.

Proof: Otherwise, let $J = \{$nonzero elements of $\mathrm{Jac}(R[\lambda])$ having smallest degree$\}$, and $J_0 = \{$leading coefficients of the element of $J\}$. Clearly $J_0 \cup \{0\}$ is an ideal of R, which we claim is nil. Indeed suppose $p \in J$. Then $\lambda p \in \mathrm{Jac}(R[\lambda])$, so there exists q in $R[\lambda]$ such that $(1 - \lambda p)q = 1$. Then $q = \lambda pq + 1$, which is the case $m = 1$ of the formula

$$q = \lambda^m p^m q + \sum_{i=0}^{m-1} \lambda^i p^i, \tag{1}$$

which we verify now by induction on m. Indeed, assume (1) holds for $m - 1$. Then

$$q = \lambda^{m-1} p^{m-1} q + \sum_{i=0}^{m-2} \lambda^i p^i = \lambda^{m-1} p^{m-1}(\lambda pq + 1) + \sum_{i=0}^{m-2} \lambda^i p^i$$

$$= \lambda^m p^m q + \sum_{i=0}^{m-1} \lambda^i p^i,$$

as desired.

Write $p = \sum_{i=0}^k r_i \lambda^i$. If $x_1 r_k x_2 = 0$ for suitable x_i in R then $\deg(x_1 p x_2) < k$; so, by definition of J, we have $x_1 p x_2 = 0$. We shall use this observation repeatedly. For example, write $q = \sum_{i=0}^t r_i' \lambda^i$ for suitable r_i' in R and take $m > t$; matching leading coefficients in (1) yields $0 = r_k^m r_t' = r_k^{m-1} r_k r_t'$, so $r_k^{m-1} p r_t' = 0$.

Hence $r_k^{m-1} r_i r_t' = 0$ for all i, so $r_k^{m-2} p r_i r_t' = 0$, implying $r_k^{m-2} p^2 r_t' = 0$. Continuing this argument shows $p^m r_t' = 0$, so $0 = \lambda^m p^m r_t' \lambda^t$ and (1) becomes

$$q = \lambda^m p^m \sum_{i=0}^{t-1} r_i' \lambda^i + \sum_{i=0}^{m-1} \lambda^i p^i.$$

As above, we now obtain $p^m r_{t-1}' = 0$; continuing in this way, finally yields $0 = r_k^m r_0' = r_k^m$ (since $q = \lambda pq + 1$ implies $r_0' = 1$). Thus we proved the leading coefficient of each element of J_0 is nil, as desired. But now $J_0 \cup \{0\} \subseteq \mathrm{Nil}(R) = 0$, proving J is empty, i.e., $\mathrm{Jac}(R[\lambda]) = 0$.   Q.E.D.

"Nakayama's Lemma"

One of the basic tools in studying a non-semiprimitive ring R is the result $\mathrm{Jac}(R)M \neq M$ for every f.g. R-module $M \neq 0$. This is known as "Nakayama's lemma," although it is generally agreed that a better attribution would be Azumaya-Krull-Jacobson. Actually any of the following assertions shall be called "Nakayama's lemma" in the sequel.

Proposition 2.5.24:   *"Nakayama's lemma." Suppose $0 \neq M \in R\text{-}\mathscr{F}imod$ and $J = \mathrm{Jac}(R)$.*

  (i) $M \neq JM$.
  (ii) *If $N \leq M$ and $M = N + JM$ then $N = M$.*
  (iii) *If M/JM is spanned by $x_1 + JM, \ldots, x_k + JM$ then $x_1, \ldots, x_k$ span M.*

Proof:

  (i) Write $M = \sum_{i=1}^t Rx_i$ with t minimal possible. If $M = JM$ then $x_t = \sum a_i x_i$ for a_i in J, so $(1 - a_t)x_t = \sum_{i=1}^{t-1} a_i x_i$; but $1 - a_t$ is invertible so $x_t \in \sum_{i=1}^{t-1} Rx_i$, contradiction.
  (ii) $M/N = J(M/N)$ so $M/N = 0$ by (i), implying $N = M$.
  (iii) Let $N = \sum_{i=1}^k Rx_i$. Then $M = N + JM$ so $N = M$ by (ii).   Q.E.D.

There are several improvements of this result. First we record one due to Schelter, also, cf., exercise 12.

Proposition 2.5.25:   *If $M \in R\text{-}\mathscr{F}imod$ and $N < M$ then there is some primitive $P \lhd R$ with $N + PM \neq M$.*

Proof: By proposition 0.2.15 there is some maximal submodule M' containing N. Then M/M' is simple so $P = \mathrm{Ann}_R M/M'$ is a primitive ideal with $PM \subseteq M'$; hence $N + PM \subseteq M' \neq M$.   Q.E.D.

The Radical of a Module

Definition 2.5.26: If $M \in R\text{-}\mathcal{M}od$ define $\text{Rad}(M) = \bigcap \{\text{maximal submodules of } M\}$, where we say $\text{Rad}(M) = M$ if M has no maximal (proper) submodules.

Viewing R in $R\text{-}\mathcal{M}od$ we have $\text{Rad}(R) = \text{Jac}(R)$ by proposition 2.5.2. Thus the radical of a module generalizes the Jacobson radical and is of considerable use in the technical study of f.g. modules, partly because of the following results. (However, our uses of $\text{Rad}(M)$ will be confined to the exercises, cf., exercises 2.8.8ff.)

Remark 2.5.27: $\text{Jac}(R)M \subseteq \text{Rad}(M)$. (Indeed, if N is a maximal submodule then $\text{Jac}(R) \subseteq \text{Ann}(M/N)$ so $\text{Jac}(R)M \subseteq N$, implying $\text{Jac}(R)M \subseteq \bigcap \{\text{maximal submodules}\}$).

This improves Nakayama's lemma because proposition 0.2.15 shows $\text{Rad}(M) < M$ for every f.g. module M.

F Supplement: Finer Results Concerning the Jacobson Radical

In this supplement we collect a wide range of lovely theorems about $\text{Jac}(R)$, including Wedderburn's principal theorem and several theorems of Amitsur. We start by comparing $\text{Jac}(R)$ with $\text{Jac}(T)$ when R is a subring of T. Our object is to compare Jacobson radicals of tensor extensions, but many results can be framed in the following general context.

Definition 2.5.28: Suppose $T \in R\text{-}\mathcal{M}od\text{-}R$. An element x of T is R-*normalizing* if $xR = Rx$. (This parallels the group-theoretic notion of the normalizer of a subgroup.) T is R-*normalizing* if T is spanned (as R-module) by a set of normalizing elements; if this set is finite and $R \subset T$ are rings we say T is a *finite normalizing extension* of R. (See exercises 13ff and 2.12.10ff.)

The obvious example of a normalizing extension is a centralizing extension; also the skew polynomial ring $R[\lambda; \sigma]$ is R-normalizing iff σ is onto. The next result is basic in studying normalizing extensions.

Proposition 2.5.29: *Suppose T is a finite normalizing extension of R spanned by R-normalizing elements $a_1, \dots, a_n$. If $M \in T\text{-}\mathcal{M}od$ is simple then M is a direct product of $\leq n$ simple modules in $R\text{-}\mathcal{M}od$.*

Proof: For any $x \neq 0$ in M we have $M = Tx = \sum_{i=1}^{n} Ra_i x$ so M is an f.g. R-module and thus has a maximal R-submodule by proposition 0.2.15. Let $N_i = \{x \in M: a_i x \in N \text{ for } 1 \leq i \leq n\}$.

Then $a_i(rx) \in Ra_i x \subseteq N$ for each x in N_i, implying $N_i \in R\text{-}\mathcal{M}od$.

We claim each M/N_i is simple in $R\text{-}\mathcal{M}od$. Indeed, if $N_i < M' \leq M$ then $a_i M'$ is an R-submodule of M *not* contained in N so $N + a_i M' = M$. For any element x in M we then have $a_i x = y + a_i x'$ for suitable y in N and x' in M', so $y = a_i(x - x')$, implying $x - x' \in N_i$ and $x \in x' + N_i \subset M'$. Therefore, $M = M'$, proving the claim.

On the other hand, $\bigcap_{i=1}^{n} N_i = \{x \in M: a_i x \in N \text{ for each } i\} = \{x \in M: Tx \subset N\}$, a T-submodule of M which is thus 0. Hence M is a direct product of the simple modules M/N_i by proposition 2.2.5.   Q.E.D.

This result is originally due to Formanek-Jategaonkar [74], cf., exercise 15.

Corollary 2.5.30: *Suppose T is a finite normalizing extension of R. Then* $\operatorname{Jac}(R) \subseteq \operatorname{Jac}(T)$. *More precisely, if P is a primitive ideal of T then $P \cap R$ is a finite intersection of primitive ideals of R; if, furthermore, $P \cap R \in \operatorname{Spec}(R)$ then $P \cap R$ is a primitive ideal of R. (Aside: See also corollary 3.4.14)*

Proof: Suppose P is a primitive ideal of T. Take a simple T-module M with $P = \operatorname{Ann}_T M$; writing $M = \prod_{i=1}^{n} N_i$ for N_i simple in $R\text{-}\mathcal{M}od$ and putting $P_i = \operatorname{Ann}_R N_i$, we have $P_1 \cap \cdots \cap P_n = \operatorname{Ann}_R M = P \cap R$. This proves the second assertion. If $P \cap R$ is prime then $P_1 \cdots P_n \subseteq P \cap R$ implies some $P_i = P \cap R$ so $P \cap R$ is primitive. The first assertion also follows, viewing $\operatorname{Jac}(R)$ as $\bigcap \{\text{primitive ideals}\}$.   Q.E.D.

This raises interest in the reverse inequality: When is $R \cap \operatorname{Jac}(T) \subseteq \operatorname{Jac}(R)$? A positive result was given in proposition 2.5.17, but the following example should prevent over-expectations.

Example 2.5.31: Fix a prime p and let $S = \{m \in \mathbb{N}: p \nmid m\}$, writing the elements of S in ascending order as $m_1 = 1, m_2, m_3$, etc. Let $S(i)$ be the submonoid of $\mathbb{N}$ generated by $m_1, \ldots, m_i$ (i.e., $S(i)$ is the set of products of powers of $m_1, \ldots, m_i$). Thus $\bigcup R_{S(i)} = R_S$, a local ring, but each $R_{S(i)}$ is semiprimitive (since each prime not dividing $m_1 \cdots m_i$ generates a maximal ideal and the intersection of these is 0).

Nevertheless, many positive results can be obtained by applying proposition 2.5.17(i).

Lemma 2.5.32: *The hypothesis of proposition 2.5.17(i) holds if R is a summand of T either in* R-$\mathcal{M}od$ *or in* $\mathcal{M}od$-R. *(Thus $R \cap \mathrm{Jac}(T) \subseteq \mathrm{Jac}(R)$ in either case.)*

Proof: By symmetry we may assume R is summand of T in $\mathcal{M}od$-R. It suffices to show $R \cap Tr = Rr$, for then r invertible in T would imply $Rr = R$ so $r^{-1} \in R$. We shall verify a slightly stronger result.   Q.E.D.

Sublemma 2.5.32′: *If T is any right R-module of which R is a summand then $R \cap TL = L$ for any left ideal L of R.*

Proof of sublemma: Write $T = R \oplus M$. Clearly $R \cap TL \supseteq L$. Conversely, if $r = \sum a_i b_i \in R$, where $a_i \in T$ and $b_i \in L$, then writing $a_i = r_i + x_i$ for r_i in R and x_i in M yields $r = \sum r_i b_i + \sum x_i b_i$; matching components in R shows $r = \sum r_i b_i \in L$.   Q.E.D.

Actually sublemma 2.5.32′ provides an easily verified criterion for the lattice of left ideals $L(R)$ to be isomorphic to a sublattice of $L(T)$, in case T is a ring. This will be one of the most-quoted results in later chapters.

Before applying these results, we should point out that Heinicke-Robson [84] have proved the following general result: If T is a finite normalizing extension of R and $R \subseteq W \subseteq T$ are rings then $\mathrm{Jac}(R) = R \cap \mathrm{Jac}(W)$. Also see exercise 13.

Proposition 2.5.33: *If T is a finite normalizing extension of R which is free in* R-$\mathcal{M}od$ *with a base containing 1, then $\mathrm{Jac}(R) = R \cap \mathrm{Jac}(T)$.*

Proof: Corollary 2.5.30 and lemma 2.5.32.   Q.E.D.

Proposition 2.5.34: *Suppose $R = \bigcup R_i$.*

(i) *If every element of R_i invertible in R is already invertible in R_i then $\mathrm{Jac}(R) \subseteq \bigcup \mathrm{Jac}(R_i)$; in particular, this holds if R is free in R_i-$\mathcal{M}od$ for each i.*

(ii) *The reverse inclusion holds if R is a finite normalizing extension of each R_i.*

Proof: (i) Any r in $\mathrm{Jac}(R)$ lies in some R_i, so $r \in \mathrm{Jac}(R_i)$ by proposition 2.5.17(i). (ii) By corollary 2.5.30.   Q.E.D.

Theorem 2.5.35: *If $R \in C$-$\mathcal{Alg}$ and H is an integral commutative C-algebra, then $\mathrm{Jac}(R \otimes 1) = (R \otimes 1) \cap \mathrm{Jac}(R \otimes H)$.*

Proof: Recall H is integral iff $C[h]$ is an f.g. C-module for each h in H. By proposition 2.5.34 we may assume H is f.g. over C, so $R \otimes H$ is a finite centralizing extension of $R \otimes 1$ and we are done by proposition 2.5.33.
$$\text{Q.E.D.}$$

One would like the stronger result that $\text{Jac}(R \otimes H) = \text{Jac}(R) \otimes H$, but this has been seen to be false in example 1.7.25. There H was an inseparable field extension, and, in fact, the following useful result holds:

Theorem 2.5.36: *Suppose R is an algebra over a field F and $H \supseteq F$ is an algebraic field extension. Then $\text{Jac}(R) \otimes H \subseteq \text{Jac}(R \otimes H)$, with equality if H is separable over F. (Tensors are over F.)*

Proof: Let $\{H_i : i \in I\}$ be the finite extensions of F contained in H. Let $J = \text{Jac}(R)$. If $x = \sum_{j=1}^t r_j \otimes h_j \in J \otimes H$ where each $r_j \in J$ then some H_i contains $h_1, \ldots, h_t$ so $x \in J \otimes H_i \subseteq \text{Jac}(R \otimes H_i)$ and x is quasi-invertible, proving $J \otimes H \subseteq \text{Jac}(R \otimes H)$, which is the first assertion.

Now assume H is separable over F, and let $\bar{R} = R/J$. Since $(R \otimes H)/(J \otimes H) \approx \bar{R} \otimes H$ (by proposition 1.7.30) it suffices to prove $\text{Jac}(\bar{R} \otimes H) = 0$, i.e., we may assume $J = 0$. Moreover, any base of H over H_i is a base of $R \otimes_F H$ over $R \otimes_F H_i$; by proposition 2.5.34 we have $\text{Jac}(R \otimes H) \subseteq \bigcup \text{Jac}(R \otimes H_i)$, so it suffices to prove each $\text{Jac}(R \otimes H_i) = 0$, i.e., we may assume $[H:F] < \infty$. Letting K be the normal closure of H, we have just proved $\text{Jac}(R \otimes H) \otimes_H K \subseteq \text{Jac}(R \otimes H \otimes_H K) = \text{Jac}(R \otimes K)$, so it suffices to prove $\text{Jac}(R \otimes K) = 0$.

K is Galois over F; letting G be the Galois group, we let $G' = \{1 \otimes \sigma : \sigma \in G\}$, a group of automorphisms of $R \otimes K$ whose fixed subring is R. Suppose $a_1, \ldots, a_n$ is a base for K over F and $z = \sum_{i=1}^n r_i \otimes a_i \in \text{Jac}(R \otimes K)$. For any given a_j and σ in G, we have $\sum_{i=1}^n r_i \otimes \sigma(a_i a_j) = \sigma(z(1 \otimes a_j)) \in \text{Jac}(R \otimes K)$ by proposition 2.5.6; summing over all σ in G and letting α_{ij} be the trace $T_{K/F} a_i a_j$, we have $\sum_i \alpha_{ij} r_i \in R \cap \text{Jac}(R \otimes K) = \text{Jac}(R) = 0$ (using theorem 2.5.35). Cramer's rule yields $d r_i = 0$ for each i, where $d = \det(\alpha_{ij}) \neq 0$. Thus each $r_i = 0$, implying $z = 0$. This proves $\text{Jac}(R \otimes K) = 0$.   Q.E.D.

Wedderburn's Principal Theorem

We are ready for a celebrated theorem of Wedderburn, which explicitly describes finite dimensional algebras. Recall a field F is *perfect* if each algebraic extension of F is separable, e.g., if F is finite or $\text{char}(F) = 0$ or F is algebraically closed.

Theorem 2.5.37: *Suppose R is a finite dimensional algebra over a perfect field F, and $N = \mathrm{Jac}(R)$. Then $R = S \oplus N$ (as a vector space over F) where S is a subalgebra of R isomorphic to R/N.*

Proof: Write $\bar{R}$ for R/N. We shall use $\bar{R}$ repeatedly to analyze R.

Case I. $N^2 = 0$ and $\bar{R}$ is split, i.e., $\bar{R} \approx \prod_{u=1}^{t} M_{n_u}(F)$ for suitable t and n_u in $\mathbb{N}$. Note N is idempotent-lifting by corollary 1.1.28. Let $z_u = (0,\ldots,0,1,0,\ldots,0)$ in $\bar{R}$, where 1 appears in the u position. Then $z_1,\ldots,z_t$ are orthogonal idempotents of $\bar{R}$ which can be lifted to orthogonal idempotents $e_1,\ldots,e_t$ of R, and $\sum e_i = 1$ (since $1 - \sum e_i$ is an idempotent of N and thus 0.) Let $R_u = e_u R e_u$. Then $\bar{R}_u = z_u \bar{R} z_u \approx M_{n_u}(F)$, so proposition 1.1.25 shows R_u also has a set of matric units $\{e_{ij}^{(u)} : 1 \le i, j \le n_u\}$. Let $S_u = \sum_{i,j} F e_{ij}^{(u)} \approx \bar{R}_u$. Then $S = \sum S_u$ is a subalgebra of R and $S \approx \prod S_u \approx \prod \bar{R}_u \approx R$. Since $\bar{R} = S/(N \cap S)$ we see $N \cap S = 0$; also $[S + N : F] = [S : F] + [N : F] = [\bar{R} : F] + [N : F] = [R : F]$, proving $S + N = R$.

Case II. $N^2 = 0$ (no assumption on $\bar{R}$). This is the only place we shall use the fact F is perfect. Let $m = [R : F]$, and let F' be the algebraic closure of F, which is separable by hypothesis. Then $[R \otimes_F F' : F'] = m$ and $\mathrm{Jac}(\bar{R} \otimes F') = \mathrm{Jac}(\bar{R}) \otimes F' = 0$, so F' is a splitting field for $\bar{R}$, and $\bar{R}$ has a splitting field K which is f.g. over F, by proposition 2.3.21; letting $R_1 = R \otimes_F K$ and $N_1 = \mathrm{Jac}(R_1) = N \otimes K$, we have $R_1 \approx S_1 \oplus N_1$ by case I, where $S_1 \approx R_1/N_1$ is a subalgebra of R_1.

Let V be an F-subspace of R complementary to N, such that $1 \in V$. Letting $t = [V : F]$, we take a base $\{x_1,\ldots,x_t\}$ of V over F with $x_1 = 1$. Note $m - t = [N : F] = [N_1 : K]$, so $[S_1 : K] = t$. Let $\{x_{t+1},\ldots,x_m\}$ be a base of N over F; then $\{x_1,\ldots,x_m\}$ is a base of R over F and thus of R_1 over K (identifying R with $R \otimes 1 \subset R_1$). Thus every element of S_1 has the form $y + \sum_{i=1}^{t} a_i x_i$ where $y \in N_1$ and the $a_i \in K$; it follows readily (by counting dimensions) that S_1 has a base $\{x_u + y_u : 1 \le u \le t\}$ where each $y_u \in N_1$.

Picking a base $\{a_1, a_2, \ldots\}$ of K over F with $a_1 = 1$, we define a balanced map $R \times K \to R$ sending $(r, \sum \alpha_i a_i)$ to $\alpha_1 r$; we thereby get an $R - R$ bimodule map $\sigma : R \otimes K \to R$ satisfying $\sigma(r \otimes \sum \alpha_i a_i) = \alpha_1 r$. Let $S = \sigma S_1$. Clearly S is spanned by the $\sigma(x_u + y_u) = x_u + \sigma y_u$ for $1 \le u \le t$, which are independent because the x_u are independent. Hence $[S : F] = t$. Moreover, using $N^2 = 0$ we have

$$(x_u + \sigma y_u)(x_v + \sigma y_v) = x_u x_v + (\sigma y_u)x_v + x_u \sigma y_v = \sigma((x_u + y_u)(x_v + y_v)) \in \sigma S_1 = S$$

for each u, v, implying S is a subalgebra of R. Clearly $\bar{S} \approx \bar{R}$ so $S \cap N = 0$ and $S \oplus N = R$.

General case. Induction on $[R:F]$. By case II we may assume $N^2 \neq 0$. Now write $\bar{R}$ for R/N^2. By induction $\bar{R} = \bar{S} + \bar{N}$ for a suitable semisimple subalgebra $\bar{S} \approx \bar{R}/\bar{N} \approx R/N$. But the preimage S of $\bar{S}$ also is a subalgebra and $\bar{S} \approx S/(N^2 \cap S)$; since N is nilpotent we see $N^2 \cap S = \text{Jac}(S)$. By induction $S = S_1 \oplus (N^2 \cap S)$ for some subalgebra $S_1 \approx \bar{S} \approx R/N$. Now

$$S_1 + N = S_1 + (N^2 \cap S) + N = S + N = R$$

and $N \cap S_1 = (N^2 \cap S) \cap S_1 = 0$, proving $R \approx S_1 \oplus N$ as desired.   Q.E.D.

Definition 2.5.38:   A finite dimensional algebra R over a field F is *separable* if $R \otimes_F K$ is semisimple Artinian for every field extension K of F.

Wedderburn's principal theorem can be obtained more generally for arbitrary separable algebras over a field, with virtually identical proof, cf., exercise 11; this is a true generalization by exercise 8. The main new ingredient needed here is the existence of separable splitting fields, proved by Koethe, cf., exercise 10. A more modern treatment of separable algebras including Wedderburn's principal theorem is given in §5.3.

F Supplement: Amitsur's Theorem and Graded Rings

We return now to Amitsur's theorem (2.5.23) on when the Jacobson radical of $R[\lambda]$ is 0. Amitsur also proved a similar result for $R(\lambda)$, and, following a lovely idea of Bergman [73], we unify these results by means of a general theorem on graded rings.

Lemma 2.5.39:   *Suppose R is a $(\mathbb{Z}/n\mathbb{Z})$-graded ring, and write R_u for the homogeneous component corresponding to u for $0 \leq u \leq n-1$. If $r = \sum_{u=0}^{n-1} r_u \in \text{Jac}(R)$ for $r_i \in R_i$ then each $nr_i \in \text{Jac}(R)$.*

Proof: Let ζ be a primitive n-th root of 1 in $\mathbb{C}$. Then $\mathbb{Z}[\zeta]$ is a free $\mathbb{Z}$-module (since the cyclotomic polynomial is monic) and thus putting $R' = R \otimes_{\mathbb{Z}} \mathbb{Z}[\zeta]$ we have R' is an f.g. free R-module by proposition 1.7.15. Let $J = \text{Jac}(R')$. Proposition 2.5.33 yields $\text{Jac}(R) = R \cap J$, so $r \in J$.

R' itself is $\mathbb{Z}/n\mathbb{Z}$-graded by putting $R'_u = R_u \otimes \mathbb{Z}[\zeta]$, and we can define an automorphism σ of R' by putting $\sigma \sum a_u = \sum \zeta^u a_u$ (for $a_u \in R'_u$ arbitrary). Then $\sigma J \subset J$ by proposition 2.5.6. Also, $\sum_{j=1}^{n} \zeta^{jk} = \sum (\zeta^k)^j = 0$ whenever $n \nmid k$, so

$$nr_u = \sum_{k=1}^{n}\left(\sum_{j=1}^{n}\zeta^{j(k-u)}r_k\right) = \sum_{j=1}^{n}\zeta^{-uj}\left(\sum_{k=1}^{n}\zeta^{kj}r_k\right) = \sum_{j=1}^{n}\zeta^{-uj}\sigma^j r \in J.$$

Hence $nr_u \in J \cap R = \mathrm{Jac}(R)$, as desired.   Q.E.D.

Theorem 2.5.40: (*Bergman*) *If R is a $\mathbb{Z}$-graded ring then $\mathrm{Jac}(R)$ is a graded ideal of R.*

Proof: Let $J = \mathrm{Jac}(R)$. Given r in J write $r = \sum r_u$ where the r_u are the homogeneous components; then there is m such that $r_u = 0$ for $|u| > m$. Take $n \ge 2m$. Since the $\mathbb{Z}$-grading induces a $\mathbb{Z}/n\mathbb{Z}$-grading (cf., remark 1.9.8), we have $nr_u \in J$ for each u. But using $(n+1)$ instead of n we have each $(n+1)r_u \in J$, so $r_u \in J$, as desired.   Q.E.D.

Lemma 2.5.41: *Suppose $r \in R$ and $r\lambda$ is left quasi-invertible in $R[\lambda]$. Then r is nilpotent.*

Proof: Suppose $\left(\sum_{i=0}^{t}a_i\lambda^i\right)(1-r\lambda) = 1$. Matching powers of λ shows $a_0 = 1$ and $a_i = a_{i-1}r$ for each $i \ge 1$, so inductively $a_i = r^i$. Thus $r^{t+1} = 0$.   Q.E.D.

Reproof of Amitsur's Theorem (2.5.23): We shall show for any ring R that $\mathrm{Jac}(R[\lambda]) = N[\lambda]$ for a suitable nil ideal N of R. Let $J = \mathrm{Jac}(R[\lambda])$ and $N = J \cap R$. Clearly $N[\lambda] \subseteq J$. Grade $R[\lambda]$ according to degree in λ. By theorem 2.5.40 if $\sum_{i=0}^{t}r_i\lambda^i \in J$ then each $r_i\lambda^i \in J$. But there is an automorphism σ of $R[\lambda]$ given by $\sigma\lambda = \lambda + 1$; thus $r_i(\lambda + 1)^i = \sigma(r_i\lambda^i) \in \sigma J = J$, so the constant term $r_i \in J \cap R = N$ for each i. This proves $J = N[\lambda]$. Finally, if $r \in N$ then $r\lambda \in J$ implying r is nilpotent by lemma 2.5.41.   Q.E.D.

Theorem 2.5.42: (*Another theorem of Amitsur*) *Suppose R is an algebra over a field F, and let F' be the field of rational fractions $F(\lambda)$. Then $\mathrm{Jac}(R \otimes F') = N \otimes F'$ for some nil ideal N of R.*

Proof: We claim F' can be $\mathbb{Z}/n\mathbb{Z}$-graded as $\bigoplus_{u=0}^{n-1}A_u$ where $A_u = \lambda^u F(\lambda^n)$ To see this, first note $A_uA_v = A_{u+v}$ (subscripts modulo n). Also $A_0,\ldots,A_{n-1}$ are independent: If $\sum_{u=0}^{n-1}\lambda^u g_u = 0$ with $g_u \in F(\lambda^n)$, then clearing denominators we may assume each $g_u \in F[\lambda^n]$ and thus has degree 0 (modulo n); the $\lambda^u g_u$ have different degrees for each u and cannot cancel, implying each $\lambda^u g_u$ must already be 0, so each $g_u = 0$, as desired. It remains to prove $\sum_{u=0}^{n-1}A_u = F'$. Clearly, $F[\lambda] \subseteq \sum_{u=0}^{n-1}A_u$, so we must show $f^{-1} \in \sum_{u=0}^{n-1}A_u$ for f in $F[\lambda]$, and we may assume f is monic. Working in the splitting field K of the polynomial

$\lambda^n - 1$ over F, we let $\lambda^n - 1 = (\lambda - \zeta_1) \cdots (\lambda - \zeta_n)$ and $h = f(\lambda)f(\zeta_1\lambda) \cdots f(\zeta_n\lambda)$. It is easy to see $h \in F[\lambda^n]$, either by arguing combinatorically or by using Galois theory; details are left to the reader. But f divides h in $K[\lambda]$, so viewing $F' \subset K(\lambda)$ canonically we have $f^{-1}h \in K[\lambda] \cap F' = F[\lambda] \subset \sum A_u$. But $h^{-1} \in F(\lambda^n) = A_0$, so $f^{-1} = (f^{-1}h)h^{-1} \in (\sum A_u)A_0 = \sum A_u$, proving the claim.

We are ready to apply the preceding theory. Write $J = \text{Jac}(R \otimes F')$ and $N = R \cap J$, viewing R as $R \otimes 1 \subset R \otimes F$. Clearly, $N \otimes F' \subset J$; to get the reverse inclusion suppose $x \in J$. We can write $x = \sum_{i=0}^t r_i \otimes \lambda^i f^{-1}$ for suitable $t \in \mathbb{N}$ and $f \in F[\lambda]$. It suffices to show $x(1 \otimes f) \in N \otimes F'$ since $x = x(1 \otimes f)(1 \otimes f)^{-1}$, so we may assume $x = \sum_{i=0}^t r_i \otimes \lambda^i$. For any $n > t$ we have $R \otimes F'$ graded by $\mathbb{Z}/n\mathbb{Z}$ with homogeneous components $R \otimes \lambda^u F(\lambda^n)$, so lemma 2.5.39 implies $nr_u \in J \cap R = N$ for each u. But using $(n + 1)$ instead of n we also have $(n + 1)r_u \in N$, so each $r_u \in N$, as desired.

To see N is nil view $F(\lambda)$ in the Laurent series ring $R((\lambda))$. For any r in N we have $(1 - r\lambda)^{-1} = \sum_{i=0}^{\infty} r^i\lambda^i \in R \otimes F'$, implying the powers of r generate a finite dimensional F-subspace of R, i.e., r is algebraic over F, and is nilpotent by corollary 2.5.18.   Q.E.D

Corollary 2.5.43: *Suppose R is an algebra over a field F and $F' = F(\lambda_i : i \in I)$ for any infinite set of commuting indeterminates. Then $\text{Jac}(R \otimes F')$ is nil.*

Proof: For $|I|$ finite this is an easy induction; for arbitrary I apply proposition 2.5.34(i).   Q.E.D.

See exercises 2.6.11ff for results on graded rings.

F Supplement: The Jacobson Radical of a Rng

Despite a determined attempt to impose 1 in ring theory, there are times when it is more convenient to deal with rings without 1. This is true in particular for certain "elementary" aspects of the theory. In this brief digression we shall sketch Jacobson's original description of his radical for $\mathscr{Rng}$ (cf., §1.5). R_0 throughout will denote a ring without 1. We need a substitute for 1.

Definition 2.5.44: A left ideal L of R_0 is *modular* if there is e in R_0 such that $r - re \in L$ for all r in R_0; e is called a *left unit (modulo L)*.

Remark 2.5.45: A modular left ideal L is proper iff $e \notin L$ (for, otherwise, $r \in re + L \subseteq L$ for all r in R_0, proving $L = R_0$). Consequently, Zorn's lemma shows any modular left ideal $L \neq R_0$ is contained in a maximal modular left ideal

$(\neq R_0)$, with the same left unit. Thus one can define $\text{Jac}(R_0) = \bigcap\{\text{maximal modular left ideals of } R_0\}$.

We can characterize $\text{Jac}(R_0)$ in terms of quasi-invertible elements, where now we say $r \in R_0$ is left *quasi-invertible* if $r + r' - r'r = 0$ for some r' in R_0. (If $1 \in R_0$ this says $(1 - r')(1 - r) = 1$, which is the definition we gave earlier.) A set is *quasi-invertible* if each of its elements is quasi-invertible.

Proposition 2.5.46: $\text{Jac}(R_0)$ *is a quasi-invertible left ideal which contains every quasi-invertible left ideal.*

Proof: Let $J = \text{Jac}(R_0)$. If $a \in J$ is not quasi-invertible then $L = \{r - ra : r \in R_0\}$ is a modular left ideal, for a is a left unit modulo L; but L is contained in a maximal modular left ideal not containing a, contradiction. Thus J is quasi-invertible. Now let J' be any quasi-invertible left ideal. We claim J' is contained in every maximal modular left ideal L. Indeed, otherwise $e \in J' + L$, where e is a left unit modulo L. Writing $e = r + a$ for r in J' and a in L, we see $a = e - r \in L$. But $r - re \in L$ by hypothesis, so $e - re \in L$. Take r' such that $r + r' - r'r = 0$. Then

$$e = e - (r + r' - r'r)e = (e - re) - r'(e - re) \in L$$

contrary to hypothesis on e. Thus $J' \subseteq \bigcap\{\text{maximal modular left ideals}\} = J$.

Q.E.D.

It is now a straightforward matter to carry over the other characterizations of $\text{Jac}(R)$ to $\mathscr{R}ng$. In particular, $\text{Jac}(R_0) \lhd R_0$ and is left-right symmetric. The reason behind such a project is to be able to focus on the radical itself as a rng; exercise 21ff presents Sasiada's example of a simple radical rng.

Having seen that modular ideals are useful, we would like to know that they are abundant enough.

Proposition 2.5.47: *If L_1 is a maximal modular left ideal and L_2 is any modular left ideal then $L_1 \cap L_2$ is modular.*

Proof: This is obvious unless $L_2 \nsubseteq L_1$; thus $L_1 + L_2 = R_0$. Let e_i be a left unit modulo L_i for $i = 1, 2$. Writing $e_i = a_i + b_i$ for $a_i \in L_1$ and $b_i \in L_2$, we claim $e = a_2 + b_1$ is a left unit modulo $L_1 \cap L_2$. Indeed, for any r in R_0 we have

$$r - re = r - r(a_2 + b_1) = r - re_1 + r(a_1 - a_2) \in L_1$$

and, likewise, $r - re \in L_2$, as desired.　　Q.E.D.

Corollary 2.5.48:   *Any finite intersection of maximal modular left ideals is modular.*

Proof: Induction applied to proposition 2.5.47.   Q.E.D.

Galois Theory of Rings

With these modest beginnings, one can obtain some rather deep results concerning group actions on rings. Given a rng R_0 and a finite group G of automorphisms on R_0, we write R_0^G for $\{r \in R_0 : \sigma r = r$ for all σ in $G\}$. Of course, R_0 may have 1, but we find it convenient not to require this. Write $t(r)$ for $\sum_{\sigma \in G} \sigma r$. We follow Puczylowski [84].

Lemma 2.5.49:   *If L is a maximal modular left ideal of R_0 then there is e in R_0 for which $|G|r - rt(e) \in L$ for all r in R_0.*

Proof: Let e be a left unit modulo $L' = \bigcap_{\sigma \in G} \sigma L$, which is modular by corollary 2.5.48. Then $r - re \in L'$, so applying σ and replacing r by $\sigma^{-1} r$ shows $r - r\sigma e \in L'$; summing over σ yields $|G|r - rt(e) \in L'$.   Q.E.D.

Lemma 2.5.50:   *If $t(R_0)$ is quasi-invertible then $|G|R_0$ is quasi-invertible.*

Proof: Let L be a maximal modular left ideal of the rng $|G|R_0$ and take $e = |G|e_0$ as in lemma 2.5.49 *applied to the rng* $|G|R_0$. Thus for all r in $|G|R_0$ we have $|G|(r - rt(e_0)) = |G|r - rt(e) \in L$, implying $|G|^3 R_0 \subseteq L$ since $t(e_0)$ is quasi-invertible. Thus $(|G|R_0)^2 \subseteq L$ so $|G|R_0 \subseteq L$ since L is modular. This proves $|G|R_0 = \text{Jac}(|G|R_0)$, as desired.   Q.E.D.

Theorem 2.5.51:   *(Montgomery [76]-Martindale [78])* $|G|\text{Jac}(R^G) \subseteq \text{Jac}(R)$ *whenever G is a finite group of automorphisms acting on a ring R.*

Proof: Let $J = \text{Jac}(R^G)$. Taking $R_0 = RJ$ we have $t(R_0) = t(R)J \subseteq R^G J = J$, a quasi-invertible left ideal of R_0. Hence $|G|R_0$ is quasi-invertible by lemma 2.5.50 and is obviously a left ideal of R, implying $|G|R_0 \subseteq \text{Jac}(R)$ as desired.
                                                                                       Q.E.D.

A related theorem is the theorem of Bergman-Isaacs, that if G is a finite group of automorphisms on a rng R_0 and $R_0^G = 0$ then $|G|R_0$ has a nontrivial nilpotent ideal. This follows at once from the following result, which also is amenable to Puczylowski's techniques:

Theorem 2.5.52:   *(Bergman-Isaacs) If G is a finite group of automorphisms acting on a rng R then there is $n = n(G)$ such that $(|G|\text{Ann}t(R)R)^n = 0$.*

Proof: Let $A = |G| \operatorname{Ann} t(R)R$. We prove the result by means of a sequence of embeddings.

Step 1. $A \subseteq \operatorname{Jac}(R)$ by lemma 2.5.50, since $t(\operatorname{Ann} t(R)R) = (\operatorname{Ann} t(R))t(R) = 0$.

Step 2. A is nil by Amitsur's theorem (2.5.23) since $A \subseteq |G| \operatorname{Ann} t(R[\lambda])R[\lambda] \subseteq \operatorname{Jac}(R[\lambda])$ by step 1.

Step 3. A is nilpotent. Indeed let X be a set of noncommuting indeterminates of the same cardinality as A and write $X = \{X_a : a \in A\}$. Then $A\{\{X\}\} \subseteq |G| \operatorname{Ann} t(R\{\{X\}\})$ is nil by step 2. But in the formal power series ring $R\{\{X\}\}$, we see the element $\sum_{a \in A} aX_a$ is nilpotent, i.e., $\left(\sum_{a \in A} aX_a\right)^n = 0$ for some n, and checking coefficients we see $A^n = 0$.

Step 4. $A^n = 0$ where n depends only on G (not on R). Indeed, otherwise, for each $m > 0$ there would be a ring R_m with G acting on R_m such that $(|G| \operatorname{Ann} t(R_m))^m \neq 0$. But then taking $R = \prod_{m \in \mathbb{N}} R_m$ we see $A = \prod |G| \operatorname{Ann} t(R_m)$ so $A^m \neq 0$ for each m, contrary to step 3.   Q.E.D.

Corollary 2.5.53: *If $R^G = 0$ then $|G|R$ is nilpotent.* (*Indeed $t(R) \subseteq R^G = 0$.*)

The embedding technique of this proof is obviously quite useful and probably has much unused potential. We return to it in exercises 2.12.20ff (Passman's "primitivity machine"). Further results on fixed rings under group actions are given in exercises 2.6.13ff and exercise 6.2.1.

The results of this discussion assume their strongest form when $|G|$ is invertible in R (e.g., char $(R) = 0$); then theorem 2.5.51 implies $\operatorname{Jac}(R^G) \subseteq \operatorname{Jac}(R)$, and Bergman-Isaacs says if $R_0^G = 0$ then R_0 is nilpotent. On the other hand, one may want characteristic-free results. Then one must add an assumption, such as G is *outer* on R, which classically means that the restriction of the automorphisms in G to $Z(R)$ are all distinct. Such a theory will be needed in §7.2, where some of the classical results are presented; a sweeping theory of automorphisms due to Kharchenko is found in Montgomery [80B].

§2.6 Nilradicals

For the purposes of this book we define a *radical* as the intersection of a certain class of ideals; for example, the Jacobson radical is the intersection of the primitive ideals. It is often difficult to reduce to the case $\operatorname{Jac}(R) = 0$, and so it is useful to have smaller radicals to refine our study of rings, especially in light of theorems to be proved below on prime rings. In this section we consider radicals which are nil ideals. One candidate is the radical arising from prime ideals, called the "prime radical" or "lower nilradical;" another is the "upper nilradical," "which is easier to describe, as the "largest" nil ideal, but which presents technical difficulties discussed in 2.6.33ff.

The following observation holds if we fix a class $\mathcal{R}$ of rings and define $\text{Rad}(R) = \bigcap \{A \lhd R : R/A \in \mathcal{R}\}$.

Remark 2.6.0: The proof of proposition 2.5.1' goes through for the radical "Rad." Thus if $\text{Rad}(R) \supseteq A \lhd R$ then $\text{Rad}(R/A) = \text{Rad}(R)/A$; if $\text{Rad}(R/A) = 0$ then $A \supseteq \text{Rad}(R)$.

Although the situation for noncommutative rings is considerably more complicated than in the commutative case (for which every finite subset is nilpotent), nil ideals do satisfy certain pleasant properties.

Lemma 2.6.1: *If $A \lhd R$ is nil and S/A is a nil subset of R/A then S is nil.*

Proof: Suppose $s \in S$. Then some $s^u \in A$ and is thus nilpotent, implying s is nilpotent. Q.E.D.

Proposition 2.6.2: *Any ring has a unique maximal nil ideal.*

Proof: By Zorn's lemma there exists some maximal nil ideal N. We claim N contains every nil ideal B. Indeed $(B + N)/N$ is nil, so $B + N$ is nil by lemma 2.6.1, implying $B + N = N$, so $B \subseteq N$ as desired. Q.E.D.

Definition 2.6.3: $\text{Nil}(R)$ is the unique maximal nil ideal of R.

Remark 2.6.4: If $A \lhd R$ then $(\text{Nil}(R) + A)/A \subseteq \text{Nil}(R/A)$, equality holding if A is nil. In particular, $\text{Nil}(R/\text{Nil}(R)) = 0$.

In analogy to $\text{Jac}(R)$ we would like to display $\text{Nil}(R)$ as a radical.

Definition 2.6.5: A ring R is *strongly prime* if R is prime with no nonzero nil ideals. $P \lhd R$ is *strongly prime* if R/P is strongly prime.

Remark 2.6.6: (To be generalized shortly) if S is a submonoid of R and $P \lhd R$ is maximal with respect to $P \cap S = \varnothing$ then P is a prime ideal. (Indeed, if $A_i \supset P$ for $i = 1, 2$ then there is s_i in $A_i \cap S$, so $s_1 s_2 \in A_1 A_2 \cap S$ implying $A_1 A_2 \nsubseteq P$.)

Proposition 2.6.7: $\text{Nil}(R) = \bigcap\{strongly\ prime\ ideals\ of\ R\}$.

Proof: Let $\mathscr{P} = \{$strongly prime ideals of $R\}$. If $P \in \mathscr{P}$ then $\text{Nil}(R/P) = 0$, implying $\text{Nil}(R) \subseteq P$. Conversely, we shall show $\bigcap\{P \in \mathscr{P}\}$ is a nil ideal and is thus $\text{Nil}(R)$. To see this we shall show that for any $a \in R$ which is *not* nilpotent, there is some P in $\mathscr{P}$ not containing a. Indeed, $S = \{a^i : i \geq 0\}$ is a submonoid of R, and by Zorn's lemma there is some ideal P maximal with respect to $P \cap S = \varnothing$.

We claim $P \in \mathscr{P}$. Indeed P is a prime ideal by remark 2.6.6. If R/P had a nonzero nil ideal B/P then $B \supset P$ so $B \cap S$ would contain some a^k; but then some power of a^k would be in P, contrary to $P \cap S = \varnothing$. This proves the claim.   Q.E.D.

Note: An examination of the above proof shows that $\text{Nil}(R)$ is also the intersection of those strongly prime ideals P_y satisfying the additional property that for each ring $R_y = R/P_y$ there is some non-nilpotent element r_y such that every nonzero ideal of R_y contains a power of r_y. This condition is somewhat technical, but is useful in implementing certain proofs.

Corollary 2.6.7': $\text{Nil}(R) = 0$ iff R is a subdirect product of strongly prime rings.

Two results motivate us especially to study $\text{Nil}(R)$:

(i) $\text{Nil}(R) \subseteq \text{Jac}(R)$ by corollary 1.1.28.
(ii) If $\text{Nil}(R) = 0$ then $\text{Jac}(R[\lambda]) = 0$ by theorem 2.5.23.

On the other hand, we have seen in §2.5 that $\text{Jac}(R)$ is nil for several classes of rings.

Nilpotent Ideals and Nilradicals

Definition 2.6.13:   Given a subset S of a ring define $S^1 = S$ and, inductively, $S^n = S^{n-1}S$. We say S is *nilpotent* of *index n* if $S^n = 0$ and $S^{n-1} \neq 0$.

Whereas in a *commutative* ring every finite nil subset is nilpotent, this is hardly true in general, as evidenced by the following easy example:

Example 2.6.14:   Let $R = M_2(F)$ and $S = \{e_{12}, e_{21}\}$.

    This definition leads us to examine how far a nil ideal is from being nil-potent, with the hope at times of closing the gap. Accordingly, we define a *nilradical* to be a radical contained in Nil(R) which contains every nilpotent ideal of R. Nil(R) is obviously the largest possible nilradical and is therefore called the *upper nilradical*. On the other hand, we can define the *prime radical* of R, denoted prime rad(R), to be the intersection of all prime ideals of R.

Proposition 2.6.15:   Prime rad(R) *is a nilradical of* R.

Proof: If $P \vartriangleleft R$ is prime and $N \vartriangleleft R$ is nilpotent then some $N^j = 0 \subseteq P$ implying $N \subseteq P$. Thus $N \subseteq$ prime rad(R). On the other hand, every strongly prime ideal is prime, so prime rad(R) $\subseteq \bigcap\{$strongly prime ideals of $R\} =$ Nil(R) by proposition 2.6.7.   Q.E.D.

    We shall show the prime radical is the smallest possible nilradical, and is therefore also called the *lower nilradical*. Let me first take the opportunity to state a personal preference for the upper nilradical, which is often much easier to use and which is certainly easier to derive than the lower nilradical. Never-theless, the lower nilradical has played an important historical role in the proofs of many theorems.

    Our first step in discussing the prime radical is to generalize remark 2.6.6. A subset S of R-$\{0\}$ is called an *m-set* if s_1, s_2 in S imply $s_1 r s_2 \in S$ for some r in R. Obviously, the complement of any prime ideal is an *m*-set.

Lemma 2.6.16:   (i) *The complement of an m-set contains a prime ideal;* (ii) *If R has no nonzero nilpotent ideals then R is semiprime.*

Proof: (i) Let S be an *m*-set. By Zorn's lemma there exists $P \vartriangleleft R$ maximal with respect to $P \cap S = \varnothing$; we claim P is prime. Indeed if $P \subset A_i \vartriangleleft R$ for $i = 1, 2$ then each A_i contains some s_i in S, so S has an element $s_1 r s_2 \in A_1 A_2$ implying $A_1 A_2 \nsubseteq P$.

    (ii) For each $r \neq 0$ in R we shall build an *m*-set containing r and be done by (i). Indeed, let $s_1 = r$ and inductively given s_i in S pick nonzero s_{i+1} in $s_i R s_i$. (If $s_i R s_i = 0$ then $R s_i R$ would be a nilpotent ideal, contrary to hypothe-sis. We claim that $S = \{s_1, s_2, \ldots\}$ is the desired *m*-set. Indeed, we claim $S \cap s_i R s_j \neq 0$ for all i, j, which is immediate for $i = j$. Thus we may assume $i < j$ (for a parallel argument works if $j < i$). By induction on $(j - i)$ there is some r' in R with $s_{i+1} r' s_j \in S$. Writing $s_{i+1} = s_i r_i s_i$ we get $s_i(r_i s_i r')s_j = s_{i+1} r' s_j \in S$, as desired.   Q.E.D.

Theorem 2.6.17: *The following are equivalent for a ring R: (i) R is semiprime; (ii) R has no nilpotent left ideals $\neq 0$; (iii) R has no nilpotent ideals $\neq 0$; (iv) if $A \lhd R$ and $A^2 = 0$ then $A = 0$.*

Proof: In view of remark 2.2.10 and lemma 2.6.16(ii) it only remains to prove (iv) $\Rightarrow$ (iii). But if $A^n = 0$ for $A \lhd R$ then $(A^{n-1})^2 = 0$ so $A^{n-1} = 0$, and continuing by induction on n we see $A = 0$. Q.E.D.

This characterization of semiprime rings is so fundamental we shall use it without further reference.

To continue our investigation we need to make the following definition by transfinite induction:

$$N_0(R) = 0, \; N_1(R) = \sum\{\text{nilpotent ideals of } R\},$$

$$N_\alpha(R) = \{r \in R: r + N_\beta(R) \in N_1(R/N_\beta(R))\} \qquad \text{for } \alpha = \beta^+,$$

$$N_\alpha(R) = \bigcup_{\beta < \alpha} N_\beta(R) \qquad \text{for } \alpha \text{ a limit ordinal.}$$

For example, $N_2(R)/N_1(R) = N_1(R/N_1(R))$. The $N_\alpha(R)$ form an increasing chain which, once it stops growing, remains constant for all the succeeding ordinals. Thus, putting $L(R) = N_\alpha(R)$, where α is any ordinal of cardinality $> |R|$, we have $L(R) = N_{\alpha'}(R)$ for all $|\alpha'| > |R|$.

Remark 2.6.18: By transfinite induction applied to remark 2.6.0, every nilradical contains $L(R)$.

Theorem 2.6.19: *(Levitzki) $L(R)$ is the prime radical of R (and is thus a nilradical contained in every nilradical).*

Proof: Let N be the prime radical of R. Then $L(R) \subseteq N$ by remark 2.6.18. But $R/L(R)$ has no nonzero nilpotent ideals, by definition of $L(R)$, so by lemma 2.6.16, $0 = \text{prime rad}(R/L(R)) = N/L(R)$, proving $N = L(R)$. Q.E.D.

It is convenient to know at what ordinal α we already have $N_\alpha(R) = L(R)$, i.e., how many steps are needed to reach the lower nilradical. Exercise 3 gives a general construction where we need arbitrarily large α to reach $L(R)$. On the other hand, $N_2(R) = L(R)$ for large classes of rings (as we shall see), and, in fact, one can make the following observation:

Remark 2.6.20: If $N_1(R)$ is nilpotent then $N_1(R) = L(R)$. (Indeed, for any nilpotent ideal $N/N_1(R)$ of $R/N_1(R)$ we see N is necessarily nilpotent, so $N \subseteq N_1(R)$ and $N/N_1(R) = 0$, proving $N = N_1(R)$.)

The Nilradical of Noetherian Rings

We shall see now that the nilradicals of a left Noetherian ring coincide; in fact every nil left or right ideal is nilpotent. The proof uses the following fundamental connection between nil left and right ideals.

Remark 2.6.21: If $(ra)^t = 0$ then $(ar)^{t+1} = a(ra)^t r = 0$. Consequently if Ra is nil then aR is nil.

Lemma 2.6.22: *If R satisfies ACC(two-sided ideals) then the lower nilradical is nilpotent.*

Proof: By remark 2.6.20 we need only show $N_1(R)$ is nilpotent. But by hypothesis {nilpotent ideals of R} has a maximal member, which is just $N_1(R)$.   Q.E.D.

Theorem 2.6.23 (Levitzki): *Every nil left or right ideal of a left Noetherian ring is nilpotent.*

Proof: Factoring out the lower nilradical (which we just saw is nilpotent), we could assume R is semiprime, so it suffices to prove the following stronger result.

Proposition 2.6.24: *Suppose R satisfies ACC on all chains of left ideals of the form $\mathrm{Ann}_R r$.*

  (i) *Every nonzero nil right ideal T contains a nonzero nilpotent right ideal;*
  (ii) *If R is semiprime then R has no nonzero nil left or right ideals.*

·**Proof:** (i) Take $0 \neq a \in T$ with $\mathrm{Ann}_R a$ maximal possible. If $r \in R$ with $(ar)^n \neq 0$ then $\mathrm{Ann}((ar)^n) = \mathrm{Ann}\, a$. (Indeed $\mathrm{Ann}(ab) \supseteq \mathrm{Ann}\, a$ for all b in R, but $\mathrm{Ann}\, a$ was chosen maximally.) But $ar \in T$ so there is $n \geq 1$ with $(ar)^n \neq 0$ and $(ar)^{n+1} = 0$. Hence $ar \in \mathrm{Ann}((ar)^n) = \mathrm{Ann}\, a$ so $ara = 0$. This proves $aRa = 0$, so aR is nilpotent.

(ii) Otherwise R has a nonzero nil left or right ideal T; picking $a \neq 0$ in T we have Ra nil or aR nil (respectively), so aR is nil by remark 2.6.21. By (i) R has a nilpotent right ideal, contrary to R being semiprime.   Q.E.D.

This result previews the Goldie theorems. Also a similar argument given in exercise 7 shows that under these hypotheses the lower nilradical contains every nil left or right ideal.

Bounded Index

Definition 2.6.25: A nil subset N of a ring R has *bounded index* $\leq t$ if $a^t = 0$ for all a in N. $N(R) = \{a \in R: Ra \text{ is nil of bounded index}\}$.

Although $N(R)$ will not be seen to be an ideal until further in this discussion, we can see the definition is left-right symmetric, by remark 2.6.21. $N(R)$ is important to us because the following result links it to the prime radical; the elegant proof is due to A. Klein.

Proposition 2.6.26: (i) If L is a nonzero nil left ideal of R of bounded index n and $a \in L$ with $a^{n-1} \neq 0$ then $(Ra^{n-1}R)^2 = 0$.
  (ii) $N_1(R) \subseteq N(R) \subseteq \text{Prime rad}(R)$; in particular, R is semiprime iff $N(R) = 0$.

Proof: (i) For $r \in R$ arbitrary write $r' = ra^{n-1} \in L$. Then $r'a = 0$ and $(r')^n = 0$, so $0 = (r' + a)^n = \sum_{i=1}^{n-1} a^i(r')^{n-i} = (1 + \hat{r})a^{n-1}r'$ where $\hat{r} \in aR$ is nilpotent, so $1 + \hat{r}$ is invertible and $0 = a^{n-1}r' = a^{n-1}ra^{n-1}$, proving $(Ra^{n-1}R)^2 = 0$, as desired.
  (ii) If $a \in N_1(R)$ then Ra is nilpotent so $a \in N(R)$, proving the first inequality. We shall conclude by proving every nil left ideal L of bounded index is in the prime radical I. Passing to R/I we may assume R is semiprime. But then $L = 0$ by (i).   Q.E.D.

Proposition 2.6.26(ii) provides a useful test for semiprime rings, which we shall need later.

Theorem 2.6.27: (*Amitsur*) Given a ring R, let I be an infinite set of cardinality $\geq |R|$, and put $R' = R^I$, writing (r_i) for the element of R' whose i-component is $r_i \in R$ for all i in I. Identify R as a subring of R' under the "diagonal injection" $r \to (r_i)$ with each $r_i = r$. Then $N(R) = \text{Nil}(R') \cap R$. Consequently, $N(R) \lhd R$ and there is an injection $R/N(R) \to R'/\text{Nil}(R')$.

Proof: For any $a \in R \cap \text{Nil}(R')$ take $y = (r_i) \in R'$ such that $\{r_i : i \in I\} = R$. Then $(ya)^t = 0$ for some t, implying Ra is nil of bounded index $\leq t$; thus $a \in N(R)$ proving $R \cap \text{Nil}(R') \subseteq N(R)$. Conversely, if $a \in N(R)$ then Ra is nil of some bounded index t. Each component of $R'a$ is in Ra so $R'a$ is also nil of bounded index $\leq t$, implying $R'a \subseteq \text{Prime rad}(R')$ by proposition 2.6.26. Therefore

$$N(R) \subseteq R \cap \text{Prime rad}(R') \subseteq R \cap \text{Nil}(R') \subseteq N(R),$$

proving equality holds at every stage. Now the canonical homomorphism $R \to R'/\mathrm{Nil}(R')$ has kernel $R \cap \mathrm{Nil}(R') = N(R)$, yielding the last assertion.

Q.E.D.

Derivations and Nilradicals

For later reference let us see how derivations apply to nilradicals.

Proposition 2.6.28: *If R is a $\mathbb{Q}$-algebra and δ is a derivation then $\delta(\mathrm{Nil}(R)) \subseteq \mathrm{Nil}(R)$ and $\delta(\mathrm{Prime\ rad}(R)) \subseteq \mathrm{Prime\ rad}(R)$.*

Proof: Suppose $a \in \mathrm{Nil}(R)$. For arbitrary r_{1i}, r_{2i} in R we have

$$\delta^n\left(\left(\sum r_{1i}ar_{2i}\right)^n\right) = \sum_u \delta^n(h_{u1} \cdots h_{un})$$

where each h_{uj} has the form $r_{1i}ar_{2i}$. But $\delta(r_{1i}ar_{2i}) = r_{1i}a\delta r_{2i} + r_{1i}(\delta a)r_{2i} + (\delta r_{1i})ar_{2i} \in r_{1i}(\delta a)r_{2i} + \mathrm{Nil}(R)$, so applying the same idea to each $\delta^n(h_{u1} \cdots h_{un})$ by iterating Leibniz' formula (1.6.29) we see

$$\delta^n\left(\left(\sum r_{1i}ar_{2i}\right)^n\right) \in m\left(\sum r_{1i}(\delta a)r_{2i}\right)^n + \mathrm{Nil}(R) \qquad \text{for suitable } m > 0.$$

Now $\sum r_{1i}ar_{2i} \in \mathrm{Nil}(R)$ so $\left(\sum r_{1i}ar_{2i}\right)^n = 0$ for some n, implying $m\left(\sum r_{1i}(\delta a)r_{2i}\right)^n \in \mathrm{Nil}(R)$ and is thus nilpotent. Hence $\sum r_{1i}(\delta a)r_{2i}$ is nilpotent. Since the r_{1i}, r_{2i} are arbitrary we see $R(\delta a)R$ is nil, proving $\delta(\mathrm{Nil}(R)) \subseteq \mathrm{Nil}(R)$.

To show Prime rad(R) is invariant, we view Prime rad as the lower radical, built by transfinite induction (cf., theorem 2.6.19). The same argument as above shows that if $\alpha = \beta^+$ then $\delta(N_\beta(R)) \subseteq N_\alpha(R)$, yielding the desired result. Q.E.D.

Nil Subsets

Often the question arises as to when a nil subset of a ring is nilpotent. Our approach to this subject will be through the structure theory, starting with simple Artinian rings and a theorem of Jacobson.

Remark 2.6.29: Suppose $M \in R\text{-}\mathcal{M}od$ and S, N are respective subsets of R, M. If $S^iN = S^{i+1}N$ then $S^iN = S^jN$ for all $j > i$. (Indeed, by induction on j we have $S^iN = S^{j-1}N$; then $S^jN = SS^{j-1}N = SS^iN = S^{i+1}N = S^iN$.)

Call a subset S of R *weakly closed* if for each s_1, s_2 in S we have some v in $\mathbb{Z}$ (depending on s_1 and s_2) such that $s_1s_2 + vs_2s_1 \in S$. This encompasses the cases $v = 0$ (in which case S is a multiplicative semigroup), $v = -1$ (Lie multiplication) and $v = +1$ (Jordan multiplication). We shall call $s_1 s_2 + v s_2 s_1$ the *weak product* of s_1 and s_2 in S.

Proposition 2.6.30: (*Jacobson*) *Suppose S is a nil weakly closed subset of $R = M_n(D)$ where D is a division ring. Then $S^n = 0$.*

Proof: Take a minimal left ideal L of R. Then $[L:D] = n$ viewing L in $\mathcal{M}od$-D. We claim $S_0^n = 0$ for any nilpotent subset S_0 of R. Indeed, otherwise $S_0^n L \neq 0$ but some $S_0^j L = 0$ by assumption, so remark 2.6.29 implies $S_0^{i+1} L \subset S_0^i L$ for each $i \leq n$. Thus $[L:D] > [S_0 L:D] > [S_0^2 L:D] > \cdots$ so $[S_0^n L:D] = 0$, yielding $S_0^n L = 0$ and proving the claim.

It follows {nilpotent subsets of S} has a maximal member S_0 which is nonempty since any singleton is a nilpotent subset. Also $S_0^n = 0$. Let $V = S_0 L \subset L$ and $m = [V:D]$, so $0 < m < n$. Put $S_1 = \{s \in S : sV \subseteq V\}$. We work inductively on n. Since S_1 acts as a nil weakly closed set of transformations in $\operatorname{End} V_D \approx M_m(D)$ we have by induction $S_1^m V = 0$. On the other hand, S_1 acts on the quotient space L/V (over D) by the action $s(a + V) = sa + V$, and $[L/V:D] = n - m < n$, so $S_1^{n-m} L \subseteq V$ (again by induction), and hence $S_1^n L = S_1^m S_1^{n-m} L \subseteq S_1^m V = 0$. Since $S_0 \subseteq S_1$, we have $S_0 = S_1$ by the maximality of S_0, i.e., $s S_0 L \not\subseteq S_0 L$ for all s in $S - S_0$. We shall obtain a contradiction by assuming $S \neq S_0$.

Let $s_1 \circ s_2$ denote the "weak product" $s_1 s_2 + v s_2 s_1$ in S described above. If there exists s in $S - S_0$ such that $s \circ S_0 \subseteq S_0$ then for every element s_0 in S_0 and a in L we have $s(s_0 a) = (s \circ s_0)a - v s_0(sa) \in S_0 L$ for suitable v in $\mathbb{Z}$, proving $s \in S_1$, a contradiction. Thus we may assume $s \circ S_0 \not\subseteq S_0$ for every s in $S - S_0$. In particular, take s_1 in $S - S_0$ and s_{01} in S_0 such that $s_2 = s_1 \circ s_{01} \notin S_0$; inductively, given s_i in $S - S_0$, take s_{0i} in S_0 such that $s_{i+1} = s_i \circ s_{0i} \notin S_0$. Then s_{2n+1} is a sum of ordinary products of length $2n + 1$ of permutations of $s_1, s_{01}, s_{02}, \ldots, s_{0,2n}$; i.e., each term contains a consecutive string of elements of S_0 of length n and is thus 0. Hence $s_{2n+1} = 0$, contrary to $s_{2n+1} \notin S_0$. **Q.E.D.**

Note that some restriction was necessary on the nil set S since $\{e_{12}, e_{21}\}$ is a nil subset of $M_2(D)$ with $e_{12}e_{21} = e_{11}$ idempotent. On the other hand, if S is indeed weakly closed then (in the above notation) $L \supset SL \supset S^2 L \supset \cdots \supset S^t L = 0$ for some $t \leq n$; we can take a base of $S^{t-1}L$ (over D) and extend it to a base of $S^{t-2}L$ and then to $S^{t-3}L$, etc. until we reach a base of L with respect to which S is in strictly lower triangular form. Thus we have a generalization of Engel's theorem from Lie algebra.

We are ready to apply structure theory.

Theorem 2.6.31: (*Jacobson*) *Suppose R has a nilpotent ideal N such that R/N is semisimple Artinian.* (*In particular, this holds if R is left Artinian.*) *Then there exists k such that* $S^k = 0$ *for every nil, weakly closed subset S of R.*

Proof: Write $N = \text{Jac}(R)$. Then $N^t = 0$ for some t and $R/N \approx \prod_{i=1}^{k} M_{n_i}(D_i)$ for suitable n_i. Let $n = \max\{n_1, \ldots, n_k\}$. Writing S_i for the image of S in $M_{n_i}(D_i)$ we have $S_i^n = 0$ for each i, by proposition 2.6.30, implying $S^n \subseteq N$ and $S^{nt} = 0$.   Q.E.D.

Theorem 2.6.31 hints that we may be interested in finding a criterion for a ring R to be a subring of a suitable left Artinian ring T; then any nil, weakly closed subset S of R would *a fortiori* be a subset of T and thus nilpotent. It is also useful to know when multiplicative subsets of a ring are nil. The following result will be generalized later for Pl-rings.

Proposition 2.6.32: *Suppose R is a finite dimensional algebra over a field F. If S is a multiplicatively closed subset each of whose elements is a sum of nilpotent elements then S is nilpotent.*

Proof: By induction on $n = [R:F]$. Let $R_0 = F + FS$, the F-subalgebra generated by S. If $R_0 \subset R$ we are done by induction, so we may assume $R_0 = R$. In particular, FS is an ideal of R, which is maximal by a dimension count. If $\text{Jac}(R) \neq 0$ then $FS/\text{Jac}(R)$ is nilpotent in $R/\text{Jac}(R)$ by induction, so FS is nilpotent and again we are done. Thus we may assume $\text{Jac}(R) = 0$, so R is semisimple Artinian; checking the simple components, we may assume R is simple. Letting K be a splitting field of R, we can replace R by $R \otimes_{Z(R)} K$ and thus assume $R \approx M_n(K)$. But every nilpotent element has trace 0, so every element of FS has trace 0, implying $FS \lhd R$. Since R is simple, we have $FS = 0$, so $S = 0$.   Q.E.D.

Remark 2.6.32': $S^n = 0$ by a dimension count, where $n = [R:F]$.

G Supplement: Koethe's Conjecture

Having seen that the unique largest quasi-invertible ideal contains every quasi-invertible left ideal, we are led to the following analog for nil ideals:

Conjecture 2.6.33: (*Koethe's Conjecture*) *The upper nilradical contains every nil left ideal.*

Although this conjecture has been verified for many different classes of rings, it is believed *not* to hold by most ring theorists, perhaps because it has withstood the efforts of several brilliant mathematicians. There are many equivalent formulations, which we shall discuss now. First note that Koethe's conjecture holds iff LR is nil for every nil left ideal L of R.

Remark 2.6.34: If L is a nil left ideal of R and $r \in R$ then Lr is also a nil left ideal (for if $a \in L$ then $ra \in L$ so $(ra)^m = 0$ for some m, implying $(ar)^{m+1} = a(ra)^m r = 0$).

Theorem 2.6.35: *The following assertions are equivalent:*

  (i) *Koethe's conjecture holds.*
  (ii) *The sum of two nil left ideals is necessarily nil.*
  (iii) $\text{Nil}(M_n(R)) = M_n(\text{Nil}(R))$ *for all rings R and for all n.*
  (iv) $\text{Jac}(R[\lambda]) = \text{Nil}(R)[\lambda]$ *for all rings R, where λ is a commuting indeterminate.*

Proof: Write $N = \text{Nil}(R)$. (i) $\Rightarrow$ (ii) If L_1, L_2 are nil left ideals then $L_1, L_2 \subseteq N$ so $L_1 + L_2 \subseteq N$ is nil.

(ii) $\Rightarrow$ (iii) $\text{Nil}(M_n(R)) = M_n(A)$ for some $A \lhd R$. But Ae_{11} is nil, so A is nil. Thus $\text{Nil}(M_n(R)) \subseteq M_n(N)$. It remains to show $M_n(N)$ is nil. We use (ii), which implies inductively that any finite sum of nil left ideals is nil. $M_n(N) = \sum_{j=1}^n (\sum_{i=1}^n Ne_{ij})$ and each $\sum_{i=1}^n Ne_{ij}$ is nil (since $(\sum_{i=1}^n a_{ij}e_{ij})^k = (\sum a_{ij}e_{ij})a_{jj}^{k-1}$).

(iv) $\Rightarrow$ (i) Suppose L is a nil left ideal of R. Let R_1 be the ring obtained by adjoining 1 formally to L (viewed as a ring without 1). A typical element of R_1 is then (m, a) where $m \in \mathbb{Z}$ and $a \in L$, which is nilpotent iff $m = 0$. Hence $L = \text{Nil}(R_1)$ so by (iv) we have $L[\lambda] = \text{Jac}(R_1[\lambda])$. By Lemma 2.5.3 we see $L[\lambda]$ remains quasi-invertible when viewed as a left ideal of $R[\lambda]$, implying $L[\lambda] \subseteq \text{Jac}(R[\lambda]) = (\text{Nil } R)[\lambda]$, proving $L \subseteq \text{Nil}(R)$.   Q.E.D.

It remains to prove (iii) $\Rightarrow$ (iv). This follows at once from the following result of Krempa-Amitsur.

Proposition 2.6.36: *Let R be an arbitrary ring and $N = \text{Nil}(R)$. If $M_n(N)$ is nil then every polynomial of $N[\lambda]$ having degree $\leq n$ is quasi-invertible.*

Proof: We want to prove every polynomial $p = 1 + \sum_{i=0}^n a_i \lambda^i$ is left invertible where $a_i \in N$. Since $a_0 \in \text{Jac}(R)$ we have $(1 + a_0)$ invertible; replacing p by $(1 + a_0)^{-1}p$ we may assume $a_0 = 0$. Let $q = 1 - \sum_{i=1}^\infty b_j \lambda^j$ be the inverse of p in the ring of formal power series $R[[\lambda]]$, cf. proposition 1.2.27. Then each $b_j = \sum_{i=1}^{n} b_{j-i} a_i$, where we formally put $b_0 = 1$ and $b_u = 0$ for $u < 0$. Let A be the $n \times n$ matrix $\sum_{i=1}^{n-1} e_{i+1,i} + \sum_{i=1}^{n} a_{n+1-i} e_{in}$. By induction we see $(0, \ldots, 0, 1)A^j = (b_{j-n+1}, \ldots, b_{j-1}, b_j)$ for each j in $\mathbb{N}$. But the matrix A^n has entries only in N and so is nilpotent by hypothesis. Thus $A^t = 0$ for some t, and hence $b_j = 0$ for all $j \geq t$, thereby proving $q \in R[\lambda]$.

Thus we proved every polynomial of $N[\lambda]$ having degree $\leq n$ is left quasi-invertible; right quasi-invertibility is proved analogously.   Q.E.D.

§2.7 Semiprimary Rings and Their Generalizations

In this section we generalize left Artinian rings by isolating some of their properties, mostly in connection with chain conditions on certain classes of left or right ideals. As we shall see, these rings have certain easily discovered properties which cast strong light on the structure of Artinian rings themselves. Our starting point is the nilpotence of the Jacobson radical of a left Artinian ring (theorem 2.5.16).

Definition 2.7.1: R is a *semilocal ring* if $R/\mathrm{Jac}(R)$ is semisimple Artinian. A semilocal ring R is *semiprimary* if $\mathrm{Jac}(R)$ is nilpotent.

Hopkins and Levitzki proved independently that left Artinian rings are left Noetherian. Their result can now be seen to be an easy application of semiprimary rings:

Theorem 2.7.2: *A ring R is left Artinian iff R is left Noetherian and semiprimary.*

Proof: Every left Artinian ring is semiprimary, so it remains to prove the following claim:

Claim. A semiprimary ring R is left Noetherian iff R is left Artinian.

Proof of Claim. Put $J = \mathrm{Jac}(R)$ and take n such that $J^n = 0$.

We shall show if R is left Artinian or left Noetherian then R has a composition series in R-$\mathcal{M}od$ and thus is both left Artinian and left Noetherian. To build the composition series we start with $R > J > J^2 > \cdots > J^n = 0$; clearly, it suffices to show each factor $M_i = J^{i-1}/J^i$ has a composition series (where $R = J^0$). Since $JM_i = 0$ we can view M_i in $\bar{R}$-$\mathcal{M}od$ where $\bar{R} = R/J$, so M_i is a direct sum of simple $\bar{R}$-modules, which can also be viewed as simple R-modules. If R is left Noetherian (resp. Artinian) then so is J^{i-1}, viewed as R-submodule, and hence so is M_i. Thus M_i is a *finite* direct sum of simple submodules, and consequently has finite composition length, as desired.

Q.E.D.

The rest of this section deals with elementary properties of semiprimary rings, although results are obtained more generally for semiperfect rings. Certain key results, however, involve homological dimension, cf., exercises 5.2.40ff and Auslander [55]; the groundwork is laid in exercises 16ff. One useful fact is that left ideals are "eventually idempotent," cf., exercise 17.

Chain Conditions on Principal Left Ideals

Semiprimary rings can actually be characterized in terms of a chain condition on principal left ideals; to see this we first need some preliminaries about Artinian rings.

Remark 2.7.3: Suppose R is semisimple Artinian and $f: aR \to bR$ is an isomorphism of right ideals of R (viewed as right modules) with $fa = b$. Then $a = ub$ for some invertible element u of R. (Indeed, aR and bR have respective complements N_1, N_2 in R as right R-module. By the Jordan-Holder theorem there is an isomorphism $g: N_1 \to N_2$, so $h = f \oplus g: R \to R$ is an isomorphism. Let $u = h^{-1}1$. Then $(h1)u = hu = 1$ and $u(h1) = h^{-1}h1 = 1$, proving u is invertible; also $a = h^{-1}b = (h^{-1}1)b = ub$.)

Remark 2.7.4: Suppose R is semisimple Artinian and $Ra = Rb$. Then $a = ub$ for some invertible element u of R. (First assume a is an idempotent e. Then $b \in Re$ so $be = b$. Left multiplication by b thus gives an onto map $f: eR \to bR$. Moreover, $\ker f = 0$ since if $0 = f(er) = ber = br$ then $0 = Rbr = Rer$. Remark 2.7.3 then yields $e = ub$ as desired.

In general, Ra is a summand of R so there is an idempotent e with $Re = Ra = Rb$. Thus $e = u_1 a = u_2 b$ by the special case, so $a = u_1^{-1} u_2 b$.)

When studying chains of principal left ideals one should realize that if $Ra > Rb$ are principal left ideals then $b = a'a$ for suitable a' in R, so $Rb = Ra'a$. Thus any chain of n principal left ideals can be put in the form

$$Ra_1 > Ra_2 a_1 > \cdots > Ra_n \cdots a_1.$$

In particular, if $\overline{Ra_1} > \cdots > \overline{Ra_n \cdots a_1}$ is a chain in a homomorphic image $\bar{R}$ of R then $Ra_1 > \cdots > Ra_n \cdots a_1$, leading to the following observation:

Remark 2.7.5: Any chain condition on principal left ideals of R is transferred to all homomorphic images of R. (By "chain condition" we mean some restriction on the chains, such as ACC, DCC, or bounded length.)

The other key observation we need involves the Jacobson radical.

Remark 2.7.6: If $a \in \mathrm{Jac}(R)$ and $Rr = Rar$ then $r = 0$. (For $r = xar$ for some x in R, so $(1 - xa)r = 0$, implying $r = 0$ since xa is quasi-invertible.)

Theorem 2.7.7: *R is semiprimary iff every chain of principal left ideals of R has length $\leq m$ for some m. More precisely, write $J = \mathrm{Jac}(R)$ and $\bar{R} = R/J$.*

(i) *If every chain of principal left ideals of R has length $\leq m$ then $J^m = 0$ and $\bar{R}$ has length $\leq m$ as $\bar{R}$-module (and is thus semisimple Artinian).*
(ii) *If $J^n = 0$ and $\bar{R}$ has composition length k then we can take $m \leq k^{n+1}$.*

Proof:

(i) For any $a_1, \dots, a_m$ in J we have $R \geq Ra_1 \geq Ra_2a_1 \geq \cdots \geq Ra_m \cdots a_1 \geq 0$, so equality must hold at some stage, i.e., $Ra_{i+1}a_i \cdots a_1 = Ra_i \cdots a_1$ for some i, so $a_i \cdots a_1 = 0$ by remark 2.7.6.

It remains to show $\bar{R}$ has length $\leq m$. In view of remark 2.7.5 we may assume $J = 0$, i.e., $\bar{R} = R$. Note that R has at most m orthogonal idempotents $e_1, e_2, \dots$ since otherwise $R > R(1 - e_1) > R(1 - e_1 - e_2) > \cdots$ is a chain of length $> m$. Thus $\mathrm{soc}(R)$ is a sum of at most m minimal left ideals. Moreover, every left ideal of R contains a minimal left ideal, since every minimal principal left ideal is a minimal left ideal (for if $L < Ra$ then taking $b \in L$ one has $Rb < Ra$). Thus R is semisimple Artinian by corollary 2.3.11.

(ii) This is the harder direction, and we prove it in a sequence of steps, the first two of which are applicable to arbitrary semilocal rings.

Step 1. If $L_1 \geq L_2$ are principal left ideals with $\bar{L}_1 = \bar{L}_2$ then there are elements r_1, r_2 of R with $\bar{r}_1$ idempotent such that $\overline{r_2 r_1} = \bar{r}_1$, $L_1 = Rr_1$, and $L_2 = Rr_2r_1$.

Proof of Step 1. Write $L_1 = Ra$. Then $\overline{Ra} = \bar{R}\bar{e}$ for some idempotent $\bar{e}$ of $\bar{R}$; so $\bar{e} = \overline{u}\overline{a}$ for some invertible $\bar{u}$ by remark 2.7.4. Hence u is invertible in R lemma 2.5.5). Let $r_1 = ua$. Then $L_1 = Rr_1$ and $\bar{r}_1 = \bar{e}$ is idempotent. Write $L_2 = Rb$ and $b = rr_1$ for suitable r in R. Then $\bar{R}\bar{r}_1 = \bar{L}_2 = \bar{R}\bar{b}$ so $\bar{r}_1 = \bar{v}\bar{b}$ for some invertible v in R; taking $r_2 = vr$ yields $Rr_2r_1 = Rvb = Rb = L_2$, and $\overline{r_2 r_1} = \bar{v}\bar{b} = \bar{r}_1$.

Step 2. We call a chain $Rr_1 > Rr_2r_1 > \cdots > Rr_j \cdots r_1$ *admissible if* $\overline{r_i r_{i-1}} = \overline{r_{i-1}}$ for each $i \geq 1$. In this case, putting $L_i = Rr_i \cdots r_1$, there is a sequence $a_1, \ldots, a_j$ in J such that $L_i \subseteq Ra_i \cdots a_1 + L_j$ for each $i \leq j$.

Proof of Step 2. Let $a_i = r_i - r_j \cdots r_i$ for $i \leq j$. Then $\overline{a_i} = \overline{r_i} - \overline{r_j \cdots r_i} = \overline{r_i} - \overline{r_i} = 0$ so $a_i \in J$. Moreover, in evaluating $a_i \cdots a_1$ we have a sum of terms each ending in $r_j \cdots r_1$ except $r_i r_{i-1} \cdots r_1$. Thus $r_i \cdots r_1 \in a_i \cdots a_1 + Rr_j \cdots r_1$, so $L_i \subseteq Ra_i \cdots a_1 + L_j$.

Step 3. Every chain $L_1 > \cdots > L_t$ for $t \geq k^n$ has an admissible subchain of length $n - 1$.

Proof of Step 3. Clearly $\overline{L_2} \neq \overline{R}$ since if $L_2 = Ru$ with $\overline{u}$ invertible then u is invertible. Hence there are $k^n - 1$ left ideals $\overline{L_2}, \overline{L_3}, \ldots$ each having length $0, 1, \ldots,$ or $k-1$, implying there is a subchain $L_1' > \cdots > L_s'$, where $\overline{L_1'} = \cdots = \overline{L_s'}$ and $s = k^{n-1}$. Write $L_i' = Rr_i \cdots r_1$ and $b_i = r_i \cdots r_2$. Then letting $L_i'' = Rb_i$ we have $L_2'' > \cdots > L_s''$. Continue inductively using step 1.

Conclusion of Proof. Step 2 shows we cannot have an admissible chain of length n, so step 3 shows there is no chain of length k^{n+1}.   Q.E.D.

Note: With care one could lower the bound for m a bit; I think the sharpest bound is unknown.

Corollary 2.7.8:   *If $J^n = 0$ and R/J has (composition) length k then every chain of cyclic R-modules has length $\leq k^{n+1}$.*

Proof: Suppose $M_1 > M_2 > \cdots$ is a chain of   cyclic modules; putting $M = Rx_i$ and letting $x_{i+1} = r_{i+1} x_i$ we have $R > Rr_2 > Rr_3 r_2 > \cdots$.
                                              Q.E.D.

One can translate results on chains of cyclic modules to chains of f.g. modules by means of remark 1.1.19 and the following observation:

Remark 2.7.9:   If R is semiprimary then $M_t(R)$ is semiprimary for each t. More explicitly, if $J^n = 0$ and $\text{length}(R/J) = k$ then $M_t(J)^n = 0$ and $M_t(R)/M_t(J) \approx M_t(R/J)$; moreover, $\text{length}(M_t(R/J)) = tk$.

Corollary 2.7.10:   *If $J^n = 0$ and R/J has length k then every chain of R-modules, each spanned by $\leq t$ elements, has length $\leq (tk)^{n+1}$.*

Proof: If $M_1 > M_2 > \cdots$ is a chain of modules spanned by $\leq t$ elements then $M_1^{(t)} > M_2^{(t)} > \cdots$ is a chain of cyclic $M_t(R)$-modules, so apply corollary 2.7.8 to $M_t(R)$.   Q.E.D.

Example 2.7.11: If M is a $D-E$ bimodule where D and E are division rings then $\begin{pmatrix} D & M \\ 0 & E \end{pmatrix}$ is semiprimary. For example $\begin{pmatrix} \mathbb{Q} & \mathbb{R} \\ 0 & \mathbb{Q} \end{pmatrix}$ is semiprimary, but neither left nor right Artinian.

Passing from R to eRe and Back

It is easy to transfer many properties from R to eRe (and likewise to $(1-e)R(1-e)$), as illustrated in the next observation.

Lemma 2.7.12: *The following properties pass from R to eRe, where $e \in R$ is idempotent: (i) left Artinian; (ii) left Noetherian; (iii) primitive; (iv) simple Artinian; (v) semisimple Artinian; (vi) semilocal.*

Proof:

(i) Suppose $L_1 > L_2 > \cdots$ are left ideals in eRe. Then $RL_1 > RL_2 > \cdots$ are left ideals in R. (Note $RL_i \neq RL_{i+1}$ since $eRL_i = (eRe)L_i = L_i \neq L_{i+1} = eRL_{i+1}$.)

(ii) Dual to (i).

(iii) If $M \in R\text{-}\mathcal{Mod}$ is faithful simple then eM clearly is faithful in $(eRe)\text{-}\mathcal{Mod}$ and is simple by proposition 1.1.15.

(iv) By (i) and (iii).

(v) If $R \approx \prod_{i=1}^{t} R_i$ where each R_i is simple Artinian then $eRe \approx \prod_{i=1}^{t} eR_i e$; so we are done by (iv).

(vi) Write $\bar{R}$ for $R/\text{Jac}(R)$ which is semisimple Artinian. By proposition 2.5.14, $eRe/\text{Jac}(eRe) \approx \overline{eRe}$ is semisimple Artinian by (v).   Q.E.D.

We are also interested in passing from eRe and $(1-e)R(1-e)$ to R.

Lemma 2.7.13: *Suppose e is an idempotent of R and $N \lhd R$. N is nilpotent iff eNe and $(1-e)N(1-e)$ are nilpotent.*

Proof: ($\Rightarrow$) is clear. Conversely assume $(eNe)^m = 0$ and $((1-e)N(1-e))^n = 0$; we shall show $N^{m+n} = 0$. Indeed, suppose $a_i \in N$ for $1 \leq i \leq m+n$. Then putting $t = m + n$

$$a_1 \cdots a_t = (e + (1 - e))a_1(e + (1 - e))a_2 \cdots (e + (1 - e))a_t(e + (1 - e)).$$

So it suffices to prove $e_0 a_1 e_1 a_2 e_2 \cdots a_t e_t = 0$ for all $e_i \in \{e, 1 - e\}$.

Either $e_i = e$ for at least $m + 1$ values of i, or $e_i = 1 - e$ for at least $n + 1$ values of i. Assuming the former, we can rewrite $e_0 a_1 e_1 \cdots a_t e_t$ as $e_0 a'_0 e a'_1 e \cdots a'_m e a'_{m+1}$, where for $1 \le u \le m$ each a'_u is a nonempty string of the form $a_{i_u + 1} e_{i_u + 1} \cdots a_{i_{u+1}} \in N$; hence $e_0 a_1 e_1 \cdots a_t e_t \in e_0 a'_0 (eNe)^m a'_{m+1} = 0$, as desired.   Q.E.D.

Proposition 2.7.14:   *The following properties hold in R iff they hold both in eRe and $(1 - e)R(1 - e)$, where e is idempotent:*

   (i) *semisimple Artinian in case R is semiprimitive (cf., example 2.7.11)*
   (ii) *semilocal*
   (iii) *semiprimary*

Proof:

   (i) ($\Rightarrow$) is lemma 2.7.12(v). To prove ($\Leftarrow$) note R has a complete set of orthogonal primitive idempotents (putting together those from eRe and $(1 - e)R(1 - e)$), so R is semisimple Artinian by corollary 2.3.12.
   (ii) Pass to $R/\mathrm{Jac}(R)$, recalling $\mathrm{Jac}(eRe) = e\mathrm{Jac}(R)e$.
   (iii) By (ii) and lemma 2.7.13.   Q.E.D.

An immediate consequence of this result is the following basic fact about endomorphism rings.

Corollary 2.7.15:   *If M is an f.g. module over a semisimple Artinian ring R then $E = \mathrm{End}_R M$ is a semisimple Artinian.*

Proof: Write M as a direct sum $\bigoplus_{i=1}^{t} M_i$ of simple R-modules. Take $f \in E$. For some i we see $fM_i \ne 0$ and thus is a simple submodule of M isomorphic to M_i (as in Schur's lemma); taking a complement M' of fM we define g to be 0 on M' and to be f^{-1} on fM. Now $(1 - gf)M_i = 0$, proving $f \notin \mathrm{Jac}(E)$; since this is true for any f we conclude $\mathrm{Jac}(E) = 0$. Letting e_i be the projection of M onto M_i we know each $e_i E e_i$ is a division ring, so E is semisimple Artinian.   Q.E.D.

This fact and many of its generalizations (to be discussed later) can also be obtained as immediate consequences of the Morita theory of Chapter 4.

Semiperfect Rings

Much of the theory of Artinian rings comes from lifting idempotents and the information which they contain. The properties which we need are pinpointed in the next definition.

Definition 2.7.16:   A semilocal ring R is *semiperfect* if $\mathrm{Jac}(R)$ is idempotent-lifting.

The usages "semiperfect" and "semiprimary" unfortunately do not fit our general meaning of "semi" given in §2.2, and it may be worth giving the mathematical etymology of these words. In definition 2.8.31 below we define a projective cover for a module. Bass [60] defined a ring R to be *perfect* if every module has a projective cover; R is *semiperfect* if every f.g. module has a projective cover. Our definitions of perfect (definition 2.7.31 below) and semiperfect are equivalent, respectively, to these definitions, although this shall only be seen in exercises 2.8.28 and 2.8.22.

The terminology concerning "primary" and "semiprimary" is hazier. Jacobson [64B] used "semiprimary" where we use "semilocal," and he used "primary" for a ring R such that $R/\mathrm{Jac}(R)$ is simple Artinian. As time passed, "semiprimary" took on the added connotation that $\mathrm{Jac}(R)$ is nilpotent. According to the usage of this book, we shall have the following kinds of rings in increasing generality:

left Artinian, semiprimary, perfect, semiperfect, semilocal

As we shall see later in §2.7 and in §2.8, perfect rings are nice because of their various characterizations, which enable one to study them using a wide variety of techniques from ring and module theory. On the other hand, most examples of perfect rings which one encounters are already Artinian. The class of *semiperfect* rings contains many non-Artinian examples, including all local domains (since there are no nontrivial idempotents to lift).

Nevertheless, many important properties of Artinian rings are obtained in this setting, as we shall see now (through theorem 2.7.30) and later in theorem 2.9.18 (which relies on an added ingredient, the Krull-Schmidt theorem).

Lemma 2.7.17:   *Suppose $A \lhd R$ and $A \subseteq \mathrm{Jac}(R)$, and $e \in R$ is idempotent. Write $\bar{R} = R/A$.*

(i) *If $L \leq Re$ and $\bar{L} + \bar{R}(\overline{1-e}) = \bar{R}$ then $L = Re$.*

(ii) *If e is a primitive idempotent and A is idempotent-lifting then $\bar{e}$ is a primitive idempotent of $\bar{R}$.*

(iii) *If R is semiperfect and e is a primitive idempotent and $L \le Re$ with $L \not\subseteq \mathrm{Jac}(R)$ then $L = Re$.*

Proof: (i) Let $L_1 = L + R(1 - e)$. If $\bar{L}_1 = \bar{R}$ then $1 \in \bar{L}_1$ so L_1 has an invertible element by lemma 2.5.5 and thus $1 \in L_1$; hence $e \in L_1 e = L$, implying $L = Re$.

(ii) Otherwise, $\bar{e} = \bar{e}_1 + \bar{e}_2$ for orthogonal idempotents $\bar{e}_1, \bar{e}_2 \ne 0$. Lift $\bar{e}_1$ to an idempotent e_1 orthogonal to $(1 - e)$. Then $e_1 e = e = ee_1$ but $e \ne e_1$ since $\bar{e} \ne \bar{e}_1$. Hence $e = e_1 + (e - e_1)$ is a sum of orthogonal idempotents, contrary to hypothesis.

(iii) Put $A = \mathrm{Jac}(R)$, which by hypothesis, is idempotent-lifting. Then $0 = \bar{L} \le \bar{R}e$ and $\bar{e}$ is primitive by (ii), so $\bar{L} = \bar{R}e$ and $\bar{L} + \bar{R(1 - e)} = \bar{R}$; we conclude by applying (i).   Q.E.D.

Lemma 2.7.18: *Suppose e is a primitive idempotent of R and R is semiperfect; then eRe is a local ring.*

Proof: $\bar{R} = R/\mathrm{Jac}(R)$ is semisimple Artinian and $\bar{e}$ is a primitive idempotent by lemma 2.7.17(ii), so $\bar{R}e$ is a minimal left ideal, implying $\overline{eRe}$ is a division ring by proposition 2.1.21.   Q.E.D.

Semiperfect rings have two key ingredients—one is lemma 2.7.18, and the other is the existence of a *complete set* of primitive idempotents, by which we mean orthogonal primitive idempotents $e_1, \ldots, e_t$ with $\sum_{i=1}^{t} e_i = 1$.

Remark 2.7.19: If $A \lhd R$ is idempotent-lifting then any complete set $\{\bar{e}_1, \ldots, \bar{e}_t\}$ (in R/A) of primitive idempotents can be lifted to orthogonal idempotents $e_1, \ldots, e_t$ of R by proposition 1.1.25; obviously, each e_i is a primitive idempotent, and $\{e_1, \ldots, e_t\}$ is a complete set since $1 - \sum_{i=1}^{t} e_{ii}$ is an idempotent of A which is thus 0. In particular, every semiperfect ring has a complete set of primitive idempotents.

Example 2.7.19':

(i) A commutative semiperfect ring C is merely a direct product of local rings. Indeed a complete set of primitive idempotents of C/J lifts up to C; but these are central since C is commutative, so provide the desired ring decomposition.

(ii) If R is f.g. (as module) over a local Noetherian ring C then R is semilocal and $(\mathrm{Jac}\, R)^k \subseteq (\mathrm{Jac}\, C)R$ for suitable k. Indeed let $J = \mathrm{Jac}(C)$. Then for every simple R-module M one has $JM \ne M$ by Nakayama so $JM = 0$ implying J annihilates every simple R-module; thus $J \subseteq \mathrm{Jac}(R)$. But R/JR is f.g. as module over the Artinian ring C/J, so R/JR and thus $R/\text{Jac}(R)$ are Artinian. Furthermore, $\text{Jac}(R)/JR = \text{Jac}(R/JR)$ is nilpotent so $(\text{Jac } R)^k \subseteq JR$ for some k.

There is a rich theory in the literature in case C is complete; in this case R is semiperfect, and one should expect much of this theory to hold for arbitrary complete semiperfect rings (with respect to the $\text{Jac}(R)$-adic topology). Thus there are significant examples of semiperfect Noetherian rings which are not Artinian.

Recapitulating, we have the following result:

Proposition 2.7.20: *A semiperfect ring R has a complete set of primitive idempotents $e_1, \ldots, e_t$ such that each $e_i R e_i$ is a local ring.*

This condition characterizes semiperfect rings, as we shall see in theorem 2.8.40. It is extremely useful when applied in conjunction with the Krull-Schmidt theorem, cf., theorem 2.9.18 below.

Proposition 2.7.21: *If R is semiperfect and $R/\text{Jac}(R) \approx M_n(D)$ then $R \approx M_n(T)$ for some local ring T such that $T/\text{Jac}(T) \approx D$. (Rings having this property are sometimes called quasi-local.)*

Proof: In view of proposition 1.1.25 we can lift a set of matric units of $R/\text{Jac}(R)$ to a set of matric units of R, so $R \approx M_n(T)$ for some T. But $T \approx e_{11} R e_{11}$ is local by lemma 2.7.18, whose proof shows $T/\text{Jac}(T) \approx D$.
$$\text{Q.E.D.}$$

Example 2.7.22: A commutative semilocal ring which is not semiperfect. Let n be a product of distinct prime numbers $p_1 \cdots p_t$ and let $\varphi: \mathbb{Z} \to \bar{\mathbb{Z}} = \mathbb{Z}/n\mathbb{Z}$ be the canonical homomorphism. Let $S = \{m \in \mathbb{Z} : m \text{ is relatively prime to } n\}$, so that the elements of $\bar{S}$ are invertible. $S^{-1}\mathbb{Z}$ is semilocal. Indeed φ extends to a surjection $\varphi_S: S^{-1}\mathbb{Z} \to \bar{\mathbb{Z}}$ given by $\varphi_S(s^{-1}r) = \bar{s}^{-1}\bar{r}$, and $\ker \varphi_S = nS^{-1}\mathbb{Z}$ is quasi-invertible, so $\ker \varphi_S \subseteq \text{Jac}(S^{-1}\mathbb{Z})$; equality holds since the Chinese Remainder Theorem shows $\bar{\mathbb{Z}} \approx \prod_{i=1}^t (\mathbb{Z}/p_i\mathbb{Z})$ which is semisimple Artinian. Then $\bar{\mathbb{Z}}$ has nontrivial idempotents for $t > 1$, but $S^{-1}\mathbb{Z} \subset \mathbb{Q}$ has no nontrivial idempotents and thus is not semiperfect.

Structure of Idempotents of Rings

We shall now study the structure of the left module Re when e is idempotent and R is semilocal. This yields useful information for semiperfect R which is important even when R is Artinian.

Remark 2.7.23: Suppose $e \in R$ is idempotent. Any R-module map $f: Re \to M$ is given by right multiplication by the element $fe = efe \in eM$. In particular, $f(Re) \subseteq ReM$. (Indeed $f(re) = f(re^2) = (re)fe$ so $fe = efe$.)

Proposition 2.7.24: *If e, e' are idempotents of R then $\mathrm{Hom}(Re, Re') \approx eRe'$ as additive groups.*

Proof: Define $\psi: eRe' \to \mathrm{Hom}(Re, Re')$ by letting $\psi(ere')$ be right multiplication by ere'. Then ψ is a group homomorphism which is onto by remark 2.7.23; if $ere' \in \ker \psi$ then $0 = e(ere') = ere'$ proving ψ is 1:1. Q.E.D.

Proposition 2.7.25: *Suppose e_1, e_2 are idempotents of R. Then $Re_1 \approx Re_2$ iff there are elements e_{12} in $e_1 Re_2$ and e_{21} in $e_2 Re_1$ such that $e_{12}e_{21} = e_1$ and $e_{21}e_{12} = e_2$.*

Proof: ($\Rightarrow$) Let $f: Re_1 \to Re_2$ be the given isomorphism and $g = f^{-1}$. Taking $e_{12} = e_1 f e_1$ and $e_{21} = e_2 g e_2$ as in remark 2.7.23, we have $e_1 = g f e_1 = e_1 e_{12} e_{21} = e_{12}e_{21}$ and $e_2 = f g e_2 = e_2 e_{21} e_{12} = e_{21}e_{12}$.

($\Leftarrow$) Define $f: Re_1 \to Re_2$ by $f(re_1) = re_{12}$ and $g: Re_2 \to Re_1$ by $g(re_2) = re_{21}$; then $gf(re_1) = re_{12}e_{21} = re_1$ and $fg(re_2) = re_{21}e_{12} = re_2$ so f and g are inverses.   Q.E.D.

This result has several nice consequences.

Corollary 2.7.26: *Suppose e_1, e_2 are idempotents of R. $Re_1 \approx Re_2$ iff $e_1 R \approx e_2 R$.*

Proof: The condition of proposition 2.7.25 is left-right symmetric. Q.E.D.

Our other application involves the Jacobson radical.

Lemma 2.7.27: *Suppose e, e' are idempotents of R, and $J = \mathrm{Jac}(R)$. If $f: Re \to Re'$ is a map then letting $^-$ denote the canonical image in $\bar{R} = R/J$ we have a map $\bar{f}: \overline{Re} \to \overline{Re'}$ given by $\bar{f}(\overline{re}) = \overline{f(re)}$; if f is an isomorphism then $\bar{f}$ is an isomorphism in $\bar{R}$-$\mathcal{M}od$.*

Proof: First note that if $\overline{re} = 0$ then $re \in J$ and $f(re) = (re)fe \in J$; hence $\bar{f}$ is a well-defined map. If f^{-1} exists then $\overline{f^{-1}} = \bar{f}^{-1}$ implying $\bar{f}$ is an isomorphism.   Q.E.D.

Proposition 2.7.28:   *Notation as in lemma 2.7.27. $Re \approx Re'$ in R-$\mathcal{M}od$ iff $\overline{Re} \approx \overline{Re'}$ in $\overline{R}$-$\mathcal{M}od$.*

Proof: ($\Rightarrow$) is the lemma. ($\Leftarrow$) By proposition 2.7.25 there are elements a in eRe' and b in $e'Re$ with $\overline{ab} = \overline{e}$ and $\overline{ba} = \overline{e'}$. Lemma 2.5.5 implies ab is invertible in eRe and ba is invertible in $e'Re'$. Hence the map $f_a: Re \rightarrow Re'$ given by $f_a x = xa$ is 1:1 and onto, and thus is an isomorphism.   Q.E.D.

To appreciate these results, suppose R is semilocal and has a complete set of primitive idempotents $e_1, \ldots, e_t$. (For example, this holds if R is semi-perfect). Then by the Peirce decomposition $R = \bigoplus e_i Re_j$, and each $e_i Re_i$ is local. Thus much of the structure of R hinges on the $e_i Re_j$; in particular, R is a direct product of local rings if the $e_i Re_j = 0$ for all $i \neq j$.

Definition 2.7.29:   A set of orthogonal primitive idempotents $\{e_1, \ldots, e_k\}$ is called *basic* if for every primitive idempotent e' we have $Re' \approx Re_i$ for exactly one e_i, $1 \leq i \leq k$. In this case $e = \sum_{i=1}^{k} e_i$ is called a *basic idempotent* and eRe is called a *basic subring of R. R* is a *basic ring* if 1 is a basic idempotent.

Theorem 2.7.30:   *Every semiperfect ring R has a basic idempotent e, and $eRe/\text{Jac}(eRe)$ is a direct product of division rings.*

Proof: Take a complete set of primitive idempotents $\{e_1, \ldots, e_t\}$; reordering it we get some $k \leq t$ such that for any $j > k$ we have $Re_j \approx Re_i$ for suitable $1 \leq i \leq k$. Then $\{e_1, \ldots, e_k\}$ is a basic set. Indeed, for any primitive idempotent e', we have $\overline{Re'}$ isomorphic to some $\overline{Re_j}$ in $\overline{R}$ for suitable $j \leq t$ (since $\overline{R}$ is semi-simple Artinian), so $Re' \approx Re_j$ by proposition 2.7.28, and $Re_j \approx Re_i$ for some $i \leq k$. The remainder of the proposition is clear.   Q.E.D.

Besides being much easier to handle, basic subrings are useful because R-$\mathcal{M}od$ is equivalent (as a category) to T-$\mathcal{M}od$ if T is a basic subring of R. This important result can be proved in a few lines, but we would like to save the proof to introduce Morita's theorems, cf., example 4.1.11.

H Supplement: Perfect Rings

Definition 2.7.31:   *R* is *right perfect* if it satisfies the DCC on principal left ideals.

This definition is a very natural generalization of semiprimary rings, in view of theorem 2.7.7, but the use of "right" may seem strange (and for this reason Bjork calls these "left coperfect" in his papers). We can understand the terminology by turning to the history of perfect rings. They were brought to light by Kaplansky [50], as part of a program to study rings satisfying abstract chain conditions, and various *ad hoc* results were obtained during the next ten years. The breakthrough came in Bass [60], who found many different characterizations, using varied tools from ring theory. Since then algebraists have tried to extend the theory of Artinian rings to this broader setting, but this project has slowed as it has become clear that many of the recent strides in the theory of Artinian rings (cf., §2.9) simply do not apply to non-Artinian perfect rings.

In the next few pages we shall give several elementary characterizations of perfect rings, largely parallel to the previous results on semiprimary rings, so the proofs will be abbreviated; in the next section we shall give the important link to projective covers. The key characterization here is in terms of the Jacobson radical, and needs another definition.

Definition 2.7.32: A subset $S \subset R$ is *right T-nilpotent* (also called *vanishing*) it for every sequence $s_1, s_2, \ldots$ in S there is suitable n with $s_n \cdots s_1 = 0$.

Theorem 2.7.33: R *is right perfect iff* R *is semilocal with* $\mathrm{Jac}(R)$ *right T-nilpotent.*

Proof: ($\Rightarrow$) Almost word-for-word the same as that of theorem 2.7.7(i) and is left for the reader.

($\Leftarrow$) Otherwise, there is an infinite chain of principal left ideals $L_1 > L_2 > \cdots$; starting far enough along we may assume $\overline{L_1} = \overline{L_2} = \cdots$. Pick an infinite chain with $\overline{L_1}$ having maximal possible length. We shall use Step 1 of the proof of theorem 2.7.7(ii). Writing $L_1 = Rr_1$ with $\overline{r_1}$ idempotent, we can write $L_i = L_i'' r_1$ with $L_2'' > L_3'' > \cdots$. Then $\overline{L_j''} = \overline{L_{j+1}''} = \cdots$ for some j, so starting at j we may assume $\overline{L_2''} = \overline{L_3''} = \cdots$. Now, by hypothesis, length $\overline{L_2''} =$ length $\overline{L_1} =$ length $\overline{L_2}$; right multiplication by $\overline{r_1}$ gives an epic $\overline{L_2''} \to \overline{L_2}$ which is thus an isomorphism. Writing $L_2'' = Rr_2$ we thus have $\overline{r_2 r_1} = \overline{r_2 u_2}$ for some invertible u_2, by remark 2.7.3. On the other hand, we could choose r_2 such that $\overline{r_2 r_1} = \overline{r_1}$ (by Step 1), so $\overline{r_1} = \overline{r_2 u_2}$. Iterating this procedure yields a new chain $L_1 > L_2 > \cdots$ such that $L_i = Rr_i \cdots r_1$ with $\overline{r_i \cdots r_1} = \overline{r_1} = \overline{r_i u_i}$ for suitable invertible u_i.

Write $\overline{r_1} = \overline{e}$ for e idempotent (recalling any nil ideal is idempotent lifting), and let $a_i = r_i - r_i \cdots r_1 e u_i^{-1}$. Then $\overline{a_i} = 0$ so $a_i \in \mathrm{Jac}(R)$ and hence there is n with

$$0 = a_n \cdots a_1 \in r_n \cdots r_1 + r_n \cdots r_1 eR.$$

Thus $r_n \cdots r_1 = r_n \cdots r_1 er'$ for some r' in R. Letting $s_i = e - r_i \cdots r_1$ for $i \geq n$ we get $r_i \cdots r_1 = (1 - s_i)er'$. But $\bar{s}_i = 0$ so $s_i \in \mathrm{Jac}(R)$ and $Rr_i \cdots r_1 = Rer'$ independently of i; consequently, $L_i = Rer' = L_{i+1}$ for all $i \geq n$.   Q.E.D.

Other characterizations can be obtained easily by bringing in idempotents.

Lemma 2.7.34: *Suppose e is an idempotent of R, and $N \lhd R$. N is right T-nilpotent iff eNe and $(1 - e)N(1 - e)$ are T-nilpotent.*

Proof: ($\Leftarrow$) Otherwise, we have an infinite sequence $a_1, a_2, \ldots$ with each $a_n \cdots a_1 \neq 0$. Thus we can choose e_i in $\{e, 1 - e\}$ for each i such that $a_n e_n \cdots a_1 e_1 \neq 0$ for each n. We may assume an infinite number of e_i are e, but then one easily builds a sequence from eNe which is not T-nilpotent (parallel to lemma 2.7.13), contrary to hypothesis.   Q.E.D.

Proposition 2.7.35: *Suppose e is an idempotent. R is right perfect iff eRe and $(1 - e)R(1 - e)$ are right perfect. In particular, if R is right perfect then $M_n(R)$ is right perfect.*

Proof: Combine proposition 2.7.14(ii) with lemma 2.7.34.   Q.E.D.

There is an *ascending* chain condition discovered by Jonah [70].

Theorem 2.7.36: *The following properties are satisfied by any right perfect ring:*

  (i) DCC *on cyclic modules;*
  (ii) DCC *on modules spanned by $\leq t$ elements;*
  (iii) ACC *on principal right ideals;*
  (iv) ACC *on cyclic submodules of any given right module;*
  (v) ACC *on submodules spanned by $\leq t$ elements of any given right module.*

Proof:

  (i) As in corollary 2.7.8.
  (ii) Apply remark 1.1.19 to proposition 2.7.35, to reduce to (i).
  (iii), (iv), (v). We prove (v), which implies the other two properties. Suppose $M \in \mathcal{M}od\text{-}R$ and has a chain $M_1 \leq M_2 \leq M_3 \leq \cdots$ of submodules of M, with each M_i spanned by $\leq t$ elements. Let $N = \bigcup_{i=1}^{\infty} M_i \leq M$, and let $J = \mathrm{Jac}(R)$. Then we write $^-$ for the canonical image in $\bar{N} = N/NJ$, which can be viewed as a module over the semisimple Artinian ring R/J.

If n is the (composition) length of R/J then each $\bar{M}_i$ has length $\leq tn$ as (R/J)-module, and thus the chain $\bar{M}_1 \leq \bar{M}_2 \leq \bar{M}_3 \leq \cdots$ must terminate.

But, clearly, $\bar{N} = \bigcup \bar{M}_i$ so some $\bar{M}_k = \bar{N}$, implying $M_k + NJ = N$. To complete the proof we need to conclude $M_k = N$, which we get by applying the following variant of Nakayama's lemma:

Lemma 2.7.37:   *If J is right T-nilpotent and $N \leq M$ in $\mathcal{M}od$-R with $N + MJ = M$ then $N = M$.*

Proof: Passing to M/N we may assume $N = 0$. Inductively, we can construct a sequence $a_1, a_2, \ldots$ of J with $Ma_n \cdots a_1 \neq 0$ for each n, for if $Ma_{n-1} \cdots a_1 \neq 0$ then $0 \neq MJa_{n-1} \cdots a_1$ since $MJ = M$. But this contradicts the presumed T-nilpotence of J.   Q.E.D.

The statements (i)–(v) are actually each equivalent to the ring being perfect, c.f., exercise 9. Moreover, Bjork has proved that theorem 2.7.36(ii) can be strengthened to the DCC on *all* f.g. modules, c.f., exercise 5; in fact, the DCC on cyclic modules formally implies the DCC on f.g. modules. A detailed account is given in Faith [76B]. Unfortunately, the notion of perfect is *not* left-right symmetric, as seen in the following example:

Example 2.7.38:   Let M be a right vector space over a field having a well-ordered base $x_1, x_2, x_3, \ldots$ and let R be the F-subalgebra of End M_R generated by $J = \{ f \in \text{soc}(\text{End } M_F): fx_i \in \sum_{j<i} x_j F \}$. (This is an infinite analogue of triangular matrices.) Clearly, $J \lhd R$ and $R/J \approx F$; moreover, J is T-nilpotent since given $a_1 \in J$ we have $a_1 M \subseteq \sum_{i=1}^{n} x_i F$ for suitable n, and then, clearly, $J^n a_1 = 0$. Hence R is right perfect. On the other hand, if we take $a_i = e_{i,i+1}$ then $a_1 \cdots a_n = e_{1,n+1} \neq 0$ for each n, so the left-handed version of T-nilpotence fails.

§2.8 Projective Modules (An Introduction)

In this section we introduce projective modules, a categorically defined notion which can often be used in place of free modules. Projective modules play an important role in much of the structure theory, and, in particular, they bear on the kinds of rings discussed in the last few sections (which is why they are introduced here).

Projective Modules

Definition 2.8.1:   An M-module P is *projective* if it satisfies the following *lifting property*: For any map $f: P \to N$ and any epic map $h: M \to N$ there is a map $\hat{f}: P \to M$ such that $f = h\hat{f}$. We then say f *lifts* to $\hat{f}$.

Clearly this definition could be formulated for any abelian category in terms of the diagram

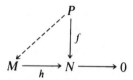

where here the dotted arrow completing the diagram is *not* uniquely defined.

Remark 2.8.2:   Every free module P is projective. (Indeed, suppose $\{x_i : i \in I\}$ is a base of P; select y_i in M such that $hy_i = fx_i$ and define $\hat{f}: P \to M$ such that $\hat{f}x_i = y_i$ for all i. Then $fx_i = h\hat{f}x_i$ for each i, implying $f = h\hat{f}$.)

In fact, projective modules are "close" to being free, in view of the following basic characterization:

Proposition 2.8.3:   *The following are equivalent for an R-module P*:

(i) *P is projective.*
(ii) *P is a summand of a free module.*
(iii) *Every epic $h: M \to P$ splits.*
(iv) *If $h: M \to N$ is epic then the induced map $\mathrm{Hom}(P, M) \to \mathrm{Hom}(P, N)$ (given by $g \to hg$) is onto.*

Proof: (i) $\Leftrightarrow$ (iv) is a restatement of the definition.

(i) $\Rightarrow$ (iii) Take $N = P$ and $f = 1_P$ in definition 2.8.1. Then $h\hat{f} = 1$ so h splits.
(iii) $\Rightarrow$ (ii) Take a free module M of large enough cardinality.
(ii) $\Rightarrow$ (i) By remark 2.8.2 and the following observation:

Q.E.D.

Lemma 2.8.4: $P = \bigoplus_{i \in I} P_i$ *is projective iff each P_i is projective.*

Proof: Let $\mu_i: P_i \to P$ be the canonical monic, and $\pi_i: P \to P_i$ be the projection for each i. Suppose $h: M \to N$ is epic.

($\Rightarrow$) Given $f: P_i \to N$ we have $f\pi_i: P \to N$ which lifts to some $g: P \to M$. Then $f = (f\pi_i)\mu_i = hg\mu_i$, so f lifts to $g\mu_i$.

($\Leftarrow$) Given $f: P \to N$ we lift each $f\mu_i$ to $g_i: P_i \to M$ and then we have $g: P \to M$ such that $g_i = g\mu_i$ for each i; hence $f\mu_i = hg\mu_i$ for each i, proving $f = hg$.

Q.E.D.

Remark 2.8.4':   ("Eilenberg's trick") If P is projective then there is a *free* module F such that $P \oplus F$ is free. (Indeed, if $P \oplus Q = F$ is free then adding on another free, if necessary, we may assume F is *not* f.g. and so

$$F \approx F^{(\mathbb{N})} \approx P \oplus Q + \cdots \approx P \oplus (Q \oplus P) \oplus (Q \oplus P) + \cdots$$

$$\approx P \oplus F^{(\mathbb{N})} \approx P \oplus F.)$$

This innocuous remark will be useful in the study of "large" projective modules, but one should note that F as constructed was *not* f.g. The condition that $P \oplus F$ is free for some f.g. free F is stronger than P projective and will be studied in §5.1.

Remark 2.8.5: Every f.g. projective module is a summand of an f.g. free module. In fact, if $P = \sum_{i=1}^{n} Rx_i$ then the proof (iii)⇒(ii) of proposition 2.8.3 shows P is a summand of $R^{(n)}$.

Corollary 2.8.6: $P \in R\text{-}\mathcal{M}od$ *is cyclic and projective iff* $P = Re$ *for some idempotent e of R.*

Proof: (⇒) P is a summand of R by remark 2.8.5, so viewing P as a left ideal of R we have $P = Re$ where e is the image of 1 under the projection $R \to P$.
(⇐) Re is a summand of R, so is projective. Q.E.D.

Projective Versus Free

Example 2.8.7: Every minimal left ideal L of $R = M_n(D)$ is cyclic and projective, but is not free if $n > 1$ since $[L:D] = n$ but $[R:D] = n^2$.

Example 2.8.8: If $m = \prod_{i=1}^{t} a_i \in \mathbb{Z}$ where the a_i are powers of prime integers with $a_1, \ldots, a_t$ pairwise relatively prime, then $\mathbb{Z}/m\mathbb{Z} \approx \prod_{i=1}^{t} a_i\mathbb{Z}$ by the Chinese Remainder Theorem, yielding examples of projective modules which are not free when $t > 1$ (seen by counting the number of elements).

A deeper example of a nonfree projective module is given in exercise 1. Nevertheless, there are certain rings R over which all projective modules are in fact free, and we give three important instances.

Example 2.8.9: Projective modules over local rings are free. This is obtained in exercise 2.9.2, but there is a straightforward proof for f.g. projective modules, which, anyway, is the case of greatest interest. In fact, we shall show now if $P = \sum_{i=1}^{n} Rx_i$ with n minimal possible then $\{x_1, \ldots, x_n\}$ is a base of P. Let $J = \text{Jac}(R)$ and let F be the free R-module with base $\{e_1, \ldots, e_n\}$. Let $\pi: F \to P$ be the map given by $\pi e_i = x_i$ for $1 \le i \le n$. We claim ker $\pi \le JF$. Otherwise, there is $x = \sum_{i=1}^{n} r_i e_i \in \ker \pi$ with some $r_i \notin J$; we may as well assume $r_1 \notin J$, so r_1 is invertible and $0 = \pi x = \sum r_i \pi e_i = \sum r_i x_i$, implying $x_1 = -\sum_{i=2}^{n} r_1^{-1} r_i x_i$, contrary to the minimality of n. Having proved the claim, we note $\pi: F \to P$ splits by proposition 2.8.3(iii); viewing P as a summand of F we have $F = P \oplus \ker \pi \leq P + JF$ so $F = P$ by "Nakayama's lemma."

Despite the categorical preference for projective modules, one often finds it easier to deal with free modules, and the second sentence of a proof involving a projective module P often is, "Take P' such that $P \oplus P'$ is free." For example, if $R \subset T$ and P is a projective R-module, then $T \otimes_R P$ is a projective T-module. (Proof: If $P = R^{(n)}$ then $T \otimes R^{(n)} \approx (T \otimes R)^{(n)} \approx T^{(n)}$ is free. In general, take P' such that $P \oplus P'$ is free. Then $(T \otimes P) \oplus (T \otimes P') \approx T \otimes (P \oplus P')$ is free.)

This is the key observation in utilizing example 2.8.9, since if T is local then $T \otimes P$ is free; we shall pursue this approach further in the discussions of localizations.

Example 2.8.10:   Projective modules over the polynomial ring $F[\lambda_1, \ldots, \lambda_t]$ are free for any t. This is a famous theorem of Quillen and Suslin, which will be presented in §5.1. For the time being, we shall consider easy special cases of this result. For $t = 0$, every module over a division ring is free, so the first nontrivial case is $t = 1$, to be discussed now in a more general setting.

Hereditary Rings

Definition 2.8.11:   A ring R is (left) hereditary if every left ideal is projective.

Example 2.8.12:   Suppose R is a PLID. Then any left ideal has the form Rx for x in R, but there is a module isomorphism $\varphi_x: R \to Rx$ given by $\varphi_x r = rx$, so every left ideal is free in $R\text{-}\mathcal{M}od$. In particular, every PLID is hereditary.

Example 2.8.13:   The ring of upper triangular matrices over a division ring is hereditary, as seen easily from example 1.1.9. Likewise, the ring of lower triangular matrices over a division ring is hereditary.

Triangular matrix rings are the motivating example in the study of hereditary rings and play an important role in their structure theory, cf. exercise 5.2.20, as well as in the construction of counterexamples. Exercise 5 gives a left but not right hereditary ring.

The purpose of the present discussion is to describe the basic connection between hereditary rings and projective modules, which will lead to the homological description of hereditary rings in Chapter 5. However, it is difficult to study hereditary rings in full generality, and usually additional properties are imposed, such as left Noetherian or semiprimary or both. Some of the best examples of such research are Robson [72], Fuelberth-Kuzmanovich [75], and Chatters-Jondrup [83]. A fruitful generalization of hereditary rings is PP-rings, c.f., exercise 3.2.20ff. Also see exercises 4ff.

Theorem 2.8.14: *Suppose R is hereditary. Every submodule of a free module F is isomorphic to a direct sum of left ideals, the number of summands being at most the cardinality of a base of F.*

Proof: Let $\{x_i : i \in I\}$ be a base of F, and take a well-ordering on I. For any i in I let $F_i = \bigoplus_{j<i} Rx_j < F$, and define $\pi_i : Rx_i + F_i \to R$ to pick off the coefficient of x_i (i.e., $\pi_i(rx_i + \sum_{j<i} r_j x_j) = r$).

Now suppose $M < F$, and write $M_i = M \cap F_i$ for each i in I. Let f_i denote the restriction of π_i to M_i. Then $\ker f_{i+1} = M_i$; putting $L_i = f_{i+1} M_{i+1}$ yields the exact sequence $0 \to M_i \to M_{i+1} \to L_i \to 0$. L_i is projective by hypothesis so $M_{i+1} \approx L_i \oplus M_i$. Continuing by induction, we get $M_\alpha \approx \bigoplus_{\beta < \alpha} L_\beta$ for any ordinal α, so we are done taking α to be the ordinal of I. Q.E.D.

Corollary 2.8.15: *R is hereditary iff every submodule of a projective module is projective.*

Proof: ($\Rightarrow$) Any submodule of a projective module is obviously a submodule of a free module and is thus a direct sum of projectives, which by lemma 2.8.4 is itself projective.

($\Leftarrow$) Every left ideal is a submodule of R and is thus projective. Q.E.D.

Corollary 2.8.16: *Every projective module over a PLID is free. In particular, if F is a free module with base of cardinality α and $M \leq F$ then M is free with base of cardinality $\leq \alpha$.*

Proposition 2.8.17: *Suppose R is hereditary and P, Q are projective modules. If $f \in \mathrm{Hom}(P, Q)$ then $\ker f$ is a summand of P. In particular, if P has no nonzero summands then every nonzero map from P to Q is monic.*

Proof: Since R is hereditary $fP < Q$ is projective, so $P \to fP \to 0$ splits, i.e., $P \approx fP \oplus \ker f$. Q.E.D.

Dual Basis Lemma

So far we have no concrete way of determining projectivity in terms of elements. To rectify this we need to generalize another property of free modules.

Remark 2.8.18: Suppose $\{e_i : i \in I\}$ is a base of a free R-module F. Defining the projections $\pi_i : F \to R$ by $\pi_i(\sum r_j e_j) = r_i$, we see for any y in F that $\pi_i y = 0$ for almost all i in I, and $y = \sum (\pi_i y) e_i$.

This elementary property actually characterizes projective modules, as we shall now see.

Definition 2.8.19:   A *dual base* of a module P is a set of pairs $\{(x_i, f_i): i \in I\}$ where $x_i \in P$ and $f_i: P \to R$ is a map satisfying the following condition for each x in P: $f_i x = 0$ for almost all i in I, and $x = \sum_{i \in I}(f_i x)x_i$.

Obviously this equation shows each $x \in \sum Rx_i$, so the x_i span P, but there is a much more decisive result.

Proposition 2.8.20:   (*"Dual basis lemma"*)

(i) *Suppose P is projective and $\{x_i: i \in I\}$ span P. Then there are $f_i: P \to R$ such that $\{(x_i, f_i): i \in I\}$ is a dual base. (In particular, every projective module has a dual base.)*

(ii) *Conversely, if P has a dual base then P is projective.*

Proof: In each direction we assume $\{x_i: i \in I\}$ spans P, and F is a free module with a base $\{e_i: i \in I\}$ and projections $\pi_i: F \to R$ as in remark 2.8.18. Define $\varphi: F \to P$ by $\varphi e_i = x_i$ for each i in I.

(i) Since P is projective we have $\psi: P \to F$ with $\varphi\psi = 1_P$. Put $f_i = \pi_i \psi$ for each i in I. Then for every x in P we have

$$x = \varphi\psi x = \varphi\left(\sum(\pi_i\psi x)e_i\right) = \varphi\left(\sum(f_i x)e_i\right) = \sum(f_i x)\varphi e_i = \sum(f_i x)x_i$$

as desired.

(ii) We shall show P is a summand of F by finding $\psi: P \to F$ with $\varphi\psi = 1_P$. Indeed, define ψ by $\psi x = \sum(f_i x)e_i$ for each x in P. Then $\varphi\psi x = \sum(f_i x)\varphi e_i = \sum(f_i x)x_i = x$.   Q.E.D.

Flat Modules

Let us note one more important property of free modules which is shared by projective modules.

Definition 2.8.21:   An R-module M is *flat* if whenever $f: N_1 \to N_2$ is monic in $\mathscr{M}\!od\text{-}R$ we have $f \otimes 1: N_1 \otimes_R M \to N_2 \otimes_R M$ is monic.

Remark 2.8.22:   Obviously R is flat since $N \otimes_R R \approx N$ naturally.

Proposition 2.8.23: $M = \bigoplus M_i$ is flat iff each M_i is flat.

Proof: Suppose $f: N_1 \to N_2$ is monic. Then using proposition 1.7.15 we can view $f \otimes 1: N_1 \otimes M \to N_2 \otimes M$ as the composition

$$N_1 \otimes \left(\bigoplus M_i\right) \approx \bigoplus (N_1 \otimes M_i) \to \bigoplus (N_2 \otimes M_i) \approx N_2 \otimes \left(\bigoplus M_i\right),$$

so we identify $\ker(f \otimes 1_M)$ as $\bigoplus \ker(f \otimes 1_{M_i})$ which is 0 iff each $f \otimes 1_{M_i}$ is monic.   Q.E.D.

Corollary 2.8.23': *Every projective module is flat.*

Proof: Every free module is a direct sum of copies of R and so is flat; a projective module is a summand of a free module and thus is flat.   Q.E.D.

Of course, nonflat modules exist; the $\mathbb{Z}$-module $M = \mathbb{Z}/2\mathbb{Z}$ is seen to be nonflat when confronted with the monic $\mathbb{Z} \to \mathbb{Q}$.

Proposition 2.8.24: *If* $0 \to F_1 \to M \to F_2 \to 0$ *is exact with* F_1, F_2 *flat, then* M *is flat.*

Proof: Suppose $f: N_1 \to N_2$ is monic. We want to show $f \otimes 1_M$ is monic. Looking inside $N_1 \otimes M$ let K denote the submodule $N_1 \otimes F_1$. The composition

$$N_1 \otimes M \xrightarrow{f \otimes 1} N_2 \otimes M \to N_2 \otimes F_2$$

has kernel containing K, so by proposition 1.7.30 we have a natural map

$$N_1 \otimes F_2 \approx (N_1 \otimes M)/K \to N_2 \otimes F_2$$

which corresponds to $f \otimes 1_{F_2}$. By hypothesis this is monic, so we conclude $\ker(f \otimes 1_M) \subseteq K$. But $\ker(f \otimes 1_{F_1}) = 0$ by hypothesis, so we conclude $\ker(f \otimes 1_M) = 0$.   Q.E.D.

Remark 2.8.25: If every f.g. submodule of M is flat then M is flat. (Indeed, otherwise there is $f: N_1 \to N_2$ monic and some element $z = \sum_{i=1}^{t} x_i \otimes y_i \neq 0$ in $\ker(f \otimes 1_M)$; then $M' = \sum_{i=1}^{t} R y_i$ is not flat since $0 \neq z \in \ker(f \otimes 1_{M'})$.)

Further study of flat modules requires the use of injectives and will be handled in §2.11.11ff. Important homological properties of flat modules will be discussed in chapter 5.

Schanuel's Lemma and Finitely Presented Modules

One of the most important results concerning projective modules is the following fact, which lies at the heart of homology theory.

Proposition 2.8.26: (*Schanuel's lemma*) *If* $0 \to K_i \to P_i \overset{g_i}{\to} M \to 0$ *are exact with* P_i *projective for* $i = 1, 2$ *then* $P_1 \oplus K_2 \approx P_2 \oplus K_1$.

Proof: Identify K_i with $\ker g_i$. Lifting g_1 to a map $h: P_1 \to P_2$ satisfying $g_1 = g_2 h$, define $f: P_1 \oplus K_2 \to P_2$ by $f(x, y) = hx - y$. Note f is epic (for if $x' \in P_2$ there is x in P_1 with $g_1 x = g_2 x'$ implying $g_2(hx - x') = g_1 x - g_2 x' = 0$, so $hx - x' \in K_2$ and $x' = f(x, hx - x')$). Thus f is split so $P_1 \oplus K_2 \approx P_2 \oplus \ker f$. Noting $x \in K_1$ iff $0 = g_1 x = g_2(hx)$ iff $hx \in K_2$, we see $\ker f = \{(x, y): hx = y \in K_2\} = \{(x, hx): x \in K_1\} \approx K_1$.   Q.E.D.

The first among many applications of Schanuel's lemma is to f.g. modules.

Definition 2.8.27: A module M is *finitely presented* if there is a map $f: F \to M$ with F f.g. free, such that $\ker f$ is also f.g.

Equivalently, M is finitely presented iff there are f.g. free modules F_i with $F_2 \overset{h}{\to} F_1 \to M \to 0$ exact. (Indeed, this condition clearly implies $M \approx F_1/hF_2$ is finitely presented; conversely, if $M \approx F/K$ with K f.g. then taking an epic $h: F_2 \to K$ with F_2 f.g. free we have $F_2 \overset{h}{\to} F \to M \to 0$ exact). Such a sequence will be called a *finite presentation* of M.

Examples 2.8.28:

(i) If R is left Noetherian then every f.g. R-module M is finitely presented. (Indeed take $f: F \to M$ epic with F f.g. free. Then $\ker f$ is f.g. since F is a Noetherian R-module.)

(ii) Every f.g. projective module is finitely presented. (Indeed if $\pi: R^{(t)} \to P$ is epic then $R^{(t)} \approx P \oplus \ker \pi$ so $\ker \pi$ is also an image of $R^{(t)}$ and thus is f.g..)

Proposition 2.8.29: *If* N *is finitely presented and* $f: M \to N$ *is epic with* M *f.g. then* $\ker f$ *is f.g..*

Proof: First assume M is projective and take an exact sequence $0 \to K \to P \to N \to 0$ with P free and K, P f.g.. Schanuel's lemma shows $\ker f$ is a summand of $M \oplus K$ and is thus f.g..

For general M take epic $\pi: F \to M$ with F f.g. free. Then $\ker(f\pi)$ is f.g. by what we just saw. But $\ker f = \pi(\ker f\pi)$. (Indeed, $\pi(\ker f\pi) \leq \ker f$; conversely, if $\pi x \in \ker f$ then $x \in \ker(f\pi)$). Thus $\ker f$ is f.g..   Q.E.D.

H Supplement: Projective Covers

Often when studying a module M we pass to a projective module P above it, i.e., there is an epic $P \to M$. We shall now describe a situation for which a unique P "closest" to M can be found, which casts a strong light on idempotent-lifting, and perfect and semiperfect rings.

Definition 2.8.30: A submodule M' of M is *small* or *superfluous* (written $M' \ll M$) if for any $N < M$ we have $N + M' < M$.

We usually verify the contrapositive, i.e., given $N + M' = M$ we prove $N = M$. Small submodules play a hidden role in "Nakayama's lemma," which says $\text{Jac}(R)M \ll M$ for every M in R-$\mathcal{F}imod$. Moreover, small left ideals tie in with the Jacobson radical as follows: $L \ll R$ iff L is contained in every maximal left ideal (seen by checking the definition), iff $L \leq \text{Jac}(R)$. Thus $\text{Jac}(R)$ is a small left ideal which contains every small left ideal.

Definition 2.8.31: A *projective cover* of a module M is an epic $f: P \to M$ such that P is projective and $\ker f \ll P$.

Part of the definition is easy to satisfy, since M is a homomorphic image of a suitable free module. Our next effort will be in analyzing the key condition that $\ker f$ is small, which leads to an improvement of Schanuel's lemma in this case: Projective covers are unique up to isomorphism (proposition 2.8.34 below).

Remark 2.8.32: If $f: M \to N$ is a map and $M_1 \leq M$ with $fM_1 = fM$ then $M = M_1 + \ker f$. (Indeed, if $x \in M$ then $fx = fx_1$ for suitable x_1 in M_1, and then $x - x_1 \in \ker f$.)

Lemma 2.8.33: *Suppose M has a projective cover $f: P \to M$. For any epic g: $Q \to M$ with Q projective, we can write $Q = P' \oplus P''$ with $P' \approx P$ and $P'' \leq \ker g$, such that the restriction of g to P' is a projective cover for M.*

Proof: Since Q is projective we can lift $g: Q \to M$ to $h: Q \to P$ such that $g = fh$. In particular, $f(hQ) = gQ = M$ so $hQ + \ker f = P$ by remark 2.8.32, implying $hQ = P$ since $\ker f$ is small. But P is projective so h splits, i.e., we can identify P with some summand P' of Q and write $Q = P \oplus P''$ where h is the projection onto P. Let K be the kernel of the restriction of g to P. Since $gP = fhP = fhQ = M$ it remains to show $K \ll P$. Suppose $N + K = P$. Then $gN = gP = M$ so $hN = P$ as shown above. But $hN = N$ so $N = P$, proving K is small.   Q.E.D.

Proposition 2.8.34:   (*Uniqueness of projective cover*) *If* $f_i: P_i \to M_i$ *is a projective cover of* M_i *for* $i = 1, 2$ *and* $\varphi: M_1 \to M_2$ *is an isomorphism then there is an isomorphism* $\varphi': P_1 \to P_2$ *with* $f_2\varphi' = \varphi f_1$.

Proof: Apply lemma 2.8.33, taking $f = f_2$, $g = \varphi f_1$, $P = P_2$, and $Q = P_1$; we need only show the epic map $h: Q \to P$ obtained there is an isomorphism, for $\varphi f_1 = g = fh$, and we would then take φ' to be h. But $P'' = \ker h$ is small in Q (for $P'' \leq \ker g \ll P_1 = Q$) and $P' + P'' = Q$, implying $P'' = 0$, as desired.   Q.E.D.

Example 2.8.35:   A module need not have a projective cover. Indeed, suppose M is a cyclic but *not* projective module over a ring R which has no nonzero small left ideals. Then M has no projective cover (since, otherwise, taking $Q = R$ in lemma 2.8.33 and letting $f: R \to M$ be the canonical onto map, we see there is a projective cover $g': P' \to M$ where P' is a left ideal of R, but $\ker g' \ll R$ implies $\ker g' = 0$ so $M \approx P'$ is projective, contrary to assumption). In particular $\mathbb{Z}/2\mathbb{Z}$ has no projective cover as a $\mathbb{Z}$-module. This idea is carried further in exercise 20, in which it is shown for R semiprimitive that only the projective R-modules have projective covers.

Lemma 2.8.36:   *A cyclic module* M *has a projective cover iff* $M \approx Re/A$ *for some idempotent* e *and left ideal* $A \leq \mathrm{Jac}(R) \cap Re$. *In this case, the canonical map* $f: Re \to M$ *is a projective cover.*

Proof: ($\Leftarrow$) $\ker f = Ae \subseteq \mathrm{Jac}(R)Re \ll Re$ by "Nakayama's lemma."
   ($\Rightarrow$) Let $g: R \to M$ be the canonical onto map. Then $R \approx P' \oplus P''$ under the notation of lemma 2.8.33 so $P' \approx Re$ for some idempotent e and $M \approx Re/A$ where $A \ll Re$, so $A \subseteq \mathrm{Jac}(R)$.   Q.E.D.

Proposition 2.8.37:   *Suppose* $A \triangleleft R$ *and* $A \subseteq \mathrm{Jac}(R)$. *A is idempotent-lifting iff every summand of the R-module* R/A *has a projective cover.*

Proof: ($\Rightarrow$) $\bar{R} = R/A$ has the same lattice of submodules in R-$\mathcal{M}od$ or R/A-$\mathcal{M}od$ so any summand can be written as $\overline{Re}$ where $\bar{e} \in \bar{R}$ is idempotent;

take $e \in R$ idempotent. Then $\overline{Re} \approx Re/A \cap Re = Re/Ae$ has a projective cover by the lemma.

($\Leftarrow$) Suppose $\bar{e} \in \bar{R} = R/A$ is idempotent. The $\overline{Re}$ is a summand of $\bar{R}$ and thus has a projective cover P which by the lemma can be written in the form Re/A' where e is an idempotent over $\bar{e}$.   Q.E.D.

We are ready for the key results of this discussion, tying projective covers in with idempotent-lifting and thus with semiperfect rings.

Remark 2.8.38:   Suppose there is an epic $\psi: M \to N$ with $\ker \psi \ll M$. Any projective cover $f: P \to N$ lifts to a projective cover $g: P \to M$. (Indeed, f lifts to a map $g: P \to M$ with $f = \psi g$; then $gP + \ker \psi = M$ by remark 2.8.32 so g is epic since $\ker \psi$ is small. Moreover, $\ker g \le \ker f \ll P$.)

Lemma 2.8.39:

   (i) *If $K_1 \ll M$ and $K_2 \ll M$ then $K_1 + K_2 \ll M$.*
   (ii) *If $K_i \ll M_i$ for $1 \le i \le n$ then $K_1 \oplus \cdots \oplus K_n \ll M_1 \oplus \cdots \oplus M_n$.*
   (iii) *If $f_i: P_i \to M_i$ are projective covers for $1 \le i \le n$ then $(f_1, \ldots, f_n)$: $\bigoplus_{i=1}^{n} P_i \to \bigoplus_{i=1}^{n} M_i$ is a projective cover.*

Proof:

   (i) If $K_1 + K_2 + N = M$ then $K_2 + N = M$ so $N = M$.
   (ii) Inductively one may assume $n = 2$. Write $M = M_1 \oplus M_2$; by (i) it suffices to show $K_i \ll M$ for $i = 1, 2$. But if $N + K_1 = M$ then $(N \cap M_1) + K_1 = M_1$ so $N \cap M_1 = M_1$, implying $M_1 \le N$ and $N \ge N + K_1 = M$.
   (iii) Clearly $(f_1, \ldots, f_n)$ is epic, and its kernel is $\bigoplus \ker f_i$, which is small by (ii).   Q.E.D.

Theorem 2.8.40:   *R is a semiperfect ring iff R has a complete set of orthogonal primitive idempotents $e_1, \ldots, e_n$ such that each $e_i Re_i$ is a local ring. Moreover, in this case every f.g. R-module has a projective cover.*

Proof: ($\Rightarrow$) is proposition 2.7.20. ($\Leftarrow$) R is semilocal by proposition 2.7.14. Let $J = \text{Jac}(R)$. We need to show J is idempotent-lifting, so in view of proposition 2.8.37 we need merely prove the last assertion, that every f.g. R-module M has a projective cover. M/JM is an f.g. module over the semi-simple Artinian ring $\bar{R} = R/J$, and $\bar{e}_1, \ldots, \bar{e}_n$ are a complete set of orthogonal idempotents which are all primitive since $\overline{e_i Re_i} \approx (e_i Re_i)/\text{Jac}(e_i Re_i)$ is a division ring. Thus $M/JM \approx \bigoplus_{\text{finite}} \overline{Re_{u_i}}$ for suitable u_i, where each $\overline{Re_{u_i}} \approx Re_{u_i}/Je_{u_i}$ is simple in $\bar{R}\text{-}\mathcal{M}od$ and thus in $R\text{-}\mathcal{M}od$. Now M/JM has a projective cover by lemmas 2.8.36 and 2.8.39(iii), implying M has a projective cover by remark 2.8.38.   Q.E.D.

Note that theorem 2.8.40 shows for a semiperfect ring that every f.g. module has a projective cover. This condition characterizes semiperfect rings (exercise 22) and actually can be weakened slightly (exercise 23).

§2.9 Indecomposable Modules and LE-Modules

We have seen in §2.4 that every completely reducible module can be written "uniquely" as a direct sum of simple modules. In this section we study modules having a weaker form of decomposition and apply this theory to modules having composition series and to semiperfect rings.

The reason this section follows the section on projective modules is that one important module which we would like to decompose is the free R-module or, more specifically, R itself; in this case all the summands are obviously projective.

Definition 2.9.1:   An R-module M is *indecomposable* if M cannot be written as a direct sum of two proper submodules. A *decomposition* of M is a direct sum $\bigoplus_{i \in I} M_i = M$ for M_i indecomposable. M is *completely decomposable (of Krull-Schmidt length $|I|$)* if M has a decomposition as above.

Our ultimate goal is to study the completely decomposable modules by means of the indecomposables. At first blush this task would seem near impossible, although amazing progress has been made in recent years. We defer the difficult project of determining the indecomposables and, for the time being, try to identify completely decomposable modules; one easy case is when M has a composition series.

The Krull-Schmidt Decomposition

Proposition 2.9.2: (*Existence of Krull-Schmidt decomposition*) *if M has a composition series of length $\leq t$ then M is completely decomposable of Krull-Schmidt length $\leq t$.*

Proof: (Induction on t) If M is indecomposable we are done. So assume $M = M_1 \oplus M_2$ for proper submodules M_1, M_2. Then $M > M_1 > 0$ can be refined to a composition series of length $\leq t$ so M_1 and $M_2 \approx M/M_1$ each have composition series of respective lengths $\leq t_1, t_2$ with $t_1 + t_2 = t$. By induction M_1, M_2 are completely decomposable of length $\leq t_1, t_2$, respectively, so $M = M_1 \oplus M_2$ is completely decomposable of length $\leq t_1 + t_2 = t$.

$$\text{Q.E.D.}$$

Proposition 2.9.3:

(i) *Suppose* $M = M_1 \oplus M_2$, *and let* e_i *be the projection of* M *onto* M_i *for* $i = 1, 2$ *(i.e.,* $e_i(x_1, x_2) = x_i$). *Then* e_1, e_2 *are orthogonal idempotents in* $\text{End}_R M$ *and* $e_1 + e_2 = 1$.

(ii) *Conversely, if* e_1, e_2 *are orthogonal idempotents in* $\text{End}_R M$ *with* $e_1 + e_2 = 1$ *then* $M_i = e_i M$ *are submodules of* M *with* $M = M_1 \oplus M_2$.

Proof: Each step is straightforward; details are left to the reader.   Q.E.D.

Example 2.9.4: Take $M = R$ so that $R \approx \text{End}_R R$ by the regular representation. R is a direct sum of n left ideals iff 1 is a sum of n orthogonal idempotents.

In this way, we see proposition 2.9.3 is a rather sweeping observation which also has the following immediate application.

Corollary 2.9.5: *M is indecomposable iff* $\text{End}_R M$ *has no nontrivial idempotents (which is true if* $\text{End}_R M$ *is local, by remark 2.5.13).*

The criterion that $\text{End}_R M$ be local will turn out to be the key to this discussion. To see why, we need another fundamental result, called *Fitting's lemma.*

Lemma 2.9.6: *Suppose* $f \in \text{End}_R M$. *If* M *is Noetherian and* f *is epic then* f *is an isomorphism.*

Proof: $\ker f \leq \ker f^2 \leq \cdots$ so $\ker f^n = \ker f^{n+1}$ for some n. But for any x in $\ker f$ we can write $x = f^n y$ for suitable y in M, so $0 = fx = f^{n+1} y$ implying $0 = f^n y = x$; hence $\ker f = 0$ and f is monic.   Q.E.D.

Proposition 2.9.7: *("Fitting's lemma") Suppose M has a composition series of length n, and* $f \in \text{End}_R M$. *Then* $M = M_1 \oplus M_2$, *where* $M_1 = f^n M$ *and* $M_2 =$

ker f^n. *Each M_i is invariant under f; moreover, f restricts to an isomorphism $f_1: M_1 \to M_1$ and a nilpotent map $f_2: M_2 \to M_2$.*

Proof: Note $M \geq fM \geq f^2 M \geq \cdots$ and $0 \leq \ker f \leq \ker f^2 \leq \cdots$; since M has length n we see at once $f^n M = f^{n+1}M = \cdots$ and $\ker f^n = \ker f^{n+1} = \cdots$. Since $f^n M = f^n (f^n M)$ we have $M = f^n M + \ker f^n$ by remark 2.8.32. For $x \in f^n M \cap \ker f^n$ writing $x = f^n y$ we have $f^{2n} y = f^n x = 0$ so $y \in \ker f^{2n} = \ker f^n$ implying $x = f^n y = 0$; this proves $M = f^n M \oplus \ker f^n$. Now $fM_1 = M_1$ so f_1 is an isomorphism by lemma 2.9.6; obviously $f^n M_2 = 0$.   Q.E.D.

Corollary 2.9.8: *If M has a composition series and is indecomposable then $\mathrm{End}_R M$ is a local ring.*

Proof: Given f in $\mathrm{End}_R M$, we have either $M_1 = M$ or $M_2 = M$ in Fitting's lemma (because M is indecomposable), so f is either an isomorphism or is nilpotent. Thus we are done once we can prove the following result:

Proposition 2.9.9: *Suppose R is a ring each of whose elements is invertible or nilpotent. Then R is a local ring.*

Proof: We need to show if $a_1 + a_2 = 1$ then a_1 or a_2 is invertible, by remark 2.5.12. If a_1 is not invertible then a_1 is nilpotent by assumption, so $a_2 = 1 - a_1$ is invertible by remark 1.1.27.   Q.E.D.

Krull-Schmidt decompositions enable us to generalize corollary 2.7.15, as in the next interesting result.

Theorem 2.9.10: *If M has a composition series of length n then $\mathrm{End}_R M$ is a semiprimary ring and $\mathrm{Jac}(\mathrm{End}_R M)^n = 0$.*

Proof: Write $M = \bigoplus_{i=1}^n M_i$ where the M_i are indecomposable, and let $e_i \in \mathrm{End}_R M$ be the projection onto M_i for $1 \leq i \leq n$. Then the e_i are orthogonal primitive idempotents, and $e_i(\mathrm{End}_R M)e_i \approx \mathrm{End}_R(e_i M) = \mathrm{End}_R M_i$ is local by corollary 2.9.8. Hence $\mathrm{End}_R M$ is semilocal, by proposition 2.7.14; it remains to show $J^n = 0$ where $J = \mathrm{Jac}(\mathrm{End}_R M)$. Suppose $f \in J$. By Fitting's lemma we can write $M = M_1 \oplus M_2$ where $M_1 = f^n M$ and $M_2 = \ker f^n$; and f restricted to M_1 has an inverse g, which we extend to an endomorphism of M by putting $gx = 0$ for all x in M_2. Then $(1 - gf)M_1 = 0$. But $f \in J$ implies $1 - gf$ is invertible. Hence $0 = M_1 = f^n M$, so we see J is nil of bounded index n.

We shall actually prove the apparently stronger assertion that every nil (multiplicative) semigroup of $\text{End}_R M$ is nilpotent; then $J^k = 0$ for some $k > 0$ and if $J^n \neq 0$ we have $M > JM > J^2 M > \cdots > J^n M > 0$ contrary to the Jordan-Holder theorem. (In fact, our assertion is equivalent to the nilpotence of J, by theorem 2.6.31) So assume S is a nil multiplicative subgroup of $\text{End}_R M$. If $J = 0$ then we are done by theorem 2.6.31. Otherwise, $\text{End}_R M$ has a nilpotent ideal $A \neq 0$ by proposition 2.6.26, and $0 < AM < M$ so AM has length $< n$. S acts as a nil semigroup of endomorphisms on AM so by induction $S^v AM = 0$ for some $v > 0$. Likewise, $S^u(M/AM) = 0$ for some $u > 0$, so $S^{uv} M \subseteq S^v AM = 0$, as desired.   Q.E.D.

Uniqueness of the Krull-Schmidt Decomposition

We turn to the uniqueness of the Krull-Schmidt decomposition, a fairly delicate issue which requires careful analysis.

Definition 2.9.11: M is an LE-*module* if $\text{End}_R M$ is local (i.e., M has a *local endomorphism* ring). We saw that every LE-module M is indecomposable, the converse holding when M has a composition series. In fact, LE-modules provide a fairly general situation in which it is possible to prove both the Krull-Schmidt theorem and a generalization due to Azumaya. Following Beck [78] we obtain very precise information concerning decompositions of sums of LE-modules. Write J_M for $\text{Jac}(\text{Hom}_R(M, M))$.

Lemma 2.9.12:   *Suppose* $f: M \to N$ *and* $g: N \to M$ *are maps in* R-$\mathcal{M}od$. $1_N + fg$ *is an isomorphism iff* $1_M + gf$ *is an isomorphism.*

Proof: $(\Rightarrow)$ Noting $f(1_M + gf) = f + fgf = (1_N + fg)f$ and, likewise, $(1_M + gf)g = g(1_N + fg)$, one sees easily that $1 - g(1_N + fg)^{-1}f$ is the inverse of $1_M + gf$. The opposite direction is analogous.   Q.E.D.

Proposition 2.9.13:   *Suppose* $M, N \in$ R-$\mathcal{M}od$ *and* M *is LE. If* $f: M \to N$ *and* $g: N \to M$ *satisfy* $fg \notin J_N$ *then* ghf *is invertible for some* $h: N \to N$. *In particular,* f *is split monic and* g *is split epic.*

Proof: By hypothesis, hfg is not quasi-invertible for some $h: N \to N$, so $1_N - hfg$ is not an isomorphism; lemma 2.9.12 implies $1_M - ghf$ is not an isomorphism so $ghf \notin J_M$. Since M is LE we conclude ghf is an isomorphism, as desired, and the second assertion follows from proposition 1.4.10.
                                                                          Q.E.D.

To utilize this result fully we need an easy property of summands.

Proposition 2.9.14: *Suppose $M = K \oplus N$ and e_K is the projection of M onto K. $M = L \oplus N$ (for $L \le M$) iff e_K restricts to an isomorphism $L \to K$.*

Proof: ($\Leftarrow$) Clearly $L \cap N = 0$. If $x \in M$ then $e_K x = e_K y$ for some y in L, so $x = y + (x - y) \in L + N$, proving $M = L \oplus N$.

($\Rightarrow$) Let f be the restriction of e_K to L. Then $\ker f = \{a \in L : e_K a = 0\} = L \cap N = 0$. Note f is onto since for any x in K we can write $x = a + b$ for a in L and b in N, so then $x = e_K x = e_K a + e_K b = e_K a$.   Q.E.D.

Theorem 2.9.15: ("*Exchange Property*") *Suppose $M = N \oplus N' = \bigoplus_{i \in I} M_i$ where each M_i is an LE-module. Then some M_i is isomorphic to a summand of N. Moreover, if N is indecomposable then we can pick i such that $N \approx M_i$ and $M = (\bigoplus_{j \ne i} M_j) \oplus N = M_i \oplus N'$.*

Proof: Let $\pi: \bigoplus_{i \in I} M_i \to N$ be the projection onto N and let $\pi_i : M_i \to N$ be the restriction of π to M_i; let $e_i : N \to M_i$ be the restriction of the projection $M \to M_i$. Pick $x \ne 0$ in N and write $x = \sum_{u=1}^{t} x_u$ for x_u in M_{i_u}. Let $f = \sum \pi_{i_u} e_{i_u} \in \operatorname{Hom}(N, N)$. Then $x = \pi x = \sum_{u=1}^{t} \pi_{i_u} x_u = \sum_{u=1}^{t} \pi_{i_u} e_{i_u} x = f x$ so $(1 - f)x = 0$; hence f is not quasi-invertible and $f \notin J_N$. Hence some $\pi_i e_i \notin J_N$. By proposition 2.9.13 this π_i is split monic, so $M_i \approx \pi_i M_i$ is a summand of N; likewise, e_i is split epic.

If N is indecomposable then e_i and π_i are both isomorphisms, so applying proposition 2.9.14 each time yields $M = (\bigoplus_{j \ne i} M_j) \oplus N$ and $M = M_i \oplus N'$.
                                                                            Q.E.D.

Definition 2.9.16: Two decompositions $M = \bigoplus_{i \in I} M_i = \bigoplus_{j \in J} N_j$ are equivalent if $I = J$ and there is a 1:1 correspondence $\sigma : I \to J$ with $M_i \approx N_{\sigma i}$ for all $i \in I$.

Theorem 2.9.17: (*Wedderburn-Krull-Schmidt-Azumaya*) *If M is a direct sum of LE-modules then all decompositions of M are equivalent.*

Proof: Given an indecomposable summand K of M let $I(K) = \{i \in I : M_i \approx K\}$ and $J(K) = \{j \in J : N_j \approx K\}$. Clearly it suffices to prove $|I(K)| = |J(K)|$ for each K. By symmetry we need only show $|I(K)| \le |J(K)|$; note by theorem 2.9.15 that $I(K) \ne \varnothing$ iff $J(K) \ne \varnothing$.

Case I. $n = |J(K)|$ is finite. For convenience write $J(K) = \{1, \ldots, n\}$. By theorem 2.9.15 we have some i such that $M_i \approx N_1$ and $M \approx (\bigoplus_{i' \ne i} M_{i'}) \oplus N_1$. Thus $\bigoplus_{i' \ne i} M_{i'} \approx M/N_1 \approx \bigoplus_{j \ne 1} N_j$, we conclude by induction on n (having removed one module from each of $I(K)$ and $J(K)$).

Case II. $|J(K)|$ is infinite. We resort to the trick of theorem 0.3.2, suitably modified. Given j in $J(K)$ write $I'(j) = \{i \in I(K): M = M_i \oplus (\bigoplus_{j' \neq j} N_{j'})\}$. $I(K) = \bigcup_{j \in J(K)} I'(j)$ by theorem 2.9.15, so it suffices to prove each $I'(j)$ is finite since then $|I(K)| \leq \sum_{j \in J(K)} |I'(j)| \leq |\aleph_0||J(K)| = |J(K)|$.

Pick $x \neq 0$ in N_j, and write $x = \sum x_i$ where each $x_i \in M_i$. Since $M = \bigoplus M_i$ we see almost all $x_i = 0$. But if $x_i = 0$ then x lies in the kernel of the restriction of the canonical projection $M \to M_i$ to N_j, so $i \notin I'(j)$ by proposition 2.9.14. Thus $I'(j)$ is finite, as desired.   Q.E.D.

Applications of the Krull-Schmidt Theorem to Semiperfect Rings

In 2.7.23-2.7.30 we initiated the study of a semilocal ring in terms of a complete set of primitive idempotents. The results become much sharper when we apply the finite case of theorem 2.9.17, usually called the *Krull-Schmidt* theorem. Note that these results are interesting even for Artinian rings.

Theorem 2.9.18: *Suppose R is semiperfect.*

(i) *There is an LE-decomposition $R = \bigoplus_{i=1}^{t} Re_i$ where $e_1, \ldots, e_t$ is a complete set of primitive idempotents.*

(ii) *Every summand of R is a finite direct sum of submodules isomorphic to the Re_i, and consequently every idempotent is a sum of orthogonal primitive idempotents (which can be expanded to a complete set of primitive idempotents).*

(iii) *If $f_1, \ldots, f_{t'}$ is another complete set of primitive idempotents of R then $t = t'$, and there is some invertible u in R such that $u^{-1} e_i u = f_i$ for $1 \leq i \leq t$ (where the f_i have been rearranged such that $Re_i \approx Rf_i$).*

Proof:

(i) is a restatement of proposition 2.7.20.

(ii) By example 2.9.4 we can write any summand L of R in the form Re for e idempotent. The exchange property shows $L = \bigoplus L_i$ where each $L_i \approx Re_i$ for suitable i; but each L_i is a summand of R so $L_i = Re_i'$ where clearly the e_i' are primitive and orthogonal.

(iii) $R = \bigoplus Re_i = \bigoplus Rf_i$ so by Krull-Schmidt, $t = t'$ and the f_i can be rearranged to yield an isomorphism $\psi_i: Re_i \to Rf_i$. Then $\bigoplus \psi_i: \bigoplus Re_i \to \bigoplus Rf_i$ is an isomorphism of R as R-module and is thus given by right multiplication by some invertible u in R. For r_i in R we thus have $f_i = \psi_i(r_i e_i) = r_i e_i u$ for $1 \leq i \leq t$. Then $f_i u^{-1} = r_i e_i$, so for $1 \leq j \leq t$ we get

$$f_j u^{-1} e_i u = r_j e_j e_i u = \delta_{ij} r_i e_i u = \delta_{ij} f_i$$

implying $u^{-1} e_i u = (\sum_{j=1}^{t} f_j) u^{-1} e_i u = f_i$, as desired.　　　Q.E.D.

The modules Re_i of theorem 2.9.18 are projective, being summands of R, and are called the *principal indecomposable* modules of R, for obvious reasons.

Corollary 2.9.19:　*Suppose $R = M_n(T)$ where T is a local ring and $\{e_{ij}: 1 \leq i, j \leq n\}$ is a set of matric units of R. If $\{e'_{ij}: 1 \leq i, j \leq n\}$ is another set of matric units of R then there is some invertible u in R with $u^{-1} e_{ij} u = e'_{ij}$ for all i, j.*

Proof:　From the theorem we have u_0 in R with $u_0^{-1} e_{ii} u_0 = e'_{ii}$ for $1 \leq i \leq n$, since $\{e_{11}, \ldots, e_{nn}\}$ is a complete set of primitive idempotents. Let $u = \sum_{i=1}^{n} e_{i1} u_0 e'_{1i}$ and $u' = \sum_{i=1}^{n} e'_{i1} u_0^{-1} e_{1i}$. Then

$$uu' = \sum e_{i1} u_0 e'_{11} u_0^{-1} e_{1i} = \sum e_{i1} e_{11} e_{1i} = \sum e_{ii} = 1$$

and, likewise, $u'u = 1$, so $u' = u^{-1}$. Moreover,

$$u^{-1} e_{ij} u = u' e_{ij} u = e'_{i1} u_0^{-1} e_{1i} e_{ij} e_{j1} u_0 e'_{1j} = e'_{i1} e'_{11} e'_{1j} = e'_{ij}$$

as desired.　　　Q.E.D.

This has the following surprising consequence.

Corollary 2.9.20:　*Suppose $R = M_n(T)$ where T is a commutative local ring. Any ring injection $f: R \to R$ fixing T is an inner automorphism.*

Proof:　If $\{e_{ij}: 1 \leq i, j \leq n\}$ is a set of matric units then so is $\{fe_{ij}: 1 \leq i, j \leq n\}$, so $fe_{ij} = u^{-1} e_{ij} u$ for some u in R, and the result follows at once.
　　　　　　　　　　　　　　　　　　　　　　　　　　Q.E.D.

We can also characterize the indecomposable projective modules.

Proposition 2.9.21:　*Suppose R is semiperfect. An R-module P is indecomposable projective iff P is principal indecomposable. Consequently R has only a finite number of indecomposable projective modules.*

Proof:　Let $\{e_1, \ldots, e_t\}$ be a complete set of orthogonal primitive idempotents. ($\Rightarrow$) P is a summand of some free R-module, which by theorem 2.9.18 is a sum of copies of the Re_i; the exchange property shows $P \approx Re_i$ for some i. The converse is clear. Consequently R has at most t indecomposable projectives.　　　Q.E.D.

Applying theorem 2.9.18 to theorem 2.8.40 we see that a module M has a finite LE-decomposition iff $\text{End}_R M$ is semiperfect, and this has become a standard tool in generalizing the Krull-Schmidt theorem to more general classes of rings.

Decompositions of Modules Over Noetherian Rings

Let us pause to see how these results fit into the general study of f.g. R-modules. When R is left Artinian then R is also left Noetherian, so every f.g. R-module is Artinian and Noetherian and thus has a composition series; hence this theory applies in its entirety.

When R is left Noetherian but not Artinian, every f.g. R-module M is Noetherian and thus has a decomposition of finite length. (Indeed, let M_1 be a maximal summand of M, and inductively let M_{i+1} be a maximal summand of M_i; writing N_{i+1} for the complement of M_{i+1} in M_i we have $N_1 < N_1 \oplus N_2 < \cdots$ in M, so $M = N_1 \oplus \cdots \oplus N_t$ for some t.) This raises the hope of studying R-$\mathscr{F}imod$ in terms of the indecomposable modules. When R is a commutative principal ideal domain this has been done classically, including the "uniqueness" part of Krull-Schmidt (cf., Jacobson [85B; §3.9]); however, in general, there are easy counterexamples:

Example 2.9.22: (Failure of Krull-Schmidt for Noetherian rings) If L_1, L_2 are left ideals of R with $L_1 + L_2 = R$ then $L_1 \oplus L_2 \approx R \oplus (L_1 \cap L_2)$ since the epic $L_1 \oplus L_2 \to R$ given by $(x_1, x_2) \to x_1 - x_2$ splits (because R is projective as R-module). If we can take L_1, L_2 *not* principal then these two decompositions are inequivalent. A concrete example: Take $R = \mathbb{Z}[\sqrt{-5}]$ and $L_1 = \langle 3, 2 + \sqrt{-5} \rangle$ and $L_2 = \langle 3, 2 - \sqrt{-5} \rangle$.

See the unabridged edition for a much fuller discussion of Krull-Schmidt.

C Supplement: Representation Theory

The main underlying problem is the classficiation of the f.g. indecomposable modules of a left Artinian Ring, for then by the Krull Schmidth theorem we would "know" all the modules. Of course, one could consider this problem for even wider classes. Nevertheless, as we shall see, even the Artinian question is exceedingly complex.

When R is semisimple Artinian every indecomposable module is simple, by theorem 2.4.9, and there is precisely one isomorphism class of simple module for each irreducible component, by remark 2.3.8'. Thus we have only a finite number of nonisomorphic simple modules.

Definition 2.9.24: R has *finite representation type* (f.r.t) if it has only a finite number of f.g. indecomposable modules (up to isomorphism).

Thus every semisimple Artinian ring has f.r.t., and, moreover, we know how to determine all its indecomposables. (Other examples are given in exercises 20, 22.) Hence by Maschke's theorem (8.1.10' below) every group algebra of a finite group over a field F of characteristic 0 has f.r.t. When $\text{char}(F) \neq 0$ the following sort of example arises.

Example 2.9.25: Let $G = (\mathbb{Z}/2\mathbb{Z})^{(2)} = \{1, g, h, gh\}$ and let F be an algebraically closed field of characteristic 2. Then $F[G]$ is commutative of dimension 4 over F. The homomorphism $F[G] \to F$ given by $g \mapsto 1$ and $h \mapsto 1$ has kernel $N = Fa + Fb + Fc$ where $a = g - 1 = g + 1$, $b = h + 1$, and $c = ab = gh + g + h + 1$; clearly $N^2 = Fc$ and $N^3 = 0$, so $F[G]$ is local with radical N. Fc is the unique minimal ideal, so any ideal of dimension 2 over F is indecomposable. Any such ideal has the form $M_{\alpha\beta} = F(\alpha a + \beta b) + Fc$ for suitable α, β in F, and we shall prove there are $|F|$ of these by showing $M_{\alpha\beta} \approx M_{\alpha'\beta'}$ iff $\alpha^{-1}\alpha' = \beta^{-1}\beta'$. Indeed ($\Leftarrow$) is clear for then $F(\alpha a + \beta b) = F(\alpha' a + \beta' b)$. To prove ($\Rightarrow$) suppose there is an isomorphism $\varphi: M_{\alpha\beta} \to M_{\alpha'\beta'}$. Then $\varphi(\alpha a + \beta b) = \gamma(\alpha' a + \beta' b)$ for some γ in F, so

$$\beta\varphi c = \varphi(a(\alpha a + \beta b)) = a\gamma(\alpha' a + \beta' b) = \beta'\gamma c \qquad \text{and}$$

$$\alpha\varphi c = \varphi(b(\alpha a + \beta b)) = b\gamma(\alpha' a + \beta' b) = \alpha'\gamma c,$$

so $\beta^{-1}\beta'\gamma c = \alpha^{-1}\alpha'\gamma c$ implying $\beta^{-1}\beta' = \alpha^{-1}\alpha'$ as desired.

These examples led to the general study of the indecomposable modules of finite dimensional algebras. A sketch of the rudiments of this theory is given on pp. 249–260 of the unabridged edition.

§2.10 Injective Modules

In category theory one often is led to dualize important concepts with the hope that the new notions will also turn out to be useful. This strategy has been quite successful at times. (The reader should be warned, however, that any strategy followed blindly can lead to disaster, and many pages in the literature have been wasted in this manner.) The concept dual to projective modules is in fact one of the most important in module theory.

Definition 2.10.1: An R-module E is *injective* if it satisfies the following property dual to the lifting property. For any map $f: N \to E$ and any monic h: $N \to M$ there is a map $\hat{f}: M \to E$ such that $f = \hat{f}h$, i.e., $\hat{f}$ completes the following diagram (*not* uniquely):

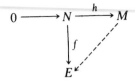

$$0 \longrightarrow N \overset{h}{\longrightarrow} M$$

In this case we say $\hat{f}$ *extends* f.

Remark 2.10.2: E is injective iff for every monic $h: N \to M$ the map $h^{\#}$: $\mathrm{Hom}(M, E) \to \mathrm{Hom}(N, E)$ given by $g \mapsto gh$ is onto.

Remark 2.10.3: Any map $f: R \to E$ is given by right multiplication by an element of E (namely $f1$). In particular, if E is injective and L is a left ideal of R then any map $f: L \to E$ is given by right multiplication by some element of E (seen by first extending f to a map $\hat{f}: R \to E$). This property actually characterizes injective modules, as we now see.

Proposition 2.10.3′: ("*Baer's criterion*") E is injective iff for every left ideal L of R, each map $g: L \to E$ can be extended to a map $\hat{g}: R \to E$.

Proof: ($\Rightarrow$) *a fortiori.* ($\Leftarrow$) Suppose $f: N \to E$ and $N \leq M$; we want to extend f to M. Let $\mathscr{L} = \{(N', f'): N \leq N' \leq M$ and f is the restriction of f' to $N\}$. We make $\mathscr{L}$ a poset by putting $(N', f') \leq (N'', f'')$ if $N' \leq N''$ and f' is the restriction of f'' to N'. Clearly $\mathscr{L}$ is then inductive, and thus has some maximal (N_0, f_0). We shall conclude by proving $N_0 = M$.

Otherwise, there is x in $M - N_0$. Let $L = \{r \in R: rx \in N_0\} \leq R$, and define $g: L \to E$ by $ga = f_0(ax)$. By remark 2.10.3, g is given by right multiplication by some x' in E. Define $f_1: N_0 + Rx \to E$ by $f_1(y + rx) = f_0 y + rx'$. Then f_1 is well-defined, for if $y_1 + r_1 x = y_2 + r_2 x \in N_0 + Rx$ then $(r_2 - r_1)x = y_2 - y_1 \in N_0$ so $(r_1 - r_2) \in L$ and

$$f_1(y_1 + r_1 x) - f_1(y_2 + r_2 x) = (f_0 y_1 + r_1 x') - (f_0 y_2 + r_2 x')$$
$$= f_0(y_1 - y_2) + (r_1 - r_2)x'$$
$$= f_0(y_1 - y_2) + g(r_1 - r_2)$$
$$= f_0(y_1 - y_2 + (r_1 - r_2)x) = f_0 0 = 0.$$

Thus f_1 is a map extending f_0, contrary to the choice of f_0.   Q.E.D.

Certain properties of projective modules (e.g., proposition 2.8.3 with "free" replaced by "projective," and lemma 2.8.4) could have been formulated and proved in any abelian category $\mathscr{C}$ satisfying the axiom that for each object C

there is an epic $f: P \to C$ with P projective. (In fact, one can take P to be free but we want to stay in a categorical framework.) We shall say such a category has *enough projectives*. Then the dual results would apply automatically to injectives, provided $\mathscr{C}$ has *enough injectives* in the sense that for any object C there is a monic $g: C \to E$ with E injective. Although we shall reprove these results (2.10.13 through 2.10.15) from scratch, we see that the key must be to show every module is contained in an injective module. This is surprisingly difficult to show, but the rewards are great, since we shall end up with a theory stronger than the dual of §2.8.

Divisible Modules

In order to study injective modules further, it is convenient to introduce a related elementary concept.

Definition 2.10.4: An element x in M is *divisible* by r in R if $rx' = x$ for some x' in R. M is *divisible* by r if each element of M is divisible by r, i.e., if left multiplication by r is an epic from M to M. Finally M is a *divisible* module if M is divisible by every regular element of R.

Proposition 2.10.5:   *Every injective module E is divisible.*

Proof: Suppose $r \in R$ is regular, and $x \in E$. There is a map $f: Rr \to E$ given by $f(r'r) = r'x$, which is well-defined since $r_1 r = r_2 r$ implies $r_1 = r_2$. But f is given by right multiplication by some x' in E, so $x = fr = rx'$ proving divisibility.   Q.E.D.

There is a partial converse.

Proposition 2.10.5':   *Any divisible module over a PLID is injective.*

Proof: (By Baer's criterion). We shall extend a given map $f: L \to M$ to all of R, where $L < R$. By hypothesis $L = Ra$ for some a in R. If $a = 0$ then $f = 0$ and we are done. If $a \neq 0$ then a is regular so fa is divisible by a, and there is some x such that $ax = fa$. Fixing x, define $\hat{f}: R \to M$ by $\hat{f}r = rx$. Then $\hat{f}(ra) = rax = rfa = f(ra)$ for all r in R, as desired.   Q.E.D.

The advantage of moving back and forth from injective to divisible is that divisibility passes easily to other modules:

Remark 2.10.6: (i) Every homomorphic image of a divisible module is divisible. (ii) Every direct product of divisible modules is divisible. (iii) Every direct sum of divisible modules is divisible. (Verifications are immediate from the definition.)

Example 2.10.7: If R is an integral domain then its field of fractions Q is divisible as R-module, so Q/R is also a divisible R-module. If R is a PID then Q/R is injective by proposition 2.10.5′; in particular, $\mathbb{Q}/\mathbb{Z}$ is an injective $\mathbb{Z}$-module.

Example 2.10.8: Any $\mathbb{Z}$-module M can be embedded in an injective $\mathbb{Z}$-module. Indeed, write $M = F/K$ where F is a free $\mathbb{Z}$-module with some base $\{b_i : i \in I\}$. Then taking F' to be the free $\mathbb{Q}$-module with base $\{b_i : i \in I\}$ (i.e., $F' = F \otimes_{\mathbb{Z}} \mathbb{Q}$) and viewing $F \subset F'$, naturally we have F' divisible so F'/K is divisible (and thus injective) as $\mathbb{Z}$-module, and $M \subset F'/K$, naturally.

To proceed further we need a basic fact relating homomorphisms and tensor products. Suppose $A \in R\text{-}\mathcal{M}od\text{-}S$, $B \in S\text{-}\mathcal{M}od\text{-}T$, and $C \in R\text{-}\mathcal{M}od\text{-}W$. Using remark 1.5.18′ we have two very natural ways of forming a $T - W$ bimodule.

(i) $\operatorname{Hom}_R(A, C) \in S\text{-}\mathcal{M}od\text{-}W$ so $\operatorname{Hom}_S(B, \operatorname{Hom}_R(A, C)) \in T\text{-}\mathcal{M}od\text{-}W$.

(ii) $A \otimes_S B \in R\text{-}\mathcal{M}od\text{-}T$ so $\operatorname{Hom}_R(A \otimes B, C) \in T\text{-}\mathcal{M}od\text{-}W$.

In fact, these two constructions can be identified:

Proposition 2.10.9: (*"Adjoint isomorphism"*) Notation as above, there is an isomorphism

$$\Phi \colon \operatorname{Hom}_R(A \otimes_S B, C) \to \operatorname{Hom}_S(B, \operatorname{Hom}_R(A, C))$$

in $T\text{-}\mathcal{M}od\text{-}W$, given as follows: For f in $\operatorname{Hom}(A \otimes_S B, C)$ and b in B, define $f_b \colon A \to C$ by $f_b a = f(a \otimes b)$. Then Φf is defined by $(\Phi f)b = f_b$.

Proof: It is easy to check Φ is a $T - W$ bimodule map, using remark 1.5.18′. To construct Φ^{-1} suppose $g \colon B \to \operatorname{Hom}(A, C)$. Define $\tilde{g} \colon A \times B \to C$ by $\tilde{g}(a, b) = (gb)a$. This is balanced, so $\tilde{g}$ induces a map $A \otimes B \to C$ which we denote ψg. Clearly, ψ and Φ are inverses.   Q.E.D.

Theorem 2.10.10: *Every R-module M can be embedded in an injective R-module.*

Proof: By example 2.10.8, M can be embedded into an injective $\mathbb{Z}$-module G, which as an abelian group is a $\mathbb{Z} - \mathbb{Z}$ bimodule. Viewing $R \in \mathbb{Z}\text{-}\mathcal{M}od\text{-}R$ naturally, let $E = \text{Hom}_{\mathbb{Z}}(R, G) \in R\text{-}\mathcal{M}od$ by remark 1.5.18′. We have a series of monics $M \approx \text{End}_R(R, M) \to \text{End}_{\mathbb{Z}}(R, M) \to \text{End}_{\mathbb{Z}}(R, G) = E$, so it remains to prove the following fact:

Proposition 2.10.11: *If G is injective in $\mathbb{Z}\text{-}\mathcal{M}od$ then $E = \text{Hom}_{\mathbb{Z}}(R, G)$ is injective in $R\text{-}\mathcal{M}od$. Actually, we shall prove the following more general result:*

Proposition 2.10.12: *If G is injective in $\mathbb{Z}\text{-}\mathcal{M}od$ and F is a flat right R-module then $E = \text{Hom}_{\mathbb{Z}}(F, G)$ is injective in $R\text{-}\mathcal{M}od$ (viewing F in $\mathbb{Z}\text{-}\mathcal{M}od$ as an abelian group).*

Proof: Suppose $h: N \to M$ is monic in $R\text{-}\mathcal{M}od$. By remark 2.10.2 we should show the canonical map $h^{\#}: \text{Hom}(M, E) \to \text{Hom}(N, E)$ is epic. But $1 \otimes h: F \otimes_R N \to F \otimes_R M$ is monic since F is flat, and we have

$$\text{Hom}_R(M, E) = \text{Hom}_R(M, \text{Hom}_{\mathbb{Z}}(F, G)) \approx \text{Hom}_{\mathbb{Z}}(F \otimes_R M, G).$$

But $\text{Hom}_{\mathbb{Z}}(F \otimes M, G) \to \text{Hom}_{\mathbb{Z}}(F \otimes N, G)$ is epic by the $\mathbb{Z}$-injectivity of G; we easily translate this to $h^{\#}$ being epic. Q.E.D.

This argument can be given more precisely in terms of "exact functors," which we shall describe later. We now have "enough" injectives to dualize the projective theory and more.

Lemma 2.10.13: *Every summand of an injective module is injective.*

Proof: Suppose $E = E_1 \oplus E_2$ is injective, and let $\pi: E \to E_1$ be the projection.

Any map $f: L \to E_1$ when composed with the inclusion map $E_1 \to E$ is given by right multiplication by some x in E. But then f is given by right multiplication by πx, proving E_1 is injective by Baer's criterion. Q.E.D.

Proposition 2.10.14: *E is injective iff every monic $h: E \to M$ splits.*

Proof: ($\Rightarrow$) Taking $N = E$ and $f = 1_E$ we have $\hat{f}: M \to E$ with $1_E = \hat{f}h$, as in definition 2.10.1 proving h is split monic.

($\Leftarrow$) Let E' be an injective module containing E. Then the inclusion map $E \to E'$ splits, proving E is a summand of E' and by the lemma is injective. Q.E.D.

Remark 2.10.15: If $\{E_i : i \in I\}$ are injective then $\prod E_i$ is injective. (Indeed, given a monic $h: N \to M$ and a map $f: N \to \prod E_i$ write $f = (f_i)$ with $f_i: N \to E_i$, i.e., $fx = (f_i x)$; extending each f_i to $\hat{f_i}: M \to E_i$ let $\hat{f} = (\hat{f_i})$.)

Essential Extensions and the Injective Hull

We turn now to the dual notion of projective cover, called the *injective hull*, which turns out to be much more important (because it always exists!) We continue the discussion of large submodules, initiated in 2.4.2ff.

Definition 2.10.16: If N is a large submodule of M we say M is an *essential extension* of N, or M is *essential over* N.

Remark 2.10.17: Suppose $N < N' < M$. M is essential over N, iff M is essential over N' and N' is essential over N. (Immediate from definition 2.4.2.)

Remark 2.10.18: If N' is an essential complement of N in M then M/N' is an essential extension of N. (Indeed, $N \approx (N + N')/N' \le M/N'$. To prove N is large in M/N', suppose $N' \le K \le M$ with $(K/N') \cap N = 0$. Then $K \cap (N + N') \le N'$ so $(K \cap N) \cap (N + N') \le N \cap N' = 0$; by hypothesis, $N + N'$ is large in M so $K \cap N = 0$, implying $K = N'$ by maximality of N', so $K/N' = 0$.)

Lemma 2.10.19: *(Extension lemma) Suppose $f: M \to E$ is monic and E is injective. If N is an essential extension of M and we take $\hat{f}: N \to E$ extending f, then $\hat{f}$ also is monic.*

Proof: $M \cap \ker \hat{f} = 0$, so $\ker \hat{f} = 0$ since N is essential over M.   Q.E.D.

Theorem 2.10.20: *The following are equivalent for modules $M \le E$:*

   (i) *E is a maximal essential extension of M.*
   (ii) *E is injective and is essential over M.*
   (iii) *E is a minimal injective containing M.*

Moreover, given any module M one can find E satisfying these equivalent properties.

Proof: (i)$\Rightarrow$(ii) Take $E' \ge E$ injective and let N be an essential complement of E in E'. Then E'/N is essential over $(E + N)/N \approx E$, so the hypothesis implies equality holds, i.e., $E' = E \oplus N$ so E is injective, yielding (ii).

Next we show there exists E satisfying (i). Take E' injective containing M.

{essential extensions of M contained in E'} is inductive and non-empty (since M is essential over M) and so has a maximal element E. To see E satisfies (i) suppose $N \geq E$ is essential over M. By the extension lemma, the natural inclusion map $E \to E'$ extends to a monic $i: N \to E'$. Thus we may assume $N \leq E'$, so $E = N$ by choice of E. Hence E satisfies (i) and thus (ii), so E also is injective.

(iii) $\Rightarrow$ (i) Take E_0 maximal in {essential extensions of M contained in E}. We just saw E_0 satisfies (i) and is injective, so $E_0 = E$ by hypothesis.

(ii)$\Rightarrow$(iii) Suppose $M \leq E_0 \leq E$ with E_0 injective. Then the inclusion $E_0 \leq E$ splits so $E = E_0 \oplus N$ for some $N \leq E$; since $N \cap M = 0$ and E is essential, we conclude $N = 0$, i.e., $E = E_0$.   Q.E.D.

Definition 2.10.21:   The module E of theorem 2.10.20 is called an *injective hull* or *injective envelope* of M.

Property (ii) shows us the injective hull is the dual of the projective cover. However, whereas arbitrary projective covers do not necessarily exist (unless the ring is perfect), the injective hull of M always exists, and its use pervades ring theory. We shall now show the injective hull is unique.

Proposition 2.10.22:   *The injective hull E of M is unique up to isomorphism. In fact, if E' is another injective module containing M then the inclusion map $i: M \to E'$ extends to a monic $E \to E'$.*

Proof: The second assertion follows from lemma 2.10.19; the first assertion then follows from theorem 2.10.20 (iii) applied to E'.   Q.E.D.

Remark 2.10.23:   Writing $E(M)$ for the injective hull of M, one has the following properties:

(i) $E(N) = E(M)$ for every large submodule N of M.
(ii) $E(M) = M$ iff M is injective.
(iii) $E(M_1 \oplus M_2) = E(M_1) \oplus E(M_2)$.

(Proofs: (i) $E(M)$ is an essential extension of N, so use theorem 2.10.20(ii), which also yields (ii); to see (iii) note $E(M_1) \oplus E(M_2)$ is an essential extension of $M_1 \oplus M_2$ and is injective.)

Criteria for Injectivity and Projectivity

There are several useful equivalent conditions for a module to be injective or projective. The first one is important in the study of quotient modules and rings, and improves Baer's criterion.

Remark 2.10.24: Suppose $L \leq M$ and $f: L \to N$ is an R-module map. If L' is an essential complement of L in M then f extends to a map $\hat{f}: L \oplus L' \to N$ by $\hat{f}(x_1, x_2) = fx_1$.

Proposition 2.10.25: E is injective iff for every large left ideal L of R, every map $f: L \to E$ extends to a map $\hat{f}: R \to E$.

Proof: Apply remark 2.10.24 to Baer's criterion.   Q.E.D.

Proposition 2.10.26: E is injective if every short exact sequence $0 \to E \to M \to C \to 0$ with C cyclic, splits.

Proof: Otherwise, let E' be the injective hull of E and let E''/E be a cyclic submodule of E'/E. By hypothesis, the inclusion map $E \to E''$ splits, contrary to E' being essential over E.   Q.E.D.

Our final result along these lines links projective modules to injective modules.

Proposition 2.10.27: A module P is projective iff for each epic $h: E \to N$ with E injective, every map $f: P \to N$ can be lifted to a map $\hat{f}: P \to E$ (such that $f = h\hat{f}$).

Proof: ($\Rightarrow$) is by definition. Conversely, suppose $h: M \to N$ is epic; we want to show any map $f: P \to N$ lifts to M. Embed M in an injective module E. Then $N \approx M/\ker h$ is a submodule of $E/\ker h$; we may view $f: P \to E/\ker h$, which by hypothesis we can lift to $\hat{f}: P \to E$. But, in fact, $\hat{f}P \subseteq M$; indeed, if $x \in \hat{f}P$ then $hx \in h\hat{f}P = fP \subseteq N$ so $hx = hx'$ for some x' in M, implying $x - x' \in \ker h \subseteq M$ and thus $x \in M$. Therefore we have lifted f.   Q.E.D.

J Supplement: Krull-Schmidt Theory for Injective Modules

Injective modules have a very pleasant decomposition theory. Our first proposition is reminiscent of the Schröder-Bernstein theorem of set theory.

Lemma 2.10.28: *Suppose $M = K \oplus N$ and there is a monic $f: M \to K$. Then $\{f^i N : i \in \mathbb{N}\}$ is an independent set of isomorphic submodules of M.*

Proof: We shall show by induction that $N, fN, \ldots, f^n N$ are independent for each n. This is clear for $n = 1$. Given the induction hypothesis for n note $fN, \ldots, f^{n+1} N$ are independent since f is monic; also

$$N \cap \sum_{i=1}^{n+1} f^i N \leq N \cap fM = 0,$$

implying $N, fN, \ldots, f^{n+1} N$ are independent, as desired.   Q.E.D.

Proposition 2.10.29: *If E_1, E_2 are injective and each isomorphic to a submodule of the other then $E_1 \approx E_2$.*

Proof: We use repeatedly the fact that an injective submodule is a summand. $E_1 = E_2' \oplus N$ where $E_2 \approx E_2'$, and there is a monic $f: E_1 \to E_2'$. Thus $\{f^i N : i \in \mathbb{N}\}$ is independent. Let $A = \bigoplus_{i=0}^{\infty} f^i N \leq E_2'$; $A = fA \oplus N$. E_2' is injective and thus contains an injective hull B of fA. Hence the injective module $B \oplus N$ is an injective hull of $fA \oplus N = A$. Since $A \approx fA$ we have $B \oplus N \approx B$. Writing $E_2' = B' \oplus B$ we get

$$E_1 = E_2' \oplus N = (B' \oplus B) \oplus N = B' \oplus (B \oplus N) \approx B' \oplus B = E_2' \approx E_2.$$

Q.E.D.

Next we have the exchange property, which is treated most easily by means of a useful notion.

Definition 2.10.30: A submodule N of M is *closed* if N has no proper essential extension in M.

Lemma 2.10.31: *Any closed submodule N of an injective module E is injective.*

Proof: By theorem 2.10.20 it suffices to show N is maximal essential. But by lemma 2.10.19, any essential extension of N is isomorphic to a submodule of E and thus equals N by hypothesis.   Q.E.D.

Proposition 2.10.32: *If* $E = E_1 \oplus E_2 = E_1' \oplus E_2'$ *then there are decompositions* $E_1 = M_1 \oplus N_1$ *and* $E_2 = M_2 \oplus N_2$ *such that* $M_1 \oplus M_2 \approx E_1'$ *and* $N_1 \oplus N_2 \approx E_2'$.

Proof: Let $\pi_i : E \to E_i$ be the usual projections for $i = 1, 2$. Let $M_i = E_1 \cap E_1'$ and $M_2 = \pi_2 E_1'$. M_1 is closed in E_1, for if M_1 is large in $N \le E_1$ then clearly E_1' is large in $N + E_1'$ so $N \subseteq E_1'$ and thus $N \subseteq E_1 \cap E_1' = M_1$. Hence M_1 is injective by lemma 2.10.31, so the exact sequence $0 \to M_1 \to E_1' \to M_2 \to 0$ splits, proving $E_1' \approx M_1 \oplus M_2$.

Thus for $i = 1, 2$ we see M_i is injective and so has a complement N_i in E_i. Since $M_2 = (1 - \pi_1) E_1' \le E_1' + E_1$ we have

$$E = E_1 + M_2 + N_2 = E_1' + E_1 + N_2 = (E_1' + M_1) + N_1 + N_2 = E_1' + N_1 + N_2.$$

Let $K = E_1' \cap (N_1 + N_2)$. If we can show $K = 0$ then $N_1 \oplus N_2 \approx (E_1' + N_1 + N_2)/E_1' = E/E_1' \approx E_2'$ as desired. But $\pi_2 K \le \pi_2 E_1' \cap \pi_2 (N_1 + N_2) = M_2 \cap N_2 = 0$, and so $K \le E_1 \cap E_1' = M_1$. Hence $\pi_1 K \le M_1 \cap \pi_1 (N_1 + N_2) \le M_1 \cap N_1 = 0$, as desired. Q.E.D.

Theorem 2.10.33: *Suppose E is a finite direct sum of indecomposable injectives $E_1 \oplus \cdots \oplus E_n$. Then every decomposition of E is equivalent to this decomposition.*

Proof: Use proposition 2.10.32 in the argument of theorem 2.9.17.

Q.E.D.

§2.11 Exact Functors

In this section we continue the study of projective and injective modules by tying them in with exact sequences, using the very important idea of *exactness* of functors. This leads us to characterize various kinds of rings and modules in terms of exact sequences, especially using flat modules. The motivation springs from some easy observations about R-$\mathcal{M}od$; the interested reader will have no difficulty in seeing how to generalize to arbitrary abelian categories. A functor F is *additive* if for all maps $f, g : M \to N$ we have $F(f + g) = Ff + Fg$. Then if $M = M_1 \oplus M_2$ we must have $FM = FM_1 \oplus FM_2$. Indeed, if $\pi_i : M \to M_i$ are the projections and $\mu_i : M_i \to M$ are the canonical monics for $i = 1, 2$ satisfying $\pi_i \mu_i = 1_{M_i}$ and $\mu_1 \pi_1 + \mu_2 \pi_2 = 1_M$, then $F\pi_i F\mu_i = 1_{FM_i}$ and $F\mu_1 F\pi_1 + F\mu_2 F\pi_2 = 1_{FM}$. We have proved

Remark 2.11.1: Any additive functor $F: R\text{-}\mathcal{M}od \to W\text{-}\mathcal{M}od$ preserves split exact sequences, in the sense that if $0 \to K \xrightarrow{f} M \xrightarrow{g} N \to 0$ is split exact then so is $0 \to FK \xrightarrow{Ff} FM \xrightarrow{Fg} FN \to 0$.

What happens when the given sequence merely is exact? To handle this situation we formulate the following definition.

Definition 2.11.2: A functor $F: R\text{-}\mathcal{M}od \to W\text{-}\mathcal{M}od$ is *exact* if whenever $K \xrightarrow{f} M \xrightarrow{g} N$ is exact we have $FK \xrightarrow{Ff} FM \xrightarrow{Fg} FN$ is exact.

Piecing together this definition, we see an exact functor "preserves" the exactness of every exact sequence. However, it is convenient to focus on short exact sequences.

Proposition 2.11.3: *A functor F is exact iff whenever $0 \to K \xrightarrow{f} M \xrightarrow{g} N \to 0$ is exact then $0 \to FK \xrightarrow{Ff} FM \xrightarrow{Fg} FN \to 0$ is exact.*

Proof: ($\Rightarrow$) from the above discussion. ($\Leftarrow$) One can cut a given exact sequence $K \xrightarrow{f} M \xrightarrow{g} N$ into two sequences:

$$0 \to \ker f \to K \xrightarrow{f} fK \to 0$$

$$0 \to \ker g \to M \xrightarrow{g} gM \to 0$$

Applying F to each of these sequences shows $(Ff)FK = F(fK) = F(\ker g) = \ker(Fg)$ as desired.   Q.E.D.

This criterion is useful because it usually is *not* satisfied! This paradox will become clear when we introduce another definition:

Definition 2.11.4: F is a *left exact* functor if $0 \to K \xrightarrow{f} M \xrightarrow{g} N$ exact implies $0 \to FK \xrightarrow{Ff} FM \xrightarrow{Fg} FN$ is exact; F is *right exact* if $K \xrightarrow{f} M \xrightarrow{g} N \to 0$ exact implies $FK \xrightarrow{Ff} FM \xrightarrow{Fg} FN \to 0$ is exact.

A contravariant functor F is *left exact* if $K \xrightarrow{f} M \xrightarrow{g} N \to 0$ exact implies $0 \to FN \xrightarrow{Fg} FM \xrightarrow{Ff} FK$ is exact.

Viewing $\text{Hom}(A, \_\_)$ and $\text{Hom}(\_\_, A)$ as functors from $R\text{-}\mathcal{M}od$ to $\mathbb{Z}\text{-}\mathcal{M}od$ we have the following important examples.

Proposition 2.11.5: $\text{Hom}(A, \_\_)$ *is a left exact functor and* $\text{Hom}(\_\_, A)$ *is a left exact contravariant functor for any R-module A.*

Proof: We show the contravariant functor $F = \operatorname{Hom}(\underline{\quad}, A)$ is left exact; the other verification is analogous. Suppose $K \xrightarrow{f} M \xrightarrow{g} N \to 0$ is exact. If $h \in FN = \operatorname{Hom}(N, A)$ with $0 = (Fg)h = hg$ then $h = 0$ since g is epic, proving $0 \to FN \xrightarrow{Fg} FM$ is exact. To see $FN \xrightarrow{Fg} FM \xrightarrow{Ff} FK$ is exact we first note $(Ff)(Fg) = F(gf) = 0$.

It remains to show that if $h: M \to A$ with $(Ff)h = 0$ then $h = h'g$ for some $h': N \to A$. To obtain h' write any x in N as gy for suitable y in M, and define $h'x$ to be hy. To see h' is indeed well-defined suppose $gy' = gy = x$; then $y - y' \in \ker g = fK$ so $h(y - y') \in hfK = ((Ff)h)K = 0$, implying $hy = hy'$ as desired.   Q.E.D.

Corollary 2.11.6: $\operatorname{Hom}(P, \underline{\quad})$ *is an exact functor iff P is a projective module.* $\operatorname{Hom}(\underline{\quad}, E)$ *is an exact contravariant functor iff E is an injective module.*

Proof: Proposition 2.8.3 and remark 2.10.2, coupled with proposition 2.11.5.
                                                                            Q.E.D.

We can use these results to characterize semisimple Artinian rings in terms of exact functors.

Remark 2.11.7:   Recall a ring R is semisimple Artinian iff every short exact sequence in R-$\mathcal{M}od$ splits. It follows at once that R is semisimple Artinian iff every R-module is projective, iff every R-module is injective.

Proposition 2.11.8:   *R is semisimple Artinian iff for every ring W, every additive functor $F: R\text{-}\mathcal{M}od \to W\text{-}\mathcal{M}od$ is exact.*

Proof: ($\Rightarrow$) Every short exact sequence in R-$\mathcal{M}od$ splits, implying F is exact by remark 2.11.1 and proposition 2.11.3.

($\Leftarrow$) Take $W = \mathbb{Z}$. By hypothesis, for every R-module P, $\operatorname{Hom}(P, \underline{\quad})$ is exact so P is projective; hence R is semisimple Artinian by remark 2.11.7.
                                                                            Q.E.D.

Using $\operatorname{Hom}(\underline{\quad}, E)$ one could prove the analogous assertion for contravariant functors. One can also define functors by means of the tensor product.

Proposition 2.11.9:   *For $A \in R\text{-}\mathcal{M}od\text{-}T$ define $A \otimes \underline{\quad}: T\text{-}\mathcal{M}od \to R\text{-}\mathcal{M}od$ sending M to $A \otimes_T M$; define $\underline{\quad} \otimes A: \mathcal{M}od\text{-}R \to \mathcal{M}od\text{-}T$ by $M \to M \otimes_R A$.*

Then $A \otimes$ ___ and ___ $\otimes A$ are right exact functors, where one sends $f: M \to N$, respectively, to $1 \otimes f: A \otimes M \to A \otimes N$ and $f \otimes 1: M \otimes A \to N \otimes A$.

Proof: These are clearly functors, and it remains to prove right exactness. Suppose $K \xrightarrow{f} M \xrightarrow{g} N \to 0$ is exact; we aim to show $A \otimes K \xrightarrow{1 \otimes f} A \otimes M \xrightarrow{1 \otimes g} A \otimes N \to 0$ is right exact. (The other verification is analogous.) Clearly, $(1 \otimes g)(A \otimes M) = A \otimes gM = A \otimes N$. Likewise, $(1 \otimes f)(A \otimes K) = A \otimes fK = A \otimes \ker g$, which is $\ker(1 \otimes g)$ by proposition 1.7.30.   Q.E.D.

Corollary 2.11.10:   *The functor* ___ $\otimes A$ *of proposition 2.11.9 is exact iff A is a flat R-module.*

Proof: The missing part of exactness is precisely the definition of flatness.
                                                                                    Q.E.D.

Remark 2.11.10′:   Let us analyze this in a particular case. Let $A \lhd R$ and $F: R\text{-}\mathcal{M}od \to (R/A)\text{-}\mathcal{M}od$ be given by $M \to M/AM$. By example 1.7.21′ this functor is given by tensoring, and thus is right exact. On the other hand, for F to be exact we must have R/A flat as right R-module.

Thus we see projective, injective, and flat each play a role in making a suitable functor exact.

Flat Modules and Injectives

We shall continue the study of flat modules by relating them to injectives. The key here is the *character module* $M^{\#} = \text{Hom}_{\mathbb{Z}}(M, \mathbb{Q}/\mathbb{Z})$. Besides being injective in $\mathbb{Z}\text{-}\mathcal{M}od$, $\mathbb{Q}/\mathbb{Z}$ satisfies the following useful property:

Remark 2.11.11:   If G is an arbitrary abelian group and $g \in G$ then there is a group homomorphism $\varphi: G \to \mathbb{Q}/\mathbb{Z}$ with $\varphi g \neq 0$. (Indeed, by injectivity of $\mathbb{Q}/\mathbb{Z}$ in $\mathbb{Z}\text{-}\mathcal{M}od$ it suffices to take $G = \mathbb{Z}g$. If g has finite order t define $\varphi: G \to \mathbb{Q}/\mathbb{Z}$ by $\varphi(ng) = n/t + \mathbb{Z}$; if g has infinite order define φ by $\varphi(ng) = n/2 + \mathbb{Z}$).

Lemma 2.11.12:   *A sequence $K \xrightarrow{f} M \xrightarrow{g} N$ of maps is exact iff $N^{\#} \xrightarrow{g^{\#}} M^{\#} \xrightarrow{f^{\#}} K^{\#}$ is exact.*

Proof: ($\Rightarrow$) The contravariant functor $\text{Hom}_{\mathbb{Z}}(\text{___}, \mathbb{Q}/\mathbb{Z})$ is exact since $\mathbb{Q}/\mathbb{Z}$ is injective.

($\Leftarrow$) Take any $k \in K$. For every $\varphi: N \to \mathbb{Q}/\mathbb{Z}$ we have $\varphi(gfk) = (f^{\#}g^{\#}\varphi)k = 0$; by remark 2.11.11 we conclude $gfk = 0$. Thus $fK \subseteq \ker g$.

Now given $\varphi: M/fK \to \mathbb{Q}/\mathbb{Z}$ we define $\psi: M \to \mathbb{Q}/\mathbb{Z}$ by putting $\psi x = \varphi(x + fK)$. Then $0 = \psi f = f^{\#}\psi$ so $\psi = g^{\#}\rho = \rho g$ for some ρ in $N^{\#}$, implying $\psi(\ker g) = 0$. This proves $\varphi((\ker g)/fK) = 0$ for any such φ, implying $(\ker g)/fK = 0$ by remark 2.11.11, and $\ker g = fK$.   Q.E.D.

Theorem 2.11.13:   *An R-module F is flat iff $F^{\#}$ is injective in $\mathcal{M}od$-R.*

Proof: ($\Rightarrow$) By proposition 2.10.12.

($\Leftarrow$) Any monic $h: N \to M$ in $\mathcal{M}od$-R induces an epic $\operatorname{Hom}_R(M, F^{\#}) \to \operatorname{Hom}_R(N, F^{\#})$ which by the adjoint isomorphism (2.10.9) translates to an epic $(M \otimes F)^{\#} \to (N \otimes F)^{\#}$, which in turn provides the monic $h \otimes 1: N \otimes F \to M \otimes F$ by lemma 2.11.12.   Q.E.D.

We are finally ready for a useful intrinsic characterization of flatness.

Theorem 2.11.14:   *The following conditions are equivalent for an R-module F:*

(i) *F is flat.*

(ii) *For any right ideal I of R, the canonical map $\psi: I \otimes F \to R \otimes F$ is monic.*

(iii) *For any right ideal I which is f.g. (as right R-module), the canonical map $I \otimes F \to R \otimes F$ is monic.*

(iv) *For any right ideal I the map $\varphi: I \otimes F \to IF$ given by $\varphi(a \otimes x) = ax$ is monic (of course, φ is always epic.)*

(v) *As in (iv), for any f.g. right ideal I.*

Proof: (i) $\Rightarrow$ (ii) by definition.

(ii) $\Rightarrow$ (i) $(R \otimes F)^{\#} \to (I \otimes F)^{\#}$ is epic so the adjoint isomorphism shows $\operatorname{Hom}_R(R, F^{\#}) \to \operatorname{Hom}_R(I, F^{\#})$ is epic, i.e., $F^{\#}$ is injective by Baer's criterion, so F is flat.

(ii) $\Rightarrow$ (iii) *is a fortiori*; conversely, (iii) $\Rightarrow$ (ii) since any right ideal is a direct limit of f.g. right ideals, enabling us to use proposition 1.8.10.

(ii) $\Leftrightarrow$ (iv) $R \otimes_R F \approx F = RF$ so we consider the commutative diagram

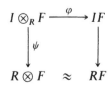

which shows ψ is monic iff φ is monic.

(iii) $\Leftrightarrow$ (v) As in (ii) $\Leftrightarrow$ (iv).   Q.E.D.

Condition (v) is sometimes called the "flat test" and can be used to provide yet another criterion for flatness, by turning to a technique of great interest in the sequel.

Proposition 2.11.15: *("The 5 lemma") Suppose we have the following commutative diagram of module maps in which the rows are exact:*

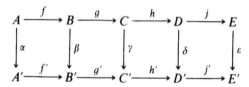

(i) *If α is epic and β and δ are monic then γ is monic.*

(ii) *If β and δ are epic and ε is monic then γ is epic.*

(iii) *If every vertical arrow but γ is an isomorphism then γ is also an isomorphism.*

Proof: The technique of proof is called *diagram-chasing*, and the proof actually shows what is needed in the hypothesis.

(i) To show γ is monic, suppose $\gamma c = 0$. Then $0 = h'\gamma c = \delta hc$; δ monic implies $c \in \ker h = gB$. Writing $c = gb$ we have $0 = \gamma c = \gamma gb = g'\beta b$ so $\beta b = f'a'$ for some a' in A'; writing $a' = \alpha a$ yields $\beta fa = f'\alpha a = f'a' = \beta b$, so $fa = b$ and $c = gb = gfa = 0$.

(ii) To show γ is epic suppose $c' \in C'$. Then $h'c' = \delta d$ for some d, and $0 = j'h'c' = j'\delta d = \varepsilon j d$, implying $jd = 0$. Hence $d = hc$ for some c. Now $h'(c' - \gamma c) = h'c' - h'\gamma c = \delta d - \delta hc = 0$ so $c' - \gamma c = g'b'$ for some b'. But $b' = \beta b$ for suitable b in B, so $\gamma(c + gb) = \gamma c + g'\beta b = \gamma c + g'b' = c'$, proving $c' \in \gamma C$.

(iii) Combine (i) and (ii).   Q.E.D.

Corollary 2.11.16: *Suppose we have the following commutative diagram in which the rows are exact, and ρ is an isomorphism.*

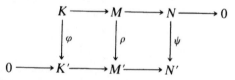

Then φ is epic iff ψ is monic.

Proof: ($\Rightarrow$) Take $A = K$, $B = M$, $C = N$, and $D = E = 0$ in proposition 2.11.15(i). ($\Leftarrow$) Take $A = B = 0$, $C = K$, $D = M$ and $E = N$ in (ii).   Q.E.D.

Proposition 2.11.17:   *Suppose F is flat and $0 \to K \to F \to M \to 0$ is exact. The following assertions are then equivalent:*

(i) M *is flat.*
(ii) $IK = K \cap IF$ *for every right ideal I of R.*
(iii) $IK = K \cap IF$ *for every f.g. right ideal I of R.*

Proof: Apply Corollary 2.11.15 to the diagram

$$
\begin{array}{ccccccc}
I \otimes K & \longrightarrow & I \otimes F & \longrightarrow & I \otimes M & \longrightarrow & 0 \\
\downarrow{\scriptstyle\varphi} & & \downarrow & & \downarrow{\scriptstyle\psi} & & \\
0 \longrightarrow K \cap IF & \longrightarrow & IF & \longrightarrow & IM & &
\end{array}
$$

noting that ψ monic is the flat test, and φ is epic iff $IK = K \cap IF$.   Q.E.D.

We continue the study of flat modules in exercises 4ff; one useful result (exercise 8) is that every finitely presented flat module is projective.

Regular Rings

Recall R is semisimple Artinian iff every R-module is projective, iff every R-module is injective. This raises the question, "When is every R-module flat?" The answer is quite easy.

Definition 2.11.18: R is a *(von Neumann) regular* ring if for each r in R there exists s such that $rsr = r$.

Note this definition is left-right symmetric, so we could interchange "left" and "right" in what follows.

Remark 2.11.19:   (i) Notation as above, rs and sr are idempotents of R. Moreover, $Rr = Rsr$ (since $Rr = Rrsr \subseteq Rsr \subseteq Rr$). Consequently, every principal left ideal is generated by an idempotent.
(ii) Since $\text{Jac}(R)$ contains no nontrivial idempotents, (i) implies $\text{Jac}(R) = 0$.

The connection with flat modules is in the next result.

Proposition 2.11.20:   *The following conditions are equivalent for a ring R:*

(i) R *is regular.*
(ii)   *Every principal right ideal is generated by an idempotent.*

(iii) *Every f.g. right ideal is generated by an idempotent.*

(iv) *Every f.g. right ideal is a summand of R.*

(v) *Every R-module is flat.*

Proof: (i) $\Rightarrow$ (ii) by remark 2.11.19 (right-hand version).

(ii) $\Rightarrow$ (iii) It suffices to show that if $e_1 R$ and $e_2 R$ are principal right ideals then $I = e_1 R + e_2 R$ is principal; by hypothesis, we may assume e_1 and e_2 are idempotent. $I = e_1 R + (e_2 - e_1 e_2)R$; write $(e_2 - e_1 e_2)R = eR$ for e idempotent. Then $e_1 e \in e_1(e_2 - e_1 e_2)R = 0$; so $e = (e - e_1)e \in (e - e_1)R$ and $e_1 = e - (e - e_1) \in (e - e_1)R$, proving $I = (e - e_1)R$, as desired.

(iii) $\Rightarrow$ (iv) Obvious.

(iv) $\Rightarrow$ (v) By theorem 2.11.14(v); if I is an f.g. right ideal then taking a complement I' of I in R we have for any module F:

$$(I \otimes F) \oplus (I' \otimes F) \approx (I \oplus I') \otimes F \approx R \otimes F \approx F = IF \oplus I'F,$$

so the restriction $I \otimes F \to IF$ is monic.

(v) $\Rightarrow$ (i) Take $F = R$, $K = Rr$, and $I = rR$ in proposition 2.11.17; then $r \in K \cap I = K \cap IF = IK = rRr$.   Q.E.D.

Corollary 2.11.21: *R is semisimple Artinian iff R is Noetherian and regular.*

Proof: ($\Rightarrow$) R is regular by (v) and is Noetherian by Levitzki-Hopkins.

($\Leftarrow$) Every right ideal is f.g. and thus a summand of R, by (iv), implying R has a composition series in $\mathcal{M}od$-R.   Q.E.D.

The next example of a regular ring is more typical.

Example 2.11.22: If an R-module M is semisimple then $\text{End}_R M$ is regular. (Indeed, given $f: M \to M$ let M_1 be a complement of $\ker f$. The restriction $f: M_1 \to fM$ is an isomorphism whose inverse $f: fM \to M_1$ can be extended to M as in remark 2.10.24. Clearly $fgf = f$.)

The abstract theory of noncommutative regular rings has been a disappointment. For example, the following two properties of commutative regular rings were hoped to hold in general but are known now to have counterexamples.

1. Is every prime regular ring primitive? (No, cf., example 8.1.40)
2. If all prime images of a semiprime ring R are regular then is R regular? (No, cf., exercise 16.)

Some of the basic properties of regular rings are given in exercises $17-32$, and regular quotient rings are discussed in §3.3. Further information can be found in Goodearl's books and Jacobson [64B].

§2.12 The Prime Spectrum

In this section we introduce $\text{Spec}(R)$, one of the main themes in ring theory. After establishing some basic properties we turn to the use of prime ideals in studying the rank of a projective module. Then we look at various non-commutative versions of the Nullstellensatz, and finally compare Spec of ring extensions.

Definition 2.12.1: Define the *prime spectrum* $\text{Spec}(R) = \{\text{proper prime ideals of } R\}$, partially ordered under set inclusion. Given $A \subseteq R$ define $\mathscr{S}(A) = \{P \in \text{Spec}(R): A \subseteq P\}$.

Jacobson [64B] introduced the *primitive spectrum*, defined analogously as $\{\text{primitive ideals of } R\}$, but the prime ideals were catapulted into prominence by Goldie's theorems (c.f., §3.2 below). Later we shall relate the primitive and prime spectrums in certain cases; first we consider the following important topology.

Remark 2.12.2: There is topology on $\text{Spec}(R)$ having $\{\mathscr{S}(A): A \subseteq R\}$ as the closed sets, as a consequence of the following facts:

(i) $\varnothing = \mathscr{S}(\{1\})$ and $\text{Spec}(R) = \mathscr{S}(\{0\})$
(ii) $\bigcap \mathscr{S}(A_i) = \mathscr{S}(\bigcup A_i)$
(iii) $\mathscr{S}(A_1) \cup \mathscr{S}(A_2) = \mathscr{S}(A_1 R A_2)$
(iv) If $A_1 \subseteq A_2$ then $\mathscr{S}(A_1) \supseteq \mathscr{S}(A_2)$

In the opposite direction define $\mathscr{I}: \text{Spec}(R) \to \{\text{Ideals of } R\}$ by $\mathscr{I}(\mathscr{S}) = \bigcap\{P \in \mathscr{S}\}$ for any $\mathscr{S}$ in $\text{Spec}(R)$. Write $\mathscr{I}(A)$ for $\mathscr{I}(\mathscr{S}(A))$.

Remark 2.12.3:

(i) $\mathscr{I}(A)$ is a semiprime ideal of R and $A \subseteq \mathscr{I}(A)$, equality holding iff A is a semiprime ideal of R.

(ii) If $A \subseteq B \subseteq \mathcal{I}(A)$ then $\mathcal{S}(A) = \mathcal{S}(B)$. Thus $\mathcal{S}(\quad)$ is determined by the semiprime ideals.

(iii) $\mathcal{S}(A) = \mathcal{S}(RAR)$ by (ii); consequently, for $A_i \triangleleft R$ we have

$$\bigcap_{i \in I} \mathcal{S}(A_i) = \mathcal{S}(\sum A_i).$$

Partially ordering the ideals of R by means of reverse inclusion, i.e., $A_1 < A_2$ iff $A_2 \subset A_1$, we see $\mathcal{S}(\quad)$ is an order-preserving function from $\{$ideals of $R\}$ to $\{$closed sets of $\mathrm{Spec}(R)\}$, whose restriction to $\{$semiprime ideals of $R\}$ is 1:1 and onto. Thus the topological properties of $\mathrm{Spec}(R)$ take an algebraic significance. It is natural to ask which topological spaces can be of the form $\mathrm{Spec}(R)$, and we can make the following reduction to R semiprime:

Remark 2.12.4: Let $\bar{R} = R/\mathrm{Prime}\ \mathrm{rad}(R)$, with $\bar{}$ denoting the canonical homomorphic image. Then $\mathcal{S}(A) \to \mathcal{S}(\bar{A})$ defines a homeomorphism from $\mathrm{Spec}(R)$ to $\mathrm{Spec}(\bar{R})$. (Indeed, $P \to \bar{P}$ gives a 1:1 correspondence of the prime ideals; moreover, $A \subseteq P$ iff $\bar{A} \subseteq \bar{P}$.)

Certain standard topological concepts have significance in this discussion.

Proposition 2.12.5: (i) $\mathrm{Spec}(R)$ *satisfies the finite intersection property.* (ii) *Every point in* $\mathrm{Spec}(R)$ *is closed iff every prime ideal of* R *is maximal.* (iii) *If* $R = R_1 \times R_2$ *then* $\mathrm{Spec}(R_1)$ *and* $\mathrm{Spec}(R_2)$ *are clopen subsets of* R. (iv) $\mathrm{Spec}(R)$ *is connected iff* $R/\mathrm{Prime}\ \mathrm{rad}(R)$ *has no central idempotents.* (v) *For* R *commutative the clopen subsets of* R *are in* 1:1 *correspondence to* $\{\mathcal{S}(e): e$ *idempotent in* $R\}$. (vi) *The closure of a set* $\{P_i: i \in I\}$ *is* $\mathcal{S}(\bigcap_{i \in I} P_i)$; *in particular,* $\{P_i: i \in I\}$ *is dense iff* $\bigcap P_i = \mathrm{Prime}\ \mathrm{Rad}(R)$.

Proof:

(i) Suppose $A_i \triangleleft R$ with $\varnothing = \bigcap_{i \in I} \mathcal{S}(A_i) = \mathcal{S}(\sum A_i)$. Then $\sum A_i$ is not contained in any maximal ideal so $1 \in \sum A_i$ implying $1 \in \sum_{u=1}^{t} A_{i_u}$ for suitable $i_1, \ldots, i_t$; thus $\bigcap_{u=1}^{t} \mathcal{S}(A_{i_u}) = \mathcal{S}(\sum A_{i_u}) = \varnothing$.

(ii) $\{P\} \in \mathrm{Spec}(R)$ is closed iff no other prime ideal contains P, so P is maximal.

(iii) Writing $\pi: R \to R_1$ for the canonical projection let $e = \pi 1 \in Z(R)$. Then $\mathcal{S}(e) \cup \mathcal{S}(1-e) = \mathcal{S}(eR(1-e)) = \mathcal{S}(Re(1-e)) = \mathcal{S}(0) = \mathrm{Spec}(R)$ and $\mathcal{S}(e) \cap \mathcal{S}(1-e) = \mathcal{S}(Re + R(1-e)) = \mathcal{S}(R) = \varnothing$; thus $\mathcal{S}(e) = \mathrm{Spec}(R_2)$ and $\mathcal{S}(1-e) = \mathrm{Spec}(R_1)$ are clopen.

(iv) We may assume $\mathrm{Prime}\ \mathrm{rad}(R) = 0$ by remark 2.12.4. ($\Rightarrow$) By (iii). ($\Leftarrow$) Suppose $\mathrm{Spec}(R)$ has clopen sets $\mathcal{S}_1$ and $\mathcal{S}_2 = \mathrm{Spec}(R) - \mathcal{S}_1$, and let $A_i = \mathcal{I}(\mathcal{S}_i)$. Then $A_1 \cap A_2 = \mathcal{I}(\mathcal{S}_1 \cup \mathcal{S}_2) = 0$ and $A_1 + A_2 = R$ (since otherwise

$A_1 + A_2$ would be contained in a proper maximal ideal $P \in \mathscr{S}_1 \cap \mathscr{S}_2 = \varnothing$.)
By the Chinese Remainder Theorem $R \approx R/A_1 \times R/A_2$.

(v) By (iv), since idempotents lift up the prime radical.

(vi) $\mathscr{S}(\bigcap_{i \in I} P_i)$ is closed by definition; on the other hand, if each $P_i \in \mathscr{S}(A)$ for suitable A then $A \subseteq P_i$ for each i and thus $A \subseteq \bigcap_{i \in I} P_i$. This proves the first assertion, and the second is an instant application.   Q.E.D.

Many results of ring theory depend on selecting the correct set of primes having intersection 0, i.e., a dense subset of Spec(R).

Corollary 2.12.5': *Suppose R is prime and $\{P_i: i \in I\}$ is dense (in Spec(R)). Then for any $0 \neq r \in R$ we have $\{P_i: r \notin P_i\}$ is dense.*

Proof: Let $A = \bigcap\{P_i: r \notin P_i\}$. Then $rA \subseteq \bigcap\{P_i: r \in P_i\} \cap A = \bigcap\{P_i: i \in I\} = 0$, so $A = 0$ since R is prime.   Q.E.D.

Remark 2.12.6: If $f: R_1 \to R_2$ is a ring homomorphism for which R_2 is a centralizing extension of fR_1 then $f^{-1}: \mathrm{Spec}(R_2) \to \mathrm{Spec}(R_1)$ is continuous.

Proposition 2.12.7 ("Prime avoidance"): *If $A, B \lhd R$ and $P_1, \ldots, P_t \in \mathrm{Spec}(R)$ with $A - B \subseteq \bigcup_{i=1}^{t} P_i$ then either $A \subseteq B$ or $A \subseteq P_i$ for some i.*

Proof: To make the argument symmetric we define $P_{t+1} = B$ (not necessarily prime) and write $A \subseteq \bigcup_{i=1}^{t+1} P_i$. We shall prove by induction on t that $A \subseteq P_i$ for some i. For $t = 0$ this is clear so assume the result is true for $t - 1$. We are done unless for each j there is $a_j \in A - \bigcup\{P_i: i \neq j\}$; then each $a_j \in P_j$. Note $a_2 R a_3 R \cdots a_t R a_{t+1} \not\subseteq P_1$ so there is some a in $a_2 R a_3 R \cdots a_t R a_{t+1} - P_1$. Then $a \in \bigcap_{j=2}^{t+1} P_j$ so $a_1 + a \notin P_j$ for each j, contrary to $a_1 + a \in A$.   Q.E.D.

There is one more fundamental result about prime ideals which will be of frequent use. Recall by remark 2.6.10 that minimal prime ideals exist.

Proposition 2.12.8: *Suppose $P_1, \ldots, P_n$ are incomparable prime ideals of R for $n > 1$ with $\bigcap_{i=1}^{n} P_i = 0$. Then $\{P_1, \ldots, P_n\}$ contains the set of minimal prime ideals of R.*

Proof: For any prime ideal P we have $P_1 \cdots P_n = 0 \subseteq P$, so some $P_i \subseteq P$.   Q.E.D.

Localization and the Prime Spectrum

Let us turn now to the effect of central localization on the prime spectrum.

Remark 2.12.9: Suppose S is a submonoid of $Z(R)$ and P is a prime ideal of R with $P \cap S = \varnothing$. Then P is S-prime. (Indeed, if $sr \in P$ then $sRr \subseteq P$ so $r \in P$.) Consequently, if $s^{-1}r \in S^{-1}P$ then $r \in P$, by lemma 1.10.9.)

Proposition 2.12.9′: *The correspondence Φ of proposition 1.10.8 induces a homeomorphism from $\{P \in \text{Spec}(R): P \cap S = \varnothing\}$ (with the topology induced from $\text{Spec}(R)$) to $\text{Spec}(S^{-1}R)$.*

Proof: Remark 2.12.9 shows that theorem 1.10.11 can be used to show Φ acts 1:1 on $\{P \in \text{Spec}(R): P \cap S = \varnothing\}$. On the other hand, if $v: R \to S^{-1}R$ is given by $vr = 1^{-1}r$ then for any P' in $\text{spec}(S^{-1}R)$ we have $v^{-1}P' \in \text{Spec}(R)$ and clearly $\Phi(v^{-1}P') = P'$, yielding an (onto) 1:1 correspondence. The homeomorphism follows at once from the following observation:

Lemma 2.12.10: *Suppose $P \in \text{Spec}(R)$ and $P \cap S = \varnothing$. Then $A \subseteq P$ iff $S^{-1}A \subseteq S^{-1}P$.*

Proof: ($\Rightarrow$) is obvious. ($\Leftarrow$) If $a \in A$ then $1^{-1}a \in S^{-1}A \subseteq S^{-1}P$ implying $a \in P$.   Q.E.D.

Define $U(a) = \{P \in \text{Spec}(R): a \notin P\} = \text{Spec}(R) - \mathscr{S}(\{a\})$, an open set. Clearly, the $U(a)$ are a base of $\text{Spec}(R)$ since any open set has the form $R - \mathscr{S}(A) = \bigcup\{U(a): a \in A\}$.

Corollary 2.12.11: $\text{Spec}(R[s^{-1}])$ *is homeomorphic to the open set $U(s)$ of* $\text{Spec}(R)$, *for any $s \in Z(R)$.*

Remark 2.12.12: If S is a submonoid of $Z(R)$ and $A \triangleleft R$ with $A \cap S = \varnothing$ then $S^{-1}R/S^{-1}A \approx \bar{S}^{-1}\bar{R}$, writing ‾ for the canonical image in R/A. (Indeed, the surjection $S^{-1}R \mapsto \bar{S}^{-1}\bar{R}$ given by $s^{-1}r \mapsto \bar{s}^{-1}\bar{r}$ has kernel $\{s^{-1}r: s'r \in A$ for some s' in $S\} = S^{-1}A$.)

Localizing at Central Prime Ideals

Definition 2.12.13: Given $P \in \text{Spec}(Z(R))$ let $S = Z(R) - P$; write R_P for $S^{-1}R$, and for any R-module M write M_P for $S^{-1}M$. (This notation may be confusing initially but is unambiguous since $0 \in P$, and we certainly would not invert 0 in a ring.) If $A \subseteq R$ write A_P for $S^{-1}A$.

The next result shows how localization gets its name.

Example 2.12.14: If C is a commutative ring and $P \in \text{Spec}(C)$ then C_P is a local ring whose unique maximal ideal is P_P, and $C_P/P_P \approx$ field of fractions of C/P. Indeed, let $S = C - P$. Then P is maximal with respect to $P \cap S = \emptyset$, implying by proposition 2.12.9' that P_P is the unique maximal ideal of C_P, so C_P is local; the last assertion is remark 2.12.12.

Proposition 2.12.15: *Let $Z = Z(R)$.*

(i) *If $M \in R\text{-}\mathcal{M}od$ then there is a monic $M \to \prod\{M_P : P$ maximal ideal of $Z(R)\}$ given by $x \to (1^{-1}x)$.*

(ii) *If M is $(f.g.)$ projective in $Z\text{-}\mathcal{M}od$ then M_P is free in $Z_P\text{-}\mathcal{M}od$ for every P in $\text{Spec}(Z)$.*

Proof:

(i) Suppose $x \in M$. Then $1 \notin \text{Ann}_Z x$, which is thus contained in a maximal ideal P of Z; hence $sx \neq 0$ for each s in $Z - P$ so $1^{-1}x \neq 0$ in M_P.

(ii) $M_P \approx Z_P \otimes M$ is projective over the local ring Z_P and is thus free (cf., example 2.8.9 and the discussion following it).   Q.E.D.

We should like to examine now what happens to free modules under central localization; the key to the puzzle lies with finitely presented modules.

Remark 2.12.16: (relying heavily on proposition 1.10.5) If M is a finitely presented module and S is a submonoid of $Z(R)$ then $S^{-1}M$ is a finitely presented $S^{-1}R$-module. (Indeed, if $f: F \to M$ is epic with F f.g. free and $\ker f$ is f.g. then f extends to an epic $\hat{f}: S^{-1}F \to S^{-1}M$ and $\ker \hat{f} = S^{-1}(\ker f)$ is f.g. in $S^{-1}R\text{-}\mathcal{M}od$; since $S^{-1}F$ is f.g. free we see $S^{-1}M$ is finitely presented.)

Proposition 2.12.17: *Suppose M is finitely presented and $S^{-1}M$ is free in $S^{-1}R\text{-}\mathcal{M}od$. Then there is s in S such that $M[s^{-1}]$ is already free in $R[s^{-1}]\text{-}\mathcal{M}od$, with a base of the same size as the base of $S^{-1}M$ over $S^{-1}R$.*

Proof: Pick a base $z^{-1}x_1, \dots, z^{-1}x_m$ of $S^{-1}M$ as $S^{-1}R$-module, where $z \in S$ and $x_i \in M$. Although not necessarily free, $M[z^{-1}]$ is spanned by $z^{-1}x_1, \dots, z^{-1}x_m$ and is finitely presented in $R[z^{-1}]\text{-}\mathcal{M}od$ by remark 2.12.16. Let F be a free $R[z^{-1}]$-module with base $e_1, \dots, e_m$ and define $f: F \to M[z^{-1}]$ by $fe_i = z^{-1}x_i$. Then $\ker f$ is f.g. by proposition 2.8.29 so we can write $\ker f = \sum_{j=1}^{t} R[z^{-1}]y_j$ for suitable y_j in F. But $S^{-1}\ker f = 0$ by proposition 1.10.5 so there are s_j in S with $s_j y_j = 0$. Writing $s = s_1 \cdots s_t z$ yields $s \ker f = 0$, so the induced map $\tilde{f}: R[s^{-1}]^{(t)} \to M[s^{-1}]$ is an isomorphism, i.e., $M[s^{-1}]$ is free.
     Q.E.D.

Corollary 2.12.18: *If M is f.g. projective over a commutative ring C then for every P in $\mathrm{Spec}(C)$ there is s in $C - P$ such that $M[s^{-1}]$ is a free $C[s^{-1}]$-module.*

Proof: M is finitely presented and M_P is free over C_P.   Q.E.D.

The Rank of a Projective Module

Define the *rank* of a free module to be the cardinality of its base. This is well-defined when R has the invariant base number, which is the case if R is commutative (c.f., corollary 1.3.27).

Proposition 2.12.19: *Suppose M is an f.g. projective module over a commutative ring C. Then there is a continuous function $f:\mathrm{Spec}(C) \to \mathbb{N}$ (under the discrete topology) given by $fP = \mathrm{rank}\, M_P$ (as C_P-module).*

Proof: We must show for any n in $\mathbb{N}$ that $f^{-1}(n)$ is open. Suppose $P \in f^{-1}(n)$, i.e., rank $M_P = n$ and take $s \in C - P$ such that rank $M[s^{-1}] = n$, using corollary 2.12.18. We claim rank $M_Q = n$ for all Q in the open set $U(s)$, i.e., whenever $s \notin Q$. Indeed, taking $S = C - Q$ we have

$$M_Q = S^{-1}M = S^{-1}(M[s^{-1}]) = S^{-1}(R[s^{-1}]^{(n)}) \approx (S^{-1}R[s^{-1}])^{(n)}$$
$$= (S^{-1}R)^{(n)} = R_Q^{(n)},$$

so rank $M_Q = n$, as desired. Thus we have shown $f^{-1}(n)$ is a union of open sets.   Q.E.D.

Corollary 2.12.20: *Notation as above, f takes on only a finite number of values in $\mathbb{N}$.*

Proof: $\{f^{-1}(n) : n \in \mathbb{N}\}$ is an open cover of $\mathrm{Spec}(C)$ and thus has a finite subcover by the *FIP* (proposition 2.12.5(i)).   Q.E.D.

Definition 2.12.21: A projective module M over a commutative ring C has *constant rank n* if rank $M_P = n$ for all P in $\mathrm{Spec}(C)$.

Theorem 2.12.22: *Suppose M is an f.g. projective module over a commutative ring C. Then C has orthogonal idempotents $e_1, \ldots, e_t$ such that each $M_i = e_i M$ is projective of constant rank over $C_i = e_i C$; also $M \approx \prod_{i=1}^{t} M_i$ and $C \approx \prod_{i=1}^{t} C_i$ canonically.*

Proof: By corollary 2.12.20, Spec(C) is covered by a finite number of disjoint clopen sets, each corresponding to the ranks of localizations of M; by proposition 2.12.5 these correspond to orthogonal idempotents of C, and the rest of the assertion is clear.   Q.E.D.

Corollary 2.12.23: *Notation as in the theorem, if C has no nontrivial idempotents then M has constant rank.*

Remark 2.12.24: If M is projective of constant rank n then for any maximal ideal $P \lhd C$ we have M/PM free over the field C/P of dimension n. (Indeed, the canonical map $M \to M/PM$ yields an epic $M_P \to M/PM$; any base of M_P gets sent to a base of M/PM by Nakayama's lemma.)

Definition 2.12.24′: A projective module of constant rank 1 is called *invertible.*

Invertible projective modules have a deep connection with algebra and geometry, since they are the natural categorical generalization of ideals of a PID (which are isomorphic to the ring itself), cf., exercises 7, 8.

K Supplement: Hilbert's Nullstellensatz and Generic Flatness

For commutative algebras $R = F[a_1, \ldots, a_t]$, Hilbert's Nullstellensatz shows the Spec topology coincides with the Zariski topology defined in §2.3. On the other hand, in §2.5 we introduced the "weak Nullstellensatz," that $\mathrm{Jac}(R/A)$ is nil for every $A \lhd R$ and noted that this property generalizes Hilbert's Nullstellensatz. Now we shall describe a different generalization and see how these properties are interrelated. These results are at the foundation of the noncommutative versions of the Hilbert Nullstellensatz, and thus are relevant to theories of noncommutative algebraic geometry. We start with a technical definition. We say a prime ideal A is a *G-ideal* if $A \neq \bigcap \{P \in \mathrm{Spec}(R): A \subset P\}$. An integral domain C is a *G-domain* if 0 is a G-ideal.

Remark 2.12.25: An integral domain C is a *G-domain* iff $C[s^{-1}]$ is a field for some $0 \neq s$ in C. (Indeed $(\Rightarrow)$ is seen by taking $0 \neq s \in \cap \{0 \neq P \in \mathrm{Spec}(C)\}$, since then $C[s^{-1}]$ has no proper maximal ideals so must be simple. $(\Leftarrow)$ Every nonzero prime ideal contains a power of s and thus s itself.)

Definition 2.12.26: A ring R is *Jacobson* if every prime ideal is the intersection of primitive ideals. An algebra R over a field F satisfies the

Nullstellensatz if $\text{End}_R M$ is algebraic over F for every simple module $M \neq 0$.

The Nullstellensatz can be studied in terms of two auxiliary notions: A C-algebra R satisfies the *strong Nullstellensatz* if $\text{Ann}_C M$ is a G-ideal of C for every simple R-module $M \neq 0$; R satisfies the *maximal Nullstellensatz* if for every simple R-module $M \neq 0$ we have $\text{Ann}_C M$ is a maximal ideal of C such that $R/\text{Ann}_R M$ satisfies the Nullstellensatz over the field $C/\text{Ann}_C M$.

An alternative formulation of the maximal Nullstellensatz is, "If P is a primitive ideal of R then $P \cap C$ is a maximal ideal of C and R/P satisfies the Nullstellensatz over the field C/P." This is seen by writing $P = \text{Ann}_R M$ for M simple.

Remark 2.12.26′: If R is Jacobson then $\text{Spec}(R)$ produces the same topology on R as the primitive spectrum. Thus the Jacobson property is very significant and is the focus of this discussion. Note that $P \in \text{Spec}(R)$ is the intersection of primitive ideals iff $\text{Jac}(R/P) = 0$, and it follows easily that R is Jacobson iff $\text{Jac}(R/A) = \text{Prime rad}(R/A)$ for every $A \lhd R$. In most classes of rings of interest to us it will turn out $\text{Nil}(R/A) = \text{Prime rad}(R/A)$, cf., theorem 2.6.23 for example. Thus the key to the discussion is verifying the "weak Nullstellensatz," that $\text{Jac}(R/A)$ is nil.

To avoid repetition of the word "Nullstellensatz" we write Null, SN, and MN respectively for the Nullstellensatz, strong Nullstellensatz, and maximal Nullstellensatz. Note by definition that MN implies SN since every maximal ideal is a G-ideal. On the other hand, MN and Null become the same if C is a field, but SN is vacuous for fields.

Remark 2.12.27: Null is quite powerful when F is algebraically closed, since then for any simple R-module M we see the division ring $\text{End}_R M$ is algebraic over the algebraically closed field F, implying $\text{End}_R M = F$. When R is primitive we can take M to be faithful; then each element of $Z(R)$ yields an endomorphism of M and thus belongs to F, proving $Z(R) = F$.

Example 2.12.28: Suppose R is an F-algebra and $|F| - 1 > [R:F]$ as cardinal numbers. Then R satisfies both the weak Nullstellensatz and Null. Indeed, the hypotheses pass to R/A for each $A \lhd R$ so we get the weak Nullstellensatz by theorem 2.5.22. On the other hand, if M is a simple R-module then $D = \text{End}_R M$ is a division algebra over F and M is a right vector space over D; since M is cyclic we have $[R:F] \geq [M:F] = [M:D][D:F]$ so $[D:F] < |F| - 1$. Consequently, for any d in D,

$\{(d - \alpha)^{-1} : \alpha \in F\}$ is F-dependent; proposition 2.5.21 implies d is algebraic over F, proving Null.

Remark 2.12.29:

(i) If C is semiprimitive (commutative) and not a field then C is not a G-domain (since $\cap\{\text{maximal ideals}\} = 0$).

(ii) If R is a C-algebra and M is a simple R-module then $\text{Ann}_C M \in \text{Spec}(C)$, analogously to remark 2.1.14.

Proposition 2.12.30:   *Suppose R is an algebra over a field F. If $R[\lambda]$ satisfies SN over $F[\lambda]$ then R satisfies Null.*

Proof: Suppose M is a simple R-module. We want to show any endomorphism $f : M \to M$ is algebraic over F. View M as an $R[\lambda]$-module by the action $(\sum r_i \lambda^i)x = \sum r_i f^i x$ for x in M. This action restricts to the given action over R, so M is a simple $R[\lambda]$-module, implying $P = \text{Ann}_{F[\lambda]} M$ is a prime G-ideal of $F[\lambda]$. But $F[\lambda]$ is semiprimitive so $P \neq 0$; taking $0 \neq p(\lambda) \in P$ we have $0 = p(\lambda)M = p(f)M$ as desired.   Q.E.D.

Duflo's approach stems from the following easy observation.

Lemma 2.12.31:   *Suppose R is an algebra over an integral domain C whose field of fractions is F, and M is a simple R-module which is faithful over C. Then M is naturally an $F \otimes_C R$-module. If M is free over $S^{-1}C$ (under this action) for a submonoid S of C then $S^{-1}C = F$.*

Proof: Given $0 \neq c \in C$ let $\hat{c} : M \to M$ denote the left multiplication map by c. By Schur's lemma $\hat{c}$ is invertible in $\text{End}_R M$, so we have the action of $F \otimes R$ on M given by $(c^{-1} \otimes r)x = \hat{c}^{-1}(rx)$ where $r \in R$ and $x \in M$; this is clearly well-defined and makes M an $F \otimes R$-module.

Now suppose M is free over $S^{-1}C$. Take a base $x_1, \ldots, x_t$. If $z \in S^{-1}C$ then $z^{-1}x_1 = \sum_{i=1}^{t} z_i x_i$ for z_i in $S^{-1}C$. Then $x_1 = \sum z z_i x_i$; matching coefficients of x_1 shows $z z_1 = 1$, so z is invertible, proving $S^{-1}C = F$.   Q.E.D.

As a special case we see if M is free over $C[s^{-1}]$ for some $0 \neq s \in C$ then $C[s^{-1}] = F$ so C is a G-domain. This leads us to the following definition.

Definition 2.12.32: R satisfies *generic flatness* over C if for every simple R-module M there is $s \neq 0$ in C such that $M[s^{-1}]$ is free over $C[s^{-1}]$.

The usual definition of generic flatness requires the condition to hold for all f.g. R-modules, but we shall only require M simple (so that lemma 2.12.31 is applicable). Several trivial observations should be noted.

Remark 2.12.33:

(i) In verifying generic flatness we may assume M is faithful over C, for if $cM = 0$ then $M[c^{-1}] = 0$ is free over $C[c^{-1}]$. In particular, C is a domain by remark 2.12.29(ii), and $M = M[c^{-1}]$ by lemma 2.12.31.

(ii) Any commutative Noetherian algebra satisfies generic flatness, by proposition 2.12.17. We shall find another important example in §6.3.

Lemma 2.12.33': *If R/PR satisfies generic flatness as C/P-algebra for each P in $\mathrm{Spec}(C)$ then R satisfies SN.*

Proof: If M is a simple R-module then $P = \mathrm{Ann}_C M \in \mathrm{Spec}(C)$ and $PR \subseteq \mathrm{Ann}_R M$; since M as R/PR-module is simple and (C/P)-faithful, we see P is a G-ideal of C by the comment after lemma 2.12.31.   Q.E.D.

In case C is Jacobson, we can conclude more.

Remark 2.12.33'': Suppose R satisfies SN over a Jacobson ring C. Then $\mathrm{Ann}_C M$ is maximal in C for every simple R-module M. (Indeed, $P = \mathrm{Ann}_C M$ is a G-ideal and is prime by remark 2.12.29(ii). Thus P is the intersection of primitive ideals of C, so P is itself a primitive ideal. Hence C/P is commutative primitive and thus a field.)

Remark 2.12.34: Suppose R is a C-algebra and view $C \subseteq R$; suppose $A \lhd R$. Then any of the hypotheses Null, SN, MN, and generic flatness pass from R as C-algebra to R/A as $C/(C \cap A)$-algebra. Indeed, any simple R/A-module M is also simple as R-module under the "change-of-rings" action, and $\mathrm{End}_R M \approx \mathrm{End}_{R/A} M$ canonically.

Lemma 2.12.35: *Suppose R satisfies SN over a Jacobson ring C, and $(R/PR)[\lambda]$ satisfies Null over C/P for every maximal ideal P of C. Then the weak Nullstellensatz holds.*

Proof: In view of remark 2.12.34, we need only show $\mathrm{Jac}(R)$ is nil; by lemma 2.5.41 it suffices to show $R[\lambda](1 - r\lambda) = R[\lambda]$ for all r in $\mathrm{Jac}(R)$. If $R[\lambda](1 - r\lambda) < R[\lambda]$ take a maximal left ideal L of $R[\lambda]$ containing $1 - r\lambda$

and let $M = R[\lambda]/L$. By remark 2.12.33'' $\mathrm{Ann}_C M$ is a maximal ideal P of C. But $\lambda \notin L$ so left multiplication by λ defines a nonzero map $\sigma: M \to M$ which is invertible by Schur's lemma. By hypothesis, σ^{-1} is algebraic so $\sigma = p(\sigma^{-1})$ for some p in $(C/P)[\lambda]$.

On the other hand, letting $x_0 = 1 + L \in M$ we see $(1 - \lambda r)x_0 = 0$ so $\lambda r x_0 = x_0$, implying $\lambda^i r^i x_0 = (\lambda r)^i x_0 = x_0$ for all i. Thus $\sigma^{-i} x_0 = r^i x_0$ for all i, implying $\sigma x_0 = p(\sigma^{-1})x_0 = p(r)x_0$ so

$$0 = (1 - r\lambda)x_0 = x_0 - r\sigma x_0 = (1 - rp(r))x_0,$$

contrary to $1 - rp(r)$ being invertible (since $r \in \mathrm{Jac}(R)$).   Q.E.D.

Theorem 2.12.36: *Suppose R is an algebra over a commutative Jacobson ring C. Consider the condition*

  (∗) *R/PR satisfies generic flatness over C/P for all P in $\mathrm{Spec}(C)$.*

Then R satisfies both the weak Nullstellensatz and MN, provided $R[\lambda_1, \lambda_2]$ satisfies (∗) as $C[\lambda_1]$-algebra.

Proof: $R[\lambda_1, \lambda_2]$ satisfies SN over $C[\lambda_1]$ by lemma 2.12.33'. Also R satisfies SN over C, seen by remark 2.12.34 by specializing $\lambda_1, \lambda_2 \to 0$. For any maximal ideal P of C we see $(R/PR)[\lambda_1, \lambda_2]$ satisfies SN over $(C/P)[\lambda_1]$, implying $(R/PR)[\lambda_2]$ satisfies Null by proposition 2.12.30. Therefore the weak Nullstellensatz holds by lemma 2.12.35.

To verify MN suppose M is a simple R-module. Then $P = \mathrm{Ann}_C M$ is maximal by remark 2.12.33'', and $R/\mathrm{Ann}_R M$ is a homomorphic image of $(R/PR)[\lambda_2]$ and thus satisfies Null.   Q.E.D.

(Note: If we assume (∗) for R we can weaken the hypothesis on $R[\lambda_1, \lambda_2]$ by restricting our attention to those prime ideals of $C[\lambda_1]$ whose intersection with C is maximal.)

Corollary 2.12.37: (*Commutative Nullstellensatz*) *If R is a commutative algebra which is finitely generated as an algebra over a field F, then R is Jacobson and satisfies Null over R.*

Proof: $R[\lambda_1, \lambda_2]$ is commutative Noetherian and thus satisfies (∗).
                                                                          Q.E.D.

Remark 2.12.37': The commutative Nullstellensatz enables us to sharpen use of ultraproducts described in theorem 1.4.15ff. Suppose we want to prove an elementary sentence φ about an integral domain C of characteristic 0 which is *not* a field, knowing (i) φ holds in characteristic p for almost all $p > 0$, and (ii) φ passes to subrings, i.e., φ is universal. In view of theorem 1.4.15 and proposition 1.4.21, we may assume C is finitely generated as $\mathbb{Z}$-algebra. By the commutative Nullstellensatz, C is semiprimitive so it has an infinite set $\{P_i : i \in \mathbb{N}\}$ of maximal ideal having intersection 0. By hypothesis, φ holds in each C/P_i, and often we can conclude by applying proposition 1.4.20. This technique is called *passing to characteristic p* and is particularly useful for group algebras and enveloping algebras, cf., Chapter 8.

The point of generic flatness is that it can often be verified more readily than the Nullstellensatz properties, and at this stage one could already prove significant generic flatness results, such as Irving [79, theorem 2]. However, we shall postpone the climax of this discussion until §8.4 and McConnell's "vector generic flatness." Incidentally when the "generic flatness" condition is dropped, some of the results fall flat on their faces, cf., exercise 5.

Comparing Prime Ideals of Related Rings

Suppose throughout $R \subseteq R'$ are rings. One of the main techniques of computing $\text{Spec}(R')$ is by comparing it to $\text{Spec}(R)$. To do this we must have a way of passing prime ideals back and forth from R to R'. This procedure is so basic that it has been investigated very carefully in the literature, leading to striking results by Heinecke-Robson [83]. Unfortunately, a presentation in this generality would obscure the simplicity of the proofs in special cases, so we shall settle on the compromise, studying centralizing extensions and treating normalizing extensions in exercises 10ff.

Remark 2.12.38:   Suppose $R' = \sum Rr_i$ where each $r_i \in C_{R'}(R)$. If $A \lhd R$ then $AR' = \sum Ar_i = \sum r_i A = R'A$ (since $A = RA = AR$); in particular $AR' \lhd R'$.

In view of remark 2.12.38 we say $R \subseteq R'$ is *ideal compatible* if $AR' \lhd R'$ for all $A \lhd R$. (Thus $R'AR' = AR'$.)

Proposition 2.12.39:   *If $R \subseteq R'$ is ideal-compatible there is a map* $\text{Spec}(R') \to \text{Spec}(R)$ *given by* $P' \to P' \cap R$. *(Likewise, if $P' \lhd R'$ is semiprime then $P' \cap R$ is a semiprime ideal of R.)*

Proof: Suppose $AB \subseteq P' \cap R$ for ideals A, B of R. Then $AR'BR' \subseteq ABR' \subseteq P'$ so $AR' \subseteq P'$ or $BR' \subseteq P'$, proving $A \subseteq P' \cap R$ or $B \subseteq P' \cap R$.   Q.E.D.

Our main concern is how much information is lost by this map, especially concerning chains of prime ideals. This leads us to study the following concepts:

Definition 2.12.40: Given $P_1 \subseteq P_2$ in $\text{Spec}(R)$, define the following possible situations:

(i) $LO(P_1) = $ "lying over" means there is P' in $\text{Spec}(R')$ with $P' \cap R = P_1$.

(ii) $GU(P_1, P_2) = $ "going up" means given any P'_1 in $\text{Spec}(R')$ lying over P we have P'_2 lying over P_2 with $P'_1 \subseteq P'_2$.

(iii) $INC(P_1) = $ "incomparability" means that one cannot have $P'_1 \subset P''_1$ in $\text{Spec}(R')$ each lying over P_1.

Similarly $GU(\_\_, P)$ means $GU(P_1, P)$ for all $P_1 \subseteq P$ in $\text{Spec}(R)$, and LO, GU, and INC mean, respectively, $LO(P)$, $GU(\_\_, P)$, and $INC(P)$ for all P in $\text{Spec}(R)$.

Given $A' \lhd R'$ and $B \lhd R$ we say an ideal B' of R' is (A', B)-*maximal* if $A' \subseteq B'$ and B' is maximal such that $B' \cap R \subseteq B$.

Lemma 2.12.41: *If $A' \cap R \subseteq B$ then (A', B)-maximal ideals exist. Furthermore any such ideal is a prime (resp. semiprime) ideal of R' if B is a prime (resp. semiprime) ideal of R.*

Proof: The first assertion is by Zorn's lemma. If P' is (A', B)-maximal and $B'_1, B'_2 \supseteq P'$ with $B'_1 B'_2 \subseteq P'$ then $(B'_1 \cap R)(B'_2 \cap R) \subseteq P' \cap R \subseteq B$; if B is prime (resp. semiprime with $B'_1 = B'_2$) then $B'_1 \cap R \subseteq B_1$ or $B'_2 \cap R \subseteq B_2$, so $B'_1 = P'$ or $B'_2 = P'$ by maximality, proving the second assertion.   Q.E.D.

Remark 2.12.42: If $R \cap R'PR' = P$ then $LO(P)$ holds. (Indeed, take P' to be $(R'PR', P)$-maximal; then $P \subseteq P' \cap R \subseteq P$ so P' lies over P).

The key to both LO and GU is the following result:

Proposition 2.12.43: *In order to prove $LO(P)$ and $GU(P_1, P)$ hold in R' over R it suffices to prove for each P' in $\text{Spec}(R')$ with $P' \cap R \subset P$ that the following weakened version of LO holds in $\overline{R'} = R'/P'$ over $\overline{R} = (R + P')/P'$:*
There is $\overline{A'} \lhd \overline{R'}$ for which $0 \neq \overline{A'} \cap \overline{R} \subseteq \overline{P}$.

Proof: To prove $LO(P)$ we take P' in $\text{Spec}(R')$ to be $(0, P)$-maximal, and claim $P \subseteq P'$. Otherwise, $\overline{P} \neq 0$ in $\overline{R'} = R'/P'$ so, by hypothesis, there is an ideal $A' \supset P'$ such that $0 \neq \overline{A'} \cap \overline{R} \subseteq \overline{P}$; then $A' \cap R \subseteq P$ contrary to choice of P'. Thus we established the claim and LO(P). The same argument shows if $P_1 \subseteq P$ and P_1' lies over P_1 then any (P_1', P)-maximal ideal of R' lies over P, proving GU(P_1, P). Q.E.D.

Remark 2.12.44: If $R \subseteq R'$ is ideal-compatible then GU implies LO. (Indeed, given P in Spec(R) take P_1' to be $(0, P)$-maximal. Then $P_1' \in$ Spec(R') so applying GU($P_1' \cap R, P$) to P_1' produces a prime ideal of R' lying over P.)

Let us now verify LO and GU in several instances.

Proposition 2.12.45: *Suppose* $R \subseteq R'$ *have a common ideal* A. *Then* LO(P) *and* GU(P_1, P) *hold under the following circumstances:*

(i) $A \nsubseteq P$.

(ii) $R \subset R'$ *is ideal-compatible and* P *is minimal prime over* $A + B$ *for a suitable ideal* $B = R \cap R'BR'$ *such that* $A \nsubseteq B$.

Proof: We verify the hypotheses of proposition 2.12.43.

(i) If $P' \cap R \subset P$ then $0 \neq \bar{A} \lhd \bar{R}' = R'/P'$. Let $\bar{A}' = \overline{APA} \neq 0$. Then $\bar{A}' \cap \bar{R} = \bar{A}' \subseteq \bar{P}$.

(ii) Passing to R/B and $R'/R'BR'$ one may assume $B = 0$. Note that if $P' \cap R \subset P$ then $A \nsubseteq P' \cap R$, since, otherwise, we contradict the minimality of P over A; hence we conclude as in (i). Q.E.D.

Remark 2.12.45': To apply (ii) we need instances when $B = R \cap R'BR'$; this is clearly the case if $B = 0$ or if B is a prime ideal *not* containing A (by (i)).

Actually a stronger incomparability result is also available.

Proposition 2.12.46: *Suppose* $R \subseteq R'$ *have a common ideal* A, *and* $A \nsubseteq P$ *for some* P *in* Spec(R). *Then* INC *holds; in fact, if* P_1', P_2' *are primes of* R' *and* $P_1' \cap R \subseteq P_2' \cap R = P$ *then* $P_1' \subseteq P_2'$.

Proof: $AP_1' \subseteq P_1' \cap A \subseteq P_1' \cap R \subseteq P_2'$ but $A \nsubseteq P_2'$ so $P_1' \subseteq P_2'$. Q.E.D.

Remark 2.12.47: If R' is a centralizing extension of R and R' is prime then $\text{Ann}_R a = 0$ for every R-normalizing element $a \neq 0$ of R'. (Indeed, let $A = \text{Ann}_R a$; then $AR'a = C_{R'}(R)AaR = 0$ implying $A = 0$ since R' is prime.)

Theorem 2.12.48: *(Bergman)* LO *and* GU *hold for all finite centralizing extensions.*

Proof: (Robson-Small [81]) in view of proposition 2.12.43 we may assume R' is prime and need only show for each $0 \neq P \in \mathrm{Spec}(R)$ there is $A' \lhd R'$ for which $0 \neq A' \cap R \subseteq P$. Note R is prime by proposition 2.12.39. Write $R' = \sum_{i=1}^{t} Ra_i$ with each $a_i \in C_{R'}(R)$ and $a_1 = 1$. We expand a_1 to a maximal independent set of the a_i which by reordering the a_i we may assume is $\{a_1, \ldots, a_m\}$. Then $T = \sum_{i=1}^{m} Ra_i = \sum_{i=1}^{m} a_i R$ is free as left or right R-module by remark 2.12.47. Let $B_i = \{r \in R : ra_i \in T\}$ and $B = \bigcap_{i=1}^{t} B_i \neq 0$. Then $PB \neq 0$ and $PBR' \lhd R'$. Let $A' = PBR' \subseteq PT \subseteq T$; sublemma 2.5.32' shows $A' \cap R \subseteq PT \cap R = P$, and $A' \cap R \supseteq PB \neq 0$, as desired.   Q.E.D.

So far the only known proofs of INC for centralizing extension involve the "central closure," a construction which we shall not encounter until §3.4; consequently, we postpone the proof of INC until theorem 3.4.13.

We can push theorem 2.12.48 further by means of the following observation.

Proposition 2.12.49:   *Suppose R' is the direct limit of subrings $R_i \supset R$ each satisfying LO, where the φ_i^j of example 1.8.2 are the inclusion maps. Then $R' \supset R$ satisfies LO.*

Proof: By remark 2.12.42 we are done unless $R \cap R'PR' \supset P$ for some P in $\mathrm{Spec}(R)$. But then writing $r = \sum_k r'_{k1} p_k r'_{k2} \notin P$ for p_i in P we see this equation holds in some R_i, implying $R \cap R_i PR_i \supset P$ contrary to LO(P) in R_i.
                                                                            Q.E.D.

Corollary 2.12.50:   *If R is an algebra over a field F then $R \otimes_F K$ satisfies LO and GU over R for any field $K \supseteq F$.*

Proof: By proposition 2.12.49 we may assume K is finitely generated. Then K is a finite algebraic extension of some purely transcendental field extension K_0 of F; by theorem 2.12.48 it suffices to show $R \otimes K_0$ satisfies LO and GU over R. By induction on the size of the transcendence base we may assume $K_0 = F(\lambda)$, i.e., we need to show $R(\lambda) \supset R$ satisfies LO and GU. But LO is clear by sublemma 2.5.32', and GU follows then from proposition 2.12.43.
                                                                            Q.E.D.

LO and the Prime Radical

One instant application is to the prime radical.

Remark 2.12.51:   If R' is an ideal-compatible extension of R then Prime

$\operatorname{rad}(R) \subseteq R \cap \operatorname{Prime} \operatorname{rad}(R')$, equality holding if LO is satisfied. (Indeed, let $N' = \operatorname{Prime} \operatorname{rad}(R')$. Then

$$R \cap N' = \bigcap \{R \cap P' : P' \in \operatorname{Spec}(R)\} \supseteq \bigcap \{P \in \operatorname{Spec}(R)\},$$

equality holding if we have LO.)

This modest observation gives us the analogue to corollary 2.5.36.

Theorem 2.12.52: *Suppose K is a field extension of F, and let $N = \operatorname{Prime} \operatorname{rad}(R)$ and $N' = \operatorname{Prime} \operatorname{rad}(R \otimes_F K)$. Identifying R with $R \otimes 1$ we have $N = R \cap N'$; furthermore, $N' = NK$ if K is a separable field extension of F.*

Proof: The first assertion follows by applying remark 2.12.51 to corollary 2.12.50. To prove the second assertion note we have proved $N' \supseteq NK$, so factoring out NK we may assume $N = 0$, i.e., R is semiprime. We want to show $R \otimes_F K$ has no nilpotent ideal $A \neq 0$. Otherwise, A has some nonzero element in RK_1 for some finite extension K_1 of F, so we may replace K by K_1 and assume K is finitely generated. Letting K_0 be a subfield generated by a transcendence base of K over F, we have $RK_0 \approx R \otimes K_0$ semiprime; replacing R, F by RK_0 and K_0, we may assume K is algebraic over F, so K is a finite separable extension. Let $\bar{K}$ be the normal closure of K. Then $R \otimes \bar{K}$ is a finite centralizing extension of $R \otimes K$, so by remark 2.12.51 it suffices to prove $R \otimes \bar{K}$ semiprime.

At this point we can copy out the proof of theorem 2.5.36 since the prime radical is invariant under automorphisms.   Q.E.D.

Corollary 2.12.53: *If R is an algebra over a field F of characteristic 0, then $\operatorname{Prime} \operatorname{rad}(R \otimes_F K) = \operatorname{Prime} \operatorname{rad}(R) \otimes K$ for any field extension K of F.*

Unfortunately if R' is a finite *normalizing* extension of R and $P \in \operatorname{Spec}(R')$ then $P \cap R$ need not be a prime ideal of R, so one might be tempted to throw up one's hands in despair. However, $P \cap R$ turns out to be a finite intersection of prime ideals of R, and corresponding weakened versions of LO and GU can be proved. INC also holds but is quite difficult. Definitive results were obtained by Heinicke-Robson [81]. Passman [81] showed how to transfer many of these questions to the primitive spectrum, and we shall use his approach in exercises 10ff. The results for normalizing extensions were generalized to the case where R' is merely a subring of a finite normalizing extension; this is called an *intermediate normalizing extension* in Heinicke-Robson [84a]

Height of Prime Ideals

Definition 2.12.54: If P is a prime ideal in R we say P has *height* 0 if P is minimal; inductively, height $(P) = 1 + \max\{\text{height}(Q): Q \subset P \text{ in Spec}(R)\}$. In other words, height (P) is the largest length of a chain $P_0 \subset P_1 \subset \cdots \subset P$ in Spec(R). (The height is sometimes called the "rank.")

Although prime ideals need not have finite height, in general, the height is a very useful measure of a prime ideal.

Example 2.12.55: If S is a submonoid of R with $P \cap S = \varnothing$ then the height of P in R equals the height of $S^{-1}P$ in $S^{-1}R$, by proposition 2.12.9'.

We bring in the height here because of its connection to LO, GU, and INC.

Remark 2.12.56: If $R \subseteq R$ is ideal-compatible then LO(P) holds for every minimal prime ideal P of R. (Indeed, if P' is $(0, P)$-maximal then $R \cap P' \subseteq P$ implies $R \cap P' = P$ since P is minimal prime.)

Proposition 2.12.57: *Suppose $R \subseteq R'$ is ideal-compatible, and suppose $P \in$ Spec(R) of height n.*

   (i) *If INC holds then* height$(P') \le n$ *for every P' in* Spec(R') *lying over P.*
   (ii) *If GU holds then there is some P' lying over P with* height$(P') \ge n$.
   (iii) *If INC and GU hold there is some P' lying over P with* height$(P') = n$.

Proof:

   (i) If $P'_0 \subset \cdots \subset P'_t = P'$ in Spec(R) then $P'_0 \cap R \subset \cdots \subset P'_t \cap R = P$ in Spec(R), so $t \le n$.
   (ii) Suppose $P_0 \subset \cdots \subset P_n$ in Spec(R), then there is P'_0 lying over P_0 and inductively we build a chain $P'_0 \subset \cdots \subset P'_n$ by applying GU(P_{i-1}, P_i) to P'_{i-1}. Thus P'_n has height $\ge n$.
   (iii) Combine (i) and (ii).   Q.E.D.

Exercises

§2.1

1. Suppose M is a simple R-module with $D = \text{End}_R M$, and $x_1, \ldots, x_n \in M$ are arbitrary. If V is a finite D-subspace of M and $x_1 \notin V$ then there exist elements $d_1 = 1, d_2, \ldots, d_n$ in D such that for any x in M there is r in $\text{Ann}_R V$ such that $rx_i = xd_i$ for $1 \le i \le n$. (Hint: Enlarge a base of V to a base of $V + \sum_{i=1}^{n} x_i D$

containing x_1 and apply the density theorem to this set.) This formulation of the density theorem is amenable to computation.

2. Suppose M is a faithful simple R-module with $D = \text{End}_R M$. If $L < R$ with $LM \neq 0$ then (as a ring without 1) L is dense in $\text{End } M_D$. In particular, this holds if $L \lhd R$.

3. Suppose R_0 is a rng. If $0 \neq L < R_0$ then $L \cap \text{Ann}'L \lhd L$. If R_0 is prime (resp. primitive) then so is the rng $L/L \cap \text{Ann}'L$. (Hint: $M/\{x \in M: Lx = 0\}$ is faithful and simple.)

4. Let I be the ideal of $F\{X_1, X_2\}$ generated by $X_1 X_2 - X_2 X_1 - X_1$. If $\text{char}(F) = 0$ then $F\{X_1, X_2\}/I$ is primitive with 0 socle.

5. Let $R = \text{End } M_D$ where M is a right vector space over D. Then $R/\text{soc}(R)$ is a prime ring. (Hint: If $a, b \in R$ do not have finite rank then find r such that arb does not have finite rank.)

6. An example of a primitive ring R such that $R/\text{soc}(R)$ is not prime. Let R be the subring of $T = \text{End } M_D$ generated by $\text{soc}(T)$ and a transformation a given by $ax_i = x_i$ if i is even and $ax_i = 0$ if i is odd. Then $aR(1 - a) \in \text{soc}(R)$ but a, $1 - a \notin \text{soc}(R)$.

7. Display the Weyl algebra explicitly as a dense subring of $\text{End } M_D$.

8. (Rieffel's short proof of the Wedderburn-Artin theorem). Viewing any left ideal L of R as R-module put $T = \text{End}_R L$ and $R' = \text{End } L_T$. For any x, y in L and f in R', note $(fx)y = f(xy)$ because right multiplication by y is an element of T. Now assume R is simple. Viewing $R \subseteq R'$ by the regular representation, conclude L is a left ideal of R' so $1 \in R = LR \leq R'$ proving $R' = R$ is simple. To prove the Wedderburn-Artin theorem note that if L is a minimal left ideal then T is a division ring and $0 \neq \text{soc } R' \lhd R'$ implies $[L:T] = n$ for some $n < \infty$, implying $R \approx M_n(T)$.

8'. If R is a prime ring with a left ideal L satisfying $LR = R$ then $R \approx \text{End } L_T$ where $T = \text{End}_R L$. (Hint: Same proof as above.)

9. Perhaps the slickest proof of the Wedderburn–Artin theorem: Suppose R is simple with minimal left ideal L. Write $1 = \sum_{i=1}^{n} a_i r_i$ for $a_i \in L$, $r_i \in R$ with n minimal. Then there is an epic $\psi: L^{(n)} \to R$ given by $(x_1, \ldots, x_n) \to \sum x_i r_i$; show $\ker \psi = 0$ because n is minimal. Hence $R \approx \text{End}_R R \approx \text{End}_R L^{(n)} \approx M_n(\text{End}_R L)$ by exercise 1.5.0.)

10. Prove directly that any domain R with a minimal left ideal Ra is a division ring. (Hint: $a \in Rra$ for any r in R.)

Primitivity of Free Products Following Lichtman [78] we shall show that the free product of two nontrivial algebras over a field is primitive provided one of them has dimension ≥ 3. This hypothesis is needed because of the following example:

11. Suppose $R_1 \approx R_2 \approx F[\mathbb{Z}/2\mathbb{Z}]$, i.e., $R_i = F + Fa_i$ where $a_i^2 = 1$. Then $R_1 \coprod R_2 \approx F[G]$ where G is the group $\langle a_1, a_2 : a_1^2 = a_2^2 = 1 \rangle$. Show $a = a_1 a_2$ has infinite period, and $\langle a \rangle \lhd G$ of index 2; furthermore, $R_1 \coprod R_2$ is not primitive. (Hint: Embed G in $GL(2, \mathbb{Q})$ by $a_1 \to e_{12} + e_{21}$ and $a_2 \to 2e_{12} + \frac{1}{2}e_{21}$. Since $\langle a \rangle$ has index 2, one can then embed $F[G]$ into $M_2(F[\langle a \rangle])$ by the regular representation. But no nontrivial group ring is simple, so using the density theorem show that $F[G]$ cannot be primitive. A faster conclusion would be via Kaplansky's theorem on polynomial identities.)

In exercises 13–16 we carry the following set-up: C is a commutative domain and R_1, R_2 are C-algebras which are free as C-modules having bases each containing 1.

For $k = 1, 2$ denote the base of R_k as $\{1\} \cup B_k$. Enumerate B_1 as $\{e_i : i \in I\}$, and B_2 as $\{f_j : j \in J\}$. Let $\mathcal{W} = \{\text{words alternating in the } e_i \text{ and } f_j\}$; then $R_1 \coprod R_2$ exists and is a free C-module with base $\mathcal{W} \cup \{1\}$. Assume $|B_1| \geq |B_2|$. Let $\mathcal{W}_k = \{\text{words of } \mathcal{W} \text{ starting with an element of } B_k\}$. For example, $f_1 e_1 \in \mathcal{W}_2$ and $e_2 f_1 e_3 \in \mathcal{W}_1$. Let $\mathcal{V} = \{\text{words of } \mathcal{W}_1 \text{ of even length}\}$, and let T be the C-subspace of $R_1 \coprod R_2$ spanned by $\mathcal{V}$.

12. T is a subalgebra of $R_1 \coprod R_2$ isomorphic to the free C-algebra (generated by the $e_i f_j$, which act as indeterminates).

13. $A \cap T \neq 0$ for every nonzero ideal A of $R_1 \coprod R_2$. (Hint: Let $0 \neq a \in A$. Multiplying on the left by e_1 and on the right by f_1, if necessary, one may assume some word in supp(a) of maximal length n is in T. Writing $e_u e_1 = \alpha_u + \sum \alpha_{uk} e_k$ and $f_u f_1 = \beta_u + \sum \beta_{uk} f_u$ for suitable α_u, β_u in C, show modulo T that $e_u(e_1 f_1) = \alpha_u f_1$ and $e_s f_i e_u(e_1 f_1) = \alpha_u e_s f_i f_1 = \alpha_u \beta_i e_s$. Thus any term of $a' = (e_1 f_1)^n a(e_1 f_1)^{n+1}$ not in T is congruent to $(e_1 f_1)^i e_1 (e_1 f_1)^j$ for suitable i, j with $j > 0$. Then $0 \neq a' e_1 f_1 - \alpha_1 \beta_1 a' \in T \cap A$.

14. Let $R = R_1 \coprod R_2$. For each $0 \neq A \lhd R$ choose an element s_A in $T \cap A$, and let $S = \{s_A : A \lhd R\}$. R is primitive if either of the following holds: (i) R is countable and $|B_1| \geq 2$. (ii) $|B_1| \geq R$. (Hint: (i) S is countable so enumerating S as $\{s_1, s_2, \ldots\}$ define $r_n = 1 + s_n e_1 f_1 (e_2 f_1)^n + f_1 s_n e_1 (f_1 e_2)^n$, and let $L = \sum_{n=1}^{\infty} R r_n$. Matching coefficients of words ending in $e_1 f_1 (e_2 f_1)^n$, one sees $1 \notin L$. But each nonzero ideal A contains some s_n, so $1 \in R r_n + A \subseteq L + A$, implying R is primitive.)

(ii) B_1 has some subset which can be listed as $\{e_s : s \in S\}$, taking $r_s = 1 + f_1 s e_s + s e_s f_1$ and $L = \sum_{s \in S} R r_s$, show as in (i) that L is comaximal with every ideal and thus R is primitive.

15. If C is a field and $|B_1| \geq 2$ then R is primitive, without any restriction on R. (Hint: It suffices to find S as in exercise 14 with S countable or $|S| \leq |B_1|$, since then one can conclude as in exercise 14(i) or (ii), respectively. Suppose $s = \sum_{k=1}^{t} \alpha_k w_k \in A \cap T$ where $w_k \in V$ are words and $\alpha_k \in C$. Replacing s by $(e_1 f_1)^{n+1} s (e_2 f_1)^{n+1} = \sum \alpha_k v_k$ where $v_k = (e_1 f_1)^{n+1} w_k (e_2 f_1)^{n+1}$ we see $v_1, \ldots, v_k$ are algebraically independent. But then some higher commutator of the v_k is in $A \cap T$, and there are at most $|B_1|$ of these if $|B_1|$ is infinite. For example, if $s = \alpha_1 w_1 + \alpha_2 w_2 + \alpha_3 w_3$ then

$$[[w_1, w_3], [w_2, w_3]] = \alpha_1^{-1}[[s, w_3], [w_2, w_3]] \in A \cap T.$$

16. For $\alpha > 2$ show example 2.1.36 is not a unique factorization domain. (Hint: property (iii).)

17. Suppose R has a sequence of idempotents $e_1, e_2, e_3, \ldots$ such that each $e_{i+1} \in e_i R e_i$. Then there is a primitive ideal P such that each $e_i \notin P$. (Hint: Take a maximal left ideal L containing $\sum R(1 - e_i)$ and let $P = \text{core}(L)$.)

§2.2

1. A ring R is *subdirectly irreducible* if $\bigcap \{\text{nonzero ideals of } R\} \neq 0$. Prove every ring is a subdirect product of subdirectly irreducible rings. (Hint: Given r in R take

$A_r \lhd R$ maximal with respect to $r \notin A_r$. Then $\bigcap\{A_r : r \in R\} = 0$ and R/A_r is sub-directly irreducible.)

2. Suppose C is a commutative ring and $f \in C[\lambda]$. If f is a product $f_1, \ldots, f_t$ of pairwise relatively prime polynomials then $C[\lambda]/\langle f \rangle \approx \prod_{i=1}^{t} C[\lambda]/\langle f_i \rangle$.

3. If A, B are comaximal ideals of R then A^m and B^n are comaximal, for any m, n in $\mathbb{N}$.

4. Suppose $A_1, \ldots, A_t$ are comaximal ideals of R. For any $r_1, \ldots, r_t$ in R and any $m_1, \ldots, m_t$ in $\mathbb{N}$ there is some r in R such that $r - r_i \in A_i^{m_i}$ for $1 \leq i \leq t$. (Hint: Apply exercise 3 to proposition 2.2.1.) This is a useful "approximation" result.

§2.3

1. (Lattice version of Jordan-Holder theorem) A lattice L with 0, 1 has a *composition chain* $a_0 = 0 < a_1 < a < \cdots < a_n = 1$ if each a_i covers a_{i-1}. If L is a modular lattice with a composition chain then every chain can be modified to a composition chain of the same length. Moreover, defining $h(a)$ as the length of a composition chain ending in a, one has $h(a) + h(b) = h(a \wedge b) + h(a \vee b)$. (Hint: Use exercise 0.0.3.)

2. If M is an R-module of length n then every $R[\lambda]$-submodule N of $M[\lambda]$ is spanned by at most n elements. Here we view $M[\lambda]$ naturally as module over $R[\lambda]$, as in exercise 1.7.6. (Hint: Viewing the elements of $M[\lambda]$ as polynomials, let $N_i = \{$leading coefficients of elements of N having degree $\leq i\} \cup \{0\}$. Then $0 \leq N_0 \leq N_1 \leq \cdots$ has at most n strict increases, so some N_k is maximal and the chain can be refined to a composition series $0 = M_0 < M_1 < \cdots < M_t$ where $t \leq n$ and $N_i = M_{\pi i}$ for some nondecreasing function $\pi: \{1, \ldots, t\} \to N$. Write $M_j = M_{j-1} + Rx_j$ for $1 \leq j \leq t$, and take p_j in N of minimal degree having leading coefficient x_j. Then $N = \sum_{j=1}^{t} Rp_j$.) Special case: If R has length $\leq n$ in $R\text{-}\mathcal{M}od$ then each left ideal in $R[\lambda]$ is spanned by $\leq n$ elements. This includes the fact that $D[\lambda]$ is a PLID for any division ring D.

3. A prime ring with a faithful module of finite length is primitive. (Hint: One of the composition factors is faithful)

4. (Amitsur-Small [78]) Suppose R is a central simple F-algebra and M is an R-module of finite composition length. If $W \subseteq \text{End}_R M$ is prime then $R' = R \otimes_F W^{op}$ is primitive. (Hint: M is a faithful R'-module since otherwise $\text{Ann}_{R'} M \supseteq R \otimes A^{op}$ for $0 \neq A \lhd W$, and then $MA = 0$, impossible. Apply exercise 3.)

5. If D is a division ring with center F such that the field $F(\lambda_1, \ldots, \lambda_t)$ is isomorphic to a subfield of $M_n(D)$ then $D[\lambda_1, \ldots, \lambda_t]$ is primitive, by exercise 4. The converse follows from an application of the Nullstellensatz, cf., exercise 8.4.1.

6. The ultraproduct of copies of a semisimple Artinian ring R is semisimple Artinian. (Hint: See exercise 1.4.7′)

7. Suppose R is a ring not necessarily with 1. If R is semiprime Artinian then R has 1 after all. (Hint: If R is prime then it is primitive, so use the density theorem. In general, show R has only a finite number of prime (and thus maximal) ideals.

§2.4

1. (Jacobson's density theorem for completely reducible modules) Suppose $M \in R\text{-}\mathcal{M}od$ is completely reducible and $T = \text{End}_R M$. For every f in $\text{End} M_T$ and any

$x_1, \ldots, x_n$ in M there is r in R with $rx_i = fx_i$ for $1 \le i \le n$. (Hint: For $n = 1$ write $M = Rx_1 \oplus N$ and let $\pi \in T$ be the projection from M to Rx_1; then $fx_1 = f(\pi x_1) = \pi f x_1$ so $fx_1 \in Rx_1$. For arbitrary n define $\bar{f} = (f, \ldots, f): M^{(n)} \to M^{(n)}$ and note for $T' = \operatorname{End}_R M^{(n)}$ that $\bar{f} \in \operatorname{End} M_{T'}$. By the previous case there is r in R with $r(x_1, \ldots, x_n) = \bar{f}(x_1, \ldots, x_n) = (fx_1, \ldots, fx_n)$.) Note that this short proof (found in Lang [65B, p. 444]) yields Jacobson's original theorem as a special case.

2. $\operatorname{soc}(M)$ is a completely reducible module which contains every completely reducible submodule of M.

3. If M is a completely reducible R-module then $R/\operatorname{Ann}_R M$ is semiprimitive. (Hint: Take the intersection of the annihilators of the simple components of M.) Conversely, if $R/\operatorname{Ann}_R M$ is semisimple Artinian then M is a completely reducible R-module.

§2.5

Characterizations of Jac(R)

1. Define the circle operation $\circ$ on R by $r_1 \circ r_2 = r_1 + r_2 - r_1 r_2 = 1 - (1 - r_1) \cdot (1 - r_2)$. Then $\circ$ is associative and $r \circ 0 = 0 \circ r = r$. $\operatorname{Jac}(R)$ is a group under $\circ$, and $r \to 1 - r$ defines a group injection $\operatorname{Jac}(R) \to \operatorname{Unit}(R)$.

2. $\operatorname{Jac}(R) = \{r \in R : r + a \text{ is invertible whenever } a \text{ is invertible}\}$. (Hint: $(r + a)^{-1} = (1 + a^{-1}r)^{-1}a^{-1}$.)

3. $\operatorname{Jac}(R) = \bigcap \{\text{cores of maximal left ideals of } R\}$. (Hint: proposition 2.1.9.)

4. A (Jordan) *inner ideal* is an additive subgroup A of R such that $ara \in A$ for all a in A and r in R. (Examples: (i) any left or right ideal is inner; (ii) the intersection of inner ideals is inner.)

   Show $\operatorname{Jac}(R) = \bigcap \{\textit{cores of maximal inner ideals of } R\}$, where the *core* of an inner ideal is the sum of all ideals contained in it. (Hint: ($\supseteq$) Every maximal left ideal is contained in a maximal inner ideal by Zorn's lemma, and their cores must be equal. ($\subseteq$) Otherwise, $\operatorname{Jac}(R) \nsubseteq \operatorname{core}(l)$ for some maximal inner ideal l, so $l + \operatorname{Jac}(R)$ is an inner ideal containing l and is thus R. Then $r + a = 1$ for suitable r in $\operatorname{Jac}(R)$ and a in l; hence a is invertible implying $l = R$.) Note that we could define R to be *inner primitive* if R has a maximal inner ideal whose core is 0. As just shown, all primitive rings are inner primitive, but the converse is false (since "inner primitive" is left-right symmetric). These ideas have been used with astounding success by E. Zelmanov in studying Jordan algebras (cf. Jacobson [81B]).

6. Let E be the infinite dimensional exterior algebra in $x_1, x_2, \ldots$ over a field F. Then $N = \sum E x_i$ is a nil ideal, and $E/N \approx F$; thus E is local. Furthermore, $(Nx_i)^2 = 0$ for all x_i in E. On the other hand, if $\sigma(\text{resp. } \delta)$ is the endomorphism (resp. derivation) of E over F given by $x_i \mapsto x_{i+1}$ then $E[\lambda; \sigma]$ and $E[\lambda; \delta]$ are prime. $E[\lambda; \sigma]$ is semiprimitive, whereas $\operatorname{Jac}(E[\lambda; \delta]) = E[\lambda; \delta]N$. (Hint: The assertions for σ are easy; for example, if $0 \ne a, b \in E[\lambda; \sigma]$ and no x_i appears in a for $i > n$ then $a\lambda^n b \ne 0$, seen by checking the "leading" term in the x_i. To prove $E[\lambda; \delta]$ is prime first note any ideal $B \ne 0$ contains a monomial $x_1 \cdots x_t$. But then $x_2 \cdots x_{2t} = \delta^t(x_1 \cdots x_t)x_{t+2} \cdots x_{2t}$, and thus for any i there is suitable $i' > i$ such

that B contains $x_{i+1}x_{i+2}\cdots x_{i'}$. Conclude $AB \neq 0$ for all ideals A, $B \neq 0$, for if $x_1\cdots x_u \in A$ then $x_1\cdots x_{u'} \in AB$.) This example is taken from Bergen-Montgomery-Passman [87], which contains further ramifications.

7. If $J_i = \mathrm{Jac}(R_i)$ for each i in I and $\mathscr{F}$ is an ultrafilter of I then $\mathrm{Jac}(\prod R_i/\mathscr{F}) = \prod J_i/\mathscr{F}$. (Hint: ($\supseteq$) by quasi-invertibility. ($\subseteq$) If $(a_i) \notin \prod J_i/\mathscr{F}$ then $\{i: a_i \notin J_i\} \in \mathscr{F}$ so $1 - (a_i)(b_i)$ is not invertible for some (b_i).)

Separable Algebras

8. If F is perfect and R is a semisimple Artinian F-algebra then R is separable. (Hint: To show $R \otimes_F K$ is semiprimitive one may assume K is finitely generated; using theorem 2.5.42 one may assume K is separable and algebraic.)

9. Suppose D is a division algebra of characteristic p with center F, and $d \in D - F$ with $d^p \in F$. Then there exists a in D with $dad^{-1} = a + 1$; in particular, $F(a)$ is *not* a purely inseparable extension of F. (Extensive hint: Let δ denote the inner derivation $a \to [d, a]$. Then $\delta^p = 0$ for $\delta^p a = [d^p, a] = 0$. Choose b with $0 \neq [d, b]$. Take n minimal with $\delta^{n+1}b = 0$ and put $c = (\delta^{n-1}b)(\delta^n b)^{-1}$. Then $\delta c = 1$, so putting $a = cd$ conclude $[d, a] = \delta a = \delta(cd) = d$ as desired.)

10. (Koëthe-Noether-Jacobson theorem) Suppose D is a noncommutative division algebra which is algebraic over its center F. Then D contains an element not in F which is separably algebraic over F. (Hint: Otherwise, each element of $D - F$ is purely inseparable over F, so we have d in $D - F$ with $d^p \in F$. But then exercise 9 yields a contradiction.) Conclude that if $[D:F]$ is finite then D has a separable splitting field. (Hint: Split off a piece of D and use induction on $[D:F]$.)

11. Prove the following improved version of Wedderburn's principal theorem: Let R be a finite dimensional algebra over a field F, and $N = \mathrm{Jac}(R)$. If R/N is separable as an F-algebra then $R \approx S \oplus N$ as R-modules, where S is a subalgebra of R isomorphic to R/N.

Normalizing Extensions (also, cf., exercises 2.12.10ff)

12. Strong version of Nakayama's lemma (Robson-Small [81], Resco [82]) Suppose M is an f.g. normalizing $R - R$ bimodule and $L < R$. If $ML = M$ then there is a in L with $M(1 - a) = 0$. Consequently, $ML < M$ if M is faithful; in particular, this holds for f.g. normalizing extensions, so the hypothesis of proposition 2.5.17 also holds. (Hint: Write $M = \sum_{i=1}^{n+1} Rx_i$ where $x_i R = Rx_i$ and formally $x_{n+1} = 0$. Then $x_j = \sum x_i a_{ij}$ for a_{ij} in L, so it suffices to prove the following:

12'. Suppose M is an $R - R$ bimodule and $L < R$ and there are $x_1, \ldots, x_t$ in M with $x_i R = Rx_i$ for $1 \leq i \leq t$ satisfying equations $x_j = \sum_{i=1}^{t} x_i a_{ij}$ for $1 \leq j \leq t$, suitable a_{ij} in L. Then for each $n \leq t$ there is a_n in L with $(\sum_{i \leq n} Rx_i)(1 - a_n) \subseteq \sum_{i > n} x_i L$. (Hint: induction on n. For $n = 1$ note $x_1(1 - a_{11}) = \sum_{i > 1} x_i a_{i1}$. In general, $x_n(1 - a_{nn})(1 - a_{n-1}) = \sum_{i < n} x_i a_{ij}(1 - a_{n-1}) + \sum_{i > n} x_i a_{ij}(1 - a_{n-1}) \subseteq \sum_{i \geq n} x_i L$ by induction. Write $r = (1 - a_{nn})(1 - a_{n-1}) \in 1 + L$ and $x_n r = \sum_{i \geq n} x_i b_i$ for b_i in L. Then $x_n(r - b_n) \in \sum_{i > n} x_i L$ so take $a_n = 1 - (1 - a_{n-1})(r - b_n)$.)

13. If T is a finite normalizing extension of R then $\mathrm{Jac}(R) = R \cap \mathrm{Jac}(T)$. (Hint: exercise 12.)

14. Suppose T is a finite normalizing extension of R and $M \in T\text{-}\mathcal{M}od$. Then any large R-submodule N of M contains a large T-submodule of M. (Hint: Write $T = \sum Ra_i$ where $Ra_i = a_iR$ and $M_i = \{x \in M : a_i x \in N\}$, a large R-submodule. Then $T(\bigcap M_i)$ is large and $T(\bigcap M_i) = \sum a_i(\bigcap M_i) \le N$.)

15. (Formanek-Jategaonkar [74]) Suppose T is a finite normalizing extension of R and $M \in T\text{-}\mathcal{M}od$.
    (i) M is Noetherian iff M is Noetherian as R-module.
    (ii) M has a composition series iff M has a composition series as R-module. (Hint: (i) ($\Leftarrow$) trivial. ($\Rightarrow$) By Noetherian induction, assume M/K is Noetherian as R-module for every $K < M$ in $T\text{-}\mathcal{M}od$. Assume M has an R-submodule N which is not f.g. as R-submodule since $N + K$ is not f.g. for any essential complement K of N, so N contains a large T-submodule N_0. But N/N_0 is f.g.; also N_0 is Noetherian in $T\text{-}\mathcal{M}od$ and thus f.g. in $R\text{-}\mathcal{M}od$. Hence N is f.g., a contradiction. (ii) By (i) and proposition 2.5.29.)

16. If $R \subseteq T \subseteq M_n(R)$ then $\operatorname{Jac}(T)^n \subseteq M_n(\operatorname{Jac}(R))$. Hint: Every simple $M_n(R)$-module has the form $M^{(n)}$ by corollary 1.1.18, so it suffices to prove $\operatorname{Jac}(T)^n M^{(n)} = 0$. But $M^{(n)}$ has composition length $\le n$ in $R\text{-}\mathcal{M}od$ and thus in $T\text{-}\mathcal{M}od$.)

16'. If $R \subseteq T$ are rings and $T \approx R^{(n)}$ as R-module then $\operatorname{Jac}(T)^n \subseteq \operatorname{Jac}(R)T$; in particular, if $\operatorname{Jac}(R) = 0$ then $\operatorname{Jac}(T)$ is nilpotent. (Hint: View $T \subseteq M_n(R)$ by the left regular representation, so any element of $\operatorname{Jac}(T)^n$ can be viewed as a matrix (r_{ij}) with r_{ij} in $\operatorname{Jac}(R)$. Writing $1 = \sum r_i x_i$ one has $r = 1r = \sum r_i r_{ij} x_j \in \operatorname{Jac}(R)T$.)

17. Suppose $S = \sum_{i \in I} Ra_i$ where the a_i are R-normalizing elements of S, and, moreover, $S \ne \sum_{i \ne 1} Ra_i$. (This can be arranged if I is finite.) If S is a simple ring then there are maximal ideals $\{M_{ij} : i, j \in I\}$ of R such that $\bigcap_{i,j} M_{ij} = 0$; in particular, if I is finite then R is a finite direct product of simple rings. (Hint: Embed $\sum_{i \ne 1} Ra_i$ in an $R - R$ sub-bimodule M of S maximal with respect to $a_1 \notin M$, and let $M_{ij} = \{r \in R : a_i r a_j \in M\} \lhd R$. The M_{ij} are maximal ideals since if $M_{ij} \subset B \lhd R$ then $a_i Ba_j + M = S$ so $a_i a_j \in a_i ba_j + M$ and $a_i(1 - b)a_j \in M$, implying $1 \in B$. Moreover, $S(\bigcap M_{ij})S \subset M \ne S$, implying $\bigcap M_{ij} = 0$. Use the Chinese Remainder Theorem at the end.)

18. Define the *Brown-McCoy* radical $BM(R) = \bigcap\{\text{maximal ideals of } R\}$. If $A \lhd BM(R)$ then $BM(R/A) = BM(R)/A$. Moreover, if $a \in BM(R)$ then $1 \in R(1 - a)R$. (Hint: Otherwise, $1 - a \in M$ for some maximal $M \lhd R$; but $a \in M$, so $1 \in M$.) We say $A \lhd R$ has the BM property if $a \in R(1 - a)R$ for all a in A. Then $BM(R)$ has the BM property and contains every ideal A having the BM property. (Hint: If $A \nsubseteq M$ for maximal $M \lhd R$ then $1 \in A + M$ so $1 - a \in M$ for some a in A, implying $a \in R(1 - a)R \subseteq M$, so $1 \in M$.) Despite its superficial similarity to the Jacobson radical, the Brown-McCoy radical has not been very useful because the condition $1 \in R(1 - a)R$ is much more unwiedly than quasi-invertibility.

18'. Hypothesis as in exercise 17, show $BM(R) \subset BM(S)$. We shall see later that $BM(R) = R \cap BM(S)$ when I is finite.

19. $N \lhd R$ is *strictly nil* if $N \otimes_C H$ is a nil ideal of $R \otimes_C H$ for every commutative C-algebra H. Show N is strictly nil iff $N[\lambda_1, \lambda_2, \ldots]$ is a nil ideal in the polynomial ring $R[\lambda_1, \lambda_2, \ldots]$ in countably many indeterminates. Prove the following (easier) variant of Amitsur's theorem: If R has no strictly nil ideals $\ne 0$ then $R[\lambda_1, \lambda_2, \ldots]$ is semiprimitive. (cf. exercise 2.6.6.)

§2.6

A ring R is *reduced* if $r^2 = 0$ implies $r = 0$.

1. If R is a reduced ring and $S \subseteq R$ then $\operatorname{Ann}_R S \vartriangleleft R$ and $R/\operatorname{Ann}_R S$ is a reduced ring.

2. (Klein [80]) Prove every reduced ring is a subdirect product of domains, by proving for each x that if P is an ideal disjoint from $\{x^i : i > 0\}$ maximal with respect to R/P reduced then R/P is a domain. (Hint: Otherwise, P is not a prime ideal; take $A, B \supset P$ with $AB \subseteq P$, and using exercise 1 conclude A and B each contain powers of x.)

3. (Amitsur [71]) For each ordinal α there is a ring R_α with $N_{\alpha+1}(R) \supset N_\alpha(R)$. In fact, the R_α are defined inductively on α, with the inductive step described as follows: Given a ring T let $M_\infty(T)$ be the ring of row-finite matrices of exercise 1.1.5 and view $T \subseteq M_\infty(T)$ via scalar matrices (i.e., $a \to a \cdot 1$). Let T' be the subring of $M_\infty(T)$ generated by its socle (i.e., $T = \{a \cdot 1 + \sum_{\text{finite}} a_{ij}e_{ij} : a, a_{ij} \in T\}$ of $M_\infty(T)$, and $T'' = \{a \cdot 1 + \sum_{\text{finite}} a_{ij}e_{ij} : a_{ij} \in \operatorname{Prime\,rad}(T)$ for $i < j\}$. Thus $T \subset T'' \subset T'$. Given $R_\beta = T$ we will take $R_{\beta^+} = T''$. Taking R_0 to be any field, this defines R for successor ordinals; for limit ordinals R_α is defined as the direct limit of $\{R_\beta : \beta < \alpha\}$.

   The desired properties of R_α are proved by examining the inductive step. Given $B \vartriangleleft T$ define $B' = \{b \cdot 1 + \sum_{\text{finite}} b_{ij}e_{ij} : b, b_{ij} \in B\} \vartriangleleft T'$; if $B \subseteq \operatorname{Prime\,rad}(T)$ then $B' \vartriangleleft T''$ and, in fact, $N_\gamma(R_\beta)' \subseteq N_\gamma(R_{\beta^+})$ for each γ. Writing N_α for $\operatorname{Prime\,rad}(R_\alpha)$, prove $N_\alpha = N_\alpha(R_\alpha)$ and $R_\alpha/N_\alpha \approx R_0$. (Extensive hint: When $\alpha = \beta^+$ prove, in fact, $N_\alpha = N_\alpha(R_\alpha) = N'_\beta$ and $R_\alpha/N_\alpha \approx R_\beta/N_\beta$ by induction, as follows: Define $\psi : R_\alpha \to R_\beta/N_\beta$ by $\psi(r \cdot 1 + \sum r_{ij}e_{ij}) = r + N_\beta$. $R_\beta/N_\beta \approx R_0$ so $N_\alpha \subseteq \ker \psi$. But any matrix in $\ker \psi$ is strictly upper triangular modulo N'_β so the left ideal it generates is nilpotent modulo N'_β. $N_\beta = N_\beta(R_\beta)$ by induction so $N'_\beta \subseteq N_\beta(R_\alpha)$ and $\ker \psi \subseteq N_\alpha(R_\alpha) \subseteq N_\alpha$; hence equality holds.)

4. If e is an idempotent of R and $N = \operatorname{Nil}(R)$ then $eNe = eRe \cap N \subseteq \operatorname{Nil}(eRe)$. Show equality would be implied by Koethe's conjecture.

5. If L is a quasi-invertible left ideal of R then $M_n(L)$ is quasi-invertible. (Hint: View $L \subseteq \operatorname{Jac}(R)$.) By adjoining 1, show Koethe's conjecture is equivalent to the assertion, "If every quasi-invertible left ideal of R is nil then the same holds for $M_n(R)$."

5'. Show Koethe's conjecture holds for algebras over uncountable fields. (Hint: theorems 2.5.22 and 2.6.35(iii).)

6. The following result provides an easier substitute for Amitsur's theorem (2.5.23) and also relates to Koethe's conjecture. Let $R' = R[\lambda_1, \lambda_2, \ldots]$ the polynomial ring in an infinite number of commuting indeterminates. $\operatorname{Jac}(R')$ is nil. (Hint: If $a = f(\lambda_1, \ldots, \lambda_t) \in \operatorname{Jac}(R')$ then $(1 - a\lambda_{t+1})^{-1}$ exists but formally is $\sum_{i=0}^\infty a^i \lambda_{t+1}^i$, implying $a^n = 0$ for some n.) Define $\operatorname{Absnil}(R) = R \cap \operatorname{Jac}(R')$, a nil ideal of R. If $\operatorname{Absnil}(R) = 0$ then $\operatorname{Jac}(R') = 0$; thus $\operatorname{Jac}(R') = (\operatorname{Absnil}(R))[\lambda_1, \lambda_2, \ldots]$. Develop a radical theory for this interesting nilradical.

7. Hypotheses as in proposition 2.6.24, show any nil left or right ideal L of R is contained in the lower nilradical N. (Hint: Otherwise, take $a \in L - N$ with $\operatorname{Ann}_R a$ maximal possible. But $\operatorname{Ann}(ara) \supset \operatorname{Ann} a$ for each r in R so $aRa \subseteq N$, proving $a \in N$, contradiction.) In particular, $N = \operatorname{Nil}(R)$.

8. (The Levitzki radical) A subset S of a ring is *locally nilpotent* if every finite subset of S is nilpotent. Prove there is a unique maximally locally nilpotent ideal of R, called $\text{Lev}(R)$, and this contains every locally nilpotent left ideal L of R. (Hint: For the first assertion note if $N \lhd R$ is locally nilpotent and $(S + N)/N$ is locally nilpotent in R/N then S is locally nilpotent. For the second assertion show every finite set $\{a_2 r_1, \ldots, a_m r_m : a_i \in L\}$ is nilpotent.)

9. An ideal A of R is locally nilpotent iff $A \otimes_Z T \subseteq \text{Nil}(R \otimes_Z T)$ for every T in $Z(R)$-$\mathcal{Alg}$ (Hint: ($\Leftarrow$) Take $T = Z\{X\}$ and consider $\sum_{i=1}^{t} a_i \otimes x_i$.)

Graded Rings

10. Suppose G is a group, written multiplicatively and $g_1, \ldots, g_{nt} \in G$ for $n, t \in \mathbb{N}$. Writing $h_i = g_1 \cdots g_i$, if $\{h_1, \ldots, h_{nt}\}$ take on at most n distinct values then there exist $0 \le u_0 \le u_1 \le \cdots \le u_t \le n$ with $g_{u_j + 1} \cdots g_{u_{j+1}} = 1$ for $0 \le j \le t - 1$. (Hint: $t + 1$ of the h_i are equal, i.e., $h_{u_0} = h_{u_1} = \cdots = h_{u_t}$.)

11. Suppose R is graded by a multiplicative group G with unit 1, and $|\text{supp } R| = n$. R_1 denotes the identity component of R. (i) If A is a graded subring (without 1) of R with $A_1 = 0$ then $A^n = 0$. (ii) If $A < R_1$ and $A^t = 0$ then $(RA)^{nt} = 0$. (iii) If R has no nilpotent graded ideals then R_1 is semiprime and $(Rr)_1 \ne 0$ for every $r \ne 0$ in R. (Hint: (i) Show $A_{g_1} \cdots A_{g_n} = 0$ for every $g_1, \ldots, g_n$ in G. This is certainly true unless $g_1 \cdots g_i \in \text{supp } R$ for each i, in which case exercise 10 is applicable with $t = 1$. (ii) Note $(RA)_1 = A$ and use exercise 10. (iii) By (ii) and (i).)

12. If R is graded by a multiplicative monoid S then $L \to RL$ gives a lattice injection from $\mathcal{L}(R_1)$ to {graded left ideals of R}; in particular, if R satisfies ACC (resp. DCC) then so does R_1. Conversely, suppose the set-up of exercise 11 (iii) holds and R_1 is semisimple Artinian. Then R is Artinian; in fact, each R_g is f.g. in R_1-$\mathcal{Mod}$. (Hint: 1 is a sum of orthogonal primitive idempotents e_i in R_1, so it suffices to show $e_i R_g e_j \approx e_i R_1 e_i$ in $e_i R e_i$-$\mathcal{Mod}$ whenever $e_i R_g e_j \ne 0$. Exercise 11(iii) shows $e_i R_g e_j R_h e_i \ne 0$ where $h = g^{-1}$; taking x_h in $e_j R_h e_i$ such that $x_h R_g e_j \ne 0$ we have a nonzero right ideal of the division ring $e_j R_1 e_j$ so $x_h R_g e_j = e_j R_1 e_j$ and $x_h y_g = e_j$ for some y_g. Conclude $e_i R_g e_j = e_i R_1 e_i y_g$.)

Group Actions

13. If G is a finite group of automorphisms on a ring R then $|G|$ Prime rad$(R^G) \subseteq$ Prime rad(R). (Hint: Apply the embeddings in the proof of theorem 2.5.52 to theorem 2.5.51.) In particular, if R has no $|G|$-torsion and is semiprime then R^G is semiprime.

14. Suppose G is a finite group of automorphisms of R, with $|G|^{-1} \in R$. Then $e = |G|^{-1} \sum_{g \in G} g$ is an idempotent of the skew group ring $R * G$, and $e(R * G)e \approx R^G$. In view of lemma 2.7.12, this exercise shows that information passed from R to $R * G$ can land in R^G.

M. Cohen found a Morita context (c.f., §4.1 below) involving R^G and $R * G$, and recently the results have been put into the very general framework of Hopf algebras (c.f., §8.4 below). The standard text is Montgomery [80B] although there are several more recent developments, c.f., Passman [83], Montgomery [84], and Blattner-

Cohen-Montgomery [86]. The next two exercises present two standard counter-examples.

15. (A prime ring without 1 having solvable group of automorphisms whose fixed subring is 0) Let F be a field of characteristic $p \neq 0$, containing a primitive n-th root ζ of 1. Let W be the free associative F-algebra $F\{X_1, X_2\}$ and $R' = M_2(W)$.

Let $w(u,v) = \zeta^u X_v \in W$ and define H as the submonoid generated by

$$\left\{ \begin{pmatrix} 1 & mw(u,v) \\ 0 & 1 \end{pmatrix} : 0 \leq m < p, \ \ 0 \leq u < n, \ \ 1 \leq v \leq 2 \right\},$$

an abelian multiplicative subgroup of R'. Letting G be the multiplicative subgroup of R' generated by H and $\begin{pmatrix} \zeta & 0 \\ 0 & 1 \end{pmatrix}$, we see H is normal in G and G/H is cyclic. Let R be the subring of R' consisting of those matrices whose entries each have constant term 0. Then conjugation by G in R' induces a group of automorphisms of R, and $R^H \subset We_{12}$ so $R^G = 0$.

16. Let $R = M_2(F)$ for a field F of characteristic 2, and let $G = \left\{ \begin{pmatrix} 1 & m \\ 0 & 1 \end{pmatrix} : 0 \leq m < p \right\}$, a cyclic group of order p, which acts on R by conjugation. Then

$$R^G = C_R\left(\begin{pmatrix} 1 & 1 \\ 0 & 1 \end{pmatrix} \right) = \left\{ \begin{pmatrix} \alpha & \beta \\ 0 & \alpha \end{pmatrix} : \alpha, \beta \in F \right\}$$

which is not even semiprime, although R is simple. Note that this example and the previous example show there is something "wrong" with inner automorphisms, and many of the positive results involve groups of automorphisms which are not even restrictions of inner automorphisms (suitably defined), c.f., the papers of Kharchenko and also Montgomery [80B].

17. If a is an invertible element of R then $C_R(a)$ is the fixed subring of the inner automorphism determined by a. This fact is used by M. Cohen [78] to link the structure of R and $C_R(a)$ after determining the structure of fixed subrings under an algebraic automorphism.

§2.7

1. The following assertions are equivalent for a module M over a left Artinian ring R: (i) M is f.g.; (ii) M is Artinian; (iii) M is Noetherian. (Hint: (ii) $\Rightarrow$ (i) Induction on the smallest t such that $J^t M = 0$ where $J = \text{Jac}(R)$. $J^{t-1}M$ is f.g. by induction, and $M/J^{t-1}M$ is an Artinian R/J-module and is thus f.g.)
2. The ring of upper triangular matrices over a field is a semiprimary ring in which central idempotents of $R/\text{Jac}(R)$ lift to idempotents which *cannot* be central.

Perfect Rings

3. In example 2.7.38 if $[M:F]$ is uncountable then $\bigcap_{i=1}^{\infty} \text{Jac}(R)^i \neq 0$ since the transformation taking x_ω to x_1 is in each $\text{Jac}(R)^i$ (where ω is the first infinite ordinal).
4. If R is (right) perfect then $\text{Jac}(R)$ is locally nilpotent. (Hint: If S is a finite subset

of Jac(R) which is not nilpotent then for some s_1 in S one has $S^k s_1 \neq 0$ for each k in $\mathbb{N}$. Continue in this way to build a non-T-nilpotent sequence).

5. (Bjork) The following assertions are equivalent: (i) R is right perfect. (ii) R has no infinite set of orthogonal idempotents, and every R-module contains a simple submodule. (iii) Every R-module satisfies DCC (f.g. submodules). Hint: (iii) $\Rightarrow$ (ii) follows instantly from the DCC on cyclic submodules. (ii) $\Rightarrow$ (i) If $a_1, a_2, \ldots \in J = \mathrm{Jac}(R)$ with $a_n \cdots a_1 \neq 0$ for each n then take $L < R$ maximal with respect to $a_n \cdots a_1 \notin L$ for each n. Take a simple submodule L_1/L of R/L. Some $a_n \cdots a_1 \in L_1$, so $L_1 = Ra_{n+1} \cdots a_1 + L$, implying $a_n \cdots a_1 \in L$ as in remark 2.7.6. This contradiction proves J is T-nilpotent; prove R/J is semisimple Artinian by corollary 2.3.11 and proposition 2.2.11) (i) $\Rightarrow$ (iii) the hardest direction. Take $N < M$ maximal with respect to satisfying DCC (f.g. submodules), and assume $N < M$. We want a contradiction. There is $x \in M - N$ with $Jx \subseteq N$ (for, otherwise, one could define a nonvanishing sequence in J by stipulating inductively $a_{i+1}(a_i \cdots a_1 x) \notin N$. There is a primitive idempotent e with $ex \notin N$; we may assume $M = N + Rex$. Let $M_1 > M_2 > \cdots$ be an arbitrary chain of f.g. submodules; to show it terminates one may assume each $M_i \not\leq N$. The strategy is to build f.g. $N_i \leq N \cap M_i$ with $N_1 > N_2 > \cdots$ and show $this$ sequence must terminate. In fact putting $N_0 = N$, $M_0 = M$, and $x_0 = ex$, we shall find $N_i \leq N_{i-1} \cap M_i$ and x_i in eN_{i-1}, such that $M_i = N_i + Rw_i$ where $w_i = \sum_{u=0}^i x_i$. Indeed, given y_i in $M_i < M_{i-1}$ one inductively has y'_{i-1} in N_{i-1} and r_1 in R with $y_i = y'_{i-1} + r_i w_{i-1}$. This implies $r_i e x_0 \notin N$ so $r_i e \notin J$. Hence $e \in Rr_i e$ (lemma 2.7.17) so $e = eb_i r_i e$ for some b_i. Take $x_i = eb_i y'_{i-1} \in eN_{i-1}$. Then $w_i = eb_i y_i \in eM_i$. Write $M_i = \sum Rx_{ij}$ and, as before, $x_{ij} = x'_{ij} + r_{ij} w_{i-1}$ where $x'_{ij} \in N_{i-1}$. Define $N_i = \sum_{j=1}^{t(i)} R(x'_{ij} - r_{ij} x_i) \subseteq N_{i-1} \cap M_i$ and note $M_i \leq N_i + \sum_j Rr_{ij} w_i \leq M_i$.) This argument can be modified to show DCC (cyclic submodules) implies DCC (f.g. submodules), as shown in Faith [76B].)

6. (Converse to theorem 2.7.35, Jonah [70]) If every right R-module satisfies the ACC on cyclic right submodules then R is right perfect. (Hint: Show for any sequence $r_1, r_2, \ldots$ in R then there is $i < j$ with $r_j \cdots r_i R = r_j \cdots r_{i+1} R$, seen as follows: Let $S_i = r_i R$ and define an equivalence $\sim$ on the disjoint union S of the S_i by putting $s_i \sim s_u$ if there is $j > \max(i, u)$ with $r_j \cdots r_{i+1} s_i = r_j \cdots r_{u+1} s_u$ (where $s_i \in S_i$ and $s_u \in S_u$). Then $M = S/\sim$ is a right module and is a special case of the $direct\ limit$; the canonical image of each S_i is a submodule M_i. Since $M_1 \leq M_2 \leq \cdots$ one has $M_i = M_{i+1}$ for some i, yielding $r_j \cdots r_{i+1} r_i \in r_j \cdots r_{i+2}(r_{i+1} R)$ as desired. Now go down the structure theory. If each $r_i \in J$ then remark 2.7.6 shows J is T-nilpotent so assume $J = 0$. An easy application of the density theorem shows any primitive image of R is simple Artinian, so, in particular, $\bigcap\{$maximal ideals of $R\} = 0$. Enumerate the maximal ideals A_i taking $A_i \not\supseteq A_1 \cap \cdots \cap A_{i-1}$ and $r_i \in A_i - \bigcap_{u=1}^{i-1} A_u$; one may assume $1 \in r_i + A_i$ for each i, so that $r_j \cdots r_i R \neq r_j \cdots r_{i+1} R$ for all $j > i$, a contradiction, unless a finite number of maximal ideals has intersection 0. Conclude by the Chinese Remainder Theorem.)

7. Every left Noetherian right perfect ring is left artinian. (Hint: Special case of exercise 8).

8. The following assertions are equivalent for a right perfect ring R with $J = \mathrm{Jac}(R)$: (i) R is left Artinian; (ii) R/J^2 is left Artinian; (iii) $J/J^2 \in R\text{-}\mathscr{F}imod$. (Hint: (iii)$\Rightarrow$(i) If $J = J^2 + \sum_{i=1}^t Ra_i$ then $J^n = J^{n+1} + \sum Ra_{i_1} a_{i_2} \cdots a_{i_n}$, so by the proof of

theorem 2.7.2 it suffices to show J is nilpotent. This is obvious unless $J > J^2 > J^3 > \cdots$, in which case for each n some $a_{i_1} \cdots a_{i_n} \notin J^{n+1}$, contrary to T-nilpotence.)

Semiperfect Rings

9. Suppose R is semiperfect, and $J = \mathrm{Jac}(R)$. An idempotent e is primitive iff Re/Je is simple in R-*Mod* iff Je is the unique maximal submodule of Re. (Hint: If N is a maximal submodule of Re then $N = Le$ where $L = R(1 - e) + N$ is a maximal left ideal of R.)

9'. If $A \lhd R$ then $\mathrm{Hom}(Re/Ae, R) \approx e\mathrm{Ann}'\, A$ in *Mod*-R for any idempotent e. (Hint: Given r in $\mathrm{Ann}'\, A$ define $\varphi_r: Re \to R$ by right multiplication by er, and note φ_r yields a map $Re/Ae \to R$.)

10. Suppose R is semilocal and $J = \mathrm{Jac}(R)$, and e, e' are primitive idempotents of R with $R \not\approx Re'$. Then any map $f{:}Re \to Re'$ satisfies $f(Re) \subseteq Je'$. (Hint: in lemma 2.7.27 show $f = 0$.)

11. Hypotheses as in exercise 10, the following are equivalent: (i) $\mathrm{Hom}(Re, Re') \neq 0$; (ii) $eRe' \neq 0$; (iii) $eJe' \neq 0$. (Hint: If $0 \neq f \in \mathrm{Hom}(Re, Re')$ then $0 \neq fe \in eJe'$.) This exercise has important applications to representation theory.

12. (Zaks [67]) Suppose R is semiperfect and has complete set of idempotents $e_1, \ldots, e_n$ such that $e_i Re_i$ is a division ring for each i. Then $R = R' + J$ where $R' = \sum\{e_i Re_j: Re_i \approx Re_j\}$ is semisimple Artinian and $J = \mathrm{Jac}(R) = \sum\{e_i Re_j: Re_i \not\approx Re_j\}$. (Hint: Although this turns out to be a special case of Wedderburn's principal theorem, we sketch Zak's easy direct proof. First note that if $Re_i \approx Re_j$ then $e_i Je_j = 0$, thus implying every map $f: Re_i \to Re_j$ is an isomorphism. Thus putting $R_u = \sum\{e_i Re_j: Re_i \approx Re_j \approx Re_u\}$, a simple ring, one concludes $R' \approx \prod R_u$ is semisimple artinian. To see $J \lhd R$ it suffices to show $e_i re_j \in J$ implies $e_i re_j r' e_k \in J$; but this is clear unless $Re_i \approx Re_k$, in which case right multiplication by $re_j r' e_k$ gives an isomorphism from Re_i to Re_k, implying $Re_i \approx Re_j$, contradiction. The rest is clear.)

13. Suppose R is a semiperfect ring with $J = \mathrm{Jac}(R)$, and e is a primitive idempotent. An R-module M with composition series has a composition factor isomorphic to the simple module Re/Je iff $eM \neq 0$.

Structure of Semiprimary Rings

16. If R is semiperfect with $\mathrm{Jac}(R)$ nil then for all $L < R$ and all $j > k$ one has $L^k = L^j + (L \cap \mathrm{Jac}(R))^k$. (Hint: Let $L_0 = L \cap \mathrm{Jac}(R)$. Then $L = L_0 + Re$ for a suitable idempotent e, by corollary 1.1.28, and an easy induction shows $L^k = ReL + L_0^k$ for all k.)

17. If R is semiprimary and $L < R$ then $L^t = L^{t+1}$ where $\mathrm{Jac}(R)^t = 0$. (Hint: Exercise 16.)

18. If e is a primitive idempotent of a semiprimary ring R then $Re > Je > J^2 e > \cdots$ refines to a composition series.

19. Given R let $R' = \prod R_i/\mathscr{F}$ where each $R_i \approx R$ and $\mathscr{F}$ is an ultrafilter. If R is semilocal then R' is semilocal (cf., exercise 2.5.7 and 2.3.6). If R is semiperfect then R' is semiperfect.

20. Notation as in exercise 19, suppose R is not left Artinian. Then R' is not left Noetherian. Indeed, if $L_1 > L_2 > L_3 > \cdots$ are left ideals of R then taking $l_k = (\prod L_{i-k})/\mathscr{F}$ where $L_u = 0$ for $u \leq 0$ show $l_1 < l_2 < \cdots$. This gives examples of semiperfect rings which are not left Noetherian. Analogously, if R is not right perfect then we can take each L_i to be principal, in which case l_k is principal and R' fails the ACC on principal left ideals.

21. Notation as in exercise 20. $M = R'/l_1$ has an onto map which is not 1:1. (Note R' is semiperfect but fails ACC on principal left ideals.)

22. (Bjork) A semiprimary ring R with a cyclic module M having an onto map which is not 1:1. Let K be a field and D be a division ring containing the free algebra $K\{X, Y\}$. The field of fractions F of $K[X]$ can be viewed in D. Then $R = \begin{pmatrix} K & D \\ 0 & F \end{pmatrix}$ is semiprimary; taking $L = \sum_{i=1}^{\infty} KYX^i$ we see $\begin{pmatrix} 0 & D/L \\ 0 & F \end{pmatrix}$ is a cyclic module spanned by $\begin{pmatrix} 0 & 0 \\ 0 & 1 \end{pmatrix}$, and right multiplication by X is the desired map.

§2.8

1. (Kaplansky) Let $C = \{$continuous real-valued functions on the interval $[0,1]\}$, a commutative subring of $\mathbb{R}^{[0,1]}$, and $P = \{f \in C : f$ vanishes at a neighborhood of $0\}$. Then $P \lhd C$. For each n in $\mathbb{N} - \{0\}$ define p_n in P by $p_n x = 0$ for $x \leq \frac{1}{n}$ and $p_n x = x - \frac{1}{n}$ for $x > \frac{1}{n}$. Then $\{p_n : n \in \mathbb{N}\}$ spans P. Find a dual base, thereby proving P is projective.

2. $\mathbb{Q}$ is *not* projective in $\mathbb{Z}$-$\mathscr{M}od$. (Hint: the only map $\mathbb{Q} \to \mathbb{Z}$ is 0.)

3. The following assertions are equivalent: (1) R is semisimple Artinian. (2) Every R-module is projective. (3) Every cyclic R-module is projective. (Hint for (3)$\Rightarrow$(1): Suppose $L < R$. Then the epic $R \to R/L$ splits, so L is a summand of R.)

4. A ring R is *semihereditary* if every f.g. left ideal of R is projective. In this case show every f.g. submodule of a projective module is projective; every f.g. projective module is isomorphic to a finite direct sum of copies of left ideals of R.

5. The ring $R = \begin{pmatrix} \mathbb{Q} & \mathbb{Q} \\ 0 & \mathbb{Z} \end{pmatrix}$ is left hereditary but not right hereditary. (Hint: If L is a left ideal then $L = e_{11}L \oplus e_{22}L$; $e_{11}L$ is a $\mathbb{Q}$-submodule of $\mathbb{Q} \oplus \mathbb{Q}$ and $e_{22}L$ is a $\mathbb{Z}$-submodule of $\mathbb{Z}$. On the other hand, the right module Re_{12} acts like $\mathbb{Q}$ as a $\mathbb{Z}$-module and is not projective by exercise 2.) A complete characterization of hereditary rings of the form of example 1.1.9 is given in Goodearl [76B, theorem 4.7].

Isomorphisms (also, cf., exercises 2.11.12ff)

7. Suppose P is an f.g. projective module over a commutative ring C, and $M \in C$-$\mathscr{M}od$. Then $\text{End}_C M \otimes \text{End}_C P \approx \text{End}_C(M \otimes P)$. (Hint: Define the balanced map $\text{End}_C M \times \text{End}_C P \to \text{End}_C(M \otimes P)$ by sending (f, g) to $f \otimes g$. This defines the desired map, which is an isomorphism for $P = C$, and thus for P f.g. free, and finally for P a summand of an f.g. free C-module.)

7'. Suppose P is an f.g. projective R-module and $M \in R\text{-}\mathcal{M}od\text{-}T$ and $N \in \mathcal{M}od\text{-}T$. Then $\operatorname{Hom}_T(M, N) \otimes_R P \approx \operatorname{Hom}_T(\operatorname{Hom}_R(P, M), N)$. (Hint: Define the balanced map $\operatorname{Hom}(M, N) \times P \to \operatorname{Hom}(\operatorname{Hom}(P, M), N)$ by $(f, x) \to f_x$ where $f_x(g) = f(gx)$. This induces a map from the tensor product which is an isomorphism for P f.g. free, and thus for P f.g. projective.)

The Radical of a Module cf., Definition 2.5.26.

8. If $N \leq \operatorname{Rad}(M)$ then $\operatorname{Rad}(M/N) = \operatorname{Rad}(M)/N$; in particular, $\operatorname{Rad}(M/\operatorname{Rad} M) = 0$. On the other hand, if $\operatorname{Rad}(M/N) = 0$ then $\operatorname{Rad} M \leq N$. (Hint: As in proposition 2.5.1'.)

9. $\operatorname{Rad} N \leq \operatorname{Rad} M$ for all $N \leq M$. (Hint: If M' is a maximal submodule of M use lemma 2.4.3(ii) to show $M' \cap N$ is either N itself or a maximal submodule.)

10. If $M' \ll M$ then M' is contained in every maximal submodule of M, so $M' \leq \operatorname{Rad}(M)$ (which we recall is the intersection of maximal submodules of M).

11. $\operatorname{Rad}(M) = \sum\{\text{small submodules of } M\}$. (Hint: Prove for $x \in M$ if Rx is *not* a small submodule of M then there exists a maximal submodule M'' of M not containing x; indeed, if $Rx + N = M$ for $N < M$ then one can take $M'' \supseteq N$ maximal with respect to $x \notin M''$. This proves $(\subseteq)$; $(\supseteq)$ is by exercise 10.)

    $\operatorname{Rad}(M)$ need not be small, cf. exercise 17.

12. $\operatorname{Rad}(M_1 \oplus M_2) = \operatorname{Rad}(M_1) \oplus \operatorname{Rad}(M_2)$ for any M_1, M_2 in $R\text{-}\mathcal{M}od$. (Hint: $(\subseteq)$ Let $N = \operatorname{Rad}(M_1 \oplus M_2)$. $N \leq N' \oplus M_2$ for every maximal submodule N' of M_1, so $N \leq \operatorname{Rad}(M_1) \oplus M_2$; likewise $N \leq M_1 \oplus \operatorname{Rad}(M_2)$. $(\supseteq)$ by exercise 9. A similar argument shows $\operatorname{Rad}(\bigoplus M_i) = \bigoplus(\operatorname{Rad}(M_i))$.)

13. If F is free in $R\text{-}\mathcal{M}od$ then $\operatorname{Rad}(F) = \operatorname{Jac}(R)F$ by exercise 12.

14. $\operatorname{Rad}(P) = \operatorname{Jac}(R)P$ for every projective module P. (Hint: Write $J = \operatorname{Jac}(R)$ and $P \oplus P' = F$ free; then apply exercises 12 and 13.)

15. $\operatorname{Rad}(P) \neq P$ for every nonzero projective module P. (Hint: Otherwise, $P = \operatorname{Rad}(P) = \operatorname{Jac}(R)P$. Take $P \oplus P' = F$ free and let $\pi: F \to P$ be the projection. Let $\{y_i : i \in I\}$ be a base of F and note $P \subseteq \operatorname{Jac}(R)F$. Hence any $x \neq 0$ can be written as $\sum a_i y_i$ for a_i in $\operatorname{Jac}(R)$, so $x = \pi x = \sum a_i(\pi y_i)$. Writing πy_i as $\sum_j b_{ij} y_j$ for b_{ij} in $\operatorname{Jac}(R)$ yields $0 = x - x = \sum_i a_i y_i - \sum_{j,i} a_i b_{ij} y_j = \sum_j (\sum_i a_i(\delta_{ij} - b_{ij})) y_j$ so $\sum_i a_i(\delta_{ij} - b_{ij}) = 0$ for each i. Hence $a(1 - (b_{ij})) = 0$ where $a = (a_i)$; since there are a finite number of b_{ij} the matrix (b_{ij}) is quasi-invertible implying $a = 0$.)

T-Nilpotence

16. Suppose M is a free right R-module with countable base $x_1, x_2, \dots$. Given $r_1, r_2, \dots$ in R let $y_i = x_i - x_{i+1} r_i$ and let $N = \sum y_i R \leq M$. Then $\{y_1, y_2, \dots\}$ is an independent set of elements. Also $x_k \in N$ iff $r_n r_{n-1} \cdots r_k = 0$ for some $n > k$; thus $N = M$ iff for each k we have suitable $n = n(k) > k$ with $r_n \cdots r_k = 0$.

17. The following are equivalent for a right ideal J of a ring R: (i) J is T-nilpotent. (ii) $MJ \neq M$ for every $M \neq 0$ in $\mathcal{M}od\text{-}R$. (iii) $MJ \ll M$ for every $M \neq 0$ in $\mathcal{M}od\text{-}R$. (iv) $MJ \ll M$ where M is a free R-module with countable base. (Hint: (i) $\Rightarrow$ (ii). If $MJ = M$ then one could inductively find $a_1, a_2, \dots$ with $Ma_n \cdots a_1 \neq 0$ for all n; (ii) $\Rightarrow$ (iii) If $N \leq M$ then $(M/N)J \neq M/N$; (iv) $\Rightarrow$ (i) By exercise 16.)

18. If every right R-module has a maximal proper submodule then $\operatorname{Jac}(R)$ is T-nilpotent. (Hint: Use (ii) $\Rightarrow$ (i) in exercise 17.)

19. (Strooker [66]) We say a map $f: M \rightarrow N$ is a *cover* if $\ker f \ll M$. Suppose $A \lhd R$ and N is an f.g. R/A-module. If $f: M \rightarrow N$ is a cover then M is f.g., and if N is R/A-projective then $M/AM \approx N$. (Hint: let $\varphi: M \rightarrow M/AM$ be the canonical map. Lift an epic $(R/A)^{(n)} \rightarrow N$ to a map $g: R^{(n)} \rightarrow M$ such that the appropriate square commutes; conclude g is epic. If N is R/A-projective then viewing N as a summand of M/AM show equality holds by moving around the square.)

Projective Covers

20. If $\text{Jac}(R) = 0$ and M has a projective cover then M is itself projective (by exercises 11 and 14.)

21. Using lemma 2.8.39 reprove any finite set of orthogonal idempotents can be lifted orthogonally.

22. R is semiperfect iff every f.g. R-module has a projective cover. (Hint: ($\Leftarrow$) Every summand of $\bar{R} = R/\text{Jac}(R)$ has a projective cover, so J is idempotent-lifting. $\bar{R}$ is semisimple Artinian by exercises 3 and 20.)

23. R is semiperfect iff every simple R-module has a projective cover. (Hint: It suffices to prove every f.g. module M has a projective cover; since $\text{Rad}(M) \ll M$ one can pass to $M/\text{Rad}(M)$ which is a finite subdirect product of simple modules; conclude using proposition 2.2.5.)

24. Suppose $P \in R\text{-}\mathcal{M}od$ is projective and $f \in \text{End}_R P$. Then $f \in \text{Jac}(\text{End}_R P)$ iff $fP \ll P$. (Hint: ($\Rightarrow$) Suppose $N + fP = P$. Show $N = P$ by proving the canonical map $\varphi: P \rightarrow P/N$ is 0 as follows: φf lifts to a map $h: P \rightarrow P$ with $(\varphi f)h = \varphi$; then $\varphi(fh - 1) = 0$ implying $\varphi = 0$. ($\Leftarrow$) For any $g: P \rightarrow P$ one has $P = (1 - fg)P + fgP$, so $1 - fg$ is onto, and thus 1_P "lifts" to a map $h: P \rightarrow P$ with $(1 - fg)h = 1$, proving fg is right quasi-invertible.)

25. Suppose $P \in R\text{-}\mathcal{M}od$ is projective and $JP \ll P$ where $J = \text{Jac}(R)$. Letting $E = \text{End}_R P$ show $\text{Jac}(E) \approx \text{Hom}_R(P, JP)$ and $\text{End}_R(P/JP) \approx E/\text{Jac}(E)$. (Hint: $JP = \text{Rad}(P)$ so $N \ll P$ iff $N \leq JP$, and exercise 24 yields the first assertion. Now there is $\psi: E \rightarrow \text{End}_R(P/JP)$ given by $(\psi f)(x + JP) = fx + JP$ where $f \in E$; $\ker \psi = \text{Jac}(E)$, and ψ is onto since any $g: P/JP \rightarrow P/JP$ yields a map $P \rightarrow P/JP$ which lifts to a map $P \rightarrow P$.)

26. The following statements about a projective R-module P are equivalent: (i) There is a projective cover $f: P \rightarrow M$ with M simple. (ii) P has a small and maximal submodule. (iii) $\text{Rad } P$ is a small and maximal submodule of P. (iv) $\text{End}_R P$ is a local ring; (v) Every maximal submodule of P is small. (Hint: (ii)$\Rightarrow$(iii) If N is maximal then $\text{Rad } P \leq N$; if, moreover, $N \ll P$ then $N = \text{Rad } P$ by exercise 10; (iii) $\Rightarrow$(iv) by exercise 24; (v)$\Rightarrow$(i) P has a maximal submodule N by exercise 15, so $P \rightarrow P/N$ is a projective cover.)

Characterizations of Right Perfect Rings

27. R is right perfect iff $R/\text{Jac}(R)$ is semisimple Artinian and each right R-module has a proper maximal submodule. (Hint: exercises 17, 18.)

28. R is right perfect iff every right module has a (right) projective cover, in which case any projective right module is isomorphic to $\bigoplus_{i \in I} e_i R$ for suitable primitive idempotents e_i of R. (Extensive hint: ($\Rightarrow$) Suppose $M \in R\text{-}\mathcal{M}od$ and $J = \text{Jac}(R)$.

$M/MJ = \bigoplus_{i \in I} e_i R/e_i J$ for primitive idempotents e_i. Let $P = \bigoplus_{i \in I} e_i R$. Exercise 15 shows $PJ \ll P$ and $MJ \ll M$. Thus $P \to M/MJ$ is a projective cover, which then lifts to a projective cover $P \to M$. In case M is already projective then $P \approx M$. ($\Leftarrow$) R is semiperfect so it suffices to show any nonzero right module $M \neq 0$ contains a maximal submodule. Let $f: P \to M$ be a projective cover. By exercise 15 there is a maximal submodule N of P, and fN is maximal in M.)

Other characterizations are given in exercise 2.9.10, 2.11.7.

Amitsur-Ware-Zelmanowitz Proof of Characterization of $\mathrm{Jac}(M_\infty(R))$ We
say a set $\{L_i : i \in I\}$ of left ideals of R is *right-vanishing* if for any sequence $a_1, a_2, \ldots$ where $a_u \in L_{i_u}$ with i_u *distinct* we have $a_1 \cdots a_n = 0$ for suitable n. Ware-Zelmanowitz [70] proved for any projective module P that any $f: P \to P$ is in $\mathrm{Jac}(\mathrm{End}_R P)$ iff P has a dual basis $\{(x_i, g_i) : i \in I\}$ such that $\{g_i f P : i \in I\}$ is a vanishing set of left ideals contained in $\mathrm{Jac}(R)$, iff *every* dual basis satisfies this property. As an application they reproved the theorem of Sexauer-Warnock [67] characterizing $\mathrm{Jac}(M_\infty(R))$ since this is $\mathrm{Jac}(\mathrm{End}_R F)$ for F countable free. We shall give this special case of their proof. Accordingly, we say a base $\{x_1, x_2, \ldots\}$ of F is *WZ* (for Ware-Zelmanowitz) *over* a map $f: F \to F$ if $\{\pi_i f F : i \in I\}$ is a vanishing set of left ideals contained in $\mathrm{Jac}(R)$, where $\pi_i: F \to R$ is given by $\pi_i x_j = \delta_{ij}$.

29. Write $E = \mathrm{End}_R F$ and $J = \mathrm{Jac}(E)$ where F is countable free. The following are equivalent for f in E: (i) $f \in J$; (ii) F has a base WZ over f; (iii) every base of F is WZ over f. (Hint: Recall from exercise 24 that $f \in J$ iff $fF \ll F$. (ii) $\Rightarrow$ (i) Let $\{x_1, x_2, \ldots\}$ be the given dual base WZ over f, and put $L_i = \pi_i f F$. Then $fx \in \sum L_i x_i$ for each x in F, so it suffices to show $\sum L_i x_i \ll F$. Suppose, on the contrary, $N + \sum L_i x_i = P$ with $N \neq P$. Then there is x in P with $0 \neq \bar{x} \in \bar{P} = P/N$, so $\bar{x} = \sum_{i \in I'} a_i \bar{x}_i$ for a_i in L_i and suitable finite $I' \subseteq I$; for each j in I' we have $\bar{x}_j = \sum_{i \in I''(j)} a_{ij} \bar{x}_i$ for some finite $I''(j) \subset I$. If $j \in I''(j)$ then $\bar{x}_j = \sum ((1 - a_{jj})^{-1} a_{ij}) \bar{x}_i$ summed over $i \in I''(j) - \{j\}$; thus we may assume $j \notin I''(j)$. Continuing in this way, for each n write $\bar{x}$ as a sum of terms of the form $a_{i_1} a_{i_2} \cdots a_{i_n} \bar{x}_{i_n}$ where the a_{i_u} are form distinct L_i; thus one could build an infinite sequence which contradicts the right-vanishing of the L_i.

    (i) $\Rightarrow$ (iii) Given any base $\{x_1, x_2, \ldots\}$ let $L_i = \pi_i f F \subseteq \mathrm{Jac}(R)$ by exercises 10, 14. It remains to show the L_i are right-vanishing. We aim to show for any sequence $a_1, a_2, \ldots$ with $a_i \in L_i$ that $a_1 \cdots a_n = 0$ for suitable n. Write $a_i = \pi_i f w_i$ where $w_i \in F$. Define $\psi: F \to F$ by the following inductive procedure. Take $m_1 = 1$ and $m_2 > 1$ such that $f w_1 \in \sum \{Rx_j : j < m_2\}$ and put $\psi x_1 = x_2$ and $\psi x_j = 0$ for all j with $m_1 < j < m_2$; inductively, given m_i take $m_{i+1} > m_i + 1$ such that $f w_u \in \sum \{Rx_j : j < m_{i+1}\}$ for each $u < m_i$ and put $\psi x_j = 0$ for all $m_i < j < m_{i+1}$ and $\psi x_{m_i} = b_i x_{i+1}$ where $b_i = a_{m_i+1} \cdots a_{m_{i+1}-1}$. Putting $b_0 = 1$ let $N = \sum_{i=1}^{\infty} R(x_i - b_{i-1} a_{m_i} x_{i+1})$. Then $a_1 x_2 = \pi_1 f w_1 x_2 = \psi f w_1$ so $x_1 = (x_1 - a_1 x_2) + a_1 x_2 \in N + \psi f F$. Inductively, each $x_i \in N + \psi f F$ so $N = F$ since $\psi f F$ is small; by exercise 16 some product $a_1 \cdots a_n$ is 0.)

30. An infinite matrix $(r_{ij}) \in \mathrm{Jac}(M_\infty(R))$, iff each $r_{ij} \in \mathrm{Jac}(R)$ and the left ideals generated by the columns of (r_{ij}) are right vanishing.

31. Extend exercise 29 to free modules of arbitrary cardinality. (Hint: (ii) $\Rightarrow$ (i) is the

same; (i)⇒(iii) is obtained by defining ψ to send all the "extra" base elements to 0.)

32. Prove the Ware-Zelmanowitz theorem as quoted above, for arbitrary projective modules P. (Hint: (ii) ⇒ (i) is almost exactly the same as before, using the g_i of the dual basis in place of π_i; (i) ⇒ (iii) seems to require writing $P \oplus P' = F$ free and defining ψ on F using the same trick of "spacing out" the base elements.)

§2.9

Countable Modules and LE-Decompositions A module is *countable* if it is spanned by a countable set of elements. (In particular, every f.g. module is countable.)

1. Suppose $M = \bigoplus_{i \in I} M_i$ and $N \le M$ where N is spanned by $\{x_j : j \in J\}$. Then there is some set $I' \subseteq I$ with $|I'| \le |J|$ such that $N \le \bigoplus_{i \in I'} M_i$. (Hint: Each x_j is in a finite sum of the M_i.)

2. (Kaplansky-Walker) If M is a direct sum of countable submodules then so is every summand. (Extensive hint: Say M is dsc if M is a direct sum of countable submodules. Write $M = \bigoplus_{i \in I} M_i$ with each M_i countable. Suppose $M = M' \oplus M''$, and consider all pairs (N', N'') satisfying $N' < M'$, $N'' < M''$ with N' and N'' each dsc, and such that $N' \oplus N''$ is a direct sum of M_i. Order such pairs by putting $(N', N'') < (P', P'')$ if N' and N'' are respective summands of P' and P'' whose complements are also dsc. Then Zorn's lemma gives a maximal pair (N', N''); write $N' \oplus N'' = \bigoplus_{i \in I'} M_i$. Replacing I by $I - I'$ one may assume $N' = N'' = 0$. Let π_1, π_2 denote the respective projections of M onto M' and M''. Now take any $M_i \ne 0$ and let $Q_1 = M_i$. Inductively, given Q_j countable, we take $N'_j = \pi_1 Q_j$ and $N''_j = \pi_2 Q_j$, which are countable, and let Q_{j+1} be a countable sum of M_i which contains $N'_j \oplus N''_j$. Then $\bigcup_{j=1}^{\infty} Q_j = \bigcup N'_j \oplus \bigcup N''_j$ with each module dsc, contrary to hypothesis.)

3. Suppose $M = M' \oplus M''$ and N is a summand of M with an LE-decomposition $N = \bigoplus_{i=1}^{n} N_i$. Then there are respective summands L', L'' of M', M'' such that $M = L' \oplus L'' \oplus N$. (Hint: By induction and proposition 2.9.14 one may assume N is an LE-module.)

4. (Crawley-Jønsson-Warfield theorem) If $M = \bigoplus_{i \in I} M_i$ is an LE-decomposition with each M_i countable then every summand M' of M is a direct sum of modules isomorphic to suitable M_i. (Hint: M' is a direct sum of countable submodules by exercise 2, so we may assume M' is countable. Let $\{x_1, x_2, \ldots\}$ be a countable spanning set for M'. Inductively one wants to find independent submodules $N_1, N_2, \ldots$ of M' such that $N_1 \oplus \cdots \oplus N_n$ is a summand of M containing $x_1 \ldots, x_n$ for each n, and, moreover, such that each N_j is a finite direct sum of submodules isomorphic to various M_i. (For then $M' = \bigoplus_{j=1}^{\infty} N_j$.) So take $N_0 = 0$ and given $N_1, \ldots, N_{n-1}$ take N'_{n-1} such that $(\bigoplus_{u=1}^{n-1} N_u) \oplus N'_{n-1} = M'$. Then $x_n = a_n + b_n$ where $a_n \in \bigoplus_{u=1}^{n-1} N_u$ and $b_n \in N'_{n-1}$. Surely $b_n \in \bigoplus_{i \in I(n)} M_i$ for some finite subset $I(n)$ of I. Let $S_n = \bigoplus_{i \in I(n)} M_i$, a finite direct sum of LE-modules, and write M'' for a complement of M'. Then exercise 3 permits us to write $M = L' \oplus L'' \oplus S_n$ for suitable respective summands L', L'' of M', M''. Put $N_n = M' \cap (S_n \oplus L'')$. Then $b_n \in N_n$ so $x_n \in \bigoplus_{u=1}^{n} N_u$. Moreover $M' = N_n \oplus L'$. Let N'' be a complement of L'' in M''. Then $S_n \approx M/(L' \oplus L'') \approx N_n \oplus N''$. But S_n has a finite LE-decomposition so N_n is a sum of various M_i.)

5. Every projective module P over a local ring R is free. (Hint: Write $P \oplus P' = \bigoplus_{i \in I} R$, *an* LE-decomposition since $\operatorname{End}_R R \approx R$ is local, and apply exercise 4.)

6. Every projective module P over a semiperfect ring R has an LE-decomposition $P = \bigoplus_{i=1}^{t} P_i$ where P_i is a direct sum of copies of Re_i and $\{e_1, \ldots, e_t\}$ is a basic set of idempotents. (Hint: This is true for R and thus for any free module, so apply exercise 4.)

14. (Swan) Failure of Krull-Schmidt for an f.g. module over a local Noetherian ring R. Let F be any field and let $A = F[x, y]/\langle x^3 - x^2 + y^2 \rangle$. Then A is a commutative domain in which the images of x and y generate a maximal ideal P. Localizing at $S = A - P$ produces a commutative local Noetherian domain $R = S^{-1}A$ whose maximal ideal we call J. Let K be the field of fractions of R and let $z = y^{-1}x \in K$. Let $T = R[z] \subset K$. T is *not* local, having maximal ideals $M_1 = J + T(z - 1)$ and $M_2 = J + T(z + 1)$. As in example 2.9.22, one has $M_1 \oplus M_2 \approx T \oplus (M_1 \cap M_2)$ as T-modules and thus as R-modules. These modules are all torsion-free and indecomposable. But if $T \approx M_i$ over R then one would have $T \approx M_i$ over T, which is impossible since M_1, M_2 require two generators as T-module. Note that the trick was to expand R to a nonlocal ring T, in which it is easier to violate uniqueness of the Krull-Schmidt decomposition.

Finite Representation Type (f.r.t.)

18. Any homomorphic image of a ring with f.r.t. has f.r.t.
19. If R is a semiperfect PLID then every indecomposable is principal indecomposable; thus any semiperfect PLID has f.r.t.
20. If G is a cyclic group then any group algebra $F[G]$ has f.r.t. (Hint: Apply exercise 19 to early results of §8.1.)
21. Suppose H is a subgroup of a finite group G, and F is a field. Given an $F[H]$-module M define the *induced* $F[G]$-module $\tilde{M}$ to be $F[G] \otimes_{F[H]} M$. Let $b_1, \ldots, b_t$ be a transversal of H in G, i.e., $[G:H] = t$. If M is an $F[G]$-module then forming $\tilde{M}$ by viewing M naturally as $F[H]$-module (by restricting scalars) show there is a map $M \to \tilde{M}$ given by $x \to \sum_{i=1}^{t} b_i \otimes b_i^{-1}x$. If t is invertible in F this is a split monic, so in this case M is a summand of $\tilde{M}$ as $F[G]$-modules.
22. (Higman [54]) Suppose H is a Sylow p-subgroup of a finite group G, and $\operatorname{char}(F) = p$. Every indecomposable $F[G]$-module is a summand of a module induced from an indecomposable $F[H]$-module. Likewise, every indecomposable $F[H]$-module is a summand of some indecomposable $F[G]$-module (viewed naturally as $F[H]$-module). Thus $F[G]$ has f.r.t. iff $F[H]$ has f.r.t.. Using exercise 20 conclude $F[G]$ has f.r.t. iff H is cyclic. (Hint: If M is an indecomposable $F[G]$-module then as $F[S]$-module $M \approx M_1 \oplus \cdots \oplus M_t$ for M_i indecomposable; thus $\tilde{M} \approx \tilde{M}_1 \oplus \cdots \oplus \tilde{M}_t$ so some $\tilde{M}_i$ is a summand of M by exercise 21 and the exchange property. This proves the first assertion, and the second assertion is analogous; now use exercise 23.)
23. If F is a field of characteristic p and G is a noncyclic p-group then $F[G]$ does *not* have f.r.t. (Hint: Show G has a normal subgroup N with $G/N \approx (\mathbb{Z}/p\mathbb{Z})^{(2)}$ and thus use exercise 22 to reduce to the case $G \approx (\mathbb{Z}/p\mathbb{Z})^{(2)}$; now proceed as in example 2.9.25.

§2.10

1. Using the pushout reprove that if every monic from E splits then E is injective. (Hint: If $N \subseteq M$ and $f: N \to E$ is a map then forming a suitable pushout M' yields a monic $E \to M'$ which splits; travelling around the square yields a map $M \to E$ extending f.)

2. $\mathbb{Q}$ is the injective hull of $\mathbb{Z}$ in $\mathbb{Z}$-$\mathcal{M}od$. If $R = \begin{pmatrix} D & D \\ 0 & D \end{pmatrix}$ for a division ring D then $M_2(D)$ is the injective hull of R in R-$\mathcal{M}od$. (Hint: It is injective by Baer and is an essential extension.)

3. A ring R is (left) hereditary iff E/M is injective for every injective R-module E and every $M \le E$. (Hint: ($\Rightarrow$) If $L < R$ then any map $f: L \to E/M$ lifts to $\hat{f}: L \to E$; now apply Baer. ($\Leftarrow$) Reverse the argument, using proposition 2.10.27.)

4. Suppose E is an injective R-module and M is an R-S-bimodule which is flat in R-$\mathcal{M}od$. Then $\text{Hom}_R(M, E)$ is injective in $\mathcal{M}od$-S. (Hint: As in proof of proposition 2.10.12.)

5. Suppose R is a finite dimensional algebra over a field F. If $M \in R$-$\mathcal{F}imod$ then the injective hull of M is in R-$\mathcal{F}imod$. (Hint: By exercise 4 we may replace R by F; but all F-modules are injective.) An f.g. left R-module E is injective iff E is a finite direct sum of duals of principal indecomposable right modules. (Hint: Use exercise 1.5.5 repeatedly.)

7. A functorial method of embedding a module into an injective. Let E be the injective hull of $\bigoplus_{L < R} R/L$. Then any R-module M has a monic into $E^{\text{Hom}(M,E)}$, cf., remark 4.2.3 below. Conclude that the direct limit of injectives is injective.

8. Suppose E is an injective module. If $\text{Ann}\{r_1, \ldots, r_n\} = 0$ for $r_1, \ldots, r_n$ in R then $\sum r_i E = E$. (Hint: Define $f: R \to R^{(n)}$ by $fr = (rr_1, \ldots, rr_n)$. For any x in E right multiplication by x can be extended to a map $g: R^{(n)} \to E$, i.e., $rx = gfr$. Then write $g = (g_1, \ldots, g_n)$ where each $g_i: R \to E$ and note $x = gf1 = \sum r_i g_i 1$.

9. The following facts hold for any indecomposable injective module E:

   (i) Every monic $f: E \to E$ is onto (Hint: fE a summand of E).
   (ii) Every nonzero submodule of E is large.
   (iii) If $f \in \text{End}_R E$ then f or $1 - f$ is an isomorphism. (Hint: $\ker(1 - f) \cap \ker f = 0$).
   (iv) $\text{End}_R E$ is local.
   (v) $E \approx E(R/L)$ for some $L < R$. (Hint: There is a nonzero map $f: R \to E$, so $R/\ker f$ is a nonzero submodule of E.

Criteria for a Ring to be Left Noetherian

10. R is left Noetherian iff the direct sum of injective R-modules must be injective. Hint: ($\Rightarrow$) Suppose $f: L \to \bigoplus E_i$ for $L \le R$. Then fL is f.g. so is contained in a finite direct sum of E_i, which is injective, so f extends to R. ($\Leftarrow$) Suppose $L_1 \le L_2 \le \cdots$ and put $L = \bigcup L_i$. Define $f: L \to \bigoplus E(R/L_i)$ by $(fa)_i = a + L_i$. Extend f to R and put $x = f1$. Then there is i such that $x_j = 0$ for all $j \ge i$, and $L/L_i = 0$ proves $L = L_i$.)

11. If R is left Noetherian then every injective left R-module E is a direct sum of indecomposable injective R-modules. (Hint: For any x in E note Rx cannot contain an infinite direct sum, so neither can $E(Rx)$, implying $E(Rx)$ has an indecomposable summand. Zorn's lemma yields a maximal set of indecomposable summands of E, whose sum is injective and thus a summand of E.) The converse is also true, by exercise 13 below.

12. (Faith-Walker theorem) R is left Noetherian iff there is a cardinal c such that every injective R-module is a direct sum of modules each spanned by at most c elements. (Hint: ($\Rightarrow$) By exercise 9 the indecomposable injective R-modules belong to $\{E(R/L): L < R\}$, a set, so take c to be the maximal cardinality. ($\Leftarrow$) One must show any direct sum $\bigoplus_{i \in I} E_i$ of injectives is injective; we may assume each E_i is spanned by $\leq c$ elements. Let $E' = \prod E_i$ injective, let E'' be the direct product of 2^c copies of E', and write $E'' = \bigoplus_{j \in J} E''_j$ where each E''_j is spanned by $\leq c$ elements. Each element of E_1 is in at most a finite number of E''_j, so $E_1 \leq \sum E''_j$ summed over a set of cardinality $\leq c$. Continuing inductively partition J into $\{J_i : i \in I\}$ such that $E_i \leq \sum \{E''_j : j \in J_i\}$. Hence E_i is a summand of $\sum E''_j$ so $E = \bigoplus E_i$ is a summand of E'' and is thus injective.)

13. If there is a set $\mathscr{S}$ of modules such that each injective R-module is a direct sum of modules isomorphic to modules in $\mathscr{S}$ then R is left Noetherian. (Hint: exercise 12.) Conclude the converse of exercise 11: If every injective left R-module is a direct sum of indecomposables then R is left Noetherian. (Hint: exercise 9(v).)

14. Any decomposition of an injective into indecomposables (if such exists) complements summands. Every indecomposable injective projective R-module M is a summand of R. (Hint: Take $0 \neq x \in M$. Then $M = E(Rx)$ is a summand of some free module $R^{(I)}$; $x \in R^{(n)}$ for some n implies $M \leq R^{(n)}$, and $n = 1$ since M is indecomposable.)

15. (Miller-Turnidge [73]) A Noetherian injective module which is not Artinian. Let M be a right vector space having countable base $\{x_i : i \in \mathbb{N}\}$ over a division ring D, and let R be the subring $\{r \in \operatorname{End} M_D : rx_k \in \sum_{i=1}^k x_i D\}$. Let e_1 be the primitive idempotent of R sending $x_1 \to x_1$ and all other $x_i \to 0$. Then Re_1 is a summand of $\operatorname{End} M_D$ and is thus injective as well as Noetherian but is not Artinian.

§2.11

0. A converse of exact: Suppose $\mathscr{C}$ is an abelian category. If $f: A \to A'$ and $g: A' \to A''$ are morphisms such that $0 \to \operatorname{Hom}(A'', C) \xrightarrow{g^\#} \operatorname{Hom}(A', C) \xrightarrow{f^\#} \operatorname{Hom}(A, C)$ is exact for all C in $\mathscr{C}$ then $A \xrightarrow{f} A' \xrightarrow{g} A'' \to 0$ is exact. (Hint: First try this for $R\text{-}\mathscr{M}od$. Note $A'' \to \operatorname{cok} g$ is 0 so g is epic. Clearly $fA \subseteq \ker g$, and equality holds because the map $A' \to \operatorname{cok} f$ factors through g.)

1. A functor F is half-exact if $0 \to A \to B \to C \to 0$ exact implies $FA \to FB \to FC$ is exact. If $F: R\text{-}\mathscr{M}od \to \mathscr{C}$ is half exact and l is a nilpotent ideal of R such that $FM = 0$ for every module M annihilated by l then $F = 0$. (Hint: Consider $0 \to l^{t+1}M \to l^tM \to l^tM/l^{t+1}M \to 0$ where $l^{t+2}M = 0$.)

2. (Osofsky [64a]) The following are equivalent: (i) R is semisimple Artinian. (ii) Every R-module is injective. (iii) Every f.g. R-module is injective. (iv) Every cyclic R-module is injective. (Hint: The only hard part is (iv) $\Rightarrow$ (i). First note by

proposition 2.11.20 that R is regular; it suffices to prove R has a finite complete set of orthogonal idempotents, which is implied by the next exercise, gleaned from Osofsky [68].)

3. Say an infinite set $\{e_i : i \in I\}$ of orthogonal idempotents satisfies the *Osofsky condition* if for each $J \subset I$ there is an element r_J in R with $e_i r_J = e_i$ for all i in J, and $r_J e_i = 0$ for all i in $I - J$.

   (i) If R is a regular ring with $R = E(R)$ then any infinite set of orthogonal idempotents satisfies the Osofsky condition. (Hint: Let π_J be the projection from R onto $E(\sum_{j \in J} Re_j)$ and let $r_J = \pi_J 1$. For $i \notin J$ let $a = r_J e_i$ and take b with $aba = a$. Note $e_j a = 0$. Then $Rab \cap \sum_{j \in J} Re_j = 0$ so $0 = \pi_J(ab) = abr_J$ implying $a = abr_J e_i = 0$.)

   (ii) If $\{e_i : i \in I\}$ satisfies the Osofsky condition and $\pi : R \to \prod Re_i$ is defined by $\pi r = (re_i)$ and $L = \ker \pi + \sum_{i \in I} Re_i < R$ then $\bar R = R/L$ is not injective as R-module. (Extensive hint: Note $L = \{r \in R : \text{almost all } re_i = 0\}$. Let $\mathcal{S}_0$ be a given infinite partition of I with each S in $\mathcal{S}_0$ infinite. By Zorn enlarge $\mathcal{S}_0$ to $\mathcal{S} \subset \mathcal{P}(I)$ maximal with respect to each S in $\mathcal{S}$ infinite and $S \cap T$ finite for all S, T in $\mathcal{S}$. Define $f : \sum_{S \in \mathcal{S}} Rr_S \to \bar R$ by the rule $fr_S = 0$ if $S \in \mathcal{S} - \mathcal{S}_0$ and $fr_S = \bar r_S$ if $S \in \mathcal{S}_0$. This is a well-defined map which does not extend to R. Indeed, if f were given by right multiplication by $\bar a$ then $r_S a = r_S + b + \sum_{u=1}^{n} r_u e_{i_u}$ for suitable b in $\ker \pi$, so $\{i \in S : e_i ae_i = e_i\}$ is infinite for each S in $\mathcal{S}_0$; form S_0 by taking one i from each of these sets. Let $r = r_{S_0}$. Obviously $S_0 \notin \mathcal{S}_0$ so $0 = fr = \bar r a$ and thus $rae_i = 0$ for almost all i. But $S_0 \cap S$ is infinite for some S in $\mathcal{S}$ (by maximality) and for all i in $S_0 \cap S$ we have $e_i rae_i = e_i ae_i = e_i \neq 0$, contradiction.)

Flat Modules

4. The direct limit of flat modules is flat. (Hint: Apply proposition 1.8.10 to the definition.) Note the following corollaries: (i) Every central localization of a flat module is flat. (ii) A module is flat iff every f.g. submodule is flat.

5. F is flat iff for every relation $\sum_{i=1}^{n} r_i x_i = 0$ (for x_i in F) there exist $y_1, \ldots, y_m$ in F for some m and r_{ij} in R with $\sum_{i=1}^{n} r_i r_{ij} = 0$ for each j and $\sum_{j=1}^{m} r_{ij} y_j = x_i$ for each i. (In other words, each relation has a reason.) (Hint: ($\Leftarrow$) by the flat test. ($\Rightarrow$) Let $I = \sum_{i=1}^{n} r_i R$ and let V be a free right R-module with base $v_1, \ldots, v_n$. Defining $f : V \to I$ by $fv_i = r_i$ and letting $K = \ker f$ one has $0 \to K \otimes F \to V \otimes F \to I \otimes F \to 0$ and $\sum v_i \otimes x_i \in \ker(f \otimes 1) = K \otimes F$. Write $\sum v_i \otimes x_i = \sum k_j \otimes y_j$ for $k_j = \sum v_i r_{ij} \in K$, and match components.)

5'. (Villamayor) Suppose F is free and $0 \to K \to F \to M \to 0$ is exact. The following are equivalent: (i) M is flat. (ii) For every x in K there is $\pi : F \to K$ with $\pi x = x$. (iii) For every $x_1, \ldots, x_n$ in K there is $\pi : F \to K$ with $\pi x_i = x_i$ for $1 \leq i \leq n$. (Hint: (i) $\Rightarrow$ (ii) Choose a basis $\{v_i\}$ of F and let $x = \sum_{i=1}^{n} r_i v_i$ and let $I = \sum r_i R$. Then $x \in K \cap IF = IK$ so writing $x = \sum r_i y_i$ define π by $\pi v_i = y_i$ for $1 \leq i \leq n$ and $\pi v_i = 0$ for all other i. (ii) $\Rightarrow$ (i) If $x \in K \cap IF$ then $x = \pi x \in IK$. (ii) $\Rightarrow$ (iii) By induction on n. Take $\pi_n : F \to K$ with $\pi_n x_n = x_n$ and apply induction to $\{x_1', \ldots, x_n'\}$ where $x_i' = x_i - \pi_n x_1$.)

6. If P is a projective module and $N \ll P$ with P/N flat then $N = 0$. (Hint: You

may assume P is free; by exercise 5' if $x \in N$ then Rx is a small summand of P so $x = 0$.) Conclude that any flat module with a projective cover is itself projective.

7. R is a left perfect ring iff every flat R-module is projective. (Hint: ($\Rightarrow$) by exercise 6. ($\Leftarrow$) It suffices to show there is no infinite descending chain $r_1 R > r_1 r_2 R > r_1 r_2 r_3 R > \cdots$ Let F be a free module with base $x_1, x_2, \ldots$ and let $N = \sum Ry_i \leq F$ where $y_i = x_i - r_i x_{i+1}$. By exercise 2.8.16 the y_i are independent; using the idea at the end of exercise 2.9.10, it suffices to prove N is a summand of F. This is true by hypothesis if F/N is flat. But $F/\sum_{i=1}^k Ry_i$ is free for each k, so apply proposition 2.11.17.)

8. An f.g. module P is projective iff P is flat and finitely presented. (Hint: ($\Rightarrow$) by §2.8. ($\Leftarrow$) One should show that if $M \to N$ is epic then $\operatorname{Hom}(P, M) \to \operatorname{Hom}(P, N)$ is epic or, equivalently, that the map of character modules $\operatorname{Hom}(P, N)^* \to \operatorname{Hom}(P, M)^*$ is monic. But $N^* \to M^*$ is monic so $N^* \otimes P \to M^* \otimes P$ is monic. It remains to show $M^* \otimes P \approx \operatorname{Hom}(P, M)^*$ canonically. Let $F_1 \to F_2 \to P \to 0$ be a presentation of P with F_1, F_2 f.g. free. Then $M^* \otimes F_i \approx \operatorname{Hom}(F_i, M)^*$ by exercise 2.8.7', so proposition 2.11.15 yields the result.) In particular, every f.g. flat module over a left Noetherian ring is projective.

9. The following are equivalent: (i) M is f.g. (ii) $\psi: (\prod N_i) \otimes M \to \prod (N_i \otimes M)$ given by $\psi((x_i) \otimes y) = (x_i \otimes y)$ is onto. (iii) The canonical map $R^I \otimes M \to M^I$ as in (ii) is onto, for any set I. (Hint: (iii) $\Rightarrow$ (i) Take the element of M^M whose component corresponding to $x \in M$ is x, and write it as the image of $\sum (r_x)_i \otimes x_i$. Then $M = \sum Rx_i$).

9'. M is finitely presented iff the canonical map $R^I \otimes M \to M^I$ given by $(r_i) \otimes x \to (r_i x)$ is an isomorphism for each I. Hint: Consider $0 \to K \to F \to M \to 0$ and apply exercise 7 to K in the diagram

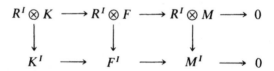

$$
\begin{array}{ccccccc}
R^I \otimes K & \longrightarrow & R^I \otimes F & \longrightarrow & R^I \otimes M & \longrightarrow & 0 \\
\downarrow & & \downarrow & & \downarrow & & \\
K^I & \longrightarrow & F^I & \longrightarrow & M^I & \longrightarrow & 0
\end{array}
$$

Coherent Rings and Modules A module M is *coherent* if every f.g. submodule is finitely presented. A ring R is *(left) coherent* if R is coherent as R-module. Examples of coherent rings include (i) Noetherian rings; (ii) semihereditary rings.

10. (Chase's theorem) The following are equivalent: (i) R is coherent. (ii) Every product of flat right R-modules is flat (iii). F^I is a flat right module for every free right module F and every set I. (iv) Every finitely presented R-module is coherent. (Hint: (i) $\Rightarrow$ (iii) Apply theorem 2.11.13. If $\sum_{i=1}^t x_i r_i = 0$ where $x_i \in F^I$ and $r_i \in R$ then take $0 \to K \to R^{(t)} \to \sum_{i=1}^t Rr_i \to 0$ with K finitely presented. (iii) $\Rightarrow$ (ii) Suppose $M_i : i \in I$ are flat and take F free with $0 \to K_i \to F \to V_i \to 0$ exact. Then $0 \to \prod K_i \to \prod F \to \prod V_i \to 0$ where the middle term is flat by assumption. Conclude with the flat test. (ii) $\Rightarrow$ (iv) If $N < M$ is f.g. then the canonical map $(R^I) \otimes N \to N^I$ is monic because this is true for M using exercise 9'. (iv) $\Rightarrow$ (i) obvious.)

11. An exact sequence $0 \to K \xrightarrow{f} M \to N \to 0$ is *pure* if it remains exact upon tensoring by any right module H. (Compare this to the definition of flat.) In this case fK is

called a *pure submodule* of M. Note that N is redundant since the crux of the issue is whether $1_H \otimes f : H \otimes K \to H \otimes M$ is monic. Every split short exact sequence is pure, and every summand of M is a pure submodule. A submodule K of M is pure iff $IK = K \cap IM$ for every right ideal I of R. An alternate necessary and sufficient condition for K to be a pure submodule of M: Given a_j in K for $1 \leq j \leq t$ and given r_{ij} in R, if all $a_j \in \sum_{i=1}^{n} r_{ij}M$ then all $a_j \in \sum_{i=1}^{n} r_{ij}K$. For M flat we have $K < M$ pure iff M/K is a flat module.

Localization and Module Properties

12. An example where $S^{-1}\mathrm{Hom}_R(M,N) \not\approx \mathrm{Hom}_{S^{-1}R}(S^{-1}M, S^{-1}N)$: Take $R = \mathbb{Z}$, $S = \mathbb{Z} - \{0\}$, $M = \mathbb{Q}$, $N = \mathbb{Z}$. (compare with exercise 2.8.7) This motivates the following chain of ideas leading to exercise 15 below.

13. Suppose R is a C-algebra, and K is a commutative C-algebra. Any R-module is to be viewed as C-module, and tensors are to be taken over C. If M is a finitely presented R-module and N is a $K \otimes R$-module then there is an isomorphism $\mathrm{Hom}_R(M,N) \to \mathrm{Hom}_{K \otimes R}(K \otimes M, N)$ sending f to $\bar{f}$ where $\bar{f}(a \otimes x) = afx$. (Hint: This is true for $M = R$ and thus M free; pass down a finite presentation using the "five lemma.")

14. Assumptions as in exercise 13 suppose, furthermore, K is a flat C-module. Then $K \otimes \mathrm{Hom}_R(M,N) \approx \mathrm{Hom}(M, K \otimes N)$ given by $a \otimes f \to f_a$ where $f_a x = a \otimes fx$. (Hint: as in exercise 13.)

15. Assumptions as in exercise 14 then $K \otimes \mathrm{Hom}_R(M,N) \approx \mathrm{Hom}_{K \otimes R}(K \otimes M, K \otimes N)$. (Hint: $K \otimes \mathrm{Hom}_R(M,N) \approx \mathrm{Hom}_R(M, K \otimes N) \approx \mathrm{Hom}_{K \otimes R}(K \otimes M, K \otimes N)$ by exercises 13, 14). Two applications: (i) C is a field and K is a field extension of C. (ii) $K = S^{-1}C$ for a suitable submonoid S of $Z(R)$.

15′. Suppose E is an f.g. injective right T-module, $M \in R\text{-}\mathcal{M}od\text{-}T$, and N is a finitely presented R-module. Then $\mathrm{Hom}_T(M,E) \otimes_R N \approx \mathrm{Hom}_T(\mathrm{Hom}_R(N,M),E)$. (Hint: Use the same map as in exercise 2.8.7′ and, as in the previous exercises, we may assume N is free.)

Regular Rings

16. (Kaplansky) Let F be a field and let R be the set of sequences of 2×2 matrices which are eventually upper triangular, i.e., $R = \{(a_i) \in M_2(F) : a_i \text{ is upper triangular for all } i > i_0, \text{ for suitable } i_0\}$. R is a semiprime ring, and is *not* regular (for $x = (e_{12})$ has no y with $xyx = x$). On the other hand, any prime homomorphic image of R is isomorphic to F or to $M_2(F)$ and thus is regular. (Hint: Write e_i for the element of R whose only nonzero component is the identity matrix in the i position. Suppose P is a given prime ideal of R. If $e_i \notin P$ and $r \in R$ then $(r - re_i) \cdot Re_i = 0$ so $r - re_i \in P$, implying R/P is a homomorphic image of $Re_i \approx M_2(F)$ and thus $R/P \approx Re_i$. On the other hand, if each $e_i \in P$ then R/P is commutative regular.)

17. Suppose R is regular. Show (i) eRe is regular for every idempotent e of R. (ii) $M_n(R)$ is regular. (Hint: (i) easy. (ii) Use flat modules; alternatively one can prove it by direct computation, for it is enough to assume $n = 2$.)

18. R is regular iff every f.g. submodule M of a projective module P is a summand (Hint: ($\Leftarrow$) Take $P = R$. ($\Rightarrow$) We may assume P is free, so assume $P = R^{(n)}$. Then $\operatorname{Hom}(R^{(n)}, M)$ is a left ideal of $M_n(R)$ and thus is a summand; thus $M \approx \operatorname{Hom}_R(R, M)$ is projective in $R\text{-}\mathcal{M}od$. This argument previews Morita equivalence).

19. R is regular iff every finitely presented module is projective. (Hint: exercise 8.)

20. The following are equivalent: (i) R is regular. (ii) R/L is a flat module for every left ideal L of R. (iii) R/Rr is flat for every r in R. (iv) $L \cap l = lL$ for every left ideal L and right ideal l. (Hint: (i) $\Rightarrow$ (ii) $\Rightarrow$ (iii) is clear and (ii) $\Leftrightarrow$ (iv) by a criterion for flatness. Likewise, (iv) $\Rightarrow$ (i) since $Rr \cap rR = rRr$.)

21. Suppose R is regular and $E \in R\text{-}\mathcal{M}od$ is injective as $R/\operatorname{Ann}_R E$-module. Then E is injective as R-module. (Hint: Let $A = \operatorname{Ann}_R E$. Suppose $\varphi: L \to E$ is a map. If $r \in L \cap A$ taking $r = rr'r$ yields $\varphi r = rr'\varphi r = 0$; the map $\bar{\varphi}: (L + A)/A \to E$ extends to R/A and thus φ extends.

22. Suppose R is regular, and every primitive image is simple Artinian. Then each ideal $A \neq 0$ of R contains a nonzero central idempotent. In particular, if R is prime then R itself is simple Artinian. (Hint: Say $\{e_{ij} : 1 \leq i, j \leq n\}$ is a set of partial $n \times n$ matric units if $e_{ij}e_{uv} = \delta_{ju}e_{iv}$; we do not require $\sum_{i=1}^{n} e_{ii} = 1$. Given such a set of partial $n \times n$ matric units one can show that if $e_{11}Re_{11}$ has zero-divisors or if $\sum e_{ii}$ is not central then there is a new set of $(n + 1) \times (n + 1)$ partial matric units $\{e'_{ij} : 1 \leq i, j \leq n + 1\}$ such that $e'_{11} \in e_{11}Re_{11}$, cf., Jacobson [64B, p.238] for the explicit computation which relies on exercise 17. Starting with any idempotent $0 \neq e \in A$, inductively build a set of partial $n \times n$ matric units $\{e_{ij}^{(n)} : 1 \leq i, j \leq n\}$ such that $e_{11}^{(n+1)} \in e_{11}^{(n)}Re_{11}^{(n)} \subseteq A$. In view of exercise 2.1.17, this procedure cannot continue indefinitely. Thus for some n we have $\sum_{i=1}^{n} e_{ii}^{(n)}$ is central, and $e_{11}^{(n)}Re_{11}^{(n)}$ is a reduced ring D. Since each $e_{ii}^{(n)} = e_{i1}^{(n)}e_{1i}^{(n)} \in A$ this proves the first assertion. To see the second assertion note a prime ring has no central idempotent, so $R \approx M_n(D)$; D is a regular domain and thus is a division ring.)

§2.12

1. Suppose S is a submonoid of $Z(R)$ and $I \lhd R$ with $I \cap S = \varnothing$. Then $\bar{S}^{-1}\bar{R} \approx S^{-1}R/S^{-1}I$ where $^-$ denotes the image in R/I. In particular, $\bar{S}^{-1}\bar{R}$ is simple if I is maximal such that $I \cap S = \varnothing$.

2. If S is a submonoid of $Z(R)$ then $S^{-1}R$ is a flat R-module. (Hint: If $f: N \to M$ is monic then $\hat{f}: S^{-1}N \to S^{-1}M$ is monic.) Conclude that there is an exact functor $F: R\text{-}\mathcal{M}od \to S^{-1}R\text{-}\mathcal{M}od$ given by $FM = S^{-1}M$.

3. The R-module $\bigoplus\{R_P : P \in \operatorname{Spec}(Z(R))\}$ is faithfully flat. If M_P is flat for every maximal ideal P of $Z(R)$ then M is flat.

4. Suppose $f: M \to N$ is a map of R-modules. Then f is monic (resp. epic) if the induced map $\hat{f}: M_P \to N_P$ is monic (resp. epic) for each maximal ideal P of $Z(R)$.

Jacobson rings (also see exercises 8.4.13 and 8.4.10)

5. (Pearson-Stephenson) Example of a skew polynomial extension of a commutative local (but not Noetherian) Jacobson ring which is prime but not semiprimi-

tive (and thus not Jacobson). Let $C = F[\Lambda]/A$ where F is a field, $\Lambda = \{\lambda_i : i \in \mathbb{Z}\}$ is an infinite set of commuting indeterminates, and A is the ideal generated by all monomials $\lambda_{i(1)} \cdots \lambda_{i(t)}$ where $t \geq 2$, $i(1) \leq \cdots \leq i(t)$, and $i(t) - i(1) \leq t^2$. Then $\lambda_i \mapsto \lambda_{i+1}$ defines an automorphism σ of $F[\Lambda]$ and so gives an automorphism of C. Then the maximal ideal M of C consisting of the images of nonconstant polynomials is nil (because $\lambda_i^2 \in A$); hence M is the unique prime ideal of C, and C is local Jacobson. Let $T = C[\mu; \sigma]$ where μ is a new (skew) indeterminate. T is prime but not semiprimitive. (Hint: To show T is prime suppose as in exercise 1.6.11 that $0 \neq A \lhd C$ and $B \lhd (C, \sigma)$ with $AB = 0$. Then $a\sigma^m b = 0$ for all a in A and b in B. But checking leading terms under the lexicographic order shows $a\sigma^m b \neq 0$ for m large enough unless $a = 0$ or $b = 0$. On the other hand, the image z_0 of λ_0 in C generates a nil left ideal in T because $\left(\left(\sum_{i=0}^t c_i \mu^i\right) z_0\right)^t = 0$.) Incidentally, an Ore extension of a left Noetherian Jacobson ring is Jacobson as well as left Noetherian, cf., exercise 8.4.10.

6. Any finite centralizing extension of a Jacobson ring is Jacobson. (Hint: One may assume both rings are prime and use corollary 2.5.30.)

Invertible Projectives

7. The following are equivalent for M in C-$\mathcal{M}od$: (i) M is invertible projective. (ii) M is f.g. projective and $\operatorname{End}_C M \approx C$. (iii) M is finitely presented and $M_P \approx C_P$ for every maximal ideal P of C. (iv) There is an isomorphism $\operatorname{Hom}_C(M, C) \otimes M \to C$ given by $f \otimes x \to fx$. (v) $N \otimes M \approx C$ for some N. (Hint: (i) $\Rightarrow$ (ii) $\Rightarrow$ (iii) $\Rightarrow$ (iv) Localize and use exercise 4 where appropriate; (iv) $\Rightarrow$ (v) obvious; (v) $\Rightarrow$ (i) count ranks by localizing.)

8. Using exercise 7 conclude {isomorphism classes of invertible projective modules over a commutative ring C} form an abelian group under tensor products, which is called the *Picard group*, written $\operatorname{Pic}(C)$. If H is a commutative C-algebra then there is a group homomorphism $\operatorname{Pic}(C) \to \operatorname{Pic}(H)$ given by $M \to M \otimes_C H$. $\operatorname{Pic}(C)$ is trivial iff every invertible projective is free.

9. If C is a central subring of a prime ring R then $C \cap Rc = Cc$ for any c in C. Consequently, LO holds for height 1 primes of C.

Passman's Approach to Normalizing Extensions The next exercises sketch the approach of Passman [81] to the question of comparing prime ideals of finite normalizing extensions. The approach is quicker than that of Heinicke-Robson, but fails to capture that elusive INC, which is proved by Heinicke-Robson. The basic construction is of a suitable large primitive ring. Write $C\{\{X\}\}$ for the ring of formal power series in a set of noncommuting indeterminates X, and write $*$ for the free product.

10. Suppose R is a prime C-algebra. Let X be a set of noncommuting indeterminates of cardinality $= \max(\mathbb{N}_0, |R|)$; let $T = R\{\{X\}\}$, let Y be a set of noncommuting indeterminates of cardinality $|T|$, and let $\tilde{R} = T\{Y\}$. Then $\tilde{R}$ is primitive. (Hint: As in exercise 2.1.14. Namely, index X by the elements of R, i.e., as $\{X_r : r \in R\}$; and noting $|\tilde{R}| = |T|$ write $Y = \{Y_w : w \in \tilde{R}\}$. Letting $a = \sum_{r \in R} X_r r$

show $L = \sum_{0 \neq w \in \tilde{R}} \tilde{R}(1 + awaY_w)$ is a proper left ideal comaximal with every non-zero ideal. To show $L \neq \tilde{R}$ one supposes $1 \in L$ and reaches a contradiction by taking leading terms in Y and then lowest terms in X.)

11. Given $A \lhd R$ write $\tilde{A}$ for the set of elements of $\tilde{R}$ whose coefficients are all in a, i.e., $(A * C\{\{X\}\}) * C\{Y\}$. Then $\tilde{A} \cap R = A$ and $(R/A)^{\sim} \approx \tilde{R}/A$; $(\bigcap A_i)^{\sim} = \bigcap \tilde{A}_i$; if A is a prime (resp. semiprime) ideal of R then $\tilde{A}$ is a primitive (resp. semiprimitive) ideal of $\tilde{R}$.

12. If B is annihilator ideal of $\tilde{R}$ then $(B \cap R)^{\sim} \subseteq B$, and if $P \in \operatorname{Spec}(\tilde{R})$ is an annihilator ideal then $P \cap R \in \operatorname{Spec}(R)$. (Hint: If $r \in B \cap R$ then r annihilates every coefficient of every element of $\operatorname{Ann}'B$, implying $(B \cap R)^{\sim} \subseteq \operatorname{Ann}(\operatorname{Ann}'B) = B$.)

Now that Passman's "primitivity machine" has been set up we need some results on normalizing extensions and simple modules. In exercises 13–18 assume $R' = \sum_{i=1}^{n} Ra_i$ where each a_i is R-normalizing.

13. $\tilde{R}' = \sum_{i=1}^{n} \tilde{R}a_i$ and the a_i are $\tilde{R}$-normalizing. If $A \lhd R'$ with $A \cap R = 0$ then $\tilde{A} \cap \tilde{R} = 0$ in $\tilde{R}'$.

14. If L is a maximal left ideal of R then $R'/R'L$ has composition length $\leq n$ as R-module. (Hint: $(Ra_i + R'L)/R'L$ is an image of the module $(R + R'L)/R'L \approx R/(R \cap R'L)$ which is simple or 0.)

15. (Bit-David and Robson) If $A \lhd R$ and is large as $R - R$ bimodule (i.e., A intersects all ideals nontrivially) then there is $A' \lhd R'$ with $0 \neq A' \cap R \subseteq A$. (Hint: Take A' an $R - R$ bisubmodule of R' such that $A' \oplus R$ is a large bisubmodule of R'. Then $M = A' \oplus A$ is large in R'. For each i, j one sees $\{r' \in R': a_i r' a_j \in M\}$ is a large bisubmodule of R', so their intersection I is a large bisubmodule. But $I = \{r' \in R': R'r'R' \subseteq M\}$ so $I \subseteq M$ and $0 \neq I \cap R \subseteq A$.)

16. **Cutting Down.** If $P' \in \operatorname{Spec}(R')$ then $P' \cap R$ is an intersection of $\leq n$ prime ideals. (Hint: One may assume R' is prime. Then $\tilde{R}$ has primitive ideals $P_1, \ldots, P_k$ for some $k \leq n$ with $\bigcap P_i = 0$ by corollary 2.5.30; each $P_i = \operatorname{Ann}_{j \neq i} \cap P_j$ so $P_i \cap R \in \operatorname{Spec}(R)$ by exercise 12.) In particular if R' is prime then 0 is a finite intersection of prime ideals of R.

17. LO generalized. If $P \in \operatorname{Spec}(R)$ then there exists P' in $\operatorname{Spec}(R')$ such that $P' \cap R \subseteq P$ and P is minimal in $\mathscr{S}(P' \cap R)$. This generalizes LO since if $P' \cap R$ were in $\operatorname{Spec}(R)$ we would have $P = P' \cap R$. (Hint: As proposition 2.12.43 one may assume R' is prime and $A' \cap R \nsubseteq P$ for all $0 \neq A' \lhd R'$. By exercise 15 there is $0 \neq B \lhd R$ with $P \cap B = 0$. "Cutting down" shows there are $P_1, \ldots, P_k$ in $\operatorname{Spec}(R)$ with $\bigcap P_i = 0$; we may assume each P_i is minimal prime. But some $P_i \nsupseteq B$ so $PB = 0 \subseteq P_i$ implies $P \subseteq P_i$ so P is minimal.)

18. There are at most n distinct P_i' in $\operatorname{Spec}(R')$ for which any $P \in \operatorname{Spec}(R)$ is minimal in $\mathscr{S}(P_i' \cap R)$. (Hint: Using the primitive machine one may assume P is primitive. Write $P = \operatorname{Core}(L)$ for L maximal and let $M = R'/PR'$ which by exercise 14 has composition length $\leq n$. Each P_i can be displayed as an annihilator of a suitable composition factor; however, the reader should be warned this argument relies on lemma 2.5 of Passman [81] which takes two pages to prove.) Passman [81] contains related results and examples, and it would be interesting to see if INC could be proved by these methods, to shorten Heinicke-Robson [81].

19. Prime $\operatorname{rad}(R) = R \cap \operatorname{Prime} \operatorname{rad}(R')$. (Hint: either directly or exercise 2.5.16.)

3 Rings of Fractions and Embedding Theorems

The main theme of this chapter is embedding a ring into a nicer ring which can be studied more easily. One familiar example from undergraduate study is embedding an integral domain into its field of fractions; another example is embedding a commutative Noetherian domain into a local Noetherian domain. In fact, these are both examples of "localization," a term which has evolved considerably during the last 30 years. Our first objective will be to find a larger ring in which a given set S of elements becomes invertible; surely this implies S is regular, but possibly other conditions may be required on S as well. When S is central we have "central localization," which was encountered already in §1.8 and §2.11. A slightly weaker condition is the "Ore condition," which ensures that localization can be carried out similarly to the commutative procedure, and Ore localization suffices for most applications in ring theory.

Ore localization, also called "classical localization," is studied in depth in §3.1. The most successful application of Ore localization is Goldie's theorems, proved in §3.2, which characterized those prime and semiprime rings having semisimple Artinian rings of fractions. In particular, every prime Noetherian ring satisfies Goldie's conditions, and the reader may wish to pass directly from Goldie's theorem to §3.5, which contains the basic theory of Noetherian rings.

271

Goldie's theorem leads naturally to several generalizations. On the one hand, it is of interest to know which rings can be embedded into Artinian rings, since many properties of Artinian rings pass to their subrings. The remainder of §3.2 is spent on questions of this sort, including Small's theorem.

On the other hand, one may wish to find non-Ore localization procedures which generalize Goldie's theorem and apply to wider classes of rings. One such theory is Johnson's theory of nonsingular rings, which is presented in some detail in §3.3, including Goodearl's theorems; however, this topic is not pursued in later chapters. Johnson's theory has been generalized further to an abstract localization theory, sketched in §3.4; one important application is Martindale's ring of quotients and, in particular, his central closure, which is an indispensible tool in parts of the structure theory.

§3.1 Classical Rings of Fractions

In this section we consider general conditions under which a given sub-monoid S of elements of R can be made invertible, cf., definition 1.10.1. Often we shall require S to be *regular*, i.e., every element of S is regular in R.

Recall $s \in R$ is *regular* if $rs \neq 0$ and $sr \neq 0$ for all nonzero r in R. Hopefully there will be no confusion with our other usage of "regular"—"regular ring" means von Neumann regular, cf., definition 2.11.18.

Definition 3.1.1: Suppose S is a (multiplicative) submonoid of R. Q is a (left) *ring of fractions of R with respect to S* if the following two conditions are satisfied: (1) There is a ring homomorphism $v: R \to Q$ such that vs is invertible for all s in S, with ker $v = \{r \in R: sr = 0 \text{ for some } s \text{ in } S\}$, and (2) every element of Q has the form $s^{-1}r$ where $s \in S$ and $r \in R$. (Technically, this should be written $(vs)^{-1}vr$, but we shall write $s^{-1}r$ abusing the notation slightly.)

Important Special Case. If $S = \{\text{all regular elements of } R\}$ and if $\varphi: R \to Q$ is an injection then Q is called the *classical ring of fractions of R*, and R is called an *order* in Q.

Classical rings of fractions need not exist, and our first objective will be to find necessary and sufficient conditions for constructing rings of fractions.

Definition 3.1.2: A (left) *denominator set* (also called *Ore set*) is a submonoid S of R satisfying the following two conditions:

(i) For any s_1 and S and r_1 in R there exist s_2 in S and r_2 in R such that $s_2 r_1 = r_2 s_1$.

(ii) If $rs = 0$ for r in R, s in S, then there is s' in S with $s'r = 0$.

Proposition 3.1.3: *If R has a ring of fractions with respect to S then S is a denominator set.*

Proof: We check the conditions of definition 3.1.2.

(i) Given s_1, r_1 we have $(1^{-1}r_1)(s_1^{-1}1) = s^{-1}r$ for suitable s, r; then $1^{-1}sr_1 = 1^{-1}rs_1$ implies $sr_1 - rs_1 \in \ker v$ so for some s' in S we have $0 = s'(sr_1 - rs_1) = s'sr_1 - s'rs_1$; let $s_2 = s's$ and $r_2 = s'r$.

(ii) If $rs = 0$ then $0 = (1^{-1}rs)(s^{-1}1) = 1^{-1}r$, implying $s'r = 0$ for some s' in S.   Q.E.D.

Before considering the converse, let us discuss two important special cases. First, any submonoid S of $Z(R)$ is obviously a denominator set, and the corresponding ring of fractions $S^{-1}R$ exists and was already constructed three different ways (construction 1.10.2, construction 1.10.3, and exercise 1.10.1).

On the other hand, suppose S is regular. Then condition (ii) is automatic, so we need only check (i). In case $S = \{$all regular elements of $R\}$ we call (i) the *Ore condition,* and R is called (left) Ore; for example, any PLID is Ore by remark 1.6.19. The proof that Ore rings have classical rings of fraction follows the lines of the general proof presented below, but with the simplification that one does not worry about zero divisors.

We are ready now to proceed to the construction of a ring of fractions with respect to an arbitrary denominator set. As usual there are two approaches—the "brute force" construction which follows one's naive intuition but involves verifications at every step of the way, and a slick construction using direct limits. The slick construction yields a much more encompassing result (cf., theorem 3.4.25ff) but leaves us without a "feel" for computing in the ring of fractions. Consequently, we shall wade through the brute force construction, and the reader who is so inclined can view it as a corollary of theorem 3.4.25 below.

Theorem 3.1.4: *Suppose S is a denominator set.*

(i) *$S \times R$ has an equivalence $\sim$ defined as follows: $(s_1, r_1) \sim (s_2, r_2)$ if there exist r, r' in R such that $rr_1 = r'r_2$ and $rs_1 = r's_2 \in S$. Write $S^{-1}R$ for the set of equivalence classes, and write $s^{-1}r$ for the equivalence class of (s, r).*

(ii) *If $s \in S$ and $as \in S$ then $(as)^{-1}(ar) = s^{-1}r$ in $S^{-1}R$.*

(iii) *To check a function f on $S^{-1}R$ is well-defined it suffices to prove f has the same value on $s_1^{-1}r_1$ and $(rs_1)^{-1}(rr_1)$ where $rs_1 \in S$.*

(iv) $S^{-1}R$ *has a natural ring structure (given in the proof), which is a ring of fractions for R with respect of S.*

Proof:

(i) Clearly $\sim$ is reflexive and symmetric; to prove transitivity we need the following technical lemma, which also cleans up the other sticky points in proving the theorem:

Lemma: *Suppose $(s_1,r_1) \sim (s_2,r_2)$ and $as_1 = a's_2 \in S$ for a, a' in R. Then there is b in R with $bar_1 = ba'r_2$ and $bas_1 = ba's_2 \in S$.*

Proof of lemma: There are r,r' in R satisfying $rr_1 = r'r_2$ and $rs_1 = r's_2 \in S$, by hypothesis. Take r'', s'' such that

$$s''(rs_1) = r''(as_1).$$

Then $(s''r - r''a)s_1 = 0$ so there is s_1' in S satisfying $s_1'(s''r - r''a) = 0$, i.e.,

$$s_1's''r = s_1'r''a.$$

Now $s_1'r''(a's_2)=s_1'r''as_1=s_1's''rs_1=s_1's''r's_2$ so $(s_1'r''a' - s_1's''r')s_2 = 0$. Hence for some s_2' we have

$$s_2's_1'r''a' = s_2's_1's''r'.$$

Let $b = s_2's_1'r''$ and $s = s_2's_1's''$. Then $ba' = sr'$ and

$$ba = s_2'(s_1'r''a) = s_2'(s_1's''r) = sr.$$

Hence $bar_1 = srr_1 = sr'r_2 = ba'r_2$. Also, by hypothesis, $bas_1 = ba's_2$ and is in S since $bas_1 = s_2's_1'r''(as_1) = s_2's_1's''(rs_1) \in S$.   Q.E.D. for lemma.

For transitivity of $\sim$ suppose $(s_1,r_1) \sim (s_2,r_2)$ and $(s_2,r_2) \sim (s_3,r_3)$. For $i = 1, 2$ we take a_i, a_i' in R with $a_ir_i = a_i'r_{i+1}$ and $a_is_i = a_i's_{i+1} \in S$. Since S is a denominator set we have s in S and a in R with $sa_2 = aa_1'$; then $(aa_1)s_1 = (aa_1')s_2 = (sa_2)s_2$ and $(sa_2)s_2 = s(a_2s_2) \in S$. Applying the lemma to aa_1 and sa_2 we get b in R with $b(aa_1)r_1 = b(sa_2)r_2$ and $b(aa_1)s_1 = b(sa_2)s_2 \in S$. Then

$$(baa_1)r_1 = bsa_2r_2 = (bsa_2')r_3 \quad \text{and} \quad (baa_1)s_1 = bsa_2s_2 = (bsa_2')s_3 \in S.$$

Thus $\sim$ is transitive, and so is an equivalence.

(ii) follows at once from definition of $\sim$.

(iii) is immediate for if $rr_1 = r'r_2$ and $rs_1 = r's_2 \in S$ then, by hypothesis,

$$f(s_1^{-1}r_1) = f((rs_1)^{-1}rr_1) = f((r's_2)^{-1}r'r_2) = f(s_2^{-1}r_2).$$

(iv) We shall define, respectively, the product and sum of $s_1^{-1}r_1$ and $s_2^{-1}r_2$. Although not needed formally the case $s_1 = r_2 = 1$ is very instructive, for if we take r in R, s in S with $sr_1 = rs_2$ then "intuitively" one would want $(1^{-1}r_1)(s_2^{-1}1) = r_1s_2^{-1} = s^{-1}r$. This trick for switching s_2 past r_1 motivates the formal definitions:

$(s_1^{-1}r_1)(s_2^{-1}r_2) = (as_1)^{-1}rr_2$ where $r, a \in R$ are chosen such that $as_1 \in S$ and $ar_1 = rs_2$

$(s_1^{-1}r_1) + (s_2^{-1}r_2) = (as_1)^{-1}(ar_1 + rr_2)$ where r, a are chosen such that $as_1 \in S$ and $as_1 = rs_2$.

We shall show multiplication is well-defined in three steps.

Step I. Independent of choice of r, a. By definition 3.1.2(i) there are r_0 in R and a_0 in S for which $a_0r_1 = r_0s_2$, and we shall prove $(as_1)^{-1}rr_2 = (a_0s_1)^{-1}r_0r_2$ for each choice of r, a. Taking r' in R and s' in S satisfying $r'a_0 = s'a$ we have

$$s'(rs_2) = s'(ar_1) = (s'a)r_1 = r'a_0r_1 = r'r_0s_2,$$

so $(s'r - r'r_0)s_2 = 0$. Hence $s(s'r - r'r_0) = 0$ for some s in S, so $ss'r = sr'r_0$. Moreover, $sr'a_0s_1 = ss'as_1 \in S$ so by (ii) we see indeed

$$(a_0s_1)^{-1}r_0r_2 = (sr'a_0s_1)^{-1}sr'r_0r_2 = (ss'as_1)^{-1}ss'rr_2 = (as_1)^{-1}rr_2.$$

Step II. Well-defined in first argument. By (iii) it suffices to show $((bs_1)^{-1}br_1) \cdot (s_2^{-1}r_2) = (s_1^{-1}r_1)(s_2^{-1}r_2)$ whenever $bs_1 \in S$. But taking r, a such that $a(br_1) = rs_2$ and $a(bs_1) \in S$ we have

$$((bs_1)^{-1}br_1)(s_2^{-1}r_2) = (abs_1)^{-1}rr_2 = (s_1^{-1}r_1)(s_2^{-1}r_2)$$

by definition since $(ab)s_1 \in S$ and $(ab)r_1 = rs_2$.

Step III. Well-defined in second argument. As in Step II, if $bs_2 \in S$ then taking r, a such that $as_1 \in S$ and $ar_1 = rbs_2$ we have

$$(s_1^{-1}r_1)((bs_2)^{-1}br_2) = (as_1)^{-1}rbr_2 = (s_1^{-1}r_1)(s_2^{-1}r_2)$$

viewing $ar_1 = (rb)s_2$.

This proves multiplication is well-defined; addition is well-defined in the same way, and the ring verifications are now routine, cf., exercise 2. Finally note $r_1 \in \ker v$ iff $1^{-1}r_1 = 1^{-1}0$, which occurs when there are r, r' in R such that $rr_1 = r'0 = 0$ and $r = r' \in S$, proving $S^{-1}R$ is indeed a ring of fractions of R with respect to S.   Q.E.D.

Properties of Fractions

Remark 3.1.5: Suppose S is a regular denominator set. Since the canonical homomorphism $v: R \to S^{-1}R$ has kernel 0, we can identify $R \subseteq S^{-1}R$ in which all elements of S become invertible.

Now we shall show the ring of fractions of R with respect to S is unique, by characterizing $S^{-1}R$ as a universal in terms of remark 3.1.5. Given $A \subset R$ write $s^{-1}A$ for $\{s^{-1}a : s \in S \text{ and } a \in A\}$.

Theorem 3.1.6: *If S is a denominator set then the ring of fractions $S^{-1}R$ (together with $v: R \to S^{-1}R$) is the universal of definition 1.10.1. In particular, if $f: R \to T$ is a homomorphism with fs invertible for all s in S then f extends uniquely to a homomorphism $\bar{f}: S^{-1}R \to T$; moreover, $\bar{f}$ is given by*

$$\bar{f}(s^{-1}r) = (fs)^{-1}fr \tag{1}$$

and $\ker \bar{f} = S^{-1} \ker f$.

Proof: Any homomorphism $\bar{f}$ extending f must satisfy (1), so we shall show (1) defines a homomorphism. First note for $s_1 \in S$ and $rs_1 \in S$ that $f(rs_1)^{-1}frfs_1 = 1$ so

$$(fs_1)^{-1} = f(rs_1)^{-1}fr.$$

Now $\bar{f}$ is well-defined by theorem 3.1.4(ii) since

$$f(rs_1)^{-1}f(rr_1) = f(rs_1)^{-1}frfr_1 = (fs_1)^{-1}fr_1.$$

To show $\bar{\ }$ is a homomorphism take r, a as in the definition of sum and product in the proof of theorem 3.1.4, where a is chosen in S (which we can do by the dominator set condition).

$$\bar{f}((s_1^{-1}r_1)(s_2^{-1}r_2)) = f(as_1)^{-1}f(rr_2) = (fs_1)^{-1}(fa)^{-1}frfr_2$$

$$= (fs_1)^{-1}fr_1(fs_2)^{-1}fr_2 = \bar{f}(s_1^{-1}r_1)\bar{f}(s_2^{-1}r_2).$$

$$\bar{f}(s_1^{-1}r_1 + s_2^{-1}r_2) = f(as_1)^{-1}f(ar_1 + rr_2) = (fs_1)^{-1}(fa)^{-1}(fafr_1 + frfr_2)$$

$$= (fs_1)^{-1}fr + (fs_1)^{-1}(fa)^{-1}frfr_2$$

$$= \bar{f}(s_1^{-1}r_1) + (fs_1)^{-1}fs_1(fs_2)^{-1}fr_2$$

$$= \bar{f}(s_1^{-1}r_1) + \bar{f}(s_2^{-1}r_2).$$

To prove the last assertion first note $S^{-1}\ker f \subseteq \ker \bar{f}$, so it suffices to show if $0 = \bar{f}(s^{-1}r) = (fs)^{-1}fr$ then $s^{-1}r \in S^{-1}\ker f$. But this is obvious since $fr = 0$. Q.E.D.

Corollary 3.1.7: *Suppose $S_1 \subseteq S_2$ are denominator sets of R. Then there is a canonical homomorphism $S_1^{-1}R \to S_2^{-1}R$ under which we may view $S_2^{-1}R$ as $(S_1^{-1}S_2)^{-1}(S_1^{-1}R)$. Moreover, if $1^{-1}s$ is invertible in $S_1^{-1}R$ for every s in S_2 then $S_1^{-1}R \approx S_2^{-1}R$ canonically. In particular, if every regular element of R is invertible then R is its own classical ring of fractions.*

Since every regular element of a left Artinian ring is invertible, we see that every left Artinian ring is its own classical ring of fractions. If we want to study rings in terms of the classical ring of fractions, the natural procedure is to start with a given kind of left Artinian ring and to characterize orders in this kind of ring. For example, those rings which are orders in fields are the commutative domains. Likewise, we have the following basic result due to Ore:

Proposition 3.1.8: *R is an order in a division ring iff R is an Ore domain.*

Proof: ($\Rightarrow$) by proposition 3.1.3, noting that every nonzero element of R is invertible in a larger ring and hence is regular in R; ($\Leftarrow$) is theorem 3.1.4, taking $S = R - \{0\}$ (since $(s^{-1}r)^{-1}$ is then $r^{-1}s$). Q.E.D.

Thus any PLID is an order in a division ring but need *not* be a right order in a division ring, in view of example 1.6.26. Nevertheless, proposition 3.1.8 is one of the cornerstones of ring theory, along with its generalization by Goldie to orders in semisimple Artinian rings, to be discussed in §3.2. We conclude this section with some useful observations concerning any denominator set S of R.

Remark 3.1.9: For any s in S and r_i in R we have $s^{-1}r_1 + s^{-1}r_2 = s^{-1}(r_1 + r_2)$, by an easy calculation.

Lemma 3.1.10: *(Common denominator) Given $q_1,\ldots,q_n$ in $S^{-1}R$ one has $r_1,\ldots,r_n$ in R and s in S with $q_i = s^{-1}r_i$ for $1 \le i \le n$.*

Proof: Inductively write $q_i = s^{-1}r_i$ for $1 \le i \le n-1$ and write $q_n = s_n^{-1}r_n$. Then picking r' in R and s' in S with $s's_n = r's$ we have $(s's_n)^{-1}(s'r_n) = q_n$ and $(s's_n)^{-1}(r'r_i) = (r's)^{-1}r'r_i = s^{-1}r_i = q_i$ for $1 \le i \le n-1$. Q.E.D.

Proposition 3.1.11: *If L is a left ideal of R and $q_1,\ldots,q_n \in (S^{-1}R)L$ then there are $a_1,\ldots,a_n$ in L and s in S such that $q_i = s^{-1}a_i$ for $1 \le i \le n$. (In particular, $(S^{-1}R)L = S^{-1}L$.)*

Proof: Write $q_i = \sum_{j=1}^{t(i)} q_{ij}b_{ij}$ for $q_{ij} \in S^{-1}R$ and $b_{ij} \in L$. Then we can find a common s such that $q_{ij} = s^{-1}r_{ij}$ for all i,j (suitable r_{ij} in R). Hence $q_i = \sum_{j=1}^{t(i)} s^{-1}r_{ij}b_{ij} = s^{-1}\sum r_{ij}b_{ij}$ so we are done by taking $a_i = \sum_{j=1}^{t(i)} r_{ij}b_{ij}$.
$$\text{Q.E.D.}$$

Corollary 3.1.12: $L \cap S \ne \varnothing$ *iff* $(S^{-1}R)L = S^{-1}R$.

Proof: ($\Rightarrow$) If $s \in L \cap S$ then $1 = s^{-1}s \in S^{-1}L$. ($\Leftarrow$) If $(S^{-1}R)L = S^{-1}R$ then $1^{-1}1 = s^{-1}a$ for some s in S and a in L, implying $s'(s - a) = 0$ for some s' in S, and $sa = s's \in L \cap S$.   Q.E.D.

For future reference we list several properties that pass from R to $S^{-1}R$ when S is a denominator set.

Proposition 3.1.13: *If R is a left Noetherian ring then so is $S^{-1}R$.*

Proof: Any left ideal of $S^{-1}R$ has the form $S^{-1}L$ where $L < R$. If $L = \sum_{i=1}^{t} Ra_i$ then $S^{-1}L = \sum_{i=1}^{t}(S^{-1}R)a_i$ is f.g., proving $S^{-1}L$ is left Noetherian.   Q.E.D.

This can also be seen as part of a more general lattice correspondence (exercise 5).

Proposition 3.1.14: *If $I \lhd R$ is nilpotent and $S^{-1}I \lhd S^{-1}R$ then $S^{-1}I$ is nilpotent.*

Proof: We shall show for any sequence $s_1^{-1}a_1, s_2^{-1}a_2,\ldots$ in $S^{-1}I$ that there are s_i' in S and a_i' in I such that for each n we have $s_n^{-1}a_n \cdots s_1^{-1}a_1 = (s_n')^{-1}a_n' \cdots a_1'$; it follows at once that $I^n = 0$ implies $(S^{-1}I)^n = 0$. We find the s_i' and a_i' by induction. Suppose we have found them for all $i \le n - 1$; then $s_{n-1}^{-1}a_{n-1} \cdots s_1^{-1}a_1 = (s_{n-1}')^{-1}a_{n-1}' \cdots a_1'$. Let $x = (s_n^{-1}a_n)(s_{n-1}')^{-1}1 \in S^{-1}I$ by hypothesis. Then $x = (s_n')^{-1}a_n'$ for suitable s_n' in S and a_n' in I so

$$s_n^{-1}s_{n-1}^{-1} \cdots s_1^{-1}a_1 = s_n^{-1}a_n(s_{n-1}')^{-1}a_{n-1}' \cdots a_1' = x1^{-1}a_{n-1}' \cdots a_1'$$

$$= (s_n')^{-1}a_n'a_{n-1}' \cdots a_1'$$

as desired.   Q.E.D.

Proposition 3.1.15:   *If S is regular and R is prime (resp. semiprime) then $S^{-1}R$ is prime (resp. semiprime).*

Proof: We prove the result for prime; for semiprime we would take $r_1 = r_2$ and $s_1 = s_2$. View $R \subseteq S^{-1}R$. If $s_1^{-1}r_1 R s_2^{-1}r_2 = 0$ then $0 = s_1 s_1^{-1}r_1(rs_2)s_2^{-1}r_2 = r_1 r r_2$ for all r in R; thus $r_1 = 0$ or $r_2 = 0$.   Q.E.D.

Proposition 3.1.16:   *Suppose R is both left and right Ore, having left (resp. right) ring of fractions Q_1 (resp. Q_2). Then there is an isomorphism $\varphi: Q_1 \to Q_2$ given by $\varphi(s^{-1}r) = (1s^{-1})(r1^{-1})$.*

Proof: φ exists and is 1:1 by theorem 3.1.6 applied to the canonical injection $R \to Q_2$. On the other hand, φ^{-1} can be constructed symmetrically, implying φ is an isomorphism.   Q.E.D.

Localizations of Modules

It shall be useful to extend several of these results to modules.

Definition 3.1.17:   Suppose S is a left denominator set for R, and $M \in R\text{-}\mathcal{M}od$. Then define $S^{-1}M = S^{-1}R \otimes_R M$.

Remark 3.1.18:   Any element of $S^{-1}M$ has the form $s^{-1} \otimes x$ for s in S and x in M. (Indeed, by lemma 3.1.10 we can rewrite any element $\sum s_i^{-1} r_i \otimes x_i$ of $S^{-1}M$ in the form $\sum s^{-1} r_i' \otimes x_i = \sum s^{-1} \otimes r_i' x_i = s^{-1} \otimes \sum r_i' x_i$.)

There is an alternate way of describing $S^{-1}M$, more in line with our earlier results. Namely, take the set of equivalence classes of $S \times M$ under the equivalence $(s_1, x_1) \sim (s_2, x_2)$ iff there are r, r' in R with $rx_1 = r'x_2$ and $rs_1 = r's_2 \in S$. Writing the equivalence class of (s, x) as $s^{-1}x$ we can define scalar multiplication over $S^{-1}R$ by defining $(s_1^{-1}r_1)(s_2^{-1}x) = (ss_1)^{-1}rx$ where $r \in R$, $s \in S$ are chosen such that $sr_1 = rs_2$. Addition of the $s^{-1}x$ defined as in theorem 3.1.4, we can duplicate the proof of theorem 3.1.4 to prove that these operations provide a module structure which we temporarily call M_S.

Proposition 3.1.19: $S^{-1}M \approx M_S$ *under the isomorphism given by $s^{-1} \otimes x \to s^{-1}x$.*

Proof: Analogous to that of proposition 1.10.18. There is a balanced map $S^{-1}R \times M \to S^{-1}M$ given by $(s^{-1}r, x) \to s^{-1}(rx)$, yielding the desired map $S^{-1}R \otimes M \to M_S$. It suffices to prove its inverse, given by $s^{-1}x \to s^{-1} \otimes x$, is well-defined, i.e., if $s_1^{-1}x_1 = s_2^{-1}x_2$ then we want $s_1^{-1} \otimes x_1 = s_2^{-1} \otimes x_2$. Indeed, take r, r' in R with $rx_1 = r'x_2$ and $rs_1 = r's_2 \in S$. Then

$$s_1^{-1} \otimes x_1 = (rs_1)^{-1}r \otimes x_1 = (rs_1)^{-1} \otimes rx_1 = (r's_2)^{-1} \otimes r'x_2$$

$$= (r's_2)^{-1}r' \otimes x_2 = s_2^{-1} \otimes x_2$$

as desired.   Q.E.D.

An alternate proof is given in exercise 4.

Theorem 3.1.20:   *Define the localization functor* $F: R\text{-}\mathcal{M}od \to S^{-1}R\text{-}\mathcal{M}od$ *to be* $S^{-1}R \otimes_R -$. *Then* F *is an exact functor.*

Proof: We know F is a right exact functor, so it remains to show that if $f: M \to N$ is monic then $1 \otimes f$ is monic. Suppose $s^{-1} \otimes x \in \ker(1 \otimes f)$, i.e., $s^{-1} \otimes fx = 0$. In view of proposition 3.1.19, this means $s^{-1}fx = 1^{-1}0$, i.e., there are r, r' in R with $rfx = 0$ and $rs = r' \in S$. But $f(rx) = rfx = 0$ implies $rx = 0$, yielding $s^{-1} \otimes x = 0$ by the next remark.   Q.E.D.

Remark 3.1.21:   By proposition 3.1.19 we see $s^{-1} \otimes x = 0$ iff $s^{-1}x = 0$, i.e., there exist r, r' in R with $rx = 0$ and $rs = r' \in S$. In particular, $S^{-1}M = 0$ iff each $1 \otimes x = 0$, iff for each x in M there is s in S such that $sx = 0$.

§3.2 Goldie's Theorems and Orders in Artinian Quotient Rings

The main object in studying rings of fractions is to relate rings to "nicer" rings. Since any left Artinian ring is already its own classical ring of fractions, we are led to consider which rings are orders in simple and semisimple Artinian rings. Goldie [58, 60] classified these rings; in particular, every semi-prime Noetherian ring is an order in a simple Artinian ring (also, cf., Lesieur-Croisot [59]). Goldie's theorems thereby provided a framework unifying Noetherian rings with commutative rings, and the impact on ring theory has been immense. As Goldie's conditions involve left annihilators and direct sums of left ideals, these aspects of the structure theory have taken on an interest in their own right, and we shall begin by familiarizing ourselves with them.

Annihilators

Definition 3.2.1: $L \subset R$ is a *left annihilator* if $L = \text{Ann}(S)$ for some subset S of R, i.e., $L = \{r \in R: rs = 0\}$. Similarly, we define $\text{Ann}'\, S = \{r \in R: Sr = 0\}$ and call such a set a *right annihilator* of R. The terminology $\ell(S)$ and $\imath(S)$ are often used in the literature for $\text{Ann}(S)$ and $\text{Ann}'(S)$.

Remark 3.2.2:

   (i) Every left annihilator is a left ideal of R, and every right annihilator is a right ideal of R.

   (ii) If $S_1 \subseteq S_2$ then $\text{Ann } S_1 \supseteq \text{Ann } S_2$.

   (iii) $S \subseteq \text{Ann}(\text{Ann}' S)$ and $S \subseteq \text{Ann}'(\text{Ann } S)$.

   (iv) If $A = \text{Ann } S$ then $A = \text{Ann}(\text{Ann}' A)$. (Indeed, ($\subseteq$) is by (iii); on the other hand, $S \subseteq \text{Ann}'(\text{Ann } S) = \text{Ann}' A$ so $A \supseteq \text{Ann}(\text{Ann}' A)$ by (ii).)

   (iv') If $A = \text{Ann}' S$ then $A = \text{Ann}'(\text{Ann } A)$. (Analogous to (iv)).

   (v) $\bigcap \text{Ann } S_i = \text{Ann}(\sum S_i)$

   (vi) $\sum \text{Ann } S_i \subseteq \text{Ann}(\bigcap S_i)$ (but equality does *not* necessarily hold, as evidenced by $\mathbb{Z}$, in which $\text{Ann}(\bigcap_{m \neq 0} m\mathbb{Z}) = \text{Ann } 0 = \mathbb{Z}$, but $\sum \text{Ann}(n\mathbb{Z}) = \sum 0 = 0$.)

   (vii) $\sum \text{Ann } S_i = \text{Ann}(\bigcap S_i)$ if the S_i are right annihilators. (Indeed, let $A_i = \text{Ann } S_i$; then $S_i = \text{Ann}' A_i$ by (iv'), so (v) yields $\text{Ann}(\bigcap S_i) = \text{Ann}(\text{Ann}'(\sum A_i)) = \sum A_i = \sum \text{Ann } S_i$.)

   (viii) If R is a subring of T and $S \subset R$ then $\text{Ann}_R S = R \cap \text{Ann}_T S$

   (ix) If $\text{Ann } x = \text{Ann } x^2$ then $Rx \cap \text{Ann } x = 0$ (for if $rx \in \text{Ann } x$ then $rx^2 = 0$ so $rx = 0$.)

   These easy facts have far-reaching consequences and are used so often that we shall refer to them merely as (i) through (ix)

Proposition 3.2.3: $\mathscr{L} = \{left\ annihilators\}$ *is a complete sublattice of* $\{left$ *ideals of* $R\}$ *and is isomorphic to the dual of* $\{right\ annihilators\}$ *under the correspondence* $A \rightarrow \text{Ann}' A$.

Proof: The first assertion follows at once from (v) and (vii). The second assertion follows from (ii) and (iv).   Q.E.D.

Corollary 3.2.4: *If* R *is a subring of* T *then there is a lattice injection* $\{left\ annihilators\ in\ R\} \rightarrow \{left\ annihilators\ in\ T\}$ *given by* $A \rightarrow \text{Ann}_T \text{Ann}'_R A$.
     Q.E.D.

   Incidentally, the second assertion of proposition 3.2.3 is purely lattice-theoretic and can be given in a much more general context.

ACC (Ann)

Our interest lies in the case when $\{$left annihilators$\}$ satisfies the ascending chain condition, which we shall call ACC(Ann). This condition is quite tractable.

Remark 3.2.5: If T satisfies ACC(Ann) then every subring of T satisfies ACC(Ann), by corollary 3.2.4. In particular, every subring of a left Noetherian ring satisfies ACC(Ann).

Remark 3.2.6: If R is a semiprime ring with ACC(Ann) then R has no non-zero nil left ideals by proposition 2.6.24.

Lemma 3.2.7: *If R is semiprime with* ACC(Ann) *and* $0 \neq r \in R$ *then the left ideal* Ann r *is not large.*

Proof: Let $J = \{r \in R: \text{Ann}\, r \text{ is large}\}$, clearly a right ideal of R since Ann $ra \supseteq$ Ann r for all a in R. We need to prove $J = 0$; by remark 3.2.6 it suffices to prove J is nil. Well for any z in J we have Ann $z \subseteq$ Ann $z^2 \subseteq \cdots$ implying Ann $z^n =$ Ann $z^{n+1} = \cdots$ for some n (depending on z). Putting $x = z^n$ we then have Ann $x =$ Ann x^2 so $Rx \cap$ Ann $x = 0$ by (ix); but $x \in J$ so $Rx = 0$, proving $z^n = 0$.   Q.E.D.

Proposition 3.2.8: *If R is semiprime with* ACC(Ann) *and Rr is large then* Ann$'\, r = 0$.

Proof: If $a \in$ Ann$'\, r$ then Ann $a \supseteq Rr$ is large so $a = 0$ by the lemma.
                                                                                        Q.E.D.

Digression 3.2.9: Proposition 3.2.3 shows that ACC(Ann) is equivalent to the DCC(Ann$'$). This useful observation can be used to prove for example that any primitive ring R satisfying ACC(Ann) and having nonzero socle is simple Artinian. (Indeed, take an orthogonal set of rank 1 idempotents $e_1, e_2, \ldots$ and observe that Ann$'\, e_1 >$ Ann$'\{e_1, e_2\} > \cdots$ shows that the rank is bounded, implying by the density theorem that R is simple Artinian, cf., remark 2.1.26.)

Goldie Rings

Definition 3.2.10: R satisfies $ACC \oplus$ if R does *not* have an infinite independent set of left ideals. R is (left) *Goldie* if R satisfies ACC(Ann) and ACC $\oplus$. More generally, we say a module M satisfies ACC $\oplus$ if M does not have an infinite independent set of submodules.

Any semisimple Artinian ring R has finite length in $R\text{-}\mathcal{M}od$ and thus satisfies ACC $\oplus$ as well as ACC(Ann) and so is left Goldie. The condition

ACC $\bigoplus$ is rather strong, permitting us to relate regular elements to large left ideals; this turns out to be one of the key aspects of Goldie's theorem.

Proposition 3.2.11: *Suppose* $\operatorname{Ann} r = 0.$ *(i) If* $L \cap Rr = 0$ *then* $\{Lr^i : i \in \mathbb{N}\}$ *is a set of independent left ideals. (ii) If* R *satisfies* $\operatorname{ACC} \bigoplus$ *then* Rr *is a large left ideal of* R.

Proof: (i) Otherwise, take $\sum_{i=m}^{n} a_i r^i = 0$ with $a_i \in L$ and $a_m \neq 0$; then $\sum_{i=0}^{n-m} a_{i+m} r^i \in \operatorname{Ann} r^m = 0$ so $a_m = -\sum_{i=1}^{n-m} a_{i+m} r^i \in L \cap Rr = 0$, contradiction.
  (ii) By (i) we have $L \cap Rr \neq 0$ for every $0 \neq L < R$.   Q.E.D.

Proposition 3.2.12: *If* R *is semiprime Goldie and* $\operatorname{Ann} r = 0$ *then* r *is regular*.

Proof: Rr is large, so proposition 3.2.8 implies $\operatorname{Ann}' r = 0$; since we are given $\operatorname{Ann} r = 0$ we conclude r is regular.   Q.E.D.

Proposition 3.2.13: *Every large left ideal* L *of a semiprime Goldie ring* R *contains a regular element*.

Proof: We have just seen it suffices to find r in L with $\operatorname{Ann} r = 0$. We start with r_1 non-nilpotent such that $\operatorname{Ann} r_1$ is maximal possible. Then $\operatorname{Ann} r_1^2 = \operatorname{Ann} r_1$ so $Rr_1 \cap \operatorname{Ann} r_1 = 0$ by (ix). Continuing inductively, suppose we have chosen non-nilpotent $r_1, \ldots, r_k$ in L such that $Rr_1, \ldots, Rr_k$, and A_k are independent left ideals, where $A_k = L \cap \operatorname{Ann} r_1 \cap \cdots \cap \operatorname{Ann} r_k$. If $A_k \neq 0$ then A_k is not nil so we can pick non-nilpotent r_{k+1} in A_k with $\operatorname{Ann} r_{k+1}$ maximal possible. Then $Rr_{k+1} \cap \operatorname{Ann} r_{k+1} = 0$ so $Rr_1, \ldots, Rr_{k+1}$, and $A_{k+1} = L \cap \operatorname{Ann} r_1 \cap \cdots \cap \operatorname{Ann} r_{k+1}$ are clearly independent. We have thereby established an inductive procedure, which must terminate because R satisfies $\operatorname{ACC}(\bigoplus)$, i.e., for some k we have $A_k = 0$ so $\bigcap_{i=1}^{k} \operatorname{Ann} r_i = 0$ since L is large. Let $r = \sum_{i=1}^{k} r_i$. If $0 = ar = \sum ar_i$ then each $ar_i = 0$ (since the Rr_i are independent) so $a \in \bigcap \operatorname{Ann} r_i = 0$, proving $\operatorname{Ann} r = 0$.   Q.E.D.

To prove the next result we need the following special case of proposition 3.3.0 below (which can also be verified directly): Given a left ideal L of R and $r \in R$ define $Lr^{-1} = \{a \in R : ar \in L\}$; if L is large then Lr^{-1} is large.

Theorem 3.2.14: *(Goldie's second theorem) The following assertions are equivalent for a ring* R:

(i) R *is semiprime Goldie*.

(ii) *A left ideal of R is large iff it contains a regular element.*

(iii) *R is Ore and its (classical) ring of fractions is semisimple Artinian.*

Proof: (i) $\Rightarrow$ (ii) by 3.2.11 and 3.2.13.

(ii) $\Rightarrow$ (iii) First we show R is Ore. Given $a, r \in R$ with a regular we need to find a', r' with a' regular and $a'r = r'a$. Ra is large, so Rar^{-1} is also large (in R) and thus has a regular element a'; then $a'r \in Ra$ as desired.

Let Q be the ring of fractions of R. We shall prove Q is semisimple Artinian by showing Q has no large left ideals other than Q itself (cf., theorem 2.4.9). Well suppose $L < Q$ is large. Then $L \cap R$ is large in R. (Indeed, if $0 \neq r \in R$ then $L \cap Qr \neq 0$ so there is $a_1^{-1}r_1$ in Q with $0 \neq a_1^{-1}r_1r \in L$ and then $r_1r \in L \cap R$.) Hence $L \cap R$ has a regular element which thus is invertible in Q, proving $L = Q$.

(iii) $\Rightarrow$ (i) R has ACC(Ann) by remark 3.2.5. Next we claim if L is a large left ideal of R then L has a regular element. Indeed, let $Q = S^{-1}R$ be the classical ring of fractions. Then QL is large in Q (since for any $s^{-1}r \in Q$ one has $0 \neq L \cap Rr \subseteq QL \cap Qs^{-1}r$). Hence $QL = Q$, so L has a regular element by Corollary 3.1.12.

To show R is semiprime suppose $N \lhd R$ with $N^2 = 0$ and let $L < R$ be an essential complement of N, i.e., $L \cap N = 0$ and $L + N$ is large (cf., proposition 2.4.4). By the previous paragraph there is a regular element a in $L + N$, and $Na \subseteq NL \subseteq L \cap N = 0$; since a is regular we have $N = 0$ proving R is semiprime.

It remains to prove R satisfies ACC $\bigoplus$. This follows at once from the following lemma:

Lemma 3.2.15:   *Suppose R is an order in Q. If $\{L_i : i \in I\}$ is an independent set of left ideals of R then $\{QL_i : i \in I\}$ is an independent set of left ideals of Q.*

Proof: If $\sum_{u=1}^{t} q_u = 0$ for q_u in QL_{i_u} then writing $q_u = s^{-1}r_u$ for $1 \leq i \leq t$ with $r_u \in L_{i_u}$ (by proposition 3.1.11) we have $\sum r_u = 0$ so each $r_u = 0$.   Q.E.D.

The most important direction of Goldie's second theorem is (i) $\Rightarrow$ (iii) since it enables us to study R through its ring of fractions. Condition (ii) is also very useful since it contains the technical information needed in many proofs about Goldie rings.

Theorem 3.2.16:   *(Goldie's first theorem) R is prime Goldie iff R is an order in a simple Artinian ring Q.*

Proof: $(\Rightarrow)$ Q is semisimple Artinian by theorem 3.2.14. If Q has simple components $Q_1 \neq Q_2$ then $(Q_1 \cap R)(Q_2 \cap R) = 0$, contrary to R prime; hence Q is simple Artinian.

$(\Leftarrow)$ R is Goldie, so we need to show R is prime. Suppose $0 \neq A \lhd R$. Then $0 \neq QAQ \lhd Q$ so $1 = \sum_{i=1}^{t} q_{i1} a_i q_{i2}$ for suitable q_{ij} in Q and a_i in A. Write $q_{i1} = s^{-1} r_{i1}$ for $1 \leq i \leq t$ with r_{i1}, $s \in R$ and s regular. Then $s = \sum_{i=1}^{t} r_{i1} a_i q_{i2} \in AQ$, implying $(\text{Ann}_R A)s = 0$. But s is regular so $\text{Ann}_R A = 0$, proving R is prime. Q.E.D.

Goldie Rank and Uniform Dimension

Of course, once Goldie's theorem was established the Goldie conditions became objects of study in their own right, so we digress a little to introduce the relevant notions.

R has *uniform* (or *Goldie*) dimension n if there is no set of $n + 1$ independent left ideals of R, with n minimal possible. R *satisfies* $\text{ACC}_n(\text{Ann})$ if every chain of left annihilators has length $\leq n$. R has *Goldie rank n* if R has uniform dimension n and satisfies $\text{ACC}_n(\text{Ann})$.

Remark 3.2.17: Suppose $R \subseteq T$. If T satisfies $\text{ACC}_n(\text{Ann})$ then R satisfies $\text{ACC}_n(\text{Ann})$ by corollary 3.2.4.

Proposition 3.2.18: *If R is semiprime Goldie then there is n in $\mathbb{N}$ such that R has Goldie rank n.*

Proof: The classical ring of fractions Q of R has a composition series of some length n, so Q satisfies $\text{ACC}_n(\text{Ann})$ and has uniform dimension $\leq n$; these properties pass to R by remark 3.2.17 and lemma 3.2.15. Q.E.D

Remark 3.2.18: Suppose R is prime Goldie, and let $Q = M_n(D)$ be its simple Artinian ring of fractions. Obviously n is the Goldie rank of Q, which also is the Goldie rank of R. This easy formulation of the Goldie rank of a prime ring is used repeatedly in Chapter 8.

$\text{ACC}(\text{Ann})$ does *not* imply $\text{ACC}_n(\text{Ann})$ for suitable n, cf., exercises 8, 9. However, see proposition 3.2.19.

The uniform dimension of a ring has a very useful generalization to modules, whose basic properties we shall record here for later use. The *uniform* (or *Goldie*) dimension of a module M is the largest n (if it exists) for which there is a set of n independent submodules; a module of uniform dimension 1 is called *uniform*.

In other words, M is uniform iff no submodule of M can be written as a direct sum of two proper nonzero submodules. It follows at once that every submodule of a uniform module is large and is uniform as a module itself. Let us use uniform modules to characterize the uniform dimension.

Proposition 3.2.19:   *Suppose M satisfies ACC $\bigoplus$. Then M has some uniform dimension n. More precisely,*

(i) *Every submodule N of M contains a uniform submodule.*
(ii) *There are independent uniform submodules $U_1, \ldots, U_n$ of M such that $\sum_{i=1}^{n} U_i$ is large in M.*
(iii) *The number n of (ii) is uniquely determined and is the uniform dimension of M.*

Proof:

(i) We are done unless N is *not* uniform, so $N \geq N_0 \oplus N'$ for suitable independent submodules N_0 and N'; continuing with N' we have an inductive procedure to build an infnite set of independent submodules, contrary to hypothesis unless at some stage we reach a uniform.

(ii) By (i) we have a uniform $U_1 \leq M$. Inductively, given $U_1, \ldots, U_t$ independent uniform submodules of M let M_t be an essential complement of $U_1 \oplus \cdots \oplus U_t$. If $M_t \neq 0$ it contains a uniform U_{t+1}; this gives us an inductive procedure for finding an infinite set of independent uniforms, contrary to hypothesis, so some $M_n = 0$; i.e., $U_1 \oplus \cdots \oplus U_n$ is large in M.

(iii) Assume, on the contrary, M has independent submodules $V_1, \ldots, V_{n+1}$. Choosing the V_i such that $V_i \in \{U_1, \ldots, U_n\}$ for $1 \leq i \leq u$ with u maximal possible, we have $u \leq n$. Let $V = \sum_{i=1}^{n} V_i$. If $V \cap U_j = 0$ then $\{V_1, \ldots, V_n, U_j\}$ would be a set of independent submodules, contrary to choice of u. Thus $V \cap U_j \neq 0$ for each j, implying V is large in $\bigoplus U_j$ by proposition 2.4.5'(ii), and thus V is large in M, contrary to $V_{n+1} \cap V = 0$.   Q.E.D.

Note. (Cohn) This proof could be streamlined by noting that {uniform modules} have a weak dependence relation (cf., remark 0.3.3) given by $V \in_{\text{dep}} \mathcal{S}$ if $V \cap \sum \{U \in \mathcal{S}\} = 0$.

Digression 3.2.19:   An alternate proof of proposition 3.2.19 actually yields stronger results. The idea is to translate these notions to the injective hull $E(M)$, viewed as the maximal essential extension of M, viz. theorem 2.10.20.

U is uniform iff $E(U)$ is indecomposable. Furthermore, $N \leq M$ implies $E(N)$ is a summand of $E(M)$, so the uniform dimension of M is precisely the number of summands in a decomposition of $E(M)$, which is unique by theorem 2.10.33.

This approach makes available other invariants, most notably Ann $E(U_i)$. We return to the uniform dimension in exercises 3.3.18ff and in §3.5.

Annihilator Ideals in Semiprime Rings

Historically, of course, Goldie's first theorem was proved first; there is an interesting argument (due to Herstein) that it implies Goldie's second theorem, based on annihilators; the key points also have other applications.

Lemma 3.2.20: *Suppose R is semiprime and $A \lhd R$. The following are then equivalent for $B \lhd R$:*

(i) $B = \mathrm{Ann}\, A$.
(ii) *B is maximal (as ideal) such that $A \cap B = 0$.*
(iii) *B is a maximal essential complement of A (as left ideal).*
(iv) $B = \mathrm{Ann}'\, A$.

Proof: (i) $\Rightarrow$ (ii) $(A \cap B)^2 \subseteq BA = 0$, so $A \cap B = 0$. If $A \cap B' = 0$ for $B' \lhd R$ then $B'A \subseteq A \cap B' = 0$ so $B' \subseteq \mathrm{Ann}\, A = B$.

(ii) $\Rightarrow$ (i) Clearly, $\mathrm{Ann}\, A \lhd R$ since $A \lhd R$, so $A \cap \mathrm{Ann}\, A = 0$ as above. But $BA \subseteq A \cap B = 0$ so $B \subseteq \mathrm{Ann}\, A$, implying $B = \mathrm{Ann}\, A$ by maximality of B.

(ii) $\Leftrightarrow$ (iv). analogous to (i) $\Leftrightarrow$ (ii) above.

(ii), (iv) $\Rightarrow$ (iii) Suppose $(A + B) \cap Rr = 0$. Then $ARr \subseteq A \cap Rr = 0$ implying $Rr \subseteq \mathrm{Ann}'\, A = B$ so $Rr \subseteq B \cap Rr = 0$. This proves $A + B$ is large, and we are given $A \cap B = 0$.

(iii) $\Rightarrow$ (ii) by definition of essential complement.   Q.E.D.

One consequence of this lemma is that Ann and Ann' mean the same when applied to ideals of a semiprime ring.

Proposition 3.2.21: *The following assertions are equivalent for R semiprime and $A \lhd R$:*

(i) *A is large as a left ideal.*
(ii) *A is large as a right ideal.*
(iii) *$A \cap I \neq 0$ for every $0 \neq I \lhd R$.*

   (iv) Ann $A = 0$

   (v) Ann$' A = 0$.

Proof: (iii) $\Rightarrow$ (iv) $\Rightarrow$ (i) is a special case of the lemma, and (i) $\Rightarrow$ (iii) is trivial. (iii) $\Rightarrow$ (v) $\Rightarrow$ (ii) $\Rightarrow$ (iii) follows by left-right symmetry of (iii).   Q.E.D.

In view of these two results we are led to consider annihilators of two-sided ideals and formally define an *annihilator ideal* to be an ideal which is a left annihilator. We aim for the surprising result that in a semiprime ring ACC(annihilator ideals) is equivalent to having a finite number of prime ideals with intersection 0.

Remark 3.2.22: If A is an annihilator ideal of a semiprime ring R then $A = \text{Ann}(\text{Ann } A)$ so A is the left annihilator of an ideal. (Immediate by lemma 3.2.20.)

Proposition 3.2.23: *If $M \in R\text{-}\mathcal{M}od$ and if the lattice $\{\text{Ann } N: 0 \neq N < M\}$ has a maximal member P then P is a prime ideal of R.*

Proof: Suppose $A \supseteq P$, $B \supseteq P$ are ideals with $AB \subseteq P$. Then $ABN = 0$ for some $0 \neq N < M$ so $A(BN) = 0$. But $BN \leq N$ so $A = P$ by hypothesis unless $BN = 0$, in which case $B = P$.   Q.E.D.

Remark 3.2.23': In a semiprime ring, each annihilator ideal is a semiprime ideal. (Indeed if $B^2 \subseteq \text{Ann } A$ then $(B \cap A)^3 = 0$ so $B \cap A = 0$ implying $B \subseteq \text{Ann } A$.)

Theorem 3.2.24:

   (i) *If R is semiprime with ACC(annihilator ideals) then R has a finite set $P_1, \ldots, P_n$ of prime ideals with $P_1 \cap \cdots \cap P_n = 0$. (In fact, $P_1, \ldots, P_n$ are precisely the maximal annihilator ideals.)*

   (ii) *Conversely, if $\bigcap_{i=1}^{n} P = 0$ for prime ideals $P_1, \ldots, P_n$ then R is semiprime with ACC(annihilator ideals). In fact, every annihilator ideal then is an intersection of some of the P_i, so, in particular, every chain of annihilator ideals has length $\leq n$.*

Proof: (i) Let $\mathcal{P} = \{$maximal annihilator ideals$\}$. Then all members of $\mathcal{P}$ are prime ideals by proposition 3.2.23. We shall see $\mathcal{P}$ is finite with intersection 0. Take distinct $P_1, P_2, \ldots$ in $\mathcal{P}$ noting $P_i \not\subseteq P_j$ for all $i \neq j$, and let $B_k =$

$\mathrm{Ann}(\bigcap_{i=1}^{k} P_i)$. Then $\mathrm{Ann}\ P_{k+1} \nsubseteq B_k$ (for, otherwise, $(\mathrm{Ann}\ P_{k+1})P_1 \cdots P_k = 0 \subseteq P_{k+1}$, implying $\mathrm{Ann}\ P_{k+1} \subseteq P_{k+1}$ and $(\mathrm{Ann}\ P_{k+1})^2 = 0$; thus $\mathrm{Ann}\ P_{k+1} = 0$, contrary to remark 3.2.22).

Hence $B_k \subset B_{k+1}$ for each k, so by ACC(annihilator ideals) we cannot have an infinite number of B_k, proving $\mathscr{P}$ has a finite number of members $P_1, \ldots, P_n$. Let $A = \bigcap_{i=1}^{n} P_i$. If $A \neq 0$ then $B_n = \mathrm{Ann}\ A$ implies $B_n \subseteq P_i$ for some i; rearranging the indices we may assume $B_n \subseteq P_n$. Then $(B_n(\bigcap_{i=1}^{n-1} P_i))^2 \subseteq B_n A = 0$ implying $B_n \subseteq B_{n-1}$, contrary to what we saw above. Hence $A = 0$, as desired.

(ii) Let $I = \{1, \ldots, n\}$. We shall show every annihilator ideal A has the form $\bigcap_{i \in I'} P_i$ for suitable $I' \subset I$, from which all the assertions follow at once. Let $B = \mathrm{Ann}\ A$ and $I' = \{i \in I : A \subseteq P_i\}$ and $I'' = \{i \in I : B \subseteq P_i\}$. Since $AB = 0 \subseteq P_i$ for each i we have $I = I' \cup I''$. Then $A = \mathrm{Ann}\ B \supseteq \mathrm{Ann}(\bigcap \{P_i : i \in I''\}) \supseteq \bigcap \{P_i : i \in I'\} \supseteq A$, so equality holds at every stage.   Q.E.D.

Remark 3.2.25:   The last sentence of the proof shows that $\mathrm{Ann}\ P_i = \bigcap_{j \neq i} P_j$.

Corollary 3.2.26:   *If R is semiprime with* ACC(annihilator ideals) *then the maximal annihilator ideals are precisely the minimal prime ideals of R, and these are finite in number.*

Proof: By theorem 3.2.24 and proposition 2.12.8 each minimal prime is a maximal annihilator ideal. But each maximal annihilator ideal P contains a minimal prime which then must be P itself.   Q.E.D.

Theorem 3.2.27:   *Suppose R has a finite set $P_1, \ldots, P_n$ of minimal prime ideals with intersection 0, and each R/P_i is prime Goldie with ring of fractions Q_i. Then R is Goldie with ring of fractions isomorphic to $\prod_{i=1}^{t} Q_i$.*

Proof: Take $0 \neq r_i \in \bigcap_{j \neq i} P_j$ for each j. Put $R_i = R/P_i$. Letting $^-$ denote the canonical image in R_i we have $\bar{r}_i \neq 0$; thus $\overline{Rr_iR}$ is large in R_i and by Goldie's theorem contains a regular element $\bar{a}_i$, with $a_i \in \bigcap_{j \neq i} P_j$. Note $a = \sum_{i=1}^{n} a_i$ is regular.

Consider the injection $R \to \prod_{i=1}^{n} R_i$ given by $r \to (r + P_1, \ldots, r + P_n)$, so that a_i corresponds to an element whose only nonzero component is in R_i. If $b \in R$ is regular then $\bar{b} = b + P_i$ is regular in R_i (for if $\bar{b}r = 0$ then $bra_i = 0$ so $ra_i = 0$, implying $\bar{r} = 0$; likewise, if $\overline{rb} = 0$ then $\bar{r} = 0$). Thus regular elements of R are invertible in $Q = \prod_{i=1}^{n} Q_i$, where each Q_i is the simple Artinian ring of fractions of R_i. Using the previous injections to view $R \subseteq \prod R_i \subseteq \prod Q_i = Q$, we claim Q is the ring of fractions of R.

Indeed, we must show that any $q \in Q$ has the form $b^{-1}r$ where $b \in R$ with b regular. Write $q = \sum_{i=1}^{n} q_i$ where $q_i \in Q_i$ and write $q_i = \bar{b}_i^{-1}\bar{r}_i$ with $\bar{b}_i$ regular in $\bar{R}_i$. Take $b = \sum_{i=1}^{n} a_i b_i$ and $r = \sum_{i=1}^{n} a_i r_i$. Clearly, b is regular and $q = b^{-1}r$ in Q, by componentwise verification.   Q.E.D.

We are led to an analogue of the Chinese Remainder Theorem after a useful preliminary about idempotents.

Lemma 3.2.28: *Suppose R is a semiprime ring and A is an ideal which is a summand of R as left module. Then $A = Re$ for a central idempotent e, and $R \approx Re \times R(1 - e)$ as rings.*

Proof: $A = Re$ for some idempotent e (since A is a summand of R) and $\text{Ann}' A = \text{Ann} A \lhd R$ by lemma 3.2.20. But $\text{Ann}' A = (1 - e)R$ so $(1 - e)Re = 0$ and also $eR(1 - e) \subseteq A(1 - e) = 0$ since $A \lhd R$. Thus $R = eRe + (1 - e)R(1 - e)$ by the Pierce decomposition, yielding $R \approx R_1 \times R_2$ where $R_1 = eRe$ and $R_2 = (1 - e)r(1 - e)$. Since $e, 1 - e$ are the respective unit elements of R_1, R_2 we see they are central in R.   Q.E.D.

Theorem 3.2.29:   *A semiprime ring R is isomorphic to a finite direct product of prime images iff R satisfies ACC(Annihilator ideals) and every maximal annihilator ideal has the form Re for e idempotent (and central by the lemma).*

Proof: ($\Rightarrow$) If $R \approx \prod_{i=1}^{n} R/P_i$ then $\bigcap_{i=1}^{n} P_i = 0$ so the P_i are the maximal annihilator ideals by theorem 3.2.25, and $P_i = Re_i$ for $e_i = (1, \ldots, 1, 0, 1, \ldots, 1)$ where 0 appears in the i position.

($\Leftarrow$) Letting $P_1, \ldots, P_n$ be the maximal annihilator ideals of R we have $\bigcap_{i=1}^{n} P_i = 0$. By the Chinese Remainder Theorem it remains to show $P_i + P_j = R$ for each $i \neq j$. By hypothesis, we can write $P_i = Re_i$ and $P_j = Re_j$. Then $1 - e_i \in \text{Ann} P_i \subseteq P_j$ (cf., remark 3.2.25) so $1 = e_i + (1 - e_i) \in P_i + P_j$ as desired.   Q.E.D.

Corollary 3.2.30:   *A semiprime Goldie ring R is a direct product of prime Goldie rings iff every maximal annihilator ideal can be written in the form Re for a suitable idempotent e.*

Proposition 3.2.31:   *Suppose R satisfies ACC(annihilator ideals) and $A, B \lhd R$ with $A \nsubseteq B$. If $A - B \subseteq$ a finite union of annihilator ideals then A is contained in some annihilator ideal, so, in particular, $\text{Ann}' A \neq 0$.*

Proof: $A - B \subseteq$ finite union of maximal annihilators; each of which is prime, so apply proposition 2.12.7.   Q.E.D.

Digression: Orders in Semilocal Rings

Motivated by Goldie's theorem one might wish to characterize orders in left Artinian rings or, more specifically, Noetherian orders in Artinian rings. This problem was solved independently by two prominent ring theorists, Robson [67] and Small [66], each then at the start of his career. We shall introduce the relevant concepts in the text but leave most of the proofs for the exercises. The underlying idea is to pass down the nilradical; more generally, we deal with arbitrary $I \lhd R$ and let S be a submonoid of R.

Definition 3.2.32: S is a *denominator set* for I if the following two conditions are satisfied:

(i) For any $s_1 \in S$ and $r_1 \in I$ there exist s_2 in S and r_2 in I with $s_2 r_1 = r_2 s_1$.
(ii) If $rs = 0$ for r in I, s in S, then there is s' in S with $s'r = 0$.

Remark 3.2.33: Suppose S is a denominator set of R. Then S is a denominator set for I iff $S^{-1}I \lhd S^{-1}R$. (Indeed (ii) holds *a fortiori*, and (i) is equivalent to $(1^{-1}r)(s^{-1}1) \in S^{-1}I$ for all r in I.) In case $S = \{$regular elements$\}$ this property is called *I-quorite* in the literature.

We aim to generalize remark 2.12.12 and thus consider the following three properties, where $^-$ denotes the image in $\bar{R} = R/I$:

(1) S is a denominator set of R.
(2) S is a denominator set for I.
(3) $\bar{S}$ is a denominator set of $\bar{R}$.

Weakening definition 3.2.32(i) by means of exercise 2, one can easily prove

Proposition 3.2.34: ($=$*exercise* 3) *If two of the properties* (1), (2), (3) *hold then the third also holds, and* $\bar{S}^{-1}\bar{R} \approx S^{-1}R/S^{-1}I$.

Now suppose (1) holds and each element of S is regular. Let $Q = S^{-1}R$, and view $R \subseteq Q$ canonically. Given $A \lhd Q$ we put $I = R \cap A \lhd R$; letting $\bar{Q} = Q/A$ we have an injection $\bar{R} \to \bar{Q}$. Moreover, $\bar{S}$ is invertible in $\bar{Q}$ and thus regular in $\bar{R}$, implying $\bar{Q} = \bar{S}^{-1}\bar{R}$. Thus (1) and (3) hold, so (2) also holds.

In particular, when $S = \{$all regular elements of $R\}$ we have shown $\bar{s}$ is regular in $\bar{R}$ for all regular s in R, leading us to consider the converse.

Definition 3.2.35: R satisfies the *regularity condition (over I)* if $\bar{r}$ regular in $\bar{R}$ implies r is regular in R. When I is not specified we take $I = \mathrm{Nil}(R)$.

Theorem 3.2.36: $(=exercise~4)$ R *is contained naturally in a semilocal ring of fractions* $Q = S^{-1}R$, *for suitable* $I \lhd R$ *and* $S = \{r \in R : \bar{r}$ *is regular in* $R/I\}$, *iff the following conditions are satisfied*:

  (i) $\bar{R}$ *is semiprime Goldie.*
  (ii) *Every s in S is regular.*
  (iii) *S is a denominator set for I.*

Moreover, in this case $S^{-1}I = \mathrm{Jac}(Q)$.
  Two important consequences of this result:

Robson's theorem: R *is an order in a semilocal ring* Q *iff there is* $I \lhd R$ *such that* (i) R *satisfies the regularity condition over* I; (ii) $\bar{R}$ *is semiprime Goldie, and* (iii) $\{$*regular elements of* $R\}$ *is a denominator set for* I.

Small's theorem: *A left Noetherian ring is an order in a left Artinian ring iff it satisfies the regularity condition.*

  Robson's theorem follows at once from theorem 3.2.36 taking $S = \{$regular elements of $R\}$. Small's theorem requires a further argument, and is proved in exercise 3.5.3 using more machinery. The recent literature contains several other characterizations of when Noetherian rings are in orders in Artinian rings. when Noetherian rings are orders in Artinian rings.

Embedding of Rings

Among other things Goldie's theorem says any semiprime left Noetherian ring can be embedded in a simple Artinian ring. This already has interesting consequences, as we shall soon see, and we are led to consider procedures of embedding rings into "nicer" rings. In particular, we shall present a general procedure of embedding a prime ring into a primitive ring and shall consider the problem of embedding rings into matrix rings over commutative rings.

Universal Sentences and Embeddings

The theme pervading this discussion is that any universal sentence holding in a ring R' also holds in each subring R. Thus to verify such a property on R we might do best by finding an appropriate overring. One example is the sentence $\forall xy(xy = 1 \Rightarrow yx = 1)$ which defines *weakly* 1-*finite* (cf., definition 1.3.30); likewise, weakly n-finite could be defined by an appropriate universal sentence, for each n.

Theorem 3.2.37:   *Every* subring R *of a left Noetherian ring is weakly finite.*

Proof: We may assume R itself is left Noetherian, and we want to prove R is weakly n-finite for any given n, i.e., that $M_n(R)$ is weakly 1-finite. $M_n(\mathrm{Nil}(R))$ is nilpotent since $\mathrm{Nil}(R)$ is nilpotent. Since invertibility passes up the Jacobson radical we may replace R by $R/\mathrm{Nil}(R)$ which is semiprime and thus is an order in a semisimple Artinian ring R'. But then $M_n(R) \subseteq M_n(R')$ which is semisimple Artinian and thus 1-finite, as desired.   Q.E.D.

Using Small's theorem we could improve this result, but the statement would become rather technical. A more subtle use of universal sentences asserts the nilpotence of a given subset.

Theorem 3.2.38:   *If R is a subring of a left Noetherian ring then there is some k such that $S^k = 0$ for every nil, weakly closed subset S of R (cf., proposition 2.6.30).*

Proof: The nilradical N of R is nilpotent, so we may pass to R/N, which is an order in a semisimple Artinian ring, so we are done by theorem 2.6.31.
      Q.E.D.

There are more general nil $\Rightarrow$ nilpotent theorems, cf., exercises 32ff, but this result shows the elegance of a very powerful method. For the rest of the section we shall collect some embedding theorems but shall also pay attention as to the manner of embedding which, although irrelevant in the above applications, can be of supreme importance in other applications.

Embedding into Primitive and Semiprimitive Rings

A certain embedding procedure was outlined preceding theorem 2.5.23, and we elaborate it here for further reference, because of its importance in the structure theory:

Definition 3.2.39:   A primitive ring R is *closed* if it has a faithful simple module M such that $F = \mathrm{End}_R M$ is a field.

Definition 3.2.40:   A *central embedding* is an injection $R \to R'$ with respect to which R' is a central extension of R.

Construction 3.2.41:   A central embedding of a primitive ring R into a closed primitive ring. (This will rely heavily on centralizers.) Let M be a faithful R-module, and $D = \operatorname{End}_R M$. Taking any maximal subfield F of D let $T = \operatorname{End} M_F$. Then $R \subseteq \operatorname{End} M_D \subseteq T$ naturally. On the other hand, we can inject $F \to T$ under the action $\alpha \to \rho_\alpha$ where $\rho_\alpha : M \to M$ is given by $\rho_\alpha x = x\alpha$. Since $F \subseteq C_T(R)$ we have the subring RF of T; we claim RF is a closed primitive ring, dense in $\operatorname{End} M_F$.

Indeed, RF acts on M by $(r\alpha)x = (rx)\alpha$ for r in R, α in F, and x in M, whereby we see M is an RF-module which is simple (since any proper submodule would be an R-submodule and thus 0), and, moreover, $RF \subseteq \operatorname{End} M_F$ shows M is faithful. But $F \subseteq \operatorname{End}_{RF} M \subseteq \{d \in D : x(\alpha d - d\alpha) = 0 \text{ for all } x \text{ in } M \text{ and } \alpha \text{ in } F\} = C_D(F) = F$ by proposition 0.0.6, so $F = \operatorname{End}_{RF} F$, proving the claim.

Thus RF is the desired closed primitive ring containing R, and we would like to add one more observation.

Remark 3.2.42: $Z(RF) = F$. (Indeed $Z(RF) \subseteq C_T(R) \subseteq D$ so $Z(RF)$ is a commutative subring of D containing F, implying $Z(RF) = F$.)

Embedding sequence 3.2.43:   Embedding a semiprime ring R into a direct product of endomorphism rings of vector spaces over fields.

Step I. R can be embedded in a ring R_1 which is a homomorphic image of a direct product of copies of R, with $\operatorname{Nil}(R_1) = 0$. (Proof: theorem 2.6.27, since $N(R) = 0$.)

Step II. $R_2 = R_1[\lambda]$ is semiprimitive by Amitsur's theorem.

Step III. R_2 is a subdirect product of primitive images R_{3i}.

Step IV. Each R_{3i} can be centrally embedded in a closed primitive ring R_{4i}.

Step V. Each R_{4i} is a dense subring of the endomorphism ring of a vector space M_i over a field F_i, by construction 3.2.41.

The nice feature of this embedding is that at each step I through IV we only took direct products, homomorphic images, and central embeddings, so sentences preserved under these operations would pass from R to its overring. This will be important when we study polynomial identities. In case R is prime there is a further embedding which we can apply.

Theorem 3.2.44: (*Amitsur* [67]) *If R is a prime ring contained in a semiprimitive ring R' then R can be embedded into a primitive ring R" that satisfies all sentences holding in each primitive homomorphic image of R'.*

Proof: Write R' as a subdirect product of $\{R_i : i \in I\}$ and introduce the following filter on I: Writing r_i for the canonical image of r in R_i under the composition $R \to R' \to R_i$, let $I_r = \{i \in I : r_i \neq 0\}$. Then $I_r \neq \varnothing$ for each $r \neq 0$. Moreover, for any nonzero a, b in R we have $arb \neq 0$ for some r in R, so $I_a \cap I_b \supseteq I_{arb}$. Thus $\{I_r : r \neq 0\}$ are the base of a filter $\mathscr{F}_0$. Let $\mathscr{F}$ be any ultrafilter containing $\mathscr{F}_0$; let $R"$ be the ultraproduct $(\prod_{i \in I} R_i)/\mathscr{F}$. Then $R"$ satisfies any elementary sentence satisfied by all the R_i. Moreover, since being a faithful simple module is a property which can be described elementarily, we see $M" = (\prod_{i \in I} M_i)/\mathscr{F}$ is a faithful simple $R"$-module. Thus $R"$ is primitive, as desired.   Q.E.D.

Remark 3.2.45: Amitsur' proof, based on a proof by M. Rabin of an earlier theorem of A. Robinson, actually yields more: If $\mathscr{C}$ is a class of rings defined by elementary sentences then any prime ring which is a subring of a direct product of rings from $\mathscr{C}$, can be embedded in a ring from $\mathscr{C}$.

Corollary 3.2.46: *The first three steps of 3.2.43 combined with theorem 3.2.44 provides an embedding of a prime ring R into a primitive ring R".*

Of course, there is a more straightforward embedding—let $S = Z(R) - \{0\}$ for any prime ring R. Then $R' = S^{-1}R$ is an algebra over the field $F = S^{-1}Z(R)$ so $R \subseteq R' \subseteq \operatorname{End} R'_F$ via the regular representation. However, one has lost all hold on the elementary sentences of R in this procedure.

L Supplement: Embeddings into Matrix Rings

It is useful to know when a ring can be embedded into matrices over a commutative ring. We start with an interesting counterexample of Bergman. We say R is quasi-local if $R/\operatorname{Jac}(R)$ is simple Artinian.

Remark 3.2.47: If e is an idempotent of a quasi-local ring R then $ze \neq 0$ for all $z \neq 0$ in $Z(R)$. (Indeed, $\operatorname{Ann} z \lhd R$ so $\operatorname{Ann} z \subseteq \operatorname{Jac}(R)$ which has no trivial idempotents.)

Example 3.2.48: (Bergman) A finite ring R which *cannot* be embedded in $M_n(C)$ for any n in $\mathbb{N}$ and any commutative ring C. Let $G = \mathbb{Z}/p\mathbb{Z} \oplus \mathbb{Z}/p^2\mathbb{Z}$, a group of order p^3, and let $R = \mathrm{Hom}_{\mathbb{Z}}(G, G,)$. Let π_i be the projection of G to the i-th component for $i = 1, 2$; then π_1 and π_2 are idempotent with $\pi_1 + \pi_2 = 1$, so $R = \pi_1 R \pi_1 \oplus \pi_1 R \pi_2 \oplus \pi_2 R \pi_1 \oplus \pi_2 R \pi_2$ by the Peirce decomposition. It is easy to see each of these components is a cyclic abelian group with respective generators $\pi_1, \pi_{12}, \pi_{21}$, and π_2, where $\pi_{12} : \mathbb{Z}/p^2\mathbb{Z} \to \mathbb{Z}/p\mathbb{Z}$ is given by $\pi_{12}(1 + p^2\mathbb{Z}) = 1 + p\mathbb{Z}$, and $\pi_{21} : \mathbb{Z}/p\mathbb{Z} \to \mathbb{Z}/p^2\mathbb{Z}$ is given by $\pi_{21}(1 + p\mathbb{Z}) = p + p^2\mathbb{Z}$; these generators have respective order p, p, p, and p^2, so $|R| = p^5$.

Suppose we had $R \subseteq M_n(C)$. Viewing C as scalar matrices, localize at some maximal ideal $P \subseteq \mathrm{Ann}_C(p\pi_2)$. Then $1^{-1}p\pi_2 \neq 0$ in the quasi-local ring $M_n(C_P)$. In particular, $1^{-1}p \neq 0$ and $1^{-1}\pi_1 \neq 0$ since $p\pi_2 = \pi_{21}\pi_1\pi_{12}$. But $(1^{-1}p)(1^{-1}\pi_1) = 1^{-1}p\pi_1 = 0$, contrary to remark 3.2.47.

L Supplement: Embedding into Division Rings

Let us now turn to the famous problem of embeddings into division rings and, more generally, into left Artinian rings. The question of whether every domain can be embedded in a division ring arose in van der Waerden [49B]. Remark 3.2.45 shows any domain which is a subring of a direct product of division rings is in fact embeddible into a division ring. However, Mal'cev [37] came up with the following counterexample:

Example 3.2.49: A domain which cannot be embedded in a division ring. Let $R = \mathbb{Q}\{X_1, \ldots, X_4, Y_1, \ldots, Y_4\}/A$ where the ideal A is generated by $X_1Y_1 + X_2Y_3, X_3Y_1 + X_4Y_3$, and $X_3Y_2 + X_4Y_4$. Let x_i, y_i denote the respective images of X_i, Y_i in R. To check R is a domain, suppose $f(x, y)g(x, y) = 0$ in R. We may assume f, g are homogeneous in total degree and that each monomial of f ends in suitable x_i and each monomial of g begins in suitable y_j; then checking the various cases shows that not all the products of monomials cancel, contradiction.

On the other hand, the relations from A imply

$$\begin{pmatrix} x_1 & x_2 \\ x_3 & x_4 \end{pmatrix}\begin{pmatrix} y_1 & y_2 \\ y_3 & y_4 \end{pmatrix} = \begin{pmatrix} 0 & a \\ 0 & 0 \end{pmatrix} \qquad \text{where } a = x_1 y_2 + x_2 y_4. \qquad (1)$$

If R were embeddible in a division ring D then multiplying both sides on the left by

$$\begin{pmatrix} 1 & -x_1 x_3^{-1} \\ 0 & 1 \end{pmatrix} \quad \text{and on the right by} \quad \begin{pmatrix} 1 & -y_3^{-1} y_4 \\ 0 & 1 \end{pmatrix} \quad \text{yields}$$

$$\begin{pmatrix} 0 & x'_2 \\ x_3 & x_4 \end{pmatrix} \begin{pmatrix} y_1 & y'_2 \\ y_3 & 0 \end{pmatrix} = \begin{pmatrix} 0 & a \\ 0 & 0 \end{pmatrix} \text{ in } M_2(D)$$

for suitable x'_2, y'_2 in D, and this is impossible.

It turns out that the multiplicative monoid $R - \{0\}$ cannot be embedded in a group, so Mal'cev refined the question as follows: If $R - \{0\}$ is embeddible in a group then is R embeddible in a division ring? Counterexamples were found in 1966 independently by Bokut [69], Bowtell [67], and Klein [67]. Bokut's example was a monoid algebra, and he based his proof on a theory he had developed concerning embeddibility of monoid algebras.

L Supplement: Embedding into Artinian Rings

The next question one might ask is, "Which rings are embeddible into simple Artinian rings?" Necessary conditions certainly include "weakly finite," nilpotence of nil subrings, by theorems 3.2.37 and 3.2.38, and bounded length on chains of left (or right) annihilators.

Schofield [85B] found a lovely characterization of these rings in terms of "rank functions" of maps and modules, which we describe here very briefly. Suppose $\psi: R \to T$ is a ring homomorphism from R to a simple Artinian ring $T = M_n(D)$. Viewing T as $T - R$ bimodule by means of ψ, we can define a *rank function* $\rho: R\text{-}\mathcal{F}imod \to \mathbb{Z}$ by $\rho M = t$ where $T \otimes_R M$ has length t in $T\text{-}\mathcal{M}od$, i.e., $T \otimes_R M \approx N^{(t)}$ where N is the simple T-module. Then ρ satisfies the following properties:

(i) $\rho R = n$.
(ii) $\rho(M_1 \oplus M_2) = \rho M_1 + \rho M_2$.
(iii) If $M'' \to M \to M' \to 0$ is exact then $\rho M' \leq \rho M \leq \rho M'' + \rho M'$ (since the functor $T \otimes \_$ is right exact).

(Actually Schofield "normalizes" ρ by dividing through by ρR.) Note that if T is a division ring then $n = 1$ so $\rho R = 1$, and Cohn's theory can be modified to show any rank function ρ satisfying $\rho R = 1$ arises from a homomorphism from R to a division ring. More generally, Schofied shows (for R an algebra over a field F) that any rank function ρ with $\rho R = n$ gives rise to a homomorphism $R \to M_n(D)$ for a suitable division ring D. He achieves this by means of the free product $R \coprod M_n(F)$ which by proposition 1.1.3 is isomorphic to $M_n(R')$ for some R'. Indeed, the rank function ρ lifts to a rank function on $M_n(R')$-$\mathcal{M}od$ and thus on R-$\mathcal{M}od$ by theorem 1.1.17, whose proof, in fact, shows $\rho R' = n/n = 1$. Hence there is a homomorphism from R' to a division ring D, yielding the composition

$$R \to R \coprod M_n(D) \approx M_n(R') \to M_n(D').$$

Examples:

(i) If R is prime left Noetherian then its ring of fractions $Q(R)$ is simple Artinian, and the ensuing rank function ρ is called the *reduced rank*, one of the key tools of Noetherian theory in §3.5.

(ii) If R is left Noetherian then any P in Spec(R) gives rise to the composition $R \to R/P \to Q(R/P)$; the ensuing rank function is essentially Stafford's important $\hat{g}$ function defined before remark 3.5.64 (unabridged edition).

(iii) If R is left Artinian then the composition length of an f.g. module defines a rank function; it follows that any left Artinian algebra over a field is embeddible into a simple Artinian algebra. (For R semiprime this is an easy exercise.) Note that the ring of example 3.2.48 cannot be embedded in a simple Artinian ring because there is an element annihilated by p^2 but not by p. Thus the condition of being an algebra over a field is necessary in Schofield's theorem. Westreich [87] has generalized the Cohn-Malcolmson construction.

On the other hand, Small constructed a left Noetherian ring failing to have minimal left annihilators and thus not embeddible in a left (or right) Artinian ring, cf., exercise 40. Such a construction is impossible for left and right Noetherian rings, but recently Dean-Stafford [86] have shown a certain image of the universal enveloping algebra of the Lie algebra SL(2, $\mathbb{C}$) is not embeddible into an Artinian ring, and Dean conjectures that every enveloping algebra of a finite dimensional semisimple Lie algebra has a nonembeddible image.

§3.3 Localization of Nonsingular Rings and Their Modules

In this section we shall reexamine classical localization in a more idealtheoretic manner, which will permit us to bypass some of the computations while at the same time preparing the ground for a new construction, Johnson's ring of quotients of a nonsingular ring. This ring has several very nice features, being self-injective, (von Neumann) regular, and quickly yields a proof of Goldie's theorem (exercise 14). Also the ideals are totally ordered by Goodearl's theorem (exercise 41; also see exercises 51, 52, 53). This is rather impressive, since most of the rings we shall study are nonsingular. Furthermore, the obstruction to being nonsingular is an ideal, called the *singular ideal.* Nonetheless, the Johnson-Goodearl theory has been difficult to apply, because a primitive, self-injective, regular ring whose ideals are totally ordered still is a far cry from semisimple Artinian! For example, it need not be weakly finite, cf., example 1.3.33.

The philosophy of this section is to use large left ideals in place of regular elements (cf., theorem 3.2.14 (i) ⇔ (ii)), and it will be useful to deal more generally with large submodules. Accordingly, we shall be extending some of the results of §2.4. Given R-modules $N \leq M$ and given $x \in M$, define $Nx^{-1} = \{r \in R : rx \in N\}$.

The tools of this section lead naturally to techniques of "hard" module theory—closed submodules, self-injective rings, quasi-Frobenius rings; these are treated in the exercises.

Proposition 3.3.0: *Suppose N is a large submodule of M. (i) Nx^{-1} is a large left ideal of R for each x in M. (ii) If $f: M' \to M$ is any map then $f^{-1}N$ is a large submodule of M'.*

Proof: Clearly (ii) implies (i), taking $f: R \to M$ to be right multiplication by x. To prove (ii), suppose $N' < M'$ with $N' \cap f^{-1}N = 0$. Then $fN' \cap N \leq f(N' \cap f^{-1}N) = 0$ so $fN' = 0$; hence $N' \leq f^{-1}N$ so $N' = N' \cap f^{-1}N = 0$.
$$\text{Q.E.D.}$$

Johnson's Ring of Quotients of a Nonsingular Ring

Construction 3.3.1:   Let $\mathscr{F} = \{\text{large left ideals}\}$, viewed as a sublattice of the dual lattice of $\mathscr{L}(R)$, i.e., $L_1 \leq L_2$ iff $L_1 \supseteq L_2$. Given M in $R\text{-}\mathscr{M}od$ define $M_i = \text{Hom}_R(L_i, M)$ for each L_i in $\mathscr{F}$; for $L_i \supseteq L_j$ define $\varphi_i^j: M_i \to M_j$ by sending a map $f: L_i \to M$ to its restriction $f: L_j \to M$. Let $\hat{M}$ denote $\varinjlim(M_i; \varphi_i^j)$ together with the canonical maps $\mu_i^j: M_i \to \hat{M}$.

Remark 3.3.2:   In view of theorem 1.8.7 we can describe $\hat{M}$ explicitly as an R-module, as follows. Let $\mathscr{S} = \{\text{pairs } (L, f) \text{ where } L \in \mathscr{F} \text{ and } f: L \to R \text{ is an } R\text{-module map}\}$. Define the equivalence $\sim$ on $\mathscr{S}$ by saying $(L_1, f_1) \sim (L_2, f_2)$ if f_1 and f_2 agree on some L in $\mathscr{F}$ contained in $L_1 \cap L_2$, i.e., their restrictions to L are the same. Then $\hat{M}$ is the set of equivalence classes $\mathscr{S}/\sim$, made into an R-module under the following operations, writing $[L, f]$ for the equivalence class of (L, f):

$$[L_1, f_1] + [L_2, f_2] = [L_1 \cap L_2, f_1 + f_2].$$

$$r[L_1, f_1] = [L_1 r^{-1}, rf_1] \qquad \text{where the map } rf_1 \text{ is given by}$$
$$(rf_1)a = f_1(ar).$$

(This is reminiscent of remark 1.5.18′.) The verifications of these assertions are an easy consequence of the following observation:

Remark 3.3.3: $\mathscr{F} = \{\text{large left ideals of } R\}$ is a filter on $\mathscr{L}(R)$ satisfying the following extra properties (by proposition 3.3.0):

  (i) If $L \in \mathscr{F}$ and $r \in R$ then $Lr^{-1} \in \mathscr{F}$.
  (ii) More generally, if $L_1, L_2 \in \mathscr{F}$ and $f: L_1 \to R$ is a map then $f^{-1}L_2 \in \mathscr{F}$.

Before using remark 3.3.2 we should like to relate M to $\hat{M}$ given in the next result.

Remark 3.3.4: Modifying the proof of proposition 1.5.19 we can define a canonical map $\varphi_M: M \to \hat{M}$ by the rule $\varphi_M x = [R, \rho_x]$ where $\rho_x: R \to M$ is right multiplication by x. Then $\ker \varphi_M = \{x \in M: \text{Ann}_R x \in \mathscr{F}\}$. (Indeed, $[R, \rho_x] = 0$ iff ρ_x is 0 on a large left ideal L i.e., $Lx = 0$, iff $\text{Ann } x$ is large.)

This leads us to the following fundamental definition.

Definition 3.3.5: The *singular submodule* $\text{Sing}(M) = \{x \in M: \text{Ann}_R x \text{ is large}\}$. (Thus $\text{Sing}(M) = \ker \varphi_M$). M is *singular* if $\text{Sing}(M) = M$; M is *nonsingular* if $\text{Sing}(M) = 0$.

This definition should be compared with a related notion. We say a module M is *torsion* if $\text{Ann}_R x$ contains a regular element of R for each x in M; M is *torsion-free* if $\text{Ann}_R x$ does *not* contain a regular element of R for each x in M. For R semiprime Goldie we see by Goldie's theorem that $\text{Ann}_R x$ contains a regular element iff $x \in \text{Sing}(M)$; thus we see for such R that torsion = singular and torsion–free = nonsingular. *A* deeper explanation can be found via torsion theory, which is described in §3.4.

Remark 3.3.6: $f(\text{Sing}(M)) \leq \text{Sing}(N)$ for any map $f: M \to N$ (for if $Lx = 0$ then $Lfx = f(Lx) = 0$). In particular, if $M \leq N$ then $\text{Sing}(M) \leq \text{Sing}(N)$.

Let us now consider the important case where $M = R$.

Theorem 3.3.7: $\hat{R}$ *is a ring, under the multiplication*

$$[L_1, f_1][L_2, f_2] = [f_1^{-1}L_2, f_2 f_1].$$

The canonical map $\varphi_R: R \to \hat{R}$ is then a ring homomorphism whose kernel is $\text{Sing}(R)$, i.e., we have a natural ring injection $R/\text{Sing}(R) \to \hat{R}$.

Proof: Note that $f_1^{-1}L_2 \in \mathscr{F}$ by remark 3.3.3(ii). Since $\mathscr{F}$ is closed under finite intersections we can verify the ring properties of $\hat{R}$ on a large left ideal contained in all the domains of the f_i. Since φ_R was already seen to be a module map it remains to verify $\varphi_R(ab) = \varphi_R a \varphi_R b$ for all a, b in R, i.e., that $[R, \rho_a][R, \rho_b] = [R \rho_{ab}]$. By definition

$[R, \rho_a][R, \rho_b] = [\rho_a^{-1}R, \rho_b\rho_a] = [R, \rho_b\rho_a] = [R, \rho_{ab}]$ as desired.   Q.E.D.

We call $\hat{R}$ the *Johnson ring of quotients.*

Corollary 3.3.8: $\text{Sing}(R) \lhd R.$ (*This can also be proved directly using remark 3.3.3.*)

In view of this theorem we want to obtain many examples of nonsingular rings.

Lemma 3.3.9:   *If R satisfies ACC(Ann) then $\text{Sing}(R)$ is nil.*

Proof: Given $r \in \text{Sing}(R)$ take $n \in \mathbb{N}$ with $\text{Ann}(r^n)$ maximal. Let $a = r^n \in \text{Sing}(R)$. Then $\text{Ann}\, a = \text{Ann}\, a^2$ so remark 3.2.2 (ix) shows $Ra \cap \text{Ann}\, a = 0$, implying $Ra = 0$, i.e., $r^n = 0$.   Q.E.D.

In fact, under this hypothesis $\text{Sing}(R)$ is nilpotent, cf., exercise 6. However, a nilpotent ideal need *not* be in $\text{Sing}(R)$, as we shall see.

Example 3.3.10:

(i) Any semiprime ring with ACC(Ann) is nonsingular by lemma 3.3.9 coupled with proposition 2.6.24. This enables us to deduce Goldie's theorem from these results (cf., exercises 12–14).

(ii) A commutative ring C is semiprime iff $\text{Sing}(C) = 0$. (Proof: ($\Rightarrow$) If $0 \neq c \in C$ then $(Cc \cap \text{Ann}\, c)^2 \subseteq (\text{Ann}\, c)Cc = 0$ so $Cc \cap \text{Ann}\, c = 0$. $\text{Ann}\, c$ is *not* large. Conversely, if $c^2 = 0$ and I is an essential complement of Cc then $Ic \subseteq I \cap Cc = 0$ so $\text{Ann}\, c \supseteq I + Cc$ is large.)

Example 3.3.11:   The semiprimary ring $R = \begin{pmatrix} F & F \\ 0 & F \end{pmatrix}$ is nonsingular, for any field F. (Proof: Lemma 3.3.9 shows $\text{Sing}(R) \subseteq \text{Jac}(R) = Fe_{12}$, but $\text{Ann}\, e_{12} = Fe_{12} + Fe_{22}$ which is not large, so $\text{Sing}(R) = 0$.)

The Injective Hull of a Nonsingular Module

We shall now describe $\hat{M}$ and $\hat{R}$ much more concisely, using the injective hull $E(M)$.

Remark 3.3.12:   If $N \leq M$ is large then M/N is a singular module. Conversely, if M is nonsingular and M/N is singular then N is large in M. (Proof: ($\Rightarrow$) For any x in M proposition 3.3.0 shows Nx^{-1} is a large left ideal; but $(Nx^{-1})x \leq N$ so $x + N \in \text{Sing}(M/N)$. ($\Leftarrow$) For any x in M we have $Lx \leq N$ for a large left ideal L, so $0 \neq Lx \subseteq N \cap Rx$.)

Remark 3.3.13: If M is nonsingular and $h: K \rightarrow M$ has $\ker h$ large in K then $h = 0$. (Indeed, $K/\ker h$ is a singular module isomorphic to hK, so $hK \leq \text{Sing}(M) = 0$.

Lemma 3.3.14: *Suppose M is nonsingular.* (i) *The injective hull $E(M)$ is also nonsingular.* (ii) *If $h: K \rightarrow E(M)$ with $\ker h$ large in K then $h = 0$.*

Proof:

(i) $M \cap \text{Sing } E(M) \subseteq \text{Sing}(M) = 0$ so $\text{Sing } E(M) = 0$, since M is large in $E(M)$.

(ii) Apply remark 3.3.13 to $E(M)$.   Q.E.D.

Proposition 3.3.15: *If M is a nonsingular module then $\hat{M} \approx E(M)$ canonically.*

Proof: Let us define the isomorphism $\varphi: E(M) \rightarrow \hat{M}$. Given x in $E(M)$ we have $L = Rx^{-1}$ large in R by proposition 3.3.0, so define $\varphi x = [L, \rho_x]$ where ρ_x is right multiplication by x. Surely φ is a map and is epic since any $f: L \rightarrow M$ (with L large in R) can be viewed as a map $f: L \rightarrow E(M)$ which, by remark 2.10.3 is given by right multiplication by some x in $E(M)$. Finally φ is monic by lemma 3.3.14.   Q.E.D.

Theorem 3.3.16: $\hat{R} \approx \text{End}_R E(R)$ *as rings, for any nonsingular ring R.*

Proof: First note as modules that there is an isomorphism $\psi: \text{End}_R E(R) \rightarrow \text{Hom}_R(R, E(R))$ which sends a map $f: E(R) \rightarrow E(R)$ to its restriction $f: R \rightarrow E(R)$; indeed ψ is epic since $E(R)$ is injective, and ψ is monic by lemma 3.3.14(ii). Thus we have a composite module isomorphism

$$\text{End}_R E(R) \rightarrow \text{Hom}_R(R, E(R)) \approx E(R) \rightarrow \hat{R},$$

by proposition 3.3.15. But this composite preserves the ring multiplication, as one sees instantly by comparing definition 1.5.6 to theorem 3.3.7.   Q.E.D.

Carrying Structure of R to $\hat{R}$

Having constructed $\hat{R}$ twice for R nonsingular, we shall study how to transfer structure from R to $\hat{R}$ and then shall examine some of the special properties of $\hat{R}$. First note by proposition 3.3.15 that $\hat{R}$ is an *essential ring extension* of R, in the sense that $\hat{R}$ is a ring containing R which is essential as R-module.

Proposition 3.3.17: *Suppose Q is any essential ring extension of R. If R is prime (resp. semiprime) then Q is prime (resp. semiprime).*

Proof: Suppose $A, B \lhd Q$ with $AB = 0$. Then $(A \cap R)(B \cap R) = 0$ so $A \cap R$ or $B \cap R = 0$, implying $A = 0$ or $B = 0$.   Q.E.D.

Further information requires a more careful analysis.

Proposition 3.3.18: *Suppose $M, N \in \hat{R}\text{-}\mathcal{M}od$ and $\mathrm{Sing}(N) = 0$. (i) Any R-module map $f: M \to N$ is also an $\hat{R}$-module map. (ii) If N is a summand of M in $R\text{-}\mathcal{M}od$ then N is a summand of M in $\hat{R}\text{-}\mathcal{M}od$.*

Proof:

(i) Fix x in M and define $h: \hat{R} \to N$ by $hq = f(qx) - qfx$. Then h is an R-module map with $hR = 0$, so $h = 0$ by remark 3.3.13.

(ii) The projection $\pi: M \to N$ is an $\hat{R}$-module map, by (i), so $M \approx N \oplus \ker \pi$ in $\hat{R}\text{-}\mathcal{M}od$.   Q.E.D.

Proposition 3.3.19: *Suppose $\mathrm{Sing}(R) = 0$. (i) A left ideal L of $\hat{R}$ is large iff $L \cap R$ is large in R; (ii) $\hat{R}$ is a nonsingular ring.*

Proof:

(i) ($\Rightarrow$) by proposition 3.3.0 (where $f: R \to \hat{R}$ is the inclusion map). ($\Leftarrow$) $\hat{R}$ is an essential extension of R and thus of $L \cap R$ in $R\text{-}\mathcal{M}od$, so $\hat{R}$ is an essential extension of L in $R\text{-}\mathcal{M}od$ and *a fortiori* in $\hat{R}\text{-}\mathcal{M}od$.

(ii) Immediate from (i).   Q.E.D.

Proposition 3.3.20: *If M is a nonsingular R-module then $\hat{M}$ is a nonsingular injective $\hat{R}$-module under the action*

$$[L_1, f][L_2, h] = [f^{-1}L_2, hf]$$

$$\text{for } f: L_1 \to R \quad \text{and} \quad h: L_2 \to M \text{ where } L_1, L_2 \in \mathcal{F}.$$

Proof: Note $f^{-1}L_2 \in \mathcal{F}$ by remark 3.3.3(ii). The argument of theorem 3.3.7 shows $\hat{M}$ is an $\hat{R}$-module. To see $\mathrm{Sing}(\hat{M}) = 0$ suppose $h: L \to M$ has $\mathrm{Ann}_{\hat{R}}[L, h]$ large in $\hat{R}$. Then $L' = \mathrm{Ann}_R[L, h]$ is large in R by proposition 3.3.19(i). For any a in $L' \cap L$ we have some large left ideal $L'' < La^{-1}$ satisfying $0 = h(L''a) = L''ha$; thus $ha \in \mathrm{Sing}(M) = 0$. This proves $h(L' \cap L) = 0$ so $[L, h] = 0$ by definition.

It remains to show $\hat{M}$ is injective as $\hat{R}$-module. Otherwise, $\hat{M}$ has a proper essential extension $\hat{N}$. But $\hat{M}$ is a summand of $\hat{N}$ as R-modules (since

$\hat{M} \approx E(M))$ and thus as $\hat{R}$-modules by proposition 3.3.18(ii), contradiction.

Q.E.D.

Let us push this result a bit further.

Theorem 3.3.21: *Suppose* $\text{Sing}(R) = 0$. *There is an exact functor* $F_J: R\text{-}\mathcal{M}od$ $\to \hat{R}\text{-}\mathcal{M}od$ *given by* $F_J M = \hat{N}$, *where* $N = M/\text{Sing}(M)$ *is nonsingular. Thus* F_J *sends R-modules to nonsingular injective* $\hat{R}$-*modules.*

Proof: Write $M_0 = \text{Sing}(M)$. First we want to show $N = M/M_0$ is non-singular. Write $\text{Sing}(N) = M'/M_0$. For any x in M' we take an essential complement L of $\text{Ann}_R x$ in R. Right multiplication by x provides an isomorphism $L \to Lx$, so $\text{Sing}(Lx) \approx \text{Sing}(L) = 0$. Thus $Lx \cap M_0 = 0$. But then

$$Lx \approx Lx/(Lx \cap M_0) \approx (Lx + M_0)/M_0 \subseteq \text{Sing}(N)$$

is singular so $Lx = 0$, implying $L = 0$. Hence $\text{Ann } x$ is large so $x \in M_0$, proving $\text{Sing}(N) = M_0/M_0 = 0$.

Now $\hat{N}$ is nonsingular and injective in $\hat{R}\text{-}\mathcal{M}od$ by proposition 3.3.20, so we have F_J acting on R-modules. Recalling $\hat{N} \approx E(N)$ we shall now define F_J acting on a map $f: M \to M'$. Indeed, the composition

$$M \xrightarrow{f} M' \to M'/\text{Sing}(M') = N'$$

has kernel containing $\text{Sing}(M)$ since $f(\text{Sing}(M)) \subseteq \text{Sing}(M')$, so we have an induced map $\bar{f}: N \to N'$. Now $\bar{f}$ composed with the injection $N' \to E(N')$ gives a map $N \to E(N')$ which (since $E(N')$ is injective) extends to a map $\hat{f}: E(N) \to E(N')$; $\hat{f}$ is unique in view of lemma 3.3.14(ii). (Indeed, if $\hat{f}_1$ and $\hat{f}_2$ both extend f then $\hat{f}_1 - \hat{f}_2$ has kernel containing N and is thus 0). We denote $\hat{f}$ as $F_J f$.

Given $f: M \to M'$ and $f': M' \to M''$ the argument of the last paragraph shows $F_J(f'f) = F_J f' F_J f$ (since they agree on N), proving F_J is a functor.

To prove F_J is exact consider the short exact sequence $0 \to M \xrightarrow{f} M' \xrightarrow{f'} M'' \to 0$, i.e., f is monic and f' is epic and $fM = \ker f'$. Putting $g = F_J f$ and $g' = F_J f'$ we want to show g is monic, g' is epic, and $g(E(N)) = \ker g'$ where (as above) $N = M/\text{Sing}(M)$ and $N' = M'/\text{Sing}(M')$ and $N'' = M''/\text{Sing}(M'')$. Write $\bar{f}: N \to N'$ and $\bar{f}': N' \to N''$ for the maps induced from f and f', respectively, and for ease of notation view $M \leq M'$ via f.

First note $\text{Sing}(M) = M \cap \text{Sing}(M')$ so $\bar{f}$ is also monic and $0 = \ker \bar{f} = N \cap \ker g$, implying $\ker g = 0$ and g is monic. In particular, $g(E(N))$ is the in-jective hull of gN and so is a maximal essential extension. On the other hand, $g'g = F_J(f'f) = F_J(0) = 0$ so $g(E(N)) \subseteq \ker g'$; to prove equality it suffices to show $\ker g'$ is essential over $gN = \bar{f}N$. Since $\ker g'$ is essential over $N' \cap \ker g' = \ker \bar{f}'$ we want $\ker \bar{f}'$ to be essential over $\bar{f}N$. But $\ker \bar{f}' = f'^{-1}(\text{Sing } M'')/\text{Sing } M'$ so $(\ker \bar{f}')/\bar{f}N$ is a homomorphic image of

$$f'^{-1}(\text{Sing } M'')/fM \approx f'^{-1}(\text{Sing } M'')/\ker f' \approx \text{Sing } M'',$$

a singular module. Since $\ker \bar{f}' \le N'$ is nonsingular we conclude by remark 3.3.12 that $\bar{f}N$ is large in $\ker \bar{f}'$ proving $g(E(N)) = \ker g'$.

It remains to show g' is epic. Note $g'N' = \bar{f}'N' = N''$ (since f' is epic), which is large in $E(N'')$, so it suffices to show $g'(E(N'))$ is injective. But the sequence $0 \to \ker g' \to E(N') \to g'(E(N')) \to 0$ splits since we saw above $\ker g' = g(E(N)) \approx E(N)$ is injective. Hence $g'(E(N'))$ is a summand of $E(N')$ and is injective, as desired.   Q.E.D.

The real reason that $\hat{R}$ is useful is that its structure is so nice, as we begin to see in the next result.

Definition 3.3.22:   A ring R is *self-injective* if R is injective as R-module.

Theorem 3.3.23:   *If R is nonsingular then $\hat{R}$ is a nonsingular, regular, self-injective ring.*

Proof: $\hat{R}$ is nonsingular and self-injective by proposition 3.3.20 (taking $M = R$).To prove $\hat{R}$ is regular suppose $[L, f] \in \hat{R}$, i.e., $f: L \to R$ is a map with L a large left ideal. Take an essential complement N of $\ker f$ in L; then $f: N \to fN$ is an isomorphism. Letting N' be an essential complement of fN in R we define $g: fN + N' \to R$ by $g(fx + x') = x$. For any x in N and y in $\ker f$ we have

$$fgf(x + y) = fgfx = f(gfx) = fx = f(x + y),$$

implying $(fgf - f)(N + \ker f) = 0$. Thus

$$[L, f][fN + N', g][L, f] = [L, f]$$

proving R is regular.   Q.E.D.

§3.4 Noncommutative Localization

As beautiful as Goodearl's results of the last section are, we are left with the problem that rings need not be nonsingular. There is a way of circumventing this problem by using "dense" submodules instead of "large submodules," cf., definition 3.4.1. However, a construction of Martindale dealing with two-sided ideals instead of left ideals has proved more useful for prime rings and has several important consequences for the prime spectrum (cf., theorem 3.4.13ff).

All of these constructions can be gathered under a single umbrella construction, due to Gabriel [62], which we describe briefly at the end. Gabriel's theory has developed into an entire subject, discussed in depth in Golan [86B].

Digression: The Maximal Ring of Quotients

Our first two constructions do away with the singular ideal. The first construction, due to Findlay-Lambek and Utumi, works for any ring and is called the *maximal ring of quotients.* Recall that in the construction of $\hat{R}$ we had $[L, f] = 0$ iff ker f is large. In this case f induces a map $\bar{f}: L/\ker f \to R$, so in order to avoid this contingency we introduce the following definition.

Definition 3.4.1: N is a *dense* R-submodule of M if $\operatorname{Hom}_R(M'/N, M) = 0$ for all $N \leq M' \leq M$. (In other words, if $f: M' \to M$ is a map and $N \leq \ker f$ then $f = 0$.)

Remark 3.4.2: Every dense submodule N of M is large. (Indeed if $K \cap N = 0$ then the projection $\pi: K + N \to K$ has kernel N, so $\pi = 0$ implying $K = 0$.) On the other hand, a large submodule need not be dense (exercise 1). However, a large submodule of a *nonsingular* module is dense, by remark 3.3.13.

Example 3.4.3: If $M = \mathbb{Q}$ then every $\mathbb{Z}$-submodule N of M is dense, because M/N is torsion and M is torsion-free. For this reason, we say M is a *rational extension* of N when N is dense in M.

Proposition 3.4.4: $\mathscr{F}_D = \{$*dense submodules of* $M\}$ *is a filter; moreover, if* $L < R$ *is dense and* $r \in R$ *then* Lr^{-1} *is also a dense left ideal.*

Proof: If $N < M$ is dense and $N \leq N' \leq M$ then N' is dense in M, for if $N' \leq M' \leq M$ and $f: M' \to M$ with $fN' = 0$ then $fN = 0$ so $f = 0$. To show $\mathscr{F}_D$ is a filter it remains to show $N \cap N'$ is dense if N and N' are dense (in M). Suppose $N \cap N' \leq M' \leq M$ and $f: M' \to M$ with $f(N \cap N') = 0$. Then we have an induced map $\bar{f}: (N \cap M')/(N \cap N') \to M$ which we combine with the isomorphism $((N \cap M') + N')/N' \approx N \cap M'/N \cap N'$ to produce a map $((N \cap M') + N')/N' \to M$ which must be 0 by hypothesis. Hence $f(N \cap M') = 0$. But now we get an induced map $(N + M')/N \approx M'/(N \cap M') \to M$ which is 0 so $fM' = 0$, i.e., $f = 0$.

To prove the second assertion suppose $Lr^{-1} \leq L' \leq R$ and $f: L' \to R$ is a map with $f(Lr^{-1}) = 0$. Then we have a map $g: (L'r + L) \to R$ given by $g(ar + b) = fa$ for a in L' and b in L; g is well-defined since if $ar + b = 0$ then $ar = -b \in L$ so $a \in Lr^{-1}$ and $fa = 0$. But $gL = 0$ so $g = 0$ (since L is dense); hence $f = 0$. (This entire argument is generalized in exercise 3.)

$$\text{Q.E.D.}$$

Armed with this result we form the following ring of quotients.

Construction 3.4.5: Let $\mathscr{F} = \{\text{dense left ideals}\}$, and let $\mathscr{S} = \{\text{pairs } (L, f)$ where $L \in \mathscr{F}$ and $f: L \to R$ is an R-module map$\}$. Define the equivalence $\sim$ on $\mathscr{S}$ by $(L_1, f_1) \sim (L_2, f_2)$ if f_1 and f_2 agree on $L_1 \cap L_2$, and form $Q_{\max}(M)$ to be the set of equivalence classes $\mathscr{S}/\sim$, made into an R-module under the following operations (where $[L, f]$ denotes the equivalence class of (L, f)):

$$[L_1, f_1] + [L_2, f_2] = [L_1 \cap L_2, f_1 + f_2].$$

$$r[L_1, f_1] = [L_1 r^{-1}, rf_1] \qquad \text{where } rf_1 \text{ is given by } (rf_1)a = f_1(ar).$$

(Note that this is analogous to remark 3.3.2, since here f_1 and f_2 agree on a dense submodule L of $L_1 \cap L_2$ iff $(f_1 - f_2)L = 0$, iff $f_1 - f_2 = 0$ on $L_1 \cap L_2$. Hence this construction could be viewed as a direct limit as in construction 3.3.1.)

Theorem 3.4.6: $Q_{\max}(R)$ is a ring, under multiplication

$$[L_1, f_1][L_2, f_2] = [f_1^{-1}L_2, f_2 f_1].$$

The canonical map $\varphi_R: R \to Q_{\max}(R)$ given by $\varphi_R r = [R, \rho_r]$ (where ρ_r is right multiplication by r) is a ring injection.

Proof: Analogous to theorem 3.3.7; clearly $r \in \ker \varphi_R$ iff $[R, \rho_r] = [0, \rho_r]$ iff $r = 1r = 0r = 0$.   Q.E.D.

Key results about dense submodules are given in exercise 1–9. Note $Q_{\max}$ generalizes Johnson's ring of quotients, in view of remark 3.4.2.

The Martindale-Amitsur Ring of Quotients

As pleasing as the maximal ring of quotients is, there is the severe drawback that in practice it is hard to identify the dense left ideals of an arbitrary ring. A more successful alternative (for applications) was discovered by Martindale

[69] for prime rings and later generalized by Amitsur [72a] for semiprime rings. The key observation here is that a (two-sided) ideal A of a semiprime ring is large as left ideal iff Ann $A = 0$ (by proposition 3.2.21). In particular, every ideal of a prime ring is large, for its annihilator is 0; consequently, we can inject every prime ring into a very useful ring of quotients, described as follows:

Construction and theorem 3.4.7:   *Suppose R is a semiprime ring. Let $\mathcal{F} = \{large\ 2\text{-}sided\ ideals\ of\ R\}$, and let $\mathcal{S} = \{(A, f)\ where\ A \in \mathcal{F}\ and\ f: A \rightarrow R\ is\ an R\text{-}module\ map\}$. Define the equivalence $\sim$ on $\mathcal{S}$, by $(A_1, f_1) \sim (A_2, f_2)$ iff f_1 and f_2 agree on some L in $\mathcal{F}$ contained in $L_1 \cap L_2$. Denoting the equivalence class of (A, f) by $[A, f]$ we can form the ring $Q_0(R) = \{equivalence\ classes\ of\ \mathcal{S}/\sim\}$; made into a ring under the operations*

$$[A_1, f_1] + [A_2, f_2] = [A_1 \cap A_2, f_1 + f_2] \qquad and$$

$$[A_1, f_1][A_2, f_2] = [(A_1 \cap A_2)^2, f_2 f_1].$$

There is an injection $\varphi: R \rightarrow Q_0(R)$ given by $\varphi r = [R, \rho_r]$, where ρ_r is right multiplication by r.

Proof: $Q_0(R)$ is a direct limit of abelian groups (under addition) and so is an abelian group under the designated addition. Multiplication is defined because Ann$(A_1 \cap A_2)^2 = 0$ by iteration, implying $(A_1 \cap A_2)^2 \in \mathcal{F}$, and $f_1(A_1 \cap A_2)^2 \subseteq (A_1 \cap A_2) f_1(A_1 \cap A_2) \subseteq A_1 \cap A_2$ (on which f_2 is defined); clearly multiplication is well-defined. The map $\varphi: R \rightarrow Q_0(R)$ is a ring homomorphism, as in theorem 3.3.7, and $r \in \ker \varphi$ iff $Ar = 0$ for some $A \in \mathcal{F}$, iff $r = 0$, by proposition 3.2.21.   Q.E.D.

This construction is fast and efficient, and provides a theory of localization. However, we continue the present discussion with an observation in a different direction. Define the *extended centroid* of R to be the subring $\{[A, f] \in Q_0(R)$: $\mathcal{F}: A \rightarrow R$ is an $R - R$ bimodule map$\}$.

Remark 3.4.8:   The extended centroid of R is a regular ring containing $Z(R)$ and which is contained in $Z(Q_0(R))$. (Indeed, if $f: A \rightarrow R$ is a bimodule map then for all $[A_1, f_1]$ in $Q_0(R)$ and all a, a' in $A \cap A_1$ we have

$$ff_1(aa') = f(af_1 a') = (fa)f_1 a' = f_1((af)a') = f_1 f(aa'),$$

proving $[A, f] \in Z(Q_0(R))$. Clearly, $\rho_z: R \rightarrow R$ is a bimodule map for any z in Z. The extended centroid is regular by the same type of argument as in theorem 3.3.23: Given a bimodule map $f: A \to R$ let $B = \text{Ann}(fA) \lhd R$ and define $g: B + fA \to R$ by $g(b + fa) = a$. Then

$$[A, f][B + fA, g][A, f] = [A, f]$$

by a simple calculation.   Q.E.D.

Definition 3.4.9: The *central closure* $\hat{R}$ of a semiprime ring R is the subring $R\hat{Z}$ of $Q_0(R)$, where $\hat{Z} = Z(Q_0(R))$. R is *centrally closed* if $R = R\hat{Z}$ in $Q_0(R)$. The *normal closure* of R is RN where $N = \{q \in Q_0(R): q \text{ normalizes } R\}$.

In this text we shall only deal with the central closure, but the normal closure is used in an analogous way to prove results about normalizing extensions. Note $\hat{R}$ is an essential central extension of R, since $Q_0(R)$ is essential over R. We shall see that centrally closed rings behave much like central simple algebras. Before examining how this is so, let us note that if R is simple then $\hat{Z} = Z(R)$ and $\hat{R} = R$. Indeed, if $f: A \to R$ is a nonzero bimodule homomorphism then $A = R$ and $f1 \in Z(R)$ since $rf1 = fr = f1r$; hence $[A, f]$ can be identified with $f1$.

Proposition 3.4.10: *The central closure $\hat{R}$ of R is itself centrally closed, and its extended centroid is $\hat{Z}$ which is also $Z(\hat{R})$.*

Proof: Let Z' be the extended centroid of R. We have a natural injection $\varphi: \hat{Z} \to Z'$ defined by the regular representation; equivalently $\varphi[A, f] = [\hat{R}A\hat{R}, \hat{f}]$ where $\hat{f}: \hat{R}A\hat{R} \to \hat{R}$ is given by $\hat{f}\sum_i \hat{r}_{i1} a_i \hat{r}_{i2} = \sum \hat{r}_{i1}(fa_i)\hat{r}_{i2}$. If we could show φ were onto then we could identify $\hat{Z}$ with Z' so $\hat{R} = \hat{R}\hat{Z}$ is centrally closed and $\hat{Z} \subseteq Z(\hat{R}) \subseteq Z' = \hat{Z}$, which would prove the proposition.
   Given any large $\hat{A} \lhd \hat{R}$ and an $\hat{R} - \hat{R}$ bimodule map $\hat{f}: \hat{A} \to \hat{R}$ we let $A = \hat{A} \cap \hat{f}^{-1}(R)$ and $f: A \to R$ be the restriction of f to A. Since $\hat{A}$ and $\hat{f}^{-1}R$ are large in $\hat{R}$ we see A is large in $\hat{R}$ and thus in R, so $[A, f]$ is admissible, and clearly $\varphi[A, f] = [\hat{A}, \hat{f}]$, proving φ is onto.   Q.E.D.

When R is prime its central closure $\hat{R}$ is prime by proposition 3.3.17, and $Z(\hat{R})$ is a field; furthermore, we can generalize certain important properties of the tensor product from theorem 1.7.27 and corollary 1.7.28.

Theorem 3.4.11: *Suppose R is prime and centrally closed. If $R' = RW$ where W is a $Z(R)$-subalgebra of $C_{R'}(R)$ without R-torsion then every nonzero ideal of R' contains an ideal $I_1 I_2 \neq 0$ where $I_1 \lhd R$ and $I_2 \lhd W$. Consequently $R' \approx R \otimes_{Z(R)} W$ canonically if R' is prime.*

Proof: First we claim any $0 \neq A \lhd RW$ contains an element $rw \neq 0$. Indeed, take $0 \neq a = \sum_{i=1}^{t} r_i w_i \in A$ with t minimal; then the w_i are clearly $Z(R)$-independent. We claim $t = 1$. Otherwise, define $f: Rr_1 R \to Rr_2 R$ by

$$f\left(\sum_j a_j r_1 b_j \right) = \sum a_j r_2 b_j \qquad \text{for } a_j, b_j \text{ in } R.$$

If some $\sum_j a_j r_1 b_j = 0$ then writing s_i as $\sum a_j r_i b_j$ in R we have $a' = \sum_{i=2}^{t} s_i w_i = \sum_j a_j a b_j \in A$ so a' must be 0 by minimality of t; thus each $s_i = 0$ and, in particular, f is well-defined. Now $[Rr_1 R, f] = z$ for some z in $\hat{Z}$ and

$$a = r_1 (w_1 + z w_2) + \sum_{i=3}^{t} r_i w_i,$$

contrary to the minimality of t. Hence $t = 1$, as claimed.

Now $0 \neq RrRWwW = RWrwRW \subseteq A$, proving the first assertion. Also note the claim applied to $R \otimes W$ shows any nonzero ideal of $R \otimes W$ contains some $r \otimes w \neq 0$.

Now we prove the last assertion by showing the canonical surjection $\varphi: R \otimes W \to R'$ given by $\varphi(r \otimes w) = rw$ is monic; otherwise, taking $0 \neq r \otimes w$ in $\ker \varphi$ we get $0 = W\varphi(r \otimes w)R = WrwR = rR'w$ so $r = 0$ or $w = 0$, contrary to $r \otimes w \neq 0$.   Q.E.D.

Corollary 3.4.12: *Suppose R is prime and centrally closed, and $R' = RF$ with $F = Z(R')$ a field. Then R' is prime and centrally closed.*

Proof: If $A'B' = 0$ for nonzero ideals A', B' of R' then taking $0 \neq A, B \lhd R$ with $AF \subseteq A'$ and $BF \subseteq B'$ we have $AB = 0$ in R, impossible; thus R' is prime. To prove R' is centrally closed suppose $f: A' \to R'$ is a bimodule map and $A' \lhd R'$. By theorem 3.4.11 we have some element $0 \neq a \in A' \cap R$. Let $fa = \sum r_i \alpha_i$ where $\{\alpha_i : i \in I\}$ is a $Z(R)$-base of F. Defining $f_i: RaR \to R$ via $f_i a = r_i$ we see $[A', f] = \sum [RaR, f_i] \alpha_i \in Z(R)F \subseteq F$, as desired. (The f_i are well-defined since $R' \approx R \otimes F$.)   Q.E.D.

We now have the tools at hand to use the central closure effectively.

Theorem 3.4.13: *(Bergman) Any finite centralizing extension R' of R satisfies INC. In fact, if $P' \in \operatorname{Spec}(R')$ and $P' \subset B' \lhd R'$ then $B' \cap R \supset P' \cap R$.*

Proof: Passing to R'/P' and $R/(P' \cap R)$ we may assume R and R' are prime with $P' = 0$. Write $R' = \sum_{i=1}^{t} Rr_i'$ with $r_i' \in C_{R'}(R)$. Let $\hat{R}, \hat{R}'$ denote the respective central closures of R and R'. We claim that $\hat{R} \subseteq \hat{R}'$ canonically. Since $R \subseteq \hat{R}'$ it suffices to view any $[A, f]$ in the extended centroid of R as an element of the extended centroid of R'. The obvious way is to define f': $\sum Ar_i' \to R'$ by $f'(\sum a_i r_i') = \sum (fa_i) r_i'$. This is well-defined, for if $\sum a_i r_i' = 0$ then for any a in A we have

$$\sum (fa_i) r_i' a = \sum fa_i a r_i' = \sum a_i (fa) r_i' = \left(\sum a_i r_i'\right) fa = 0,$$

implying $\sum (fa_i) r_i' \in \text{Ann}_{R'} AR' = 0$. Furthermore, f' is an $R' - R'$ bimodule map for if $r' \in R'$ then writing $r_i' r' = \sum_j r_{ij} r_j'$ for suitable r_{ij} in R we have

$$f'\left(\sum a_i r_i'\right) r' = \sum (fa_i) r_i' r' = \sum_{i,j} (fa_i) r_{ij} r_j' = \sum f(a_i r_{ij}) r_j' = f'\left(\sum a_i r_i' r'\right)$$

and $r' f'(\sum a_i r_i') = f'(\sum r' a_i r_i')$ by an analogous argument.

Let $F = Z(\hat{R}')$. Then $\hat{R}F$ is centrally closed by corollary 3.4.12. But every ideal of $\hat{R}F$ intersects $\hat{R}$ nontrivially by theorem 3.4.11 and thus intersects R nontrivially; replacing R, R' and B', respectively, by $\hat{R}F$, $\hat{R}'$ and $\hat{R}'B'$, we may assume R and R' are centrally closed with the same center F.

Shrink $\{r_1', \ldots, r_t'\}$ to an F-independent set which we can call $\{r_1', \ldots, r_m'\}$, and let $W = \sum_{i=1}^m Fr_i'$. W is closed under multiplication. Indeed, if $r_i' r_j' = \sum_u r_{iju} r_u'$ then for each r in R we get $0 = [r, r_i' r_j'] = \sum_u [r, r_{iju}] r_u'$ implying $[r, r_{iju}] = 0$ since $R' \approx R \otimes C_{R'}(R)$; hence each $r_{iju} \in Z(R) = F$ as desired. Consequently W is a prime finite dimensional F-algebra and so is simple Artinian.

Now $R' \approx R \otimes_F W$ by theorem 3.4.11, so $B' \cap R \neq 0$ by theorem 1.7.27. Q.E.D.

Let us summarize these results.

Theorem 3.4.13':

(i) *Any finite centralizing extension R' of R satisfies LO, GU, and INC.*

(ii) *If K is any algebraic field extension of F then $R \otimes_F K$ as an extension of R satisfies LO, GU, and INC.*

Proof:

(i) Restatement of theorem 2.12.48 and theorem 3.4.13.

(ii) LO and GU have already been shown in corollary 2.12.50 without any assumptions on K. To check INC suppose $P_1 \supset P_2$ in $\text{Spec}(R \otimes K)$ lie over P. Then taking $x \in P_1 - P_2$ we see $x \in R \otimes_F K_0$ for some finitely generated extension K_0 of F; hence $[K_0 : F] < \infty$ and thus $R' = R \otimes_F K_0$ is a finite centralizing extension of R. But $R' \cap P_1 \supset R' \cap P_2$ in $\text{Spec}(R')$ by proposition 2.12.39.   Q.E.D.

Corollary 3.4.14: *If R' is a finite centralizing extension of R and $P' \in$ Spec(R') then P' is a primitive ideal of R' if $P' \cap R$ is a primitive ideal of R.*

Proof: Passing to R/P' we may assume $P' = 0$. Suppose L is a maximal left ideal having core 0. We claim $R'L$ is a proper left ideal of R'. Indeed, if $R'L = R'$ then using the notation of the proof of theorem 2.12.48 we have $B = BR'L \subseteq TL$ so $B \subseteq R \cap TL = L$ by sublemma 2.5.32', contrary to core$(L) = 0$. Having proved the claim we see R' is primitive by proposition 2.1.11 unless $A' + R'L < R'L$ for some $A' \lhd R'$. But then $(A' \cap R) + L < R$ so $A' \cap R \subseteq L$, contrary to $A' \cap R \neq 0$ by the theorem.   Q.E.D.

Corollary 3.4.15: *If R' is a finite centralizing extension of R and P' is a G-ideal of R' then $P' \cap R$ is a G-ideal of R.*

Proof: Let $B' = \bigcap\{$prime ideals of R' containing $P'\} \supset P'$. Then $P' \cap R \subset B' \cap R \subseteq \bigcap\{$prime ideals of R containing $P' \cap R\}$ since LO holds.
                                                                                Q.E.D.

We close the discussion with a fairly explicit description for primitive rings.

Example 3.4.16: Suppose a primitive ring R is displayed as a dense subring of End M_D where M is a faithful simple R-module and $D = $ End$_R M$. Given an $R - R$ bimodule map $f: A \to R$ with $A \lhd R$ we define $\bar{f}: M \to M$ by $\bar{f}(ax) = (fa)x$ for a in A and x in M. This is well-defined since if $a_1 x_1 = a_2 x_2$ then $0 = (fA)(a_1 x_1 - a_2 x_2) = A((fa_1)x_1 - (fa_2)x_2)$ so $(fa_1)x_1 = (fa_2)x_2$. But $AM = M$ so $\bar{f}$ is defined on all of M, and clearly $\bar{f} \in D$. Furthermore, $\bar{f} \in Z(D)$ by inspection, so we have an injection $\varphi: \hat{Z} \to Z(D)$ given by $\varphi[A, f] = \bar{f}$. On the other hand, multiplication by any element z of $Z(D)$ yields a map $\rho_z: M \to M$ so we can view $Z(D) \subseteq $ End M_D, and then we have $\hat{R} \subseteq RZ(D) \subseteq $ End M_D. Thus by corollary 3.4.12 we see $RZ(D)$ is centrally closed.

When soc$(R) \neq 0$ we must have $\hat{Z} = Z(D)$ since $\rho_z$soc$(R) = $ soc(R); thus in this case we get $\hat{R} = RZ(D)$. I do not know a specific example where $\hat{Z} \neq Z(D)$ but suspect one could be constructed by taking a simple ring (e.g., a skew polynomial ring) whose center is "too small."

C Supplement: Idempotent Filters and Serre Categories

Having several examples of quotients at our disposal we should like now to find general principles which tie them together. There are several places where one could start; the filter of left ideals used to build the quotient ring seems as good a place as any. In an attempt to give some of the flavor of the subject we shall skip some details.

Definition 3.4.17: A filter $\mathcal{F}$ of left ideals of R is *topologizing* if $Lr^{-1} \in \mathcal{F}$ for every L in $\mathcal{F}$ and every r in R. A topologizing filter is *idempotent* if $L' \in \mathcal{F}$ whenever $L'a^{-1} \in \mathcal{F}$ for all a in some L in $\mathcal{F}$.

The reason for the word "topologizing" is that every such filter defines a topological ring structure on R, via a neighborhood system at 0. Let us describe our previous examples in this context.

Example 3.4.18:

(i) Let S be a denominator set of R. Then $\mathcal{F} = \{L < R : L \cap S \neq \varnothing\}$ is an idempotent filter. (Indeed if $s \in L \cap S$ and $s' \in L' \cap S$ then there is some s_2 in S and r_2 in R with $s_2 s = rs' \in L \cap L' \cap S$, proving $\mathcal{F}$ is a filter. Likewise, if $s \in L \cap S$ and $r \in R$ then taking s_2 in S and r_2 in R with $s_2 r = r_2 s$ we have $s_2 \in Lr^{-1}$, proving $\mathcal{F}$ is topologizing. Finally suppose $L \in \mathcal{F}$ and $L'a^{-1} \in \mathcal{F}$ for all a in L. Taking $a \in L \cap S$ and $s \in L'a^{-1} \cap S$ gives $sa \in L' \cap S$, proving $\mathcal{F}$ is idempotent.

(ii) If $\operatorname{Sing}(R) = 0$ then $\mathcal{F} = \{$large left ideals$\}$ is an idempotent filter. Indeed, $\mathcal{F}$ is a topologizing filter by remark 3.3.3. To prove $\mathcal{F}$ is idempotent suppose $L'r^{-1}$ is large for all r in R. We want L' is large. But if $L \cap Rr = 0$ then $(Lr^{-1})r \subseteq L \cap Rr = 0$ implying $r \in \operatorname{Sing}(R) = 0$.

(iii) $\{$Dense left ideals of $R\}$ is an idempotent filter, as can be seen readily from exercise 4. Details are left to the reader.

(iv) For Martindale-Amitsur's ring of quotients left $\mathcal{F} = \{$left ideals of R containing a large two-sided ideal$\}$. $\mathcal{F}$ is clearly a filter and is topologizing since if $I \lhd R$ then $Ir \subseteq I$ so $I \subset Lr^{-1}$ for every r in R and every left ideal $L \supset I$.

We can use topologizing filters to mimic the results of §3.3.

Definition 3.4.19: A *radical* is a left exact functor Rad: $R\text{-}\mathcal{M}od \to R\text{-}\mathcal{M}od$ satisfying the following properties for all modules M, N:

(i) $\operatorname{Rad} M \leq M$.
(ii) If $f \in \operatorname{Hom}(M, N)$ then $\operatorname{Rad} f$ is the restriction of f to $\operatorname{Rad} M$. (Thus $f(\operatorname{Rad} M) \subseteq \operatorname{Rad} N$).
(iii) $\operatorname{Rad}(M/\operatorname{Rad} M) = 0$.

Rad is a *torsion* (or *hereditary*) radical if $N \cap \operatorname{Rad} M = \operatorname{Rad} N$ for all $N \leq M$.

For example, $\text{Rad}(M)$ considered in §2.5 is a radical (*not* torsion), but the point of view here is different based on the following results. Given a filter $\mathscr{F}$ define $\mathscr{T}_{\mathscr{F}}(M) = \{x \in M : \text{Ann}_R x \in \mathscr{F}.\}$

Proposition 3.4.20:   *If $\mathscr{F}$ is an idempotent filter then $\mathscr{T}_{\mathscr{F}}$ is a torsion radical.*

Proof: (i) and (ii) of definition 3.4.19 are clear. To see (iii) let $N = \mathscr{T}_{\mathscr{F}}(M)$ and suppose $x + N \in \mathscr{T}_{\mathscr{F}}(M/N)$. Then $Lx \subseteq N$ for some L in $\mathscr{F}$. Let $B = \text{Ann } x$. For any a in L we have $ax \in N$ so $Ba^{-1} \in \mathscr{F}$, implying $B \in \mathscr{F}$ and $x \in N$. Hence $\mathscr{T}_{\mathscr{F}}$ is a radical, which is clearly torsion.   Q.E.D.

Example 3.4.21:   Let us consider the torsion radicals corresponding, respectively, to the first three (idempotent) filters of example 3.4.18.

   (i) $\mathscr{T}_{\mathscr{F}}(M) = \{x \in M : sx = 0 \text{ for some } s \text{ in } S\}$, called the *torsion submodule* of M (with respect to S). This justifies the use of the name "torsion."
   (ii) $\mathscr{T}_{\mathscr{F}}(M) = \{x \in M : \text{Ann } x \text{ is large}\} = \text{Sing}(M)$.
   (iii) $\mathscr{T}_{\mathscr{F}}(M) = \{x \in M : \text{Ann } x \text{ is dense}\} = 0$.

Example 3.4.22:   When $\mathscr{F}$ is as in example 3.4.18(iv) then $\mathscr{T}_{\mathscr{F}}(M) = \{x \in M : Ix = 0 \text{ for some large ideal } I \text{ of } R\}$. Even though this need not be a torsion radical $\mathscr{T}_{\mathscr{F}}(R) = 0$ for R semiprime.

Construction 3.4.23:   Given a topologizing filter $\mathscr{F}$ define $Q_{\mathscr{F}}(M) = \varinjlim\{\text{Hom}(L, M/\mathscr{T}_{\mathscr{F}}(M)) : L \in \mathscr{F}\}$ for every R-module M, under the system analogous to construction 3.3.1; i.e., putting $\bar{M} = M/\mathscr{T}_{\mathscr{F}}(M)$ let $M_i = \text{Hom}(L_i, \bar{M})$, and for $L_i \supseteq L_j$ define $\varphi_i^j : M_i \to M_j$ by sending a map $f : L_i \to \bar{M}$ to its restriction $f : L_j \to \bar{M}$.

   $Q_{\mathscr{F}}(M)$ can be explicitly described as an R-module as follows: Writing (L, f) for a map $f : L \to \bar{M}$ define the equivalence $\sim$ on $\mathscr{S} = \{\text{Hom}(L, \bar{M}) : L \in \mathscr{F}\}$, by $(L_1, f_1) \sim (L_2, f_2)$ iff there is $L \subseteq L_1 \cap L_2$ in $\mathscr{F}$ with $L \leq \ker(f - g)$; then $Q_{\mathscr{F}}(M) = \mathscr{S}/\sim$.

   Thus $M \to Q_{\mathscr{F}}(M)$ defines a functor from $R\text{-}\mathscr{M}od$ to $R\text{-}\mathscr{M}od$, called the *localization functor* with respect to $\mathscr{F}$. An alternate description in exercise 13 yields many of its important functorial properties.

Remark 3.4.24:   (as in remark 3.3.2) If $\mathscr{F}$ is a topologizing filter then $Q_{\mathscr{F}}(M)$ is an R-module under the operations

$$[L_1, f_1] + [L_2, f_2] = [L_1 \cap L_2, f_1 + f_2].$$

$$r[L, f] = [Lr^{-1}, fr]$$

where $fr : Lr^{-1} \to M$ is defined by $(fr)a = f(ar)$.

There is a map $M \to Q_{\mathcal{F}}(M)$ sending x to right multiplication by x; the kernel is $\mathcal{T}_{\mathcal{F}}(M)$.

Theorem 3.4.25: $Q_{\mathcal{F}}(R)$ *is a ring, under multiplication.*

$$[L_1, f_1][L_2, f_2] = [f_1^{-1}L_2, f_2 f_1],$$

and the map $R \to Q_{\mathcal{F}}(R)$ of remark 3.4.24 is a ring homomorphism.

Proof: As in theorem 3.3.7.   Q.E.D.

This result unifies Theorems 3.1.4, 3.3.7, 3.4.6, and 3.4.7, in view of examples 3.4.21 and 3.4.22. Note that it gives us a functor from R-$\mathcal{M}od$ to $Q_{\mathcal{F}}(R)$-$\mathcal{M}od$.

§3.5 Left Noetherian Rings

In this section we generalize the classical theory of commutative Noetherian rings to a noncommutative setting. The flavor is entirely different from the results surrounding the Faith-Walker theorem (exercises 9-13 of §2.10) which described left Noetherian rings in terms of direct sums of injectives. Instead we shall analyze left Noetherian rings in terms of the Noether-Jacobson-Levitzki structure theory, also relying heavily on Goldie's theorems. *Assume until definition 3.5.24 that R is a left Noetherian ring.* By *Noetherian* we mean "left and right Noetherian." We have already obtained the following basic facts:

1. Any homomorphic image of R is left Noetherian as R-module (by proposition 0.2.19) and thus as a ring itself; any localization of R with respect to a denominator set is left Noetherian (proposition 3.1.13).
2. Every f.g. R-module is a Noetherian module (corollary 0.2.21); consequently, every submodule of an f.g. module is f.g.
3. Nil(R) is nilpotent (theorem 2.6.23); more generally, any nil, weakly closed subset is nilpotent (theorem 3.2.38).
4. Suppose R is semiprime. Then R is Goldie and thus an order in a semisimple Artinian ring (Goldie's theorem). Furthermore, there is a finite number of prime ideals with intersection 0, and these are the minimal prime ideals; equivalently, these are the maximal annihilator ideals (theorem 3.2.24 and proposition 3.2.26).

5. In general, R has a finite set of prime ideals whose product is 0, taken with suitable repetitions. (Indeed, this is true for $R/\text{Nil}(R)$ by #4, so apply #3.)

We shall also make heavy use of *Noetherian induction*: When proving an assertion about a Noetherian module M we may assume the assertion holds in every proper homomorphic image of M; indeed, if M were a counter-example we could take $N < M$ maximal for which M/N is a counterexample, and then replace M by M/N. Likewise, when proving an assertion about R we may assume the assertion holds in R/A for every nonzero ideal A of R.

Digression: There is a converse of fact 2, due to Formanek-Jategaonkar [74] and generalizing earlier work of Eakin and Nagata: If R is a commutative subring of a left Noetherian ring which is f.g. as R-module then R is left Noetherian. This cannot be generalized to arbitrary noncommutative rings, as seen by taking R to be the subring $\begin{pmatrix} \mathbb{Z} & \mathbb{Q} \\ 0 & \mathbb{Q} \end{pmatrix}$ of $M_2(\mathbb{Q})$. See exercise 2.5.15 for another result of theirs in a similar vein.

Much of the material of the first third of this section is drawn from Chatters-Hajarnavis [80B].

Constructing Left Noetherian Rings

First we shall see that the class of left Noetherian rings is large enough to be interesting.

Remark 3.5.1: If the ring $T \supseteq R$ is f.g. as R-module then T also is a left Noetherian ring (since T is a Noetherian R-module and thus *a fortiori* a Noetherian T-module). In particular, $M_n(R)$ is left Noetherian.

The other elementary way of constructing commutative Noetherian rings is by means of the Hilbert basis theorem, which says the polynomial ring $C[\lambda]$ is Noetherian if C is Noetherian. Of course, the free algebra $F\{X_1, X_2\}$ over a field F is *not* left Noetherian, but there is the following result:

Proposition 3.5.2: (*Hilbert Basis Theorem generalized*) *Suppose W is gener-ated as a ring by R and an element a such that $R + aR = R + Ra$. Then W is also left Noetherian. (In particular, this holds if a is R-normalizing.)*

Proof: We aim to show every left ideal L of W is f.g. Define $L_i = \{r \in R:$ there is $\sum_{u=0}^{i} a^u r_u$ in L with $r_i = r$ and all $r_u \in R\}$, and write $\tilde{w}(r)$ for such an element $\sum a^u r_u$. Despite having written the a^u on the left, we see each L_i is a left ideal since $R + aR = R + Ra$. $L_0 \subseteq L_1 \subseteq L_2 \cdots$, so for some m we have $L_i = L_{i+1}$ for all $i \geq m$. Let us write $L_i = \sum_{j=1}^{t(i)} Rs_{ij}$ and let $L' = \sum_{i=0}^{m} \sum_{j=1}^{t(i)} W\tilde{w}(s_{ij})$. $L' \leq L$ by definition of the $\tilde{w}(r)$.

We claim $L' = L$. Indeed we shall prove by induction on k that given any $w = \sum_{u=0}^{k} a^u r_u$ in L (with r_u in R) we have $w \in L'$. First suppose $k \leq m$. Writing $r_k = \sum r_{kj} s_{kj}$ and noting $a^k r_{kj} \in r'_{kj} a^k + \sum_{v<k} a^v R$ for suitable r'_{kj} in R (since $aR \subseteq Ra + R$) we see $w - \sum r'_{kj} \tilde{w}(s_{kj}) \in L$ has degree $< k$ and so is in L' by induction on k. Thus $w \in L'$ as desired.

Hence we may assume $k > m$. Now $r_k \in L_k = L_m$. As in the previous paragraph r_k appears as the leading coefficient of an element w' of L' of degree m; thus r_k is the "leading coefficient" of $a^{k-m}w'$, and $w - a^{k-m}w'$ has degree $< k$. By induction $w - a^{k-m}w' \in L'$, implying $w \in L'$.   Q.E.D.

Corollary 3.5.3. *The Ore extension $R[\lambda; \sigma, \delta]$ also is left Noetherian, provided σ is an automorphism. (The assertion may fail if σ is not onto.)*

The Reduced Rank

In studying noncommutative Noetherian rings one would rather not face the awesome hurdle of noncommutative localization, even in prime Noetherian rings. An attractive alternative introduced by Goldie [64] is the *reduced rank*, which came to the fore later when used in the proof of several important theorems. We start by recalling from §3.1 that if M is an R-module and S is a left denominator set for R then we can form the $S^{-1}R$-module $S^{-1}M = S^{-1}R \otimes_R M$.

Definition 3.5.4. Suppose M is an f.g. module over a left Noetherian ring R. Define the *reduced rank* $\rho(M)$ as follows:

Case (i). If R is semiprime then $\rho(M)$ is the (composition) length of $S^{-1}M$ as $S^{-1}R$-module, where $S = \{$regular elements of $R\}$.
Case (ii). $N = \mathrm{Nil}(R)$ is nilpotent of index $t > 1$. If $NM = 0$ then viewing M naturally as R/N-module, take $\rho(M)$ to be its reduced rank as R/N-module. In general, by induction on i such that $N^i M = 0$ define $\rho(M)$ as $\rho(NM) + \rho(M/NM)$. Explicitly $\rho(M) = \sum_{i=1}^{t} \rho(M'_i)$ where $M'_i = N^{i-1}M/N^i M$.

Remark 3.5.5: If $x_1, \ldots, x_t$ span M as R-module then $1 \otimes x_1, \ldots, 1 \otimes x_t$ span $S^{-1}M$ as $S^{-1}R$-module (notation as in (i)); since $S^{-1}R$ is semisimple Artinian by Goldie's theorem we see that case (i) makes sense.

Proposition 3.5.6: ρ *is additive, i.e., if $K \leq M$ then $\rho(M) = \rho(K) + \rho(M/K)$.*

Proof: First assume R is semiprime, i.e., case (i) holds. By theorem 3.1.20 the sequence $0 \to S^{-1}K \to S^{-1}M \to S^{-1}(M/K) \to 0$ is exact, so length$(S^{-1}M) =$ length$(S^{-1}K) +$ length$(S^{-1}(M/K))$ as $S^{-1}R$-module, as desired.

Next we consider case (ii), i.e., $N = \text{Nil}(R)$ is nilpotent. The argument is rather computational. By induction on the smallest j such that $N^j M = 0$, we note we are done by case (i) if $j = 1$ since M is then an R/N-module. We shall appeal repeatedly to Noetherian induction, and assume the assertion holds for every proper homomorphic image of M. By definition

$$\rho(M) = \rho(NM) + \rho(M/NM) \qquad \text{and } \rho(K) = \rho(NK) + \rho(K/NK).$$

By induction

$$\rho(NM) = \rho(NK) + \rho(NM/NK) = \rho(K) - \rho(K/NK) + \rho(NM/NK)$$

so

$$\rho(M) = (\rho(K) - \rho(K/NK) + \rho(NM/NK)) + \rho(M/NM).$$

If $NK \neq 0$ then applying Noetherian induction on M/NK yields

$$\rho(M) = \rho(K) - \rho(K/NK) + \rho(M/NK) = \rho(K) + \rho(M/K).$$

Thus we may assume $NK = 0$. But then case (i) applies to K and $\rho(K) = \rho(K \cap NM) + \rho(K/K \cap NM)$. By induction

$$\rho(NM) = \rho(K \cap NM) + \rho(NM/K \cap NM)$$

$$= \rho(K) - \rho(K/K \cap NM) + \rho(NM/K \cap NM).$$

Thus

$$\rho(M) = (\rho(K) - \rho(K/K \cap NM) + \rho(NM/K \cap NM)) + \rho(M/NM)$$

$$= \rho(K) - \rho(K/K \cap NM) + \rho(M/K \cap NM)$$

$$= \rho(K) + \rho(M/K)$$

by Noetherian induction unless $K \cap NM = 0$. But now $K \approx (K + NM)/NM \leq M/NM$, so $\rho(M/NM) = \rho(K) + \rho(M/K + NM)$ and

$$\rho(M) = (\rho(K) + \rho(M/K + NM)) + \rho(NM)$$

$$= \rho(K) + \rho(M/K + NM) + \rho(K + NM/K) = \rho(K) + \rho(M/K)$$

by Noetherian induction.   Q.E.D.

This proof could be formulated directly without Noetherian induction, cf., Chatters-Hajarnavis [80B]. Nevertheless, the proof presented here has a certain inevitability. It is useful to know when $\rho(M) = 0$.

Proposition 3.5.7: $\rho(M) = 0$ *iff for each x in M there is an r in $\operatorname{Ann}_R x$ such that $r + N$ is regular in R/N. (Notation as in definition 3.5.4 (ii).)*

Proof: Induction on the smallest j such that $N^j M = 0$. If $j = 1$ then we may assume $N = 0$ and case (i) holds. But then $\rho(M) = 0$ iff $S^{-1} \otimes M = 0$, iff (by remark 3.1.21) $S \cap \operatorname{Ann}_R x \neq \varnothing$ for each x in M, as desired.

For $j > 1$ we see using case (ii) that $N^{i-1}M/N^i M$ has reduced rank 0 for each $i \leq j$; by case (i) for each x in $N^{i-1}M$ there is s_i in S such that $s_i x \in N^i M$. But then for any x in M we have suitable $s_1, \ldots, s_j$ in S with $s_j \cdots s_1 x = 0$, as desired.   Q.E.D.

Digression 3.5.7: *Noetherian orders.* Although our immediate interest in the reduced rank is in proving the principal ideal theorem, other important applications are given in exercises 1, 2, and 3. These include a direct proof of Small's characterization of left Noetherian orders in left Artinian rings. One of the main tools in studying such rings is the *Artinian radical* $A(R)$ defined below in 3.5.19. Another method is studying those prime ideals consisting of zero divisors; P. F. Smith showed these are precisely the minimal prime ideals iff R is an order in an Artinian ring. Small-Stafford [82] have fitted Smith's result into a general theory of regularity modulo prime ideals. Stafford [82a] has shown that a Noetherian ring R is its own classical ring of fractions iff $\operatorname{Ann} A(R) \cap \operatorname{Ann}' A(R) \subseteq \operatorname{Jac}(R)$; in particular R must be semilocal.

Quite recently interest has returned to the theory of Noetherian rings which are embeddible into Artinian rings. We discussed this topic in §3.2; Small has observed that the reduced rank ρ yields a "Sylvester rank function" in the sense of Schofield [85B] and thus a homomorphism from an arbitrary left Noetherian ring R to a simple Artinian ring, whose kernel N_0 is $\{r \in R : \rho(R/Rr) = \rho(R)\}$. In view of proposition 3.5.6, $N_0 = \{r \in R : \rho(Rr) = 0\}$; since N_0 is f.g. as R-module we get $\rho(N_0) = 0$, so, in particular, $N_0 \subseteq \operatorname{Nil}(R)$. If R is also right Noetherian then N_0 is f.g. as right R-module and thus annihilates an element regular modulo $\operatorname{Nil}(R)$. In certain instances $N_0 = 0$ (e.g., when R is "Krull homogeneous," cf., after proposition 3.5.46 below).

The Principal Ideal Theorem

One of the key tools in the study of $\text{Spec}(C)$ for a commutative Noetherian domain C is the *Principal Ideal Theorem*, which states that any prime ideal minimal over a nonzero element c is a minimal prime ideal of C. In other words, if $c \in P$ and $0 \neq P_1 \subset P$ then $c \in P_1$. For a while the following example of Procesi discouraged attempts to generalize this result to noncommutative Noetherian rings:

Example 3.5.8: Let $R = M_2(\mathbb{Z}[\lambda])$ and $r = \begin{pmatrix} 2 & 0 \\ 0 & \lambda \end{pmatrix}$, a regular element. Then $P = RrR = 2R + \lambda R$ is a prime ideal since $R/P \approx M_2(\mathbb{Z}/2\mathbb{Z})$, but P is not a minimal prime since $\lambda R \in \text{Spec}(R)$.

Nevertheless, Jategaonkar [74a] proved a noncommutative principal ideal theorem, which we present below. Jategaonkar's proof was a masterpiece in the application of abstract torsion theory to a concrete problem about rings. On the positive side this was a triumph for torsion theory, and recently Jategaonkar has used his theory to extend this theorem still further. However, hardly anyone understood the proof, as evidenced by the long interval between Jategaonkar's results, and most ring theorists breathed a sigh of relief when Chatters-Goldie-Hajarnavis-Lenagan [79] recast the proof in terms of the reduced rank.

Lemma 3.5.9: *Suppose $a \in R$ is a normalizing element and $P \in \text{Spec}(R)$ with $a \notin P$. If $ra \in P$ then $r \in P$.*

Proof: $rRa = raR \subseteq P$ so $r \in P$.   Q.E.D.

Theorem 3.5.10: *(Principal ideal theorem) Suppose R is prime (left Noetherian) and $a \neq 0$ is a normalizing element of R. If P is a prime ideal of R minimal over a then P is a minimal (nonzero) prime ideal of R.*

Proof: Suppose on the contrary, there is P_1 in $\text{Spec}(R)$ with $0 \subset P_1 \subset P$. Since R is prime P_1 contains a regular element s. Also note a is regular by lemma 3.5.9. Let $L_k = a^{-k}Rs = \{r \in R : a^k r \in Rs\}$, a left ideal since a is normalizing. Then $L_0 \leq L_1 \leq L_2 \leq \cdots$ so for some m we have $L_k = L_{k+1}$ for all $k \geq m$. Replacing a by a^m we may assume $m = 1$. Thus $L_1 = L_2$. Consequently, if $a^2 r \in Rs$ then $ar \in Rs$.

Let $M = (Ra + Rs)/Ra^2$, viewed naturally as an R/Ra^2-module, and let $M_1 = (Ra^2 + Rs)/Ra^2 \le M$. We claim $\rho(M/M_1) = 0$, i.e., $\rho(M) = \rho(M_1)$. To prove this claim we use proposition 3.5.6. Let $M' = Ra/Ra^2$, $M'_1 = (Ra^2 + Rsa)/Ra^2$, and $M''_1 = (Ra^2 + Ras)/Ra^2$. Right multiplication by a yields an epic $M \to M'_1$ with kernel M', so $\rho(M) = \rho(M') + \rho(M'_1)$. On the other hand, there is a ring automorphism $\psi: R \to R$ given by $\psi r = ara^{-1}$. Since $\psi(Ra^2) = aRa = Ra^2$ we see that ψ induces a ring automorphism of R/Ra^2, which sends M'_1 (as a subset of R/Ra^2) to M''_1. Thus $\rho(M'_1) = \rho(M''_1)$. We can prove the claim by finding an epic $f: Ra \to M_1/M''_1$ having kernel Ra^2, for then f would yield an isomorphism $M' \to M_1/M''_1$ implying $\rho(M_1) = \rho(M') + \rho(M''_1) = \rho(M') + \rho(M'_1) = \rho(M)$, as desired. Our candidate for f is given by $f(ra) = rs + (Ra^2 + Ras)$. Clearly f is onto and $Ra^2 \subseteq \ker f$, so it remains to show that $\ker f \subseteq Ra^2$. If $ra \in \ker f$ then $rs \in Ra^2 + Ras = a^2R + Ras$; writing $rs = a^2 r_1 + r_2 as$ we get $a^2 r_1 \in Rs$ so $ar_1 \in Rs$ by the first paragraph. Writing $ar_1 = r's$ we have $rs = ar's + r_2 as$ so $r = ar' + r_2 a \in Ra$ and $ra \in Ra^2$; hence $\ker f \subseteq Ra^2$, proving the claim.

Write $^-$ for the canonical image in $\bar{R} = (R/Ra^2)/\mathrm{Nil}(R/Ra^2)$, a semiprime Goldie ring. By proposition 3.5.7 applied to M/M_1 there is some element r of R such that $\bar{r}$ is regular in $\bar{R}$ and $ra \in Ra^2 + Rs$. Write $ra = r_1 a^2 + r_2 s$ for r_i in R. Then $(r - r_1 a)a = r_2 s \in P_1$ implying $r - r_1 a \in P_1 \subset P$. Hence $r \in P$, so $\bar{P}$ is a minimal prime ideal of $\bar{R}$ containing the regular element $\bar{r}$, contrary to $\bar{P}$ being an annihilator ideal (fact 4). This proves the theorem.   Q.E.D.

Corollary 3.5.11: *Any prime ideal P minimal over a normalizing element in R has height ≤ 1.*

Proof: Otherwise $P \supset P_1 \supset P_2$; pass to R/P_2.   Q.E.D.

This result can be improved still further, with Ra replaced by any invertible ideal, cf., Chatters-Hajarnavis [80B, p. 45–50].

Height of Prime Ideals

For commutative Noetherian rings one proves directly from the principal ideal theorem that every prime ideal has finite height. Although no such sweeping result is available for the noncommutative case, we do have a good partial result.

Lemma 3.5.12:   *Suppose $P \in \mathrm{Spec}(R)$ contains a normalizing element a of R.*

Then any chain of prime ideals $P = P_0 \supset P_1 \supset \cdots \supset P_t$ can be modified to a chain $P = P_0 \supset P'_1 \supset \cdots \supset P'_{t-1} \supset P_t$ with $a \in P'_{t-1}$.

Proof: Choose $P = P_0 \supset P'_1 \supset \cdots \supset P'_{t-1}$ such that $a \in P'_k$ with k maximal possible. We are done if $k = t - 1$, so assume $k \leq t - 2$. Let $P'_t = P_t$. In the prime ring $\bar{R} = R/P'_{k+2}$ we have $\bar{P}'_k = \bar{P}'_{k+1} \supset 0$. Thus $\bar{P}'_{k+1}$ is *not* minimal over $\bar{a}$, by theorem 3.5.10, so there is $\bar{P}''_{k+1} \subset \bar{P}'_k$ with $\bar{a} \in \bar{P}''_{k+1}$. Replacing P_{k+1} by P''_{k+1} yields a contradiction.   Q.E.D.

Theorem 3.5.13:   (R is left Noetherian) *Suppose $P \in \operatorname{Spec}(R)$ and $\bar{P}$ contains a nonzero normalizing element of $\bar{R}$ for every prime homomorphic image $\bar{R}$ of R in which $\bar{P} \neq 0$. Then P has finite height.*

Proof: By Noetherian induction we may assume for any nonzero ideal $A \subset P$ that P/A has finite height in R/A. Passing to $R/\operatorname{Nil}(R)$ we may assume R is semiprime, for this does not change the prime spectrum. Since R has only a finite number of minimal prime ideals we may assume furthermore R is prime. Take $0 \neq a \in P$ with a normalizing in R. Then $\bar{P} = P/Ra$ has some finite height t in $\bar{R} = R/Ra$. But this implies height$(P) \leq t + 1$ in R, for, otherwise, given a chain $P = P_0 \supset \cdots \supset P_{t+2}$, we could assume $a \in P_{t+1}$ by lemma 3.5.12, yielding $\bar{P} \supset \bar{P}_1 \supset \cdots \supset \bar{P}_{t+1}$ in $\bar{R}$, a contradiction.   Q.E.D.

It will turn out that this condition can indeed be verified in several important cases (such as PI-rings).

Artinian Properties of Noetherian Rings: Prelude to Jacobson's Conjecture

Let us now introduce Artinian techniques, i.e., we consider ideals which are Artinian as left modules.

Lemma 3.5.17:   *Suppose $A \lhd R$ is a minimal (two-sided) ideal which also is Artinian as a right R-module. Then $R/\operatorname{Ann}_R A$ is a simple Artinian ring.*

Proof: Let $P = \operatorname{Ann}_R A$. $P \in \operatorname{Spec} R$ (for if B_1, B_2 are ideals properly containing P then $0 \neq B_2 A \subseteq A$ so $B_2 A = A$ and thus $B_1 B_2 A \neq 0$ implying $B_1 B_2 \not\subseteq P$ as desired). Hence $\bar{R} = R/P$ is prime and is an order in a simple

Artinian ring Q. Write $A = \sum_{i=1}^{t} a_i R$ and let us also identify A naturally as an f.g. $\bar{R}$-module. Then $ra = 0$ iff $\bar{r}a = 0$.

Let $S = \{r \in R : \bar{r}$ is regular in $\bar{R}\}$ and $A' = \{a \in A : \bar{S} \cap \text{Ann } a \neq \varnothing\}$. Then $A' \lhd R$ (seen either by noting $A' = \text{Sing}(A)$ or by applying the Ore property directly). Hence $A' = 0$ or $A' = A$. But $A' \neq A$ for, otherwise, there are s_i in S with $s_i a_i = 0$ for $1 \leq i \leq t$; using the Ore condition we could find s in S with $sa_i = 0$ for all i, implying $sA = 0$ contrary to $\bar{s} \neq 0$ in $\bar{R} = R/P$. Thus $A' = 0$.

Now for any s in S we have $A \supseteq sA \supseteq \cdots$. Since A is right Artinian we have $s^i A = s^{i+1} A$ for some i, implying $A = sA$ by the preceding paragraph. Thus A naturally becomes a Q-module, implying $Q = \bar{R}$ by lemmas 3.5.14 and 3.5.15. Thus $\bar{R}$ is simple Artinian.   Q.E.D.

Proposition 3.5.18: *Suppose $A \lhd R$ is Artinian as a right R-module. Then $R/\text{Ann}_R A$ is a left Artinian ring, so, in particular, A is Artinian as a left R-module.*

Proof: Take $A_0 = A$ and inductively given $A_i \lhd R$ take $A_{i+1} \lhd R$ maximal with respect to $A_{i+1} \subset A_i$. (This is possible since A_i is a left submodule of R and is thus Noetherian.) Then $A_0 \supset A_1 \supset \cdots$ so this chain must terminate when viewed as *right* modules. In other words, some $A_t = 0$. Then A_i/A_{i+1} is a minimal ideal of R_i/A_{i+1}, so by lemma 3.5.17 there is a maximal ideal P_i in R such that $P_i A_i \subseteq A_{i+1}$, and, furthermore, R/P_i is simple Artinian. Let $B = \bigcap_{i=0}^{t-1} P_i$. Then R/B is semisimple Artinian by the Chinese Remainder Theorem. But $B^t A \subseteq P_{t-1} \cdots P_0 A = 0$ so $B^t \subseteq \text{Ann}_R A$. Consequently, $R/\text{Ann}_R A$ is a semiprimary left Noetherian ring which is thus left Artinian.

Now A is an f.g. module over $R/\text{Ann}_R A$ and is thus Artinian, and so is Artinian as R-module.   Q.E.D.

Digression 3.5.19: (Also see exercise 12). These considerations lead one to define the *Artinian radical* $A(R)$ to be the sum of all the left ideals of R which are Artinian (as R-modules). If $L < R$ is Artinian then so is Lr for any r in R, so $A(R) \lhd R$. Since R is left Noetherian we see $A(R)$ is f.g. and is thus Artinian itself. In other words, $A(R)$ is Artinian and contains all Artinian left ideals. In this context proposition 3.5.18 shows this definition is left-right symmetric for R Noetherian, i.e., $A(R)$ is Artinian as a right R-module and contains all Artinian right ideals. The Artinian radical is one of the central features of Chatters-Hajarnavis [80B]. (Also see Stafford [82a] in this regard.)

Remark 3.5.20: If R is Noetherian and $\bar{R}$ is a left Artinian image of R then $\bar{R}$ is Artinian. (Indeed, $\bar{R}$ is semiprimary and thus right Artinian by theorem 2.7.2.)

The interplay between Noetherian and Artinian is quite enlightening.

Lemma 3.5.21:   *If $A, B \lhd R$ and R/A and R/B are Artinian then R/AB is Artinian.*

Proof: B/AB is an f.g. module over R/A and so is Artinian, and thus is Artinian as R/AB-module. Hence $R/AB > B/AB > 0$ refines to a composition series of R/AB as module over itself.   Q.E.D.

We are ready to introduce the key Ginn-Moss prime ideal.

Lemma 3.5.22:   *(Ginn and Moss) Suppose R is Noetherian and $M \in R\text{-}\mathcal{M}od$ such that $R/\text{Ann}_R M$ is not Artinian. Then there is $P \in \text{Spec}(R)$ such that $P = \text{Ann } N$ for some $N \leq M$ and R/P is not Artinian.*

Proof: Take $P \lhd R$ maximal with respect to R/P not Artinian but $P = \text{Ann } N$ for suitable $N \leq M$. We are done unless P is not prime, i.e., $AB \subseteq P$ for ideals $A, B \supset P$. Then $ABN = 0$; replacing A by $\text{Ann}_R(BN)$ we have R/A Artinian by assumption on P. Now $\bar{B} = B/AB$ is an f.g. R/A-module and is thus Artinian. Hence $\bar{B}$ is Artinian as module over $\bar{R} = R/AB$ by proposition 3.5.18. Let $\bar{I} = \text{Ann}'_{\bar{R}} \bar{B}$, i.e., $BI = AB$. The right-handed version of proposition 3.5.18 shows R/I is Artinian. But $BIN = 0$ so putting $I' = \text{Ann}(IN) \supseteq B \supset P$ we see R/I' is Artinian. By lemma 3.5.21, $R/I'I$ is Artinian. But $I'I \subseteq \text{Ann } N = P$, so R/P is Artinian, as desired.   Q.ED.

Jacobson's Conjecture

Let $J = \text{Jac}(R)$. The celebrated Krull Intersection Theorem states $\bigcap_{i \in \mathbb{N}} J^i M = 0$ for every f.g. module M when R is commutative Noetherian, and one of the major projects in ring theory has been to verify this for various classes of noncommutative rings. *Jacobson's conjecture* is that $\bigcap_{i \in \mathbb{N}} J^i = 0$ for an arbitrary Noetherian ring R. Jacobson's conjecture implies the Krull intersection theorem in the commutative theory, since $R \oplus M$ can be viewed as a Noetherian ring by taking multiplication $(r_1, x_1)(r_2, x_2) = r_1 r_2 + r_1 x_2 + r_2 x_1$. An attempt to "noncommutativize" this trick by means of upper triangular matrices actually leads to a famous counterexample.

Example 3.5.23:   (Herstein) Let $P = 2\mathbb{Z}$, $C = \mathbb{Z}_P$, and $R = \begin{pmatrix} \mathbb{Q} & \mathbb{Q} \\ 0 & C \end{pmatrix}$. In view of example 1.1.9 R is left Noetherian. On the other hand, $J = \begin{pmatrix} 0 & \mathbb{Q} \\ 0 & P_P \end{pmatrix}$, seen easily because $R/Qe_{12} \approx Q \times C$ implies $J/Qe_{12} = P_P e_{22}$. Hence $e_{12} = (2^{-k}e_{12})(2e_{22})^k \in J^{k+1}$ for each k, implying $0 \neq e_{12} \in \bigcap_{i \in \mathbb{N}} J^i$.

The status of Jacobson's conjecture for (left and right) Noetherian rings is still open. However, there is an additional condition which guarantees the conjecture.

Definition 3.5.24: *R is left bounded* if every large left ideal contains a two-sided ideal. *R is almost bounded* if a large submodule of a faithful f.g. module is necessarily faithful. *R is left fully bounded* (resp. *almost fully bounded*) if every prime homomorphic image of *R* is left bounded (resp. almost bounded).

Almost fully bounded is equivalent to the *strong second layer condition* in Jategaonkar [86B, §8.13], which is the standard terminology.

Proposition 3.5.25: *Every left bounded prime ring R is almost bounded. Every left fully bounded ring is almost fully bounded.*

Proof: We verify the first assertion, which formally implies the second. Suppose $M = \sum_{i=1}^{n} Rx_i$ is faithful, and $N \leq M$ is large. By proposition 3.3.0 each Nx_i^{-1} is a large left ideal which thus contains a two-sided ideal A_i. Let $A = A_1 \cdots A_n \neq 0$. Then $AM \subseteq \sum ARx_i \subseteq N$, so $(\operatorname{Ann} N)A \subseteq \operatorname{Ann} M = 0$, implying $\operatorname{Ann} N = 0$ since *R* is prime.   Q.E.D.

Every commutative ring is left and right fully bounded, and we shall see in §6.1 that this hypothesis passes to rings with polynomial identity. Jategaonkar [74] verified Jacobson's conjecture for left and right fully bounded Noetherian rings. This result was subsequently improved by Schelter [75] and by Cauchon [76], who removed "right fully bounded" from the hypothesis. Later Jategaonkar [81] proved the result for almost fully bounded rings. This is a substantial advance because virtually all of the Noetherian rings of interest today are almost fully bounded, whereas many are not left fully bounded. On the other hand, the almost fully bounded case is much more complicated than the left fully bounded case. We present the easier result here; Jategaonkar's stronger version is given in the unabridged version (pp. 433ff). For a very thorough treatment, the reader is referred to Jategaonkar [86B].

Note that a left bounded prime Goldie ring satisfies the property that if $s \in R$ is regular then Rs contains a nonzero ideal.

Remark 3.5.26: If *R* is primitive and left bounded and satisfies $\operatorname{ACC}(\oplus)$ then *R* is simple Artinian. (Indeed let *L* be a maximal left ideal of *R* having core 0. Then *L* is not large, so has a nonzero essential complement *L'*. Clearly *L'* is a minimal nonzero left ideal of *R*, implying $\operatorname{soc}(R) \neq 0$. By hypothesis soc($R$) is a sum of a finite number of minimal left ideals since soc(R) is completely reducible; hence R is semisimple Artinian by remark 2.1.26.)

Lemma 3.5.27: *Suppose R is almost fully bounded Noetherian with primitive images Artinian. (In particular, this would hold if R is fully bounded, by proposition 3.5.25 and remark 3.5.26.) Then every f.g. module M with a unique minimal nonzero submodule M_0 is Artinian.*

Proof: Replacing R by $R/\text{Ann}\,M$, we may assume that M is a faithful R-module. Suppose R is *not* Artinian. By lemma 3.5.22 we have $P \in \text{Spec}(R)$ such that $\bar{R} = R/P$ is *not* Artinian and $P = \text{Ann}_R M'$ for some $0 \neq M' < M$. Since $M' \neq 0$ we see $M_0 \leq M$, so M_0 is a large submodule of M'. But M' is a faithful module over R/P so, by hypothesis, M_0 is faithful as well as simple as $\bar{R}$-module, implying $\bar{R}$ is primitive and thus Artinian, contradiction.
$$\text{Q.E.D.}$$

Theorem 3.5.28: *(Jategaonkar-Schelter-Cauchon)* $\bigcap_{i \in \mathbb{N}} J^i M = 0$ *for every f.g. module M, whenever R is an almost fully bounded Noetherian ring whose primitive images are Artinian; in particular, it holds for fully bounded Noetherian rings.*

Proof: It suffices to show that for every x in M there is some $m \in \mathbb{N}$ with $x \notin J^m M$. By Zorn's lemma there is $N < M$ maximal with respect to $x \notin N$; thus it suffices to prove $J^m M \leq N$. This certainly holds by "Nakayama's lemma" if M/N has finite composition length. On the other hand, every nonzero submodule of M/N contains $Rx + N$, so passing from M to M/N we are in position to apply lemma 3.5.27.   Q.E.D.

Digression: The key idea in this proof was clearly the reduction to lemma 3.5.27, so we would like to know whether anything was lost in this transition. If R is semilocal and Jacobson's conjecture holds then indeed the conclusion of lemma 3.5.27 is true; indeed, suppose $\bigcap J^i M = 0$. Then $M_0 \nleq J^m M$ for some m, so $J^m M = 0$. Thus M is an f.g. module over R/J^m, a semiprimary Noetherian ring which is thus Artinian, implying M is Artinian.

On the other hand, the conclusion of lemma 3.5.27 can fail even when Jacobson's conjecture holds, should primitive images of R not be Artinian. Here is a counterexample, over any field F of characteristic 0.

Example 3.5.29: (Musson) Let R_0 be the polynomial ring $F[\mu]$ and R be the Ore extension $R_0[\lambda; 1, \delta]$ where $\delta\mu = \mu$, i.e., $\lambda\mu - \mu\lambda = \mu$. R is Noetherian by corollary 3.5.3, and $\mu R = R\mu$ by inspection. Let $M = R/L$ where $L = R(\mu - 1)(\lambda - 1)$, and $M_0 = Rx_0$ where $x_0 = (\lambda - 1) + L \in M$. We shall show every nonzero submodule of M contains M_0, but M is not Artinian. In fact letting $^-$ denote the image in $\bar{R} = R/L$ we shall show that M_0 and $R\mu^i\bar{1}$ are the only nonzero submodules of M. We start with some easy computations.

(i) $\mu x_0 = x_0$ since $\mu(\lambda - 1) - (\lambda - 1) = (\mu - 1)(\lambda - 1) \in L$.

(ii) $[\lambda, \mu^i] = i\mu^i$ obtained by iterating $\lambda\mu = \mu\lambda + \mu$.

(iii) $\bar{\lambda} = \bar{1} + x_0$.

(iv) $x_0 = \mu^i x_0 = \mu^i\bar{\lambda} - \mu^i\bar{1} = (\lambda\mu^i\bar{1} - i\mu^i\bar{1}) - \mu^i\bar{1} = (\lambda - i - 1)\mu^i\bar{1}$.

In particular, $M_0 \subseteq R\mu^n\bar{1}$ for each n. Since $R\mu^{n+1}\bar{1} \subseteq R\mu^n\bar{1}$ it suffices to prove the following two assertions:

(1) For any $x \neq 0$ in M we have $Rx = M_0$ or $Rx = R\mu^n\bar{1}$ for suitable n.
(2) $\mu^n\bar{1} \notin R\mu^{n+1}\bar{1}$ for each n.

Indeed, (1) would imply M_0 and the $R\mu^n\bar{1}$ are the only submodules of M and (2) would show that M_0 is the only Artinian submodule, although it is large in all the submodules.

To prove (1), note $\mu\lambda \equiv \mu + \lambda - 1 \equiv \mu + x_0$ modulo L; hence x has the form $x' + \sum_{i=n}^{t} \alpha_i\mu^i\bar{1}$ for suitable x' in M_0.

First assume each $\alpha_i = 0$; then $Rx' \subseteq M_0$, so we want to show $x_0 \in Rx'$. But $x' = \sum_{i=0}^{m} \beta_i\lambda^i x_0$ for suitable β_i in F, by (i). Furthermore, for $i \geq 1$ we have inductively $(1 - \mu)^i\lambda^i x_0 = i!x_0$ (verfication left for the reader); on the other hand, $(1 - \mu)x_0 = x_0 - \mu x_0 = 0$ so $(1 - \mu)^i\lambda^j x_0 = 0$ for $j < i$. Consequently, $(1 - \mu)^m x' = m!\beta_m x_0$ so $x_0 \in Rx'$ as desired.

Thus we may assume some $\alpha_i \neq 0$; choosing n, t appropriately, we have $\alpha_n \neq 0$ so each $\mu^i\bar{1} \in R\mu^n\bar{1}$ and thus $x \in R\mu^n\bar{1}$. To prove (1) it remains to prove $\mu^n\bar{1} \in Rx$; in fact, we see the proofs of both (1) and (2) reduce to proving the following claim (where for (2) we take $x = 0$):

If $x = x' + \sum_{i=n}^{t} \alpha_i\mu^i\bar{1}$ with $x' \in M_0$ and $\alpha_n = 1$ then $\mu^n\bar{1} \in Rx$.

We prove this by induction on t. Multiply both sides by $\lambda - t - 1$; noting by (iv) that $(\lambda - i - 1)\mu^i\bar{1} \in M_0$ we have

$$(\lambda - t - 1)x \in M_0 + \sum_{i=n}^{t-1} \alpha_i(i - t)\mu^i\bar{1}$$

and by induction we have $\mu^n \bar{1} \in R(\lambda - t - 1)x \subseteq Rx$, as needed.

Musson's example is not a counterexample to Jacobson's conjecture, since it is primitive (by exercise 2.1.4)! On the other hand, I do not know of any counterexample to the conclusion of lemma 3.5.27 when R is Noetherian with all primitive images Artinian.

D Supplement: Noetherian Completion of a Ring

In this supplement we do not assume R is Noetherian. Theorem 3.5.28 gave us a criterion for $\bigcap_{i \in \mathbb{N}} J^i = 0$ where $J = \text{Jac}(R)$. Then we can form the J-adic completion $\hat{R}$. (This notation has no connection to §3.3.) We shall examine conditions under which $\hat{R}$ also is left Noetherian; this would be useful since the module theory of complete Noetherian rings is richer than that of Noetherian rings in general. We present the results in the more general setting of rings R with filtration over $(\mathbb{Z}, +)$. Not only does this include the A-adic completion (or, more generally, the case where $R(0) = R$, cf., example 1.8.14(ii) ff.) but also enveloping algebras (as we shall see in §8.3.)

Definition 3.5.30:   Suppose R is a ring with filtration over $(\mathbb{Z}, +)$. Define the *associated graded ring* $G(R)$ to be the additive group $\bigoplus_{i \in \mathbb{Z}} R(i)/R(i + 1)$, made into a ring by defining multiplication

$$(r_i + R(i + 1))(r'_j + R(j + 1)) = r_i r'_j + R(i + j + 1)$$

where $r_i \in R(i)$ and $r'_j \in R(j)$.

If $M \in R\text{-}\mathcal{M}od$ the *associated graded module* is $\bigoplus_{i \in \mathbb{Z}} R(i)M/R(i+1)M$, made into a module over the associated graded ring by means of scalar multiplication

$$(r_i + R(i + 1))(x_j + R(j + 1)M) = r_i x_j + R(i + j + 1)M$$

where $r_i \in R(i)$ and $x_j \in R(j)M$.

An important special case is the A-adic filtration; then we write $G_A(R)$ for the associated graded ring $\bigoplus_{i \in \mathbb{N}} A^i/A^{i+1}$, and $G_A(M)$ for the associated graded module. Of course, there is a functor $R\text{-}\mathcal{M}od \to G_A(R)\text{-}\mathcal{M}od$ sending M to $G_A(M)$.

Proposition 3.5.31:   *Suppose R is a ring with filtration over $(\mathbb{Z}, +)$. (i) If R is complete and $M \in R\text{-}\mathcal{M}od$ such that $\bigcap_{i \in \mathbb{N}} R(i)M = 0$ and $G(M)$ is f.g. as $G(R)$-module, then M is f.g. (ii) If $R(0) = R$ then $G(R) \approx G(\hat{R})$.*

Proof:

(i) Write $M = \sum_{u=1}^{m} G(R)x_u$, where we may take the x_u homogeneous, i.e., $x_u = y_u + R(n_u + 1)M$ for suitable n_u in $\mathbb{N}$ and y_u in $R(n_u)M$. For any given $0 \neq y \in M$ there is a smallest n for which $y \notin R(n + 1)M$; looking in $G(M)$ we have $y' = y - \sum r_{nu}y_u \in R(n + 1)M$ for suitable r_{nu} in $R(n)$; likewise, $y' - \sum_u r_{n+1,u}y_u \in R(n + 2)M$, for suitable $r_{n+1,u}$ in $R(n + 1)$. Continuing *ad infinitum* we see for each u that $\sum_{m \geq n} r_{mu} \in \hat{R} = R$ and $y - \sum_{u=1}^{m} (\sum_{m \geq n} r_{mu})y_u \in \bigcap_{m > n} R(m)M = 0$, proving $y_1, \ldots, y_m$ generate M.

(ii) From example 1.8.14 we have $R \cap \hat{R}(n) = R(n)$, so $R(n) \cap \hat{R}(n + 1) = R(n + 1)$; also $\hat{R}(n) = R(n) + \hat{R}(n + 1)$.

Now

$$\hat{R}(n)/\hat{R}(n + 1) = (R(n) + \hat{R}(n + 1))/\hat{R}(n + 1) \approx R(n)/(R(n) \cap \hat{R}(n + 1))$$

$$= R(n)/R(n + 1),$$

proving (ii).   Q.E.D.

Corollary 3.5.32: *Suppose R is a ring with filtration over $(\mathbb{Z}, +)$, and $G(R)$ is left Noetherian.*

(i) *If $R(1) = 0$ then R is left Noetherian.*

(ii) *If $R(0) = R$ then $\hat{R}$ is left Noetherian.*

Proof:

(i) R is complete (cf., example 1.8.14(i)), so apply the proposition with $M = R$.

(i) If $L < \hat{R}$ then $G(L)$ is a left ideal of $G(\hat{R}) \approx G(R)$ so is f.g., implying L is f.g. in $\hat{R}$.   Q.E.D.

Our immediate interest is the special case of (ii) for the A-Adic completion: If $G_A(R)$ is left Noetherian then $\hat{R}$ also is left Noetherian. Thus we are led to ask when $G_A(R)$ is left Noetherian. This is an easy application of the Hilbert Basis Theorem for R commutative, and with a little more effort we can obtain a slightly more general result.

Proposition 3.5.33: *Suppose $A = \sum_{u=1}^{t} Rz_i$ with each z_i in $Z(R)$. If R/A is left Noetherian then $G_A(R)$ and $\hat{R}$ are left Noetherian.*

Proof: Write $S = \{z_1, \ldots, z_t\}$. Then $A = RS$ and, inductively, $A^i = RS^i$ for each i. But $R_1 = (R/A)[\lambda_1, \ldots, \lambda_t]$ is left Noetherian by the Hilbert Basis

Theorem, and there is a surjection $R_1 \to G_A(R)$ sending λ_u^i to the element $z_u^i + A^{i+1}$ in A^i/A^{i+1}, so $G_A(R)$ is also left Noetherian.   Q.E.D.

We shall now prove a rather sweeping result of McConnell [79] based on work of Roseblade. We say $r \in R$ is *central modulo* an ideal A if $r + A \in Z(A)$.

Definition 3.5.34: An ideal A of R is *polycentral* if there is a chain $A = A_0 \supset A_1 \supset \cdots \supset A_{t+1} = 0$ of ideals of R such that $A_i = A_{i+1} + \sum R a_{iu}$ for suitable elements $\{a_{iu}: 1 \leq u \leq t(i)\}$ each central modulo A_{i+1}. If we only have the weaker condition that the $a_{iu} + A_{i+1}$ normalize R/A_{i+1} we say A is a *polynormal* ideal.

These definitions embrace several examples from group algebras and enveloping algebras to be pursued in §8.4. To study polycentral ideals we construct the ring $\tilde{R}$, defined to be the subring of $R[\lambda]$ generated by $A_i \lambda^v$ for all $0 \leq i \leq t$ and all $1 \leq v \leq 2^i$. Then $\tilde{R}$ is $\mathbb{N}$-graded (according to the powers of λ) since the generators of $\tilde{R}$ are homogeneous elements of $R[\lambda]$. Write $\tilde{R}_{t+1} = R$ and for each $i \leq t$ let $\tilde{R}_i$ be the subring of $\tilde{R}$ generated by $\tilde{R}_{i+1}$ and all $\{A_i \lambda^v: 1 \leq v \leq 2^i\}$; thus $\tilde{R} = \tilde{R}_0$.

Lemma 3.5.35: $\tilde{R}_i$ *is generated by* $\tilde{R}_{i+1}$ *and a set* $S = \{a_{ij}\lambda^v: 1 \leq j \leq t(i),$ $1 \leq v \leq 2^i\}$ (notation as in definition 3.5.34); furthermore, $[s, s'] \in \tilde{R}_{i+1}$ and $[s, \tilde{R}_{i+1}] \subseteq \tilde{R}_{i+1}$ for all s, s' in S.

Proof: The first assertion is immediate. To prove the next assertion take $s = a_{ij}\lambda^v$ and $s' = a_{ij'}\lambda^{v'}$ in S; then $[s, s'] \in A_{i+1}\lambda^{v+v'} \subseteq \tilde{R}_{i+1}$ since $v + v' \leq 2^i + 2^i = 2^{i+1}$. It remains to show $[s, \tilde{R}_{i+1}] \subseteq \tilde{R}_{i+1}$. Write $a = a_{ij}$ and $T = \tilde{R}_{i+1}$.

Note $[s, R] = [a, R]\lambda^v \subseteq A_{i+1}\lambda^v \subseteq T$. But T is spanned by R and products of length m of generating elements for any $m > 0$, i.e., terms of the form $x = b_1 \cdots b_m \lambda^n$ where each $b_u \in A_{u'}$ for suitable $u' \geq i + 1$ and $m \leq n \leq \sum 2^{u'}$. Thus it suffices to show for each x of this form that $[s, x] \in T$. Now

$$[s, x] = \sum_{k=1}^{m} b_1 \cdots b_{k-1}[s, b_k]b_{k+1} \cdots b_m \lambda^n,$$

so it suffices to show each $b_1 \cdots b_{k-1}[a, b_k]b_{k+1} \cdots b_m \lambda^{v+n} \in T$. Since $b_k \in A_{k'} = A_{k'+1} + \sum R a_{k'j}$ for $a_{k'j}$ central modulo $A_{k'+1}$, we can break this further into two cases: $b_k \in A_{k'+1}$ and $b_k \in R a_{k'j}$ for some j.

Case I. If $b_k \in A_{k'+1}$ then $[a, b_k] \in A_{k'+1}$ so replacing k' by $k' + 1$ we have $n + v \leq \sum 2^{u'}$ as desired, and we are done.

Case II. If $b_k = r a_{k'j}$ for suitable r in R then $[a, b_k] = [a, r a_{k'j}] = [a, r] a_{k'j} + r[a, a_{k'j}] \in A_{i+1} A_{k'} + A_{k'+1}$.

By case I we can handle the part from $A_{k'+1}$, so it remains to show $b_i \cdots b_{k-1} A_{i+1} A_{k'} b_{k+1} \cdots b_m \lambda^{n+v} \subseteq T$. Take any elements b'_k from A_{i+1} and b'_{k+1} from $A_{k'}$. Now $b_1 \cdots b_{k-1} b'_k b'_{k+1} b_{k+1} \cdots b_m \lambda^{v+n} \in T$ is clear if we replace m by $m+1 \leq v+n \leq 2^{i+1} + \sum_{u'=1}^{m} 2^{u'}$. This disposes of case II, proving the lemma.   Q.E.D.

Theorem 3.5.36: *If R has a polycentral ideal A such that R/A is left Noetherian then the A-adic completion $\hat{R}$ is left Noetherian.*

Proof: We modify the proof of proposition 3.5.33. Let $A = A_0 \supset \cdots \supset A_{t+1} = 0$ and $\tilde{R}$ and $\tilde{R}_i$ be as in definition 3.5.34ff; and write $\tilde{R} = \sum_{i \in \mathbb{N}} B_i \lambda^i$ where $B_0 = R$, $B_1 = A$, and each $B_i \lhd R$. Since $\tilde{R}$ is $\mathbb{N}$-graded we have each $B_i B_j \subseteq B_{i+j}$; thus $A^i \subseteq B_i$. Also note if $i > 2^{kt}$ then $B_i \subset A^k$ since the highest power of λ available in the generators of $\tilde{R}$ is 2^t.

Let R'' be the graded ring associated to the filtration $R > B_1 > B_2 > \cdots$ of R. There is a canonical graded ring surjection $\Phi : \tilde{R} \to R''$ sending $B_i \lambda^i$ to B_i / B_{i+1} obtained by cancelling λ.

We claim $\Phi \tilde{R}$ is left Noetherian. Indeed, $\Phi \tilde{R}_{t+1} = \Phi R = R/A$ is left Noetherian by hypothesis, so it suffices to prove that if $\Phi \tilde{R}_{i+1}$ is left Noetherian then $\Phi \tilde{R}_i$ is also left Noetherian. Write $T = \Phi \tilde{R}_{i+1}$ and note from lemma 3.5.35 that $\Phi \tilde{R}_i$ is generated by T and elements $s_1, \ldots, s_m$ such that $[s_i, T] \subseteq T$ and $[s_i, s_j] \in T$ for each i, j. Write T_k for the subring of $\Phi \tilde{R}_i$ generated by T and $s_1, \ldots, s_k$. We shall show by induction that if T_{k-1} is left Noetherian then T_k is left Noetherian; clearly this would conclude the proof of the claim.

A typical element of T_{k-1} is a product of terms of the form $r_1 \cdots r_m$ where each $r_i \in T \cup \{s_1, \ldots, s_{k-1}\}$. Then $[s_k, r_1 \cdots r_m] = \sum_j r_1 \cdots r_{j-1} [s_k, r_j] r_{j+1} \cdots r_m$ and each $[s_k, r_j] \in T$, so $[s_k, r_1 \cdots r_m] \in T_{k-1}$. Hence $[s_k, T_{k-1}] \subseteq T_{k-1}$, so T_k is left Noetherian by proposition 3.5.2, establishing the claim.

Combining the assertions we see R'' is left Noetherian. We want to conclude $\hat{R}$ is left Noetherian, just as in corollary 3.5.32. But this is seen by copying out the proof of proposition 3.5.33 using R'' instead of $G_A(R)$, in view of the fact $A^i \subseteq B_i \subseteq A^k$ for all $i > 2^{kt}$ established earlier.   Q.E.D.

Krull Dimension for Noncommutative Left Noetherian Rings

The "localization" theory for Noetherian rings has been a rich source both of research and of headaches. The main difficulty is to determine when a prime ideal P is *left localizable*, i.e., when {elements of R modulo P} is a left denominator set. There has been considerable progress in the last five years, to be discussed at the end of this section and in §6.3, but localization remains a very delicate procedure.

Fortunately, the reduced rank can serve as a substitute as we have already seen. Coupling the reduced rank with the noncommutative "Krull dimension," to be defined presently, one can often bypass localization altogether. We shall look at the rudiments of Krull dimension and then study two applications, due to Stafford-Coutinho and to Jategaonkar, which illustrate how to obtain noncommutative versions of important theorems.

One of the fundamental ties from commutative ring theory to geometry is Krull's definition of the *dimension* of a local Noetherian ring as the height of its maximal ideal, i.e., the length of a maximal chain of prime ideals. There are several candidates generalizing this definition to noncommutative rings. Of course the first candidate is the obvious one—the maximal length of a chain of prime ideals (if this exists); this is the commutative ring theorist's definition of Krull dimension, which we shall call the *little Krull dimension*, and is useful for certain classes of noncommutative rings (such as PI-rings, cf. §6.3). Various module-theoretic definitions have proved more successful for noncommutative Noetherian rings. Unfortunately there is a problem in terminology—since these dimensions were considered by Gabriel and generalize Krull dimension, the appellation "Gabriel-Krull" could be applied to each of them; worse yet, there is another dimension introduced by Gelfand-Kirillov (to be considered in Chapter 6) which is now called GK dimension! Ideally the dimension considered here would honor Rentschler-Gabriel [67] who invented it, and Gordon-Robson [73] who developed the comprehensive theory. However, the title of Gordon and Robson's treatise, and the name almost universally used, is "Krull dimension."

Definition 3.5.39: (We are *not* assuming R is left Noetherian.) The *Krull dimension* of M, abbreviated as K-dim M, is defined as follows (via transfinite induction): K-dim$(0) = -1$, and, inductively, K-dim M is the smallest ordinal α (possibly infinite) such that in each descending chain $M = M_0 \geq M_1 \geq M_2 \geq \cdots$ one must have

$$\text{K-dim}(M_{i-1}/M_i) < \alpha \qquad \text{for almost all } i.$$

Let us write out this definition explicitly for low values of α.

Example 3.5.40: K-dim $M = 0$ iff M is Artinian. K-dim $M \leq 1$ iff there is no infinite descending chain $M = M_0 > M_1 > M_2 \ldots$ such that each factor M_{i-1}/M_i is *not* Artinian.

Krull dimension is possibly the principal dimension used today in general noncommutative ring theory. However, the reader should be aware that its one-sidedness makes it a bit awkward in handling rings which are not fully bounded.

As with most other dimensions, we define the *Krull dimension* of a ring R to be its K-dim as R-module. Modules need not have Krull dimension. In fact, if F is a field then $M = F^{(\mathbb{N})}$ is an F-module; let $M' = M^{(\mathbb{N})}$, and let M_i be the submodule of M' consisting of all vectors whose first i components are 0. Then clearly $M' = M_0 > M_1 > \cdots$ is an infinite descending chain with each factor isomorphic to M, so we would have K-dim $M' >$ K-dim M. But $M' \approx M$; consequently, M cannot have Krull dimension. Similarly, any infinite direct product of copies of a field is a ring which does not have Krull dimension. Nevertheless, Krull dimension exists in many important cases, as we shall soon see.

Remark 3.5.40′: If K-dim $M > \alpha$ then one can find an infinite chain $M = M_0 > M_1 > \ldots$ such that K-dim $(M_{i-1}/M_i) \geq \alpha$ for *each i.*

Remark 3.5.40″: If $M_2 \leq M_1 \leq M$ and $N \leq M$, then there is an exact sequence
$$0 \to M_1 \cap N / M_2 \cap N \to M_1/M_2 \to (M_1 + N)/(M_2 + N) \to 0.$$
(Indeed, define $f: M_1 \to (M_1 + N)/(M_2 + N)$ by $fx = x + (M_2 + N)$. Then ker $f = M_1 \cap (M_2 + N) = M_2 + (M_1 \cap N)$, so f induces a map $M_1/M_2 \to (M_1 + N)/(M_2 + N)$ having kernel $(M_2 + M_1 \cap N)/M_2 \approx (M_1 \cap N)/(M_2 \cap N)$.)

Proposition 3.4.41: K-dim $M = \sup\{$K-dim $N,$ K-dim $M/N\}$ *for any* $N \leq M$, *provided either side exists.*

Proof: The inequality "$\geq$" is clear since any descending chain from N or from M/N can be transferred to M.

To prove "$\leq$" we modify the proof of proposition 0.2.19. Suppose $M > M_1 > M_2 > \cdots$ is an infinite chain. We need to show that if K-dim $N \leq \alpha$ and K-dim $M/N \leq \alpha$ then K-dim $M \leq \alpha$, i.e., almost all factors M_i/M_{i+1} have K-dim $< \alpha$. Letting $\bar{M}_i$ denote $(M_i + N)/N$ we have $\bar{M} \geq \bar{M}_1 \geq \bar{M}_2 \geq \cdots$, so by hypothesis almost all of the factors $\bar{M}_i/\bar{M}_{i+1}$ have K-dim $< \alpha$. Likewise, almost all $(M_i \cap N)/(M_{i+1} \cap N)$ have K-dim $< \alpha$. By induction on α, we see by remark 3.5.40″ that K-dim $M_i/M_{i+1} < \alpha$.   Q.E.D.

It follows that K-dim $R = \max \{$K-dim $M : M \in R\text{-}\mathscr{F}\!imod\}$ if either side exists, so K-dim R is categorically defined, cf., §4.1. For later use let us apply remark 3.5.40″ to proposition 3.5.41.

Corollary 3.5.41′:   *If* $M_2 \leq M_1 \leq M$ *and* $N \leq M$ *then* K-dim $M_1/M_2 =$ max $\{K\text{-dim}(M_1 \cap N)/(M_2 \cap N), K\text{-dim}(M_1 + N)/(M_2 + N)\}$.

Remark 3.5.41″:   In analogy to proposition 3.5.41, we shall call a dimension on modules *exact* if dim $M = \max\{\dim K, \dim M/K\}$ for all $N \leq M$. In such a case one readily concludes dim $M \leq \max\{\dim K, \dim N\}$ for every exact sequence $K \to M \to N$. (Indeed, any map $g: M \to N$ gives rise to the exact sequence $0 \to \ker g \to M \to gM \to 0$, so dim $M = \max\{\dim(\ker g), \dim gM\}$. But $gM \leq N$, and ker g is the image of K in M.)

Proposition 3.5.42:

(i) *If* K-dim $M/N < \alpha$ *for all* $0 \neq N < M$ *then* K-dim M *exists and* $\leq \alpha$.

(ii) *If* M *is a Noetherian module then* K-dim M *exists.*

Proof: (i) If $M = M_0 > M_1 > M_2 > \cdots$ is an infinite chain then each K-dim $M_{i-1}/M_i \leq$ K-dim $M/M_i < \alpha$, so K-dim $M \leq \alpha$.

(ii) By Noetherian induction we may assume that K-dim(M/N) exists for each $N < M$. Let $\alpha = \sup\{K\text{-dim}(M/N): N \leq M\}$. By (i) K-dim $M \leq \alpha^+$.

$$\text{Q.E.D.}$$

Thus the Krull dimension forges yet another link from Artinian to Noetherian and has proved useful in low dimensions by providing classes of Noetherian rings in which a richer theory can be developed, as indicated in exercise 36–38. The following lemma provides a useful induction procedure.

Lemma 3.5.43:   *Suppose* $s \in R$ *is left regular. If* K-dim $R = \alpha$ *then* K-dim$(R/Rs) < \alpha$. *In particular, if* $\alpha = 1$ *then* R/Rs *is an Artinian module.*

Proof: $R > Rs > Rs^2 > \cdots$ is a descending series each of whose factors $Rs^i/Rs^{i+1} \approx R/Rs$, so by definition $\alpha > K\text{-dim}(R/Rs)$.   Q.E.D.

Remark 3.5.44:

(i) K-dim $R/A \leq$ K-dim R for all $A \triangleleft R$.

(ii) Suppose R is left Noetherian. K-dim $R = K$-dim R/P for some P in Spec(R).

Proof:

(i) by the change-of-rings functor from $R/A\text{-}\mathcal{M}od$ to $R\text{-}\mathcal{M}od$.

(ii) If $P_1 \cdots P_k = 0$ for prime ideals $P_1, \ldots, P_k$ of R then

$$R > P_k \geq P_{k-1}P_k \geq \cdots \geq 0$$

is a chain each of whose factors is an f.g. R/P_i-module for some i and thus has K-dim $\leq$ K-dim R/P_i, so we are done by proposition 3.5.41.   Q.E.D.

Proposition 3.5.45:   *If $P \in \mathrm{Spec}(R)$ and $P \subset A$ then*

$$\text{K-dim}(R/A) < \text{K-dim}(R/P).$$

Proof: Passing to R/P we may assume $P = 0$. Then A has a regular element s by Goldie's theorem, and K-dim$(R/P) \leq$ K-dim$(R/Ra) <$ K-dim R by lemma 3.5.43.   Q.E.D.

Corollary 3.5.45′:   *Suppose K-dim $R = n < \infty$. There are only a finite number of prime ideals P with K-dim $R/P = n$, and any such P is a minimal prime.*

Proof: By corollary 3.2.26 it is enough to prove P is minimal. But if $P' \subset P$ in $\mathrm{Spec}(R)$ then K-dim $R/P <$ K-dim $R/P' \leq n$, contrary to hypothesis.   Q.E.D.

Proposition 3.5.46:   *If R is prime left Noetherian then K-dim $L =$ K-dim R for every nonzero left ideal L of R.*

Proof: $LR = \sum_{\text{finite}} Lr_i$ for suitable r_i in R. But $LR \lhd R$ and thus has a regular element s. Since $R \approx Rs$ we get

K-dim $R =$ K-dim $Rs \leq$ K-dim $LR \leq \max\{$K-dim $Lr_i\} \leq$ K-dim L.   Q.E.D

The requirement that all nonzero submodules have the same K-dim occurs in the recent literature; such modules are called Krull *homogeneous*. There also is a very important connection between K-dim and torsion for modules over prime rings.

Proposition 3.5.47:   *Suppose R is a prime ring and M is an f.g. module. M is torsion iff K-dim $R >$ K-dim M.*

Proof: Let $\alpha =$ K-dim M. ($\Rightarrow$) Write $M = \sum_{i=1}^{t} Rx_i$. Clearly, some Rx_i has K-dim α. But $Rx_i \approx R/\mathrm{Ann}\, x_i$ so we are done by lemma 3.5.43 since $\mathrm{Ann}\, x_i$ has a regular element.

($\Leftarrow$) Suppose, on the contrary, that M is not torsion. By theorem 3.2.13 there is x in M for which $\mathrm{Ann}\, x$ is not large and thereby misses some left ideal L. Then $L \approx Lx$, so, in view of proposition 3.5.46, we have K-dim $R =$ K-dim $L =$ K-dim $Lx \leq \alpha$, contradiction.   Q.E.D.

Other Krull Dimensions

Before delving into the study of K-dim, let us introduce other noncommutative generalizations of the Krull dimension and see how they compare to K-dim.

Definition 3.5.48:

(i) The *little Krull dimension* $k \dim(R)$ is the length of a maximal chain of prime ideals of R; we permit chains of any ordinal length.

(ii) The *Gabriel-Krause* dimension of R, called "classical" Krull by Gordon-Robson and denoted cl K-dim(R), is defined inductively as follows: Take $\text{Spec}_{-1} = \varnothing$ and $\text{Spec}_\alpha = \{P \in \text{Spec}(R): \text{ Each prime ideal properly}$ containing P is in $\text{Spec}_\beta(R)$ for some $\beta < \alpha\}$. Thus $\text{Spec}_0(R) = \{$maximal ideals of $R\}$. Then cl K-dim(R) is the smallest ordinal α such that $\text{Spec}_\alpha(R) = \text{Spec}(R)$. (Thus $P \in \text{Spec}_\alpha(R)$ iff cl K-dim$(R/P) \leq \alpha$.)

Remark 3.5.48':   When cl K-dim(R) is finite it is clearly the length of a maximal chain of prime ideals and thus equals $k \dim R$.

To compare $k \dim(R)$ with K-dim R we need a couple of preliminary results about prime ideals, which in exercises 31–35 will be seen to hold for all rings having K-dim.

Proposition 3.5.49:   *If R is left Noetherian then $k \dim R \leq$ cl K-dim $R \leq$ K-dim(R).*

Proof: Let $\alpha =$ K-dim(R). First we prove cl K-dim$(R) \leq \alpha$ by showing $P \in \text{Spec}_\alpha(R)$ for every prime ideal P. Indeed, this is clear if P is maximal. On the other hand, if $P \subset P'$ in $\text{Spec}(R)$ then $\alpha \geq$ K-dim$(R/P) >$ K-dim$(R/P') \geq$ cl K-dim(R/P') by induction on K-dim, implying $P' \in \text{Spec}_\beta(R)$ for some $\beta < \alpha$, so $P \in \text{Spec}_\alpha(R)$ by definition.

It remains to show $k \dim(R) \leq$ cl K-dim(R). We shall show by induction on β that if $\text{Spec}(R)$ has a descending chain of length β then cl K-dim $R \geq \beta$. If $\beta = \gamma^+$ then the chain ends in some $P \supset P'$. But cl K-dim$(R/P) \geq \gamma$ by induction, so cl K-dim $R \geq \gamma^+ = \beta$, as desired. Thus we may assume β is a limit ordinal. But for every $\gamma < \beta$ we have a chain of length γ so by induction cl K-dim$(R) \geq \gamma$ for each $\gamma < \beta$, i.e., cl K-dim$(R) \geq \beta$ as desired.
                                                                                    Q.E.D.

Of course if R is commutative Noetherian then the finite height of prime ideals shows $k \dim(R) \leq \omega$. Yet Gordon-Robson [73] give an example of a commutative Noetherian ring with arbitrarily large infinite K-dim. Thus the next proposition shows that cl K-dim can also take on any ordinal value.

Lemma 3.5.50: *Suppose R is a left bounded, prime left Noetherian ring of K-dim α. For any $\beta < \alpha$ there is P in $\mathrm{Spec}(R)$ with K-dim$(R/P) = \beta$.*

Proof: There is an infinite chain $R = L_0 \geq L_1 \geq L_2 \geq \cdots$ with each $L_k < R$ such that K-dim$(L_{k-1}/L_k) \geq \beta$ for each β. By proposition 3.2.19 there is some k such that R does *not* have k independent left ideals. It follows at once there is $i \leq k$ such that L_i is large in L_{i-1}. Let L' be an essential complement of L_{i-1} in R. Then $L_i + L'$ also is large in R and thus contains an ideal $I \neq 0$. Then

$$\beta \leq \text{K-dim}(L_{i-1}/L_i) = \text{K-dim}((L_{i-1} + L')/(L_i + L'))$$

$$\leq \text{K-dim}(R/(L_i + L'))$$

$$\leq \text{K-dim}(R/I) < \alpha.$$

By induction on α there is some prime ideal P/I of R/I such that $\beta = $ K-dim$((R/I)/(P/I)) = $ K-dim(R/P).   Q.E.D.

Proposition 3.5.51: *Suppose R is fully bounded left Noetherian. Then cl K-dim$(R) = $ K-dim R.*

Proof: In view of proposition 3.5.49 we need only show "$\geq$". Let $\alpha = $ K dim(R). Choosing $P_0 \in \mathrm{Spec}(R)$ with K-dim$(R/P_0) = \alpha$ it suffices to prove cl K-dim$(R/P_0) \geq \alpha$, so we may pass to R/P_0 and assume R is prime. We induct on α. For any $\beta < \alpha$ there is $P \in \mathrm{Spec}(R)$ with K-dim$(R/P) = \beta$; by induction $\beta \leq $ cl K-dim(R/P). Thus cl K-dim $R \geq \sup\{\beta \leq \alpha\} = \alpha$.   Q.E.D.

Corollary 3.5.52: *If R is fully bounded left Noetherian and $k \dim(R)$ is finite then $k \dim(R) = $ cl K-dim$(R) = $ K-dim R.*

Example 3.5.53: Suppose R is the Weyl algebra $A_1(F)$ over a field F. Then cl K-dim$(R) = 0$ since R is simple, but K-dim $R \geq 1$ since R is not Artinian. Thus we need the hypothesis of proposition 3.5.51.

In fact K-dim $R = 1$, cf., exercise 25, although K-dim $F[\lambda, \mu] = 2$, cf., exercise 23. Thus K-dim is not preserved under passage to graded algebras.

McConnell [84a] computes K-dim $A_n(T)$ for a wide range of rings T including uncountable fields.

Critical Submodules

Definition 3.5.54:   A module $M \neq 0$ is *critical* if K-dim $M/N <$ K-dim M for all $0 \neq N < M$.

Proposition 3.5.55:   *If $M \neq 0$ has Krull dimension then M has a critical submodule.*

Proof: Let $\alpha =$ K-dim M. Clearly, it is enough to show some submodule M' has a critical submodule, so by induction on α one may assume K-dim $M' = \alpha$ for all $M' < M$. Taking $M_0 = M$ we build $M = M_0 > M_1 > \cdots$ inductively as follows. If M_i is noncritical take $0 \neq M_{i+1} < M_i$ such that K-dim$(M_i/M_{i+1}) = \alpha$. By definition of K-dim M, this process must terminate at some M_i, so M_i is critical.   Q.E.D.

Remark 3.5.56:   Suppose M is critical of K-dim α. Every submodule of M is also critical of K-dim α. (Indeed, if $0 \neq M' \leq M$ then K-dim $M' = \alpha$ by proposition 3.5.41; now for all $0 \neq N \leq M'$ one has K-dim$(M'/N) \leq$ K-dim $M/N < \alpha$.)

We say an ideal A of R is a *partial annihilator* of M if $A =$ Ann N for some $0 \neq N \leq M$. An *assassinator* of M is a maximal partial annihilator of M. Any module over a left Noetherian ring clearly has an assassinator, which is a prime ideal by proposition 3.2.23.

Remark 3.5.57:   Suppose M is an f.g. module over a left Noetherian ring R. Then M has a critical cyclic submodule N whose annihilator P is an assassinator of M, so, in particular, $P \in \mathrm{Spec}(R)$. (Indeed, let P be an assassinator of M. Then $P =$ Ann N for some $N \leq M$. Replacing N by a critical submodule we may assume N is critical; now in view of remark 3.5.56 we may replace N by any nonzero cyclic submodule.)

Question 3.5.93:   (Krull symmetry) Is the left and right K-dim the same for any Noetherian ring?

This question seems to have survived the vicious attack by Stafford [85a] on conjectures in Noetherian ring theory.

N Supplement: Gabriel Dimension

We can define a dimension which is better behaved categorically than K-dim.

Definition 3.5.94:   Define the *Gabriel dimension* G-dim M and α-simple
modules by induction on α, as follows:

  G-dim $0 = 0$.
  M is α-*simple* if G-dim $M/N < \alpha$ for all $0 \neq N \leq M$.
  G-dim $M \leq \alpha$ if every image of M has a nonzero β-simple submodule for
some $\beta \leq \alpha$.

  For example, the 1-simple modules are just the simple modules, so
G-dim $M = 1$ iff every image of M contains a simple submodule. For M
Noetherian this clearly implies M has a composition series, but also any completely reducible module has G-dim 1. Thus a module may have G-dim 1 but
fail to have K-dim! On the other hand, when K-dim M exists then the two
dimension nearly coincide; a discrepancy of 1 arises since G-dim starts with
0 whereas K-dim starts with -1.

Proposition 3.5.95:   *If* K-dim $M = \alpha$ *then* G-dim M *exists and* G-dim $M \leq$
K-dim $M + 1$, *equality holding when* M *is Noetherian.*

Proof: Induction on α. If $\alpha = -1$ then $M = 0$ so G-dim $M = 0$. In general,
let M_0 be some image of M. It suffices to prove M_0 has a β-simple submodule
for some $\beta \leq \alpha + 1$. Let M_1 be a critical submodule of M_0. We are done
unless M_1 does *not* have G-dim $\leq \alpha + 1$. But for every $0 \neq N < M_1$ we have
K-dim $M_1/N < \alpha$ so G-dim $M_1/N \leq \alpha$ by induction, implying M_1 is $(\alpha + 1)$-
simple, as desired.
  In case M is Noetherian let $\alpha' = $ G-dim M. We need to show $\alpha \leq \alpha' - 1$.
Take a β-simple submodule M_0 for $\beta \leq \alpha'$. By Noetherian induction we may
assume K-dim$(M/M_0) = $ G-dim$(M/M_0) - 1 \leq \alpha' - 1$, so it suffices to prove
K-dim $M_0 \leq \alpha' - 1$. But for any $0 \neq N < M_0$ we see $\alpha' > $ G-dim$(M_0/N) \geq$
K-dim$(M_0/N) + 1$ by induction on G-dim, so K-dim $M_0 \leq \alpha' - 1$ by propo-
sition 3.5.42(i).   Q.E.D.

  Thus the Gabriel dimension yields nothing new for Noetherian modules.
For M non-Noetherian one can prove $\alpha \leq$ G-dim $M \leq \alpha + 1$ whenever
K-dim $M = \alpha$ exists, since then M has finite uniform dimension (exercise 28).
Thus Gabriel dimension should be viewed as a possible way of extending the
Krull dimension theory to much more general settings.

Definition 3.5.96:   The *Gabriel dimension*, or *G-dim*, of a ring R is its
Gabriel dimension as R-module.

The G-dim of a ring exists in interesting instances where K-dim does not, cf., exercises 29 and §6.3. One might expect this to stimulate work in Gabriel dimension, but thus far the flowering of the Krull dimension and Gelfand-Kirillov dimension (to be defined in §6.2) have stunted its growth. As the reader may have guessed, the definition in Gabriel [62] is categorical, dealing with Grothendieck categories. Most of the basic properties of K-dim can be extended to G-dim, cf., Gordon-Robson [73], Gordon [74] and Dahari [87]. In particular, G-dim is exact in the sense of remark 3.5.41″, and G-dim $M \leq$ G-dim R for every R-module M.

Exercises

§3.1

1. Prove directly that any Ore domain has a classical ring of quotients, which is a division ring.
2. Conclude the details of the proof of theorem 3.1.4: Well-definedness of addition (as with multiplication), associativity, and distributivity. (Hint: Use the technical lemma for associativity.)
3. Prove the analogue of theorem 1.10.11 for arbitrary denominator sets S.
4. Display the $S^{-1}M$ and M_S as universals with respect to the same functor and conclude thereby that they are isomorphic. Also display this construction as a direct limit.
5. Let $Q = S^{-1}R$ for a denominator set S. Given $M \in R\text{-}\mathcal{M}od$ and $N' \leq S^{-1}M$ in $Q\text{-}\mathcal{M}od$ define $N = \{x \in N : s^{-1}x \in N'\}$. Then $N' = S^{-1}N$; this procedure pulls chains from $\mathscr{L}(_Q S^{-1}M)$ to $\mathscr{L}(_R M)$. In particular, if M is a Noetherian (resp. Artinian) R-module then $S^{-1}M$ is a Noetherian (resp. Artinian) Q-module.
6. If $A \triangleleft R$ and $Q = S^{-1}R$ then for any $q_1, \ldots, q_t$ in QAQ we can find $a_1, \ldots, a_t$ in AQ and s in S such that $q_i = s^{-1}a_i$. This gives a way of comparing ideals of R with ideals of Q.
7. Suppose S is a left denominator set of R. Show the following:

   (i) If R is right perfect then $S^{-1}R$ is right perfect. (Hint: Given r_i in R there is s_i in S with $Rs_i r_i$ as small as possible. If $S^{-1}Rr_1 > S^{-1}Rr_2 > \cdots$ then $Rs_1 r_1 > Rs_2 r_2 > \cdots$).
   (ii) If R is semiprimary then $S^{-1}R$ is semiprimary.

Matric Units and Rings of Fractions

8. If $M_m(R) \approx M_n(R')$ where R, R' are left Ore domains then $m = n$. (Hint: $M_m(R) \subseteq M_n(D)$ for a division ring D, so counting orthogonal idempotents shows $m \leq n$.) In particular, each Ore domain has IBN.
9. Suppose R is an order in a ring Q having a set of $n \times n$ matric units. Then Q has a set E of $n \times n$ matric units such that $Es \subseteq R$ and $s'E \subseteq R$ for suitable regular elements s, s' of R. (Hint: Suppose $\{e_{ij} : 1 \leq i, j \leq n\}$ is a set of matric units of Q. Writing $e_{ij} = s^{-1}r_{ij}$ we see $se_{ij} \in R$ for all i, j. Now let $E = \{se_{ij}s^{-1} : 1 \leq i, j \leq n\}$.)

10. If R has an involution (*) then any denominator set S satisfying $S^* = S$ is also a right donominator set and $S^{-1}R$ has the involution given by $(s^{-1}r) = (1^{-1}r^*)((s^*)^{-1}1)$.

Torsion-Free Modules We say an R-module M is *torsion-free* if $ax \neq 0$ for all $x \neq 0$ in M and all regular a in R. In what follows take Q to be the ring of fractions of an Ore ring R.

11. A torsion-free R-module M is divisible iff M is a Q-module; in fact, taking y to be that element of M satisfying $ay = rx$ show $y = (a^{-1}r)x$ for $a \in R$ regular, $r \in R$, and $x \in M$.
12. R as above. Any torsion-free R-module M is contained in a Q-module M'; if M is f.g. we can take M' f.g. as Q-module. (Hint: Let $E = E(M)$ and $E_0 = \{x \in E : \operatorname{Ann} x$ contains a regular element$\}$. Then E/E_0 is torsion free and divisible, so a Q-module by exercise 11. If $M = \sum Rx_i$ then take $M' = \sum Qx_i$ in E/E_0.)
13. If R is left and right Ore then any f.g. torsion-free R-module M can be embedded in an f.g. free R-module. (Hint: By exercise 12 we can find a monic $f : M \to Q^{(t)}$; each generator of M is in $R^{(t)}s^{-1}$, so we have a monic $M \to R^{(t)}s^{-1} \approx R^{(t)}$.) This result is quite useful in Noetherian theory.

§3.2

1. If $\{P_i : i \in \mathbb{N}\}$ are distinct height 1 prime ideals in a prime ring R satisfying ACC(ideals) then $\bigcap P_i = 0$. (Hint: If $0 \neq r \in P_i$ then R/RrR has only a finite number of minimal prime ideals.)

Orders in Semilocal Rings

2. Let $\bar{R} + R/I$. S is a denominator set for R if $\bar{S}$ is a denominator set of $\bar{R}$ and S satisfies the following two conditions:

   (i) For any s_1 in S and a_1 in I there are s_2 in S and r_2 in R with $s_2 a_1 = r_2 s_1$
   (ii) As in definition 3.2.32(ii).
3. Prove proposition 3.2.34. (Hint: (1), (2) imply $S^{-1}I \lhd S^{-1}R$, and the canonical map $\bar{r} \to 1^{-1}r + S^{-1}I$ yields (3). If (2), (3) holds then (1) holds by exercise 2. Finally, if (1), (3) holds then $S^{-1}I$ is the kernel of the map $S^{-1}R \to \bar{S}^{-1}\bar{R}$ obtained from $r \mapsto \bar{1}^{-1}\bar{r}$.)
4. Prove theorem 3.2.36. (Hint: ($\Leftarrow$) exercise 3 shows S is a denominator set of R, and thus $S^{-1}R/S^{-1}I$ is the ring of fractions of $\bar{R}$. Conclude by showing $S^{-1}I \subseteq \operatorname{Jac}(S^{-1}R)$ since each $s^{-1}a$ in $S^{-1}I$ is quasi-invertible. ($\Rightarrow$) Put $A = \operatorname{Jac}(Q)$ and $I = R \cap A$. $\bar{R}$ is an order in $\bar{Q}$ which is semisimple Artinian.)
5. Suppose $I \lhd R$ and S is a denominator set of R with $S^{-1}I \lhd S^{-1}R$. If I is nil (resp. T-nilpotent) then $S^{-1}I$ is nil (resp. T-nilpotent). (Hint: As in proposition 3.1.14).
6. Suppose R is an order in a semilocal ring Q, and let $N = \operatorname{Nil}(R)$. Then Q is right perfect iff N is T-nilpotent; Q is semiprimary iff N is nilpotent. In each case N is the ideal I of theorem 3.2.36, and, in fact, $N = \operatorname{Jac}(Q) \cap R$. (Hint: ($\Rightarrow$) Take $I = \operatorname{Jac}(Q) \cap R$. Then I is nil so $I \subseteq N$, but R/I is semiprime so $I = N$. ($\Leftarrow$) by exercise 5 and proposition 3.1.15.)

8. (Kerr [84]) A commutative ring with ACC(Ann) having annihilator chains of arbitrarily long length. Let $\lambda = \{\lambda_{ij} : 1 \leq i \leq j < \infty\}$ be a set of commuting indeterminates and let $R = F[\lambda]/l$ where l is the ideal generated by all $\lambda_{ij}\lambda_{mn}\lambda_{uv}$ and all $\lambda_{ij}\lambda_{i'j}$ for $i \neq i'$. Then R is naturally graded as $R = R_0 \oplus R_1 \oplus R_2$ where the polynomials in R_i are homogeneous of degree i. For $m \leq n$ let $A_{mn} = \mathrm{Ann}\{\lambda_{in} : m \leq i \leq n\} = R_2 + \sum_{u=1}^{m-1} F\lambda_{un}$. Then $A_{1n} < A_{2n} < \cdots < A_{nn}$ is a chain of annihilators of length n. However, R satisfies ACC(Ann), seen as follows: First note every zero divisor lies in $R_1 + R_2$ by matching components. If A is any annihilator ideal with $\mathrm{Ann}\, A \supset R_2$ then take $r = r_1 + r_2 \in A$, where $r_i \in R_i$. Then $\mathrm{Ann}\, r_1 = \mathrm{Ann}\, r \supset R_2$ so $R_1 \cap \mathrm{Ann}\, r_1 \neq 0$. Let λ_{mn} (resp. λ_{uv}) be the indeterminate of highest lexicographic order appearing in r_1 (resp. in any element y of $R_1 \cap \mathrm{Ann}(r_1)$). Then $\lambda_{mn}\lambda_{uv} = 0$ so $v = n$, and thus $A \subseteq R_2 + \sum_{u=1}^{k} \sum_{k=1}^{n} F\lambda_{uk}$. Any chain of annihilators descending from A has finite length by a dimension count; hence R satisfies DCC(Ann) and taking annihilators yields ACC(Ann).

Principal Left Ideal Rings (cf., Jategaonkar [70B]

We shall say R is PLI if each left ideal is principal; more generally, R is semi-PLI if each f.g. left ideal is principal. Recall PLIDs were studied in §1.6, also, cf., corollary 2.8.16.

13. $M_n(R)$ is PLI (resp. semi-PLI) iff every left (resp. f.g. left) R-submodule of $R^{(n)}$ is spanned by $\leq n$ elements.

14. "Goldie's third theorem" The following are equivalent: (i) R is a semiprime left Goldie semi-PLI ring; (ii) $R \approx \prod_{i=1}^{n} R_i$ where each R_i is prime left Goldie semi-PLI, and the R_i are uniquely determined up to isomorphism and permutation; (iii) $R \approx \prod_{i=1}^{n} M_{n_i}(T_i)$ where each T_i is left Ore and every f.g. T_i-submodule of $T_i^{(n_i)}$ is spanned by $\leq n_i$ elements. (Extensive hint: (i) $\Rightarrow$ (ii) R is an order in $Q = \prod Q_i$; write $Q_i = Qc_i$ for c_i a central idempotent in Q_i. Write $R' = \prod Rc_i \supseteq R$; we need $R' = R$. There is regular s with $sc_i \in R$ for each i; then $R's$ is a f.g. left ideal of R so $R's = Ru$ for some u in R which must be regular. Then $s = ru$ for r in R, implying $R'r = R$. Hence $r^{-1} \in R'$ so $r^{-1} = r^{-2}r \in R'r = R$ implying $R' = R$. (ii) $\Rightarrow$ (iii) Assume $n = 1$, i.e., R is prime Goldie with ring of fractions $Q = M_n(D)$. It suffices to find a set of $n \times n$ matric units lying in R, since then exercise 13 would enable us to conclude the proof. Let E be a set of $n \times n$ matric units as in exercise 3.1.9, i.e., $Es \subseteq R$ and $s'E \subseteq R$. Then $REs = Rs_1$ for some s_1 in R which must be regular, and $s_1 = qs$ for some q in RE. Hence $RE = Rq$. Note $Er \subseteq R$ iff $qr \in R$ iff $r \in q^{-1}R$. But $E(Eq^{-1}) = Eq^{-1} \subseteq R$ so $Eq^{-1} \subseteq q^{-1}R$ and $qEq^{-1} \subseteq q(q^{-1}R) \subseteq R$ is the desired set of matric units.

15. Suppose $A, B \triangleleft R$ (for R arbitrary). If perchance $A = Ra$ and $B = Rb$ then $AB = Rab$. (Hint: ($\subseteq$) $aR \subseteq AR = A$.)

16. (Johnson's first theorem) Call R fully Goldie if each homomorphic image is left Goldie. R is fully Goldie semi-PLI iff $R \approx \prod_{i=1}^{n} R_i$ where each R_i is fully Goldie semi-PLI with $R_i/\mathrm{Nil}(R_i)$ prime. (Hint: Write $\bar{R} = R/\mathrm{Nil}(R) = \prod \bar{R}_i$, i.e., each $\bar{R}_i$ is isomorphic to an ideal of $\bar{R}$. Write $\bar{R}_i = \bar{A}_i$ where $A_i \triangleleft R$. Then $\sum A_i = R$ so $\sum A_i^u = R$ for every u in $\mathbb{N}$, and one needs to show the A^u are independent for suitable u. By induction it suffices to prove: If $\bar{R} = R/\mathrm{Nil}(R)$ is prime and $A + B = R$ with $\overline{AB} = 0$ then $A^u \cap B^u = 0$ for some u. To see this first show by

the Chinese Remainder Theorem that $AB = \text{Nil}(R) = BA$. Thus $A^n B^n = (AB)^n$. But $(AB)^n = 0$ for some n so $A^n \subseteq \text{Ann}(B^n)$. Conclude with remark 3.2.2(ix).)

Rings for which Principal Annihilators are Summands (PP Rings)

17. Every maximal left annihilator of a ring has the form $\text{Ann}\,r$ for suitable $r \neq 0$.

18. If $e \in R$ is idempotent then $\text{Ann}\,e = R(1 - e)$. (Hint: Peirce decomposition.)

19. Suppose e_1, e_2 are idempotents such that $\text{Ann}_R(e_1(1 - e_2)) = Re_3$ for some idempotent e_3. Then there is an idempotent e satisfying $Re = Re_1 \cap Re_2$. (Hint: Put $L = Re_3$. Then $R(1 - e_1) \subseteq L$ so $e = e_3 e_1$ and $e_3(1 - e_1)$ are orthogonal idempotents of L whose sum is e_3, implying $L(1 - e_1) = R(1 - e_1)$. Clearly, $Re \subseteq L \cap \text{Ann}(1 - e_1) = L \cap Re_1 \subseteq Re_2 \cap Re_1$ for if $re_1 \in L$ then $re_1(1 - e_2) = re_1(e_1(1 - e_2)) = 0$ implying $re_1 \in \text{Ann}(1 - e_2) = Re_2$. But $Re_2 \cap Re_1 \subseteq Re$ since if $r_1 e_1 = r_2 e_2$ then $r_1 e_1(e_1(1 - e_2)) = r_1 e_1(1 - e_2) = 0$ so $r_1 e_1 \in L$ and $r_1 e_1 = (r_1 e_1)e_3$, implying $r_1 e_1 = r_1 e_1^2 = ((r_1 e_1)e_3)e_1 \in Re$.)

20. We say R is a PP-*ring* if every principal left ideal is a summand of R. (This is a common generalization of semihereditary and of regular.) R is a PP-ring iff every annihilator of the form $\text{Ann}(r)$ is a summand of R, i.e., has the form Re for e idempotent. (Hint: The sequence $0 \to \text{Ann}\,r \to R \to Rr \to 0$ splits.) Prove any PP-ring has the following properties:

(i) If R is left Noetherian then R is an order in a left Artinian ring. (Hint: Verify the regularity condition. Suppose $r \in R$ with $\bar{r}$ regular. Write $\text{Ann}\,r = Re$ for e idempotent. Then $\bar{e} = 0$ so $e = 0$. Hence Rr is large. If $rr' = 0$ then $\text{Ann}\,r'$ is a summand of R containing Rr, so $\text{Ann}\,r' = R$ implying $r' = 0$.)

(ii) If R has no infinite set of orthogonal idempotents then every left annihilator $\text{Ann}\,B$ is a summand of R. (Hint: Construct a descending chain of annihilators as follows: Take any b_i in B and put $A_1 = \text{Ann}\,b_1$; given A_i pick b_{i+1} in B, if possible, such that $A_i \not\subseteq \text{Ann}\,b_{i+1}$ and put $A_{i+1} = A_i \cap \text{Ann}\,b_{i+1}$. Using exercise 19 as the inductive step show $A_i = Re_i$ for e_i idempotent, so $Re_1 > Re_2 > \cdots$ which by hypothesis must terminate.)

(iii) Under hypothesis of (ii) show any right annihilator has the form eR since it annihilates a left annihilator. Conclude R satisfies ACC(Ann) and DCC(Ann).

(iv) If R is semiprime Goldie then R is isomorphic to a finite direct product of prime Goldie rings (by theorem 3.2.29).

(v) If R is semiprime left Noetherian then R is isomorphic to a finite direct product of prime left Noetherian rings.

21. (Bergman [71]) The center of a PP ring is PP. In fact, if R is PP and $Rz = Re$ with z in $Z(R)$ and e idempotent then $e \in Z(R)$. (Hint: Completely elementary. Pick r in R and let $r' = [e, r]$. Write $Rr' = Re'$ for e' idempotent. Then $r'z = [z, r] = 0$ so $0 = e'z = ze'$ and thus $ee' = 0$. But $zr' = 0$ implies $er' = 0$ so $er = ere$. Thus $r'e' = 0$ so $0 = (e')^2 = e'$, proving $r' = 0$.)

In the same article Bergman shows the center of a hereditary ring is a Krull domain. The sharpness of this result is shown in exercise 2.8.6

22. Any uniform left ideal L of a PP ring is minimal. (Hint: If $a \in L$ then $Ra = Re$ for e idempotent; conclude $L(1 - e) = 0$.)

23. Suppose M is a torsion-free module over a semiprime Goldie ring R. M is injective iff M is divisible. (Hint: ($\Leftarrow$) By Baer we need show only that every map $f: L \to M$ extends to a map $g: R \to M$. Furthermore, one may assume L is large and thus contains a regular element a. Define $gr = ra^{-1}fa$.)

Fractional and Invertible Ideals In what follows suppose R is a subring of a ring Q. A Q-*fractional left ideal* is an R-submodule of Q; define $L^{-1} = \{q \in Q: Lq \subseteq R\}$. L is called *invertible* if $1 \in L^{-1}L$, i.e., $\sum q_i a_i = 1$ for a_i in L and q_i in L^{-1}. Use this notation below.

24. If $r \in R$ is invertible in Q then Rr is an invertible left ideal.
25. Any invertible Q-fractional left ideal is f.g. and projective as R-module. (Hint: $L = \sum (Lq_i)a_i$. Define $f_i: L \to R$ by right multiplication by q_i. Then (a_i, f_i) is a dual base.)
26. Conversely to exercise 25, if a Q-fractional left ideal L is projective in R-*Mod* and contains an element r of R invertible in Q then L is invertible. (Hint: Letting (a_i, f_i) be a dual base, define $q_i = r^{-1}f_i a_i \in Q$. Then $r = \sum (rq_i)a_i$, so $1 = \sum q_i a_i$.)
27. A semiprime Goldie ring R is hereditary iff every large left ideal is invertible with respect to the classical ring of quotients Q of R. (Hint: ($\Rightarrow$) Every left ideal is the summand of a large left ideal.)

Rings without 1 and the Faith-Utumi Theorem

28. If a division ring D is an essential ring extension of R (i.e. as R-module) then R is an Ore domain with D as its ring of fractions.
29. Define rings of fractions for rings without 1, with respect to denominator sets. Note the ring of fractions has multiplicative unit $s^{-1}s$.
30. Suppose Q is an essential ring extension of a semiprime ring R and e is an idempotent of Q such that $D = eQe$ is a division ring with $R \cap eQe \neq 0$. Then D is an essential ring extension of $R \cap eQe$ (possibly as ring without 1).
31. "Faith-Utumi theorem". Suppose R is an order in $Q = M_n(D)$. Then R contains $M_n(T)$ where T is a suitable order (without 1) of D, with respect to a suitable set of matric units. (Hint: By exercise 3.1.9 pick a set E of $n \times n$ matric units of Q with respect to which $Es \subseteq R$ and $s'E \subseteq R$. Let $L' = \{r \in R: rE \subseteq R\}$, a large left ideal of R and $L'' = \{r \in R: Er \subseteq R\}$. Let $T_0 = L''L'$, a prime subring (without 1) of R. For any $r \neq 0$ in R we have $L' \cap L'r \neq 0$ so $T_0 \cap T_0 r \neq 0$, implying R (and thus Q) is an essential ring extension of T_0. $T = T_0 \cap e_{11}Qe_{11}$ contains $e_{11}ss'e_{11} \neq 0$ and is a left order in D by exercise 28 and 30.)

Faith [73B, p. 414–416] uses exercise 31 to determine the structure of semiprime principal left ideal rings and of simple Goldie rings—if such a ring R is an order in $M_n(W)$ then $R \approx \text{End}_T M$ where T is an order in W and M is a torsion free T-module.

Nil Subsets (also, cf., exercise 3.3.18)

32. (Shock's theorem) Suppose R satisfies ACC(Ann) and N is a nil multiplicative subset of R. If N is not nilpotent then there are $r_1, r_2, \ldots$ in N with $\{Rr_i: i \in \mathbb{N}\}$

independent, also satisfying $A_0 < A_1 < \cdots$ where $A_i = \text{Ann}'(\sum_{j \geq i} R r_j)$. (Hint: Put $T_i = \text{Ann } N^i$; some T_n is maximal among the T_i. Take $r_0 = x_0 \in N - T_n$ with $\text{Ann } x_0$ maximal possible. Inductively, given $r_i \notin T_n$ put $N_i = \{x \in N : r_i x \notin T_n\} \neq \emptyset$ since $r_i N \nsubseteq T_n$, and picking x_{i+1} in N_i with $\text{Ann } x_{i+1}$ maximal, define $r_{i+1} = r_i x_{i+1} = x_0 \cdots x_{i+1}$. In the remainder, assume throughout $i < j$.

To prove the properties define $r_{ii} = 1$ and $r_{ij} = x_{i+1} \cdots x_j$. Then $r_j x_i = (r_{i-1} x_i r_{ij}) x_i$ for $i \leq j$. Since $x_i r_{ij} \in N$ is nilpotent, $\text{Ann } x_i r_{ij} x_i > \text{Ann } x_i$ so $x_i r_{ij} x_i \notin N_{i-1}$ and so $r_{i-1} x_i r_{ij} x_i \in T_n$, i.e., $r_j x_i \in T_n$; thus $r_j r_{i-1,i+n} \in r_j x_i N^n = 0$.

In particular, $r_{i,i+n+1} \in A_{i+1} - A_i$ proving $A_i < A_{i+1}$. It remains to show that the $R r_i$ are independent, i.e., if $\sum_{u=i}^{k} a_u r_u = 0$ for a_u in R then $a_i r_i = 0$. But $0 = (\sum_{u=i}^{k} a_u r_u) r_{i,i+n+1} = a_i r_i r_{i,i+n+1} + 0 = a_i r_{i+n+1}$ so $a_i r_{i-1} \in \text{Ann}(x_i r_{i,i+n+1}) = \text{Ann } x_i$ by choice of x_i, implying $0 = a_i r_{i-1} x_i = a_i r_i$ as desired.)

33. As corollaries to Shock's theorem, prove that nil multiplicative subsets of the following types of rings are necessarily nilpotent: (i) left Goldie rings (Lanski); (ii) rings satisfying both ACC(Ann) and ACC(Ann') (Levitzki; independently Herstein-Small). There does not yet seem to be an example known of a ring satisfying ACC(Ann) and having a nil multiplicative subset which is *not* nilpotent.

Embeddings into Artinian Rings

40. (Small) A left Noetherian ring which is embeddible neither in a left nor right Artinian ring. Let W be the Weyl algebra $A_1(\mathbb{C})$ and taking any nonzero left ideal L let $R = \begin{pmatrix} W & M \\ 0 & \mathbb{C} \end{pmatrix}$ where $M = W/L$. If $S \subset M$ then $\text{Ann} \begin{pmatrix} 0 & S \\ 0 & 0 \end{pmatrix} = \begin{pmatrix} \text{Ann } S & M \\ 0 & \mathbb{C} \end{pmatrix}$. It follows that R fails DCC(Ann) and thus is not embeddible in a left Artinian ring. (Hint: Otherwise, take a finite set $S \subset M$ with $0 \neq L_0 = \text{Ann}_R S$ minimal possible. Then $L_0 = \text{Ann}(S \cup \{x\})$ for any x in M so $L_0 M = 0$, implying $0 \neq L_0 \lhd W$, contrary to W simple.) Note R is *not* right Noetherian and cannot be embedded in a right Noetherian ring. This exercise shows embeddibility does not "lift" modulo the nilradical.

§3.3

0. $(N x^{-1}) x = N \cap R x$. Consequently, if $\text{Sing}(R) = 0$ and $L r^{-1}$ is a large left ideal then L is large in $L + R r$; conclude if $R r_1$ and $R r_2$ are large in a nonsingular ring R then $R r_1 r_2$ is large. This is part of the direct proof of Goldie's theorem by Zelmanowitz [69]. A more structural proof is given in exercises 12–14.

1. Any reduced ring R is nonsingular (since $\text{Ann}_R r \lhd R$).

2. Modifying triangular matrices construct a ring R where $\text{Sing}(R/\text{Sing}(R)) \neq 0$.

3. A module M is nonsingular iff $\text{Hom}(K, M) = 0$ for every singular module K.

4. The direct product of nonsingular modules is nonsingular. (Hint: Use exercise 3) The ultraproduct of nonsingular rings is a nonsingular ring.

5. If $0 \to M_1 \to M \to M_2 \to 0$ is exact with M_1, M_2 nonsingular then M is nonsingular. (Hint: exercise 3).

6. If R satisfies ACC(Ann) then $l = \text{Sing}(R)$ is nilpotent. (Hint: Assume $l^k \neq 0$ for all

k. Take a_k in I such that $a_k l^k \neq 0$ and Ann a_k is maximal such. Show Ann$(a_k a) >$ Ann(a_k) for all a in I, implying $a_k a l^k = 0$; thus $a_k l^{k+1} = 0$ so Ann$(l^{k+1}) >$ Ann(l^k) for all k, contradiction.)

7. Define $\mathrm{Sing}_2(M)$ to be that submodule N where $N/\mathrm{Sing}(M) = \mathrm{Sing}(M/\mathrm{Sing}\, M)$. Show $M/\mathrm{Sing}_2(M)$ is a nonsingular module, and $R/\mathrm{Sing}_2(R)$ is a nonsingular ring.

8. Every PP ring (c.f., exercise 3.2.20) is nonsingular. (Hint: $0 \to \mathrm{Ann}\, r \to R \to Rr \to 0$ splits.)

8'. Every regular ring is nonsingular. Every semihereditary ring is nonsingular. (Hint: Exercise 8.)

9. $\begin{pmatrix} \mathbb{Z}/2\mathbb{Z} & \mathbb{Z}/2\mathbb{Z} \\ 0 & \mathbb{Z} \end{pmatrix}$ is nonsingular, but the *right* annihilator of e_{12} is large.

10. (Osofsky [64]) A ring R whose injective hull E (in $R\text{-}\mathcal{M}od$) cannot be given a ring structure. Let $T = \mathbb{Z}/4\mathbb{Z}$ and $R = \begin{pmatrix} T & 0 \\ 2T & T \end{pmatrix} \subset M_2(T)$, and $L_i = 2Re_{2i}$ for $i = 1, 2$, and let E_i be the injective hull of L_i (so $E_i \subset E$). Since Re_{ii} is an essential extension of L_i we may assume $e_{ii} \in E_i$ for $i = 1, 2$; likewise $e_{21} \in E_1$. There is an isomorphism $f: L_2 \to L_1$ given by $f(2e_{22}) = 2e_{21}$, which extends to an isomorphism $g: E_2 \to E_1$. But $L_1 < Re_{11}$ so $E_1 < E(Re_{11})$ and is thus a summand. Letting $\pi: E(Re_{11}) \to E_1$ be the projection and $h = g^{-1}\pi$, let $x = he_{11} \in E_2$. Then $e_{11}x = x$ so $2x = 0$ (since, otherwise, there is r in R with $0 \neq 2rx = 2re_{11}x \in L_2$, impossible). If E were a ring then $0 = (ge_{22})2x = 2e_{21}x = h(2e_{21}) = 2e_{22}$, contradiction.

11. If Q is an essential ring extension of R then $\mathrm{Sing}(R) \subseteq \mathrm{Sing}(Q)$. Consequently, if Q is regular then R is nonsingular, by results of §3.3.

Closed Submodules and Goldie's Theorem

12. Show the following chain conditions on M are equivalent: (i) ACC$(\oplus)$; (ii) ACC(closed submodules); (iii) DCC(closed submodules). Hint: (i) $\Rightarrow$ (ii) by means of essential complements. (ii) $\Rightarrow$ (iii) as in exercise 0.0.13. (iii) $\Rightarrow$ (i) Given $\{N_i : i \in \mathbb{N}\}$ independent submodules build a descending chain $K_1 > K_2 > \cdots$ of closed submodules inductively by taking $K_{j+1} \geq \sum_{i=j+1}^{\infty} N_i$ an essential complement of N_j in K_j.)

13. Every left annihilator in a nonsingular ring is closed. (Hint: Suppose $A = \mathrm{Ann}_R B$ is large in A'. Then A'/A is singular so for any a in A' there is large $L < R$ with $LaB = 0$.)

14. A semiprime ring satisfying ACC$(\oplus)$ is nonsingular iff it is left Goldie. (Hint: ($\Rightarrow$) exercises 12 and 13.) In this case identify $\hat{R}$ with the classical ring of fractions of R. (Hint: $s \in R$ is regular iff Ann s is large.)

Closed Submodules

15. Suppose N is a submodule of M. If $\mathrm{Sing}(M/N) = 0$ then N is closed; the converse holds when M is nonsingular. (Hint: ($\Rightarrow$) If N is large in $N' \leq M$ then $N'/N = 0$ by remark 3.3.12. ($\Leftarrow$) Suppose $x + N \in \mathrm{Sing}(M/N)$. Then $(Rx + N)/N$ is singular, so N is large in $Rx + N$ implying $x \in N$.)

16. A prime ring has no proper closed ideals other than 0. The only proper closed ideals of $\mathbb{Z} \times \mathbb{Z}$ are $(\mathbb{Z}, 0)$ and $(0, \mathbb{Z})$.

17. If $A < R$ is closed as left ideal and R is semiprime self-injective, then $A = Re$ for some central idempotent e of R. (Hint: A is injective and thus a summand of R; apply lemma 3.2.28.)

Finite Dimensional Modules In the following exercises write $\dim(M)$ for the uniform dimension of M.

18. If $\dim(M) = n$ then the injective hull $E(M)$ is a direct product of n simple modules. (Hint: Start with the case $n = 1$ and apply injective hulls to proposition 3.2.19.) It follows that if M satisfies $\text{ACC}(\oplus)$ then $E(M)$ has a composition series, so $\text{End}_R E(M)$ is semiprimary. Using this in conjunction with theorem 2.6.31 and exercise 2.10.20 prove the following theorem of Fisher: Any nil subring of $R/\text{Sing}(R)$ is nilpotent.

19. If N is a closed submodule of M then $\dim(M) = \dim(N) + \dim(M/N)$. On the other hand, if N is a large submodule of M then $\dim(N) = \dim(M)$.

Self-Injective Rings which are Not Necessarily Regular

20. Suppose R is *right* self-injective. Then (i) $\text{Ann}(A \cap B) = \text{Ann } A + \text{Ann } B$ for all right ideals A, B of R; (ii) every f.g. left ideal is a left annihilator. (Hint: (i) Given r in $\text{Ann}(A \cap B)$ define $f: A + B \to R$ by $f(a + b) = a + (1 + r)b$. Then f is a well-defined right module map and thus is given by left multiplication by some x in R. Then $(1 + r - x)b = (x - 1)a$ for all a, b in R, implying $1 + r - x \in \text{Ann } B$ and $x - 1 \in \text{Ann } A$ so $r \in \text{Ann } A + \text{Ann } B$, as desired. (ii) Write $L = \sum_{i=1}^{t} Rx_i$ and $L' = \text{Ann}' L$. Clearly $L \subseteq \text{Ann } L'$. On the other hand, $\text{Ann } L' = \sum \text{Ann Ann}' x_i$ by (i). If $y \in \text{Ann}(\text{Ann}' x_i)$ then the map $f: x_i R \to R$ defined by $f(x_i r) = yr$ is given by left multiplication by some w in R, so $y = fx_i = wx_i \in L$.)

21. If R satisfies ACC (f.g. left ideals) then R is left Noetherian. In particular, this holds if R satisfies ACC(Ann) and every f.g. left ideal is a left annihilator.

22. Suppose R is left self-injective and right perfect. Then (i) every simple module is isomorphic to a left ideal; (ii) $\text{Ann}(\text{Ann}' L) = L$ for every left ideal L. (Hints: (i) Take a basic set of orthogonal primitive idempotents $e_1, \ldots, e_t$ of R. Any simple R-module is also an $R/\text{Jac}(R)$-module by Nakayama's lemma and thus isomorphic to a minimal left ideal of $R/\text{Jac}(R)$, so there are at most t of them. But each Re_i contains a minimal left ideal L_i and $Re_i = E(L_i)$, so $L_1, \ldots, L_t$ are nonisomorphic and are thus all the simple modules. (ii) Let $L' = L + Rx$ for x in $\text{Ann}(\text{Ann}' L)$. L is contained in a maximal proper submodule M of L'; an injection $f: L'/M \to R$ (by (i)) yields a map $\hat{f}: L' \to R$ which then can be given by right multiplication by some r in R. Then $Lr = 0$ so $r \in \text{Ann}' L$, implying $0 = L'r = fL'$, contradiction.)

Quasi-Frobenius Rings A ring is quasi-Frobenius (*QF*) if it is left and right Artinian and left and right self-injective. The structure of these rings is so rich that they have become an object of close study, especially with respect to their modules, as we shall see later on. Our main object here is to weaken the definition.

23. If R satisfies ACC(Ann) and is left *or* right self-injective then R is *QF*. (Hint: *Case I. R* is right self-injective. Then R is left Noetherian by exercises 20, 21 so $R/\text{Jac}(R)$ is left Noetherian and regular by exercise 2.10.20; hence R is semiprimary and thus left Artinian. $l = \text{Ann}'\,\text{Ann}\,l$ for each right ideal l of R by the right-hand version of exercise 22, so R is right Noetherian as well as semiprimary; and thus is right Artinian. To show R is right self-injective it suffices to show for each right ideal l that every map $f: l \to R$ is given by left multiplication by an element r of R. Do this by induction on t where $l = \sum_{i=1}^{t} a_i R$. Namely, let $l' = \sum_{i=1}^{t-1} a_i R$ and write $fa = r'a$ for $a \in l'$. Write $r'' = fa_t$; then $f(a_t r) = r''r$. But $r' - r'' \in \text{Ann}(l' \cap a_t R) = \text{Ann}(\text{Ann}'\,\text{Ann}\,l \cap \text{Ann}'\,\text{Ann}\,a_t) = \text{Ann}\,l + \text{Ann}\,a_t$, so write $r' - r'' = s' + s''$ where $s'l = 0$ and $s''a_t = 0$. Put $r = r' - s' = r'' + s''$. *Case II. R* is left self-injective. Using exercise 20 show R satisfies DCC(principal right ideals) and is thus left perfect. In fact, R is semiprimary for if we take n with $\text{Ann}\,J^n$ maximal then $J^n = 0$, seen by applying lemma 2.7.37 to a minimal submodule of $R/\text{Ann}\,J^n$. Hence every left ideal of R is a left annihilator by exercise 22, so R is left Noetherian and thus left Artinian. Done by Case I.)

24. Suppose R is *QF*. Then $\text{Ann}(\text{Ann}'\,L) = L$ for every left ideal L, so Ann' yields a 1:1 correspondence between {minimal left ideals} and {maximal right ideals}. Using injectivity show if M is a simple R-module then $\text{Hom}(M, R)$ is a simple right R-module under the natural action. Using exercise 2.7.9' show $\text{soc}(R) = \text{Ann}\,\text{Jac}(R)$. From this conclude that the left and right socles coincide. There is a bijection between principal indecomposable modules and minimal left ideals, by $Re \to \text{soc}(R)e$, whose inverse correspondence is given by means of the injective hull (c.f., hint for exercise 22).

25. If R is *QF* then an R-module is projective iff it is injective. (Hint: ($\Leftarrow$) Any indecomposable injective has the form $E(L)$ where L is a simple module and thus isomorphic to a minimal left ideal; hence $E(L)$ is a summand of R.)

26. Every module over a *QF*-ring is a submodule of a free module. (Hint: Apply exercise 25 to the injective hull.)

27. If R is a PLID and $a \neq 0$ is a normalizing element of R then R/Ra is *QF*. (Hint: R/Ra is left Noetherian and selfinjective.)

Frobenius Algebras A finite dimensional algebra R is *Frobenius* if $R \approx R^* = \text{Hom}_F(R, F)$ as R-modules (c.f., remark 1.5.18).

28. Any Frobenius algebra is *QF*. (Hint: Apply exercise 2.10.4.)

29. The following are equivalent: (i) R is Frobenius; (ii) R has a nondegenerate bilinear form $\langle , \rangle$ which is *associative* in the sense $\langle ab, c \rangle = \langle a, bc \rangle$ for all a, b, c in R; (iii) There is an F-map $f: R \to F$ whose kernel contains no nonzero left ideal. In this case $[R:F] = [L:F] + [\text{Ann}'\,L:F]$ for every left ideal L. (Hint: (i)$\Rightarrow$(ii) Given the isomorphism $\varphi: R \to R^*$ define $\langle , \rangle$ by $\langle a, b \rangle = (\varphi b)a$. (ii)$\Rightarrow$(iii) Define $f: R \to F$ by $fa = \langle a, 1 \rangle$. (iii)$\Rightarrow$(i) Define the monic $\varphi: R \to R^*$ by $(\varphi r)a = f(ar)$. Since $[R^*:F] = [R:F]$ conclude φ is also epic. The last assertion follows from $\text{Ann}'\,L = \{r \in R: \langle L, r \rangle = 0\}$.)

30. An algebra is *symmetric* if it has a symmetric associative bilinear form, c.f., exercise 29. Every symmetric algebra has IBN. Every f.d. symmetric algebra is Frobenius. Using trace maps (c.f., definition 1.3.28) one can easily show any group algebra is symmetric, as well as any f.d. semisimple algebra over a field.

Regular Self-Injective Rings and Goodearl's First Theorem

36. Suppose Q is an essential ring extension of R. The following are equivalent: (i) R is nonsingular and $Q = \hat{R}$; (ii) Q is regular and self-injective; (iii) Q is self-injective and R is nonsingular. (Hint: (ii) $\Rightarrow$ (iii) Q is nonsingular as an R-module. (iii) $\Rightarrow$ (i) The inclusion $R \to \hat{R}$ lifts to a map $Q \to \hat{R}$, which is a ring injection since the map $g: Q \to \hat{R}$ given by $gy = fxfy - f(xy)$ vanishes on R and is thus 0. But then $\hat{R}$ is an essential ring extension of Q.

In exercises 37-41 we study a regular self-injective ring Q. By exercise 36 $\text{Sing}(Q) = 0$, and $\hat{Q} = Q$.

37. Every f.g. nonsingular Q-module is injective and projective. (Hint: Take $f: Q^{(n)} \to M$ epic. Theorem 3.3.21 yields $F_J f: F_J Q^{(n)} \to F_J M$; but $F_J Q^{(n)} = Q^{(n)}$ so $F_J f = f$. Hence $\hat{M} = fQ^{(n)} = M$ is injective. But $\ker f = \ker F_J f = E(\ker f)$ is a summand of $Q^{(n)}$ so f splits.)

38. We say a module N is *subisomorphic* to a module M, written $N \lesssim M$, if there is a monic $f: N \to M$. For any nonsingular injective Q-modules M, N there is a suitable idempotent e of Q for which $eM \lesssim eN$ and $(1 - e)N \lesssim (1 - e)M$. (Hint: Let $\mathcal{S} = \{\text{triples } (M', N', f'): M' \leq M, N' \leq N, \text{ and } f': M' \to N' \text{ is an isomorphism}\}$. By the maximal principle there is a maximal such triple (M', N', f'), in the sense that f' cannot be extended further to a map $f'': M'' \to N''$ where $M' \leq M''$ and $N' \leq N''$. In particular, $M' = E(M')$ and $N' = E(N')$. Hence $M = M' \oplus J$ and $N = N' \oplus K$ for suitable nonsingular injectives J, K. Let $A = \text{Ann}_Q J$. Then Q/A is nonsingular so exercises 15, 17 show $A = Qe$ for some central idempotent e. Then $eM = eM' \approx eN' \leq eN$.

Likewise, to show $(1 - e)N \lesssim (1 - e)M$ one needs to prove $(1 - e)K = 0$. So suppose $0 \neq y \in (1 - e)K$. Then Qy is projective so $Qy \approx Qe'$ for some idempotent e'. But $e \in \text{Ann}(Qy)$ so $ee' = 0$ and $e' \in Q(1 - e)$. Hence there is $0 \neq x = e'b \in e'J$ for suitable b. Qx is projective so right multiplication by b yields a split epic $Qe' \to Qx$. Thus Qx is isomorphic to a summand of Qy, and there is a monic $g: Qx \to Qy$; $(M' + Qx, N' + gQx, f' + g)$ extends (M', N', f'), contradiction.)

39. Suppose L is a left ideal of Q, and $r \in Q$. Then $r \in LQ$ iff $Qr \lesssim L^{(n)}$ for some n. (Hint: ($\Rightarrow$) Write $r = \sum_{i=1}^{n} a_i q_i$ for $a_i \in L$ and $q_i \in Q$. Define the map $f: L^{(n)} \to Q$ by $f(b_i) = \sum b_i q_i$. Then $r \in fL^{(n)}$. Then exercise 37 yields a monic $Qr \to f^{-1}(Qr) \lesssim L^{(n)}$. ($\Leftarrow$) Qr is a summand of $L^{(n)}$ by exercise 37. Let $\mu_i: L \to L^{(n)}$ be the injection to the i-th component, and let $\pi: L^{(n)} \to Qr$ be the projection. Then $\pi\mu_i$ is given by right multiplication by suitable q_i in Q so $Qr = \sum Lq_i \leq LQ$.)

40. (Goodearl's first theorem) If $P \in \text{Spec}(Q)$ then the lattice of ideals of Q/P is totally ordered. (Hint: Given ideals $A, B \supset P$ with $B \not\subseteq A$ one needs $A \subseteq B$. Take b in $B - A$ and arbitrary a in A. Then Qa, Qb are injective so there is a central idempotent e of with $Qea \lesssim Qeb$ and $Q(1 - e)b \lesssim Q(1 - e)a$. If $e \in P$ then $Qb =$

$Q(1 - e)b + Qeb \lesssim A^{(2)}$ so $b \in AQ = A$ by exercise 39. Thus $e \notin P$. Then $1 - e \in P$ so $a \in B$ by analogy. Thus $A \subseteq B$.)

41. If R is prime nonsingular then the lattice of ideals of $\hat{R}$ is totally ordered.

§3.4

1. $\mathbb{Z}/4\mathbb{Z}$ has the large $\mathbb{Z}$-submodule $2\mathbb{Z}/4\mathbb{Z}$ which is not dense.
2. If $N \leq M$ is dense and $y \in M$ then $L = Ny^{-1}$ satisfies $\mathrm{Ann}'_M L = 0$. (Hint: $R/L \approx (N + Ry)/N$ so $\mathrm{Hom}(R/L, M) = 0$.)
3. If $N \leq M$ is dense and $f: M' \to M$ is a map with $M' \leq M$ then $f^{-1}N$ is a dense submodule of M'. (Hint: $M'/f^{-1}N \approx (N + fM')/N$.)
4. $N \leq M$ is dense iff for every nonzero x, y in M there is r in R with $rx \neq 0$ and $ry \in N$. (Hint: ($\Rightarrow$) by exercise 2: ($\Leftarrow$) Suppose there is nonzero $f: M' \to M$ with $N \leq \ker f$. Take $x = fy \neq 0$, and get a contradiction.) In other words, N is dense iff $Ny^{-1}x \neq 0$ for all x, y in M.
5. Suppose $N \leq N' \leq M$. $N \leq M$ is dense iff $N \leq N'$ and $N' \leq M$ are dense. (Hint: ($\Leftarrow$) Suppose $N \leq M' \leq M$ and $f: M' \to M$ with $fN = 0$. Restrict f to a map g: $M' \cap N' \cap f^{-1}N' \to N'$, which is 0 by hypothesis. Hence $f(M' \cap N') = 0$ by exercise 3 and conclude as in proposition 3.4.4.)
6. Given $M \in R\text{-}\mathcal{M}od$ define $E_r(M) = \bigcap\{\ker f: f \in \mathrm{End}_R(E(M)) \text{ and } fM = 0\}$. This is a rational extension of M and if N is any rational extension of M then the inclusion $M \to E_r(M)$ extends to a monic $N \to E_r(M)$, implying $E_r(M)$ has no proper rational extensions.
7. Identify $E_r(M)$ as the module constructed in theorem 3.4.6.
8. A (left) *quotient ring* of R is a ring Q which is a rational extension of R (as R-module). Show the maximal quotient ring Q is maximal in the sense that for any quotient ring Q' of R the injection $R \to Q$ extends to a ring injection $Q' \to Q$.
9. The maximal ring of quotients Q of R is self-injective iff Rx^{-1} is dense in R for each x in $E(R)$. In this case $Q \approx \mathrm{Hom}_R(E(R), E(R))$.

§3.5

In exercises 1 through 7 assume R is left Noetherian, and $S = \{r \in R: \mathrm{Ann}_R r = 0\}$ and $N = \mathrm{Nil}(R)$. We say $r \in \mathscr{C}(N)$ if $r + N$ is regular in R/N.

Reduced Rank

1. (Goldie) If $r \in R$ and $s \in S$ then there are r', s' in R with $s'r = r's$, such that $s' \in \mathscr{C}(N)$. (Hint: $R \approx Rs$ so $\rho(R/Rs) = 0$. Thus there is s' regular mod N such that $s'(r + Rs) = 0$ i.e., $s'r \in Rs$.)
2. (Djalabi) If $s \in S$ then $s \in \mathscr{C}(N)$. (Hint: Rs has an element in $\mathscr{C}(N)$ by exercise 1, implying s is regular mod N, seen by passing to the ring of fractions of R/N.)
3. (Small) A left Noetherian ring R is an order in a left Artinian ring iff it satisfies the *regularity condition*: If $r + \mathrm{Nil}(R)$ is regular in $R/\mathrm{Nil}(R)$ then r is regular in R. (Hint: S is a left denominator set by exercises 1 and 2. To show $Q = S^{-1}R$ is left Artinian note that if $L_1 < L_2 < Q$ then $S \cap \mathrm{Ann}_R(L_2/L_1) \neq \varnothing$ so $\rho((L_2 \cap R)/(L_1 \cap R)) > 0$ and $\rho(L_1 \cap R) < \rho(L_2 \cap R)$; this implies length $(Q) \leq \rho(R)$.)

4. A Noetherian ring R is a left order in a left Artinian ring Q, iff R is a right order in Q and Q is a right Artinian ring. (Hint: Small and proposition 3.1.16.)

5. Show for $R = \begin{pmatrix} \mathbb{Z} & F \\ 0 & F \end{pmatrix}$ and $F = \mathbb{Z}/2\mathbb{Z}$ that R is Noetherian but the left and right reduced ranks of N are unequal.

6. Show in exercise 5 that $Ns = N$ holds for all s in $\mathscr{C}(N)$. Generalize Robson's theorem to show that if $Ns = N$ for all s in $\mathscr{C}(N)$ then R has an idempotent e such that $eR(1 - e) = 0$ with eRe Artinian and $(1 - e)R(1 - e)$ semiprime, and conversely.

7. If R is as in exercise 6 and R/N is prime then R is prime or Artinian.

More Counterexamples of Jategaonkar

8. (Jategaonkar) Suppose T is a PLID contained in a division ring D and suppose there is an injection $\sigma: D \to T$. The skew power series ring $R = T[[\lambda; \sigma]]$ then has the following properties: (i) R is a PLID; (ii) $r \in R$ is a unit iff its constant term is a unit in T. (iii) For any a in T we have $\lambda^n = (\sigma^n a^{-1})\lambda^n a \in Ra$ for each n. (iv) Every left ideal L of R has the form $Ra\lambda^n$ where $a \in T$. (v) $Ra\lambda^n \lhd T$ iff $a\sigma^n b \in Ta$ for all b in T; (vi) Suppose $\text{Jac}(T) = Tb \neq 0$. Letting $J = \text{Jac}(R)$ one then has $J = Rb$ and $\lambda \in \bigcap_{i=1}^{\infty} J^i$. In particular, Jacobson's conjecture fails. (vii) If T is local then R is local. (Hints: (i) as in proposition 1.6.24, using terms of *lowest* degree; (ii) as in proposition 1.2.24; (iii) clear; (iv) Write $L = Rr\lambda^n$ where r has constant term $a \neq 0$, and apply (iii); (v) clear; (vi) Suppose $L = Ra\lambda$ is a maximal left ideal. If $n \geq 1$ then $n = 1$ and $Ra = R$; but $R\lambda \subset Rb$ by (iii), contradiction. Hence $n = 0$, so $L = Ra$. It follows that $Rb \subseteq J$. Hence $\lambda^n \in J$ for each n, by (iii), so $r \in J$ iff its constant term is in J. Consequently, $J = Rb$. $\lambda \in Rb^n \subseteq J^n$ for each n; (vii) clear from (vi).

9. (A local PLID in which Jacobson's conjecture fails) Suppose Λ_1, Λ_2 are countable sets of commuting indeterminates over a field F_0. Let $F = F_0(\Lambda_1)$ and let T be any localization of $F[\Lambda_2]$ in $D = F(\Lambda_2)$. Then there is an injection $\sigma: D \to F$ since $D = F_0(\Lambda_1 \cup \Lambda_2) \approx F$. Thus we take $R = T[[\lambda; \sigma]]$ and apply exercise 8. Note $R/J \approx F$ is a field.

10. In exercise 9 the commutative primary decomposition theorem fails for $J\lambda$ where $J = \text{Jac}(R)$. (Indeed, write $J = Rb$ and note $b^n \notin J\lambda$ and $\lambda \notin J\lambda$, but $J\lambda$ is not the intersection of ideals properly containing it.) Likewise, the Artin-Rees property fails for $M = R$, $N = J\lambda$, for $J^n M \cap N = N \supset J^{n-t+1}\lambda = J^{n-t}(J^t M \cap N)$ for $t < n$. (This uses the fact $J^t = Rb^t$, cf., exercise 3.2.15.)

11. Exercise 9 can be improved using example 2.1.36 (also due to Jategaonkar) Continuing with the notation there, let $S = \{f \in R_\alpha: \text{when written in standard}$ form f has some summand in $D = R_0\}$, e.g., $1 + \lambda_1\lambda_2 + \lambda_1\lambda_3 \in S$. Show every element in $R - \{0\}$ has the form sq where $s \in S$ and q is a monic monomial in standard form. Furthermore S is a left denominator set, and $S^{-1}R$ is a local ring. If we define inductively $J_\beta = \bigcap_{i=1}^{\infty} J_\gamma^i$ for $\beta = \gamma^+$ a successor ordinal and $J_\beta = \bigcap_{\gamma < \beta} J_\gamma$ for β a limit ordinal, $J_0 = \text{Jac}(S^{-1}R)$, then $\lambda_\beta \in J_\beta$ for every successor ordinal β, and thus $J \neq 0$. To show S is a denominator set suppose $s \in S$ and $r \in R - 0$. Write $Rs + Rr = Rf$, so $s = af$ and $r = bf$, and $f = a_1 s + b_1 r$. Then $(1 - aa_1)s = ab_1 r$ and $ba_1 s = (1 - bb_1)r$, so it suffices to show either $ab_1 \in S$

or $(1 - bb_1) \in S$. If $a \notin S$ then $(1 - aa_1)s \in S$ so $ab_1 \in S$. So assume $a \in S$. If $b_1 \in S$ then $ab_1 \in S$; if $b_1 \notin S$ then $1 - bb_1 \in S$.

12. (The basic theorem concerning the Artinian radical, cf., Chatters-Hajarnavis [80B, theorem 5.1]). Suppose R is a Noetherian ring which is an order in an Artinian ring, and define $A(R)$ as in digression 3.5.19. Then $A(R) = Re$ for a suitable central idempotent e of R, such that $(Re + N)/N = \operatorname{soc}(R/N)$ and $Re = \bigcap\{Rs: s \in S\} = \bigcap\{sR: s \in S\}$. $A(R/A(R)) = 0$. (Extensive hint: Let $\bar{R} = R/N$; noting $\operatorname{soc}(\bar{R})$ is an f.g. $\bar{R}$-module one has $\operatorname{soc}(\bar{R}) = \overline{Re}$ for suitable $\bar{e}$ in $Z(\bar{R})$ by lemma 3.2.28. We may assume $e \in R$ is idempotent. Write $A = A(R)$. Since $A \lhd R$ is Artinian as left (*and* as right) R-module, it follows $A = As = sA$ for each s in S. Letting $L = \bigcap\{Rs: s \in S\}$ one has $A \subset L$. Using the Ore condition show for all r in R and s in S that $Lr \subseteq Rs$, implying $Lr \subseteq L$, so $L \lhd R$. It follows easily $Ls = L$ for each s in S. This makes L a right module over $Q = S^{-1}R$.

   Claim: L is Artinian as a right R-module and $L = eL$. (The claim instantly implies $L = A$, and e is central since $L \lhd R$). Proof of claim: Replace L by $M = N^iL/N^{i+1}L$. Each N^iL is naturally a right Q-module, so M is a right Q-module. Exercise 2 implies M is an f.g. torsion-free divisible right $\bar{R}$-module and thus is naturally a right $\bar{Q}$-module. Arguing similarly to lemma 3.5.15 and 3.5.16, conclude M is indeed Artinian as right $\bar{R}$-module, and is a sum of submodules isomorphic to minimal right ideals of $\bar{R}$.)

Fully Bounded Rings and the H-Condition (Cauchon [76a])

13. (Cauchon's theorem on fully bounded rings) A ring R *satisfies the H-condition* if for every f.g. module M there is a finite set $V \subset M$ such that $\operatorname{Ann} M = \operatorname{Ann} V$. Prove the following assertion holds for any left Noetherian ring R: (i) Every homomorphic image of R is left bounded; (ii) R is fully bounded; (iii) R satisfies the H-condition. (Hint: (i) $\Rightarrow$ (ii) obvious. (iii) $\Rightarrow$ (i) Note the H-condition passes to homomorphic images of R, so it suffices to prove R is left bounded. Suppose $L < R$ is large. Write $M = R/L$ and take finite $V \subset M$ with $\operatorname{Ann} V = \operatorname{Ann} M$. But $\operatorname{Ann} V = \bigcap\{Lv^{-1}; v \in V\}$ is large. (ii) $\Rightarrow$ (iii) The hard direction. By Noetherian induction there are submodules $M_1, \ldots, M_n$ of M such that $M_1 \cap \cdots \cap M_n = 0$ and each M/M_i is uniform. It suffices to show each M/M_i has a finite subset V_i with $\operatorname{Ann}(V_i) = \operatorname{Ann}(M/M_i)$ since then taking V to be a union of representatives of V_i in M yields $\operatorname{Ann} M = \operatorname{Ann} V$. Thus one may assume M is uniform and faithful (replacing R by $R/\operatorname{Ann} M$). Let $P \lhd R$ be maximal with respect to being $\operatorname{Ann} M'$ for suitable $M' \leq M$. Then $P \in \operatorname{Spec}(R)$, and $\operatorname{Ann}'_R P$ is large as a left ideal in R. Let $M' = \{x \in M: Px = 0\}$, so $P = \operatorname{Ann} M'$. Then M' is a faithful module over the prime ring R/P, implying $\operatorname{Sing}(M) = 0$. (Otherwise, if $\operatorname{Sing}(M) = \sum_{i=1}^{t} Rx_i$ then $\operatorname{Ann}_{R/P} x_i$ contains a nonzero ideal A_i for $1 \leq i \leq t$, so $0 \neq \bigcap A_i \subseteq \operatorname{Ann} M$, contradiction.) But then $\rho(M'') = 1$ for each $M'' \leq M'$ since $\rho(M'/M'') = 0$. Now build V inductively as follows. Take v_1 arbitrarily in M. Suppose $v_1, \ldots, v_k$ have already been chosen, and let $B_k = \operatorname{Ann}_R\{v_1, \ldots, v_k\}$. We are done unless $B \supset 0 = \operatorname{Ann} M$ so $B'_k = B_k \cap \operatorname{Ann}'_R P \neq 0$. Thus there is v_{k+1} in M such that $B'_k v_{k+1} \neq 0$. $PB'_k v_{k+1} = 0$ so $B'_k v_{k+1} \leq M'$ and $\rho(B'_k v_{k+1}) = 1$. But right multiplication by v_{k+1} gives an epic $B'_k \to B'_k v_{k+1}$ having kernel B'_{k+1} so $\rho(B'_k/B'_{k+1}) = 1$. Thus the process *must* stop after $\rho(B'_1)$ steps.)

Every fully bounded left Noetherian ring is left bounded, by Cauchon's theorem.

Completions

14. (McConnell [79]) If $A \triangleleft R$ and there is a polycentral ideal B of R such that $B + A^2 = A$ and R/A is left Noetherian then the A-adic completion of R is left Noetherian. (Hint: The proof of theorem 3.5.36 must be modified considerably since one no longer has $A \subseteq B_1$ where $\tilde{R} = \sum B_i \lambda^i$. Let $B_i' = B_1 + A^i, B_1'' = B_1' = A$, and, inductively, $B_i'' = B_i' + \sum_{u=1}^{i-1} B_u'' B_{i-u}''$. Now the B_i'' provide a filtration of R, and, as in the original proof, the graded ring R'' is left Noetherian.)

15. (Hinohara [60]) Suppose $J = \text{Jac}(R)$ is f.g. as left R-module, and $\bigcap_{i \in \mathbb{N}} J^i = 0$. If R is complete and semilocal then every f.g. submodule N of any f.g. projective R-module P is closed in the J-adic topology of P, i.e., $N = \bigcap_{i \in \mathbb{N}} N + J^i P$, and, furthermore, $\bigcap_{i \in \mathbb{N}} J^i M = 0$ for every finitely presented R-module M. (Hint: It is enough to prove the first assertion, since the second follows from considering $0 \to N \to F \to M \to 0$ exact where N is f.g. and F is f.g. free. The main step is showing $\bar{N} = N + \overline{JN}$, where $\bar{N}$ denotes the closure of N; indeed, then $J^i N = J^i N + \overline{J^{i+1} N}$ for each i (in view of exercise 0.2.3) and since R is complete one passes to the limit and gets $\bar{N} = N$. To prove $\bar{N} = N + \overline{JN}$ suppose $x \in \bar{N}$. Then for any u one can find x_u in N such that $x - x_u \in J^u P$, so $x - x_u \in (Rx + N) \cap J^u P$, and one wants $(Rx + N) \cap J^u P \subseteq \overline{JN}$ for some u. Let $N' = (Rx + N + \overline{JN})/\overline{JN}$, which is an f.g. R/J-module and thus Artinian. Hence $N' \cap (J^u P + \overline{JN})/\overline{JN} = N' \cap (J^{u+1} P + \overline{JN})/\overline{JN}$ for suitable u, from which one concludes $N' \cap (J^u P + \overline{JN})/\overline{JN} = 0$.)

16. (McConnell [79]) Suppose $A \triangleleft R$ and $A = \sum_{i=1}^{t} Ra_i$ such that $a_1 R = Ra_1$ and $[a_i, R] \subseteq \sum_{u=1}^{i-1} Ra_u$ for each $i \leq t$. If R/A is Artinian then $\hat{R}$ is Noetherian. (Hint: Assume $\bigcap_{i=1}^{\infty} J^i = 0$. Let $R_1 = R/Ra_1$. By exercise 15 the kernel B of the map $\hat{R} \to \hat{R}_1$ is $\hat{R}a_1$. By symmetry also $B = a_1 \hat{R}$. But $\hat{R}_1$ is Noetherian by induction on t, and $Gr_B \hat{R}_1$ is generated by $\hat{R}_1$ and the element $a_1 + B^2$, so $Gr_B \hat{R}$ is Noetherian.)

17. Suppose $a \in R$ is regular and normalizing. Then a induces an automorphism σ on R by $ar = (\sigma r)a$. There are examples where R has a prime ideal P with $P \neq \sigma P$. Explicitly, let R_0 be any Noetherian ring with a prime ideal P_0 not fixed by an automorphism σ; let $R = R_0[\lambda; \sigma]$, $a = \lambda$, and $P = R\lambda + P_0$.

17. (McConnell [69]) Let $R_0 = C[\mu_1, \mu_2]$ where μ_i are commuting indeterminates, let σ be given by $\sigma \mu_1 = \mu_1$ and $\sigma \mu_2 = \mu_1 + \mu_2$, and let $P_0 = R_0 \mu_2$. Forming R, P as in exercise 17, one has $\bigcap_{n \in \mathbb{N}} P^n = 0$ but the P-adic completion of R is not left Noetherian.

Krull Dimension

22. Suppose λ is a commuting indeterminate over R. Viewing $M[\lambda] = R[\lambda] \otimes_R M$ as $R[\lambda]$-module prove for any Noetherian R-module M that K-dim$_{R[\lambda]}(M[\lambda]) =$ K-dim$_R(M) + 1$. (Hint: Since M is Noetherian there is a finite chain $M > M_1 > M_2 > \cdots > M_t = 0$ each of whose factors is critical. Thus we may assume M is critical. The chain $M[\lambda] > \lambda M[\lambda] > \lambda^2 M[\lambda] \cdots$ has factors isomorphic to

M, proving "$\geq$." Thus it suffices to prove "$\leq$," i.e., K-dim$_{R[\lambda]}M[\lambda]/R[\lambda]p \leq \alpha$ for any p in $M[\lambda]$, where $\alpha = $ K-dim(M). Let $x_p\lambda^n$ be the leading term of p. Given $N < M[\lambda]$ let $N_i = \{x \in M: x$ is the leading coefficient in M of a polynomial in $M[x]$ of degree $i\}$. The map $N \to N_0 + N_1\lambda + N_2\lambda^2 + \cdots$ shows K-dim$(M[\lambda]/R[\lambda]p) \leq$ K-dim$(M[\lambda]/R[\lambda]x_p\lambda^n)$. Next consider $M[\lambda] > M[\lambda]\lambda > \cdots > M[\lambda]\lambda^n \geq R[\lambda]x_p\lambda^n$. Each factor has K-dim α except possibly the last, which is $M[\lambda]\lambda^n/R[\lambda]x_p\lambda^n \approx (M/Rx_p)[\lambda]$. But K-dim$(M/Rx_p) < \alpha$ so K-dim$_{R[\lambda]}((M/Rx_p)[\lambda]) \leq \alpha$ by induction; piece the chains togethers.)

23. Suppose R is a left Noetherian ring of K-dim α. Then K-dim$(R[\lambda])$ as a ring is α^+, by exercise 22.

24. If $L_1 > L_2 > \cdots$ is a chain of left ideals of $F[\lambda_1, \lambda_2]$ such that $\bigcap_{i \in N} L_i \neq 0$ then there is n such that L_i/L_{i+1} has finite composition length for all $i > n$. (Hint: K-dim $F[\lambda_1, \lambda_2]/\bigcap L_i \leq 1$ by exercise 22.) Using the Nullstellensatz conclude each of these L_i/L_{i+1} is finite dimensional over F.

25. K-dim $\mathscr{A}_1(F) = 1$ for any field F of characteristic 0. (Hint: Let R be the Weyl algebra $\mathscr{A}_1(F)$, and show R/L is Artinian for every $0 \neq L < R$. Indeed, otherwise, there is an infinite descending chain $L_1 > L_2 > \cdots$ containing L; passing to the associated graded algebra $F[\lambda, \mu]$ apply exercise 24 to conclude L_i/L_{i+1} is finite dimensional over F for large enough i. But then R has a simple module finite dimensional over F, which is faithful since R is simple, implying R is finite dimensional over F, absurd.)

26. If S is a left denominator set for R then K-dim $S^{-1}R \leq$ K-dim R. Conclude that K-dim $K \otimes_F K = $ tr deg K/F for any field extension K of F.

27. Recall in example 2.1.36 that there is a *Jategaonkar sequence* $\{R_\beta : \beta \leq \alpha\}$ of PLIDs for any ordinal α; $R_\gamma = R_\beta[\lambda_\beta; \sigma_\beta]$ for $\gamma = \beta^+$, where $\sigma_\beta : R_\beta \to R_0$ is an injection. Show by induction that K-dim$(R_\beta) = \beta$, proving there are (noncommutative) PLIDs of arbitrarily large K-dim. An example of a commutative Noetherian domain of arbitrarily large K-dim is given in Gordon-Robson [73, Chapter 9].

Noetherian Rings of K-dim 1 (Lenagan)

36. Suppose R is Noetherian and K-dim$(R) = 1$. If the Artinian radical $A(R) = 0$ then R satisfies Small's regularity condition. (Hint: Induction on t such that $N^t = 0$, where $N = $ Nil(R). Let $T = N \cap$ Ann N. Clearly $A(R/N) = 0$. Also $A(R/\text{Ann } N) = 0$ since if $L/\text{Ann } N$ is Artinian then $La = 0$ for each a in N. Thus $A(R/T) = 0$. By induction if $s \in \mathscr{C}(N)$ then $s + T$ is regular. We want Ann $s = $ Ann$'s = 0$. For $a \in$ Ann $s \subseteq T$ define $f: R/(N + Rs) \to Ra$ by $f\bar{r} = ra$. But $R/(N + Rs) \approx (R/N)/(N + Rs/N)$ is Artinian so Ra is Artinian and thus 0, as desired.)

37. Hypotheses as in exercise 36, if $L < R$ and R/L is Artinian then L contains a regular element of R. (Hint: By exercises 12, 36 one may assume R is semiprime. Then any essential complement of L is Artinian and thus 0; hence L is large.)

38. Jacobson's conjecture holds for any Noetherian ring of K-dim 1. (Hint: Let $J = $ Jac(R) and $J' = \bigcap_{i=1}^{\infty} J'$. For any regular s in R we have R/Rs of finite length so $J' \subseteq \bigcap Rs$. Let $A = A(R)$. One is done by exercise 36 and 12 unless

$A \neq 0$. Passing to R/A shows $J' \subseteq A$; also $R/\text{Ann}_R A$ is Artinian so Ann A contains an element s regular modulo A. Take k by Fitting's lemma such that $0 = Rs^k \cap \text{Ann } s^k \supseteq J'$.

Assassinators of Modules The following exercises link injective modules to Spec(R).

39. If M is uniform then the assassinator is unique, so we can write Ass(M); in this case Ass(M) = $\{r \in R : rN = 0$ for some $0 \neq N \leq M\}$.

40. Suppose R is left Noetherian. There is an onto map Φ: {(indecomposable injective R-modules} $\rightarrow$ Spec(R), given by $E \rightarrow$ Ass(E). (Hint: If $P \in$ Spec(R) then take a uniform left ideal L of R/P and let $E = E(L)$, noting Ass(E) = Ass(L) = P.)

To see when the correspondence of exercise 40 is 1:1 we embark on a closer examination using Krull dimension.

41. Suppose R has K-dim β. For any module M there is some α which is the smallest K-dim of a nonzero submodule. (Note that every cyclic submodule has K-dim $\leq \beta$.) Define cr dim(M) to be α; show M has a critical cyclic module of K-dim α.

42. If E is injective and $P =$ Ass(E) then E has a critical cyclic module N with Ann $N = P$ and K-dim(N) = cr dim(E); thus N is an R/P-module, so cr dim(E) $\leq$ K-dim(R/P). On the other hand, equality holds if $E = E(L)$ for L as in exercise 40.

43. The following are equivalent for a let Noetherian ring R: (i) R is fully bounded. (ii) cr dim(E) = K-dim($R/$Ass E) for every indecomposable injective E. (iii) The correspondence Φ of exercise 40 is 1:1. (Hint: (i) $\Rightarrow$ (ii) Take a critical cyclic module M with $P =$ Ann $M =$ Ass E and cr dim(E) = K-dim(M) = α. Writing $M \approx R/L$ one has core(L/P) = 0 in R/P. By hypothesis L/P is not large and so has an essential complement L' in R/P. Then $L' \leq R/L \approx M$ so by proposition 3.5.46 K-dim(R/P) = K-dim(L') $\leq$ K-dim(M) = α.

(ii) $\Rightarrow$ (iii) Let $P =$ Ass(E) and $\alpha =$ cr dim(E) = K-dim($\bar{R}$) where $\bar{R} = R/P$. Take a critical cyclic submodule N of E with $P =$ Ann N and K-dim(N) = α. Let $L_1, \ldots, L_n$ be critical left ideals of $\bar{R}$ such that $T = \bigoplus L_i$ is large. Then K-dim($\bar{R}/T$) $< \alpha$ so Hom($\bar{R}/T, N$) = 0. Thus $TN \neq 0$ so some $L_i x \neq 0$ for suitable x in N; comparing K-dim shows the map $L_i \rightarrow L_i x$ is monic, i.e., N contains a copy of L_i, so $E \approx E(L_i)$. (iii) $\Rightarrow$ (ii) by exercise 42. (ii) $\Rightarrow$ (i) Assume R is prime. Suppose $L < R$ is large. Then R/L has a critical submodule L_1/L whose annihilator is Ass(R/L); inductively we have $L = L_0 < L_1 < L_2 < \cdots$ where Ann(L_{i+1}/L_i) = Ass(L_{i+1}/L_i) $\neq 0$ by hypotheses. Then some $L_n = R$ and $0 \neq$ Ann(L_1/L) $\cdots$ Ann(L_n/L_{n-1}) $\subseteq L$.)

Stafford-Coutinho Theory Assume R is left Noetherian.

44. Prove directly that if R is simple (non-Artinian) and M is f.g. Artinian then M is cyclic; in fact, if $M = \sum_{i=1}^{t} Rx_i$ then for any $r \neq 0$ there are r_i' in R such that $M = R(x_1 + \sum_{i=2}^{t} rr_i'x_i)$. (Hint: By induction reduce to the case $t = 2$. Take regular s in R with $sx_1 = 0$. If Rx_2 is simple then there is r' in R satisfying

$Rsrr'x_2 = Rx_2 = Rs(x_1 + rr'x_2)$ and thus also $x_1 \in R(x_1 + rr'x_2)$. Thus one is done unless Rx_2 is not simple. Take a simple submodule Rax_2 of M. By induction on composition length $\bar{M} = M/Rax_2$ can be written as $R(\bar{x}_1 + rr'\bar{x}_2)$ for suitable r' in R. Then $M = R(x_1 + rr'x_2) + Rax_2$ so as above $M = R(x_1 + rr'x_2 + rr''ax_2)$ for suitable r'' in R.)

45. If R is simple and K-dim $R = 1$ then for any regular s in R there is r' in R for which $R = Rsr' + Rs$. (Hint: Apply exercise 44 to $M = R/Rs$.)

These results might whet the reader's appetite in asking for the minimal number of generators of a f.g. module over a left Noetherian ring; the lovely theory of Warfield-Stafford-Coutinho is given in the unabridged edition, pp. 421ff.

Gabriel Dimension

51. Suppose G-dim $M = \alpha$. Let $N_1(M) = \sum\{\beta\text{-simple submodules}: \beta \le \alpha\}$ and inductively let $N_{\gamma^+}(M)$ be given by $N_{\gamma^+}(M)/N_\gamma(M) = N_1(M/N_\gamma(M))$, and $N_\gamma(M) = \bigcup\{N_{\gamma'}(M): \gamma' < \gamma\}$ for limit ordinals. Then $N_\alpha(M) = M$.

52. Applying induction on γ to exercise 51, show every submodule of M contains a β-simple submodule for suitable $\beta \le \alpha$. Conclude $N_1(M)$ is large in M.

53. Every nonzero submodule of an α-simple module is α-simple.

54. G-dim is exact, by exercise 52.

55. G-dim $\sum_{i \in I} M_i = \sup_{i \in I}$ G-dim M_i. (Hint: By exercise 54 assume the sum is direct; then take finite direct sum and finally show any image of $\bigoplus_{i \in I} M_i$ contains a submodule of $\bigoplus_{i \in I'} M_i$ for suitable finite $I' \in I$.

56. If G-dim $R = \alpha$ then G-dim $M \le \alpha$ for any R-module M.

4 Categorical Aspects of Module Theory

Although space limitations have cut into our categorical treatment of R-$\mathcal{M}od$, there are times when the subject demands a strictly categorical viewpoint.

§4.1 The Morita Theorems

Recall two categories $\mathscr{C}$ and $\mathscr{D}$ are *equivalent* if there are functors $F: \mathscr{C} \to \mathscr{D}$ and $G: \mathscr{D} \to \mathscr{C}$ such that $GF \simeq 1_\mathscr{C}$ and $FG \simeq 1_\mathscr{D}$ where $\simeq$ denotes a natural isomorphism of functors; in this case F and G are each called a *category equivalence*. In theorem 1.1.17 we saw there is a category equivalence R-$\mathcal{M}od \to M_n(R)$-$\mathcal{M}od$ sending M to $M^{(n)}$. Category equivalences are important because they preserve categorically-defined notions.

Remark 4.1.1: The following notions are categorically defined and thus preserved under categorical equivalence: monic, epic, kernel, cokernel, exact sequence, direct sum, summand, direct product, projective, injective, projective cover (by proposition 2.8.42), and injective hull (by duality). More generally, direct limits and inverse limits are also categorically defined.

For example $\pi: P \to M$ is a projective cover in R-$\mathcal{M}od$ iff $\pi^{(n)}: P^{(n)} \to M^{(n)}$ is a projective cover in $M_n(R)$-$\mathcal{M}od$. This idea whets one's appetite for proving that a given notion is categorically defined. This cannot always be done: "Free" is not categorically defined since R is a free R-module but $R^{(2)}$

357

is not a free $M_2(R)$-module. Sometimes it requires some subtlety to prove a notion is categorically defined, as in the case of f.g. modules.

Definition 4.1.2: An object A of a category $\mathscr{C}$ is *finite* if A cannot be written as a direct limit of proper subobjects, i.e., if we cannot write $A = \varinjlim(A_i; \varphi_i^j)$ where, in the notation of definition 1.8.3, each μ_i is monic but *not* epic.

This definition is categorical, and is useful because of the following result:

Proposition 4.1.3: *An R-module M is f.g. iff M is finite as an object in* R-$\mathscr{M}\!od$.

Proof: ($\Leftarrow$) (The contrapositive) If M is not an f.g. module then M is a direct limit of proper submodules by example 1.8.9, so is not finite in R-$\mathscr{M}\!od$. ($\Leftarrow$) Suppose $M = \sum_{i=1}^{t} Rx_i$ could be written as a direct limit of proper subobjects M_j. Then we would have suitable j_i with $x_i \in M_{j_i}$, $1 \le i \le t$, and so some M_j in the system would contain $x_1, \ldots, x_n$ and thus M, contrary to M_j being a proper submodule.   Q.E.D.

Since R is left Noetherian iff every submodule of an f.g. module is f.g., we conclude instantly that $M_n(R)$ is left Noetherian iff R is left Noetherian. Similarly, R is semiperfect iff $M_n(R)$ is semiperfect (because semiperfect is described categorically by every f.g. module having a projective cover.) More generally, we could translate many ring theoretic properties of R to any ring R' for which R'-$\mathscr{M}\!od$ and R-$\mathscr{M}\!od$ are equivalent; in this case we say R and R' are *Morita equivalent* in honor of Morita [58] who proved

Theorem 4.1.4: (*Morita's theorem*) Rings R and R' are Morita equivalent iff $R' \approx \operatorname{End}_R P$ for a suitable f.g. projective module P which is a generator in R-$\mathscr{M}\!od$.

Categorical Notions

The remainder of the section will be spent in digesting and proving Morita's theorem. To do this most expeditiously we shall recall some categorical notions along the way. A is a *generator* of R-$\mathscr{M}\!od$ if for every module M there is an epic $A^{(I)} \to M$ for a suitable set I.

Remark 4.1.5: R is a generator of R-$\mathscr{M}\!od$. If $f: N \to A$ is epic and A is a generator then N is a generator.

The following tool is useful in determining whether A is a generator. In what follows M^* denotes $\text{Hom}(M, R)$, which is in Mod-R by defining the product fr by $(fr)x = (fx)r$ for all x in M, f in M^*, and $r \in R$.

Definition 4.1.6: The *trace ideal* $T(A) = \{\sum_{\text{finite}} fa : f \in A^*, a \in A\}$. $T(A)$ is indeed an ideal of R since $r(fa) = f(ra)$ and $(fa)r = (fr)a$.

Lemma 4.1.7: *The following are equivalent for A in R-$\mathcal{M}\!od$: (i) A is a generator of R-$\mathcal{M}\!od$; (ii) $T(A) = R$; (iii) R is a homomorphic image of $A^{(n)}$ for some n.*

Proof: (i) $\Rightarrow$ (ii) Suppose $h: A^{(I)} \to R$ is epic. Thus $1 = h(a_i)$ for some (a_i) in $A^{(I)}$. Writing $\mu_i: A \to A^{(I)}$ for the canonical injection of A into the i-th component of $A^{(I)}$, we can define $f_i = h\mu_i: A \to R$ and see $1 = \sum f_i a_i \in T(A)$.
  (ii) $\Rightarrow$ (iii) Writing $1 = \sum_{i=1}^n f_i a_i$ we see the map $A^{(n)} \to R$ given by $(x_1, \ldots, x_n) \to \sum f_i x_i$ is onto.
  (iii) $\Rightarrow$ (i) $A^{(n)}$ is a generator by remark 4.1.5, clearly implying A is a generator.   Q.E.D.

Since we are interested in proving theorem 4.1.4 we introduce the following terminology.

Definition 4.1.8: A *progenerator* is an f.g. projective R-module which is a generator.

Remark 4.1.9: Suppose M is an f.g. projective module over R. If M is faithfully projective then M is a progenerator, in particular, this is the case if R is commutative and M is faithful as R-module. (Indeed $M = T(M)M$ by the dual basis theorem, so $T(M) = R$ if M is faithfully projective. To prove the second assertion suppose, on the contrary, $A \lhd R$ and $AM < M$. Taking a maximal ideal P of R containing A, note M_P is an f.g. R_P-module and $M_P = A_P M_P$ so $M_P = 0$ by "Nakayama's lemma" since R_P is local. This implies there is s in $R - P$ annihilating each of the generators of M so $sM = 0$, contrary to M faithful.)
  See exercise 2 for an f.g. projective module which is *not* a progenerator We shall now illustrate ($\Leftarrow$) of Morita's theorem in two useful special cases.

Digression: Two Examples

Example 4.1.10: Suppose $P = R^{(n)}$, a free R-module and obviously a progenerator. Let $R' = \text{End}_R P \approx M_n(R)$. Then R and R' are Morita equivalent by theorem 1.1.17. Note $P \in R\text{-}\mathcal{M}od\text{-}R'$. On the other hand, let $P' = \text{Hom}(P, R) \in R'\text{-}\mathcal{M}od\text{-}R$. In fact, P' is isomorphic to $R^{(n)}$ viewed as an R'-module via matrix multiplication, so $(P')^{(n)} \approx R'$, implying P' is a progenerator of R'. Then $R \approx \text{End}_{R'} P'$.

Example 4.1.11: Suppose R is a semiperfect ring. Then R has a complete set of orthogonal primitive idempotents $e_1, \ldots, e_n$ from which we take a basic set $e_1, \ldots, e_t$ (rearranging the indices if necessary). Recall this means for each $j \le n$ that $Re_j \approx Re_i$ for precisely one value of i between 1 and t. Let $e = \sum_{i=1}^{t} e_i$. Then, obviously, Re is a progenerator, and $\text{End}_R Re \approx eRe$ by proposition 2.1.21. But this is the basic ring of R, which is Morita equivalent to R by Morita's theorem. Let us try to see this fact directly. Indeed, define the functor $F: R\text{-}\mathcal{M}od \to eRe\text{-}\mathcal{M}od$ by putting $FM = eM$ and, given $f: M \to N$ defining $Ff: eM \to eN$ by $(Ff)ex = f(ex) = efx$. It is not difficult to prove F is a (category) equivalence by constructing the functor G in the opposite direction, but it is useful to have an intrinsic criterion for F to be an equivalence.

Recall a functor $F: \mathscr{C} \to \mathscr{D}$ is *full* (resp. *faithful*) if the map $\text{Hom}(A, B) \to \text{Hom}(FA, FB)$ given by $f \to Ff$ is epic (resp. monic) for all A, B in $\text{Ob } \mathscr{C}$. Then we have

Proposition 4.1.12: (*cf.*, Jacobson [80B, proposition 1.3]) *A functor $F: \mathscr{C} \to \mathscr{D}$ is a category equivalence iff F is faithful and full such that for each object D in $\mathscr{D}$ there is an object A in $\mathscr{C}$ with FA isomorphic to D (in $\mathscr{D}$).*

Example 4.1.13: Now we can prove directly and easily that the functor $F: R\text{-}\mathcal{M}od \to eRe\text{-}\mathcal{M}od$ of example 4.1.11 is an equivalence. Notation as in example 4.1.11, recall from proposition 2.7.25 that given $Re_i \approx Re_j$ we have e_{ij} in e_iRe_j and e_{ji} in e_jRe_i such that $e_{ij}e_{ji} = e_i$ and $e_{ji}e_{ij} = e_j$. For convenience we take $e_{ii} = e_i$. For any R-module map $f: M \to N$ we have

$$f(e_jx) = f(e_{ji}e_ie_{ij}x) = e_{ji}f(e_ie_{ij}x).$$

In particular, if $f(eM) = 0$ then $f(e_jM) = 0$ for all j, so $f = 0$, proving F is faithful. On the other hand, the action of f on each e_jM is determined by the action of f on the e_iM so F is full. It remains to show that any eRe-module N is naturally isomorphic to FM for a suitable R-module M. Take $N_i = e_iN$ for $i \le t$, and N_j to be a copy of N_i as an abelian group with a group isomorphism $\varphi_j: N_i \to N_j$ whenever $1 < j \le n$ and $Re_i \approx Re_j$. We take $M = \bigoplus_{j=1}^{n} N_j$ as an abelian group; since $R = \bigoplus_{j,k=1}^{n} e_jRe_k$ as an abelian group it suffices to define $(e_j re_k) x_v$ for all r in R and x_v in N_v, where $1 \leq j, k, v \leq n$. This is clear: for $k \neq v$ we take $e_j re_k x_v = 0$; if $k = v$ find $i, u \leq t$ with $Re_i \approx Re_j$ and $Re_u \approx Re_v$ and define

$$e_j re_v x_v = \varphi_j(e_{ij} re_{vu} \varphi_v^{-1} x).$$

We leave it to the reader to check this extends N to an R-module.

Morita Contexts and Morita's Theorems

Let us turn now to the proof of Morita's theorem, which relies on a collection of data obtained from $M \in R\text{-}\mathcal{M}od$.

Definition 4.1.14: A *Morita Context* (also called a *set of pre-equivalence data*) is a six-tuple $(R, R', M, M', \tau, \tau')$ where R, R' are rings, $M \in R\text{-}\mathcal{M}od\text{-}R'$, $M' \in R'\text{-}\mathcal{M}od\text{-}R$ and $\tau: M \otimes_R M' \to R$ and $\tau': M' \otimes_R M \to R$ are bimodule maps under which the following diagrams commute:

$$
\begin{array}{ccc}
M \otimes M' \otimes M & \xrightarrow{\;1 \otimes \tau'\;} & M \otimes R' \\
{\scriptstyle \tau \otimes 1}\downarrow & & \downarrow \\
R \otimes M & \longrightarrow & M
\end{array}
\qquad
\begin{array}{ccc}
M' \otimes M \otimes M' & \xrightarrow{\;1 \otimes \tau\;} & M' \otimes R \\
{\scriptstyle \tau' \otimes 1}\downarrow & & \downarrow \\
R' \otimes M' & \longrightarrow & M'
\end{array}
$$

It is clear from the context whether the tensors are over R or R', and so they are left undecorated. Similarly, the unlabelled arrows are the canonical isomorphisms. Note that every requirement comes in pairs, to maintain the duality between the R, M, τ and R', M', τ'. In what follows we designate arbitrary elements from their sets as follows: $r \in R$, $r' \in R'$, $x, y \in M$, $x', y' \in M'$. We write (x, x') for $\tau(x \otimes x')$, and $[x', x]$ for $\tau'(x' \otimes x)$. Then the two commutative diagrams translate to the formulas

(1) $(x, x')y = x[x', y]$.

(2) $[x', x]y' = x'(x, y')$.

Remark 4.1.15: There is a more concise way of describing a Morita context. Namely, $\begin{pmatrix} R & M \\ M' & R' \end{pmatrix}$ is a ring under the usual matrix operations, cf., example 1.1.10.

Example 4.1.16: The Morita context which concerns us is for M an R-module and $R' = \operatorname{End}_R M$. Then $M \in R\text{-}\mathcal{M}od\ R'$. Let $M' = M^* = \operatorname{Hom}_R(M, R)$, viewed as an $R' - R$ bimodule as in remark 1.5.18'. The evaluation map $\varphi: M \times M' \to R$ given by $\varphi(x, f) = fx$ is balanced over R' since $\varphi(xr', f) = f(xr') = (r'f)x = \varphi(x, r'f)$; thus we get $\tau: M \otimes M' \to R$ given by $\tau(x \otimes f) = fx$. We wrote f instead of the promised x' because of its familiarity to the reader, but now we shall continue with x'. Likewise, we shall write the action of R' on M from the right (instead of the more familiar left) to fit into the above setting. Define $\varphi': M' \times M \to R'$ by taking the action of $\varphi'(x', x)$ on M to be given by

$$y\varphi'(x', x) = \tau(y \otimes x')x.$$

Then $y\varphi'(x'r, x) = \tau(y \otimes x'r)x = \tau(y \otimes x')rx = y\varphi'(x', rx)$ so φ' is balanced and induces $\tau': M' \otimes M \to R'$ which is seen at once to be an $R' - R$ bimodule map.

Note that by definition of τ' we have $(x, x')y = \tau(x \otimes x')y = x[x', y]$, which is (1). To see (2) we note for all y that

$$y[x', x]y' = (y, x')xy' = yx'(x, y') \text{ as desired.}$$

To add insight to this computation let us return to example 4.1.10, where $M = R^{(n)}$, written as $1 \times n$ matrices, $R' = \operatorname{Hom}_R M \approx M_n(R)$, and $M' = \operatorname{Hom}(M, R)$ which we can identify with $n \times 1$ matrices over R. Now we can view τ, τ' as the usual matrix product, i.e., $\tau(x \otimes x') = xx'$ and $\tau(x' \otimes x) = x'x$. Note that τ, τ' are epic, since $M_n(R)$ is spanned by $\tau(e_i \otimes e_j')$ where $\{e_1, \ldots, e_n\}$, $\{e_1', \ldots, e_n'\}$ are the standard bases of M, M', respectively. This is the final ingredient for the decisive theorem on Morita contexts.

Theorem 4.1.17: ("*Morita I*") *Suppose* $(R, R', M, M', \tau, \tau')$ *is a Morita context with* τ, τ' *epic. Then*

(i) *M is a progenerator in R-$\mathcal{M}od$ and in $\mathcal{M}od$-R'.*

(i') *M' is a progenerator in $\mathcal{M}od$-R and R'-$\mathcal{M}od$.*

(ii) *τ, τ' are isomorphisms.*

(iii) *$M' \approx M^* = \operatorname{Hom}_R(M, R)$ under the $R' - R$ bimodule isomorphism $x' \to (\ , x')$. There are three analogous isomorphisms $M \approx \operatorname{Hom}(M', R)_R$ (under $x \to (x, \ )$), $M \approx \operatorname{Hom}_{R'}(M', R')$ (under $x \to [\ , x]$), and $M' \approx \operatorname{Hom}(M, R')_{R'}$ (under $x' \to [x', \ ]$). Note we write the subscript on the right when we have maps of right modules.*

(iv) *$R' \approx \operatorname{End}_R M$ under the regular representation. (Likewise, $R' \approx \operatorname{End} M'_R$, and $R \approx \operatorname{End}_{R'} M' \approx \operatorname{End} M_{R'}$.)*

(v) *R-$\mathcal{M}od$ and R'-$\mathcal{M}od$ are equivalent under the functors $M \otimes_{R'}$ and $M' \otimes_R$. (Likewise, $\mathcal{M}od$-R and $\mathcal{M}od$-R' are equivalent.)*

(vi) $\operatorname{Hom}(M, \underline{\quad})$ *and* $M' \otimes \underline{\quad}$ *are naturally equivalent functors from* $R\text{-}\mathcal{M}od$ *to* $R'\text{-}\mathcal{M}od$.

Proof: Because of symmetry we need only prove the first assertion in each part. Define $(\ , x')$ in M^* as the map $x \to (x, x')$. Since τ is onto we have $\sum_{i=1}^{n}(x_i, x_i') = 1$ for suitable x_i. Likewise, write $\sum_{j=1}^{m}[y_j', y_j] = 1$. The proof largely consists of substituting these expressions for 1 whenever appropriate, to wit:

(i) M is a generator, by lemma 4.1.7. To see M is f.g. projective we note $\{(y_j, f_j): 1 \le j \le m\}$ is a dual basis, where $f_j = (\ , y_j')$; indeed, for all x in M,

$$x = \sum_{j=1}^{m} x[y_j', y_j] = \sum_{j=1}^{m}(x, y_j')y_j = \sum(f_j x)y_j.$$

(ii) We need to show if $z_k \in M$ and $z_k' \in M'$ with $a = \sum_{k=1}^{t} z_k \otimes z_k' \in \ker \tau$ then $a = 0$. Indeed $0 = \tau a = \sum_{k=1}^{t}(z_k, z_k')$ so

$$a = \sum z_k \otimes z_k' \sum_{i=1}^{n}(x_i, x_i') = \sum_{i,k} z_k \otimes [z_k', x_i]x_i' = \sum_{i,k} z_k[z_k', x_i] \otimes x_i'$$

$$= \sum_{i,k}(z_k, z_k')x_i \otimes x_i' = \sum_i \left(\sum_k (z_k, z_k')\right)x_i \otimes x_i' = 0.$$

(iii) $x' \to (\ , x')$ is clearly a bimodule homomorphism, and is monic since $(\ , x') = 0$ implies $x' = \sum_{j=1}^{m}[y_j', y_j]x' = \sum y_j'(y_j, x') = 0$. On the other hand, for any $f \in M^*$ we have

$$fx = f\left(\sum x[y_j', y_j]\right) = f\left(\sum(x, y_j')y_j\right) = \sum(x, y_j')fy_j = (x, \sum y_j'fy_j)$$

implying $f = (\ , \sum y_j'fy_j)$.

(iv) Let ρ denote the regular representation, i.e., if $r' \in R'$ then $(\rho r'): M \to M$ is right multiplication by r'. Then ρ is a ring homomorphism from R' to $\operatorname{Hom}_R(M, M)^{\mathrm{op}} = \operatorname{End}_R M$. Also $\ker \rho = 0$ for if $\rho r' = 0$ then $Mr' = 0$ and $r' = \sum[y_j', y_j]r' = \sum[y_j', y_j r'] = 0$. But ρ is onto since for any f in $\operatorname{End}_R M$ we have (in analogy to (iii))

$$fx = f\left(\sum x[y_j', y_j]\right) = f\left(\sum(x, y_j')y_j\right) = \sum(x, y_j')fy_j = x\sum y_j'fy_j$$

proving $f = \rho(\sum y_j'fy_j)$.

(v) Suppose $N \in R\text{-}\mathcal{M}od$. Then $N \approx R \otimes_R N \approx (M \otimes_{R'} M') \otimes_R N \approx M \otimes_{R'}(M' \otimes_R N)$ in view of (ii). An analogous argument using τ' shows for N' in $R'\text{-}\mathcal{M}od$ that $N' \approx M' \otimes_R(M \otimes_{R'} N')$ and so the composite of the functors $M' \otimes_R \underline{\quad}$ and $M \otimes_{R'} \underline{\quad}$ in either direction is naturally equivalent to the identity. (We leave it to the reader to draw the appropriate diagram.)

(vi) This follows from the chain of isomorphisms $\text{Hom}_R(M, N) \approx \text{Hom}_{R'}(M' \otimes M, M' \otimes N) \approx \text{Hom}_{R'}(R', M' \otimes N) \approx M' \otimes N$.   Q.E.D.

Remark 4.1.17': The proof of theorem 4.1.17(i) shows more precisely that M is f.g. projective if τ' is onto; symmetrically M' is an f.g. projective R'-module if τ is onto.

Actually two types of symmetry are at work here—one is between R and R', and the other is the left-right symmetry between R-$\mathcal{M}od$ and $\mathcal{M}od$-R; this is why we wound up with four assertions for the price of one, with an impressive effect in the following result relating the ring-theoretic structures of R and R' (also, cf., exercise 6).

Proposition 4.1.18: *Notation and hypothesis of theorem* 4.1.17. *Then the lattice of ideals of R is isomorphic to the lattice of ideals of R'. In fact, the correspondence $L \to M'L$ is a lattice isomorphism from $\mathcal{L}(R)$ to $\mathcal{L}_{R'}(M')$ which restricts to a lattice isomorphism from* {*ideals of R*} *to* {$R' - R'$ *bisubmodules of M'*}.

Proof: It is enough to prove the last assertion, since applying the two kinds of symmetry observed above would also yield a lattice isomorphism {$R' - R$ bisubmodules of M'} $\to$ {ideals of R'}, as desired. $L \to M'L$ is obviously order-preserving (under set inclusion) so it remains to show there is an order-preserving inverse correspondence, $\Phi: \mathcal{L}_{R'}(M') \to \mathcal{L}(R)$, given by $N' \to (M, N') = \{\sum(z_k, a'_k): z_k \in M, z'_k \in N'\}$. Indeed, $L = RL = (M, M')L = (M, M'L) = \Phi(M'L)$ and $N' = [M', M]N' = M'(M, N') = M'\Phi N'$.   Q.E.D.

Morita contexts can be tied in with progenerators in the following nice way.

Proposition 4.1.19: *In the Morita context of example* 4.1.16, τ *is epic if M is a generator*; τ' *is epic if M is f.g. projective. Consequently, Morita 1 applies if M is a progenerator.*

Proof: τ is epic since $T(M) = R$ by lemma 4.1.7. If M is f.g. projective then the dual basis lemma gives us x_i in M and x'_i in M^* for $1 \leq i \leq t$ satisfying $x = \sum_{i=1}^{t} x_i(x'_i, x) = \sum[x_i, x'_i]x$ for all x in P; thus $\sum[x_i, x'_i] = 1_P$ proving τ' also is epic.   Q.E.D.

Proof of Morita's Theorem and Applications

Proof of Theorem 4.1.4: (i) $\Rightarrow$ (ii). Let $F: W\text{-}\mathcal{M}od \to R\text{-}\mathcal{M}od$ be a category equivalence. Then $P = FW$ is a progenerator of R, and $W \approx \operatorname{End}_W W \approx \operatorname{End}_R FW = \operatorname{End}_R P$.

(ii) $\Rightarrow$ (i) Construct the Morita context $(R, W, P, P^*, \tau, \tau')$ of example 4.1.16, noting τ, τ' are onto by proposition 4.1.19. Then R and W are Morita equivalent by Morita I.   Q.E.D.

Remark 4.1.20: Morita's theorem gives fast proofs of the following facts that previously required considerable effort:

(i) Wedderburn-Artin Theorem. If R is simple Artinian then R is a direct sum of minimal left ideals, each of which are isomorphic to each other by proposition 2.1.15. In other words, any minimal left ideal L of R is a progenerator; thus $R \approx \operatorname{End} L_D$, where $D = \operatorname{End}_R L$ is a division ring by Schur's lemma.

(ii) If M is an f.g. module over a semisimple Artinian ring R then $\operatorname{End}_R M$ is semisimple Artinian (since M is projective.)

(iii) If P is an f.g. projective module over a semiperfect ring R then $\operatorname{End}_R P$ is semiperfect.

Morita's theorem is tied more closely to matrices in exercises 7, 8. Note that this theory gives very explicit information, and often one builds a Morita context in order to pass information concretely from one ring to a Morita equivalent ring. We conclude by introducing some ideas which "spin off" from Morita I.

Remark 4.1.21: The proof of theorem 4.1.4, coupled with Morita I, shows that any categorical equivalence $F: W\text{-}\mathcal{M}od \to R\text{-}\mathcal{M}od$ is naturally isomorphic to the functor $P \otimes_W \underline{}$ where $P = FW$. This give rise to question as to which functors can in fact be given by tensors; an answer is given by Watt's theorem (exercise 4.2.8).

Exercises

§4.1

1. Any left ideal L of a simple ring R is a generator. (Hint: If $1 = \sum_{i=1}^{n} a_i r_i$ for $a_i \in L$ and $r_i \in R$ then the map $L^{(n)} \to R$ given by $(x_1, \ldots, x_n) \to \sum x_i r_i$ is epic.)
2. Example of a faithful cyclic projective module which is *not* a generator: Let

$R = \{$lower triangular matrices over a field $K\}$, and $P = Re_{11}$. Note $T(P) = P \neq R$.

3. The following are equivalent: (i) A is a generator of R-$\mathcal{M}od$. (ii) The functor Hom$(A, \_\_)$ is faithful. (iii) For any module M there is an epic $A^{(I)} \rightarrow M$ where $I =$ Hom(A,M).

4. A *cogenerator* of a category $\mathscr{C}$ is an object A such that Hom$(, A)$ is a faithful functor. Show A is a cogenerator of R-$\mathcal{M}od$ iff for any module M there is a monic $M \rightarrow A^l$ for a suitable set l. (Hint: $l = $ Hom(M, A).) Further results on cogenerators are in Anderson-Fuller [74B].

5. $\mathbb{Q}/\mathbb{Z}$ is an injective cogenerator for $\mathbb{Z}$-$\mathcal{M}od$. In general, R-$\mathcal{M}od$ has an injective cogenerator (see remark 4.2.3).

6. If R, R' are Morita equivalent then $Z(R) \approx Z(R')$. (Hint: As in theorem 4.1.17(iv) view $R \approx $ End $M_{R'}$ and $R' \approx $ End$_R M$ in End$_{\mathbb{Z}} M$, where they are centralizers and note their centers are both End$_R M \cap $ End $M_{R'}$.)

7. Suppose $e \in R$ is idempotent. Re is a progenerator of R-$\mathcal{M}od$ iff $R = ReR$. (Hint: ($\Rightarrow$) $1 \in T(Re)$ implies $1 = \sum f_i(r_i e) = \sum r_i e s_i$ where $s_i = f_i e \in R$. ($\Leftarrow$) Go backwards, where f_i is right multiplication by s_i.)

8. R and R' are Morita equivalent iff for some n there is an idempotent e of $W = M_n(R)$ such that $WeW = W$ and $eWe \approx R'$. (Hint: ($\Rightarrow$) Let P be a progenerator with $R' \approx $ End$_R P$. P is a summand of $R^{(n)}$ for suitable n. Then End$_R R^{(n)} \approx W$ and $P = R^{(n)}e$ so End$_R P \approx eWe$. Identifying R with $e_{11}We_{11}$ show $P \approx e_{11}We$ and $T(P) = e_{11}WeWe_{11}$; since $e_{11} \in T(P)$ conclude $WeW = W$. ($\Leftarrow$) Using exercise 7 note We is a progenerator of W, so W is Morita equivalent to $eWe \approx R'$.)

9. Consider the monic $\Phi: \bigoplus_{i \in \mathbb{N}} Hom(P, P_i) \rightarrow Hom(P, \bigoplus_{i \in \mathbb{N}} P_i)$ sending (g_i) to the map $x \rightarrow (g_i x)$. A projective module P is f.g. iff Φ is an isomorphism when each $P_i = P$. (Hint: ($\Leftarrow$) Use a dual base $\{(x_i, f_i): i \in I\}$ to show there is $g: P \rightarrow \bigoplus P_i$ given by $gx = ((f_i x)x_i).$)

10. (Eilenberg) If P is a progenerator for R then $P^{(\mathbb{N})} \approx R^{(\mathbb{N})}$. (Hint: by remark 2.8.4')

11. (Camillo [84]) Rings R and T are Morita equivalent iff End$_R R^{(\mathbb{N})} \approx $ End$_T T^{(\mathbb{N})}$ as rings. (Hint: ($\Rightarrow$) Write $R_{(\mathbb{N})}$ for End$_R R^{(\mathbb{N})}$. Using exercise 10 show $R_{(\mathbb{N})} \approx $ End$_R P^{(\mathbb{N})}$ $\approx ($End$_R P)_{(\mathbb{N})}$ for a progenerator P. ($\Leftarrow$) the hard direction. Let $\varphi: R_{(\mathbb{N})} \rightarrow T_{(\mathbb{N})}$ be the isomorphism; letting $e_{11}: R^{(\mathbb{N})} \rightarrow R^{(\mathbb{N})}$ denote the projection onto the first component show $(\varphi e_{11})T^{(\mathbb{N})}$ is a progenerator, using exercise 9.)

Balanced Modules and the Wedderburn-Artin Theorem

The following exercises due to Faith [67] show how Rieffel's short proof of the Wedderburn-Artin theorem (exercise 2.1.8) fits into the Morita theory. In what follows let $M \in R$-$\mathcal{M}od$, $R' = $ End$_R M$, and $R'' = $ End $M_{R'}$. We say M is *balanced* if the regular representation yields an isomorphism from R to R''. In particular, any balanced module is faithful.

12. If there is an epic $\pi: M \rightarrow R$ then M is balanced. (Hint: R is projective so π splits. Write $M = R \oplus N$. It suffices to prove any r'' in R'' is given by left multiplication by $r''(1, 0)$. Given x in M there is r' in R' such that $(1, 0)r' = x$. Viewing $\pi \in R'$ shows $r''(1, 0) = (r''(1, 0))\pi \in R$ so $r''x = (r''(1, 0))r' = r''(1, 0)((1, 0)r') = r''(1, 0)x.$)

13. If $M^{(n)}$ is balanced then M is balanced. (Hint: Look at the diagonal.)

14. M is a generator of R-$\mathcal{M}od$ iff M is a balanced R-module and an f.g. projective R'-module. (Hint: ($\Rightarrow$) There is an epic $M^{(n)} \rightarrow R$, so $M^{(n)} \approx R \oplus N$ and M is

balanced. Moreover $(R')^{(n)} \approx \mathrm{Hom}(M^{(n)}, M) \approx M \oplus \mathrm{Hom}(N, M)$. ($\Leftarrow$) If $R'^{(n)} \approx M \oplus N$ in $\mathcal{M}od$-R' then $M^{(n)} \approx \mathrm{Hom}(R', M)^{(n)} \approx R'' \oplus \mathrm{Hom}(N, M)$ so M is a generator for $R'' \approx R$.

15. If R is simple and L is a left ideal then L is an f.g. projective right module over $T = \mathrm{End}_R L$, and $R \approx \mathrm{End}\, L_T$. T is a simple ring iff L is f.g. projective as an R-module, in which case R and T are Morita equivalent. (Hint: L is a generator, by exercise 1). Use this result to refine Koh's theorem (exercise 3.3.34.) concerning simple rings having a maximal left annihilator.

5 Homology and Cohomology

One of the major tools of algebra is homology theory and its dual, cohomology theory. Although rooted in algebraic topology, it has applications in virtually every aspect of algebra, as explained in the classic book of Cartan and Eilenberg. Our goal here is to develop enough of the general theory to obtain its main applications to rings, i.e., projective resolutions, the homological dimensions, the functors $\mathcal{T}or$ and $\mathcal{E}xt$, and the cohomology groups (which are needed to study division rings).

Duality plays an important role in category theory, so we often wish to replace $R\text{-}\mathcal{M}od$ by a self-dual category, i.e., we use only those properties of $R\text{-}\mathcal{M}od$ whose duals also hold in $R\text{-}\mathcal{M}od$. For the most part, it is enough to consider abelian categories with "enough" projectives and injectives, i.e., for each object M there is an epic $P \to M$ for P projective and a monic $M \to E$ for E injective. Given a map $f: M \to N$ we define coker $f = N/fM$, the *cokernel* of f.

§5.0 Preliminaries About Diagrams

Certain easy technical lemmas concerning diagrams are used repeatedly in homology and cohomology. For the sake of further reference we collect them here. We shall assume all of the diagrams are commutative, with all

369

rows and columns exact; as usual a dotted line indicates a map to be found. Also please see the remark at the end.

(1) *The five lemma* (*repeated from proposition 2.11.15*). *In the diagram*

$$
\begin{array}{ccccccccc}
A & \xrightarrow{f} & B & \xrightarrow{g} & C & \xrightarrow{h} & D & \xrightarrow{j} & E \\
\downarrow{\scriptstyle\alpha} & & \downarrow{\scriptstyle\beta} & & \downarrow{\scriptstyle\gamma} & & \downarrow{\scriptstyle\delta} & & \downarrow{\scriptstyle\varepsilon} \\
A' & \xrightarrow{f'} & B' & \xrightarrow{g'} & C' & \xrightarrow{h'} & D' & \xrightarrow{j'} & E'
\end{array}
$$

if every vertical arrow but γ is an isomorphism then γ is also an isomorphism.
(2) *Given*

$$
\begin{array}{ccc}
M & \xrightarrow{\beta} & N \\
\downarrow{\scriptstyle f} & & \downarrow{\scriptstyle g} \\
M'' & \xrightarrow{\gamma} & N''
\end{array}
$$

then f restricts to a map $\tilde{f}$: $\ker \beta \to \ker \gamma$, and g induces a map $\bar{g}$: coker $\beta \to$ coker γ yielding the commutative diagram

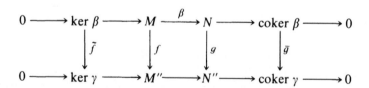

$$
\begin{array}{ccccccccc}
0 & \longrightarrow & \ker \beta & \longrightarrow & M & \xrightarrow{\beta} & N & \longrightarrow & \text{coker } \beta & \longrightarrow & 0 \\
& & \downarrow{\scriptstyle \tilde{f}} & & \downarrow{\scriptstyle f} & & \downarrow{\scriptstyle g} & & \downarrow{\scriptstyle \bar{g}} \\
0 & \longrightarrow & \ker \gamma & \longrightarrow & M'' & \longrightarrow & N'' & \longrightarrow & \text{coker } \gamma & \longrightarrow & 0
\end{array}
$$

If f and g are isomorphisms then $\tilde{f}$ and $\bar{g}$ are also isomorphisms.

Proof: If $\beta x = 0$ then $\gamma f x = g \beta x = 0$ implying $f(\ker \beta) \subseteq \ker \gamma$. On the other hand, $g(\beta M) = \gamma f M \subseteq \gamma M''$, so g induces a map $\bar{g}$: $N/\beta M \to N''/\gamma M''$. The commutativity of the ensuing diagram is obvious. The last assertion is a special case of the five lemma. To wit, adding 0 at the right so that $\bar{g}$ is in the middle of the appropriate diagram (with f at the left side) shows $\bar{g}$ is an isomorphism; adding, instead, 0 at the left so that $\tilde{f}$ is in the middle shows $\tilde{f}$ is an isomorphism.   Q.E.D.

(3) *The snake lemma. Given*

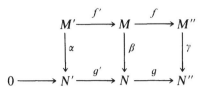

(i) $\ker \alpha \to \ker \beta \to \ker \gamma$ *is exact, where the maps are obtained as in* (2).

(ii) *If* f *is epic then there is an exact sequence*

$$\ker \alpha \xrightarrow{\tilde{f}'} \ker \beta \xrightarrow{\tilde{f}} \ker \gamma \xrightarrow{\partial} \operatorname{coker} \alpha \to \operatorname{coker} \beta \to \operatorname{coker} \gamma$$

where ∂ *is defined in the proof, and the other maps are in* (2).

Proof:

(i) The composition is certainly 0. On the other hand, if $z \in \ker \beta$ and $fz = 0$ then $z = f'x'$ for some x' in M'; then $0 = \beta z = g'\alpha x'$, so $\alpha x' = 0$ and $x' \in \ker \alpha$ as desired.

(ii) Define ∂ as follows: Given x'' in $\ker \gamma$ we take x in M with $fx = x''$; then $g\beta x = 0$ so $\beta x \in \ker g = g'N'$, yielding y' in N' with $g'y' = \beta x$, and we take $\partial x''$ to be $y' + \alpha M'$. To check ∂ is well-defined, suppose we took instead x_1 with $fx_1 = x''$ and we took y'_1 with $g'y'_1 = \beta x_1$. Then $f(x_1 - x) = 0$ so $x_1 - x \in f'M'$, and thus

$$g'(y'_1 - y_1) = \beta(x_1 - x) \in \beta f'M' = g'\alpha M'.$$

Hence $y'_1 - y_1 \in \alpha M'$ since g' is monic.

Having defined ∂, which is surely a map, we note that the hypothesis is self-dual, so it is enough to show $\ker \alpha \xrightarrow{\tilde{f}'} \ker \beta \xrightarrow{\tilde{f}} \ker \gamma \xrightarrow{\partial} \operatorname{coker} \alpha$ is exact; in view of (i) we need only show $\ker \beta \xrightarrow{\tilde{f}} \ker \gamma \xrightarrow{\partial} \operatorname{coker} \alpha$ is exact. We shall use throughout the notation of the previous paragraph. First note that if $x'' \in \tilde{f}(\ker \beta)$ then in the definition of $\partial x''$ above we could take x in $\ker \beta$, so $g'y' = \beta x = 0$ and thus $y' = 0$, proving $\partial \tilde{f} = 0$. On the other hand, if $\partial x'' = 0$ then $y' \in \alpha M'$ so writing $y' = \alpha x'$ we see $x - f'x' \in \ker \beta$ and $\tilde{f}(x - f'x') = f(x - f'x') = fx = x''$ proving $x'' \in \tilde{f}(\ker \beta)$. Q.E.D.

(4) *The nine lemma (or* 3×3 *lemma). Given*

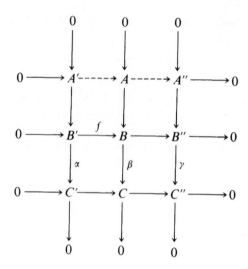

one can find dotted arrows making an exact row such that the above diagram is commutative.

Proof: Immediate from the snake lemma, since

$$\ker \alpha \xrightarrow{\tilde{f}} \ker \beta \to \ker \gamma \xrightarrow{\partial} \operatorname{coker} \alpha = 0$$

is exact and f is monic, yielding $0 \to A' \to A \to A'' \to 0$ as desired.   Q.E.D.

(5)   *The horseshoe lemma. If P' and P'' are projective then the following diagram can be "filled in" where $\mu: P' \to P' \oplus P''$ is the canonical monic and $\pi: P' \oplus P'' \to P''$ is the projection:*

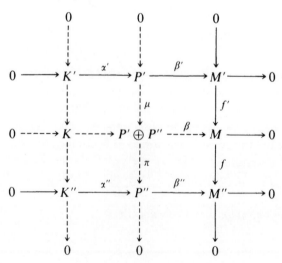

(You get the horseshoe lying on its side by erasing the dotted lines and the maps to and from 0.)

Proof: There is a map $h: P'' \to M$ lifting f, i.e., $\beta'' = fh$. Defining β by $\beta(x', x'') = f'\beta'x' + hx''$ one sees at once that the top right square is commutative. Moreover, $f\beta(x', x'') = ff'\beta'x' + \beta''x'' = 0 + \beta''\pi(x', x'')$ so the bottom right square is commutative. Taking $K = \ker \beta$ we can draw the upper row by means of the nine lemma (turned on its side). Q.E.D.

(6) *Turning the corner Given*

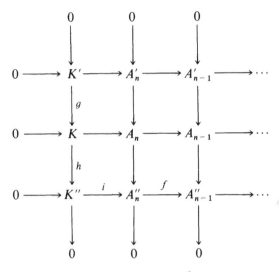

One has an exact sequence $0 \to K' \xrightarrow{g} K \xrightarrow{ih} A''_n \xrightarrow{f} A''_{n-1} \to \cdots$

Proof: $\ker f = iK'' = ihK$ since h is epic; moreover, since i is monic we have $\ker ih = \ker h = gK'$. Q.E.D.

(7) *The pullback and pushout have been introduced, respectively, as inverse and direct limits in exercises* 1.8.1 *and* 1.8.1'. *We start over again here, obtaining some obvious properties. First, given*

we define the pullback P to be $\{(a_1, a_2) \in A_1 \oplus A_2 : fa_1 = ga_2\}$, and let $\pi_i : P_i \to A_i$ be the restrictions of the projection maps from $A_1 \oplus A_2$ to A_i. Note $\ker \pi_1 = \{(0, a_2) : a_2 \in \ker g\} \approx \ker g$, yielding a commutative diagram

$$\begin{array}{ccccccc}
0 & \longrightarrow & K & \longrightarrow & P & \overset{\pi_1}{\longrightarrow} & A_1 \\
& & \downarrow{1_K} & & \downarrow{\pi_2} & & \downarrow \\
0 & \longrightarrow & K & \longrightarrow & A_2 & \underset{g}{\longrightarrow} & B
\end{array}$$

Furthermore, if g is epic then π_1 is epic for if $a_1 \in A_1$ then $fa_1 = ga_2$ for some a_2 in A_2 implying $a_1 = \pi_1(a_1, a_2)$.

Dually given

$$\begin{array}{ccc}
B' & \overset{f'}{\longrightarrow} & A'_1 \\
\downarrow{g'} & & \\
A'_2 & &
\end{array}$$

we define the pushout $P' = (A'_1 \oplus A'_2)/N$ where $N = \{f'b, -g'b) : b \in B'\}$, and let μ_1, μ_2 be the compositions of the canonical injections into $A'_1 \oplus A'_2$ with the canonical map factoring out N. Then $\operatorname{coker} \mu_1 \approx \operatorname{coker} g'$ (proof dual to the dual assertion for the pullback), and if g' is monic then μ_1 is monic.

(8) *The projective lifting lemma. If P is projective and*

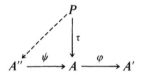

is exact with $\varphi\tau = 0$ then $\tau = \psi\sigma$ for some $\sigma : P \to A''$: (Indeed, replace A by $\psi A''$. Since $\tau P \subseteq \ker \varphi \subseteq \psi A''$ we see ψ lifts.)

Remark: Although the results of this section have been framed for modules, they are purely categorical in nature, and are applicable in many other settings. Another advantage in a more abstract treatment is that the principle of categorical duality would become available. Suppose we can prove the basic facts of this section in a class of categories that is self-dual (i.e., closed under passing to the dual category); then whenever we prove a theorem with them, we can automatically conclude the *dual theorem* obtained by reversing the arrows. Although module categories are not self-dual, there is a more general class of categories that is self-dual, yet rich enough to yield all the results of

§5.0. This is the class of *abelian categories*, discussed in depth in vol. 1 of the unabridged edition.

In brief, in an abelian category $\mathscr{C}$ we have addition of morphisms in $\text{Hom}(A, B)$ for all objects A, B, making each $\text{Hom}(A, B)$ an abelian group with composition of morphisms distributive over addition; moreover the finite direct product of objects of $\mathscr{C}$ exists as an object of $\mathscr{C}$, and every morphism f is the composite hg of an epic g and a monic h.

We shall cheat a little and claim (without proof) the dual results of theorems we prove in Chapter 5, even though we are not developing them in the general context of abelian categories.

§5.1 Resolutions and Projective and Injective Dimension

In this section we introduce two new important dimensions, the *projective dimension* and its dual, the *injective dimension*. It turns out these dimensions have immediate application to ring theory, and we shall use these to develop the motivation of the subject.

Projective Resolutions

Definition 5.1.1: A *projective resolution* $\mathscr{P}$ of a module M is an exact sequence of the form

$$\cdots \to P_n \xrightarrow{f_n} P_{n-1} \to \cdots \to P_1 \xrightarrow{f_1} P_0 \xrightarrow{\varepsilon} M \to 0$$

where each P_i is projective. We say $\mathscr{P}$ is f.g. (usually called *finite* in the literature) if each P_i is f.g.; $\mathscr{P}$ is *free* if each P_i is free. The smallest n for which $P_{n+1} = 0$ is called the *length* of the resolution $\mathscr{P}$.

Several immediate observations are available.

Remark 5.1.2:

(i) Every module M has a free resolution. Indeed, take an epic $f_0: F_0 \to M$ with F_0 free and, inductively, given $f_{i-1}: F_{i-1} \to F_{i-2}$ take an epic $f_i: F_i \to \ker f_{i-1}$. We could view $f_i: F_i \to F_{i-1}$; then $f_i F_i = \ker f_{i-1}$ so the sequence $\cdots \to F_2 \xrightarrow{f_2} F_1 \xrightarrow{f_1} F_0 \to M \to 0$ is exact.

(ii) If M is f.g. then we could take F_0 f.g. as well as free. If R is left Noetherian then $\ker f_0 \le F_0$ is f.g., and, continuing by induction, we may assume each F_i is f.g. Thus every f.g. module over a left Noetherian ring has an f.g. free resolution.

In order to study a resolution we shall repeatedly use the observation that we can "cut" an exact sequence $\cdots \to M'' \xrightarrow{f} M \xrightarrow{g} M' \to \cdots$ at g to produce exact sequences

$$\to M'' \xrightarrow{f} M \to gM \to 0 \quad \text{and} \quad 0 \to \ker g \to M \xrightarrow{g} M' \to \cdots.$$

In particular, any projective resolution $\mathscr{P}$ can be cut at f_n to yield

$$0 \to K \to P_n \xrightarrow{f_n} P_{n-1} \to \cdots \to P_0 \to M \to 0;$$

$K = \ker f_n$ is called the n-th syzygy of M. If $\mathscr{P}$ has length $n + 1$ then $K \approx P_{n+1}$ is projective. Conversely, if K is projective then we have produced a new projective resolution of length $n + 1$, which we shall compare to $\mathscr{P}$ by means of the next result.

Proposition 5.1.3: *(Generalized Schanuel's lemma) Given exact sequences*

$$0 \to K \to P_n \xrightarrow{f_n} P_{n-1} \to \cdots \to P_1 \xrightarrow{f_1} P_0 \xrightarrow{\varepsilon} M \to 0 \tag{1}$$

$$0 \to K' \to P'_n \xrightarrow{g_n} P'_{n-1} \to \cdots \to P'_1 \xrightarrow{g_1} P'_0 \xrightarrow{\varepsilon'} M \to 0 \tag{2}$$

with each P_i and P'_i projective then

$$K \oplus P'_n \oplus P_{n-1} \oplus P'_{n-2} \oplus \cdots \approx K' \oplus P_n \oplus P'_{n-1} \oplus P_{n-2} \oplus \cdots.$$

Proof: We cut (1) at f_1 to get

$$0 \to K \to \cdots \to P_2 \to P_1 \to \ker \varepsilon \to 0 \tag{3}$$

$$0 \to \ker \varepsilon \to P_0 \to M \to 0 \tag{4}$$

and cut (2) at g_1 to get

$$0 \to K' \to \cdots \to P'_2 \to P'_1 \to \ker \varepsilon' \to 0 \tag{5}$$

$$0 \to \ker \varepsilon' \to P'_0 \to M \to 0 \tag{6}$$

Let $M' = P'_0 \oplus \ker \varepsilon \approx P_0 \oplus \ker \varepsilon'$ by Schanuel's lemma (2.8.26) applied to sequences (4) and (6). Then we can modify (3), (5) to get

$$0 \to K \to \cdots \to P_2 \to P_1 \oplus P'_0 \xrightarrow{f_1 \oplus 1} M' \to 0$$

$$0 \to K' \to \cdots \to P'_2 \to P'_1 \oplus P_0 \xrightarrow{g_1 \oplus 1} M' \to 0$$

and by induction on the length we get the desired conclusion.   Q.E.D.

Definition 5.1.4: Two modules N, N' are *projectively equivalent* if there are projectives P, P' such that $N \oplus P \approx N' \oplus P'$; if, furthermore, P, P' are f.g. free then N, N' are *stably equivalent*. N is *stably free* if N is stably equivalent to a free module, i.e., $N \oplus R^{(n)}$ is free for suitable n.

Note that any module M projectively equivalent to a projective module is itself projective since M is a summand of a projective module. In particular, all stably free modules are projective. (Compare to remark 2.8.4′ and exercise 10.)

Corollary 5.1.5: *For any n the n-th syzygies of any two resolutions of M are projectively equivalent; the n-th syzygies of any two f.g. free resolutions of M are stably equivalent.*

We are ready to open an important new dimension in module theory.

Definition 5.1.6: *M* has *projective dimension n* (written $\text{pd}(M) = n$) if *M* has a projective resolution of length *n*, with *n* minimal such. In the literature the projective dimension is also called the *homological dimension*.

Remark 5.1.7: $\text{pd}(M) = 0$ iff *M* is projective.

Proposition 5.1.8: *The following are equivalent:*

(i) $\text{pd}(M) \leq n + 1$.

(ii) *The n-th syzygy of any projective resolution of M is projective.*

(iii) *Any projective resolution $\mathscr{P}$ of M can be cut at f_n to form a projective resolution of length $n + 1$.*

Proof: (i) $\Rightarrow$ (ii) By corollary 5.1.5 the *n*-th projective syzygy is projectively equivalent to a projective module and thus is itself projective; (ii) $\Rightarrow$ (iii) clear; (iii) $\Rightarrow$ (i) by definition.   Q.E.D.

Proposition 5.1.9: *The following are equivalent:*

(i) *M has a f.g. free resolution of length $\leq n + 1$.*

(ii) *The n-th syzygy of any f.g. free resolution is stably free.*

(iii) *Any f.g. free resolution can be cut and modified to form a f.g. free resolution of length $n + 1$.*

Proof: (iii) $\Rightarrow$ (i) $\Rightarrow$ (ii) is clear. To see (ii) $\Rightarrow$ (iii) note that if $P \oplus R^{(n)}$ is free and $0 \to P \to F \to F'$ is exact with F, F' free then the sequence

$$0 \to P \oplus R^{(n)} \to F \oplus R^{(n)} \to F'$$

is exact.   Q.E.D.

Elementary Properties of Projective Dimension

Let us now compare projective dimensions of modules in an exact sequence. This can be done elegantly using the functor Ext of §5.2 or in a simple but ad hoc fashion by means of the following result:

Proposition 5.1.10:   *Suppose* $0 \to M' \to M \to M'' \to 0$ *is exact. Given projective resolutions* $\mathscr{P}', \mathscr{P}''$ *of* M', M'' *respectively, we can build a projective resolution* $\mathscr{P}$ *of* M *where each* $P_n = P'_n \oplus P''_n$ *and, furthermore, letting* K'_n, K''_n *denote the n-th syzygies we have the following commutative diagram for each n, obtained by cutting* $\mathscr{P}', \mathscr{P}$, *and* $\mathscr{P}''$ *at* f'_n, f_n, *and* f''_n, *respectively:*

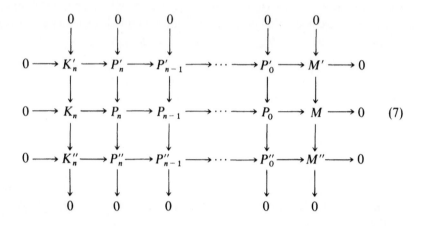

$$(7)$$

Proof: By induction on n. Suppose for $i \le n-1$ we have $P_i = P'_i \oplus P''_i$ together with the diagram

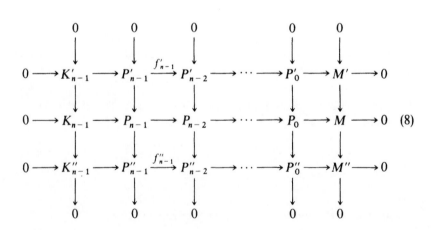

$$(8)$$

Note that $K'_{n-1} = \ker f'_{n-1} = f'_n P'_n$ and $K''_{n-1} = \ker f''_{n-1} = f''_n P''_n$. Cutting the resolutions $\mathscr{P}', \mathscr{P}''$ at f_n', f_n'' yields the diagram

$$
\begin{array}{ccccccc}
 & & & & 0 & & \\
 & & & & \downarrow & & \\
\longrightarrow & P_{n+1}' & \longrightarrow & P_n' & \longrightarrow f_n' P_n' = K_{n-1}' & \longrightarrow & 0 \\
 & & & & \downarrow & & \\
 & & & & K_{n-1} & & \\
 & & & & \downarrow & & \\
\longrightarrow & P_{n+1}'' & \longrightarrow & P_n'' & \longrightarrow f_n'' P_n'' = K_{n-1}'' & \longrightarrow & 0 \\
 & & & & \downarrow & & \\
 & & & & 0 & &
\end{array}
$$

where the column is the $n - 1$ — syzygy column from above. Now the horse-shoe lemma enables us to fill in the diagram

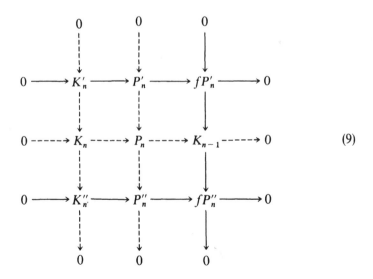

(9)

where $P_n = P_n' \oplus P_n''$, and we get (7) by pasting (9) to (8).   Q.E.D.

Corollary 5.1.11: *Suppose* $0 \to M' \to M \to M'' \to 0$ *is exact. If M', M'' have projective (resp. f.g. projective, resp. free, resp. f.g. free) resolutions of length $\leq n$ then so does M. In particular,* $\mathrm{pd}(M) \leq \max\{\mathrm{pd}(M'), \mathrm{pd}(M'')\}$*. If* $\mathrm{pd}(M') > \mathrm{pd}(M'')$ *then* $\mathrm{pd}(M) = \mathrm{pd}(M')$*.*

Proof: The first two assertions are immediate since $0 \oplus 0 \approx 0$. For the last assertion let us examine the exact sequence of n-th syzygies $0 \to K'_n \to K_n \to K''_n \to 0$ for $n = \mathrm{pd}(M'')$. This splits since K''_n is projective, so K'_n is projectively equivalent to K_n, and we apply proposition 5.1.8.   Q.E.D.

Corollary 5.1.12: *Suppose* $0 \to M' \to M \to M'' \to 0$ *is exact. Suppose* $\mathscr{P}', \mathscr{P}''$ *are projective resolutions of M', M'' and let $\mathscr{P}$ be built as in proposition 5.1.10. Let K'_m, K_m, K''_m denote the respective m^{th} syzygies. Then for any m there is an exact sequence*

$$0 \to K'_{m-1} \to K_{m-1} \to P''_{m-1} \to P''_{m-2} \to \cdots \to P''_0 \to M'' \to 0.$$

This displays K'_{m-1} as an m-th syzygy of a projective resolution of M'', if K_{m-1} is projective.

Writing n', n, n'' for the respective pd of M', M, and M'' we then have $n'' \leq \max\{n', n\} + 1$. Furthermore, $n'' = n$ if $n' < n$, and $n'' = n' + 1$ if $n' > n$.

Proof: The first assertion is obtained by "turning the corner" in (7). Everything else follows by choosing m properly. Taking $m = \max\{n', n\}$ we see K'_{m-1} and K_{m-1} are projective so $n'' \leq m + 1 = \max\{n', n\} + 1$.

If $n' < n$ then we could choose $\mathscr{P}'$ so that $K'_n = 0$; taking $m = n + 1$ we have $n'' \leq n$ and thus $n'' = n$ by corollary 5.1.11.

If $n' > n$ then taking $m = n'$ we see K_{m-1} is projective but K'_{m-1} is not, so $n'' > m$ by proposition 5.1.8, i.e., $n'' \geq n' + 1$; hence $n'' = n' + 1$ since $n'' \leq \max\{n', n\} + 1$.   Q.E.D.

Summary 5.1.13: *Suppose* $0 \to M' \to M \to M'' \to 0$ *is exact. Recapitulating corollary 5.1.12 where $n' = \mathrm{pd}(M')$, $n = \mathrm{pd}(M)$, and $n'' = \mathrm{pd}(M'')$ we have the following possibilities:*

Case I. $n < n'$*. Then* $n'' = n' + 1$*.*

Case II. $n = n'$*. Then* $n'' \leq n + 1$*.*

Case III. $n > n'$*. Then* $n'' = n$*.*

These formulas could be proved directly using induction, cf., Kaplansky [72B, p. 169]; we shall give a less ad hoc proof in §5.2.

Example 5.1.14: Suppose $f: F \to M''$ is epic with F free and M'' *not* projective. Then $\mathrm{pd}(F) = 0$ so $\mathrm{pd}(M'') = \mathrm{pd}(\ker f) + 1$. In particular, if Rr is *not* projective then $\mathrm{pd}(Rr) = \mathrm{pd}(\mathrm{Ann}\, r) + 1$.

The (Left) Global Dimension

Definition 5.1.15: The *global dimension*

$$\mathrm{gl.\,dim}(R) = \sup\{\mathrm{pd}(M): M \in R\text{-}\mathcal{M}od\}.$$

Example 5.1.16:

(i) R is semisimple Artinian iff every module is projective, iff $\mathrm{gl.\,dim}(R) = 0$.

(ii) If R is hereditary then $\mathrm{gl.\,dim}(R) \leq 1$. (Indeed, every submodule of a projective is projective, so the first syzygy is always projective, implying $\mathrm{pd}(M) \leq 1$ for any module M.)

There is another way of viewing the global dimension, by dualizing everything.

Definition 5.1.17: An *injective resolution* $\mathbb{E}$ of a module M is an exact sequence $0 \to M \xrightarrow{f_0} E_0 \xrightarrow{f_1} E_1 \to \cdots$ with each E_i injective; the n-th *cosyzygy* is $\mathrm{coker}\, f_n$. $\mathbb{E}$ has *length* n if $E_i = 0$ for all $i > n$, with n minimal such. The *injective dimension* $\mathrm{id}(M)$ (if it exists) is the smallest n for which M has an injective resolution of length n.

Dually to proposition 5.1.8 we see $\mathrm{id}(M) = n$ iff every injective resolution can be cut to a resolution of length n.

In the next section we shall see $\mathrm{gl.\,dim}(R) = \sup\{\mathrm{id}(M): M \in R\text{-}\mathcal{M}od\}$, which is easy to believe since it holds for R semisimple Artinian ($\mathrm{gl.\,dim}\, 0$) or hereditary ($\mathrm{gl.\,dim}\, 1$). Kirkman-Kuzmanovich [87] have constructed examples of Noetherian R with $\mathrm{gl.\,dim}\, R < \mathrm{gl.\,dim}\, R/\mathrm{Nil}(R)$.

Our present interest is to see the connection between R and $R[\lambda]$ for a commuting indeterminate λ. If R is a field then $R[\lambda]$ is a PID and thus hereditary, so $\mathrm{gl.\,dim}\, R[\lambda] = \mathrm{gl.\,dim}(R) + 1$. Actually, this relation holds much more generally, and we shall verify it for the skew polynomial ring $R[\lambda; \sigma]$ where R is arbitrary and σ is an automorphism of R. (In particular, one could take $\sigma = 1$ and obtain the ordinary polynomial ring.) To see this we need to introduce some new module structure.

If M is an R-module then we define $M[\lambda; \sigma]$ to be $R[\lambda] \otimes_R M$ as an additive group, i.e., the elements of $M[\lambda; \sigma]$ have the unique form $\sum \lambda^i x_i$ where $x_i \in M$. $M[\lambda; \sigma]$ is made into $R[\lambda; \sigma]$-module via scalar multiplication

$$r\lambda^j \sum \lambda^i x_i = \sum \lambda^{i+j} (\sigma^{-(i+j)} r) x_i.$$

Remark 5.1.18:   Since $R[\lambda; \sigma]$ is free over itself we see that $F[\lambda; \sigma]$ is a free $R[\lambda; \sigma]$-module for any free R-module F; consequently, any projective (resp. free) resolution $0 \to P_n \to \cdots \to P_0 \to M \to 0$ of M in $R\text{-}\mathcal{M}od$ yields a projective (resp. free) resolution $0 \to P_n[\lambda; \sigma] \to \cdots \to P_0[\lambda; \sigma] \to M[\lambda; \sigma] \to 0$ in $R[\lambda; \sigma]\text{-}\mathcal{M}od$. (This is really a fact about graded modules.) In particular, $\text{pd}_{R[\lambda;\sigma]} M[\lambda; \sigma] \leq \text{pd}_R M$; in fact, equality holds by exercise 2.

On the other hand, any $R[\lambda; \sigma]$-module M is viewed naturally as R-module by forgetting λ, and then we can form $M[\lambda; \sigma]$ as above. To distinguish between the two $R[\lambda; \sigma]$-module structures we write λx to denote the original product in M, and $\lambda \cdot x$ to denote the new product in $M[\lambda; \sigma]$. We need one more module. Define σM to be the set of formal elements $\{\sigma x : x \in M\}$ made into a module under the operations $\sigma x_1 + \sigma x_2 = \sigma(x_1 + x_2)$ (so that $\sigma M \approx M$ as abelian groups) and $r\sigma x = \sigma x'$ where x' is the product $(\sigma^{-1} r)x$ in M.

Lemma 5.1.18':   (*Hochschild's trick*) *For any $R[\lambda; \sigma]$-module M there is an exact sequence*

$$0 \to (\sigma M)[\lambda; \sigma] \to M[\lambda; \sigma] \xrightarrow{f} M \to 0$$

where $f(\lambda^i \cdot x) = \lambda^i x$.

Proof: Define $g : (\sigma M)[\lambda; \sigma] \to M[\lambda; \sigma]$ by $g(\sum \lambda^i \cdot \sigma x_i) = \sum (\lambda^i \cdot \lambda x_i - \lambda^{i+1} \cdot x_i)$. Clearly $g(\sigma M[\lambda; \sigma]) \subseteq \ker f$. Also

$$(r\lambda^j) \cdot \lambda^i \sigma x_i = \lambda^{i+j} \sigma^{-(i+j)} r \sigma x_i = \lambda^{i+j} \cdot \sigma((\sigma^{-(i+j+1)} r)x_i) \qquad \text{so}$$

$$g\left(r\lambda^j \sum_i \lambda^i \cdot \sigma x_i\right) = \sum_i (\lambda^{i+j} \cdot \lambda(\sigma^{-(i+j+1)} r)x_i - \lambda^{i+j+1} \cdot (\sigma^{-(i+j+1)} r)x_i)$$

$$= \sum_i r(\lambda^{i+j} \cdot \lambda x_i - \lambda^{i+j+1} x_i)$$

$$= r\lambda^j g\left(\sum \lambda^i \cdot \sigma x_i\right)$$

proving g is indeed a map. Moreover, we can rewrite $g(\sum \lambda^i \cdot \sigma x_i)$ as $1 \cdot \lambda x_0 + \sum_{i \geq 1} \lambda^i \cdot (\lambda x_i - x_{i-1})$. Hence $\sum \lambda^i \cdot \sigma x_i \in \ker g$ iff $x_0 = 0$ and $\lambda x_i = x_{i-1}$ for all $i \geq 1$; but one has $x_i = 0$ for large i, so $0 = x_{i-1} = x_{i-2} = \cdots$, implying g is monic.

Finally suppose $x' = \sum_{i=0}^m \lambda^i \cdot x_i' \in \ker f$. Put $x_m = 0$, $x_{m-1} = -x_m'$, and given x_i put $x_{i-1} = \lambda x_i - x_i'$. Then $g(\sum_{i \geq 0} \lambda^i \cdot \sigma x_i) = x'$, proving exactness.

$\hspace{10cm}$ Q.E.D.

We are ready for the first inequality.

Proposition 5.1.19: gl. dim $R[\lambda; \sigma] \leq$ gl. dim $R + 1$. *In fact,* $\mathrm{pd}_{R[\lambda;\sigma]} M \leq \mathrm{pd}_R \sigma M + 1$ *for any* $R[\lambda; \sigma]$-*module* M.

Proof: Let $n = $ gl. dim R. For any $R[\lambda; \sigma]$-module M we have

$$\mathrm{pd}_{R[\lambda;\sigma]} M[\lambda; \sigma] \leq \mathrm{pd}_R M \leq n,$$

and $\mathrm{pd}_{R[\lambda;\sigma]}(\sigma M)[\lambda; \sigma] \leq \mathrm{pd}_R \sigma M \leq n$. Applying corollary 5.1.12 to lemma 5.1.18' thus shows $\mathrm{pd}_{R[\lambda;\sigma]} M \leq n + 1$. $\hspace{2cm}$ Q.E.D.

Our next goal is to see, in fact, that equality holds.

Proposition 5.1.20: $\mathrm{pd}(\bigoplus_{i \in I} M_i) = \sup\{\mathrm{pd}(M_i): i \in I\}$.

Proof: Write $n_i = \mathrm{pd}(M_i)$ and $n = \mathrm{pd}(\oplus M_i)$; take projective resolutions $\mathscr{P}_i$ of M_i of respective lengths n_i, and let $\mathscr{P} = \oplus \mathscr{P}_i$, ie $\mathscr{P}$ is formed by taking the direct sum of the respective terms. Then, clearly, length $(\mathscr{P}) = \sup(n_i)$, proving $n \leq \sup(n_i)$. On the other hand, the $(n - 1)$-st syzygy of $\mathscr{P}$ is projective, so the $(n - 1)$-st syzygy of each $\mathscr{P}_i$ is a summand of a projective and thus projective, proving each $n_i \leq n$. $\hspace{2cm}$ Q.E.D.

Corollary 5.1.21: $\mathrm{pd}(M^{(I)}) = \mathrm{pd}(M)$ *for any index set* I.

Proposition 5.1.22: *Suppose* $\varphi: R \to T$ *is a ring homomorphism. Viewing any* T-*module* M *as* R-*module by means of* φ *we have*

$$\mathrm{pd}_R M \leq \mathrm{pd}_T M + \mathrm{pd}_R T.$$

Moreover, if equality holds for all M *such that* $\mathrm{pd}_T M \leq 1$ *then equality holds for all* M *having finite pd.*

Proof: Induction on $m = \text{pd}_T M$. Write $n = \text{pd}_R M$ and $t = \text{pd}_R T$; we want to show $n \le m + t$. For $m = 0$ we have M projective as T-module so $M \oplus M' = F$ for some free T-module F. This is also a direct sum as R-modules, so applying proposition 5.1.20 twice yields $n \le \text{pd}_R F = t$, as desired.

For $m > 0$ take $0 \to M' \to F \to M \to 0$ exact where F is free. Then $\text{pd}_T M' = m - 1$ by summary 5.1.13, so $\text{pd}_R M' \le (m - 1) + t$ by induction. Thus

$$n \le 1 + \max\{\text{pd}_R F, \text{pd}_R M'\} \le 1 + \max\{t, (m-1) + t\} = 1 + m - 1 + t = m + t$$

proving the first assertion. To prove the second assertion we note the same induction argument works if we have the case $m = 1$.   Q.E.D.

Remark 5.1.23: Suppose $Ra = aR$. Any M in $R\text{-}\mathcal{M}od$ has the submodule aM, and $\bigoplus M_i / a(\bigoplus M_i) \approx \bigoplus(M_i / aM_i)$. In particular, if F is a free (resp. projective) R-module then F/aF is a free (resp. projective) R/Ra-module.

Theorem 5.1.24: ("*Change of rings*") *Suppose $a \in R$ is a regular noninvertible element, and $Ra = aR$. If M is an R/Ra-module with $\text{pd}_{R/Ra} M = n < \infty$ then $\text{pd}_R M = n + 1$. In particular,* gl. dim $R \ge$ gl. $\dim(R/Ra) + 1$ *provided* gl. dim R/Ra *is finite.*

Proof: Write $\bar{R} = R/Ra$. First we show $\text{pd}_R \bar{R} = 1$. Right multiplication by a gives an isomorphism $R \to Ra$, so applying summary 5.1.13 to the exact sequence $0 \to Ra \to R \to \bar{R} \to 0$ we see $\text{pd}_R \bar{R} \le 1$. But $\bar{R}$ is not projective (for otherwise Ra would be a summand of R, implying $Ra = Re$ for an idempotent $e \ne 0$, and then $a(1 - e) = 0$, contrary to a regular.) Thus $\text{pd}_R \bar{R} = 1$.

By proposition 5.1.22 we have $\text{pd}_R M \le n + 1$, and we need to show equality for the case $n \le 1$. Assume M is a counterexample; we may assume $\text{pd}_R M \le 1$. By definition $aM = 0$. Hence M cannot be a submodule of a free R-module. In particular, M is not projective so $\text{pd}_R M = 1$. Hence $n = 1$ (since M is a counterexample). Any exact sequence of R-modules

$$0 \to M' \to F \to M \to 0 \qquad (F \text{ free}) \tag{10}$$

sends aF to $aM = 0$, so we get the exact sequence

$$0 \to M'/aF \to F/aF \to M \to 0. \tag{11}$$

But this can be read in $\bar{R}\text{-}\mathcal{M}od$, in which F/aF is free. Also we have

$$0 \to aF/aM' \to M'/aM' \to M'/aF \to 0. \tag{12}$$

Applying example 5.1.14 to (10), (11) we have $\mathrm{pd}_R\, M' = 0$ and $\mathrm{pd}_R\, M'/aF = 0$. But then M'/aM' and M'/aF are projective $\bar{R}$-modules. In particular, (12) splits, implying aF/aM' is a projective $\bar{R}$-module. Thus $M \approx F/M' \approx aF/aM'$ is projective.   Q.E.D.

Corollary 5.1.25: $\mathrm{gl.\,dim}\, R[\lambda; \sigma] = \mathrm{gl.\,dim}\, R + 1$.

Proof: Apply the theorem, with $R[\lambda; \sigma]$ and λ replacing, respectively, R and a.   Q.E.D.

(For $\mathrm{gl.\,dim}\, R = \infty$ a separate argument is needed, which is an instant application of exercise 4. In fact, infinite gl. dim has led to serious errors in the literature, cf., exercise 16 and McConnell[77].) It is also useful to have a localization result.

Proposition 5.1.26:   *If S is a left denominator set for R and $T = S^{-1}R$ then $\mathrm{pd}_T\, S^{-1}M \leq \mathrm{pd}_R\, M$ for M in R-Mod. In particular, $\mathrm{gl.\,dim}\, T \leq \mathrm{gl.\,dim}\, R$.*

Proof: Since the localization functor is exact (theorem 3.1.20), any projective resolution of M can be localized to a projective resolution of $S^{-1}M$ as T-module.   Q.E.D.

There is a right-handed version of projective dimension and thus of gl. dim, which also is 0 if $\mathrm{gl.\,dim}\, R = 0$ since "semisimple Artinian" is left-right symmetric. However, for $\mathrm{gl.\,dim}\, R = 1$ the symmetry fails because left hereditary rings need not be right hereditary, and Jategaonkar's example in §2.1 can be used to find rings of arbitrary left and right gl. dim ≥ 1. A thorough discussion of global dimension and its peculiar connection to the continuum hypothesis can be found in Osofsky [73B].

Stably Free and FFR

Definition 5.1.27:   A module M has FFR (of length n) if M has an f.g. free resolution of length n. (In particular M is f.g.)

Our interest in FFR is derived from the next result.

Lemma 5.1.28:   *If P is projective with FFR then P is stably free.*

Proof: Take an f.g. free resolution $0 \to F_n \overset{f_n}{\to} F_{n-1} \to \cdots \to F_1 \overset{f_1}{\to} F_0 \overset{\varepsilon}{\to} P \to 0$. Since ε splits we can write $F_0 = P \oplus P'$ with $f_1 F_1 = \ker \varepsilon = P'$, so P' has a f.g. free resolution $0 \to F_n \to F_{n-1} \to F_1 \to P' \to 0$ of length $n-1$. By induction P' is stably free; write $P' \oplus R^{(n)} \approx F$ for F free. Then $P \oplus F \approx P \oplus P' \oplus R^{(n)} \approx F_0 \oplus R^{(n)}$ is f.g. free, proving P is stably free.   Q.E.D.

Conversely, we have

Lemma 5.1.29: *If M has an f.g. projective resolution of length n, with each P_i stably free then M has FFR of length $\le n+1$.*

Proof: Take f.g. free F, F' such that $P_0 \oplus F' \approx F$. If $n = 0$ then $M \approx P_0$ so $0 \to F' \to F \to M \to 0$ is the desired sequence. In general, we break the resolution $0 \to P_n \to \cdots \to P_0 \to M \to 0$ at ε to get $0 \to P_n \to \cdots \to P_2 \to P_1 \to \ker f \to 0$ and $0 \to \ker f \to P_0 \to M \to 0$. By induction $\ker f$ has an f.g. free resolution $0 \to F_n \to \cdots \to F_0 \to \ker f \to 0$; piecing back together yields

$$0 \to F_n \to \cdots \to F_0 \to P_0 \to M \to 0, \qquad \text{or better yet}$$

$$0 \to F_n \to \cdots \to F_1 \to F_0 \oplus F' \to F \to M \to 0. \qquad \text{Q.E.D.}$$

Definition 5.1.30: $K_0(R) = \mathscr{F} / \mathscr{R}$, where $\mathscr{F}$ is the free abelian group whose generators correspond to the isomorphism classes of f.g. projective R-modules, and $\mathscr{R}$ is the subgroup generated by all $[P \oplus Q] - [P] - [Q]$ for P, Q f.g. projective.

Since $\sum [P_i] = [\oplus P_i]$ in $K_0(R)$, we see every element of $K_0(R)$ can be written in the form $[P] - [Q]$ for P, Q f.g. projective.

Remark 5.1.30': P is stably free iff $[P] = [R^{(k)}]$ for some k. (Proof: ($\Rightarrow$) If $P \oplus R^{(m)} \approx R^{(n)}$ then $[P] = [R^{(n-m)}]$. ($\Leftarrow$) If $[P] = [R^{(k)}]$ then $P \oplus Q \approx R^{(k)} \oplus Q$ for a suitable f.g. projective module Q; writing $Q \oplus Q' \approx R^{(m)}$ we see $P \oplus R^{(m)} \approx R^{(m+k)}$.)

In view of remark 5.1.30' we see every f.g. projective is stably free iff $K_0(R) = \langle [R] \rangle$, and we shall use this characterization from now on.

Proposition 5.1.31: *Suppose R is left Noetherian and $\operatorname{gl.dim} R < \infty$. $K_0(R) = \langle [R] \rangle$ iff every f.g. projective R-module has FFR.*

Proof: ($\Rightarrow$) by lemma 5.1.29. ($\Leftarrow$) by lemma 5.1.28.   Q.E.D.

Our principal interest here is when $K_0(R) = \langle [R] \rangle$. One way of proving this is to show that every f.g. module has FFR. Thus we might hope for an analog of corollary 5.1.25, to tell us that $K_0(R[\lambda]) = \langle [R[\lambda]] \rangle$ if $K_0(R) = \langle [R] \rangle$. Unfortunately, we lack an f.g. analog of proposition 5.1.22, so we start over again.

Proposition 5.1.32: *Suppose R is left Noetherian and* gl. dim $R < \infty$. *If $K_0(R) = \langle [R] \rangle$ then $K_0(S^{-1}R) = \langle [S^{-1}R] \rangle$ for any left denominator set S of R.*

Proof: By lemma 5.1.28 it suffices to show every f.g. projective $S^{-1}R$-module P has FFR. P is naturally an R-module; take an f.g. R-submodule M with $S^{-1}M$ maximal. If $S^{-1}M < P$ then taking x in $P - S^{-1}M$ we would have $S^{-1}(M + Rx) > S^{-1}M$, contrary to choice of M; thus $S^{-1}M = P$. But now take an FFR for M. Tensoring each term by $S^{-1}R$ yields an FFR for $S^{-1}M = P$.   Q.E.D.

This argument actually yields an epic $K_0(R) \to K_0(S^{-1}R)$, cf., exercise 12.

Proposition 5.1.33: *Suppose R is left Noetherian and $0 \to M' \to M \to M'' \to 0$ is exact. If two of the modules have FFR then so does the third.*

Proof: In view of remark 5.1.2(ii) there are f.g. free resolutions $\mathcal{P}', \mathcal{P}''$ (of possibly infinite length) of M', M'', and we form the f.g. free resolution $\mathcal{P}$ of M as in proposition 5.1.10. By lemma 5.1.29 it suffices to find some n for which the n-th syzygies K'_n, K_n, and K''_n are stably free. Recall $0 \to K'_n \to K_n \to K''_n \to 0$ is exact. If M', M'' have FFR then we may assume $\mathcal{P}', \mathcal{P}''$ are f.g., so for large enough n we get $K'_n = K''_n = 0$ and thus $K_n = 0$. Therefore, we may assume one of the two modules with FFR is M. Now the generalized Schanuel lemma implies K_n is stably free for large enough n. But M' or M'' has FFR so for n large enough we may assume $K'_n = 0$ or $K''_n = 0$; thus K_n is isomorphic to the other syzygy which is thus stably free.   Q.E.D.

Theorem 5.1.34: *(Serre's theorem) Suppose R is a left Noetherian $\mathbb{N}$-graded ring, with* gl. dim $R < \infty$. *If $K_0(R_0) = \langle [R_0] \rangle$ then $K_0(R) = \langle [R] \rangle$.*

Before presenting the proof of this important result, we should note the same proof shows $K_0(R) \approx K_0(R_0)$. This result will be generalized further in appendix A. We follow Bass [68B] and start by noting various properties of graded modules, of independent interest. Let $R_+ = \bigoplus_{n>0} R_n$, an ideal of R. Then $R_0 \approx R/R_+$. We consider the category $R\text{-}\mathcal{G}\imath\text{-}\mathcal{M}od$ of definition 1.9.1.

Remark 5.1.34': If $M \in R\text{-}\mathcal{G}\imath\text{-}\mathcal{M}od$ and $R_+M = M$ then $M = 0$. (Indeed, otherwise, take n minimal such that $M_n \neq 0$. Then $M_n \cap R_+M = 0$, contradiction.)

Remark 5.1.34'': If a projective module P happens to be in $R\text{-}\mathcal{G}\imath\text{-}\mathcal{M}od$ then there is an epic $F \to P$ in $R\text{-}\mathcal{G}\imath\text{-}\mathcal{M}od$ for F a suitable graded free $G(R)$-module (cf., remark 1.9.7), and the epic splits in $R\text{-}\mathcal{G}\imath\text{-}\mathcal{M}od$ by remark 1.9.6. In view of remark 1.9.5 we see if p.d. $M = t < \infty$ and M is graded then M has a projective resolution of length t in $R\text{-}\mathcal{G}\imath\text{-}\mathcal{M}od$.

In view of remark 5.1.34'' there is no ambiguity in talking about the graded f.g. projective R-modules, which we call $\mathcal{G}\imath\text{-}proj(R)$, viewed as a full sub-category of $R\text{-}\mathcal{G}\imath\text{-}\mathcal{M}od$.

Claim 1: *The functor $F = R \otimes_{R_0} \underline{\quad}$ (cf., remark 1.9.11) is a category equivalence $\mathcal{G}\imath\text{-}proj(R_0) \to \mathcal{G}\imath\text{-}proj(R)$.*

Proof of Claim 1: Take the functor $G = R_0 \otimes_R \underline{\quad}: R\text{-}\mathcal{G}\imath\text{-}\mathcal{M}od \to R_0\text{-}\mathcal{G}\imath\text{-}\mathcal{M}od$. Since GF is naturally equivalent to 1 it suffices to show $P \approx FGP$ (graded) for every P in $\mathcal{G}\imath\text{-}proj(R)$. Let $Q = GP \approx P/R_+P$ by example 1.7.21'; since Q is projective the epic $P \to Q$ splits in $R_0\text{-}\mathcal{M}od$, and thus in $R_0\text{-}\mathcal{G}\imath\text{-}\mathcal{M}od$ by remark 1.9.6. Hence there is a graded monic $f: Q \to P$. Now $FGP = R \otimes_{R_0} Q$ so we can define a graded map $\varphi: FGP \to P$ such that $\varphi(r \otimes x) = rfx$ for r in R and x in Q. Since G right exact and $G\varphi$ is an isomorphism we see $G(\mathrm{coker}\,\varphi) = 0$, i.e., $\mathrm{coker}\,\varphi = R_+ \mathrm{coker}\,\varphi$, implying $\mathrm{coker}\,\varphi = 0$ by remark 5.1.34'. But then φ is epic and thus split, so $G(\ker \varphi) = 0$, likewise implying $\ker \varphi = 0$. Thus φ is an isomorphism, proving claim 1.

The next part of the proof of the theorem involves two clever tricks, the first of which is very cheap. Grade $R[\lambda]$ by putting $R[\lambda]_n = \bigoplus_{i \leq n} R_i \lambda^{n-i}$. Then $R[\lambda]_0 = R_0$, so R is replaced by the larger graded ring $R[\lambda]$. The second trick is a "homogenization map" ψ_d, given by $\psi_d r = r\lambda^{d-n}$ for any r in R_n. This defines the map $\psi_d: \bigoplus_{n \leq d} R_n \to R[\lambda]_d$, so ψ_d homogenizes all elements of degree $\leq d$. Now define the functor $H: R[\lambda]\text{-}\mathcal{G}\imath\text{-}\mathcal{F}imod \to R\text{-}\mathcal{F}imod$ given by $HM = M/(1 - \lambda)M$, i.e., we specialize λ to 1. Then H is given by $R \otimes_{R[\lambda]} \underline{\quad}$, where we view R as the $R[\lambda]$-module $R[\lambda]/R[\lambda](1 - \lambda)$ as in example 1.7.21'. In particular, H is right exact.

Claim 2: *For every f.g. R-module N there is an f.g. graded $R[\lambda]$-module M with $HM = N$.*

Proof of Claim 2: Write $N \approx R^{(n)}/K$ where $K = \sum_{i=1}^{t} Rx_i$. Writing $x_i = (x_{i1}, \ldots, x_{in})$ in $R^{(n)}$, we take d greater than the maximum of the degrees of all the x_{ij}, and define $x_i' = (\psi_d x_{i1}, \ldots, \psi_d x_{in}) \in (R[\lambda]_d)^{(n)}$. Letting $K' = \sum_{i=1}^{t} R[\lambda] x_i'$, a graded submodule of $R[\lambda]^{(n)}$, we have $H(R[\lambda]^{(n)}/K') \approx H(R[\lambda]^{(n)})/HK' \approx R^{(n)}/K \approx N$ since H is right exact.

Claim 3: *H is exact.*

Proof of Claim 3: We need to show H is left exact, i.e., given a monic $f: N \to M$ in $R[\lambda]\text{-}\mathscr{G}\!\imath\text{-}\mathscr{F}\!imod$ then $Hf: HN \to HM$ is monic. But the canonical map $N \to HM = M/(1 - \lambda)M$ has kernel $N \cap (1 - \lambda)M$, so it suffices to show $N \cap (1 - \lambda)M \le (1 - \lambda)N$. Suppose $x = \sum x_n \in M$ with $(1 - \lambda)x \in N$, where the x_n are homogeneous. $x_0 = ((1 - \lambda)x)_0 \in N$, and $x_i - \lambda x_{i-1} = ((1 - \lambda)x)_i \in N$ for each $i \ge 1$, so by induction we see each $x_i \in N$. Thus $x \in N$, so $(1 - \lambda)x \in (1 - \lambda)N$ as desired.

Proof of the Theorem: Suppose N is an f.g. R-module. Write $N = HM$ for $M \in R[\lambda]\text{-}\mathscr{G}\!\imath\text{-}\mathscr{M}\!od$, by claim 2. $R[\lambda]$ is Noetherian of finite gl. dim, by proposition 5.1.19. Thus M has an f.g. projective resolution

$$0 \to P_n \to \cdots \to P_0 \to M \to 0,$$

which by remark 5.1.34'' can be taken to be graded. Then $0 \to HP_n \to \cdots \to HP_0 \to N \to 0$ is a projective resolution of N by claim 3. On the other hand, $R[\lambda]_0 = R_0$ so claim 1 shows each $P \approx R[\lambda] \otimes_{R_0} Q_i$ for suitable f.g. projective R_0-modules Q_i, which by hypothesis are all stably free. Then $HP_i \approx R \otimes_{R_0} Q_i$ are stably free. Thus $K_0(R) = \langle [R] \rangle$ by proposition 5.1.31. Q.E.D.

We now have quite a wide assortment of rings R with $K_0(R) = \langle [R] \rangle$.

Corollary 5.1.35: *Suppose R is left Noetherian, gl. dim $R < \infty$. The hypothesis "every f.g. projective is stably free" passes from R to the following rings:*

  (i) *$R[\lambda; \sigma]$ for any automorphism $\sigma : R \to R$.*
  (ii) *$R[\lambda_1, \ldots, \lambda_t]$ for commuting indeterminates $\lambda_1, \ldots, \lambda_t$.*

In particular, if R is a PLID then every f.g. projective $R[\lambda_1, \ldots, \lambda_t]$-module is stably free.

Proof:

(i) $R[\lambda; \sigma]$ is $\mathbb{N}$-graded according to degree and is left Noetherian (proposition 3.5.2) of finite gl. dim (corollary 5.1.25).

(ii) Apply (i) t times.

To see the last assertion note $K_0(R) = \langle [R] \rangle$ and gl. dim $R = 1$ since R is hereditary.   Q.E.D.

The hypothesis R left Noetherian of finite gl. dim could be weakened throughout to "Every f.g. R-module has an f.g. projective resolution of finite length," without change in proof. Such rings are called (*homologically*) *regular*.

O Supplement: Quillen's Theorem

One good theorem deserves another, and Quillen [73] proved a startling extension of Serre's theorem. Before stating Quillen's theorem let us make another definition.

Definition 5.1.36: A ring R is *filtered* if R has a chain of additive subgroups $R_0 \subseteq R_1 \subseteq \cdots$ such that $R_i R_j \subseteq R_{i+j}$ for all i, j and $\bigcup_{i \in \mathbb{N}} R_i = R$.

Note that this concept is not new to us; R is filtered iff R has a filtration over $(\mathbb{Z}, +)$ for which $R(1) = 0$. (Just take the new R_i to be the old R_{-i}). Thus example 1.8.14(i) and the results applicable to that example can be used here, and, in particular, we can form the associated graded ring $G(R)$; also note that R_0 is a subring of R. The terminology introduced here has become standard when applied to enveloping algebras, and the results to be given here are fundamental in the theory of enveloping algebras, cf., §8.4.

Quillen's Theorem: *Suppose R is filtered, such that the graded ring $G(R)$ is Noetherian and of finite gl. dim. If $G(R)$ is flat as R_0-module then the natural injection $R_0 \to R$ induces an isomorphism $\varphi: K_0(R_0) \to K_0(R)$.*

Quillen actually proved a much stronger result, that the corresponding isomorphism also holds in the "higher" K-theories. Our exposition (also, cf., the appendix and exercise 13) follows the very readable account of McConnell [85], which in turn relies on Roy [65].

The main object is to develop the homological module theory of filtered rings to the same stage that we developed the module theory of graded rings. To this end we assume throughout this discussion that R is a filtered ring; we say an R-module M is *filtered* if M has a chain of subgroups $M_0 \subseteq M_1 \subseteq \cdots$ such that $R_i M_j \le M_{i+j}$ and $\bigcup_{i \in \mathbb{N}} M_i = M$. R is itself filtered as R-module in the natural way; on the other hand, any R-module M can be *trivially filtered* by putting $M_0 = M$.

A map $f: M \to N$ of filtered R-modules will be called *filtered* if $fM_i \subseteq N_i$ for all i; f is *strictly filtered* if $fM_i = N_i \cap fM$ for all i. If N is a filtered module and $f: M \to N$ is an arbitrary map then M can be filtered in such a way that f is strictly filtered; namely, put $M_i = f^{-1}N_i$.

If M is a filtered R-module we define the associated *graded module* $G(M) = \bigoplus(M_{i+1}/M_i)$, viewed naturally as $G(R)$-module by the rule

$$(r_i + R_{i+1})(x_j + M_{j+1}) = r_i x_j + M_{i+j+1}$$

for r_i in R_i and x_j in M_j. Given x in M_i we shall write $\bar{x}$ for the corresponding element of $G(M)$.

Any filtered map $f: M \to N$ gives rise naturally to a map $G(f): G(M) \to G(N)$, so G is a functor. The point of strictly filtered maps lies in the following "exactness" feature of this functor:

If $M'' \xrightarrow{f} M \xrightarrow{g} M'$ is exact with f, g strictly filtered then

$$G(M'') \xrightarrow{G(f)} G(M) \xrightarrow{G(g)} G(M') \qquad \text{is exact.}$$

We say M is *filtered-free* if $G(M)$ is graded free over $G(R)$, cf., remark 1.9.7.

Remark 5.1.37: Any filtered-free R-module is free, for if $\{\bar{e}_i : i \in I\}$ is a base of homogeneous elements of $G(M)$ then $\{e_i : i \in I\}$ is a base of M; in fact, writing $e_i \in M_{n(i)}$ we have

$$M_k = \sum_{n(i) \le k} R_{k-n(i)} e_i.$$

Conversely, any graded free $G(R)$-module $\tilde{F}$ gives rise to a filtered-free R-module F such that $G(F) = \tilde{F}$. (Proof. ($\Rightarrow$) Clearly the above holds for $k = 0$, and thus is easily seen to hold by induction for all k, by passing to $G(M)$ at each step; likewise, any dependence would be reflected in some M_k and thus pass to $G(M)_k$. ($\Leftarrow$) Let $\{\bar{e}_i : i \in I\}$ denote a base of $\tilde{F}$ over $G(R)$ and define $F = R^{(I)}$. Labeling a base of F as $\{e_i : i \in I\}$ use the above equation to filter F; then $G(F) = \tilde{F}$ by inspection.)

Now that we want to see that our object is really "free".

Lemma 5.1.38: *Suppose M is a filtered R-module.*

(i) *If F is a filtered-free module then for every graded map $\tilde{f}: G(F) \to G(M)$ there is a filtered map $f: F \to M$ such that $G(f) = \tilde{f}$.*

(ii) *There exists a filtered-free module F and a strictly filtered epic $F \to M$.*

(iii) *If P is filtered and $G(P)$ is $G(R)$-projective then P is projective.*

Proof:

(i) Let $\{e_i : i \in I\}$ be a base of F taken as in remark 5.1.37, and define $f: F \to M$ by $\overline{f e_i} = \tilde{f} \bar{e}_i$. Then $G(f) = \tilde{f}$ by construction, and, obviously, f is filtered.

(ii) Let $\tilde{F}$ be a graded-free $G(R)$-module, together with a graded epic $\tilde{f}: \tilde{F} \to G(M)$. By remark 5.1.37 we can write $\tilde{F}$ in the form $G(F)$, so by (i) we have a filtered epic $f: F \to M$. It remains to show f is strictly filtered, i.e., $f: F_i \to M_i$ is epic for each i; we do this by induction on i, noting it is clear for $i = 0$ and thus follows for general i by applying proposition 2.11.15 to the diagram

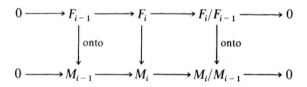

(iii) By (ii) we have an exact sequence $0 \to N \xrightarrow{f} F \xrightarrow{g} P \to 0$ where F is filtered-free and f, g are strictly filtered. Applying $G(\ )$ yields an exact sequence $0 \to G(N) \to G(F) \to G(P) \to 0$ which splits by remark 1.9.6.

We need to show g is split, i.e. $g\mu = 1_P$ for some $\mu: P \to F$. Note P_i/P_{i-1} is a summand of $G(P)$ and thus is projective; hence the epic $P_i \to P_i/P_{i-1}$ splits. But then $P_i \approx P_{i-1} \oplus (P_i/P_{i-1})$, so we can build μ inductively on i (i.e., piecing together $P_i/P_{i-1} \hookrightarrow F_i/F_{i-1}$ and $P_{i-1} \hookrightarrow F_{i-1}$ to get $P_i \hookrightarrow F_i$).   Q.E.D.

Proposition 5.1.39:   *For any filtered module M we have $\mathrm{pd}_R M \le \mathrm{pd}_{G(R)} G(M)$.*

Proof: Let $n = \mathrm{pd}_{G(R)} G(M)$. There is nothing to prove unless n is finite, so we proceed by induction on n. Lemma 5.1.38(iii) gives the result for $n = 0$. For $n > 0$ take an exact sequence $0 \to N \to F \to M \to 0$ where F is filtered-free and the maps are strictly filtered, cf., lemma 5.1.38(ii). Then we get an exact sequence $0 \to G(N) \to G(F) \to G(M) \to 0$. But $\mathrm{pd}_{G(R)} G(N) \le n - 1$ by summary 5.1.13, so $\mathrm{pd}_R N \le n - 1$ by induction, implying $\mathrm{pd}_R M \le n$ by summary 5.1.13.   Q.E.D.

Corollary 5.1.40: *If R is a filtered ring then* gl. dim $R \leq$ gl. dim $G(R)$.

Proof: Any R-module M can be filtered trivially, and then by the proposition $\text{pd}_R M \leq \text{pd}_{G(R)} G(M) \leq$ gl. dim $G(R)$. Thus gl. dim $R \leq$ gl. dim $G(R)$.

Q.E.D.

Digression: These results actually yield more. We could define the *graded* pd of a module and graded gl. dim of a ring, in terms of graded projective resolutions. Then we see at once the graded gl. dim of a graded ring equals its usual gl. dim.

We are ready for our watered-down version of Quillen's theorem.

Theorem 5.1.41: *Suppose R is filtered and $G(R)$ is Noetherian of finite* gl. dim. *If $K_0(R_0) = \langle [R_0] \rangle$ then $K_0(R) = \langle [R] \rangle$.*

Proof: R is Noetherian by corollary 3.5.32(i), and has finite gl. dim by corollary 5.1.40. We want to show that every f.g. projective R-module P is stably free. Viewing P as filtered trivially, we see $G(P)$ is f.g. over $G(R)$ and thus we can build a graded resolution of f.g. graded free modules, as in remark 5.1.34″. By hypothesis the n-th syzygy K_n is projective and is the kernel of a graded map, so is graded projective. By claim 1 (following remark 5.1.34″) there is a projective R_0-module Q_0 with $K_n \approx G(R) \otimes_{R_0} Q_0$. But Q_0 is stably free by hypothesis, so K_n is stably free as $G(R)$-module. Thus K_n is "graded" stably free by remark 5.1.34″, so just as in lemma 5.1.29 we have a graded FFR of $G(P)$ having finite length. Using lemma 5.1.38 we have the resolution in the form $0 \to G(F_n) \to G(F_{n-1}) \to \cdots \to G(P) \to 0$, and this "pulls back" to an FFR $0 \to F_n \to F_{n-1} \to \cdots \to P \to 0$ in R-$\mathcal{M}od$, as needed.

Euler Characteristic

The FFR property has an important tie to topology.

Definition 5.1.42: Suppose R has IBN. The *Euler characteristic* of a module M with an f.g. free resolution $0 \to F_n \to \cdots \to F_0 \to M \to 0$ is defined as $\chi(M) = \sum_{i=0}^{n} (-1)^i \text{rank}(F_i)$.

Remark 5.1.43: $\chi(M)$ is independent of the FFR because of the generalized Schanuel lemma. In particular, if $0 \to F_n \to \cdots \to F_0 \to 0$ is exact then $\sum (-1)^i \text{rank}(F_i) = 0$.

Remark 5.1.44:   Suppose S is a left denominator set of R and $S^{-1}R$ also
has IBN. If an R-module has FFR then $S^{-1}M$ has FFR in $S^{-1}R\text{-}\mathcal{M}od$, and
$\chi(S^{-1}M) = \chi(M)$. (Just localize the resolution, since the localization functor
is exact and preserves "free" and the rank.)

The name "Euler characteristic" comes from Euler's observation that
$\#$ faces $-$ $\#$ edges $+$ $\#$ vertices of a simplicial complex is a topological
invariant. The standard reference on properties of the Euler characteristic
is Bass [76].

Theorem 5.1.45:   *(Walker [72]) If R is a prime left Noetherian ring of finite*
gl. dim with $K_0(R) = \langle [R] \rangle$ then R is a domain.

Proof: By Goldie's theorem R is a left order in a simple Artinian ring
$S^{-1}R \approx M_n(D)$. We claim $n = 1$, which would prove R is a domain. Indeed,
$L = R \cap M_n(D)e_{11}$ is a left ideal of R which is nonzero since $M_n(D)$ is
an essential extension of R; thus $S^{-1}L \leq M_n(D)e_{11}$ so $0 < [S^{-1}L:D] \leq n$.
On the other hand, take an FFR $0 \to F_m \to \cdots \to F_0 \to L \to 0$. Then each
$S^{-1}F_i$ is a free $S^{-1}R$-module so n^2 divides $[S^{-1}F_i:D]$. Hence remark 5.1.43
shows n^2 divides $[S^{-1}L:D]$, a contradiction unless $n = 1$.   Q.E.D.

Walker's theorem is more general, with "semiprime" replacing "prime",
cf., exercise 14. However, the proof given here will be relevant when we
consider group rings later.

§5.2 Homology, Cohomology, and Derived Functors

In this section we develop (as quickly as possible) the theory of derived
functions with special emphasis on $\mathcal{T}or$ and $\mathcal{E}xt$, the derived functors of $\otimes$
and $\mathcal{H}om$. Our reason is twofold: (1) We get much deeper insight into homological dimensions; (2) tools are forged which apply to diverse subjects in the
sequel. To do this we view projective resolutions more categorically, building
homology and cohomology. We shall draw on the basic results of §5.0. To
keep the discussion explicit, we work in the category $R\text{-}\mathcal{M}od$; however, see
note 5.2.5.

Definition 5.2.1:   A (chain) *complex* $(\mathbb{A}; (d_n))$ is a sequence (not necessarily
exact) of maps

$$\cdots \to A_{n+1} \xrightarrow{d_{n+1}} A_n \xrightarrow{d_n} A_{n-1} \to \cdots$$

for all n in $\mathbb{Z}$, such that $d_n d_{n+1} = 0$ for all n. The maps d_n are called *differentiations*; when unambiguous $\mathbb{A}$ is used to denote the complex $(\mathbb{A}; (d_n))$. We call the complex $\mathbb{A}$ *positive* if $A_n = 0$ for all $n < 0$; $\mathbb{A}$ is *negative* if $A_n = 0$ for all $n > 0$. We shall also find it convenient to call a complex *almost positive* if $A_n = 0$ for all $n < -1$; in this case the map $d_0: A_0 \to A_{-1}$ has a special role and is called the *augmentation map*, designated as ε. To unify notation we shall retain d_0. There is a category $\mathscr{Comp}$ whose objects are the complexes and whose morphisms $f: \mathbb{A} \to \mathbb{A}'$ are $\mathbb{Z}$-tuples (f_i) of morphisms $f_i: A_i \to A'_i$ for each i, such that the following diagram commutes:

$$
\begin{array}{ccccccc}
\cdots \longrightarrow & A_{n+1} & \xrightarrow{d_{n+1}} & A_n & \xrightarrow{d_n} & A_{n-1} & \longrightarrow \cdots \\
& \downarrow{f_{n+1}} & & \downarrow{f_n} & & \downarrow{f_{n-1}} & \\
\cdots \longrightarrow & A'_{n+1} & \xrightarrow{d'_{n+1}} & A'_n & \xrightarrow{d'_n} & A'_{n-1} & \longrightarrow \cdots
\end{array}
$$

These morphisms f are called *chain maps*.

Example 5.2.2: Suppose $(\mathbb{A}; (d_n))$ is a complex. We can form a new complex $(\mathbb{A}'; (d'_n))$ where $A'_n = A_{n-1}$ and $d'_n = d_{n-1}$; we can view $d: \mathbb{A} \to \mathbb{A}'$ as a chain map. Note that $\mathbb{A}'$ here is almost positive iff $\mathbb{A}$ is positive.

Although originating in topology, complexes are very relevant to the study of sequences of modules. Every exact sequence is obviously a complex. In particular, any projective resolution is an almost positive complex, and any injective resolution is an almost negative complex. Although functors do not necessarily preserve exactness they *do* preserve complexes. Thus if $\mathbb{A}$ is a complex and $F: \mathscr{C} \to \mathscr{D}$ is a covariant functor then $F\mathbb{A}$ is a complex of $\mathscr{D}$ given by

$$
\cdots \to FA_{n+1} \xrightarrow{Fd_{n+1}} FA_n \xrightarrow{Fd_n} FA_{n-1} \to \cdots.
$$

There is a more concise way of describing complexes, in terms of graded objects. Suppose G is a given abelian group and M, M' are G-graded modules. A map $M \to M'$ has *degree* h for suitable h in G if $fM_g \subseteq M'_{g+h}$ for all g in G. Taking $G = \mathbb{Z}$ and $A = \bigoplus_{i \in \mathbb{Z}} A_i$ in definition 5.2.1 we see $d: A \to A$ is a map of degree -1, leading us to the following alternate definition:

Definition 5.2.3: A *complex* is a $\mathbb{Z}$-graded module A together with a map $d: A \to A$ of degree -1 satisfying $d^2 = 0$. $\mathscr{Comp}'$ is then the category of

$\mathbb{Z}$-graded modules, whose morphisms are those maps $f: A \to A'$ of degree 0 satisfying $fd = d'f$.

As we just saw, any complex $(\mathbb{A}; (d_n))$ of definition 5.2.1 gives rise to the complex $(A; d)$ of definition 5.2.3 where $A = \bigoplus A_n$ and $d = \bigoplus d_n$; and conversely. Thus $\mathscr{C}omp$ and $\mathscr{C}omp'$ are easily seen to be isomorphic categories, and we shall use the definitions interchangeably. One should note that definition 5.2.3 can be easily generalized, using different groups G in place of $(\mathbb{Z}, +)$; $G = (\mathbb{Z}^{(2)}, +)$ is a useful candidate.

Definition 5.2.4: Dually to definition 5.2.1, define a *cochain complex* $\mathbb{A}'$ to be a sequence $\cdots \leftarrow A^{n+1} \overset{d^n}{\leftarrow} A^n \overset{d^{n-1}}{\leftarrow} A^{n-1} \leftarrow \cdots$ with each $d^n d^{n-1} = 0$ where, by convention, one writes superscripts instead of subscripts. Then one can define $\mathscr{C}ocomp$ analogously.

In fact, there is an isomorphism $\mathscr{C}omp \to \mathscr{C}ocomp$ given by sending the complex $\cdots \to A_{n+1} \overset{d_{n+1}}{\longrightarrow} A_n \to \cdots$ to the cochain complex $\mathbb{A}'$ obtained by replacing n by $-n$, i.e., $(A')^n = A_{-n}$ and $(d')^n = d_{-n}$.

Note 5.2.5: If one worked with an arbitrary abelian category $\mathscr{C}$ instead of $R\text{-}\mathcal{M}od$ one could formally define the corresponding category $\mathscr{C}omp(\mathscr{C})$ of chain complexes. There is a formal isomorphism between $\mathscr{C}omp(\mathscr{C}^{op})$ and $\mathscr{C}ocomp(\mathscr{C})^{op}$ given by reversing arrows, so we see that $\mathscr{C}omp(\mathscr{C})^{op}$ and $\mathscr{C}omp(\mathscr{C}^{op})$ are isomorphic categories. Aiming for the best of both worlds, we work in $R\text{-}\mathcal{M}od$ but keep the duality in mind, often skipping dual proofs (which can be filled in easily).

Homology

We saw before that functors preserve complexes, and this is a good reason to consider complexes in preference to exact sequences. But then we should know when a complex is exact (as a sequence). Obviously the complex $\cdots \to A_{n+1} \overset{d_{n+1}}{\longrightarrow} A_n \overset{d_n}{\to} A_{n-1} \to \cdots$ is exact at A_n iff $\ker d_n / \operatorname{im} d_{n+1} = 0$, leading us to the following important definition:

Definition 5.2.6: Suppose (A, d) is a complex, notation as in definition 5.2.3. Define $Z = Z(A) = \ker d$, $B = B(A) = dA$ (the image of d), and $H = H(A) = Z/B$. B is called the *boundary* and Z the *cycle*, and H is the *homology* of A.

Remark 5.2.7: Z and B are graded submodules of A, where $Z_n = \ker d_n$ and $B_n = d_{n+1}A_{n+1}$. Moreover, since $d^2 = 0$ we have a canonical injection $i: B \to Z$, so H is a graded module and is $\operatorname{coker} i$. Dually one can define the *coboundary* $Z' = \operatorname{coker} d = A/B$, and the *cocycle* $B' = A/Z$ (the coimage of d). Then there is a canonical epic $p: Z' \to B'$, and $\ker p \approx Z/B = H$. Thus H is "self-dual." Moreover, d induces a map $\bar{d}: Z' \to Z$, and we have the exact sequence

$$0 \to H \to Z' \xrightarrow{\bar{d}} Z \to H \to 0. \tag{1}$$

In view of the above discussion we have $H \approx \bigoplus_{n \in \mathbb{N}} H_n$ where $H_n = Z_n/B_n$. *Thus $H_n = 0$ iff the complex is exact at A_n.* We write $H_n(A)$ for H_n when A is ambiguous. Actually $Z(\ )$ and $B(\ )$ can be viewed as functors. Indeed if $f: (A; d) \to (A'; d')$ is a map of complexes then f restricts to graded maps $Zf: Z \to Z'$ and $Bf: B \to B'$, by fact 5.0.2 applied to each f_n. Consequently we get a graded map $f_*: H(A) \to H(A')$.

The Long Exact Sequence

We will often want to compare the homology of different modules, especially those in an exact sequence.

Proposition 5.2.8: *The functor $Z(\ )$ is a left exact functor.*

Proof: If $0 \to A' \xrightarrow{f} A \xrightarrow{g} A''$ is exact then $ZA' \xrightarrow{Zf} ZA \xrightarrow{Zg} ZA''$ is exact by the snake lemma (5.0.3(i)); applying the same argument to $0 \to 0 \to A' \to A$ shows Zf is monic.   Q.E.D.

Remark 5.2.9: By duality Z' is a right exact functor.

Now we have all the necessary requirements for a focal result.

Theorem 5.2.10: ("*Exact homology sequence*") *If $0 \to A' \xrightarrow{f} A \xrightarrow{g} A'' \to 0$ is an exact sequence of complexes, then there is a long exact sequence*

$$\cdots \to H_{n+1}(A') \xrightarrow{f_*} H_{n+1}(A) \xrightarrow{g_*} H_{n+1}(A'') \xrightarrow{\partial} H_n(A') \xrightarrow{f_*} H_n(A) \xrightarrow{g_*} H_n(A'') \xrightarrow{\partial}$$

$$H_{n-1}(A') \to \cdots$$

where $\partial: H_{n+1}(A'') \to H_n(A')$ is obtained by applying the snake lemma to the exact sequence

$$Z'_{n+1}(A') \xrightarrow{f_{n+1}} Z'_{n+1}(A) \xrightarrow{g_{n+1}} Z'_{n+1}(A'') \longrightarrow 0$$

$$0 \longrightarrow Z_n(A') \xrightarrow{f_n} Z_n(A) \xrightarrow{g_n} Z_n(A'')$$

Proof: The rows of the sequences are exact by proposition 5.2.8 and remark 5.2.9, so we are done by the snake lemma and (1).   Q.E.D.

One could rephrase the conclusion of theorem 5.2.10 by saying the following graded triangle is exact:

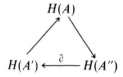

$$H(A') \xleftarrow{\partial} H(A'')$$

Cohomology

The same procedure applied to cochain complexes yields a *cohomology* functor from $\mathscr{Cocomp}(R\text{-}\mathscr{Mod})$ to $\{\mathbb{Z}\text{-graded } R\text{-modules}\}$, and theorem 5.2.10 has the following dual theorem:

Theorem 5.2.11:   (*Exact cohomology sequence*) *If* $0 \to A' \xrightarrow{f} A \xrightarrow{g} A'' \to 0$ *is an exact sequence of cochain complexes then there is a long exact sequence*

$$\to H^n(A') \xrightarrow{f^*} H^n(A) \xrightarrow{g^*} H^n(A'') \xrightarrow{\delta} H^{n+1}(A') \xrightarrow{f^*} H^{n+1}(A) \to \cdots$$

where the maps are defined as in theorem 5.2.10.

Cohomology turns out to be particularly useful in algebra, and we digress a bit to present an example of paramount importance for §7.2. First we transfer some terminology from algebras to groups.

Definition 5.2.12:   Suppose G is a group. A *G-module* is an abelian group M together with a scalar multiplication $G \times M \to M$ satisfying the axioms $g(x_1 + x_2) = gx_1 + gx_2, (g_1 g_2)x = g_1(g_2 x)$, and $1x = x$, for all g_i in G and x_i in M.

Formally we have merely written down whichever module axioms make sense. Looking at the G-module axioms from a different perspective, we could view the elements of G as maps from M to itself. In particular, any group M has the *trivial* G-module action obtained by taking $gx = x$ for all g in G. A deeper example: If G is a group of automorphisms of a field F then the multiplicative group $M = F - \{0\}$ (rewritten additively) is a G-module under the given action of G.

Example 5.2.13: Suppose M is a G-module. There is a very useful positive cocomplex $\mathbb{A}$ of abelian groups (i.e., $\mathbb{Z}$-modules): $A^n = \{\text{functions } f: G^{(n)} \to M\}$ where $G^{(0)} = \{1\}$, and $d^n: A^n \to A^{n+1}$ is defined by

$$(d^n f)(g_1, \ldots, g_{n+1}) = g_1 f(g_2, \ldots, g_{n+1}) + (-1)^{n+1} f(g_1, \ldots, g_n)$$

$$+ \sum_{i=1}^{n} (-1)^i f(g_1, \ldots, g_i g_{i+1}, \ldots, g_{n+1}). \tag{2}$$

The cohomology groups are denoted $H^n(G, M)$, and are torsion if G is a finite group, cf., exercise 2.

A key example: Any exact sequence of groups $1 \to M \to E \to G \to 1$ with M abelian determines a G-module structure on M, by the following rule: Suppose $x \in M$ and $g \in G$. View $M \subseteq E$ (and thereby use multiplicative notation). Letting h be a preimage of g in E, define gx to be $h^{-1}xh$ (which is clearly in M). Note that this action is independent of the choice of h, for if $h_1 \in Mh$ then writing $h_1 = hx_1$ we have $h_1^{-1}xh_1 = x_1^{-1}(h^{-1}xh)x_1 = h^{-1}xh$ since M is abelian.

This set-up is called a *group extension* of G by M, and there is a correspondence given in Jacobson [80B, pp. 363–366] between these group extensions and $H^2(G, M)$. In exercises 8.2.7ff. we shall see how group extensions tie in with division algebras.

Derived Functors

Proposition 5.2.14: *Suppose* $\mathbb{P} = \cdots \to P_n \xrightarrow{f_n} P_{n-1} \to \cdots \to P_0 \xrightarrow{f_0} M \to 0$ *is a projective resolution, and* $\mathbb{A} = \cdots \to A_n \xrightarrow{d_n} A_{n-1} \to \cdots \to A_0 \xrightarrow{d_0} A_{-1} \to 0$ *is an arbitrary exact sequence. Then any map* $g: M \to A_{-1}$ *can be lifted to maps* $g_i: P_i \to A_i$ *such that* $(g_n): \mathbb{P} \to \mathbb{A}$ *is a chain map, i.e., there is a commutative diagram*

$$
\begin{array}{ccccccccc}
\cdots & \longrightarrow & P_n & \xrightarrow{f_n} & \cdots & \longrightarrow & P_0 & \xrightarrow{f_0} & M & \longrightarrow & 0 \\
& & \downarrow{g_n} & & & & \downarrow{g_0} & & \downarrow{g} & & \\
\cdots & \longrightarrow & A_n & \xrightarrow{d_n} & \cdots & \longrightarrow & A_0 & \xrightarrow{d_0} & A_{-1} & \longrightarrow & 0
\end{array}
$$

Proof: $d_0: A_0 \to A_{-1}$ is epic so we can lift $gf_0: P_0 \to A_{-1}$ to a map $g_0: P_0 \to A_0$ such that

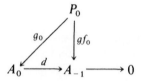

commutes, i.e., $gf_0 = d_0 g_0$. Now, inductively, suppose we have $g_{n-1}f_n = d_n g_n$. Applying 5.0.8 with $P = P_{n+1}$, $\tau = g_n f_{n+1}$, $\varphi = d_n$, and $\psi = d_{n+1}$, we see $\varphi\tau = d_n g_n f_{n+1} = g_{n-1} f_n f_{n+1} = 0$, so $g_n f_{n+1} = \tau = d_{n+1}\sigma$ for some $\sigma: P_{n+1} \to A_{n+1}$, and we are done taking $g_{n+1} = \sigma$.   Q.E.D.

Definition 5.2.15: The n-th *derived* functor $L_n T$ of a given functor T: $R\text{-}\mathcal{M}od \to R'\text{-}\mathcal{M}od$ is defined as follows: Take a projective resolution

$$\mathbb{P} = \cdots \to P_n \xrightarrow{f_n} P_{n-1} \xrightarrow{f_{n-1}} \cdots \to M \to 0$$

viewed as an almost positive complex, and put $(L_n T)M = H_n(T\mathbb{P}) = \ker Tf_n / \mathrm{im}\ Tf_{n+1}$; given $g: M \to M'$ define $(L_n T)g$ as follows: First take a projective resolution $\mathbb{A}$ of M', i.e., $A_{-1} = M'$; extend g to a chain map as in proposition 5.2.14, then apply T, and finally take the n-th homology map.

We need to prove this is well-defined, i.e., independent of the choice of projective resolution. To this end we make the following definition:

Definition 5.2.16: Suppose $(\mathbb{A}; (d_n))$ and $(\mathbb{A}'; (d_n'))$ are complexes. Two chain maps $g: \mathbb{A}' \to \mathbb{A}$ and $h: \mathbb{A}' \to \mathbb{A}$ are *homotopic* if there are $s_n: A_n' \to A_{n+1}$ for each n such that $h_n - g_n = d_{n+1}s_n + s_{n-1}d_n'$.

Proposition 5.2.17: *Homotopic chain maps induce the same maps on the homology modules.*

Proof: Suppose $g: \mathbb{A}' \to \mathbb{A}$ and $h: \mathbb{A}' \to \mathbb{A}$ are homotopic. For any cycle z in Z_n we have $h_n z - g_n z = d_{n+1}s_n z + s_{n-1}d_n' z = d_{n+1}s_n z \in B_n$, so the induced actions on H_n are equal.   Q.E.D.

One easy application of this result is a method of checking that a given sequence is indeed exact.

Proposition 5.2.18: *Suppose we are given a sequence of R-modules*

$$\mathbb{A} = \cdots \to A_n \xrightarrow{d_n} A_{n-1} \to \cdots \to A_0 \xrightarrow{d_0} A_{-1} \to 0$$

with d_0 epic and $d_0 d_1 = 0$. If there are group homomorphisms $s_i : A_i \to A_{i+1}$ such that each $s_i A_i$ spans A_{i+1} and $d_0 s_{-1} = 1_{A_{-1}}$ and $d_{i+1} s_i + s_{i-1} d_i = 1_{A_i}$ for all $i \geq 0$ then $\mathbb{A}$ is exact.

Proof: $d_i d_{i+1} s_i = d_i (1 - s_{i-1} d_i) = d_i - (1 - s_{i-2} d_{i-1}) d_i = s_{i-2} d_{i-1} d_i = 0$ by induction, so $d_i d_{i+1} = 0$ by hypothesis on s_i. Thus $\mathbb{A}$ is a complex. Now we see the chain map $1_{\mathbb{A}} : \mathbb{A} \to \mathbb{A}$ is homotopic to the zero chain map, implying the homology of $\mathbb{A}$ is 0, i.e., $\mathbb{A}$ is exact.   Q.E.D.

We are also interested in the opposite direction. To illustrate what we are aiming for, here is a partial converse to proposition 5.2.18.

Proposition 5.2.18': *Suppose* $\cdots \to P_n \overset{d_n}{\to} P_{n-1} \to \cdots \to P_0 \overset{d_0}{\to} P_{-1} \to 0$ *is an exact sequence of projective R-modules. Then there are maps* $s_i : P_i \to P_{i+1}$ *such that* $d_0 s_{-1} = 1_{P_{-1}}$ *and* $d_{i+1} s_i + s_{i-1} d_i = 1_{P_i}$ *for all* $i \geq 0$.

Proof: Since d_0 is split epic we can find s_{-1} with $d_0 s_{-1} = 1_{P_{-1}}$. Now we merely proceed by induction on i, applying 5.0(8) to the diagram

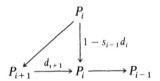

to obtain $s_i : P_i \to P_{i+1}$ satisfying $d_{i+1} s_i = 1 - s_{i-1} d_i$, or $d_{i+1} s_i + s_{i-1} d_i = 1$.
                                                                          Q.E.D.

The same idea applies to homotopy.

Lemma 5.2.19: *(Comparison lemma) Notation as in proposition 5.2.14. Suppose* $\mathbb{P}$ *is a projective resolution of an R-module M, and* $(\mathbb{A}; (d_n))$ *is exact and almost positive. If* $g : M \to A_{-1}$ *is lifted to chain maps* $(g_n), (h_n) : \mathbb{P} \to \mathbb{A}$ *as in proposition 5.2.14 then* (g_n) *and* (h_n) *are homotopic.*

Proof: Let $q_n = g_n - h_n$, which lifts $g - g = 0$. We use 5.0.8 repeatedly. First note $d_0 q_0 = (g - g) f_0 = 0$, so $q_0 P_0 \subset d_1 A_1$, and thus there is $s_0 : P_0 \to A_1$ such that $q_0 = d_1 s_0$. Inductively, suppose we have defined $s_{n-1} : P_{n-1} \to A_n$ such that $q_{n-1} = d_n s_{n-1} + s_{n-2} f_{n-1}$. We need to define s_n such that $d_{n+1} s_n = q_n - s_{n-1} f_n$. To do this it suffices to show $d_n (q_n - s_{n-1} f_n) = 0$. But $d_n q_n - d_n s_{n-1} f_n = q_{n-1} f_n - (q_{n-1} - s_{n-2} f_{n-1}) f_n = s_{n-2} f_{n-1} f_n = 0$, as desired.
                                                                          Q.E.D.

Proposition 5.2.20: *Definition 5.2.15 is well-defined.*

Proof: First $(L_n T)M$ is well-defined since for any two projective resolu-
tions $\mathbb{P}, \mathbb{P}'$ of M we have chain maps $(g_n): \mathbb{P} \to \mathbb{P}'$ and $(h_n): \mathbb{P}' \to \mathbb{P}$ lifting
1_M, implying $(h_n g_n): \mathbb{P} \to \mathbb{P}$ lifts 1_M. But the identity map $1_{\mathbb{P}}$ lifts 1_M, so
$h_n g_n$ is homotopic to 1 by lemma 5.2.19, implying $Th_n Tg_n$ is homotopic to
$T1 = 1$, and thus (Tg_n) induces an isomorphism of the homology modules,
by proposition 5.2.17. The well-definedness of $(L_n T)g$ is proved analogously.
                                                                           Q.E.D.

There is one case in which it is easy to determine $L_n T$.

Remark 5.2.21: If P is projective then $(L_n T)P = 0$ for all n. (Indeed, $0 \to$
$P \to P \to 0$ is a projective resolution where $P_0 = P$, and $0 \to TP \to TP \to 0$
is exact.)

Let us now restate theorem 5.2.10 for left derived functors.

Theorem 5.2.22: *If T is a right exact functor from $R\text{-}\mathcal{M}od$ to $R'\text{-}\mathcal{M}od$
then for any exact sequence $0 \to M' \xrightarrow{f'} M \xrightarrow{f} M'' \to 0$ there is a corresponding
long exact sequence*

$$\cdots \to (L_{n+1}T)M'' \to (L_n T)M' \xrightarrow{(L_n T)f'} (L_n T)M \xrightarrow{(L_n T)f} (L_n T)M''$$

$$\to (L_{n-1}T)M' \to \cdots \to (L_1 T)M'' \to TM' \xrightarrow{Tf'} TM \xrightarrow{Tf} TM'' \to 0.$$

Note: If T is not right exact then the conclusion becomes true when we
replace TM', TM, and TM'' by $(L_0 T)M'$, $(L_0 T)M$, and $(L_0 T)M''$, but we
shall not have occasion to use this generality. Before applying the theorem
let us discuss it briefly. As usual we note that had the results above been
stated more generally for abelian categories, the dualization would be automatic, so let us indicate briefly how this could have been done. As remarked
at the end of §5.0 the results there hold for any abelian category $\mathcal{C}$. Assuming
$\mathcal{C}$ has "enough projectives", which we recall means for any object A there is
an epic $P \to A$ with P projective, we can form projective resolutions for any
object. On the other hand, the definition of homology is applicable in any
abelian category (and actually could be formulated in terms of "graded
objects" by means of coproducts). Thus the left derived functors could be
defined for any abelian category with enough projectives, thereby yielding a
general version of theorem 5.2.22. We leave the details to the reader,
although we shall need the following consequence.

Corollary 5.2.23: *Suppose* $T: \mathcal{C} \to \mathcal{D}$ *is a right exact (covariant) functor of abelian categories, and* $\mathcal{C}$ *has enough projectives. T is exact iff the left derived functor* $L_1 T = 0$.

Proof: ($\Leftarrow$) For an exact sequence $0 \to M' \xrightarrow{f'} M \xrightarrow{f} M'' \to 0$ we have $\cdots \to (L_1 T)M \to (L_1 T)M'' \to TM' \xrightarrow{Tf'} TM \to TM'' \to 0$ exact. By hypothesis $(L_1 T)M'' = 0$ so clearly T is exact.

($\Rightarrow$) Given any M'' we could take an exact sequence with M projective; then $(L_1 T)M = 0$ and Tf' is monic by hypothesis, implying $(L_1 T)M'' = 0$ for all M''. Hence $L_1 T = 0$. $\qquad$ Q.E.D.

There are two ways to dualize the discussion; we could either switch the direction of the resolution or replace T by a contravariant functor. Let us describe these procedures respectively, assuming the underlying category has "enough injectives."

Definition 5.2.24: The *n*-th *right derived* functor $R^n T$ of the functor T is given by $(R^n T)M = H^n(T\mathbb{E})$, the *n*-th cohomology group of T applied to an injective resolution $\mathbb{E}$ of M; likewise, $(R^n T)f$ is the *n*-th cohomology map of $T(f^n)$, where (f^n) is the extension of $f: M \to M'$ to injective resolutions of M and M'.

Theorem 5.2.25: *If T is a left exact functor then for any exact sequence* $0 \to M' \to M \to M'' \to 0$ *there is a long exact sequence*

$$0 \to TM' \to TM \to TM'' \to (R^1 T)M' \to (R^1 T)M \to \cdots$$

$(R^1 T)E = 0$ *for every injective E; T is exact iff $R^1 T = 0$.*

Proof: Dual to corollary 5.2.23. $\qquad$ Q.E.D.

Definition 5.2.26: The *n*-th *right derived* functor $R^n T$ of a *contravariant* functor T is defined by $(R^n T)M = H^n(T\mathbb{P})$ where $\mathbb{P}$ is a projective resolution of M. (Note we use cohomology since $T\mathbb{P}$ is a negative complex.) Given $f: M \to M'$ we lift f to a map $(f_n): \mathbb{P} \to \mathbb{P}'$ of projective resolutions and define $(R^n T)f$ to be the *n*-th cohomology map of $T(f_n)$.

Theorem 5.2.26': *If T is a left exact contravariant functor then for any exact sequence* $0 \to M' \to M \to M'' \to 0$ *there is a long exact sequence*

$$0 \to TM'' \to TM \to TM' \to (R^1 T)M'' \to (R^1 T)M \to \cdots.$$

T is exact iff $R^1 T = 0$.

Proof: As in theorem 5.2.25.   Q.E.D.

Tor and Ext

At last we are ready to define the functors which will help us to measure homological dimensions of rings.

Definition 5.2.27: $\text{Tor}_n^R(\_\_, M)$ is defined as $L_n T: \mathcal{M}od\text{-}R \to \mathcal{A}b$ where $T = \_\_ \otimes_R M$.

$\text{Ext}_R^n(M, \_\_)$ is defined as $R^n T: R\text{-}\mathcal{M}od \to \mathcal{A}b$ where $T = \text{Hom}_R(M, \_\_)$.
$\text{Ext}_R^n(\_\_, M)$ is defined as $R^n T: R\text{-}\mathcal{M}od \to \mathcal{A}b$ where $T = \text{Hom}_R(\_\_, M)$.

We delete the "R" when the ring R is understood. These definitions make sense because $\_\_ \otimes M$ is right exact, $\text{Hom}(M, \_\_)$ is left exact, and $\text{Hom}(\_\_, M)$ is contravariant left exact. Let us review theorem 5.2.22, 5.2.25, and 5.2.26' for these particular functors.

Summary 5.2.28: If $0 \to N' \to N \to N'' \to 0$ is exact in $\mathcal{M}od\text{-}R$ or $R\text{-}\mathcal{M}od$ according to the context then there are long exact sequences

(i) $\cdots \to \text{Tor}^2(N'', M) \to \text{Tor}^1(N', M) \to \text{Tor}^1(N, M) \to \text{Tor}^1(N'', M)$

$\to N' \otimes M \to N \otimes M \to N'' \otimes M \to 0$

(ii) $0 \to \text{Hom}(M, N') \to \text{Hom}(M, N) \to \text{Hom}(M, N'') \to \text{Ext}^1(M, N')$

$\to \text{Ext}^1(M, N) \to \text{Ext}^1(M, N'') \to \text{Ext}^2(M, N') \to \cdots$

(iii) $0 \to \text{Hom}(N'', M) \to \text{Hom}(N, M) \to \text{Hom}(N', M) \to \text{Ext}^1(N'', M)$

$\to \text{Ext}^1(N, M) \to \text{Ext}^1(N', M) \to \text{Ext}^2(N'', M) \to \cdots$

There is a fundamental tie from Ext^1 to projectives and injectives.

Proposition 5.2.29: $\text{Ext}^1(P, \_\_) = 0$ iff P *is projective; dually* $\text{Ext}^1(\_\_, E) = 0$ iff E *is injective.*

Proof: We prove the first assertion. Let $T = \text{Hom}(P, \_\_)$. In view of corollary 2.11.6 we want to show $R^1 T = 0$ iff T is exact. ($\Rightarrow$) is immediate by theorem 5.2.25. ($\Leftarrow$) For any module N we can take an exact sequence $0 \to N \to E \to E/N \to 0$ with E injective, so $0 \to TN \to TE \to T(E/N) \to (R^1 T)N \to (R^1 T)E = 0$ is exact. Hence $(R^1 T)N \approx \text{coker}(TE \to T(E/N)) = 0$ since T is exact, proving $R^1 T = 0$.   Q.E.D.

There is some ambiguity in notation since $\text{Ext}^n(M, N)$ could denote either $\text{Ext}^n(M, \_)$ applied to N or $\text{Ext}^n(\_, N)$ applied to M. We have seen these are both 0 for $n = 1$ when M is projective (by proposition 5.2.29 and remark 5.2.21), thereby motivating the next theorem, that the two interpretations of $\text{Ext}^n(M, N)$ are the same. First we show this for $n = 1$.

Proposition 5.2.30: *For this result write* $\overline{\text{Ext}}(M, \_)$ *instead of* $\text{Ext}(M, \_)$. *Then* $\overline{\text{Ext}}^1(M, N) \approx \text{Ext}^1(M, N)$ *for all modules* M, N.

Proof: Take exact sequence $0 \to K \to P \to M \to 0$ and $0 \to N \to E \to K' \to 0$ with P projective and E injective. Then we have the commutative diagram

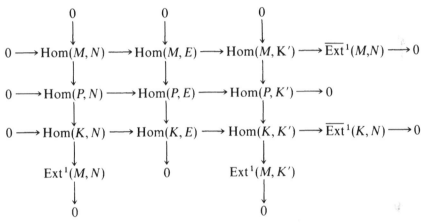

The zeros appear on the right and bottom because of remark 5.2.21 and proposition 5.2.29. Applying the snake lemma to the second and third rows gives the exact sequence

$$\text{Hom}(M, N) \to \text{Hom}(M, E) \to \text{Hom}(M, K') \to \text{Ext}^1(M, N) \to 0.$$

Thus $\text{Ext}^1(M,N) \approx \text{coker}(\text{Hom}(M,E) \to \text{Hom}(M,K')) \approx \overline{\text{Ext}}^1(M,N)$ as desired.
                                                                Q.E.D.

Remark 5.2.31: In the lower right-hand corner of the above diagram there is an isomorphism $f: \overline{\text{Ext}}^1(K, N) \to \text{Ext}^1(M, K')$, sending any x in $\overline{\text{Ext}}^1(K, N)$ to its preimage in $\text{Hom}(K, K')$ and then to the image in $\text{Ext}^1(M, K')$. (The proof is by diagram chasing. First note that if y, y' are both preimages of x in $\text{Hom}(K, K')$ then $y - y'$ has a preimage in $\text{Hom}(K, E)$ and thus in $\text{Hom}(P, E)$; travelling down the other side of the square shows $y - y'$ goes to 0 in $\text{Ext}^1(M, K')$. This proves f is well-defined, and the same analysis going in the other direction enables us to construct f^{-1}.)

In order to generalize proposition 5.2.30 to arbitrary n we need a way of shifting down.

Lemma 5.2.32: *If $\mathbb{P}$ is a projective resolution of M and K_n is the n-th syzygy for each n then $(L_{n+1}T)M \approx (L_iT)K_{n-i}$ for any right exact functor T, for each $1 \leq i \leq n$.*

Proof: By induction it suffices to prove $(L_{n+1}T)M \approx (L_nT)K_0$. But we can cut $\mathbb{P}$ at P_0 to get a projective resolution $\cdots \to P_2 \to P_1 \to K_0 \to 0$ of K_0, which we call $\mathbb{P}'$. By definition $(L_nT)K_0 = H_n(T\mathbb{P}') = H_{n+1}(T\mathbb{P}) = (L_{n+1}T)M$ as desired.   Q.E.D.

Proposition 5.2.33: *Suppose K_i is the i-th syzygy of some projective resolution of M. Then $\text{Tor}_{n+1}(\_, M)$ and $\text{Tor}_{n-i}(\_, K_i)$ are naturally isomorphic functors; also $\text{Ext}^{n+1}(\_, M)$ and $\text{Ext}^{n-i}(\_, K_i)$ are naturally isomorphic functors. Dually suppose M has an injective resolution $0 \to M \xrightarrow{f_0} E_0 \xrightarrow{f_1} E_1 \cdots$ and let $K_i' = \text{coker} f_i = E_i/f_i E_{i-1}$, called the i-th cosyzygy. Then $\text{Ext}^{n+1}(M, \_)$ and $\text{Ext}^{n-i}(K_i', \_)$ are naturally isomorphic functors.*

Proof: Apply the lemma and its analog for maps; this yields the first assertion, and the others follow in the same way.   Q.E.D.

Theorem 5.2.34: *For this result write $\overline{\text{Ext}}^n(M, \_)$ instead of $\text{Ext}^n(M, \_)$. Then $\text{Ext}^n(M, N) \approx \overline{\text{Ext}}^n(M, N)$ for all modules M, N, and all n.*

Proof: (Zaks) For $n = 0$ this is obvious and for $n = 1$ it is proposition 5.2.30. In general, take a projective resolution $\mathbb{P}$ of M, denoting the n-th syzygy by K_n, and take an injective resolution $\mathbb{E}$, denoting the n-th cosyzygy as K_n'. Applying remark 5.2.31 to the exact sequences $0 \to K_0 \to P_0 \to M \to 0$ and $0 \to K_{n-2}' \to E_{n-1}' \to K_{n-1}' \to 0$ yields $\text{Ext}^1(M, K_{n-1}') \approx \text{Ext}^1(K_0, K_{n-2}')$. Iterating remark 5.2.31 we have $\text{Ext}^1(M, K_{n-1}') \approx \text{Ext}^1(K_0, K_{n-2}') \approx \text{Ext}^1(K_1, K_{n-3}') \approx \cdots \approx \text{Ext}^1(K_{n-1}, N)$. Applying proposition 5.2.33 at both ends yields

$$\overline{\text{Ext}}^{n+1}(M, N) \approx \overline{\text{Ext}}^1(M, K_{n-1}') \approx \text{Ext}^1(M, K_{n-1}')$$

$$\approx \text{Ext}^1(K_{n-1}, N) \approx \text{Ext}^{n+1}(M, N)$$

as desired.   Q.E.D.

We are now in a position for a basic result.

Corollary 5.2.35: *The following are equivalent for a module P*:

  (i) *P is projective.*
  (ii) $\text{Ext}^1(P, \_\_) = 0$.
  (iii) $\text{Ext}^n(P, \_\_) = 0$ *for all n.*

Proof: (ii) $\Rightarrow$ (i) by proposition 5.2.29, and (i) $\Rightarrow$ (iii) by remark 5.2.21. (iii) $\Rightarrow$ (ii) is obvious.   Q.E.D.

Let us dualize all of this for Tor. Given a right module M we can define $\text{Tor}_n^R(M, \_\_)$ as $L_n T: R\text{-}\mathcal{M}od \to \mathcal{A}b$ where T is the functor $M \otimes_R \_\_$. By definition M is flat iff T is exact. Thus proposition 5.2.29 dualizes to

Proposition 5.2.36: $\text{Tor}_1(F, \_\_) = 0$ *iff F is a flat right module; analogously,* $\text{Tor}_1(\_\_, F) = 0$ *iff F is a flat left R-module.*

But any projective module P is flat, so an analogous proof to proposition 5.2.30 (but now using projective resolutions on both sides) shows $\text{Tor}_1(M, N)$ is well-defined for any right module M and left module N. The analogous use of proposition 5.2.33 enables one to prove

Theorem 5.2.37: $\text{Tor}_n(M, N)$ *is well-defined for any right module M and left module N.*

Tor has fundamental connections to tensor products, and we shall need the following one in Chapter 6:

Example 5.2.38: (i) Suppose $0 \to K \to F \to M \to 0$ is an exact sequence of R-modules with F free, and $A \triangleleft R$. Then $\text{Tor}_1^R(R/A, M)$ is canonically identified with $(K \cap AF)/AK$. Indeed, applying $\text{Tor}_1(R/A, \_\_)$ yields the exact sequence

$$0 = \text{Tor}_1(R/A, F) \to \text{Tor}_1(R/A, M) \to K/AK \to F/AF \to M/AM$$

since $K/AK \approx (R/A) \otimes K$ and $F/AF \approx (R/A) \otimes F$ by example 1.7.21'. Hence we can identify $\text{Tor}_1(R/A, M)$ with $\ker(K/AK \to F/AF) = (K \cap AF)/AK$, as needed.

  (ii) If $A, B \triangleleft R$ then $\text{Tor}_1(R/A, R/B)$ is canonically identified with $A \cap B/AB$, as seen by taking $K = B$ and $F = R$ in (i).

Digression: Ext and Module Extensions

Our objective in introducing Ext and Tor was to understand the homological dimensions better. However, there are more concrete interpretations of these functors, and we shall digress a bit to discuss Ext, leaving Tor for exercise 8ff.

Given modules K, N we say an *extension* of K by N is an exact sequence $0 \to K \xrightarrow{f} M \xrightarrow{g} N \to 0$. Sometimes we denote this extension merely as M, with f,g understood. Two extensions M, M' of K by N are *equivalent* if there is a map $\varphi: M \to M'$ such that

$$
\begin{array}{ccccccccc}
0 & \longrightarrow & K & \xrightarrow{f} & M & \xrightarrow{g} & N & \longrightarrow & 0 \\
& & \downarrow 1_M & & \downarrow \varphi & & \downarrow 1_N & & \\
0 & \longrightarrow & K & \xrightarrow{f'} & M' & \xrightarrow{g'} & N & \longrightarrow & 0
\end{array}
$$

commutes. Note by the 5 lemma that φ must be an isomorphism. The set of equivalence classes of extensions of K by N is denoted $e(N, K)$.

Proposition 5.2.39: *There is a $1 - 1$ correspondence $e(N, K) \to \mathrm{Ext}^1(N, K)$.*

Proof: Given an extension $0 \to K \to M \to N \to 0$ we shall define an element of $\mathrm{Ext}^1(N, K)$ as follows: Take a projective resolution $\mathbb{P}$ of N, and using proposition 5.2.14 build the commutative diagram

$$
\begin{array}{ccccccccccc}
\cdots & \longrightarrow & P_2 & \xrightarrow{d_2} & P_1 & \xrightarrow{d_1} & P_0 & \xrightarrow{d_0} & N & \longrightarrow & 0 \\
& & \downarrow & & \downarrow f_1 & & \downarrow f_0 & & \downarrow 1_N & & \\
\cdots & \longrightarrow & 0 & \longrightarrow & K & \longrightarrow & M & \longrightarrow & N & \longrightarrow & 0
\end{array}
\qquad (1)
$$

Now $0 = f_1 d_2 = d_2^* f_1$ so f_1 yields an element in $\ker d_2^* / \mathrm{im}\, d_1^* \in \mathrm{Ext}^1(N, K)$, and this is the element of $\mathrm{Ext}^1(N, K)$ corresponding to our original sequence. By lemma 5.2.19 and proposition 5.2.20 this is well-defined and sends equivalent extensions to the same element of $\mathrm{Ext}^1(N, K)$. Thus we have a functor $\Phi: e(N, K) \to \mathrm{Ext}^1(N, K)$.

Now we want to determine Φ^{-1}. Given a cocycle $f_1: P_1 \to K$ we want to define a suitable extension. This will be done by means of the pushout (cf., 5.0.7). We are given $0 = d_2^* f_1 = f_1 d_2$, so we have a map $\bar{f}: P_1 / d_2 P_2 \to K$, giving rise to the sequence

$$
\begin{array}{ccccccc}
0 & \longrightarrow & P_1 / d_2 P_2 & \xrightarrow{\bar{d}_1} & P_0 & \longrightarrow & N & \longrightarrow & 0 \\
& & \downarrow \bar{f} & & & & \\
& & K & & & &
\end{array}
$$

with $\bar{d}_1$ monic. Taking M to be the pushout of the upper left corner yields

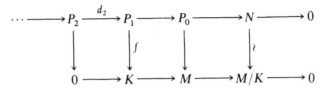

$$(2)$$

But (2) could also be rewritten as

which is the same as (1). Any extension satisfying (1) could be viewed in this manner as a pushout, so the universality of the pushout shows the extension is determined uniquely up to equivalence. Thus we have, indeed, created Φ^{-1}.   Q.E.D.

Corollary 5.2.39′: $\mathrm{Ext}^1(K, N) = 0$ *iff every extension of K by N splits.*

Proof: Every extension must be equivalent to the extension $0 \to K \xrightarrow{\mu} K \oplus N \xrightarrow{\pi} N \to 0$.   Q.E.D.

Homological Dimension and the Derived Functors

We are finally ready to apply all this machinery to homological dimension; the proofs become much clearer intuitively.

Theorem 5.2.40:   *The following are equivalent for a module M:*

(i) $\mathrm{pd}\, M \leq n$;
(ii) $\mathrm{Ext}^k(M, \_) = 0$ *for all $k > n$*;
(iii) $\mathrm{Ext}^{n+1}(M, \_) = 0$;
(iv) *The $(n - 1)$ syzygy of any projective resolution of M is projective.*

Proof: (i) $\Rightarrow$ (ii) By definition of $\mathrm{Ext}^k(\_, N)$ we see $\mathrm{Ext}^k(M, N) = 0$ for any $k > n$ since M has a projective resolution of length n. Hence $\mathrm{Ext}^k(M, \_) = 0$ by theorem 5.2.34.

(ii) $\Rightarrow$ (iii) is a fortiori.

(iii) $\Rightarrow$ (iv) Let K_{n-1} be the $(n - 1)^{\mathrm{st}}$ syzygy. Proposition 5.2.33 shows $\mathrm{Ext}^1(K_{n-1}, \_) \approx \mathrm{Ext}^{n+1}(M, \_) = 0$, so K_{n-1} is projective by corollary 5.2.35.

(iv) $\Rightarrow$ (i) By proposition 5.1.8.   Q.E.D.

One has the dual theorem for injective resolutions:

Theorem 5.2.40′: *The following are equivalent for a module N:*

(i) *N has injective dimension $\leq n$;*
(ii) $\text{Ext}^k(\underline{\quad}, N) = 0$ *for all $k > n$;*
(iii) $\text{Ext}^{n+1}(\underline{\quad}, N) = 0$;
(iv) *The $(n - 1)$ cosyzygy of every injective resolution of N is injective.*

Corollary 5.2.41: $\text{gl. dim } R = \sup\{injective\ dimensions\ of\ the\ R\text{-}modules\}$.

Proof: Each side $= \sup\{k: \text{Ext}^k(M, N) \neq 0$ for all R-modules M and $N\}$, by theorem 5.2.40(ii) and 5.2.40′ (iii).   Q.E.D.

Using Ext one can show gl. dim depends only on the cyclic modules.

Theorem 5.2.42: *(Auslander)* $\text{gl. dim } R = \sup\{\text{pd } R/L: L < R\}$.

Proof: Let $n = \sup\{\text{pd } R/L: L < R\}$; then $\text{Ext}^{n+1}(R/L, \underline{\quad}) = 0$. We need to show for any injective resolution $\mathbb{E}$ that the $(n - 1)$ cosyzygy K is injective. Well, $\text{Ext}^1(R/L, K) \approx \text{Ext}^{n+1}(R/L, N) = 0$, so $\text{Hom}(R, K) \to \text{Hom}(L, K) \to \text{Ext}^1(R/L, K) = 0$ is exact by summary 5.2.28(iii). By Baer's criterion K is injective, as desired.   Q.E.D.

Corollary 5.2.43: *R is left hereditary iff* $\text{gl. dim}(R) = 1$.

Treating Tor analogously we discover a new dimension. In view of proposition 5.2.36 it makes sense to work with flat modules.

Definition 5.2.44: A *flat resolution* of a module M is an exact sequence

$$\cdots \to F_n \overset{d_n}{\to} F_{n-1} \overset{d_{n-1}}{\longrightarrow} \cdots \to F_0 \overset{d_0}{\to} M = 0$$

where each F_n is flat. The *flat dimension of M* is the smallest n for which M has a flat resolution of length n; the *weak dimension*, written w. dim R, is the supremum of the flat dimensions of all R-modules.

Proposition 5.2.45: *For any ring* w. dim $R \leq$ gl. dim R.

Proof: Any projective resolution is a flat resolution, since projective modules are flat.   Q.E.D.

Of course w. dim $R = 0$ iff every R-module is flat, iff R is (von Neumann) regular. Since left Noetherian regular rings are semisimple Artinian, this raises the hope that w. dim $R = $ gl. dim R if R is left Noetherian. In fact this is true, and implies gl. dim is left-right symmetric for Noetherian rings. These and other basic facts about w. dim are given in exercises 13ff; of course $\mathcal{T}or$ enters heavily in the proofs.

Dimension Shifting

Dimension shifting is a rather straightforward technique which lifts properties from Ext^n to Ext^{n+1} and from Tor_n to Tor_{n+1} and thus transfers properties of Hom to Ext, and of $\otimes$ to Tor. The idea is rather simple.

Remark 5.2.46: "Dimension shifting". Suppose $K \leq M$. If $\text{Ext}^n(M, \_) = 0$ (for example, if M is projective) then $\text{Ext}^{n+1}(M/K, \_) \approx \text{Ext}^n(K, \_)$ in the sense they are naturally isomorphic functors. (Indeed, $\text{Ext}^{n+1}(M, \_) = 0$ by theorem 5.2.40, so summary 5.2.28(iii) yields the exact sequence

$$0 = \text{Ext}^n(M, N) \to \text{Ext}^n(K, N) \to \text{Ext}^{n+1}(M/K, N) \to \text{Ext}^{n+1}(M, N) = 0,$$

implying $\text{Ext}^{n+1}(M/K, N) \approx \text{Ext}^n(K, N)$ for all modules N.)
   Similarly, if $\text{Ext}^n(\_, M) = 0$ then $\text{Ext}^{n+1}(K, \_) \approx \text{Ext}^n(M/K, \_)$.
   If $\text{Tor}_n(\_, M) = 0$ then $\text{Tor}_{n+1}(M/K, \_) \approx \text{Tor}_n(K, \_)$.

The principle of dimension shifting has already been used implicitly; this simple technique produces instant results.

Remark 5.2.47: Reproof of summary 5.1.13. Suppose $0 \to M' \to M \to M'' \to 0$ is exact and $n = \text{pd } M$. $\text{Ext}^{n+1}(M, \_) = 0$, so by dimension shifting $\text{Ext}^{m+1}(M'', \_) \approx \text{Ext}^m(M', \_)$ for all $m > n$. This yields cases I and II $(n \leq n')$ at once. For case III $(n > n')$ we need a similar but modified argument. Taking $m > n'$ we have

$$0 = \text{Ext}^m(M', N) \to \text{Ext}^{m+1}(M'', N) \to \text{Ext}^{m+1}(M, N) \to \text{Ext}^{m+1}(M', N) = 0$$

for all N, so $\text{Ext}^{m+1}(M, \_) \approx \text{Ext}^{m+1}(M'', \_)$. Thus $n = n''$.

Our immediate interest in dimension-shifting is to extend certain basic properties from Hom to Ext, and from $\otimes$ to Tor.

Theorem 5.2.48: *Suppose $F: R\text{-}\mathcal{M}od^I \to R\text{-}\mathcal{M}od$ and $G: \mathcal{A}b \to \mathcal{A}b$ is a pair* *of right exact functors which commutes with* Hom *in the sense that there is a natural isomorphism* $\text{Hom}(\_\_, F(N_i)) \approx G(\text{Hom}(\_\_, (N_i)))$ *(as functors from R-Mod to Ab). Then F and G commutes with* Ext^n *for all* $n \in \mathbb{N}$, *in the same sense.*

Proof: Take an exact sequence $0 \to K \to P \to M \to 0$ with P free. Since $\text{Ext}_R^n(P, \_\_) = 0$ for all $n \geq 1$ we have the commutative diagram

$$
\begin{array}{ccccccc}
\text{Ext}^{n-1}(P, F(N_i)) & \longrightarrow & \text{Ext}^{n-1}(K, F(N_i)) & \longrightarrow & \text{Ext}^n(M, F(N_i)) & \longrightarrow & 0 \\
\downarrow & & \downarrow & & \downarrow & & \\
G(\text{Ext}^{n-1}(P, N_i)) & \longrightarrow & G(\text{Ext}^{n-1}(K, N_i)) & \longrightarrow & G(\text{Ext}^n(M, N_i)) & \longrightarrow & 0
\end{array}
$$

Identifying Ext^0 with Hom we may assume by induction on n that the first two vertical maps are isomorphisms. (In fact, for $n > 1$ the left-hand terms are 0.) By property 5.0.2 the right-hand map exists and is an isomorphism, as desired. This is all natural, by inspection.   Q.E.D.

Theorem 5.2.49: *Suppose* $F: R\text{-}Mod^{(I)} \to R\text{-}Mod$ *and* $G: Ab \to Ab$ *is a pair of left exact functors for which there is a natural isomorphism of functors* $\_\_ \otimes F(N_i)$ *and* $G(\_\_ \otimes (N_i))$. *Then there is a natural isomorphism of functors* $\text{Tor}_n(\_\_, F(N_i))$ *and* $G(\text{Tor}_n(\_\_, (N_i)))$.

Proof: Dual to theorem 5.2.48.   Q.E.D.

Corollary 5.2.50: *Suppose* N_i *are R-modules.* $\text{Ext}^n(M, \prod N_i) \approx \prod \text{Ext}^n(M, N_i)$ *for every R-module M, and* $\text{Tor}_n(M, \coprod N_i) \approx \coprod \text{Tor}_n(M, N_i)$ *for every right R-module M. (We write* $\coprod$ *instead of* $\oplus$ *to stress duality.)*

Proof: It suffices to prove $\prod$ is right exact, and dually $\coprod$ is left exact. But given $K_i \to L_i \to N_i \to 0$ exact we see at once $\coprod L_i \to \coprod N_i$ is epic with kernel $\coprod K_i$.   Q.E.D.

Let us improve this result.

Lemma 5.2.51: If I is a directed set then $\varinjlim$ is left exact, in the sense that if $(A_i'; \varphi_{ij}')$, $(A_i; \varphi_{ij})$ and $(A_i''; \varphi_{ij}'')$ are systems indexed by I and $0 \to A_i' \to A_i \to A_i''$ is exact then $0 \to \varinjlim A_i' \to \varinjlim A_i \to \varinjlim A_i''$ is exact. Indeed, let $f: \varinjlim A_i \to \varinjlim A_i''$ be the given map. Obviously, $\varinjlim A_i' \subseteq \ker f$. Conversely, suppose $(a^{(i)}) \in \ker f$. By the construction in theorem 1.8.7 this means for suitable $k \geq i$ we have $\varphi_i^k a^{(i)} \in A_k'$. But defining $a'^{(i)}$ to be $\varphi_i^k a^{(i)}$ we see easily that $(a'^{(i)})$ equals $(a^{(i)})$ in $\varinjlim A_i$, proving $\ker f = \varinjlim A_i'$.

Proposition 5.2.52: $\mathrm{Tor}_n(M, \varinjlim N_i) \approx \varinjlim \mathrm{Tor}_n(M, N_i)$ *whenever the index set is directed.*

Proof: Apply lemma 5.2.51 to theorem 5.2.49.   Q.E.D.

Remark 5.2.53: We should record some variants of these results; the proofs are analogous.

(i) $\mathrm{Ext}^n(\coprod M_i, N) \approx \prod \mathrm{Ext}^n(M_i, N)$.

(ii) $\mathrm{Tor}_n(\coprod M_i, N) \approx \coprod \mathrm{Tor}_n(M_i, N)$.

(iii) $\mathrm{Tor}_n(\varinjlim M_i, N) \approx \varinjlim \mathrm{Tor}(M_i, N)$ for any directed index set I.

Other results of this genre are given in exercises 29–35; a detailed treatment is given in the classic Cartan-Eilenberg [56B]. There are many papers in the literature which study the homology of particular rings, but we should like to point the reader to Rinehart-Rosenberg [76] and Rosenberg-Stafford [76], who by a careful use of projective resolutions calculate explicitly the homological dimensions of modules over rings of differential polynomials and over Weyl algebras.

Acyclic Complexes

Homology gives us an important connection between complexes and projective resolutions by means of the following definition:

Definition 5.2.54: A positive complex (A, d) is *acyclic* if $H_n(A) = 0$ for all $n > 0$.

Remark 5.2.55: The acyclicity condition means

$$\to A_{n+1} \xrightarrow{d_{n+1}} A_n \xrightarrow{d_n} \cdots \xrightarrow{d_1} A_0 \to 0$$

is exact at A_n for each $n > 0$. On the other hand, $H_0(A) = A_0/d_1 A_1$, so we have the exact sequence

$$\cdots \to A_{n+1} \xrightarrow{d_{n+1}} A_n \xrightarrow{d_n} \cdots \xrightarrow{d_1} A_0 \to H_0(A) \to 0.$$

(This is why we have been considering almost positive complexes). Thus we have a projective resolution of $H_0(A)$ if A_n is projective for each $n \geq 0$, or, equivalently, if $A = \bigoplus_{n \geq 0} A_n$ is projective. Reversing the argument shows a projective resolution can be viewed as a positive acyclic complex which is projective.

Proposition 5.2.56: *Suppose (A, d) and (A', d') are complexes of vector spaces over a field F. Let $\tilde{A} = A \otimes_F A'$, which is graded via $\tilde{A}_n = \bigoplus_{u+v=n} A_u \otimes_F A'_v$, and define a differentiation $\tilde{d}$ by $\tilde{d}_n(a \otimes a') = d_u a \otimes a' + (-1)^u a \otimes d'_v a'$ for $a \in A_u$ and $a' \in A'_v$. Then*

(i) *$(\tilde{A}, \tilde{d})$ is a complex.*

(ii) *If A and A' are acyclic then so is $\tilde{A}$, and $H_0(\tilde{A}) = H_0(A) \otimes H_0(A')$ canonically.*

Proof: For purposes of calculation we shall take $a \in A_u$ and $a' \in A'_v$, with $u + v = n$.

(i) $\tilde{d}_{n-1} \tilde{d}_n(a \otimes a') = \tilde{d}_{n-1}(d_u a \otimes a' + (-1)^u a \otimes d'_v a')$

$$= d_{u-1} d_u a \otimes a' + (-1)^{u-1} d_u a \otimes d'_v a' + (-1)^u d_u a \otimes d'_v a'$$

$$+ (-1)^u a \otimes d'_{v-1} d'_v a' = 0$$

since the middle two terms cancel and the end terms are each 0.

(ii) We shall apply proposition 5.2.18 to $\tilde{A}$ and $H_0(\tilde{A})$. Let $\varepsilon : A_0 \to H_0(A)$ and $\varepsilon' : A'_0 \to H_0(A')$ be the canonical maps, and $\tilde{d}_0 = \varepsilon \otimes \varepsilon'$. By proposition 5.2.18' we have $s_{-1} : H_0(A) \to A_0$ and $s_i : A_i \to A_{i+1}$ for $i \geq 0$ satisfying $\varepsilon s_{-1} = 1$, $d_1 s_0 + s_{-1} \varepsilon = 1$, and $d_{i+1} s_i + s_{i-1} d_i = 1$ for all $i > 0$. Likewise, we have $s'_{-1} : H_0(A') \to A'_0$ and $s'_i : A'_i \to A'_{i+1}$ for $i \geq 0$ satisfying $\varepsilon' s'_{-1} = 1$, $d'_1 s'_0 + s'_{-1} \varepsilon' = 1$, and $d'_{i+1} s'_i + s'_{i-1} d'_i = 1$ for all $i > 0$. Define

$$\tilde{s}_{-1} = s_{-1} \otimes s'_{-1} \quad \text{and} \quad \tilde{s}_n = s_{-1} \varepsilon \otimes s'_n + \sum_{i=0}^{n} s_i \otimes 1 \quad \text{for } n \geq 0.$$

Then $\tilde{d}_0 \tilde{s}_{-1} = 1 \otimes 1 = 1$. It remains to show $\tilde{d}_{n+1} \tilde{s}_n + \tilde{s}_{n-1} \tilde{d}_n = 1$ for all $n \geq 0$, which will be seen by evaluating it on $a_u \otimes a_v$.

Case I. $u > 1$. Then

$$(\tilde{d}_{n+1} \tilde{s}_n + \tilde{s}_{n-1} \tilde{d}_n)(a \otimes a') = \tilde{d}_{n+1} s_u a \otimes a' + \tilde{s}_{n-1}(d_u a \otimes a' + (-1)^u a \otimes d'_v a')$$

$$= d_{u+1} s_u a \otimes a' + (-1)^{u+1} s_u a \otimes d'_v a'$$

$$+ s_{u-1} d_u a \otimes a' + (-1)^u s_u a \otimes d'_v a'$$

$$= (d_{u+1} s_u + s_{u-1} d_u) a \otimes a' = a \otimes a'.$$

Case II. $u = 1$. Same calculation as in case I, except with the extra term $s_{-1}\varepsilon d_1 a \otimes s'_n a' = 0$ (since $\varepsilon d_1 = 0$).

Case III. $u = 0$.

$$(\tilde{d}_{n+1}\tilde{s}_n + \tilde{s}_{n-1}\tilde{d}_n)(a \otimes a') = \tilde{d}_{n+1}(s_{-1}\varepsilon a \otimes s'_n a' + s_0 a \otimes a') + \tilde{s}_{n-1}(a \otimes d'_n a')$$

$$= s_{-1}\varepsilon a \otimes d'_{n+1}s'_n a' + d_1 s_0 a \otimes a' - s_0 a \otimes d'_n a'$$

$$+ s_{-1}\varepsilon a \otimes s'_{n-1}d'_n a' + s_0 a \otimes d'_n a'$$

$$= s_{-1}\varepsilon a \otimes (d'_{n+1}s'_n + s'_{n-1}d'_n)a' + d_1 s_0 a \otimes a'$$

$$= (s_{-1}\varepsilon + d_1 s_0)a \otimes a' = a \otimes a'. \qquad \text{Q.E.D.}$$

Corollary 5.2.57: *If R is an F-algebra and $\mathbb{P}_k$ is a projective resolution of M_k for $k = 1,\ldots,m$ then $M_1 \otimes_F \cdots \otimes_F M_m$ has the projective resolution $\tilde{P} = \bigotimes_{k=1}^m \mathbb{P}_k$ where $\tilde{P}_n = \bigoplus_{u_1 + \cdots + u_m = n}(P_1)_{u_1} \otimes \cdots \otimes (P_m)_{u_m}$ and $\tilde{d}_n: \tilde{P}_n \to \tilde{P}_{n-1}$ is given by*

$$\tilde{d}(a_1 \otimes \cdots \otimes a_m) = \sum_{i=1}^m (-1)^{u(i)} a_1 \otimes \cdots \otimes da_i \otimes \cdots \otimes a_m$$

where each $a_i \in (A_i)_{u_i}$ and $u(i) = \sum_{k=1}^{i-1} u_k$.

Proof: View each $\mathbb{P}_k$ as a direct sum of projectives, which is thus projective; hence $\mathbb{P}$ is projective by the easy remark 5.3.0. Furthermore, $\tilde{P}$ is a projective resolution of $M_1 \otimes_F \cdots \otimes_F M_m$, by the proposition (since everything can be viewed naturally as vector spaces over F.) Q.E.D.

Some remarks about these results. Proposition 5.2.56 is a special case of the Kunneth formulas of algebraic topology. The reader's eyebrows may have been raised by the sign in the formula for the differentiation. This sign arises naturally in the definition of tensor products of morphisms of $\mathbb{Z}$-graded modules and is discussed in detail in MacLane [63B, Chapter VII, esp. pp. 190–191].

§5.3 Separable Algebras and Azumaya Algebras

As noted several times earlier, one can bypass the asymmetry between left and right modules by dealing with bimodules. On the other hand, we want the powerful techniques of module theory at our disposal, so we view $R - R$ bimodules as $R \otimes R^{\text{op}}$-modules, cf., proposition 1.7.31. Rather than taking tensors over $\mathbb{Z}$, we shall consider the more general case where R is an algebra over an arbitrary commutative ring C, and shall write R^e for $R \otimes_C R^{op}$. Furthermore, we assume throughout that M is an $R - R$ bimodule respecting the algebra structure, i.e., $cx = xc$ for all c in C, x in M. Thus we shall view M canonically as R^e-module, by the action $(r_1 \otimes r_2)x = r_1 x r_2$; also $(c \otimes 1)x = (1 \otimes c)x$ for all c in C. Now we can define the *Hochschild* homology and cohomology groups

$$H_n(R, M) = \text{Tor}_n(R, M) \quad \text{and} \quad H^n(R, M) = \text{Ext}^n(R, M)$$

where the base ring is R^e. Although there is a lovely theory developed in Hochschild's articles, Cartan-Eilenberg [56B], and the series "On the dimension of modules and algebras" in the Nagoya Journal (1950s), we focus on the special case where R is a projective R^e-module. This provides a modern and natural setting for the classical theory of separable algebras, cf., definition 2.5.28, and, in particular, for Wedderburn's principal theorem. We shall also deal with "Azumaya algebras," in preparation for Chapters 6 and 7. Our treatment follows Knus-Ojanguren [74B], Jacobson [80B], and Demeyer-Ingraham [71B].

Remark 5.3.0: If P_i are projective C-modules then $P_1 \otimes_C P_2$ is projective. (Indeed, if $F_i \approx P_i \oplus P_i'$ is free then

$$P_1 \otimes P_2 \oplus (P_1' \otimes P_2 \oplus P_1 \otimes P_2' \oplus P_1' \otimes P_2') \approx (P_1 \oplus P_1') \otimes (P_2 \oplus P_2')$$

$$\approx F_1 \otimes F_2$$

is free.)

Remark 5.3.0′: If P is a projective C-module and R is a C-algebra then $R \otimes_C P$ is a projective R-module. (For if $P \oplus P' = F$ then $R \otimes P \oplus R \otimes P' \approx R \otimes F$ is free by corollary 1.7.16.)

Remark 5.3.1: R is a cyclic R^e-module (spanned by 1), and there is an epic $p: R^e \to R$ in $R^e\text{-}\mathcal{M}od$ given by $p(\sum r_{1i} \otimes r_{2i}) = \sum r_{1i} r_{2i}$. Letting $J = \ker p$ we have the exact sequence of R^e-modules $0 \to J \to R^e \to R \to 0$. The notation J and p has become standard, and we shall use it without further ado.

Proposition 5.3.2: $J = \sum_{r \in R} R^e(r \otimes 1 - 1 \otimes r)$.

Proof: ($\supseteq$) is clear. ($\subseteq$) If $a = \sum r_{i1} \otimes r_{i2} \in J$ then $\sum r_{i1} r_{i2} = 0$ so $a = \sum_i (r_{i1} \otimes 1)(1 \otimes r_{i2} - r_{i2} \otimes 1)$ as desired.   Q.E.D.

Remark 5.3.3:   The exact sequence of remark 5.3.1 yields an exact sequence

$$0 \to \mathrm{Hom}_{R^e}(R, M) \to \mathrm{Hom}_{R^e}(R^e, M) \to \mathrm{Hom}_{R^e}(J, M) \to \mathrm{Ext}^1_{R^e}(R, M) \to 0$$

since $\mathrm{Ext}^1_{R^e}(R^e, R) = 0$.

To interpret the other groups we need some definitions. First define $M^R = \{x \in M : rx = xr \text{ for all } r \text{ in } R\}$, viewing M as $R - R$ bimodule; translated to $R^e\text{-}\mathcal{M}od$ the condition reads $(r \otimes 1)x = (1 \otimes r)x$, i.e., $Jx = 0$.

Next define $\mathrm{Der}(M) = \{\partial: R \to M : \partial(r_1 r_2) = (\partial r_1)r_2 + r_1 \partial r_2\}$, the set of *derivations of R in M*. Translated to $R^e\text{-}\mathcal{M}od$ this condition reads as $\partial(r_1 r_2) = (1 \otimes r_2)\partial r_1 + (r_1 \otimes 1)\partial r_2$. We can view $\mathrm{Der}(M)$ as C-module by defining $(c\partial)x$ to be $c(\partial x)$. For any x in M we define the *inner derivation* ∂_x by $\partial_x r = (r \otimes 1 - 1 \otimes r)x$. When $M = R^e$ we write δ for $\partial_{1 \otimes 1}$, i.e., $\delta r = r \otimes 1 - 1 \otimes r$. In fact, $\delta \in \mathrm{Der}(J)$ by proposition 5.3.2. Let us view this fact in context.

Proposition 5.3.4: $\mathrm{Hom}_{R^e}(R, M) \approx M^R$ *via* $f \mapsto f1$; $\mathrm{Hom}_{R^e}(J, M) \approx \mathrm{Der}(M)$ *via* $f \mapsto f\delta$. *Consequently, we have an exact sequence*

$$0 \to M^R \to M \to \mathrm{Der}(M) \to \mathrm{Ext}^1_{R^e}(R, M) \to 0$$

Proof: The last assertion is obtained by matching the given isomorphisms to remark 5.3.3, noting also $\mathrm{Hom}(R^e, M) \approx M$ by $f \mapsto f1$. Thus it remains to verify the two given isomorphisms.

Define $\varphi: \mathrm{Hom}_{R^e}(R, M) \to M^R$ by $\varphi f = f1$; note $f1 \in M^R$ since

$$(r \otimes 1 - 1 \otimes r)f1 = f(r - r) = f0 = 0.$$

Thus φ is a map whose inverse $M^R \to \mathrm{Hom}_{R^e}(R, M)$ is given by sending x to right multiplication by x.

Next define $\psi: \mathrm{Hom}_{R^e}(J, M) \to \mathrm{Der}(M)$ by $\psi f = f\delta$; $f\delta$ is in $\mathrm{Der}(M)$ since

$$f\delta(r_1 r_2) = f((1 \otimes r_2)\delta r_1 + (r_1 \otimes 1)\delta r_2) = (1 \otimes r_2)f\delta r_1 + (r_1 \otimes 1)f\delta r_2.$$

$\ker \psi = 0$ since $f\delta = 0$ implies $fJ = 0$ by proposition 5.3.2. It remains to show ψ is onto. Given $\partial \in \mathrm{Der}(M)$ we define $\bar{g}: R^e \to M$ by means of the balanced map $g: R \times R^{\mathrm{op}} \to M$ given by $\bar{g}(a, b) = -(a \otimes 1)\partial b$, and let f be the restriction of $\bar{g}$ to J. Then $f \in \mathrm{Hom}(J, M)$, for if $\sum a_i \otimes b_i \in J$ then

$$f((r_1 \otimes r_2)\sum a_i \otimes b_i) = \sum f(r_1 a_i \otimes b_i r_2) = -\sum (r_1 a_i \otimes 1)\partial(b_i r_2)$$

$$= -\sum ((r_1 a_i \otimes 1)(1 \otimes r_2)\partial b_i + (b_i \otimes 1)\partial r_2)$$

$$= -\sum ((r_1 a_i \otimes r_2)\partial b_i + (r_1 \sum a_i b_i \otimes 1)\partial r_2)$$

$$= -(r_1 \otimes r_2)\sum (a_i \otimes 1)\partial b_i + 0 = (r_1 \otimes r_2)f(\sum a_i \otimes b_i).$$

But $f\delta r = f(r \otimes 1 - 1 \otimes r) = -(r \otimes 1)\partial 1 + (1 \otimes 1)\partial r = 0 + \partial r$, proving $f\delta = \partial$ as desired. Thus ψ is onto.   Q.E.D.

Corollary 5.3.4′:

  (i) $\operatorname{Hom}_{R^e}(R, R) \approx Z(R)$.
  (ii) $\operatorname{Hom}_{R^e}(R, R^e) \approx \operatorname{Ann}' J$, *the right annihilator of J in R^e, under the correspondence* $f \mapsto f1$.

Proof: (i) Take $M = R$. (ii) Take $M = R^e$. Then $\operatorname{Hom}_{R^e}(R, M) \approx M^R \approx \{x \in M : (r \otimes 1)x = (1 \otimes r)x$ for all r in $R\} = \operatorname{Ann}'\{r \otimes 1 - 1 \otimes r : r \in R\} = \operatorname{Ann}' J.$
                                                                                        Q.E.D.

Remark 5.3.5:   In the exact sequence of proposition 5.3.4 the map $M \to \operatorname{Der}(M)$ is given by sending x in M to the inner derivation ∂_x. Thus $\operatorname{Ext}^1(R, M) = 0$ iff every derivation of R in M is inner.

Separable Algebras

Definition 5.3.6: R is a *separable C-algebra* if R is projective as R^e-module.

There are several nice criteria for separability.

Theorem 5.3.7:   *The following conditions are equivalent*:

  (i) *R is separable.*
  (ii) *$\operatorname{Ext}^1_{R^e}(R, M) = 0$ for every R-module M.*
  (iii) *There is an element e in $(R^e)^R$ such that $pe = 1$ where $p: R^e \to R$ is the canonical epic.*
  (iv) *The epic $p: R^e \to R$ restricts to an epic $(R^e)^R \to Z(R)$,*
  (v) *Every derivation of R in M is inner, for every module M.*

Proof: (i) $\Leftrightarrow$ (ii) by corollary 5.2.35. (i) $\Rightarrow$ (iii) p splits so there is a monic $f: R \to R^e$ for which $pf = 1_R$; take $e = f1$.
  (iii) $\Rightarrow$ (iv) $p((z \otimes 1)e) = z$ for any z in $Z(R)$.
  (iv) $\Rightarrow$ (iii) trivial. (iii) $\Rightarrow$ (i) Define $f: R \to R^e$ by $fr = (r \otimes 1)e$; we see $f((r_1 \otimes r_2)r) = f(r_1 r r_2) = (r_1 r r_2 \otimes 1)e = (r_1 r \otimes 1)((r_2 \otimes 1)e) = (r_1 r \otimes r_2)e = (r_1 \otimes r_2)fr$, proving f is a map. Hence p is split so R is a summand of R^e and thus projective.
  (ii) $\Leftrightarrow$ (v) by the exact sequence of proposition 5.3.4.   Q.E.D.

I hope no confusion will arise from the use of the symbol e both as part of the notation R^e and as an element thereof.

Remark 5.3.8: The element e of theorem 5.3.7(iii) is called a *separability idempotent* of R and is indeed idempotent since writing $e = \sum r_{j1} \otimes r_{j2}$ we have

$$e^2 = \sum (r_{j1} \otimes 1)(1 \otimes r_{j2})e = \sum (r_{j1} \otimes 1)(r_{j2} \otimes 1)e.$$
$$= \left(\sum r_{j1}r_{j2} \otimes 1\right)e = (1 \otimes 1)e = e.$$

Obviously C is separable over itself since $C \approx C^e$. $M_n(C)$ is separable since $M_n(C) \otimes M_n(C)^{\mathrm{op}} \approx M_n(C) \otimes M_n(C) \approx M_t(C)$ where $t = n^2$. Other examples are had by means of the separability idempotent, so let us record some of its properties.

Remark 5.3.9: Suppose e is a separability idempotent of R and let $Z = Z(R)$.
   (i) $eR^e \subseteq (R^e)^R$ since $(r \otimes 1)ex = ((r \otimes 1)e)x = ((1 \otimes r)e)x = (1 \otimes r)ex$ for x in R^e;
   (ii) $p(eR^e) = Z$ by (i) and theorem 5.3.7(iv); in fact

$$p(e(z \otimes 1)) = p((z \otimes 1)e) = z$$

for z in Z;
   (iii) Z is a summand of R as Z-module since $r \mapsto p(e(r \otimes 1))$ defines a projection $R \to Z$ fixing Z, by (ii);
   (iv) If $A \lhd Z$ then $A = Z \cap AR$ by (iii) and sublemma 2.5.22'

Proposition 5.3.10:

   (i) If R_i are separable C-algebras then $R_1 \times R_2$ is a separable C-algebra.
   (ii) If R_i are separable C_i-algebras where C_i are commutative C-algebras then $R_1 \otimes_C R_2$ is a separable $C_1 \otimes_C C_2$-algebra and

$$Z(R_1 \otimes R_2) = Z(R_1) \otimes Z(R_2).$$

Proof: Let e_i be a separability idempotent of R_i.

   (i) Viewing $R_1^e \times R_2^e \subset (R_1 \times R_2)^e$ canonically we see by inspection (e_1, e_2) is a separability idempotent.
   (ii) $R_1 \otimes R_2$ has the separability idempotent $e = e_1 \otimes e_2$, and thus is separable. Let $Z = Z(R_1 \otimes R_2)$. Clearly $Z(R_1) \otimes Z(R_2) \subseteq Z$. Conversely, if

$z = \sum r_{i1} \otimes r_{i2} \in Z$ then by remark 5.3.9(ii) we have

$$z = p(e(z \otimes 1)) = \sum p(e_1(r_{i1} \otimes 1) \otimes e_2(r_{i2} \otimes 1)) \in Z(R_1) \otimes Z(R_2)$$

as desired.   Q.E.D.

Example 5.3.11:

(i) The direct product of copies of C is separable.

(ii) If S is a submonoid of C and R is separable then $S^{-1}R \approx S^{-1}C \otimes_C R$ is separable (with the same separability idempotent).

(iii) If R is separable then $M_n(R) \approx M_n(C) \otimes R$ is separable.

(iv) (extension of scalars) If R is separable and C_1 is a commutative C-algebra then $C_1 \otimes_C R$ is a separable C_1-algebra.

Proposition 5.3.12: *If $f: R \to R'$ is a surjection of C-algebras and R is separable then R' is separable and $Z(R') = fZ(R)$.*

Proof: Let $p: R^e \to R$ and $p': (R')^e \to R$ be the usual epics. Let $\tilde{f} = f \otimes f: R^e \to (R')^e$. Then $fp = p'\tilde{f}$. Letting e be a separability idempotent of R, we see $\tilde{f}e$ is a separability idempotent of R', proving R' is separable. Furthermore $f(Z(R)) = fp(eR^e) = p'\tilde{f}(eR^e) = p'((\tilde{f}e)(R')^e) = Z(R')$ by remark 5.3.9(ii).
     Q.E.D.

Proposition 5.3.13: *Suppose C' is a commutative C-algebra and R is a C'-algebra.*

(i) *If R is separable over C then R is separable over C'.*

(ii) *If R is separable over C' and C' is separable over C then R is separable over C.*

Proof:

(i) The canonical epic $R \otimes_C R^{op} \to R$ factors through $R \otimes_{C'} R^{op}$, so we get a separability idempotent.

(ii) By definition 5.3.6 C' is a summand of $(C')^e = C' \otimes_C C'$. Tensoring by R^e over $(C')^e$ yields $R \otimes_{C'} R^{op}$ is a summand of R^e. But R is a summand of $R \otimes_{C'} R^{op}$ by hypothesis, so R is a summand of R^e.   Q.E.D.

To make separability "descend," we consider the following set-up: C' is a commutative C-algebra and R, R' are algebras over C, C', respectively, such that C' is a summand of R'. Write $(R')^e$ for $R' \otimes_{C'} (R')^{op}$. As C'-algebras

$(R \otimes_C R')^e \approx R^e \otimes_C (R')^e$ so any $(R \otimes R')^e$-module can be viewed naturally as R^e-module via the map $R^e \to R^e \otimes 1$. In particular, if $R \otimes R'$ is separable as C'-algebra then $R \otimes R'$ is a summand of $(R \otimes R')^e$ as module over $(R \otimes R')^e$ and thus as R^e-module. To conclude R is separable over C we merely need to show

   (i) R is a summand of $R \otimes R'$, and
   (ii) $(R \otimes R')^e$ is R^e-projective.

(i) is clear if $C \cdot 1$ is a summand of R' as C-module, since then we tensor on the left by R. On the other hand, (ii) follows if $(R')^e$ is projective as C-module since then we tensor on the left by R^e. In case R' is projective as C-module we thus have (ii) if $R' = C'$ (trivially), or if $C' = C$ by remark 5.3.0. This provides the following two important instances of descent:

Proposition 5.3.14:

   (1) *If $R \otimes_C R'$ is separable over C and R' is faithfully projective over C then R is separable over C.*

   (2) *If $R \otimes_C C'$ is separable over C' for a commutative algebra C' which is faithfully projective over C then R is separable over C.*

Proof: By proposition 2.11.29 if R' is faithfully projective over C then $C \cdot 1$ is a summand of R'. Hence (i) of the above discussion is satisfied, and (ii) also holds since $C = C'$ in (1) and $R' = C'$ in (2).   Q.E.D.

Proposition 5.3.15: *If R is separable over C and R is faithfully projective over a commutative C-algebra C' then C' is separable over C.*

Proof: R is a summand of R^e which is projective over $(C')^e$, so R is projective over $(C')^e$. But by proposition 2.11.29 C' is a summand of R as C'-module and thus as C'-bimodule since C' is commutative (so we copy out the same scalar multiplication on the right also), and thus as $(C')^e$-module. Hence C' is projective over $(C')^e$.   Q.E.D.

We are ready to justify the name "separable", also, cf., theorem 5.3.18.

Proposition 5.3.16: *Suppose R is a finite field extension of a subfield F. This field extension is separable iff R is separable as F-algebra.*

Proof: In view of propositions 5.3.13 and 5.3.15 we may assume R is a simple field extension, i.e., $R = F[r] \approx F[\lambda]/\langle f \rangle$ for some irreducible polynomial f in $F[\lambda]$.

($\Rightarrow$) Let K be a splitting field of f over F. Then $K \otimes R \approx R^{(n)}$ where $n = \deg f$, by the Chinese Remainder Theorem, cf., after remark 2.2.7, so $K \otimes R$ is separable as R-algebra, implying K is separable as F-algebra by proposition 5.3.14.

($\Leftarrow$) Suppose the field extension were not separable. Then $\text{char}(F) = p > 0$ and $R \approx L[\lambda]/\langle \lambda^n - a \rangle$ for some field $L \subset R$ and $a \in L$. Then differentiation with respect to λ yields a nontrivial derivation of R, which cannot be inner since R is commutative, so R is not separable by theorem 5.3.7(v).   Q.E.D.

Proposition 5.3.17:   *Suppose R is a separable C-algebra and is projective as C-module. Then R is f.g. as C-module.*

Proof: Take a separability idempotent $e = \sum_{\text{finite}} r_{j1} \otimes r_{j2}$. Picking a dual base $\{(x_i, f_i) : i \in I\}$ of R we have $f_i r_{j1} = 0$ for almost all i, and we claim the $x_i r_{j2}$ span R. Indeed, for any r in R we have

$$r = p((r \otimes 1)e) = p\left(\sum_j r r_{j1} \otimes r_{j2} \right) = p\left(\sum_{j,i} f_i(r r_{j1}) x_i \otimes r_{j2} \right) = \sum_{j,i} f_i(r r_{j1}) x_i r_{j2}$$

as desired.   Q.E.D.

Theorem 5.3.18:   *An algebra R over a field F is separable in the sense of this section iff it is separable in the sense of definition 2.5.38.*

Proof: ($\Rightarrow$) First we show R is semisimple Artinian. It suffices to show $\text{Hom}_R(M, \_)$ is exact for any R-module M, by corollary 2.11.6. But $\text{Hom}_F(M, \_)$ is exact. Furthermore, $\text{Hom}_F(M, N)$ is an R^e-module (i.e., $R - R$ bimodule) by remark 1.5.18; the explicit action is $((r_1 \otimes r_2)f)x = r_1 f(r_2 x)$ for f in $\text{Hom}(M, N)$. Then

$$\text{Hom}_R(M, N) = (\text{Hom}_F(M, N))^R \approx \text{Hom}_{R^e}(R, \text{Hom}_F(M, N))$$

by proposition 5.3.4. Thus $\text{Hom}_R(M, \_)$ is the composite of the exact functors $\text{Hom}_F(M, \_)$ and $\text{Hom}_{R^e}(R, \_)$ (noting R is R^e-projective), so R indeed is semisimple Artinian.

Let K be the algebraic closure of F. Then $R \otimes_F K$ is separable, so by the first paragraph is semisimple Artinian and f.d. over K by proposition 5.3.17 implying $R \otimes_F K$ is split. Now for any field extension L of F let L' be the composition of K and L; then $R \otimes L'$ is split, so $R \otimes_F L$ is semisimple Artinian. Hence R is separable in the sense of definition 2.5.38.

($\Leftarrow$) Let K be a splitting field of R. Then $R \otimes_F K$ is separable over K by example 5.3.11, implying R is separable over F.   Q.E.D.

Hochschild's Cohomology

Having seen that this notion of separable generalizes the classical notion and is more elegant, our next objective is to redo the classical theory of separable algebras in this context, in particular to recast Wedderburn's principal theorem in this setting. First we want Hochschild's explicit description of his cohomology groups.

Construction 5.3.19: Assuming R is projective as C-module, we shall construct an explicit cochain complex for any R^e-module M. Define $C^n(R, M) = \text{Hom}_C(R^{(n)}, M)$ (so $C^0(R, M) = M$) and define the cochain map $\delta^n: C^n(R, M) \to C^{n+1}(R, M)$ by

$$\delta^n f(r_1, \ldots, r_{n+1}) = r_1 f(r_2, \ldots, r_{n+1})$$

$$+ \sum_{i=1}^{n} (-1)^i f(r_1, \ldots, r_{i-1}, r_i r_{i+1}, r_{i+2}, \ldots, r_{n+1}) + (-1)^{n+1} f(r_1, \ldots, r_n) r_{n+1}$$

for r_i in R. Let $B^n = \delta^{n-1} C^{n-1}(R, M)$ and $Z^n = \ker \delta^n$.

We want to identify $H^n(R, M)$ with Z^n/B^n. To do this we must identify this cochain complex with a cochain complex arising from $\text{Hom}(\underline{\quad}, M)$ applied to a projective resolution of R in $R^e\text{-}\mathcal{M}od$.

Define $P_n = R^{\otimes(n+2)}$ for each $n \geq -1$, viewed naturally as $R - R$ bimodules (i.e., R^e-modules). Then $P_0 \approx R^e$ and the P_n are all projective C-modules by remark 5.3.0, implying $P_n \approx R^e \otimes_C P_{n-2}$ is a projective R_e-module, by remark 5.3.0'. Now by the adjoint isomorphism

$$\text{Hom}_{R^e}(P_n, M) \approx \text{Hom}_{R^e}(P_{n-2} \otimes R^e, M) \approx \text{Hom}_C(P_{n-2}, \text{Hom}_{R^e}(R^e, M))$$

$$\approx \text{Hom}_C(P_{n-2}, M) \approx C^n(R, M)$$

the last isomorphism arising from viewing an n-linear map as a balanced map. It remains to define the differentiations $\partial_n: P_n \to P_{n-1}$, by

$$\partial_n(r_1 \otimes \cdots \otimes r_{n+2}) = \sum_{i=1}^{n+1} (-1)^{i-1} r_1 \otimes \cdots \otimes r_i r_{i+1} \otimes \cdots \otimes r_{n+2}.$$

We have a diagram

$$\cdots \to P_2 \xrightarrow{\partial_2} P_1 \xrightarrow{\partial_1} P_0 \xrightarrow{P} R \to 0 \tag{$\mathbb{P}$}$$

where p is as in remark 5.3.1. Define $s_n: P_n \to P_{n+1}$ by $s_n(x_1 \otimes \cdots \otimes x_{n+2}) = 1 \otimes x_1 \otimes \cdots \otimes x_{n+2}$. Clearly $s_n P_n$ spans P_{n+1}, $ps_{-1} = 1_R$, and $\partial_{n+1}s_n + s_{n-1}\partial_n = 1$ on each P_n, by direct computation. Hence $\mathbb{P}$ is a projective resolution by proposition 5.2.18. Moreover, we have a commutative diagram

$$
\begin{array}{ccc}
\mathrm{Hom}_{R^e}(P_n, M) & \xrightarrow{\;\partial^{\#}\;} & \mathrm{Hom}_{R^e}(P_{n+1}, M) \\
\big\downarrow & & \big\downarrow \\
C^n(R, M) & \xrightarrow{\;\delta\;} & C^{n+1}(R, M)
\end{array}
$$

which yields the desired isomorphism of $H^n(R, M)$ with the cohomology group of $(C^n; \delta)$, i.e., with Z^n/B^n.

Let us look at certain relevant cocycles and coboundaries.

A map $f: R \to M$ is in B^1 iff $f = \delta^0 x$ for some $x \in C^0(R, M) \approx M$, i.e., f is the inner derivation with respect to x.

$f: R \to M$ is in Z^1 iff $r_1 f r_2 - f(r_1 r_2) + (f r_1)r_2 = 0 \qquad$ for all r_i in R,

i.e., f is a derivation of R in M. This shows $H^1(R, M)$ is the derivations modulo the inner derivations, a result implicit in the proof of proposition 5.3.4.

A map $f: R^{(2)} \to M$ is in B^2 iff there is $g: R \to M$ for which $f(r_1, r_2) = r_1 g r_2 - g(r_1 r_2) + (g r_1)r_2$ for all r_i in R. Now we have a much more intuitive proof of a strengthening of Wedderburn's principal theorem (2.5.37).

Theorem 5.3.20: *(Wedderburn's principal theorem revisited) Suppose R is a finite dimensional algebra over a field F, with $H^2(R, \_\_) = 0$. Then R has a semisimple Artinian subalgebra S for which $R \approx S \oplus \mathrm{Jac}(R)$ as vector spaces over F.*

Proof: As in the "general case" of the proof of theorem 2.5.27 we can reduce readily to the case $N^2 = 0$ where $N = \mathrm{Jac}(R)$. Let V be a complementary vector space to N in R. Then $V \approx R/N$ as vector spaces, and this would be an algebra isomorphism if V were closed under multiplication in R, implying V were semisimple Artinian, as desired. Hence we want to modify V by suitable elements of N to make V multiplicatively closed. Letting $\pi: R \to V$ be the vector space projection define $f: R^{(2)} \to N$ by

$$
f(r_1, r_2) = \pi(r_1 r_2) - \pi r_1 \pi r_2.
$$

Then

$$\delta f(r_1, r_2, r_3) = r_1 f(r_2, r_3) - f(r_1 r_2, r_3) + f(r_1, r_2 r_3) - f(r_1, r_2) r_3$$

$$= r_1 \pi(r_2 r_3) - r_1 \pi r_2 \pi r_3 - \pi(r_1 r_2 r_3) + \pi(r_1 r_2) \pi r_3$$

$$+ \pi(r_1 r_2 r_3) - \pi r_1 \pi(r_2 r_3) - \pi(r_1 r_2) r_3 + \pi r_1(\pi r_2) r_3$$

$$= (r_1 - \pi r_1)(\pi(r_2 r_3) - \pi r_2 \pi r_3) + (\pi r_1 \pi r_2 - \pi(r_1 r_2))(r_3 - \pi r_3) = 0$$

since $N^2 = 0$. Thus $f \in Z^2(R, N) = B^2(R, N)$ by hypothesis, so there is some $g: R \to N$ for which $f = \delta g$. Hence $f(r_1, r_2) = r_1 g r_2 - g(r_1 r_2) + (g r_1) r_2$. Let $S = \{\pi r + g r : r \in R\}$. $R = S + N$, so $[V:F] = [S:F]$ and $R = S \oplus N$ as vector spaces. Furthermore, $\pi r_1 g r_2 = r_1 g r_2$ since $(r_1 - \pi r_1) g r_2 \in N^2 = 0$, and, likewise, $g r_1 \pi r_2 = (g r_1) r_2$, yielding

$$(\pi r_1 + g r_1)(\pi r_2 + g r_2) = \pi r_1 \pi r_2 + g r_1 \pi r_2 + \pi r_1 g r_2 + 0$$

$$= \pi(r_1 r_2) - f(r_1, r_2) + (g r_1) r_2 + r_1 g r_2$$

$$= \pi(r_1 r_2) + g(r_1 r_2)$$

proving S is multiplicatively closed, as desired.   Q.E.D.

This also yields a direct approach to a theorem of Malcev concerning the uniqueness of S. Note first by proposition 2.5.6(i) that for any unit u in R we have $u^{-1} N u = N$ where $N = \text{Jac}(R)$, so whenever $R = S \oplus N$ we also have $R \approx u^{-1} S u \oplus N$.

Theorem 5.3.21: *Suppose R is a finite dimensional F-algebra with $S \oplus N = S' \oplus N = R$ as vector spaces, where $N = \text{Jac}(R)$ with R/N separable and $S \approx R/N \approx S'$. Then there is $a \in N$ for which $S' = (1 + a)^{-1} S(1 + a)$.*

Proof: Let $\varphi: S \to S'$ be the composition of the isomorphisms $S \approx R/N \approx S'$. Let $\bar{e} \in (R/N)^e$ be a separability idempotent of R/N, written as the image of a suitable element e of R^e (not necessarily idempotent), and let $e' \in S \otimes (S')^{\text{op}}$ be the image of $\bar{e}$ under the isomorphism $(R/N)^e \to S \otimes (S')^{\text{op}}$. Then for any s in S we have $(s \otimes 1 - 1 \otimes \varphi s) e' = 0$. Passing up to R^e and applying the canonical maps $p: R^e \to R$ and $R \to R/N$ gives $spe = (pe)\varphi s$ for all s in S. Since the image of e' in R/N is 1 we see $pe \in 1 + N$ is invertible and thus $\varphi s = (pe)^{-1} spe$ for all s in S, as desired.   Q.E.D.

This result can be formulated in the much broader scope of "inertial subalgebras," cf., Ingraham [74].

Azumaya Algebras

Separable algebras have been seen to have many properties in common with simple algebras. The next one is particularly noteworthy.

Lemma 5.3.22: *Suppose R is a separable $Z(R)$-algebra. Then any maximal ideal A has the form $A_0 R$ for suitable maximal $A_0 \lhd Z(R)$. In fact $A_0 = A \cap Z(R)$.*

Proof: Let $Z = Z(R)$ and $A_0 = A \cap Z$. By proposition 5.3.12 R/A is separable with center Z/A_0. But R/A is simple so Z/A_0 is a field and A_0 is a maximal ideal. But now $R/A_0 R$ is separable over the field Z/A_0 so is central simple by theorem 5.3.18. Hence $A_0 R$ is already a maximal ideal, implying $A_0 R = A$.   Q.E.D.

Definition 5.3.23: A C-algebra R is *Azumaya* if the following two conditions are satisfied:

(i) R is faithful and f.g. projective as C-module, and
(ii) $R^e \approx \operatorname{End}_C R$ under the correspondence sending $r_1 \otimes r_2$ to the map $r \to r_1 r r_2$.

By theorem 2.3.27 every central simple algebra is Azumaya, and this is the motivating example of our discussion. In fact, we shall see that Azumaya algebras possess properties in common with central simple algebras, because of the following connection to separability.

Theorem 5.3.24: *The following are equivalent for a C-algebra R:*
(i) R is Azumaya.
(ii) $C = Z(R)$ and R is separable over C.
(iii) *There is a category equivalence $C\text{-}\mathcal{M}od \to R^e\text{-}\mathcal{M}od$ given by $R \otimes \_\_$.*
(iv) $R^e \approx \operatorname{End}_C R$ and R is a progenerator of $C\text{-}\mathcal{M}od$.
(v) $R^e \approx \operatorname{End}_C R$ and R is a generator of $R^e\text{-}\mathcal{M}od$.
(vi) $R^e \approx \operatorname{End}_C R$ and $R^e \operatorname{Ann}' J = R^e$ (*where $\operatorname{Ann}' J$ denotes the right annihilator of J in R^e*).

Before proving this theorem let us note that conditions (i), (ii), (iii) interconnect Azumaya algebras with separability theory and Morita theory: furthermore, we shall tie Azumaya algebras to polynomial identity theory in theorem 6.1.35. Conditions (iv), (v), (vi) are more technical, and all involve $R^e \approx \operatorname{End}_C R$ plus another condition. It may well be that $R^e \approx \operatorname{End}_C R$ already implies R is Azumaya! Indeed Braun [86] has shown this is the case if R has an anti-automorphism fixing $Z(R)$, cf., exercises 6–9. Another criterion for R to be Azumaya is given in exercise 11.

Proof of Theorem 5.3.24: We show (i) $\Rightarrow$ (iv) $\Rightarrow$ (iii) $\Rightarrow$ (ii) $\Rightarrow$ (i) and (ii) $\Rightarrow$ (vi) $\Rightarrow$ (v) $\Rightarrow$ (i).

(i) $\Rightarrow$ (iv) by remark 4.1.9 (since $R^e = R \otimes R^{op}$).

(iv) $\Rightarrow$ (iii) by Morita's theorem.

(iii) $\Rightarrow$ (ii) $R = R \otimes C$ is a progenerator in R^e-$\mathcal{M}od$. By proposition 4.1.19 we can build a Morita context, thereby yielding a category equivalence in the other direction given by the functor $\operatorname{Hom}_{R^e}(R, R^e) \otimes_{R^e} \underline{\ \ }$, which by the adjoint isomorphism is naturally equivalent to $\operatorname{Hom}_{R^e}(R, \underline{\ \ })$. Thus $C \approx \operatorname{Hom}_{R^e}(R, R) \approx Z(R)$ by corollary 5.3.4′.

(ii) $\Rightarrow$ (i) $C = Z(R) \approx \operatorname{Hom}_{R^e}(R, R)$ so applying example 4.1.16 to the projective R^e-module R gives the Morita context

$$(R^e, C, R, \operatorname{Hom}_{R^e}(R, R^e), \tau, \tau') \tag{1}$$

where τ' is epic by proposition 4.1.19 since R is cyclic projective.

To conclude by Morita's theorem we need to show τ is epic; by remark 4.1.9 and proposition 4.1.19 it is enough to show R is faithfully projective over R^e. Suppose $A \lhd R^e$ and $AR = R$. We may assume A is maximal. By proposition 5.3.10 $R^e \approx R \otimes R^{op}$ is separable with center C, so $A = A_0 R^e$ for some maximal $A_0 \lhd C$, by lemma 5.3.22. Now $A_0 R = R$, which is impossible by remark 5.3.9(iv).   Q.E.D.

(ii) $\Rightarrow$ (vi) As in the proof of (ii) $\Rightarrow$ (i) we can set up the Morita context (1) with τ' epic. Thus $R^e \approx \operatorname{End}_C R$ and we need to show $R^e \operatorname{Ann}' J = R^e$. τ'epic implies every element of R^e has the form $\sum_i f_i r_i = \sum r_i f_i 1 = \sum (r_i \otimes 1) f_i 1 \in R^e \operatorname{Ann}' J$ since each $f_i 1 \in \operatorname{Ann}' J$ by corollary 5.3.4(ii); here each $f_i \in \operatorname{Hom}_{R^e}(R, R^e)$.

(vi) $\Rightarrow$ (v) Reverse the previous argument to show τ' is epic in (1), so R is a generator in R^e-$\mathcal{M}od$ by Lemma 4.1.7.

(v) $\Rightarrow$ (i) Setting up the Morita context (1) we see R is f.g. projective over $\operatorname{End}_{R^e} R \approx Z(R)$ by remark 4.1.17′.   Q.E.D.

Remark 5.3.24′: The reverse equivalence R^e-$\mathcal{M}od \to C$-$\mathcal{M}od$ is given by $M \mapsto M^R$. (Indeed, as shown in the proof of (iii) $\Rightarrow$ (ii) it is given by $\operatorname{Hom}_{R^e}(R, \underline{\ \ })$, so we appeal to proposition 5.3.4.)

Corollary 5.3.25: *If R is an Azumaya C-algebra then there is a 1:1*

correspondence between {ideals of C} and {ideals of R} given by $A_0 \mapsto A_0 R$
(and the reverse correspondence $A \mapsto A \cap C$).

Proof: Apply proposition 4.1.18, noting the R^e-submodules of R are the
ideals of R.   Q.E.D.

On the other hand, let us consider the consequences of viewing R as a
projective C-module. In view of theorem 2.12.22 we can write $C = \prod_{i=1}^{t} C_i$
where each $C_i = Ce_i$ for a suitable idempotent e_i of C, and $R_i = Re_i$ has
constant rank as projective module over Ce_i. But each R_i is Azumaya as
C_i-algebra by theorem 5.3.24(ii), and $R \approx \prod R_i$, so we have proved

Proposition 5.3.26:   *Every Azumaya algebra is a finite direct product of
Azumaya algebras of constant rank (as projective module over the center).*

One further reduction is worth noting, for this is what gives Azumaya
algebras the connection to the original definition of Azumaya [50]. We say
R is *proper maximally central of rank n* over a subring C of $Z(R)$ if R is a free
C-module of rank n and $R^e \approx \operatorname{End}_C R$.

Proposition 5.3.27:   *Suppose R is Azumaya over C of constant rank n.*

   (i) R_P *is proper maximally central over C_P of rank n, for every $P \in \operatorname{Spec}(C)$.*
   (ii) *R/P' is central simple of dimension n over $C/(P' \cap C)$, for every maximal
ideal P' of R. In particular, n is square.*

Proof:

   (i) By proposition 5.3.10(ii) $R_P \approx C_P \otimes_C R$ is separable over C_P with
$Z(R_P) = C_P$. C_P is local so R_P is free of rank n by definition 2.12.21.
   (ii) Let $P = P' \cap C$. Then $P' = PR$ by lemma 5.3.22, so $P'_P = P_P R_P$. Thus
by Nakayama's lemma the image of a base of R_P (over C_P) must be a base of
R_P/P'_P over C_P/P_P, the center of R_P/P'_P by proposition 5.3.12. We conclude
by remark 2.12.12, which shows $R_P/P'_P \approx R/P'$ and $C_P/P_P \approx C/P$.   Q.E.D.

The converse results are in Knus-Ojanguren [74B, pp. 79 and 95], and
reduce the study of Azumaya algebras to the local case.

Exercises

§5.1

1. Schanuel's lemma fails for flat modules. (Hint: Let F_0 be any localization of $\mathbb{Z}$;

take F free with $f: F \to F_0/\mathbb{Z}$ epic. Then $0 \to \mathbb{Z} \to F_0 \to F_0/\mathbb{Z} \to 0$ but $F_0 \oplus$ ker $f \not\approx \mathbb{Z} \oplus F$ unless F_0 is projective.)

2. $\mathrm{pd}_{R[\lambda]} M[\lambda] = \mathrm{pd}_R M$. (Hint: corollary 5.1.21, viewing $M[\lambda]$ as R-module.)

3. If each R-module has finite pd then gl. dim $R < \infty$. (Hint: proposition 5.1.20.)

4. (I. Cohen) If R is a subring of T such that R is a summand of T in R-$\mathcal{M}od$-R then gl. dim $R \leq$ gl. dim $T + \mathrm{pd}_R(T)$. In particular, if T is projective as R-module then gl. dim $R \leq$ gl. dim T. (Hint: Writing $T = R \oplus N$ as bimodule yields $\mathrm{Hom}_R(T, M) = M \oplus \mathrm{Hom}_R(N, M)$ for any R-module M. Let $M' = \mathrm{Hom}_R(T, M)$. Then $\mathrm{pd}_R M \leq \mathrm{pd}_R M'$. Apply proposition 5.1.22 to M'.)

5. gl. dim $M_n(R) =$ gl. dim R. (Hint: Morita.)

6. If R is left Noetherian and $0 \to M' \to M \to M'' \to 0$ is exact with two of the modules having Euler characteristic then so does the third, and $\chi(M) = \chi(M') + \chi(M'')$.

7. If $M > M_1 > \cdots > M_t = 0$ and each factor has FFR then M has FFR.

8. Fields [70] Prove proposition 5.1.19 under the weaker assumption $\sigma: R \to R$ is an injection. (Hint: Consider σR-modules.) Thus gl. dim $R \leq$ gl. dim $R[\lambda; \sigma] \leq 1 +$ gl. dim R. There are examples for strict inequality at either stage. Furthermore, these results hold for $R[\lambda; \sigma, \delta]$.

K_0-Theory

11. Define more generally $K_0(R\text{-}\mathcal{F}imod)$ to be the free abelian group whose generators are isomorphism classes of f.g. R-modules, modulo the relation $[M] = [M'] + [M'']$ if there is an exact sequence $0 \to M'' \to M \to M' \to 0$. If R is left Noetherian of finite gl. dim show $K_0(R\text{-}\mathcal{F}imod) \approx K_0(R)$ canonically. (Hint: Split an f.g. projective resolution into short exact sequences.)

12. Using exercise 11 and the proof of proposition 5.1.32 show there is a canonical epic $K_0(R\text{-}\mathcal{F}imod) \to K_0(S^{-1}R\text{-}\mathcal{F}imod)$ whose kernel is generated by those modules with S-torsion.

13. Prove that φ of Quillen's theorem is epic. (Hint: Let R' be the "Rees subring" $\sum_{n \in \mathbb{N}} R_n \lambda^n$ of $R[\lambda]$, which takes the place of $R[\lambda]$ used in proving Serre's theorem. Grading R' according to degree in λ one has $R_0 = (R')_0$. But the summands R_{n+1}/R_n of $G(R)$ are flat as R_0-modules so each $R_0, R_1, \ldots$ is flat. Hence R is flat; also R' is flat since $R'_n \approx R_n$ is flat. On the other hand, R' is also filtered by $(R')_n = R_n[\lambda] \cap R'$; the associated graded ring is isomorphic to $G(R)[\lambda]$ under the grade $G(R)[\lambda]_n = \bigoplus_{i \leq n} R_i \lambda^{n-i}$ so is left Noetherian with finite gl. dim. Thus these properties pass to R'. Now factor φ as $K_0(R) \to K_0(R') \to K_0(R)$, the first first arrow of which is an isomorphism by Serre's theorem. It remains to show the second arrow is onto, seen just as in the proof of Serre's theorem by viewing R as R'-module via specializing λ to 1 and applying the functor H given by $HM = M/(1 - \lambda)M$.)

14. (Walker [72]) If R is semiprime Noetherian and gl. dim $R < \infty$ and $K_0(R) = \langle [R] \rangle$ then R is a domain. (Hint: Use the uniform dimension instead of the Euler characteristic in the proof of theorem 5.1.45.)

15. State and prove explicitly the dual to Schanuel's lemma.

Change of Rings

16. Theorem 5.1.24 can fail if gl. dim R/Ra is infinite. (Example: $R = \mathbb{Z}$ and $a = 4$.)

17. Suppose a is a normalizing element of R and M is a module with $ax \neq 0$ for all $x \neq 0$ in M. Then any exact sequence $0 \to K \xrightarrow{g} L \xrightarrow{f} M \to 0$ yields an exact sequence (of R/Ra-modules) $0 \to K/aK \to L/aL \xrightarrow{\bar{f}} M/aM \to 0$. (Hint: $\ker \bar{f} = (K + aL)/aL \approx K/(K \cap aL)$ so it suffices to show $K \cap aL = aK$. But if $ax \in K$ then $0 = f(ax) = afx$.)

18. Second "change of rings" theorem. Hypotheses as in exercise 17 we have $\mathrm{pd}_{R/aR} M/aM \leq \mathrm{pd}_R M$. (Hint: Induction on $n = \mathrm{pd}_R M$. For $n > 0$ take $0 \to K \to F \to M \to 0$ exact, with F free. Then $\mathrm{pd}_R K/aK \leq n - 1$ by induction; conclude using exercise 17.)

19. Third "change of rings" theorem. Hypotheses as in exercise 17, if, moreover, R is left Noetherian, M is f.g. and $a \in \mathrm{Jac}(R)$, then $\mathrm{pd}_{R/aR} M/aM = \mathrm{pd}_R M$. (Hint: Using an induction procedure analogous to that of exercise 18, one is left with the case $\mathrm{pd}_{R/aR} M/aM = 0$. Take $0 \to K \to F \to M \to 0$ with F free. By exercise 17 one has $0 \to K/aK \to F/aF \to M/aM \to 0$ which thus splits. It suffices to show $K \oplus M$ is free; replacing M by $K \oplus M$ assume M/aM is free. But any base $x_1 + aM, \ldots, x_t + aM$ of M/aM lifts to a base of M. Indeed, the x_i span by Nakayama; if $\sum r_i x_i = 0$ take such r_i with Rr_1 maximal and note each $r_i \in Ra = aR$, and writing $r_i = ar_i'$ conclude $r_1 \in Rar_1'$, contrary to $a \in \mathrm{Jac}(R)$.) A more homological approach taken by Strooker [66] is given in Exercise 5.2.24ff.

20. If gl. dim $/R \leq 2$ then R is a PP ring. (Hint: By exercise 3.1.20 it suffices to show Ann a is projective for any a in R. But $0 \to$ Ann $a \to R \to R \to R/Ra \to 0$ is exact where the third arrow is given by right multiplication by a.)

21. Generalizing the argument of exercise 20, show that if gl. dim $R \leq 2$ then every left annihilator of R is projective. This result is used by Faith [73B] to characterize simple Noetherian rings of gl. dim ≤ 2; in fact, Bass [60] characterized Noetherian rings of gl. dim ≤ 2 along these lines.

22. (Zaks [71]) Suppose L is a left ideal containing $P_1 \cdots P_t$ such that each $P_i \lhd R$ is f.g. projective as left module and each R/P_i is simple Artinian. Then L is projective. (Hint: Suppose L is not projective. Let $A = P_1 \cdots P_t$. R/A is Artinian, so one may take L maximal with respect to not projective, and also there is $L < L' \leq R$ with L'/L simple. Hence $\mathrm{Ann}(L'/L)$ is some P_i. But $\mathrm{pd}_R R/P_i \leq 1$ so pd $L'/L \leq 1$. L' is projective; hence L is projective.)

23. (Zaks [71]) Suppose R is left bounded, all proper images are Artinian, and all maximal ideals are projective as left modules. Then R is left hereditary. (Hint: If $L < R$ then exercise 22 shows $L \oplus L'$ is projective where L' is an essential complement of L) This is one of a host of his results about when a ring must be left hereditary.

24. If R is left Noetherian then the direct limit of modules of injective dimension $\leq n$ has injective dimension $\leq n$.

§5.2

1. The *ring of dual numbers* $\tilde{R}$ is defined as $R[\lambda]/\langle \lambda^2 \rangle$ (also, cf., exercise 1.9.7).

Any R-module M can be viewed as $\tilde{R}$-module since R is a homomorphic image of $\tilde{R}$. Thus $Z(\ ) \approx \text{Hom}_{\tilde{R}}(R, \_)$ which also proves left exactness of Z.

Group Cohomology

2. If $t = |G|$ is finite then each $H^n(G, M)$ is t-torsion (notation as in example 5.2.13). (Hint: Let $h(g_1, \ldots, g_{n-1}) = \sum_{g \in G} f(g_1, \ldots, g_{n-1}, g)$. Summing over (2) of example 5.2.13, as g_n runs over g yields $\sum_{g \in G}(d^n f)(g_1, \ldots, g_n) = (-1)^{n+1} t f(g_1, \ldots, g_n) - (d^n h)(g_1, \ldots, g_n)$. If $d^n f = 0$ then $t f \in B^n$ as desired.

3. Let G be a group. Define $P_0 = \mathbb{Z}[G]$ and inductively $P_{n+1} = P_n \otimes \mathbb{Z}[G]$, a free $\mathbb{Z}[G]$-module with base $\{1 \otimes g_1 \otimes \cdots \otimes g_n : g_i \in G\}$. Also define maps $d_n : P_n \to P_{n-1}$ by $d_n(1 \otimes g_1 \otimes \cdots \otimes g_n) = g_1 \otimes g_2 \otimes \cdots \otimes g_n + (-1)^n \otimes g_1 \otimes \cdots \otimes g_{n-1} + \sum_{1 \leq i \leq n-1} (-1)^i \otimes g_1 \otimes \cdots \otimes g_i g_{i+1} \otimes \cdots \otimes g_n$. Then (P_n, d_n) is a projective resolution of $\mathbb{Z}$ as $\mathbb{Z}[G]$-module with the trivial G-action. (Hint: Define $s_{-1} : \mathbb{Z} \to P_0$ by $s_{-1} 1 = 1$ and $s_n : P_n \to P_{n-1}$ by $s_n(g_0 \otimes \cdots \otimes g_n) = 1 \otimes g_0 \otimes \cdots \otimes g_n$, and use proposition 5.2.18.) Conclude that $H^n(G, M) \approx \text{Ext}^n_{\mathbb{Z}[G]}(\mathbb{Z}, M)$ for any $\mathbb{Z}[G]$-module M by looking at the isomorphisms $\text{Hom}(P_n, M) \approx H^n(G, M)$.

4. Prove in detail the assertions made after theorem 5.2.22 concerning derived functors of Abelian categories, to obtain corollary 5.2.23.

Basic Properties of $\mathcal{T}or$ (also, cf., Exercises 23ff.)

5. Use Tor to prove quickly the following facts about an exact sequence $0 \to M' \to M \to M'' \to 0$: (1) If M', M'' are flat then M is flat; (2) If M'' is flat the sequence is pure.

6. Suppose $0 \to K \to F \to N \to 0$ with F flat then $\text{Tor}_n(M, N) \approx \text{Tor}_{n+1}(M, K)$.

7. If M is a $T\text{-}\mathcal{M}od\text{-}R$ bimodule then $\text{Tor}^R_n(M, \_)$ can be viewed as a functor from $R\text{-}\mathcal{M}od$ to $T\text{-}\mathcal{M}od$. Verify similar assertions for $\text{Tor}(\_, M)$ and for Ext.

$\mathcal{T}or$ as Torsion Module

8. Define the *torsion submodule* $tM = \{x \in M : \text{Ann}_R x \text{ contains a regular element}\}$. If R is left Ore then $tM \leq M$ and $t(M/tM) = 0$; we say M is *torsion* if $tM = M$. If R is semiprime Goldie then $tM = \text{Sing}(M)$.

In Exercises 9 through 12 we assume R is left Ore with ring of fractions Q, and M is an arbitrary R-module.

9. $Q \otimes_R tM = 0$. Conclude $\text{Tor}_1(Q/R, tM) \approx tM$. (Hint: Apply $\text{Tor}(\_, tM)$ to $0 \to R \to Q \to Q/R \to 0$, noting Q is R-flat.)

10. $\text{Tor}_n(Q/R, M) = 0$ for all $n \geq 2$. (Hint: As in exercise 9, noting Q and R are flat.)

11. The functors $t(\ )$ and $\text{Tor}_1(Q/R, \_)$ are naturally isomorphic. (Hint: Apply $\text{Tor}(Q/R, \_)$ to the exact sequence $0 \to tM \xrightarrow{i} M \to M/tM \to 0$ to get $0 \to \text{Tor}_1(Q/R, tM) \xrightarrow{i_*} \text{Tor}_1(Q/R, M) \to 0$ by exercises 9, 10. Conclude by composing i_* with the isomorphism $tM \to \text{Tor}_1(Q/R, tM)$ of exercise 9.)

12. If R is an integral domain then $\text{Tor}_n(M, N)$ is torsion for all R-modules M, N

and for all $n \geq 1$, and $\text{Tor}_0(M, N)$ also is torsion if N is torsion. (Hint: First assume N is torsion. Then $\text{Tor}_0(M, N) \approx M \otimes N$ is torsion. For $n \geq 1$ apply $\text{Tor}(\_\_, N)$ to a sequence $0 \to K \to F \to M \to 0$ with F free and thus flat. Finally for arbitrary N apply $\text{Tor}(M, \_\_)$ to $0 \to tN \to N \to N/tN \to 0$ to get $\text{Tor}_n(M, tN) \xrightarrow{f} \text{Tor}_n(M, N) \xrightarrow{g} \text{Tor}_n(M, N/tN)$ exact, and note $\ker f$ and $\text{im}\, g$ are torsion.

Weak Dimension and Gl. Dim

13. In analogy to the syzygy of a projective resolution define the n-th *yoke* of a flat resolution of M to be $\ker d_n$ and prove the following are equivalent: (i) M has a flat resolution of length n. (ii) $\text{Tor}_k(\_\_, M) = 0$ for all $k \geq n + 1$. (iii) $\text{Tor}_{n+1}(\_\_, M) = 0$. (iv) The $(n - 1)$ yoke of any flat resolution of M is flat.

14. Show w. dim R is the same as the right-handed version, by applying theorem 5.2.37 to the right-handed version of exercise 13. In this way w. dim is "better" than gl. dim, which is not left-right symmetric.

15. If R is left Noetherian then pd M = the flat dimension of M, for any f.g. module M. (Hint: Suppose M has flat dimension n. We need to find a projective resolution of length n. Take any f.g. projective resolution. This is also a flat resolution so the $(n - 1)$ syzygy K is flat. But K is f.g. and thus finitely presented, and so K is projective by exercise 2.11.8.)

16. If R is left Noetherian then gl. dim R = w. dim R. (Hint: By theorem 5.2.42 one need only check the f.g. modules, but this is exercise 15.) Consequently, gl. dim R is left-right symmetric for Noetherian rings. For non-Noetherian rings one can sometimes find a bound for the difference of the left and right gl. dim, as shown in Osofsky [73B, p. 57].

17. (i) A module F is flat iff $\text{Tor}_1(R/I, F) = 0$ for every right ideal I. (Hint: Use $0 \to I \to R \to R/I \to 0$.) (ii) w. dim($R$) is the supremum of the flat dimensions of the cyclic modules. (Hint: Dual of Auslander's theorem.)

18. The following are equivalent for a ring R: (i) Every left ideal is flat. (ii) w. dim $R \leq 1$. (iii) Every right ideal is flat. Consequently, if R is left or right semihereditary then w. dim(R) ≤ 1 and every submodule of a flat module is flat.

Global Dimension for Semiprimary and/or Hereditary Rings

19. (Eilenberg-Nagao-Nakayama [56]) Suppose R is hereditary and $A \lhd R$ such that $A^n = A^{n+1}$. Then gl. dim $R/A \leq 2n - 1$. (Hint: Let $B/A \lhd R/A$. The series $B \geq A \geq AB \geq A^2 \geq A^2B \geq \cdots$ yields an exact sequence

$$\cdots \to AB/A^2B \to A/A^2 \to B/AB \to B/A \to 0.$$

Each term has the form P/AP where $P \lhd R$ and is thus projective, so P/AP is projective as R/A-module. Since $A^n/A^{n+1} = 0$ conclude pd(B/A) $\leq 2n - 1$ as desired.) Using exercise 2.7.17 conclude every homomorphic image of a hereditary semiprimary ring has finite global dimension, bounded by twice the index of nilpotence of the radical.

22. (Auslander [55]) If R is semiprimary and not semisimple Artinian then gl. dim $R =$ $\mathrm{pd}_R R/J = \mathrm{pd}_R J + 1$ where $J = \mathrm{Jac}(R)$. (Hint: Let $n = \mathrm{pd}_R R/J$ and suppose $M \in R\text{-}\mathcal{M}od$. Every simple R-module is a summand of R/J so $\mathrm{Ext}_R^n(N, M) = 0$ for all simple N, implying $\mathrm{Ext}_R^n(N, M) = 0$ for all N satisfying $JN = 0$. Thus $\mathrm{Ext}_R^n = 0$ by exercise 2.11.1 proving the first equality.) Auslander also showed this is the weak global dimension of R, thereby achieving left-right symmetry.

27. Suppose R is left Noetherian and a is a regular normalizing element of R with $a \in \mathrm{Jac}(R)$. If gl. dim $R/Ra = n < \infty$ then gl. dim $R = n + 1$. (Hint: It suffices to show pd $M \le n + 1$ for every f.g. module M. Take $0 \to K \to F \to M \to 0$ with F f.g. free. Then $\mathrm{pd}_R K = \mathrm{pd}_{R/aR} K/aK \le n$ by exercise 5.1.19, so $\mathrm{pd}_R M \le n + 1$.)

§5.3

1. If $\partial \in \mathrm{Der}(M)$ and $f: M \to N$ is a map in $R^e\text{-}\mathcal{M}od$ then $f\partial \in \mathrm{Der}(N)$, so we have a commutative diagram each of whose vertical arrows arises from f:

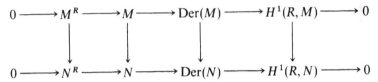

2. A map $f: J \to M$ extends to a map $R^e \to M$ iff $f\partial$ is an inner derivation of R in M).

3. If G is a group of order n and $1/n \in C$ then the group algebra $C[G]$ is separable over C. (Hint: $\sum_{\sigma \in G}(1/n)\sigma \otimes \sigma^{-1}$ is a separability idempotent.)

6 Rings with Polynomial Identities and Affine Algebras

In this chapter we shall treat two theories each of which in recent years has served as a way of attracting directly some of the techniques of commutative algebras to study suitably conditioned noncommutative algebras. Both theories have their roots in specific questions outside of "pure algebra" but have blossomed spectacularly in recent years. The two subjects complement each other beautifully and often intertwine both in hypotheses and in mathematical history, so we shall treat them together.

§6.1 is a resume of "elementary" theory of *rings with polynomial identity* (PI), focusing on central polynomials and their most direct applications. In §6.2 we introduce *affine algebras* and the Gelfand-Kirillov dimension, which perhaps has become the most prominent dimension of recent research; the springboard is the Golod-Shafarevich counterexample to Kurosch's problem. In §6.3 we study the structure of affine PI-algebras, including the positive answer to Kurosch's problem using Shirshov's techniques; one of the key features is Schelter's important "trace ring" construction which has been the arena of many significant recent advances.

§6.1 Rings with Polynomial Identities

Suppose one were to look for the most naive way possible to generalize the property of commutativity of rings. One might begin by observing that the commutative law for R reads "$X_1 X_2 = X_2 X_1$ for all substitutions of X_1, X_2 in R," or, equivalently, $X_1 X_2 - X_2 X_1$ is "identically" 0 in R. Similarly, we could view the associative law as $(X_1 X_2)X_3 - X_1(X_2 X_3)$ vanishing identically, and likewise for the distributive law. Thus it is natural to examine formal polynomials, say as element of $\mathbb{Z}\{X\}$, and see what happens when some polynomial vanishes for all substitutions for R. Some immediate examples:

Example 6.1.1:

(i) Any Boolean ring satisfies the identity $X_1^2 - X_1$.

(ii) Any finite field having m elements satisfies the identity $X_1^m - X_1$ by Fermat's little theorem.

(iii) The algebra R of upper triangular $n \times n$ matrices over a commutative ring satisfies the identity $[X_1, X_2][X_3, X_4] \cdots [X_{2n-1} X_{2n}]$.

These examples are not nearly as important as two deeper examples, finite dimensional algebras and algebraic algebras of bounded degree, which will be discussed a bit later. There is another obvious example—any algebra of characteristic p satisfies the identity pX_1. In order to exclude such a huge class of algebras, we want to dismiss such an identity as "trivial." We are ready for a formal definition. Throughout $X = \{X_1, X_2, \ldots\}$ is a countable set of non-commuting indeterminates over a commutative ring C, and a *polynomial* is an element f of $C\{X\}$, written $f(X_1, \ldots, X_m)$ to indicate that no X_i occurs in f for $i > m$. But now f can be viewed as a function $f: R^{(m)} \to R$ acting by substitution, i.e., $f(R^{(m)}) = \{f(r_1, \ldots, r_m): r_i \in R\}$.

When m is not of concern to us we shall write $f(R)$ in place of $f(R^{(m)})$. It is also convenient to deal with $f(R)^+$, defined as the additive subgroup of R generated by $f(R)$. Obviously, $f(R) = 0$ iff $f(R)^+ = 0$.

Remark 6.1.1': If f is a polynomial and σ is a C-algebra automorphism of R then $f(\sigma r_1, \ldots, \sigma r_m) = \sigma f(r_1, \ldots, r_m)$; consequently, $f(R)$ and $f(R)^+$ are invariant under all automorphisms of R.

Definition 6.1.2: f is an *identity* of R if $f(R) = 0$; an identity f is a *polynomial identity* if one of the monomials of f of highest (total) degree has coefficient 1. R is a PI-*algebra* if R satisfies a polynomial identity; R is a PI-*ring* if R satisfies a polynomial identity with $C = \mathbb{Z}$.

The reason for the somewhat technical definition of polynomial identity is to exclude identities such as $[X_1, X_2] + 2[X_1^2, X_2]$, since one usually wants to work with the monomials of highest degree.

The subject of most of this chapter is the "qualitative" PI-theory, i.e., the structure theory that can be developed for a ring on the mere assumption that it satisfies a polynomial identity. PI-rings were introduced by Dehn [22] who was searching for an algebraic framework for axiomatic Desargiuan geometry. He was missing a key ingredient, the use of inverses, and his project receded into oblivion until Amitsur [66] brought it to a successful completion. With a few exceptions mostly of historical interest, PI-rings remained in a state of suspended animation until Jacobson, Kaplansky, and Levitzki became interested in them, largely in connection with Kurosch's problem (to be discussed in §6.2). When Kaplansky proved every primitive PI-algebra is simple and finite dimensional over its center, the PI-theory began to breathe anew, especially in the hands of Amitsur. Another breakthrough came when Posner [60] proved every prime PI-ring is Goldie, yielding an instance where Goldie's theory could be applied to non-Noetherian rings. Deep applications were found to Azumaya algebras, geometry, and division algebras. But the best was yet to come. In response to an old problem of Kaplansky, Formanek [72] and Razmyslov [73] constructed polynomials on arbitrary $n \times n$ matrices which are not identities yet take only scalar values. Although the question had at first been considered mostly as a curiosity, it soon became clear that these "central polynomials" provided enough commutativity to permit direct application of standard techniques from the commutative theory, and the entire PI-theory was reworked from the bottom up to give central polynomials their proper dominant role.

During the 1970s researchers began to turn to basic matrix theory techniques to study PI-rings and expanded the rings under consideration by adjoining coefficients of the characteristic polynomials of various elements. On the one hand, this tied PI-theory to classical invariant theory, and brought in the "quantitative" PI-theory; on the other hand, Schelter and others used this technique to prove striking new results about prime PI-rings and later to develop the rudiments of a noncommutative algebraic geometry, cf, the papers of Artin-Schelter.

This section concerns that part of the PI structure theory which is obtained as a direct consequence of the existence of central polynomials for matrices. In the later sections we shall obtain the other basic PI-structure theorems, concentrating on the deeper theory of prime PI-rings. As hinted in the previous paragraphs, the structure theory hinges on what can be said about matrices, in particular on the two most basic properties of matrices—the determinant and the Hamilton-Cayley theorem. The determinant is linear and alternating in its rows; to study the analogous situation for polynomials we introduce the notation $f(X_i \mapsto h)$ to denote that h is substituted throughout for X_i: an alternate longer notation is $f(X_1, \ldots, X_{i-1}, h, X_{i+1}, \ldots, X_m)$.

Definition 6.1.3: A polynomial f is *linear in* X_i if X_i occurs exactly once (of degree 1) in every monomial of f; f is *t-linear* if f is linear in $X_1, \ldots, X_t$. f is *t-alternating* if $f(X_i \mapsto X_j) = 0$ for all $1 \le i < j \le t$. A polynomial which is t-linear and t-alternating is called *t-normal*.

For example $[X_1, X_2] = X_1 X_2 - X_2 X_1$ is 2-normal. The value of t-linear polynomials lies in the following observation:

Remark 6.1.4: If $f(X_1, \ldots, X_m)$ is t-linear then

$$f\left(\sum c_{i1} r_{i1}, \ldots, \sum c_{it} r_{it}, r_{t+1}, \ldots, r_m\right)$$

$$= \sum_{i_1, \ldots, i_t} c_{i_1 1} \cdots c_{i_t t} f(r_{i_1 1}, \ldots, r_{i_t t}, r_{t+1}, \ldots, r_m)$$

for all c_{ij} in C and r_{ij} in R. In particular, if B spans R as C-module then $f(R^{(m)}) = f(B^{(t)} \times R^{(m-t)})$. Likewise, if f is 1-linear then $f(R)^+$ is a C-submodule of R.

Example 6.1.5: The 3-linear polynomial $[[X_1, X_2], X_3]$ is an identity of each exterior algebra, seen by checking monomials.

Proposition 6.1.6: *If* $f(X_1, \ldots, X_m)$ *is t-normal and R is spanned by fewer than t elements over C then f is an identity of R.*

Proof: Suppose $R = CB$ where $|B| \le t - 1$. Then $f(b_1, \ldots, b_t, r_{t+1}, \ldots, r_m) = 0$ by definition whenever the $b_i \in B$, since two b_i must be the same. Consequently $f(B^{(t)} \times R^{(m-t)}) = 0$, so f is an identity by remark 6.1.4.

It is a simple enough matter to construct normal polynomials.

Definition 6.1.7: Define the *Capelli polynomial*

$$C_{2t}(X_1, \ldots, X_{2t}) = \sum_{\pi \in \text{Sym}(t)} (\text{sg } \pi) X_{\pi 1} X_{t+1} X_{\pi 2} X_{t+2} \cdots X_{\pi(t-1)} X_{2t-1} X_{\pi t} X_{2t}$$

and the *standard polynomial*

$$S_t(X_1, \ldots, X_t) = C_{2t}(X_1, \ldots, X_t, 1, \ldots, 1) = \sum_{\pi \in \text{Sym}(t)} (\text{sg } \pi) X_{\pi 1} \cdots X_{\pi t}$$

Proposition 6.1.8: C_{2t} and S_t are t-normal and thus are identities of any C-algebra spanned by $<t$ elements.

Proof: The terms in $C_{2t}(X_i \mapsto X_j)$ (or $S_t(X_i \mapsto X_j)$) subdivide into pairs of the same monomials appearing with opposite sign (one corresponding to a permutation π and the other corresponding to π composed with the transposition (ij) and thus having opposite sign), so the polynomial is sent to 0.   Q.E.D.

For our final example we say R is *integral of bounded degree n* if every element of R is integral of degree $\leq n$ over C, with n minimal such.

Example 6.1.9:   If R is integral of bounded degree $\leq n$ and $f(X_1,\ldots,X_m)$ is n-normal then $S_n([X_1^n, X_2],\ldots,[X_1, X_2])$ is a 2-variable identify of R.

Indeed, for any r_i in R we can write $r_1^n = \sum_{i=0}^{n-1} c_i r_1^i$ so $[r_1^n, r_2] = \sum_{i=1}^{n-1} c_i[r_1^i, r_2]$ implying $S_n([r_1^n, r_2], \ldots, [r_1, r_2]) = 0$ by remark 6.1.4, so f is an identity.

As a special case, consider $M_n(C)$ over an arbitrary commutative ring C. On the one hand, taking $t = n^2 + 1$ we see S_t and C_{2t} are identiities of $M_n(C)$ by proposition 6.1.8. On the other hand, by the Hamilton-Cayley theorem (which holds for arbitrary commutative rings, cf., Jacobson [85B, page 203]) we see $M_n(C)$ is integral of bounded degree n so $S_n([X_1^n, X_2], \ldots, [X_1, X_2])$ is an identity of $M_n(C)$.

Given these basic examples of PI-rings, we want to see how broad the class of PI-rings really is.

Remark 6.1.10:

(i) If f is an identity of R then f is also an identity of every homomorphic image of R and of every subalgebra of R.

(ii) If f is an identity of R_i for each i in I then f is an identity of $\prod\{R_i : i \in I\}$, and thus of any subdirect product of the R_i. (Indeed (i) is *a fortiori* and (ii) is by checking components and then applying (i).)

A polynomial $f(X_1, \ldots, X_m)$ is *multilinear* if X_i has degree 1 in each monomial of f, for each $1 \leq i \leq m$. Multilinear identities play a special role because of the next observation (compare with theorem 6.1.44).

Proposition 6.1.11: *Any multilinear identity f of R is an identity of each central extention R' of R.*

Proof: If $R' = RZ$ then remark 6.1.4 shows $f(R') = f(R)Z = 0$.   Q.E.D.

To utilize this result we need a method of *multilinearizing* polynomial identities. Given $f(X_1, \ldots, X_m)$ define $\Delta_i f(X_1, \ldots, X_{m+1})$ for $1 \leq i \leq n$ by

$$\Delta_i f = f(X_i \mapsto (X_i + X_{m+1})) - f - f(X_i \mapsto X_{m+1}).$$

For example, if $f = X_1^2 X_2$ then $\Delta f = (X_1 + X_3)^2 X_2 - X_1^2 X_2 - X_3^2 X_2 = X_1 X_3 X_2 + X_3 X_1 X_2$ and $\Delta_2 f = X_1^2(X_2 + X_3) - X_1^2 X_2 - X_1^2 X_3 = 0$.

The multilinearization process only works on the assumption that any X_i appearing in f appears in *all* the monomials of f; such a polynomial will be called "blended." All the polynomials we wrote down thus far are blended, and exercise 1 shows any identity is a sum of blended identities, so this assumption is very mild.

Remark 6.1.12: Suppose $f(X_1, \ldots, X_m)$ is blended. Then $\Delta_i f$ is blended (in $X_1, \ldots, X_{m+1}$). Furthermore, any monomial h in f of degree $d \geq 1$ in X_i produces $2^d - 2$ monomials in $\Delta_i f$, each having the same coefficient as h and of degree $< d$ in X_i. If f is an identity then $\Delta_i f$ is identity. (Indeed, the last assertion is obvious; furthermore, $h(X_i \mapsto X_i + X_{m+i})$ produces 2^d monomials, all of which are blended except h and $h(X_i \mapsto X_{m+i})$, and these two drop out by definition of Δ_i.)

Remark 6.1.13: Using remark 6.1.12 one can turn any blended polynomial identity into a multilinear polynomial identity of the same degree. (Indeed, take a monomial h of highest (total) degree having coefficient 1. If h is linear in each X_i then f is multilinear since f is blended. If h is not linear in X_i then $\Delta_i h$ is blended and has smaller degree in X_i; repeating the procedure we eventually reach a multilinear identity. Note that all monomials of smaller degree than h have vanished in the multilinearization procedure.)

To see how this works let us consider the identities of example 6.1.1.

(i) $\Delta_1(X_1^2 - X_1) = \Delta_1(X_1^2) = X_1 X_2 + X_2 X_1$ which is thus an identity of every Boolean algebra.

(ii) Applying Δ_1 $(m - 1)$ times to X_1^m yields $\sum_{\pi \in \text{Sym}(m)} X_{\pi 1} \cdots X_{\pi m}$, which is a multilinear identity of any field of m elements and is called the *symmetric identity*.

(iii) $[X_1 X_2] \cdots [X_{2n-1}, X_{2n}]$ is already multilinear.

Central Polynomials and Identities of Matrices

We want now to examine identities of matrices more carefully, for this will pay handsome dividends in structure theoretical applications. If x_1 and x_2 are in $M_2(C)$ then $[x_1, x_2]$ has trace 0 so $[x_1, x_2]^2$ is a scalar matrix by the Hamilton-Cayley theorem. This led Kaplansky to ask whether arbitrary $n \times n$ matrix rings have nonidentities taking on scalar values. We shall see soon the answer is "yes." (Also see exercise 22.)

Definition 6.1.14: $f(X_1, \ldots, X_m)$ is a *central polynomial* for R, otherwise called *R-central*, if $0 \neq f(R) \subseteq Z(R)$, i.e., if f is *not* an identity of R but $[X_{m+1}, f]$ is an identity of R.

Example 6.1.15: $[X_1, X_2]^2$ is $M_2(C)$-central for any commutative ring C, as we just saw. $[X_1, X_2]$ is a central polynomial for the exterior algebra, in view of example 6.1.5.

Our aim is to construct multilinear $M_n(C)$-central polynomials. We follow the exposition of Rowen [80B] (based on Razmyslov-Bergman-Amitsur) since the ideas also apply to §6.3. The study of identities and central polynomials of matrices are closely related. Our first concern is to distinguish $M_n(C)$ from $M_{n-1}(C)$ by means of their identities.

Proposition 6.1.16: For $t = n^2$, the Capelli polynomial C_{2t} is not an identity of $R = M_n(H)$ for any commutative ring H (although $C_{2(t+1)}$ is an identity, as noted earlier). In fact $C_{2t}(R)^+ = R$.

Proof: Order the matric units $\{e_{ij}: 1 \leq i, j \leq n\}$ lexicographically on the subscripts, i.e., $e_{11} < e_{12} < \cdots < e_{1n} < e_{21} < \cdots < e_{nn}$, and write r_k for the k-th matric unit on this list. Let us evaluate $C_{2t}(r_1, \ldots, r_t, r_1, \ldots, r_t)$. Taking π in $\mathrm{Sym}(t)$ let $a_\pi = r_{\pi 1} r_1 r_{\pi 2} r_2 \cdots r_{\pi t} r_t$. Then $r_1 r_{\pi 2} r_2 = e_{11} r_{\pi 2} e_{12}$ so $a = 0$ unless $r_{\pi 2} = e_{11}$; likewise, $r_2 r_{\pi 3} r_3 = e_{12} r_{\pi 3} e_{13}$ is 0 unless $r_{\pi 3} = e_{21}$. Continuing in this way we have precisely one choice of $r_{\pi 2}, \ldots, r_{\pi t}$ for a to be nonzero. Since e_{n1} has not yet been selected we take $r_{\pi 1} = e_{n1}$; then $a_\pi = e_{nn}$ for this particular π, and all other $a_\pi = 0$; proving $C_{2t}(r_1, \ldots, r_t, r_1, \ldots, r_t) = \pm e_{nn}$. By symmetry each $e_{ii} \in C_{2t}(R)$. For $i \neq j$ we have $(1 + e_{ij})^{-1} = 1 - e_{ij}$ so $(1 + e_{ij})^{-1} e_{ii}(1 + e_{ij}) = e_{ii} + e_{ij} \in C_{2t}(R)$ by remark 6.1.1'. Hence each $e_{ij} \in C_{2t}(R)^+$, proving $C_{2t}(R)^+ = R$.   Q.E.D.

Remark 6.1.16': For any field F, elements $\{b_i: 1 \leq i \leq n^2\}$ of $M_n(F)$ form a base iff there are $\{r_i: 1 \leq i \leq n^2\} \subset M_n(F)$ such that $C_{2t}(b_1,\ldots,b_t,r_1,\ldots,$ $r_t) \neq 0$ where $t = n^2$. (Indeed by remark 6.1.4 we may assume the b_i are matric units, so apply the proposition.)

Digression 6.1.17: The standard polynomial S_{2n} actually is an identity of $M_n(C)$ for any n. This celebrated theorem of Amitsur-Levitzki is proved in exercise 10 as a fairly direct consequence of the Hamilton-Cayley theorem, translated to traces by means of "Newton's formulas." Actually Razmyslov [74] has proved that *all* identities of $n \times n$ matrices can be obtained from the Hamilton-Cayley theorem by using traces formally in PI-theory. Nevertheless, S_{2n} has the smallest degree of all the identities of $M_n(C)$, as we see now.

Remark 6.1.18: No polynomial $f \neq 0$ of degree $\leq 2n - 1$ is an identity of $M_n(C)$. (Indeed, by multilinearizing we could assume $f = \sum \alpha_\pi X_{\pi 1} \cdots X_{\pi m}$ with $\alpha_{(1)} \neq 0$; then $f(e_{11}, e_{12}, e_{22}, \ldots) = \alpha_{(1)} e_{11} e_{12} \cdots \neq 0$.)

One might have hoped $S_n([X_1^n, X_2], \ldots, [X_1, X_2])$ to be the minimal 2-variable identity, but for $n \geq 8$ the polynomial identity $S_{2n}(X_1, X_2, X_1X_2,$ $X_2X_1, X_1^2X_2, \ldots)$ has lower degree.

Our key to finding central polynomials for $M_n(C)$ is the following lemma which mimics the known properties of determinants.

Proposition 6.1.19: *Suppose* $R \in C\text{-}\mathscr{A}lg$, $x_1, \ldots, x_t$ *are arbitrary elements of* R, *and* $T: \sum_{i=1}^t Cx_i \to \sum_{i=1}^t Cx_i$ *is a map (in* $C\text{-}\mathscr{M}od$). *Viewing* T *as an image of a* $t \times t$ *matrix* T' (*cf., theorem 1.5.13*) *we have the following formulas, for any* t-*normal polynomial* $f(X_1, \ldots, X_k)$:

(i) $(\det T')f(x_1, \ldots, x_k) = f(Tx_1, \ldots, Tx_t, x_{t+1}, \ldots, x_k)$,

(ii) $\det(\lambda - T')f(x_1, \ldots, x_k) = f(\lambda x_1 - Tx_1, \ldots, \lambda x_t - Tx_t, x_{t+1}, \ldots, x_k)$

for a commuting indeterminate λ (*viewed as a scalar* $t \times t$ *matrix*)

(iii) *If* $\lambda^t + \sum_{i=1}^t (-1)^i c_i \lambda^{t-i}$ *is the characteristic polynomial of* T' *then each*

$$c_i f(x_1, \ldots, x_k) = \sum f(T^{j_1}x_1, \ldots, T^{j_t}x_t, x_{t+1}, \ldots, x_k)$$

summed over $(j_1, \ldots, j_t)$ *in* $\{0, 1\}^{(t)}$ *such that* $j_1 + \cdots + j_t = i$;

(iv) $(\operatorname{tr} T')f(x_1, \ldots, x_k) = \sum_{i=1}^t f(x_1, \ldots, x_{i-1}, Tx_i, x_{i+1}, \ldots, x_k)$.

Proof:

(i) Write $Tx_j = \sum_{i=1}^{t} c_{ij}x_i$. Using the t normality of f we have

$$f(Tx_1,\ldots, Tx_t, x_{t+1},\ldots, x_k) = f\left(\sum_{i=1}^{t} c_{i1}x_i,\ldots; \sum_{i=1}^{t} c_{it}x_i, x_{t+1},\ldots, x_k\right)$$

$$= \sum_{\pi \in \text{Sym}(t)} c_{\pi 1,1} \cdots c_{\pi t,t} f(x_{\pi 1},\ldots, x_{\pi t}, x_{t+1},\ldots, x_k)$$

$$= \sum_{\pi} (\text{sg } \pi) c_{\pi 1,1} \cdots c_{\pi t,t} f(x_1,\ldots, x_k) = (\det T') f(x_1,\ldots, x_k)$$

(ii) Follows from (i), working in $R[\lambda]$ and using $\lambda 1 - T$ in place of T (where 1 is the identity map).

(iii) Match coefficients of λ^{t-i} in (ii).

(iv) Take $i = 1$ in (iii).   Q.E.D.

Theorem 6.1.20:   *There is a multilinear polynomial which is $M_n(H)$-central for every commutative ring H.*

Proof: Put $t = n^2$, and write

$$\sum_{i=1}^{t} C_{2t}(X_1,\ldots, X_{i-1}, X_{2t+1}X_iX_{2t+2}, X_{i+1},\ldots, X_{2t})$$

$$= \sum_{j=1}^{m} \sum_{i=1}^{t} h_{ij1}X_{2t+1}X_iX_{2t+2}h_{ij2}$$

for suitable m and multilinear monomials h_{ij1}, h_{ij2} (in $X_1,\ldots, X_{i-1}, X_{i+1},\ldots,$ X_{2t}). Pick arbitrary $a,b,r_1,\ldots, r_{2t}$ in $M_n(H)$; viewing $M_n(H)$ as an n^2-dimensional module over H having base e_{ij}, $1 \le i,j \le n$, we define the map $T: M_n(H) \to M_n(H)$ given by $Tx = axb$. Writing $a = (a_{ij})$ and $b = (b_{ij})$ we have $Te_{ij} = \sum_{u,v=1}^{n} a_{ui}b_{jv}e_{uv}$, whose coefficient of e_{ij} is $a_{ii}b_{jj}$. Hence as $n^2 \times n^2$ matrix, T has trace $\sum_{u,v=1}^{n} a_{ii}b_{jj} = \text{tr}(a)\text{tr}(b)$ so by proposition 6.1.19(iv), putting $w_{iju} = h_{iju}(r_1,\ldots, r_{i-1}, r_{i+1},\ldots, r_{2t})$ we have

$$\text{tr}(a)\text{tr}(b)C_{2t}(r_1,\ldots, r_{2t}) = \sum_{i=1}^{t} C_{2t}(r_1,\ldots, r_{i-1}, ar_ib, r_{i+1},\ldots, r_{2t})$$

$$= \sum_{j=1}^{m} \sum_{i=1}^{t} w_{ij1}ar_ibw_{ij2}.$$

Put $w = C_{2t}(r_1,\ldots, r_{2t})$. Taking traces of both sides yields

$$\text{tr}(\text{tr}(a)\text{tr}(w)b) = \text{tr}(a)\text{tr}(w)\text{tr}(b) = \text{tr}(\text{tr}(a)\text{tr}(b)w) = \sum_{i,j} \text{tr}(w_{ij1}ar_ibw_{ij2})$$

$$= \sum_{i,j} \text{tr}(w_{ij2}w_{ij1}ar_ib) = \text{tr}\left(\sum_{i,j} w_{ij2}w_{ij1}ar_i\ b\right),$$

implying $\mathrm{tr}((\mathrm{tr}(a)\,\mathrm{tr}(w) - \sum_{i,j} w_{ij2} w_{ij1} a r_i)b) = 0$ for all b; nondegeneracy of the trace yields $\mathrm{tr}(a)\,\mathrm{tr}(w) - \sum_{i,j} w_{ij2} w_{ij1} a r_i = 0$, i.e., $\sum w_{ij2} w_{ij1} a r_i$ equals the scalar $\mathrm{tr}(a)\,\mathrm{tr}(w)$. By proposition 6.1.16 we can pick $r_1,\ldots,r_{2t}$ such that $\mathrm{tr}(a)\,\mathrm{tr}(w) \neq 0$. Thus we define $f = \sum_{j=1}^{m} \sum_{i=1}^{t} h_{ij2} h_{ij1} X_{2t+1} X_i$, which takes only scalar values and is not an identity, i.e., f is $M_n(H)$-central.   Q.E.D.

Actually we want an n^2-normal central polynomial, but this is rather easy to find.

Definition 6.1.21: $g_n = f(C_{2n^2}(X_1,\ldots,X_{2n^2})X_{2n^2+1},\ X_{2n^2+2},\ldots,\ X_{2n^2+2t+1})$ where $f(X_1,\ldots,X_{2t+1})$ is the central polynomial for $n \times n$ matrices we found in the theorem.

Corollary 6.1.22: g_n is an n^2-normal polynomial which is $M_n(C)$-central for all commutative rings C.

Proof: Clearly, g_n is n^2-normal and is either $M_n(C)$-central or an identity. But f is not an identity, so g_n is not an identity, by proposition 6.1.17.   Q.E.D.

We carry the notation g_n throughout this chapter and shall denote its degree as m.

Digression 6.1.23:   We made little effort to limit the degree, and indeed there are multilinear polynomials of considerably lower degree. Formanek's central polynomials have degree n^2, but this is not the best possible. Also Formanek [86] verified a central polynomial of degree $2n^2$ (conjectured by Regev) which is of considerable importance in the quantitative theory, cf., §6.4.

Halpin [83] has simplified Razmyslov's original construction of a central polynomial. Let us outline the main ideas, which are of considerable interest. We say f is a *weak identity* of $M_n(C)$ if f vanishes on all matrices of trace 0. The connection between identities and weak identities is found in the *Razmyslov transformations* T_k and T'_k defined on multilinear polynomials as follows: It is enough to define them for multilinear monomials $h = h_1 X_k h_2$; define $T_k h = h_2 X_k h_1$ and $T'_k h = h_2 h_1 = T_k h(X_k \mapsto 1)$. One verifies at once

(i) $T_k^2 f = f$.

(ii) $T_k f = (T'_k(X_{m+1} f))(X_{m+1} \mapsto X_k)$.

(iii) $T_k[X_{m+1}, f] = (T_k f)(X_k \mapsto [X_k, X_{m+1}])$.

By a careful accounting of the diagonal contributions, one can show if a multilinear polynomial f is a weak identity of $M_n(C)$ then $T'_k f$ is scalar-valued for all substitutions of matrices of trace 0.

Applying (ii), one sees f is an identity iff $T_k f$ is an identity. Using commutators, it is now easy to show that any weak identity of degree d which is *not* an identity produces a central polynomial of degree $\leq 2d$. The obvious such weak identity is the Capelli polynomial C_{2n^2}. However, noting the trace is the second coefficient of the Hamilton-Cayley polynomial, one sees the multilinearization of

$$S_{n-1}([X_1^n, X_2], [X_1^{n-2}, X_2], [X_1^{n-3}, X_2], \ldots, [X_1, X_2])$$

is a weak identity which is not an identity, cf., exercise 11.

Here is another result used in verifying identities and central polynomials.

Remark 6.1.23′: Suppose R is central simple over an infinite field F. To prove a continuous map $f: R^{(m)} \to T$ satisfies $f(r, r_2, \ldots, r_m) = 0$ for all r in R (where $r_2, \ldots, r_m$ are fixed) it suffices to show this holds for a Zariski open subset. Using proposition 2.3.35 we may assume $\deg r = n$ and the roots of the minimal polynomial are distinct. Thus r is diagonalizable in $M_n(\bar{F}) \approx R \otimes \bar{F}$, where $\bar{F}$ is the algebraic closure of F. Suppose $f(ara^{-1}, ar_2 a^{-1}, \ldots, ar_m a^{-1}) = af(r, r_2, \ldots, r_m)a^{-1}$ (e.g., f is a polynomial or a trace polynomial). Diagonalizing r, it *suffices to assume r is diagonal with distinct eigenvalues* (in $M_n(\bar{F})$).

Structure Theory for PI-Rings

We can now build a structure theory for semiprime PI-rings. The cornerstone is Kaplansky's theorem, that any primitive PI-ring R is simple and finite dimensional over the center. The proof is twofold. First one notes R is simple Artinian by Jacobson's density theorem, and then one shows the underlying division ring is finite dimensional by means of some splitting technique. The splitting technique used here is a special case of example 2.12.28, spelled out for the reader's convenience.

Proposition 6.1.24: *Suppose D is a division algebra over an algebraically closed field F and D has a (possibly infinite) base over F of cardinality $<$ $|F| - 1$. Then $D = F$. (More generally, if F is not algebraically closed then D is algebraic over F.)*

Proof: If $d \in D$ were not algebraic over F then $\{(d - \alpha)^{-1}: \alpha \in F\}$ is F-independent by proposition 2.5.21, contrary to hypothesis. Thus $F[d]$ is a finite field extension of F, so $d \in F$, i.e., $D = F$. Q.E.D.

We say R has PI-*degree* n if R satisfies all multilinear identities of $M_n(\mathbb{Z})$ and g_n (of corollary 6.1.22) is R-central. This was called PI-*class* n in Rowen [80B] to avoid confusion wih the degree of a minimal polynomial identity, which is $2n$ by the Amitsur-Levitzki theorem, but the terminology PI-degree has become standard.

Theorem 6.1.25: (*Kaplansky's Theorem*) *Suppose R is a primitive ring satisfying a polynomial identity f of degree d. Then R has some PI-degree $n \leq [d/2]$, and $R \approx M_t(D)$ for a division ring D (unique up to isomorphism) with $n^2 = [R: Z(R)] = t^2[D: Z(D)]$.*

Proof: We may assume f is multilinear and $X_d \cdots X_1$ has nonzero coefficient α in f. Let M be a faithful simple R-module, and $D = \operatorname{End}_R M$. We claim $R \approx M_t(D)$ for some $t \leq d$. Otherwise, taking any x_1 in M we take r_1 such that $r_1 x_1 \notin x_1 D$, and put $x_2 = r_1 x_1$; inductively, given $r_1, \ldots, r_{i-1}$ and $x_1, \ldots, x_{i-1}$, take r_i such that $r_i x_j = 0$ for all $j < i$ and $r_i x_i \notin \sum_{j < 1} x_i D$, and put $x_i = r_i x_i$. Then $f(r_1, \ldots, r_d)x_1 = \alpha r_d \cdots r_1 x_1 = \alpha x_d \neq 0$.

Let $F = Z(D) = Z(R)$ and take an algebraically closed field K of cardinality $> 1 + [R:F]$ (possibly infinite). Then $R_1 = R \otimes_F K$ is a simple K-algebra and satisfies the identity by proposition 6.1.11, so as above $R_1 \approx M_n(D_1)$ for some n and some K-division algebra D_1 with $[D_1:K] \leq [R_1:K] = [R:F] < |K| - 1$. Hence $D_1 = K$ by proposition 6.1.24 so $R_1 \approx M_n(K)$ and $n \leq [d/2]$ by remark 6.1.18. But then $n = [R:F] = t^2[D:F]$, and clearly R has PI-degree n.   Q.E.D.

To utilize Kaplansky's theorem most effectively we need an embedding procedure into nicer rings. We say R is *embeddible in $n \times n$ matrices* if there is an injection from R into $\prod_{k<n} M_k(H_k)$ where H_k is commutative.

Theorem 6.1.26: *If R satisfies a polynomial identity of degree d then $R/N(R)$ is embeddible in $n \times n$ matrices for $n = [d/2]$. (Recall $N(R)$ from definition 2.6.25.) In fact, each H_k can be taken as a direct product of fields.*

Proof: In the notation of theorem 2.6.27 we have an injection of $R/N(R)$ into R_1 with $\operatorname{Nil}(R_1) = 0$. Multilinearizing we can pass the PI to R_1. But $R_1[\lambda]$ is semiprimitive by Amitsur's theorem 2.5.23, so can be injected into a direct product of primitive rings, each of which by Kaplansky's theorem is simple of dimension $\leq n^2$ over its center. Splitting each of these primitive components to inject it into $M_k(F_{ik})$ for a suitable field F_{ik} and $k \leq n$, we conclude by taking $H_k = \prod_i F_{ik}$.   Q.E.D.

This embedding procedure, foreshadowed in §2.6, is extremely useful. We give the principal application here and leave the others for exercises 14, 15. (Note that an arbitrary PI-ring need not be embeddible in matrices, by example 3.2.48. We pursue this question further in §6.3.)

Theorem 6.1.27: *Suppose R satisfies a PI of degree d. If R is semiprime then* $\text{Nil}(R) = 0$. *In general, for every nil weakly closed subset A of R we have* $A^{[d/2]} \subseteq N(R)$.

Proof: Let $\bar{A}$ be the image of A in $R/N(R)$. By theorem 6.1.26 $R/N(R)$ is embeddible in $n \times n$ matrices for $n = [d/2]$, so $\bar{A}^n = 0$, i.e., $A^n \subseteq N(R)$, proving the second assertion. In particular, $\text{Nil}(R)^n \subseteq N(R) = 0$ for R semiprime, so $\text{Nil}(R) = 0$.   Q.E.D.

Although exercise 12 shows this theorem is sharp insofar as n is concerned, one can replace $N(R)$ by $L_1(R)$, the sum of the nilpotent ideals, as expounded in Rowen [80B, theorem 1.6.36(i) and exercise 1.6.6]; consequently the nil-radical is reached in only two steps. Instead of pursuing this matter further, we push on to a central result.

Lemma 6.1.27′: *If f is an identity or central polynomial of $M_n(C)$ then f is an identity of $M_{n-1}(C)$.*

Proof: Matrices $x_i = \sum_{i,j=1}^{n-1} c_{ij}e_{ij}$ can be reread as elements of $M_n(C)$ so $f(x_1, \ldots, x_m)$ is some scalar α; but $\alpha e_{nn} = f(x_1, \ldots, x_m)e_{nn} = 0$ so $\alpha = 0$.   Q.E.D.

Theorem 6.1.28: *Every semiprime PI-ring R has PI-degree n for suitable n, and every ideal of R intersects the center nontrivially.*

Proof: Let d be the degree of a PI of R; let $Z = Z(R)$ and $0 \neq A \triangleleft R$. We shall show R has PI-degree $n \leq [d/2]$, and $A \cap Z \neq 0$.

Case I. R is semiprimitive. R is a subdirect product of primitive $\{R_i : i \in I\}$, and by Kaplansky's theorem each R_i is central simple over $Z(R_i)$ of degree $n_i \leq [d/2]$. Let $\pi : R \to R_i$ denote the canonical projection and let $A_i = \pi A_i$. Each A_i is an ideal of R_i so $A_i = 0$ or $A_i = R_i$. Let $I' = \{i \in I : A_i \neq 0\}$ and $n = \max\{n_i : i \in I'\}$. Then $g_n(A_i) \subseteq Z(R_i)$ for $i \in I'$ and $g_n(A_i) = 0$ for $i \notin I'$ by lemma 6.1.27′. But $g_n(A_i) \neq 0$ for i such that $n_i = n$, so $0 \neq g_n(A) \subseteq A \cap Z$. Furthermore, taking $A = R$ we have each $A_i = R_i$ so $I' = I$ and g_n is R-central, proving R has PI-degree n.

Case II. R semiprime. Then $Nil(R) = 0$ by theorem 6.1.27 so $R[\lambda]$ is semi-primitive by Amitsur's theorem. Now case I is applicable. $0 \neq A[\lambda] \cap Z[\lambda] = (A \cap Z)[\lambda]$, proving $A \cap Z \neq 0$; also $R[\lambda]$ has some PI-degree n, implying R also has PI-degree n.   Q.E.D.

Corollary 6.1.29: *If R is semiprime PI and $Z(R)$ is a field F then R is a central simple F-algebra.*

Proof: Every nonzero ideal of R contains a unit in F, so R is simple, and Kaplansky's theorem is applicable.   Q.E.D.

Theorem 6.1.30: *If R is a prime PI-ring and $S = Z(R) - \{0\}$ then $S^{-1}R$ is central simple over $S^{-1}Z(R)$ of degree n, where $n = $ PI-deg(R).*

Proof: Let $Z = Z(R)$. $S^{-1}R$ is prime by proposition 2.12.9′ and has PI-degree n since $S^{-1}R$ is a central extension of R. But proposition 1.10.13 shows $Z(S^{-1}R) = S^{-1}Z$, a field, so $S^{-1}R$ is central simple by corollary 6.1.29.
                                                                        Q.E.D.

Corollary 6.1.31: *If R is a semiprime PI-ring then $Sing(R) = 0$.*

Proof: If L is a large left ideal of R and $Lz = 0$ for z in $Z(R)$ then $z \in$ Ann $L = 0$. Thus $Z(R) \cap Sing(R) = 0$, implying $Sing(R) = 0$.   Q.E.D.

Digression: We now see the theory of nonsingular rings is applicable to semiprime PI-rings. An alternate proof of theorem 6.1.28 comes by means of rings without 1, cf., exercises 18–20, and shows, furthermore, that $L < R$ is large iff $L \cap Z(R)$ is large in Z (exercise 21), leading to an easy description of the maximal quotient ring, cf., Rowen [74]. Continuing in the general structure theory, let us exploit the n^2-normality of g_n.

Lemma 6.1.32: *Suppose R has PI-degree n. For each $r_1, \ldots, r_{m+1}$ in R we have*

$$g_n(r_1, \ldots, r_m) r_{m+1} = \sum_{i=1}^{t} (-1)^{i+1} g_n(r_{m+1}, r_1, \ldots, r_{i-1}, r_{i+1}, \ldots, r_m) r_i$$

where $t = n^2$.

Proof: Define $\tilde{g}(X_1, \ldots, X_{m+1})$ as $\sum_{i=1}^{t+1} (-1)^i g_n(X_1, \ldots, X_{i-1}, X_{i+1}, \ldots, X_{m+1}) X_i$. Then $\tilde{g}$ is $(t+1)$-alternating by inspection since $\tilde{g}(X_i \mapsto X_j)$ has exactly two nonzero parts, which appear with opposite signs. Hence $\tilde{g}$ is an identity of R, so $\tilde{g}_n(r_{m+1}, r_1, \ldots, r_m) = 0$ yielding the desired equation.   Q.E.D.

Theorem 6.1.33:   *Suppose R has PI-degree n. If there are elements $r_1, \ldots, r_m$ in R for which $g_n(r_1, \ldots, r_m) = 1$ then R is a free $Z(R)$-module with base $r_1, \ldots, r_t$ where $t = n^2$.*

Proof: Let $Z = Z(R)$. By lemma 6.1.32 we have $r = \sum_{i=1}^{t} (-1)^{i+1} g(r, r_1, \ldots, r_{i-1}, r_{i+1}, \ldots, r)r_i \in \sum Z r_i$ for each r in R, so $r_1, \ldots, r_t$ span R. If $\sum_{i=1}^{t} z_i r_i = 0$ for z_i in Z then for each $j \leq t$ we have

$$0 = g_n\left(r_1, \ldots, r_{j-1}, \sum_{i=1}^{t} z_i r_i, r_{j+1}, \ldots, r_m\right)r_j$$

$$= \sum_{i=1}^{t} g_n(r_1, \ldots, r_{j-1}, z_i r_i, r_{j+1}, \ldots, r_m)r_j$$

$$= g_n(r_1, \ldots, z_j r_j, \ldots, r_m) = z_j$$

proving $r_1, \ldots, r_t$ are independent.   Q.E.D.

The Artin-Procesi Theorem

The last result hints R may be Azumaya over $Z(R)$, leading us to another major result; the equivalence (i') $\Leftrightarrow$ (iv) below is called the *Artin-Procesi theorem*.

Lemma 6.1.34:   *Suppose R has a multilinear central polynomial g. Then $g(R)^+$ is an ideal of $Z(R)$; furthermore, if $g(R)R = R$ then $g(R)^+ = Z(R)$.*

Proof: Let $Z = Z(R)$. If $z \in Z$ then $zg(r_1, \ldots, r_m) = g(zr_1, \ldots, r_m) \in g(R)$, proving $g(R)^+$ is an ideal of Z. Suppose $g(R)R = R$. If $g(R)^+ \neq Z$ then take a maximal ideal P of Z containing $g(R)$ and let $\bar{R} = R_P/B$ where B is a maximal ideal of R_P. Then $1 \in g(R_P)R_P$ and $g(R_P) \subseteq g(R)_P \subseteq Z_P$, implying g is $\bar{R}$-central. Since $\bar{R}$ is simple we see $g(\bar{R})^+ = Z(\bar{R})$, so

$$(Z_P + B)/B \subseteq Z(\bar{R}) = g(\bar{R})^+ = (g(R_P)^+ + B)/B \subseteq (Z_P + B)/B$$

so equality holds and $1 = s^{-1}z + b$ for suitable z in $g(R)^+$, s in $Z - P$, and b in B. Thus $b = 1 - s^{-1}z \in B \cap Z_P \subseteq P_P$ since Z_P is local, implying the absurdity $1 \in P_P$. The only way out is $g(R)^+ = Z$.   Q.E.D.

For the purpose of the next result we say an identity f of $M_n(\mathbb{Z})$ is *critical* if f is not an identity of $M_{n+1}(F)$ for any field F.

Theorem 6.1.35:   *The following conditions are equivalent:*

(i) *R satisfies all the multilinear identities of $M_n(\mathbb{Z})$, and each critical multilinear identity f of $M_{n-1}(\mathbb{Z})$ is not satisfied by any homomorphic image of R.*
(i') *R satisfies all the multilinear identities of $M_n(\mathbb{Z})$, and there is some multilinear identity f of $M_{n-1}(\mathbb{Z})$ not satisfied by any homomorphic image of R.*
(ii) *R has PI-degree n and $1 \in g_n(R)R$.*
(iii) *R has PI-degree n and $1 \in g_n(R)^+$.*
(iv) *R is Azumaya of constant rank n^2.*

Proof: Let $Z = Z(R)$. (i) $\Rightarrow$ (i') is obvious. (i') $\Rightarrow$ (ii) Otherwise, take a maximal ideal P containing $g_n(R)R$. Then g_n is an identity of R/P, implying R/P has PI-degree $< n$, and thus R/P satisfies f, contradiction.

(ii) $\Rightarrow$ (iii) by lemma 6.1.34.

(iii) $\Rightarrow$ (iv) Write $1 = \sum_{i=1}^{k} g_n(r_{i1}, \ldots, r_{im})$. First we show R is an f.g. projective Z-module by means of the dual basis lemma. Indeed, define $f_{ij}: R \to Z$ by

$$f_{ij}r = (-1)^{j+1} g_n(r, r_{i1}, \ldots, r_{i,j-1}, r_{i,j+1}, \ldots, r_{im}).$$

Then lemma 6.1.32 implies $r = \sum_{i=1}^{k} \sum_{j=1}^{n^2} (f_{ij}r)r_{ij}$, so $\{(r_{ij}, f_{ij}): 1 \leq j \leq n^2\}$ is the desired dual basis. $\text{rank}(R) = n^2$ by theorem 6.1.33, since $g_n(R_P)$ contains an invertible element of Z_P for any P in $\text{Spec}(Z)$.

Clearly R is faithful over Z since $1 \in R$, so it suffices to prove the canonical homomorphism $\psi: R \otimes R^{op} \to \text{End}_Z R$ is an isomorphism. To this end write $g_n = \sum_u h_{u1} X_1 h_{u2}$ for suitable polynomials $h_{uv}(X_2, \ldots, X_m)$. Let $a_{iju} = (-1)^{j+1} h_{u1}(r_{i1}, \ldots, r_{i,j-1}, r_{i,j+1}, \ldots, r_{im})$ and $b_{iju} = h_{u2}(r_{i1}, \ldots, r_{i,j-1}, r_{i,j+1}, \ldots, r_{im})$.

To show ψ is onto let $\beta \in \text{End}_Z R$. We shall show $\beta = \psi(\sum_{i,j,u} a_{iju} \otimes b_{iju}\beta r_{ij})$. Indeed,

$$\beta r = \beta(1r) = \beta\left(\sum_i g_n(r_{i1}, \ldots, r_{im})r\right)$$

$$= \beta\left(\sum_{i,j}(-1)^{j+1} g_n(r, r_{i1}, \ldots, r_{i,j-1}, r_{i,j+1}, \ldots, r_{im})r_{ij}\right)$$

$$= \sum_{i,j}(-1)^{j+1} g_n(r, r_{i1}, \ldots, r_{i,j-1}, r_{i,j+1}, \ldots, r_{im})\beta r_{ij}$$

$$= \sum_{i,j,u} a_{iju} r b_{iju}\beta r_{ij}.$$

as desired.

To show ψ is 1:1 suppose $\sum_v x_v \otimes y_v \in \ker \psi$ for suitable x_v in R, y_v in R^{op}.

$$\sum x_v \otimes y_v = \sum_{i,v} g_n(r_{i1},\dots,r_{im})x_v \otimes y_v$$

$$= \sum_{i,j,v}(-1)^{j+1}g_n(x_v,r_{i1},\dots,r_{i,j-1},r_{i,j+1},\dots,r_{im})r_{ij}\otimes y_v$$

$$= \sum_{i,j,v}(-1)^{j+1}r_{ij}\otimes g_n(x_v,r_{i1},\dots,r_{i,j-1},r_{i,j+1},\dots,r_{im})y_v$$

$$= \sum_{i,j,u,v} r_{ij}\otimes(a_{iju}x_v b_{iju})y_v$$

$$= \sum_{i,j} r_{ij}\otimes \sum_u a_{iju}\left(\sum_v x_v b_{iju}y_u\right) = 0$$

since $\sum_v x_v b_{iju}y_v = (\sum x_v \otimes y_v)b_{iju} = 0$. Thus $\ker \psi = 0$.

(Once upon a time this was the "hard direction" of the Artin-Procesi theorem; see exercise 13 for an even shorter proof.)

(iv) $\Rightarrow$ (i) First we claim no homomorphic image R/A of R satisfies f. Indeed, otherwise, taking P maximal containing A we see R/P satisfies f; but R/P is central simple over its center and splitting by a maximal subfield we see f is an identity of $M_{n'}(F)$ for some field F, where n' is the PI-degree of R/P. Since f is critical we have $n' < n$, so R/P has dimension $\leq (n-1)^2$ over its center $Z/(Z \cap P)$, contrary to proposition 5.3.27.

It remains to show R satisfies all identities of $M_n(\mathbb{Z})$. In view of proposition 2.12.15(i) we may assume Z is local, with maximal ideal P; then PR is the unique maximal ideal of R, and $\bar{R} = R/PR$ is simple of dimension n^2 over Z/P, by proposition 5.3.27.

We need to quote the following fact from the theory of central simple algebras, to be proved below as proposition 7.1.22. There are elements r_j, a in R for which $\{\bar{a}^{i-1}\bar{r}_j: 1 \leq i,j \leq n\}$ are a base for $\bar{R}$ over $\bar{Z}$, where $\bar{}$ denotes the canonical image in $\bar{R}$. Now $\{a^{i-1}r_j: 1 \leq i,j \leq n\}$ span R over Z by Nakayama's lemma, implying $r_1,\dots,r_n$ span R as module over the commutative ring $Z[a]$. The regular representation enables us to embed R into $\mathrm{End}_{Z[a]} R$, which by theorem 1.5.13 is a homomorphic image of a subring of $M_n(Z[a])$; since $Z[a]$ is a homomorphic image of the free commutative ring $\mathbb{Z}[\lambda]$ we see R satisfies all multilinear identities of $M_n(\mathbb{Z}[\lambda])$ and thus of $M_n(\mathbb{Z})$.   Q.E.D.

The critical identity of $M_{n-1}(\mathbb{Z})$ usually used in (i') is S_{2n-2}, in view of the Amitsur-Levitzki theorem. The Artin-Procesi theorem is a very explicit computational tool for studying Azumaya algebras of rank n, and actually provides a new invariant, the minimal number k for which $1 = \sum_{i=1}^{k} z_i$ for suitable z_i in $g_n(R)$. If $k = 1$ then R is free over $Z(R)$, by theorem 6.1.33.

Remark 6.1.35':   Examining the proof of lemma 6.1.32 and theorem 6.1.35, we see that the condition in (iii) that R has PI-degree n could be weakened to "$\tilde{g}$ is an identity of R."

The n-Spectrum

One way to achieve $1 \in g_n(R)$ is by localization.

Corollary 6.1.36:   *Suppose R has PI-degree n with center Z and $s \in g_n(R)$. Then $R[s^{-1}]$ is Azumaya of rank n^2 over $Z[s^{-1}]$, and, in fact, is free over $Z[s^{-1}]$.*

Proof: $1 \in g_n(R[s^{-1}])$ so lemma 6.1.34 shows $Z(R[s^{-1}]) = g_n(R[s^{-1}])^+ \subseteq Z[s^{-1}] \subseteq Z(R[s^{-1}])$; thus equality holds throughout. Now $R[s^{-1}]$ is Azumaya by theorem 6.1.35 and is free by theorem 6.1.33.   Q.E.D.

This provides a most powerful tool, since it is applicable whenever $g_n(R) \nsubseteq P$. Accordingly we make the following definition.

Definition 6.1.37:   Suppose $Z = Z(R)$. $\text{Spec}_n(R) = \{P \in \text{Spec}(R): g_n(R) \nsubseteq P\}$ and $\text{Spec}_n(Z) = \{P \in \text{Spec}(Z): g_n(R) \nsubseteq P\}$.

$g_n(R)$ is often called the *Formanek center*; the primes in $\text{Spec}_n(R)$ are called *identity-faithful* or *regular*. The following theorem and subsequent results have more elementary proofs in Rowen [80B, p. 76–78].

Theorem 6.1.38:   *Suppose $P \in \text{Spec}_n(Z(R))$. There is a unique P' in $\text{Spec}(R)$ lying over P, and P' contains every ideal A of R for which $A \cap Z(R) \subseteq P$. In other words, R_P is local with maximal ideal $P_P R_P$.*

Proof: Pass to R_P, which is Azumaya and thus local with maximal ideal $P_P R_P$, and translate this back to R.   Q.E.D.

Corollary 6.1.39:   LO, GU, *and* INC *hold from $Z(R)$ to R in* Spec_n.

This permits us to build an inductive procedure on the PI-degree, and there is also a pleasant correspondence of maximal ideals.

Lemma 6.1.40:   *If $P' \in \operatorname{Spec}_n(R)$ is a maximal ideal of R then $Z(R) \cap P'$ is a maximal ideal of $Z(R)$.*

Proof: Let $Z = Z(R)$ and $\bar{R} = R/P'$, a simple ring. Then $1 \in g_n(\bar{R})^+$ by theorem 6.1.34 so

$$Z(\bar{R}) = g(\bar{R})^+ = \overline{g(R)^+} \subseteq \bar{Z} \subseteq Z(\bar{R})$$

implying $Z(\bar{R}) = \bar{Z} \approx Z/(P' \cap Z)$.   Q.E.D.

These results show that the "difficult" part of $\operatorname{Spec}(R)$ is the complement of the n-spectrum. Nevertheless, there are some interesting results, most notably the Bergman-Small "additivity principle" to be discussed in §6.3.

The Algebra of Generic Matrices

We shall now deal with a PI-ring which is a very important "test ring." Given a commutative ring C we let $\Lambda = \{\lambda_{ij}^{(k)} : 1 \leq i, j \leq n, k \in \mathbb{N}\}$ be a set of commuting indeterminates over C, and let Y_k denote the matrix $(\lambda_{ij}^{(k)})$ in $M_n(C[\Lambda])$. Each entry of Y_k is a different indeterminate; accordingly, we call Y_k a *generic matrix* and define the algebra of *generic matrices*, denoted $C_n\{Y\}$, to be the C-subalgebra of $M_n(C[\Lambda])$ generated by the Y_k.

Proposition 6.1.41:   *If C is an integral domain then the algebra of generic matrices $C_n\{Y\}$ is prime.*

Proof: Let K be the field of fractions of $C[\Lambda]$ and let $R' = C_n\{Y\}K \subseteq M_n(K)$. Clearly the Capelli polynomial of degree $2n^2$ is nonvanishing on generic matrices (seen by specializing the indeterminates), so $[R':K] \geq n^2$. But this means $R' = M_n(K)$; hence $M_n(K)$ is a central extension of $C_n\{Y\}$, which thus is prime by proposition 2.12.39 (taking $P' = 0$).   Q.E.D.

The significance of $C_n\{Y\}$ lies in the following result:

Proposition 6.1.42:

(i) *There is a canonical surjection $\varphi: C\{X\} \to C_n\{Y\}$ given by $\varphi X_i = Y_i$, and $\ker \varphi = \{$identities of $M_n(C[\Lambda])\}$.*

(ii) *$C_n\{Y\}$ is free in the class $\mathscr{C}$ of algebras satisfying the identities of $M_n(C[\Lambda])$.*

Proof:

(i) φ exists because $C\{X\}$ is free. If $f(X_1, \ldots, X_m)$ is an identity of $M_n(C[\Lambda])$ then $f(Y_1, \ldots, Y_m) = 0$ so $f \in \ker \varphi$. To finish the proof of (i) it suffices to prove if $f(Y_1, \ldots, Y_m) = 0$ then $f(X_1, \ldots, X_m)$ is an identity of $M_n(C[\Lambda])$. Given any matrices $a_k = (c_{ij}^{(k)})$ in $M_n(C[\Lambda])$ we could extend the map $\lambda_{ij}^{(k)} \mapsto c_{ij}^{(k)}$ to a homomorphism $C[\Lambda] \to C[\Lambda]$ and thus to a homomorphism $M_n(C[\Lambda]) \to M_n(C[\Lambda])$ sending $Y_k = (\lambda_{ij}^{(k)})$ to a_k; then $f(a_1, \ldots, a_m)$ is the image of $f(Y_1, \ldots, Y_m) = 0$.

(ii) This argument will be repeated more generally in §6.4. Suppose $R \in \mathscr{C}$. Given any map $Y_i \mapsto r_i$ for $i = 1, 2, \ldots$ we have a homomorphism $\psi: C\{X\} \to R$ sending $X_i \mapsto r_i$ for each i. We shall show $\ker \psi \supseteq \ker \varphi$, for then we can factor ψ through $C_n\{Y\}$ to get the desired homomorphism $\bar\psi: C_n\{Y\} \to R$ satisfying $\bar\psi Y_i = r_i$, and clearly $\bar\psi$ is unique such. If $f \in \ker \varphi$ then f is an identity of $M_n(C[\Lambda])$ by (i), so $f(r_1, \ldots, r_m) = 0$ since $R \in \mathscr{C}$, implying $f \in \ker \psi$.   Q.E.D.

The Identities of a PI-Algebra

Proposition 6.1.42 heightens our interest in which rings satisfy *all* the identities of $M_n(C[\Lambda])$. Accordingly we define $\mathscr{L}(R) = \{\text{identities of } R\}$, and we say C-algebras R and T are PI-*equivalent* if $\mathscr{L}(R) = \mathscr{L}(T)$. Let us start by asking when R and $R \otimes_C H$ are PI-equivalent, recalling proposition 6.1.11. Now we must require some restriction on C, as evidenced by the example that $\mathbb{Z}/2\mathbb{Z}$ satisfies the identity $f = X_1^2 - X_1$ but no other field satisfies f.

Lemma 6.1.43:   *Suppose C is an infinite integral domain, and the polynomial $p \in C[\lambda_1, \ldots, \lambda_k]$ satisfies $p(c_1, \ldots, c_k) = 0$ for all c_i in C. Then $p = 0$.*

Proof: This is well-known, but we review it here. The case $k = 1$ follows at once from the fact that any nonzero $p(\lambda_1)$ has only a finite number of zeroes in the field of fractions of C, so some $p(c) \neq 0$. In general, $p = \sum_{i=0}^t p_i(\lambda_1, \ldots, \lambda_{k-1})\lambda_k^i$ and for each $\mathbf{c} = (c_1, \ldots, c_{k-1})$ in $C^{(k-1)}$ write $p_{\mathbf{c}} = \sum_{i=0}^t p_i(c_1, \ldots, c_{k-1})\lambda_k^i \in F[\lambda_k]$. Then as seen above $p_{\mathbf{c}} = 0$ so $p_{\mathbf{c}} = 0$ for each i and all $\mathbf{c}$ in $C^{(k-1)}$, implying each $p_i = 0$ by induction on k; hence $p = 0$.   Q.E.D.

Theorem 6.1.44:   *Suppose R is torsion-free over an infinite integral domain C. Then R is PI-equivalent to $R \otimes_C H$ for any torsion-free commutative C-algebra H.*

Proof: Identify R with $R \otimes 1 \subseteq R \otimes H$. Then every identity of $R \otimes H$ is an identity of R, so it remains to show every identity f of R is an identity of $R \otimes H$. Let F be the field of fractions of C. Replacing H by $H \otimes_C F$ we may assume H is an F-algebra. $R \otimes F$ has a base $\{r_i : i \in I\}$ over F, where each $r_i \in R$. Given $x_1, \ldots, x_m$ in $R \otimes H$ write $x_j = \sum_i r_i \otimes a_{ij}$ where the $a_{ij} \in H$. Then $f(x_1, \ldots, x_m) = \sum_i r_i \otimes p_i(\mathbf{a})$ where each $p_i \in C[\lambda_1, \ldots, \lambda_m]$ and $p_i(\mathbf{a})$ is the evaluation at various a_{ij}. If all our $a_{ij} \in C$ then each $x_1, \ldots, x_m \in R$ so $f(x_1, \ldots, x_m) = 0$ and thus each $p_i(\mathbf{a}) = 0$; therefore, each $p_i = 0$ by lemma 6.1.43. But then $p_i(\mathbf{a}) = 0$ for arbitrary a_{ij} in H, implying $f(x_1, \ldots, x_m) = 0$ for all $x_1, \ldots, x_m$ in $R \otimes H$.   Q.E.D.

Noetherian PI-Rings

The subject of Noetherian PI-rings relies heavily on the following observation, which permits us at once to lift many of the results directly from §3.5.

Proposition 6.1.48:   *Every PI-ring R is left and right fully bounded.*

Proof: We may assume R is prime, and by symmetry need to show any large left ideal L contains a two-sided ideal. But letting $S = Z(R) - \{0\}$ we show $S^{-1}L$ is a large left ideal of the simple Artinian ring $Q = S^{-1}R$, so $S^{-1}L = Q$ and thus some $s \in S \cap L$. Now $0 \neq Rs \lhd R$, and clearly every nonzero ideal of a prime ring is large.   Q.E.D.

Theorem 6.1.49:   *If R is a Noetherian PI-ring then $\bigcap_{i \in \mathbb{N}} \operatorname{Jac}(R)^i M = 0$ for all f.g. R-modules M.*

Proof: By theorem 3.5.28, noting the hypotheses hold by proposition 6.1.48 and Kaplansky's theorem.   Q.E.D.

Theorem 6.1.50:   *In a left Noetherian PI-ring every prime ideal has finite height.*

Proof: By theorem 3.5.13, noting that every nonzero ideal of a prime PI-ring has a nonzero central element which thus is normalizing.   Q.E.D.

To determine when R is Noetherian we can use the following property of arbitrary prime rings of PI-degree n.

Proposition 6.1.51:   *Suppose R is prime of PI-degree n. Then any left ideal L*

is isomorphic to a $Z(R)$-submodule of a free $Z(R)$-module M of rank $\leq n^2$ contained in L. Explicitly there is $z \neq 0$ in $Z(R)$ for which $zL \subseteq M \subseteq L$.

Proof: Let $Z = Z(R)$ and take k maximal possible for which there are $r_1, \ldots, r_m$ such that $z = g_n(r_1, \ldots, r_m) \neq 0$ with k of the r_i from L. Passing to the ring of fractions of R we see the r_i are Z-independent; relabeling the r_i from L as $a_1, \ldots, a_k$ we take $M = \sum_{i=1}^{k} Za_i \approx Z^{(k)}$. It remains to show $zL \subseteq M$. But lemma 6.1.32 yields

$$za = \sum_{j=1}^{t} (-1)^{j+1} g_n(a, r_1, \ldots, r_{j-1}, r_{j+1}, \ldots, r_m) r_j$$

and the j-th summand is 0 by maximality of k unless $r_j \in L$, implying each nonzero summand is in M.   Q.E.D.

Corollary 6.1.52:   *(Formanek) Suppose R is prime PI. If $Z(R)$ is Noetherian then R is f.g. as $Z(R)$-module and hence is Noetherian. If $Z(R)$ has Krull dimension α then R has Krull dimension $\leq \alpha$.*

Proof: R is a submodule of an f.g. $Z(R)$-module.   Q.E.D.

A cute variant of this idea yields *Cauchon's theorem*, that any semiprime PI-ring with ACC(ideals) is Noetherian, cf., exercise 24.

Formanek's result raised the hope that the structure of a prime Noetherian PI-ring is closely tied to the structure of its center. Although this is true in certain special cases, e.g., when R hereditary (cf., Robson-Small [74]), the general situation is the opposite; the center of a prime Noetherian PI-ring can be rather different. Perhaps the most efficient counterexample (i.e., most pathology for least work) is exercise 27. In desperation it was asked whether a prime PI-ring must be f.g. as module over a suitable commutative subring. Building on examples by Cauchon and Bergman, Sarraille [82] produced a counterexample, cf., exercise 6.3.2.

If R is Noetherian PI then $K\text{-dim}(R) \leq \text{gl.dim } R$ when the latter exists. This was proved more generally for fully bounded Noetherian rings under the restriction that the center contains an uncountable domain, by Brown-Warfield [84]; Goodearl-Small [84] finished the result for Noetherian PI-rings by passing to the Laurent series ring $R((\lambda))$, whose center $Z((\lambda))$ is uncountable. To accomplish this reduction they had to show $\text{gl.dim } R((\lambda)) = \text{gl.dim } R$ and $K\text{-dim } R \leq K\text{-dim } R((\lambda))$. The latter fact is easy since there is a lattice injection $\mathscr{L}(R((\lambda))) \to \mathscr{L}(R)$ given by $L \mapsto L'$, where $L' = \{a \in R : a$ is the lowest order coefficient of some element of L (viewed as Laurent series)$\}$.

Vamos [77] showed the completion $\hat{R}$ of a semilocal Noetherian PI-ring R is Noetherian, and $\hat{R} \otimes$ __ is an exact functor which on $R\text{-}\mathscr{F}imod$ is naturally equivalent to the I-adic completion.

Much recent work on Noetherian PI-rings has focused on the "trace ring" $T(R)$, as described in 6.3.45ff of the unabridged version.

§6.2 Affine Algebras

Definition 6.2.1: R is an *affine C-algebra* if R is generated *as algebra* by a finite number of elements $r_1, \ldots, r_t$; in this case we write $R = C\{r_1, \ldots, r_t\}$. (In the literature C is often required to be a field.)

Commutative affine algebras over fields are obviously Noetherian, and we record the following useful result.

Noether (–Bourbaki) Normalization Theorem: (*cf.*, *Bourbaki* [72B, theorem v.3.1]) *Suppose $A_1 \subseteq \cdots \subseteq A_k$ are ideals in a commutative affine algebra C over a field F. There is a transcendence base $c_1, \ldots, c_m$ of C for which C is integral over $F[c_1, \ldots, c_m]$, and such that there are $n(1) \le n(2) \le \cdots \le n(k)$ such that $A_i \cap F[c_1, \ldots, c_m] = \sum_{u=1}^{n(i)} Cc_u$ for $1 \le i \le k$.*

On the other hand, the free algebra in two noncommuting indeterminates is far from Noetherian and so affine algebras offer an alternative to Noetherian rings in building a general geometric-flavored ring theory. We shall start with some general results concerning affine algebras and then delve into the Gelfand-Kirillov dimension.

Remark 6.2.2: If R is an affine C-algebra and $A \lhd R$ then R/A is affine over $C/(C \cap A)$.

Recall from theorem 2.5.22 that the weak Nullstellensatz holds for affine algebras over uncountable fields, so the starting point of our inquiry might be whether this is true in general, i.e., if $\mathrm{Jac}(R)$ is nil for R affine over a field. This question was open for a long time but was settled negatively by Beidar using a very easy method which we describe below now (with some Small modifications). *Note f.g. always means as module* in what follows.

Example 6.2.3: (Affinization) If T is an affine C-algebra and L is a f.g. left ideal of T then

$$R = \begin{pmatrix} C + L & T \\ L & T \end{pmatrix}$$

is affine over $C \cdot 1$, and $Z(R) \subseteq C + L$. Indeed if $T = C\{a_1, \ldots, a_t\}$ and $L = \sum_{i=1}^{n} Ta'_i$ for a'_i, a_j in T then $R = C\{e_{11}, e_{12}, a'_i e_{21}, a_j e_{22} : 1 \leq i \leq n, 1 \leq j \leq t\}$.

Note $Z(R)$ can be very poorly behaved, and even not affine, e.g., take $T = \mathbb{Q}[\lambda]$ and $L = T\lambda$. A related instance was given in exercise 6.1.28. This example also is used in exercises 3, 4. However, we want to see what happens when T is the free associative ring.

Example 6.2.4: (Failure of the Nullstellensatz for affine algebras.) Build R as in example 6.2.3, where C is a countable field F, $T = F\{X_1, X_2\}$, and $L = TX_2$. Then R has a prime ideal P for which R/P is right Noetherian affine over F but $\mathrm{Jac}(R/P) \neq 0 = \mathrm{Nil}(R/P)$. To see this let

$$T' = \left\{ \begin{pmatrix} a & 0 \\ 0 & a \end{pmatrix} : a \in F + L \right\} \approx F + L = F\{X_1^i X_2 : i \in \mathbb{N}\},$$

which is isomorphic to the free algebra in countably many indeterminates. Let H be a countable Noetherian local integral domain, such as the localization of $F[\lambda]$ at the prime ideal λ. Then $H \approx T'/A$ for some A in $\mathrm{Spec}(T')$. $R = \begin{pmatrix} T' & T \\ L & T \end{pmatrix}$; hence $RAR = \begin{pmatrix} A & AT \\ LA & TAT \end{pmatrix}$ is contained in an ideal P of R maximal with respect to $P \cap T' = A$, and $P \in \mathrm{Spec}(R)$ by remark 2.12.42. Let $\bar{R} = R/P$ and let e denote the image of e_{11} in R/P. Then $e\bar{R}e \approx T'/A \approx H$ so $0 \neq \mathrm{Jac}(e\bar{R}e) = e\,\mathrm{Jac}(\bar{R})e$ by proposition 2.5.14, proving $\mathrm{Jac}(\bar{R}) \neq 0$.

$\bar{R}$ is left Noetherian since right multiplication by the image of $X_2 \cdot 1$ gives a lattice injection from the left ideals $\mathcal{L}(\bar{R})$ to $\mathcal{L}(M_2(H))$. In particular, $\mathrm{Nil}(\bar{R}) = 0$ since $\bar{R}$ is prime and left Noetherian.

In practice many affine algebras do satisfy the weak Nullstellensatz and more, as we shall see in §6.3 and §6.4, and there is no known answer to **Question 6.2.4':** If R is Noetherian and affine over a field then is $\mathrm{Jac}(R)$ necessarily nil? (An affirmative answer would imply R is Jacobson by passing to homomorphic images.)

Incidentally, the relationship between affine and Noetherian is mysterious: it is not even known if every affine subalgebra of the Weyl algebra $\mathcal{A}_1(F)$ is Noetherian.

Let us try to salvage something concrete from the wreckage of example 6.2.4. First we note that there are cases when the center is affine.

Proposition 6.2.5: *(Artin-Tate) If C is commutative Noetherian and R is an affine C-algebra and Z is a C-subalgebra of $Z(R)$ over which R is fg. then Z is also affine.*

Proof: Write $R = C\{r_1, \ldots, r_t\}$ and $R = \sum_{k=1}^{t} Zx_k$ for suitable x_k in R. Then we have elements z_{ijk}, z'_{uk} in Z for which

$$x_i x_j = \sum_k z_{ijk} x_k \qquad \text{and} \qquad r_u = \sum_k z'_{uk} x_k.$$

Let $Z_1 = C[z_{ijk}; z'_{uk} : 1 \le i, j, k \le m \text{ and } 1 \le u \le t]$, which is commutative affine and thus Noetherian. Then one can check at once $R = \sum_{k=1}^{m} Z_1 x_k$ is f.g. over Z_1, so R is a Noetherian Z_1-module and thus its submodule Z is f.g. over Z_1, implying Z is affine. Q.E.D.

Other results verifying when an algebra is affine are given in exercises 1, 2; a sweeping result can be found in Lorenz [84].

Kurosch's Problem and the Golod-Shafarevich Counterexample

Having seen that affine algebras exist in abundance, we might wonder whether something can be said if extra restrictions are imposed.

Problem 6.2.6: (Kurosch's problem) If R is affine over a field F and algebraic over F then is R finite dimensional as a vector space over F?

This famous problem has two key special cases:

Kurosch's Problem for Division Rings: If a division algebra D is affine and algebraic over F then is D central simple? (If so then D is f.g. over $Z(D)$, implying $Z(D)$ is affine by Artin-Tate and thus $[Z(D):F] < \infty$ by corollary 6.3.2 below, and thence $[D:F] < \infty$.)

Levitzki's Problem: If N is a nil, affine F-algebra without 1 then is N nilpotent? This is cleary a special case of Kurosch, since adjoining 1 formally would give an affine algebraic algebra R and if $[R:F] < \infty$ then $N = \text{Nil}(R)$ is nilpotent.

There is a cute reduction of Kurosch's problem to the prime case, cf., exercise 5. Kurosch for division rings is still open in general, but has a positive solution for F uncountable, by proposition 6.1.24; Levitzki's problem, however, is false even for F uncountable, by a celebrated example of Golod [64] and Shafarevich. Recently Bergman [89] has studied local finiteness (and the Golod-Shafarevich example) in terms of the Jacobson radical of extension rings, and sharper examples have recently appeared. Let us study the original Golod-Shafarevich example.

Main Computational Lemma 6.2.7:   *Grade   the   free   associative   algebra*
$F\{X_1,\ldots,X_m\}$ *by the (total) degree. Let A be an ideal generated as an ideal*
by a countably infinite set S of homogeneous elements of degree ≥ 2. Writing
$S_k = \{$*elements of S of degree k*$\}$*, suppose there are numbers $m_k \geq |S_k|$ for*
which the coefficients of the power series

$$p(\lambda) = \left(1 - m\lambda + \sum_{k=2}^{\infty} m_k \lambda^k\right)^{-1}$$

are all non-negative. Then $R = F\{X_1,\ldots,X_m\}/A$ is infinite dimensional over F.

Proof: (Vinberg)   We   write   power   series $\sum a_j \lambda^j \leq \sum b_j \lambda^j$ to   indicate   that
each $a_j \leq b_j$. Let $n_j = [R_j : F]$. The aim of the proof is to show the series
$\sum_{j \in \mathbb{N}} n_j \lambda^j \geq p(\lambda)$. This would finish the proof since p has infinite support, i.e.,
an infinite number of nonzero coefficients. (Indeed, if p had finite support
then $(1 + \sum_{k \geq 2} m_k \lambda^k)p$ would have infinite support since all the coefficients
of p are non-negative; but $(1 - m\lambda + \sum_{k \geq 2} m_k \lambda^k)p = 1$, so $-m\lambda p$ must have
infinite support, implying p has infinite support.)

Write T for $F\{X_1,\ldots,X_m\}$, and T_j for the homogeneous part of degree j.
Then $T_1 = \sum F X_i$ and $TT_1 = \bigoplus_{j > 0} T_j$ so $T = TT_1 \oplus F$. Let B_j be a vector
subspace of T_j complementary to A_j, and $B = \bigoplus B_j$. Then $T = A \oplus B$ so

$$A = TST = TSTT_1 + TSF = AT_1 + TS = AT_1 + (A + B)S = AT_1 + BS.$$

Taking   homogeneous   parts   yields $A_k = A_{k-1}T_1 + \sum_j B_{k-j}S_j$;   computing
dimensions (noting $[A_j : F] + [B_j : F] = [T_j : F] = m^j$ and $[B_j : F] = [R_j : F] = n_j$)
we have

$$m^k - n_k \leq (m^{k-1} - n_{k-1})m + \sum_j n_{k-j}d_j$$

where $d_j = |S_j|$, implying

$$0 \leq n_k - n_{k-1}m + \sum_j n_{k-j}d_j \leq n_k - n_{k-1}m + \sum_j n_{k-j}m_j$$

for   all $k \geq 1$. Letting $h(\lambda) = (\sum_j n_j \lambda^j)(1 - m\lambda + \sum_u m_u \lambda^u)$ our   inequality
shows $h \geq 1$. But then $\sum_j n_j \lambda^j = hp = p + (h-1)p \geq p$ as desired, since
$(h-1)p$ has only non-negative coefficients.   Q.E.D.

As   final   preparation   we   need   a   standard   observation   about   counting.

Remark 6.2.8:   The   number   of   ways   of   selecting n objects   from q allowing
repetitions   is $\dbinom{q+n-1}{n}$ (Indeed,   since $\dbinom{q+n-1}{n}$ is   the   number   of

ways of selecting n from $q + n - 1$ without repetition, we wish to find a 1:1 correspondence between these selections and selections of n from q with repetitions. Arranging selections in ascending order, we define the "bottom" of a selection of n from $q + n - 1$ to be those numbers $\leq q$. We shall conclude by showing for any t that the number of selections of n from $q + n - 1$ having a particular bottom $(i_1, \ldots, i_t)$ is the same as the number of selections of n from q which are $i_1, \ldots, i_t$ with repetitions. Indeed, in the first case we are selecting $n - t$ objects from the $n - 1$ objects $q + 1, \ldots, q + n - 1$ without repetition, and in the latter case we are selecting $n - t$ objects from the t objects $i_1, \ldots, i_t$ with repetition, which by induction on n has $\binom{t + n - t - 1}{n - t} = \binom{n-1}{n-t}$ possibilities, so indeed these numbers are the same).

Theorem 6.2.9: (Golod [66]) *Given any field F and any $m \geq 2$ there is an infinite dimensional graded F-algebra R_0 without 1, generated as algebra without 1 by m elements, such that every subset of $<m$ elements of R_0 is nilpotent. In particular, R_0 is nil and is a counterexample to Levitzki's and Kurosch's problem. Furthermore $\bigcap_{n \in \mathbb{N}} R_0^n = 0$.*

Proof: Since we want to use lemma 6.2.7 we first look for a general instance for which the coefficients of the power series $(1 - m\lambda + \sum_{k=2}^{\infty} m_k \lambda^k)^{-1}$ are non-negative. We claim this holds for

$$m_k = \varepsilon^2 (m - 2\varepsilon)^{k-2} \qquad \text{where } 0 < 2\varepsilon < m.$$

Indeed we compute

$$1 - m\lambda + \sum_{k \geq 2} \varepsilon^2 (m - 2\varepsilon)^{k-2} \lambda^k = 1 - m\lambda + \frac{\varepsilon^2 \lambda^2}{1 - (m - 2\varepsilon)\lambda}$$

$$= \frac{((1 - m\lambda)(1 - (m - 2\varepsilon)\lambda) + \varepsilon^2 \lambda^2)}{1 - (m - 2\varepsilon)\lambda}$$

$$= \frac{(1 - (m - \varepsilon)\lambda)^2}{1 - (m - 2\varepsilon)\lambda},$$

so its reciprocal is

$$\frac{1 - (m - 2\varepsilon)\lambda}{(1 - (m - \varepsilon)\lambda)^2} = (1 - (m - 2\varepsilon)\lambda)(1 + (m - \varepsilon)\lambda + (m - \varepsilon)^2 \lambda^2 + \cdots)^2$$

$$= (1 - (m - 2\varepsilon)\lambda)\left(1 + \sum_{k \geq 1} (k + 1)(m - \varepsilon)^k \lambda^k\right)$$

$$= 1 + \sum ((k + 1)(m - \varepsilon)^k - (m - 2\varepsilon)k(m - \varepsilon)^{k-1})\lambda^k$$

$$= 1 + \sum_{k \geq 1} (m - \varepsilon)^{k-1}(m + (k - 1)\varepsilon)\lambda^k$$

and the coefficients are visibly positive.

Next we shall construct inductively a set S of homogeneous elements $\{s_1, s_2, \ldots\} \subset T$ having the following properties:

(i) There are $\leq \varepsilon^2(m - 2\varepsilon)^{k-2}$ elements of S having degree k, for each k;
(ii) If $f_1, \ldots, f_{m-1}$ are arbitrary in T having constant term 0 then $\{f_1, \ldots, f_{m-1}\}^n \subseteq \langle S \rangle$ for suitable $n = n(f_1, \ldots, f_{m-1})$.

Clearly $R_0 = T_1 T / \langle S \rangle$ would be the desired algebra. Assume by induction that for given k_0 there is $d = d(k_0)$ for which homogeneous $s_1, \ldots, s_d$ of degree k_0 have been selected and such that whenever each f_i has degree $\leq k_0$ we have $\{f_1, \ldots, f_{m-1}\}^n \subseteq \langle s_1, \ldots, s_d \rangle$ for suitable n depending on $f_1, \ldots, f_{m-1}$. We want to select the next bunch of s_j, each of degree $> k_0$. To do this formally write

$$f_i = \sum \alpha_{iw} w \qquad \text{for } 1 \leq i \leq m - 1$$

where $\alpha_{iu} \in F$ and w runs over the words in $X_1, \ldots, X_m$ having degree between 1 and k_0. Now any product of length n (to be determined) of the f_i has total degree $\leq nk_0$ and is a linear combination of homogeneous elements whose coefficients are products of the α_{iu}. Taking $s_{d+1}, s_{d+2} \cdots$ to be these homogeneous elements we formally have satisfied condition (ii).

It remains to verify (i), which we claim is automatic if n is large enough. First of all taking $n > d$ we see that each of the new s_j have degree $> d$; the number of possible new s_j is at most the number of commutative monomials in the α_{iw}, and this can be estimated as $\leq d'd''$ where

d' is the number of commutative monomials having length n in the α_{1w};
$d'' = (m - 1)^n$ is the number of possible ways to replace a product $\alpha_{1w_1} \cdots \alpha_{1w_n}$ by $\alpha_{i_1 w_1} \cdots \alpha_{i_n w_n}$ where $1 \leq i_u \leq m - 1$ for each u.

Note d' is the number of ways of choosing n from

$$q = (m - 1)(m + m^2 + \cdots + m^{k_0})$$

(the number of words of degree $\leq k_0$) allowing repetitions, which by remark 6.2.8 is

$$\binom{q + n - 1}{n} \leq (q + n - 1)^{q-1}.$$

Let $c = (m - 2\varepsilon)/(m - 1)$. Taking $\varepsilon < \frac{1}{2}$ we may assume $c > 1$. For sufficiently large n we have $(q + n - 1)^{q-1} \leq \varepsilon^2 c^n/(m - 2\varepsilon)^2$ since the exponential rises faster than the polynomial; thus for such n we see

$$d'd'' \leq (q + n - 1)^{q-1}(m - 1)^n \leq (\varepsilon^2 c^n/(m - 2\varepsilon)^2)(m - 1)^n$$

$$= \varepsilon^2(m - 2\varepsilon)^n/(m - 2\varepsilon)^2 = \varepsilon^2(m - 2\varepsilon)^{n-2}.$$

Since the new monomials all have degree $\geq n > k_0$ we see (i) holds, as desired.   Q.E.D.

Note 6.2.9':   The algebra of theorem 6.2.9 is considerably more than a counterexample to Kurosch's problem. For example, taking $m = 4$ we get a counterexample R_0 for which $M_2(R_0)$ also is nil. If we are interested in finding the quickest counterexample to Kurosch's problem we could require merely that each s_j has a power in $\langle S \rangle$; thus one can work with one j at a time and streamline the above proof considerably, cf., Herstein [69B] which produces a quick application of lemma 6.2.7 for F countable and $m = 3$.

Growth of Algebras

The crux of the Golod-Shafarevich example is that S was "sparse" enough that the algebra R although nil grew exponentially in terms of the generators. Thus it makes sense to require that our algebras do not grow so fast. This leads us to the following situation. For the remainder of this section *we assume R is an algebra over a field F. If $S = \{r_1, \ldots, r_m\}$ we write $F\{S\}$ for $F\{r_1, \ldots, r_m\}$.*

Definition 6.2.10:   Suppose $R = F\{S\}$. Let $V_n(S)$ be the subspace $\sum_{k=0}^{n} FS^k$, and define the *growth function* $G_S(n)$ of R by

$$G_S(n) = [V_n(S):F].$$

We say R has *exponential growth* if $G_S(n) \geq t^n$ for some $t > 1$ and all $n \in \mathbb{Z}$; otherwise, R has *subexponential growth*. R has *polynomially bounded growth* if $G_S(n) \leq cn^t$ for suitable c, t in $\mathbb{N}$, for all n in $\mathbb{N}$.

Examples 6.2.11:

(i) If R is the free algebra $F\{S\}$ where $S = \{X_1, \ldots, X_m\}$ then $G_S(n) = \sum_{k=0}^{n} m^k$ so R has exponential growth.

(ii) If $R = F[\lambda_1, \ldots, \lambda_m]$ then as noted in remark 6.2.8 we have $\binom{m+n-1}{n}$ elements in $\{\lambda_1, \ldots, \lambda_m\}^n$, and thus $G_S(n) = \sum_{k=0}^{n} \binom{m+k-1}{k} = \binom{m+n}{n}$

$\left(\text{since } \dbinom{m+n}{n} = \dbinom{m+n-1}{n} + \dbinom{m+n-1}{n-1}\right.$ and apply induction to

$\left.\dbinom{m+n-1}{n-1}\right)$. But $\dbinom{m+n}{n} = \dfrac{(n+m)\cdots(n+1)}{m\cdots 1} \leq 2n^m$ whenever $n > 1$. Con-

sequently $G_S(n)$ has polynomially bounded growth. Note for $m \geq 1$ and $n > 4$ that $G_S(m) \leq n^m$.

(iii) The Weyl algebra $\mathbb{A}_n(F)$ "grows" at the same rate as $F[\lambda_1, \ldots, \lambda_n]$, so $G_S(n) \leq 2n^m$.

It would be rather embarrassing if the type of growth changed according to the generating set, so we need the following fact.

Remark 6.2.12: If $R = F\{S\} = F\{S'\}$ for S' finite then $S' \subseteq S^{n'}$ for some n'. (Each element of S' lies in suitable $V_{n'}(S)$ so take the maximal such n'.) Consequently $G_{S'}(n'n) \geq G_S(n)$ for all n, thereby implying exponential, subexponential, and polynomially bounded growth are each well-defined.

This remark leads us to the following definition.

Definition 6.2.13: Suppose R, R' are affine F-algebras where $R = F\{S\}$ and $R' = F\{S'\}$. R' grows *at least as fast* as R, written $G(R) \leq G(R')$, if there are positive integers n_0, n_1 for which $G_S(n) \leq n_1 G_{S'}(n_0 n)$ for all n in $\mathbb{N}$. We say R and R' have the *same* growth if $G(R') \leq G(R)$ and $G(R') \leq G(R)$.

In view of remark 6.2.12 this definition is independent of the choice of generating sets S and S'. Thus we have the following observations.

Remark 6.2.14:

(i) If R is a homomorphic image of R' then $G(R) \leq G(R')$. In particular, if $R = F\{r_1, \ldots, r_m\}$ then $G(R) \leq G(F\{X_1, \ldots, X_m\})$.

(ii) If R is an affine subalgebra of R' then $G(R) \leq G(R')$. (Indeed, if S, S' are generating sets of R, R', respectively, then replace S' by $S \cup S'$ and the assertion is obvious.)

(iii) If $R = F\{S\}$ and $R' = F\{S'\}$ then $R \otimes_F R'$ has generating set $S'' = (S \otimes 1) \cup (1 \otimes S')$, so $V_n(S) \otimes V_n(S') \subseteq V_{2n}(S'') \subseteq V_{2n}(S) \otimes V_{2n}(S')$. In particular, if $[R':F] < \infty$ then $R \otimes_F R'$ and R have the same growth. As a special case R and $M_n(R) \approx R \otimes_F M_n(F)$ have the same growth.

(iv) If $R \subseteq R'$ and R' is f.g. as R-module then R' and R have the same growth, by (iii) and the regular representation of R' as an F-subalgebra of $M_t(R)$.

(v) The Noether normalization theorem shows if C is F-affine commutative of transcendence degree t then C has the same growth as $F[\lambda_1, \ldots, \lambda_t]$.

(vi) Viewing the $\{V_n(S): n \in \mathbb{N}\}$ as a filtration of R, we see by comparing dimensions that R has the same growth as the associated graded algebra.

There are domains having subexponential but not polynomially bounded growth, cf., exercise 8.3.14. Having various examples at our disposal, let us now show that subexponential growth has important structural implications.

Proposition 6.2.15: *(Jategaonkar) Suppose R is a domain not necessarily with 1, and R does not contain a copy of the free algebra $F\{X_1, X_2\}$ without 1. (In particular, this latter hypothesis holds when R has subexponential growth). Then R is left and right Ore.*

Proof: Given $0 \neq r_1, r_2$ in R take $f \in F\{X_1, X_2\}$ of minimal total degree such that $f(r_1, r_2) = 0$. Writing $f = f_1(X_1, X_2)X_1 + f_2(X_1, X_2)X_2$ we have $f_i(r_1, r_2) \neq 0$ for $i = 1, 2$ by assumption on i, so $f_1(r_1, r_2)r_1 = -f_2(r_1, r_2)r_2 \neq 0$, proving $Rr_1 \cap Rr_2 \neq 0$. The right Ore condition is proved analogously. Q.E.D.

This does not generalize to arbitrary prime algebras of subexponential growth, cf., exercise 7. However, Irving-Small [83] do have a generalization.

Corollary 6.2.16: *Suppose R is an affine prime F-algebra having subexponential growth. If R has a left ideal L for which Ann$'$ L is a maximal right annihilator then L is uniform.*

Proof: Let $I = \text{Ann}' L$. Then $I \cap L \lhd L$ as rng, and $\bar{L} = L/(I \cap L)$ is a domain; indeed, if $a_1, a_2 \in L - (I \cap L)$ then $La_1 \neq 0$ implying $(La_1)a_2 \neq 0$ (for, otherwise, Ann$'$ $La_1 \supset I$) and so $a_1 a_2 \notin I \cap L$. By the proposition $\bar{L}$ is left and right Ore. To prove L is uniform suppose $0 \neq L_1, L_2 < L$. Then their images $\bar{L}_1$ and $\bar{L}_2$ in $\bar{L}$ are nonzero since $L_i L \neq 0$ (for R is prime). Hence $\bar{L}_1$ and $\bar{L}_2$ have a common element $\bar{a} \neq 0$, and $0 \neq La \subseteq L_1 \cap L_2$. Q.E.D.

Theorem 6.2.17: *(Irving-Small) Suppose R is an affine, semiprime F-algebra satisfying ACC on left and right annihilators. If R has subexponential growth then R is Goldie.*

Proof: By theorem 3.2.27 we may assume R is prime. But ACC(right annihilators) implies DCC(left annihilators). Let L be a minimal left annihilator. By corollary 6.2.16 L is uniform. R is prime so $\text{Ann } L = 0$ and consequently there are $a_1, \ldots, a_n$ in L for which $\text{Ann}\{a_1, \ldots, a_n\} = 0$. The map $r \mapsto (ra_1, \ldots, ra_n)$ gives us a monic $R \to \bigoplus_{i=1}^{n} Ra_i \subseteq L^{(n)}$ implying R has finite uniform ($=$ Goldie) dimension. Since R has ACC(Ann) by hypothesis we conclude R is Goldie.   Q.E.D.

(The hypothesis ACC on *left* annihilators is superfluous, cf., exercise 8; however, some chain condition is needed in light of exercise 7.) Having whetted our appetites for studying subexponential growth and, in particular, polynomially bounded growth, we want a more precise measure of growth.

Gelfand-Kirillov Dimension

Definition 6.2.18:   The *Gelfand-Kirillov* dimension of an affine algebra R, written GK dim(R), is $\overline{\lim}_{n \to \infty} \log_n(G_S(n)) = \overline{\lim}_{n \to \infty}(\log G_S(n)/\log n)$.

The point of this definition is that if $G_S(n) \leq cn^t$ then GK dim(R) $\leq \overline{\lim}_{n \to \infty}(\log_n c + t) = t$. Thus one sees R has finite GK dim if R has polynomially bounded growth. Similarly we have

Remark 6.2.18': $\overline{\lim} \log_n G_S(n) = \overline{\lim} \log_n G_S(n_0 n)$, for fixed $n_0 > 0$, since $\log_n G_S(n_0 n) = \log G_S(n_0 n)/\log n = (\log G_S(n_0 n)/\log n_0 n)((\log n_0 n)/(\log n))$ and $(\log n_0 n)/\log n = 1 + (\log n_0/\log n) \to 1$ as $n \to \infty$. Consequently if $G(R) \leq G(R')$ then GK dim $R \leq$ GK dim R'.

When R is not affine we could still define GK dim(R) as sup$\{$GK dim(R'): R' is an affine subalgebra of $R\}$. A related notion (nowadays called GK tr deg) was introduced by Gelfand-Kirillov [66] to show that the Weyl algebras $A_n(F)$ and $A_m(F)$ have nonisomorphic rings of fractions for $n \neq m$ and to frame the Gelfand-Kirillov conjecture (cf., §8.3). Within a decade GK dim was seen to be a general tool in ring theory. A thorough, readable treatment of GK dim is given in Krause-Lenagan [85B], which is the basis of much of the discussion here.

By example 6.2.11 we see that $\text{GK}(F[\lambda_1, \ldots, \lambda_m]) = m$, and $\text{GK}(A_m(F)) = 2m$. On the other hand, K-dim $F[\lambda_1, \ldots, \lambda_m] = m$ by exercise 3.5.23, but K-dim $A_m(F) = m$ by exercise 3.5.25. See exercise 9 for a worse example.

Remark 6.2.19:   If $R \subseteq R'$ or R is a homomorphic image of R' then GK dim(R) $\leq$ GK dim(R') by remark 6.2.14(i), (ii).

Remark 6.2.20: If R' is f.g. as module over a subring R then GK dim$(R') =$ GK dim(R) by remark 6.2.14(iv). This simple observation has the following consequences:

(i) If C is commutative and F-affine then GK dim $C = \operatorname{tr} \deg(C/F)$ by Noether normalization and example 6.2.11(ii).

(ii) GK dim$(M_n(R)) = $ GK dim(R).

(iii) If R is locally finite over F then GK$(R) = 0$. (Indeed one may assume R is affine and thus finite dimensional over F, so GK dim$(R) = $ GK dim$(F) = 0$.)

As typified by these results, the GK dimension is usually difficult to compute from scratch but is very amenable because closely related rings often have the same GK dimension. Another instance is

Proposition 6.2.21: *If R is a subdirect product of $R_1,\ldots,R_t$ then* GK dim $R = \max\{$GK dim$(R_i): 1 \le i \le t\}$.

Proof: "$\ge$" is clear by remark 6.2.19, and to prove "$\le$" we note $R \subseteq \prod_{i=1}^{t} R_i$, so it suffices to assume $R = \prod R_i$. Let S_i be a generating set for R_i, with $1 \in S_i$ for each i. Then $S = S_1 \times \cdots \times S_n$ is certainly a generating set for R, and $V_n(S) = \prod V_n(S_i)$; writing $G_i(n)$ for the generating function of S_i in R_i we have

$$G_S(n) = \sum_{i=1}^{t} G_i(n) \le t \max\{G_i(n): 1 \le i \le t\}.$$

Since $\log_n t \to 0$ we see GK dim$(R) = \max$ GK dim(R_i), as desired.

Q.E.D.

Proposition 6.2.22: GK dim$(R[\lambda]) = 1 + $ GK dim(R).

Proof: Given $S = \{1, r_1, \ldots, r_t\}$ let $S' = S \cup \{\lambda\} \subset R[\lambda]$; then $V_{2n}(S') \supset \sum_{i=0}^{n} V_n(S)\lambda^i$ so in view of remark 6.2.18' we see

$$\text{GK dim}(R[\lambda]) \ge \overline{\lim} \log_n(n+1)G_S(n) = 1 + \overline{\lim} \log_n G_S(n)$$

proving $(\ge)$.

To prove $(\le)$ note that any affine subalgebra T of $R[\lambda]$ is contained in $R_1[\lambda]$ where R_1 is the affine subalgebra of R generated by the coefficients of the generators of T. Thus we may assume R is affine. Now taking S, S' as above we have $V_n(S') \subset \sum_{i=0}^{n} V_n(S)\lambda^i$ so

$$\text{GK dim}(R[\lambda]) \le \overline{\lim} \log_n(n+1)G_S(n) = \overline{\lim} \log_n(n+1) + \overline{\lim} \log_n G_S(n)$$

$$= 1 + \text{GK dim}(R). \quad \text{Q.E.D.}$$

Remark 6.2.22':   The same proof shows that $\text{GK dim}(T) \geq 1 + \text{GK dim}(R)$ whenever there is x in T such that $\{x^i : i \in \mathbb{N}\}$ are R-independent.

The opposite inequality is trickier and can fail for Ore extensions; cf., theorem 8.2.16 and exercise 8.3.13. However, it holds for differential polynomial extensions which are affine, cf., exercise 10.

Proposition 6.2.23: $\text{GK dim}(S^{-1}R) = \text{GK dim}(R)$ *for any regular submonoid S of $Z(R)$.*

Proof: Any finite subset of $S^{-1}R$ can be written in the form $s^{-1}r_1, \ldots, s^{-1}r_t$, and thus $s^n\{s^{-1}r_1, \ldots, s^{-1}r_t\}^n \subseteq \{r_1, \ldots, r_t\}^n$, from which we see at once that R grows at least as fast as $S^{-1}R$. Thus $\text{GK dim}(S^{-1}R) \leq \text{GK dim}(R)$, and equality follows by remark 6.2.14(ii) since $R \subseteq S^{-1}R$ (because S is regular).
$$\text{Q.E.D.}$$

On the other hand, this result does not hold for arbitrary Ore sets S, in view of Makar-Limanov's result that the division algebra of fractions of $A_1(F)$ does not have GK dim, cf., example 7.1.46.

Proposition 6.2.24:   *Suppose an ideal A of R contains a left regular element a. Then $\text{GK dim}(R) \geq 1 + \text{GK dim}(R/A)$.*

Proof: Take any finite set S of R containing 1 and a. Let $V'_n = A \cap V_n(S)$ and let V''_n be a complementary subspace of V'_n in $V_n(S)$. Since $V''_n \cap Ra = 0$ we see

$$V''_n, aV''_n, \ldots, a^n V''_n$$

are independent subspaces all contained in $V_{2n}(S)$, so

$$G_S(2n) \geq n[V''_n : F] = nG_{\bar{S}}(n)$$

where $\bar{S}$ is the image of S in R/A. Thus by remark 6.2.18'

$$\text{GK dim}(R) \geq \overline{\lim} \log_n G_S(2n) \geq \overline{\lim} \log_n(nG_{\bar{S}}(n))$$

$$= 1 + \overline{\lim} \log_n(G_{\bar{S}}(n)) = 1 + \text{GK dim}(R/A). \qquad \text{Q.E.D.}$$

Theorem 6.2.25:   *Suppose the prime images of R are left Goldie. (In particular, this is true if R is PI or if R is left Noetherian.) Then*

$$\text{GK dim}(R) \geq \text{GK dim}(R/P) + \text{height}(P) \qquad \text{for any P in $\text{Spec}(R)$.}$$

Proof: For any chain $P = P_0 \supset P_1 \supset \cdots \supset P_m$ we need to show GK dim$(R) \geq$ GK(R/P) + m, which would certainly be the case if GK$(R/P_i) \geq$ GK(R/P_{i-1}) + 1 for each i; but this latter assertion follows from proposition 6.2.24 since P_{i-1}/P_i is a nonzero ideal of the prime Goldie ring R/P_i and thus contains a regular element.   Q.E.D.

Note: It follows that if GK dim(R/P_1) = GK dim(R/P_2) for P_1, P_2 in Spec(R) then P_1 and P_2 are incomparable.

Corollary 6.2.25′: *Under hypothesis of theorem 6.2.25, the little Krull dimension of R is at most GK dim(R).*

Proposition 6.2.26 GK dim$(R_1 \otimes R_2) \leq$ GK dim(R_1) + GK dim(R_2).

Proof: We may assume R_1, R_2 are affine. Then the inequality follows from remark 6.2.14(iii).   Q.E.D.

A useful related result involves changing the base field.

Remark 6.2.26′: If K is a field extension of F then GK dim $R =$ GK dim $R \otimes_F K$ (as K-algebra) for any affine F-algebra R. (Indeed, if S generates R then $S \otimes 1$ generates $R \otimes_F K$, and $G_S(n) = \sum_{k \leq n}[FS^k:F] = \sum_{k \leq n}[FS^k \otimes K:K] = G_{S \otimes 1}(n)$, so appeal to the definition.)

One may wonder, "When does equality hold in proposition 6.2.26?", i.e., when does $\alpha_1 + \alpha_2 \leq$ GK dim$(R_1 \otimes R_2)$ when $\alpha_i =$ GK dim R_i? The hitch in proving equality in general via remark 6.2.14(iii) is that when taking a sequence $n_1, n_2 \ldots$ such that $\log G_{S_2}(n_i)/\log n_i \to \alpha_2$, one need not have $\log G_{S_1}(n_i)/\log n_i \to \alpha_1$. Of course, this occurs if also $\alpha_1 = \underline{\lim} \log G_{S_1}(n)/\log n$, which we shall temporarily call the *lower* GK-dim of R_1. In certain cases the GK-dim and lower GK-dim agree and are integervalued; for example, when R_1 is either commutative (remark 6.2.20(i)), affine PI (theorem 6.3.41), or the homomorphic image of an enveloping algebra of a finite dimensional Lie algebra (remark 8.3.37). Also see exercise 6.3.24.

Indeed, it is not easy to come up with an example for which GK dim R is not an integer, but Borho-Kraft [76] showed for any real α between 2 and 3 that there is a homomorphic image of $F\{X_1, X_2\}/\langle X_2^3 \rangle$ which has GK dim α. In view of proposition 6.2.22 GK dim R thus can take on any real value ≥ 2.

Warfield [84] showed equality need not hold in proposition 6.2.26, and

Krempa-Okninski [86] extend Warfield's example to an algebra R whose GK-dim and lower GK-dim take on arbitrary values $\alpha > \alpha' \geq 2$. For GK-dim ≤ 2, however, the story is completely different, as we shall now see.

Remark 6.2.27: The GK dimension of R cannot be strictly between 0 and 1; in fact, $GK(R) \geq 1$ if R is not locally finite over F. (Indeed, by remark 6.2.20(iii) we may assume R is affine but not finite dimensional over F. Then $V_n(S) \neq R$ for any n, where S is a generating set; hence $V_n(S) \subset V_{n+1}(S)$ for each n, implying $G_S(n) \geq n$ and thus $GK(R) \geq 1$.)

Bergman showed that the GK dimension also cannot be strictly between 1 and 2; his proof was given in the unabridged text.

Returning to Kurosch's problem, we see that if GK dimension = 0 or 1 then Kurosch's problem has a trivial solution. Thus we are led to ask,

Question 6.2.35: If R is algebraic over F and $GK \dim(R) < \infty$ then is $GK(R) = 0$?

The GK Dimension of a Module

Although the GK dimension is used predominantly for algebras, it is also applicable to modules.

Definition 6.2.36: If R is affine and $M = \sum_{i=1}^{t} Rx_i \in R\text{-}\mathscr{F}imod$ then $GK \dim(M) = \overline{\lim}_{n \to \infty} \log_n [\sum V_n(S) x_i : F]$, notation as in definition 6.2.10. More generally, for any module M over an F-algebra R define $GK \dim M = \sup\{GK \text{ dimension of f.g. submodules over affine subalgebras of } R\}$.

It is not difficult to show this definition is well-defined. Clearly $GK \dim(R)$ as R-module reduces to the definition of $GK \dim(R)$ as algebra, and the GK module dimension has ridden on the shoulders of the algebraic-theoretic dimension. Its main disadvantage in general is that is not *exact*, i.e., one does not always have the formula $GK \dim M = \max\{GK \dim(N), GK \dim(M/N)\}$ for $N < M$, cf., exercise 13. We do have the following partial results.

Remark 6.2.37: $GK \dim M \geq \max\{GK \dim(N), GK \dim(M/N)\}$ for any $N < M$, as an immediate consequence of the definition. Furthermore, $GK \dim M = GK \dim(M^{(t)})$ for any t since taking $x_i = (0, \ldots, 1, \ldots, 0)$

$$GK \dim M^{(t)} = \overline{\lim} \log_n \left[\sum_{i=1}^{t} V_n(S) x_i : F \right] = \overline{\lim} \log_n (t [V_n(S) x_1 : F])$$

$$= \overline{\lim}(\log_n t + \log_n[V_n(S)x_1 : F] = \text{GK dim } M.$$

In particular, for any f.g. R-module M we have GK dim $M \leq$ GK dim $R^{(t)} =$ GK dim R (for suitable t).

§6.3 Affine PI-Algebras

We start this section by verifying that for affine PI-algebras the general questions raised in §6.2 actually have positive solutions. In fact, using Shirshov's positive solution to Kurosch's problem we can describe the structure of prime affine PI-algebras extremely precisely, leading to many important results such as the catenarity of affine PI-algebras and the nilpotence of the Jacobson radical. Several techniques obtained thereby, most notably the use of the trace ring, are applicable for arbitrary prime PI-rings, and we shall note their impact on the Bergman-Small theorem and on localization (but without proof). Often the base ring C is required to be Noetherian and/or Jacobson; this is certainly the case when C is a field.

The Nullstellensatz

There are several ways of proving the Nullstellensatz for affine PI-algebras, as we shall see; possibly the fastest method is Duflo's proof using generic flatness. In order to achieve the result when the base ring is arbitrary Jacobson we need another version of the Artin-Tate Lemma.

Proposition 6.3.1: *Suppose R is C-affine and is f.g. free as Z-module with a base containing 1, for suitable $Z \subseteq Z(R)$. Then Z is affine.*

Proof: Let $R = C\{r_1, \ldots, r_n\}$ and let $b_1 = 1, \ldots, b_t$ be a base of R as Z-module. As in the proof of proposition 6.2.5 write $b_i b_j = \sum_{k=1}^{t} z_{ijk} b_k$ and $r_u = \sum_k z_{uk} b_k$, and let $C' = C[z_{ijk}, z_{uk} : 1 \leq i, j, k \leq t; 1 \leq u \leq n]$. Then C' is affine. Moreover, $\sum_{k=1}^{t} C' b_k$ is a subalgebra of R containing each r_i and so equals R. It follows $C' = Z$, for if $\sum c'_k b_k = z \in Z$ then matching coefficients of 1 shows $z = c'_1 \in C'$, as desired.   Q.E.D.

Corollary 6.3.2: *Suppose R is C-affine primitive PI. Then C is a G-domain, i.e., there is $s \neq 0$ in C for which $C[s^{-1}]$ is a field F; furthermore, $[R : F] < \infty$.*

Proof: Let F be the field of fractions of C. By Kaplansky's theorem R is finite dimensional over $Z(R)$, which is affine over F by proposition 6.3.1. Write $Z(R) = F[z_1, \ldots, z_t]$.

We claim any field of the form $F[z_1, \ldots, z_t]$ is algebraic over F. This is seen by induction on the transcendence degree over F. Suppose, on the contrary, z_t is transcendental over F, and let F_1 be the field of fraction of $F[z_1, \ldots, z_{t-1}]$ taken inside $Z(R)$. Then $Z(R) = F_1[z_t]$ is a field, implying z_t is algebraic over F_1 (for, otherwise, $F_1[z_t] \approx F_1[\lambda]$ is not a field) and thus $[Z(R):F_1] < \infty$ implying F_1 is affine by proposition 6.3.1. But tr $\deg(F_1) <$ tr deg $Z(R)$ so by induction F_1 is algebraic over F, yielding the claim.

Thus $[Z(R):F] < \infty$ so $[R:F] < \infty$. Hence F is C-affine by proposition 6.3.1. Write $F = C[s^{-1}c_1, \ldots, s^{-1}c_t]$ for s, c_i in C; then $F = C[s^{-1}]$ as desired.   Q.E.D.

Theorem 6.3.3: (*Amitsur-Procesi*) *Suppose R is an affine PI-algebra over a commutative Jacobson ring C. Then the generic flatness hypotheses of theorem 2.12.36 are satisfied, and consequently*

   (i) *R is Jacobson.*
   (ii) *R satisfies the maximal Nullstellensatz.*
   (iii) *If P is a primitive ideal of R then $C \cap P$ is a maximal ideal of C, and R/P is simple and finite dimensional over $C/C \cap P$.*

Proof: (Duflo[73]) Since any image of $R[\lambda]$ is also affine PI, we need only to show that R/PR satisfies generic flatness over C/P for every P in Spec(C); passing to P/PR over C/P we may assume C is a domain. Let M be a simple R-module; by remark 2.12.33(i) we may assume M is faithful over C. But then $R/\text{Ann}_R M$ is C-affine, so $C[s^{-1}]$ is a field for some s, by corollary 6.3.2. $M[s^{-1}]$ is free over this field, so we have verified generic flatness.

Now conditions (i) and (ii) follows from theorem 2.12.36. The first part of (iii) comes from remark 2.12.33″, taking P the annihilator of a suitable simple module. But then R/P is affine over the field $F = C/C \cap P$ and is primitive PI so we are done by corollary 6.3.2.   Q.E.D.

Another proof of theorem 6.3.3 comes from the powerful "trace ring" techniques of the latter part of this section; a far-reaching generalization of McConnell [84a] is discussed in §8.4.

The weak Nullstellensatz shows Jac(R) is nil for an affine PI-algebra R over a field, so the next question is whether Nil(R) is nilpotent. The story of this question involves most of the important ideas of affine algebras. Razmyslov [74] (and later Schelter who had been unaware of Razmylov's work) proved the nilpotence of the radical for homomorphic images of prime affine PI-algebras. To obtain the maximal value from this result we need an affine version of generic matrices. Let $F_n\{Y_1, \ldots, Y_m\}$ denote the subalgebra of the algebra of generic matrices $F_n\{Y\}$.

Remark 6.3.4: $F_n\{Y_1, \ldots, Y_m\}$ is F-affine and PI, and is prime by the argument of proposition 6.1.41. If R satisfies all the identities of $M_n(F[\Lambda])$ then R is a homomorphic image of $F_n\{Y_1, \ldots, Y_m\}$, by proposition 6.1.42(ii).

Thus Razmyslov's theorem (once proved) yields the nilpotence of the nilradical for every affine PI-algebra which satisfies the identities of $n \times n$ matrices. Since virtually all affine PI-algebras are visibly seen to satisfy the identities of $n \times n$ matrices, this formulation is quite satisfactory; furthermore, Lewin proved the converse, that nilpotence of the nilradial implies satisfaction of the identities of $n \times n$ matrices.

Razymslov also showed in characteristic 0 that it is enough to assume the affine algebra satisfies a Capelli identity. Kemer showed that every affine PI-algebra of characteristic 0 indeed satisfies a Capelli identity, thereby verifying the nilpotence of the radical of arbitrary affine PI-algebras in characteristic 0. Nevertheless, the PI community was relieved when Braun proved the nilpotence of the nilradical in complete generality, in arbitrary characteristic, using a different technique.

Shirshov's Theorem

Having established an interesting converse to our question of nilpotence of the radical, we return to our casual pursuit of the original question (or better, its solution). There is a surprising connection to Kurosch's problem, which has a positive solution for affine PI-algebras; in fact, the solutions by Kaplansky and Levitzki were the first in the long list of major theorems in PI theory. There is a fast structural solution, cf., exercise 12, but we shall use a solution of Shirshov [57] based on combinatoric techniques on words which only requires the algebraicity of a finite (albeit huge) number of elements; this becomes very important in the applications. We start with the word monoid M in $\{1, \ldots, t\}$, cf. example 1.2.4, given the (partial) lexicographic order of 1.2.8, 1.2.9. "w" always will denote a word. We say w' is a *subword* of w if there are words v', v'', (possibly blank) for which $w = v'w'v''$; w' is *initial* if v' is blank.

Note words $w_1 \neq w_2$ are incomparable if one is an initial subword of the other; otherwise they are comparable. We use an ingenious idea of Below. Write $|w|$ for the length of w.

Lemma 6.3.18: *Suppose* $|w| = n$. *Write* $w(t)$ *for the tth position in w, $1 \leq t \leq n$, and w'_i for $w(i)w(i + 1)\ldots w(n)$. Then for any $i < j$, either*

w'_i and w'_j are comparable or else w'_i has the form $u^k v$ where $|u| = j - i$, v is an initial subword of u, and $k = [(n - i)/(j - i)]$.

Proof.   Note $w'_i = uw'_j$ where $u = w(i)w(i + 1)...w(j - 1)$. We are done unless w'_i and w'_j are incomparable, so w'_j is also an initial subword of w'_i. Take k maximal such that u^k is an initial subword of w'_i, and write $w'_i = u^k v$. Then

$$|v| = |w'_i| - |u^k| = |w'_j| - |u^{k-1}|$$

If $|v| \geq |u|$, we conclude $|w'_j| \geq |u^k|$. Hence, u^k is an initial subword of w'_j, implying u^{k+1} is an initial subword of w'_i, contrary to the choice of k. Hence $|v| < |u|$, so $w'_j = u^{k-1}v$ is an initial subword of u^k, implying v is an initial subword of u.   Q.E.D.

Definition 6.3.19.   A word w is *m-decomposable* (with *m-decomposition* $w_0 w_1 ... w_m w_{m+1}$) if it can be written in the form $w_0 w_1 ... w_m w_{m+1}$ where $w_1, ..., w_m$ are subwords such that $w_{\pi 1}... w_{\pi m} < w_1 ... w_m$ for any nonidentity permutation π of $w_1, ..., w_m$.

This certainly will be the case if $w_1 > ... > w_m$. The following simplification of Shirshov's combinatorial approach is due to Belov.

Lemma 6.3.20.   *Assume $|w| > m$. Any word of the form $\tilde{w} = v_1 w v_2 w ... v_m w v_{m+1}$ (where w appears m times) is either m-decomposable or else has a subword u^k where $|u| \leq m$ and $k = [|w|/m] - 1$.*

Proof.   Apply the notation of lemma 6.3.18 to $\tilde{w}$; we are done unless $w'_1, ..., w'_m$ are pairwise comparable. (Recall these are the terminal subwords of w of respective lengths $|w|, ..., |w| - m + 1$.) Suppose $w'_{\rho 1} > ... > w'_{\rho m}$ for a suitable permutation ρ of $\{1, ..., m\}$. Write $w''_{\rho i} = w(1)w(2)...w(\rho i - 1)$, i.e.,

$$w = w''_{\rho i} w'_{\rho i}$$

Then $\tilde{w} = v_1 w''_{\rho 1} w'_{\rho 1} v_2 w''_{\rho 2} w'_{\rho 2} ... v_m w''_{\rho m} w'_{\rho m} v_{m+1}$. Setting

$$w_0 = v_1 w'_{\rho 1}, \; w_1 = w'_{\rho 1} v_2 w''_{\rho 2}, \; w_2 = w'_{\rho 2} v_3 w''_{\rho 3}, ... \; w_{m+1} = w'_{\rho m} v_{m+1},$$

we see $\tilde{w} = w_0 w_1 ... w_m w_{m+1}$ is the desired m-decomposition.   Q.E.D.

Theorem 6.3.23.   *(Belov's version of Shirshov's Theorem.) Suppose $R = C\{r_1, ..., r_t\}$ satisfies a PI of degree m, and suppose all monomials in the r_i*

of length $\leq m$ are algebraic of degree $\leq n$. Then R is f.g. as C-module, spanned by the products in the r_i of length $\leq n' = m^2(n + 1)t^{m(n + 1)}$.

Proof. It suffices to show that any word w in the r_i of length $> n'$ is spanned by smaller words (corresponding to the lexicographic order), for then we can conclude by induction (on length of word and on order). Partition w into $mt^{m(n + 1)}$ subwords each of length $m(n + 1)$. The number of distinct words of length $m(n + 1)$ is $t^{m(n + 1)}$, so m of these must be the same, and lemma 6.3.20 is applicable.

If w is m-decomposable, then the polynomial identity (multilinearized) yields w in the span of words of smaller order.

If w has a subword u^k with $|u| \leq m$ and $k = [m(n + 1)/m] - 1 = n$, then the algebraicity hypothesis yields w in the span of words of smaller order. Thus we are done in either case.   Q.E.D.

Schelter noted that the same combinatorial argument acually gives a stronger result, cf., exercise 13 and Rowen [80B, theorem 4.2.8]. Shirshov proved the following stronger result than theorem 6.3.23.

Theorem 6.3.24: *(Shirshov) Suppose $R = C\{r_1,\ldots,r_t\}$ satisfies a multilinear polynomial identity $f(X_1,\ldots,X_m)$. Let $w_1,\ldots,w_{m'}$ denote the words of length $\leq m$, and let $\bar{w}$ denote the image of a word w in R, substituting r_i for each occurrence of i. There is a number $k = k(m,t)$ such that for any word w of length $\geq m^2(m + 1)t^{m(m + 1)}$ we have*

$$\bar{w} \in \sum C\bar{v}_{i_1} \cdots \bar{v}_{i_k}$$

where each $i_j \in \{1,\ldots,m'\}$, v_i denotes a power of w_i, and each word $v_{i_1} \cdots v_{i_k}$ is a rearrangement of the letters of w.

Proof: Let $\beta = m^2(m + 1)t^{m(m + 1)}$ and $k = (m')^2 m\beta$. Proof is by induction on the order of w (cf., example 1.2.18). In particular, we may assume w does not have an m-decomposable subword since then we could use the identity f to express $\bar{w}$ in terms of lower order monomials, just as in the proof of theorem 6.3.23 case (i). Write

$$\bar{w} = \bar{v}_{i_1} \cdots \bar{v}_{i_d}$$

where each v_i denotes a power of w_i and each $i_j \in \{1,\ldots,m'\}$, such that d is minimal possible, temporarily we call d the *breadth* of w. We are done unless w has breadth $> k$, so, in particular, length$(w) > k$.

Claim. Any subword v of w of breadth $> \beta$ contains a subword $\tilde{w}^m \tilde{w}'$ where $\tilde{w},\tilde{w}'$ have length $\leq m$ and each of $\tilde{w},\tilde{w}'$ is not an initial subword of the other.

Proof of claim. *A fortiori* v cannot have an m-decomposable subword, so by lemma 6.3.20 v contains a subword $\tilde{w}^m$ with length$(\tilde{w}) \leq m$. We choose $\tilde{w}$ such that the initial subword $v'\tilde{w}^m$ has length(v') minimal possible. Write $v = v'\tilde{w}^u v''$ for $u \geq m$, such that $\tilde{w}$ is not an initial subword of v''. We are done unless v'' is an initial subword of $\tilde{w}$. Write $\tilde{w} = v''a$ for a suitable subword a, and note length$(v'') <$ length$(\tilde{w}) \leq m$. Now

$$v = v'v''(av'')^u.$$

Since length$(av'') =$ length$(\tilde{w}) \leq m$ we see breadth$(v'v'') \geq$ breadth$(v) - 1 \geq \beta$, so again lemma 6.3.20 shows $v'v''$ has a subword $\hat{w}^m$ for suitable $\hat{w}$ of length $\leq m$. But then $v'v''$ has an initial subword $a'\hat{w}^m$ where length$(a') \leq$ length$(v') +$ length$(v'') - m <$ length (v'); since $a'\hat{w}^m$ is also an initial subword of v we have contradicted the mimimality of length (v').

Having established the claim we have a subword $\tilde{w}^m\tilde{w}'$ where neither $\tilde{w}$ nor $\tilde{w}'$ is initial in the other. But the number of choices for the pair $(\tilde{w}, \tilde{w}')$ is $\leq (m')^2$; since $k = (m')^2 m\beta$ we can apply the claim successively to the remaining subwords to assure that some $\tilde{w}^m\tilde{w}'$ recurs in m distinct places. Thus w has a subword

$$\tilde{w}^m\tilde{w}'a_1\tilde{w}^m\tilde{w}'a_2 \cdots \tilde{w}^m\tilde{w}'a_{m-1}\tilde{w}^m\tilde{w}'.$$

If $\tilde{w} > \tilde{w}'$ this has an m-decomposable subword

$$(\tilde{w}^m\tilde{w}'a_1\tilde{w})(\tilde{w}^{m-1}\tilde{w}'a_2\tilde{w}^2) \cdots (\tilde{w}\tilde{w}');$$

if $\tilde{w} < \tilde{w}'$ we use instead the m-decomposable subword

$$(\tilde{w}'a_1\tilde{w}^{m-1})(\tilde{w}\tilde{w}'a_2\tilde{w}^{m-2}) \cdots (\tilde{w}^{m-2}\tilde{w}'a_{m-1}\tilde{w})(\tilde{w}^{m-1}\tilde{w}').$$

In either case w has an m-decomposable subword, so as noted above, we are done.   Q.E.D.

This result is rather technical and was slow in being incorporated into the theory. However, it has some far-reaching consequences. The first one is due to Berele, using a different proof.

Theorem 6.3.25: *Any affine PI-algebra over a field has finite Gelfand-Kirillov dimension.*

Proof: (Amitsur; Drensky) Take $S = \{\bar{w}_1, \ldots, \bar{w}_{m'}\}$ of theorem 6.3.24. In calculating $G_S(n)$ of definition 6.2.10 for n large we see $V_n(S)$ is spanned by the $\bar{v}_{i_1} \cdots \bar{v}_{i_k}$. One could formally replace i_j by j and obtain $\bar{v}_1 \cdots \bar{v}_k$, in this way creating a monomial which could be viewed in the commutative polynomial ring $C[\lambda_1, \ldots, \lambda_n]$. Since at most k^k monomials $\bar{v}_{i_1} \cdots \bar{v}_{i_k}$ correspond to $\bar{v}_1 \cdots \bar{v}_k$, we see by example 6.2.11(ii) that

$$G_S(n) \leq k^k \binom{k + n}{n} \leq k^k n^k = (kn)^k.$$

Hence the GK-dimension is bounded by

$$\lim_{n \to \infty} \log_n((kn)^k) = \lim k(1 + \log_n k) = k. \qquad \text{Q.E.D.}$$

The Theory of Prime PI-Rings

Many problems about an arbitrary PI-ring R can be reduced to the prime case by means of the structure theory. Furthermore, prime PI-rings occupy a position of prominence because of

Remark 6.3.26: If R satisfies the identities of $n \times n$ matrices then R is a homomorphic image of the algebra of generic matrices, which is prime (cf., proposition 6.1.41 and 6.1.42).

Thus the theory of prime PI-rings should bear heavily on the PI-theory in general. So far we have several powerful tools at our disposal.

Summary 6.3.27: Suppose R is a prime PI-ring with center Z, and $S = Z - \{0\}$. Then

(i) R has some PI-degree n (theorem 6.1.28)

(ii) $S^{-1}R$ is central simple over $S^{-1}Z$ of degree n (theorem 6.1.30)

(iii) There exists z in Z for which $R_{<z>}$ is a free $Z_{<z>}$-module with a base of n^2 elements, and $R_{<z>}$ is Azumaya over $Z_{<z>}$. (In fact, any $z \neq 0$ in $g_n(R)$ will do; corollary 6.1.36.)

Despite these results, $Z(R)$ can still be rather poorly behaved, as we shall see when studying Noetherian PI-rings, and even localizing at a single element can be too crude a tool to work with certain delicate questions. There is one more construction which can be used in almost every situation.

Construction 6.3.28: (*Schelter's Trace Ring*) *Let a C-algebra R be prime of PI-degree n and $Q = S^{-1}R$ where $S = Z(R) - \{0\}$. We shall expand R to a certain subring $T(R)$ of Q satisfying the following properties:*

(i) $T(R) = RC'$ *for a suitable C-subalgebra C' of $Z(Q)$; in particular, $T(R)$ is prime.*

(ii) $Rg_n(R) \neq 0$ is a common ideal of R and $T(R)$. In fact $C'g_n(R) = g_n(R)^+$.

(iii) $T(R)$ is integral over C'.

(iv) If R is C-affine then C' is C-affine and $T(R)$ is f.g. over C'; in particular, when C is Noetherian C' and $T(R)$ are Noetherian.

To construct $T(R)$ we first embed R in $M_n(F)$ by splitting Q, and taking a base $b_1, \ldots, b_k$ of $M_n(F)$ where $k = n^2$ we view $R \subseteq M_k(F)$ by the regular representation. Writing $\hat{r}$ for the $k \times k$ matrix corresponding to r we have the Hamilton-Cayley polynomial $\det(\lambda \cdot 1 - \hat{r}) = \sum (-1)^i \alpha_i \lambda^{k-i}$ for suitable α_i, $0 \leq i \leq k$, where $\alpha_0 = 1$. We shall call these α_i the *characteristic coefficients* of r. Note α_i are independent of the choice of the base $b_1, \ldots, b_k$. Furthermore, choosing the base $b_1, \ldots, b_k$ in Q it is easy to see that each $\alpha_i \in Q$.

Let C' be the C-subalgebra of Q generated by the characteristic coefficients of "enough" elements of R, and let $T(R) = RC' \subseteq Q$. "Enough" means "all" unless R is affine, i.e., $R = C\{r_1, \ldots, r_t\}$ in which case the monomials in $r_1, \ldots, r_t$ of length $\leq \beta(t, 2n^2, d)$ are "enough" (where d is the degree of a PI of R). We now have (i) and shall prove (iii), (iv), and (ii).

Proof of (iii): Any element a of $T(R)$ has the form $\sum_{i=1}^{u} a_i c_i$ where $a_i \in R$ and $c_i \in C'$. Hence $a \in C'\{a_1, \ldots, a_u\}$. But all the monomials in the a_i are in R and thus integral over C' since C' contains enough characteristic coefficients. Thus a is integral over C', by theorem 6.3.23.

Proof of (iv): As in proof of (ii), since now each a_i can be taken to be a monomial in $r_1, \ldots, r_t$, and C' contains enough characteristic coefficients by construction; the last assertion follows from elementary properties of Noetherian rings.

Proof of (ii): It suffices to show for any characteristic coefficient α of r and any $b_1, \ldots, b_m$ in R that $\alpha g_n(b_1, \ldots, b_m) \in g_n(R)^+$ where m is the degree of g_n. This is obvious unless $z = g_n(b_1, \ldots, b_m) \neq 0$, in which case $b_1, \ldots, b_k$ is a base of $M_n(F)$, seen by applying theorem 6.1.33 to $z^{-1} b_1, \ldots, b_k$. Thus we could use $b_1, \ldots, b_t$ as our base in forming $\hat{r}$. By proposition 6.1.19(iii) we have

$$\alpha g_n(b_1, \ldots, b_n) \in g_n(R)^+, \text{ proving (ii).} \qquad \text{Q.E.D.}$$

This construction was called the *characteristic closure* in Rowen [80B], but the misnomer *trace ring* has stuck since for Q-algebras all the characteristic coefficients can be obtained from the traces. Technically our ring $T(R)$ might be called the n^2-trace ring since we took the n^2-Hamilton-

Cayley polynomial; Amitsur-Small [80] analogously built the n-trace ring using the n-Hamilton-Cayley polynomial. One advantage of the n^2-trace ring is that proposition 6.1.19 is itself a useful tool; for example, it could be used in the proof of exercise 15, due to Amitsur, which implies that if $T(R)$ is defined functorially by adjoining all characteristic coefficients of elements of R then $T(T(R)) = T(R)$.

When R is a $\mathbb{Q}$-algebra the n-trace ring and n^2-trace ring are the same. A detailed proof of this fact would rely on the reduced trace (cf., definition 2.3.32 ff), but essentially boils down to the observation that if a is an $n \times n$ matrix then the trace of the $n^2 \times n^2$ matrix $\begin{pmatrix} a & & \\ & \ddots & \\ & & a \end{pmatrix}$ is $n\operatorname{tr}(a)$, and $n^{-1} \in \mathbb{Q}$.

Remark 6.3.29: $T(R)$ is isomorphic to an R-submodule of R. (Indeed, take $0 \neq z \in g_n(R)$ and note $T(R)z \subseteq Rg_n(R) \subseteq R$, so right multiplication by z embeds $T(R)$ into R.)

In order to utilize $T(R)$ we must be able to pass prime ideals from R to $T(R)$ and back, so the following observations are crucial:

Theorem 6.3.30: *Suppose a PI-ring R is integral over $C \subseteq Z(R)$. Then GU, LO and INC hold from C to R.*

Proof: Any C-affine subalgebra of R is f.g. by theorem 6.3.23, so satisfies GU and LO by theorem 2.12.48 and INC by theorem 3.4.13. Hence $C \subseteq R$ satifies GU and LO by proposition 2.12.49. To prove INC suppose $P_1 \subset P_2$ in $\operatorname{Spec}(R)$ lie over the same prime ideal of C. Passing to R/P_1 and $C/(P_1 \cap C)$ we may assume $P_1 = 0$ and $P_2 \cap C = 0$.

In particular, $Z(R)$ is a domain. Take an affine subalgebra Z' of $Z(R)$ containing some $0 \neq z \in P_2 \cap Z(R)$. Then $Z'z \cap C = 0$ so $Z'z$ is contained in a prime ideal of Z' lying over 0, by remark 2.12.42. This contradicts INC for Z' over C. Q.E.D.

Theorem 6.3.31: *Suppose R is prime of PI-degree n.*

(i) *LO(P) and GU(___, P) hold from R to $T(R)$, for all P in $\operatorname{Spec}_n(R)$ and for all P minimal over a prime in $\operatorname{Spec}_n(R)$ (i.e., for which there is P_0 in $\operatorname{Spec}_n(R)$ satisfying height $(P/P_0) = 1$ in R/P_0.)*

(ii) *The strong version of INC(P) holds from R to $T(R)$ for all P in $\operatorname{Spec}_n(R)$, in the sense that if $P'_1, P'_2 \in \operatorname{Spec} T(R)$ with $P'_1 \cap R \subseteq P'_2 \cap R = P$ then $P'_1 \subseteq P'_2$.*

(iii) *LO, GU, and INC hold from C' to $T(R)$.*

Proof:

(i) Let $A = Rg_n(R)$ and apply proposition 2.12.45 to 6.3.28(i),(ii).

(ii) By proposition 2.12.46, as in (i).

(iii) Theorem 6.3.30 is applicable by 6.3.28(iii).   Q.E.D.

Thus many questions about $\text{Spec}(R)$ can be translated to $T(R)$, which can be a suitable arena for computation even when R is not affine. One of the key computations, due to Braun [85], is a Hamilton-Cayley type expression for $T(R)$ which passes to certain homomorphic images; Braun uses this in a short proof of

Theorem (Bergman-Small): *Suppose R is prime of PI-degree n, and $P \in \text{Spec}(R)$. If $m = PI\text{-degree}(R/P)$ then $n - m$ is a sum of PI-degrees (with possible repetitions) of simple homomorphic images of R.*

In particular, if R is local and P is the unique maximal ideal then $m \mid n$. The reduction to $T(R)$ is easy, and is given in exercise 17. We shall use $T(R)$ here to polish off much of the theory of affine PI-algebras and then discuss its use for Noetherian rings. Let us start by reproving the weak Nullstellensatz.

Remark 6.3.32: If $r \in R$ is regular and is algebraic over a central subring Z then $Z \cap Rr \neq 0$. (Indeed, take $\sum_{i=0}^{t} z_i r^i = 0$ with $z_t \neq 0$, t minimal such, and let $r' = \sum_{i=1}^{t} z_i r^{i-1} \neq 0$ by minimality of t. Then $0 \neq -r'r = z_0 \in Z \cap Rr$, as desired.)

Application 6.3.33: Easy proof of the weak Nullstellensatz for affine PI-algebras over Jacobson C: It is enough to assume R is prime and to prove $\text{Jac}(R) = 0$. Using construction 6.3.28(iv) we see C' is Jacobson, by the commutative Nullstellensatz, so $\text{Jac}(C') = 0$. By proposition 2.5.33 we have $0 = C' \cap \text{Jac}(T(R))$; but $Z(R)$ is integral over C' and so $Z(R) \cap \text{Jac}(T(R)) = 0$ by remark 6.3.32 (since all of its elements are central and thus regular). Hence $\text{Jac}(T(R)) = 0$ by theorem 6.1.28.

Now $g_n(R)\text{Jac}(R) \subseteq \text{Jac}(R)$ so is quasi-invertible and is an ideal of $T(R)$ by 6.3.28(ii), implying $g_n(R)\text{Jac}(R) = 0$. Since R is prime we conclude $\text{Jac}(R) = 0$.   Q.E.D.

Theorem 6.3.34: *(Schelter) If R is affine PI over a Noetherian ring C then R satisfies ACC(prime ideals)*

Proof: Take a chain $P_1 \subset P_2 \subset \cdots$, and take m for which $PI\text{-deg}(R/P_m)$ is minimal. Replacing R by R/P_m and replacing the original chain by the chain

$P_{m+1}/P_m \subset P_{m+2}/P_m \subset \cdots$ we may assume R is prime of some PI-degree n, and each $P_i \in \mathrm{Spec}_n(R)$. But by theorem 6.3.31(i) $T(R)$ has an infinite ascending chain of prime ideals, contrary to $T(R)$ being Noetherian by 6.3.28(iv).

Q.E.D.

Of course when C is a field the previous result follows from theorems 6.2.25 and 6.3.25. We shall obtain a stronger result in corollary 6.3.36', to follow.

Nilpotence of the Jacobson Radical

Theorem 6.3.35: (*Razmyslov*) *Suppose R is affine over a Noetherian Jacobson domain C. (In particular, one could take C to be a field.) If R satisfies all the identities of $M_n(C[\Lambda])$ where Λ is an infinite set of commuting indeterminates then* $\mathrm{Jac}(R)$ *is nilpotent.*

Proof: Since R is a homomorphic image of an affine algebra of generic matrices, which is prime by remark 6.3.4, this follows at once from the weak Nullstellensatz and the following result:

Theorem 6.3.36: (*Schelter*) *Suppose R is prime PI and affine over a Noetherian ring C. If $A \lhd R$ then* $\mathrm{Nil}(R/A)$ *is a finite intersection of prime ideals of R/A and is nilpotent in R/A.*

Proof: In view of theorem 6.3.34 we may apply Noetherian induction to the prime images of R and thereby assume the theorem is true in R/P for every $0 \neq P \in \mathrm{Spec}(R)$. Pass to $T(R)$, which is Noetherian by 6.3.28(iv). Let $n = \mathrm{PI\ deg}(R)$. Then $g_n(R)A \lhd T(R)$ by 6.3.28(ii), so there are a finite number of prime ideals $Q_1, \ldots, Q_t$ of $T(R)$ minimal over $g_n(R)A$. Since $(\bigcap Q_i)/g_n(R)A = \mathrm{Nil}(T(R)/g_n(R)A)$ is nilpotent, we have $(\bigcap Q_i)^k \subseteq g_n(R)A$ for some k.

Let $P_i = Q_i \cap R \neq 0$. Then $R/(P_i + A)$ is an image of R/P_i so there are a finite set of prime ideals $P_{i1}, \ldots, P_{im}$ each containing $P_i + A$, for which $(\bigcap P_{ij})^{k'} \subseteq P_i + A$; technically m and k' depend on P_i but there are only a finite number of the P_i, so we can take m and k' to be the same for all i. Now

$$\left(\left(\bigcap_{j=1}^m P_{1j} \right)^{k'} \cdots \left(\left(\bigcap_{j=1}^m P_{tj} \right)^{k'} \right)^k \subseteq ((P_1 + A) \cdots (P_t + A))^k \subseteq g_n(R)A \subseteq A, \right.$$

implying at once $\mathrm{Nil}(R/A) = \left(\bigcap_{i=1}^t \bigcap_{j=1}^m P_{ij} \right)/A$ and is nilpotent.   Q.E.D.

Corollary 6.3.36': *Any semiprime PI-algebra affine over a Noetherian ring C has a finite number of minimal prime ideals and satisfies* ACC(*semiprime ideals*).

Proof: This is shown directly in Rowen [80B, §4.5], but our proof here is based on a reduction to theorem 6.3.34. By exercise 3.5.34, it is enough to prove 0 is a finite intersection of prime ideals (for by passing to homomorphic images we may then assume any semiprime ideal is a finite intersection of prime ideals.) Since C has only finitely many minimal primes $P_1, \ldots, P_t$ we may pass to C/P_i and assume C is prime. But by corollary 6.1.47 R is now a homomorphic image of the generic matrix algebra, which is a prime ring, so Schelter's theorem applies.   Q.E.D.

We now aim for the nilpotence of the radical in general.

Lemma 6.3.37: *(Latyshev) Suppose $R = C\{r_1, \ldots, r_t\}$ is PI. If $N \lhd R$ is nil and f.g. as two-sided ideal, then N is nilpotent.*

Proof: Writing $N = \sum_{i=1}^k Ra_i R$, we incorporate $\{a_1, \ldots, a_k\}$ into the generating set $\{r_1, \ldots, r_t\}$ for R. We shall appeal to theorem 6.3.24, with its notation. Let n be the largest index of nilpotency of {monomials in $r_1, \ldots, r_t$ of length $\leq m$ which contains some a_i}. (Certainly these are all elements of N and are thus nilpotent.) Let $n' = \max\{\beta(t, m, m), kmn + 1\}$ where $k = k(m, t)$ was obtained in theorem 6.3.24.

We shall prove $N^{n'} = 0$ by showing any monomial $\bar{w}$ in $N^{n'}$ is in fact 0. By theorem 6.3.24 we may assume $\bar{w} = \bar{v}_{i_1} \cdots \bar{v}_{i_k}$, and n' elements of N appear in $\bar{v}_{i_1} \cdots \bar{v}_{i_k}$. But $n' > kmn$ so one of these $\bar{v}_i$ contains $> mn$ elements of N. On the other hand, theorem 6.3.24 also says $v_i = w_i^u$ for some u and $u \geq n$ since length $w_i \leq m$. Hence $\bar{v} = \bar{w}_i^u = 0$.   Q.E.D.

Remark 6.3.37': Latyshev's result already shows that Nil(R) is the sum of nilpotent ideals of R since $\text{Nil}(R) = \sum_{a \in \text{Nil}(R)} RaR$.

Remark 6.3.38: Suppose R is affine over Noetherian C and f.g. as $Z(R)$-module. Then R is Noetherian so Nil(R) is nilpotent. (Indeed $Z(R)$ is affine over C by Artin-Tate (proposition 6.2.5) so is Noetherian by the Hilbert basis theorem; thus R is Noetherian.)

Theorem 6.3.39: *(Braun [84]) If C is a commutative Noetherian ring then Nil(R) is nilpotent for any PI-algebra $R = C\{r_1, \ldots, r_t\}$.*

Proof: Let C' be the localization of the polynomial ring $C[\lambda]$ at the monoid generated by all $\lambda^i - \lambda^j$ for $i > j$. Since Nil(R) is locally nilpotent we may replace R by $R \otimes C'$ and C by C', and thereby assume each homomorphic image of C has an infinite number of distinct elements (the images of the powers of λ).

Let us collect some reductions which will be needed throughout. Write $N = \mathrm{Nil}(R)$ and $Z = Z(R)$. We reformulate the theorem as:

For any $A \lhd R$ and any semiprime ideal B of R such that $B/A = \mathrm{Nil}(R/A)$ we have $B^k \subseteq A$ for suitable k (depending on A, B).

Suppose this assertion is false. Take a counterexample R for which R/N has minimal PI-degree; by corollary 6.3.36' there is a semiprime ideal B maximal with respect to $B/A = \mathrm{Nil}(R/A)$ and $B^k \nsubseteq A$ for all k. Passing to R/A we may assume $A = 0$ and $B = N$.

Reductions: (1) (Amitsur) By the last paragraph, if $B \supset N$ is a semiprime ideal and $A \lhd R$ with B/A nil then $B^k \subseteq A$ for some k.

(2) If A is a non-nil ideal then taking $B = A + N$ in (1) we see $(A + N)^k \subseteq A$ for suitable k.

(3) If $z + N$ is central in R/N then we may assume $z \in Z$. (Indeed, each $[z, r_i] \in N$ so $N_1 = \sum_{i,j=1}^{t} R[r_i, z]R$ is nilpotent by Latyshev's lemma; we obtain the desired reduction when we replace R by R/N_1.)

(4) If $z \in Z - N$ then it is enough to prove the theorem for $R/\mathrm{Ann}\, z$. (Indeed, suppose $N^{k(1)} \subseteq \mathrm{Ann}\, z$. By reduction 2 we have $N^{k(2)} \subseteq Rz$ for some $k(2)$, so $N^{k(1)+k(2)} \subseteq (\mathrm{Ann}\, z)Rz = 0$.)

Case I. (Folklore). R/N is commutative. Applying reduction 3 to each of $r_1, \ldots, r_t$ in turn, we may assume each $r_i \in Z(R)$, i.e., R is commutative, so we are done by remark 6.3.38.

Case II. R/N is not prime, i.e., there are ideals $A_1, A_2 \supset N$ with $A_1 A_2 \subseteq N$. Since R/N is semiprime we can take z_j for $j = 1, 2$ such that $0 \neq z_j + N \in (A_j/N) \cap Z(R/N)$. By reduction 3 we may assume each $z_j \in Z$ so by reduction 2 there are numbers $k(1), k(2)$ such that $N^{k(j)} \subseteq Rz_j$. But $z_1 z_2 \in A_1 A_2 \subseteq N$ so $0 = (z_1 z_2)^m = z_1^m z_2^m$ for some m, and we are done by reductions 1 and 4.

Case III. Thus we may assume R/N is prime of PI degree > 1. Write $\bar{\ }$ for the canonical image in $\bar{R} = R/N$. Then $\bar{C} \subseteq Z(\bar{R})$ is an integral domain, so $N \cap C = \mathrm{Nil}(C)$ is nilpotent. Replacing R by $R/\mathrm{Nil}(C)R$ we may assume C is an integral domain, which as noted above is infinite. Our object is to prove "enough monomials" in $r_1, \ldots, r_t$ are integral over Z to permit us to apply Shirshov's theorem (6.3.23), for then we would be done by remark 6.3.38.

In the above paragraph it is important to note "enough" is finite. Thus we want a procedure of modifying R such that a given monomial in $r_1, \ldots, r_t$ is integral over Z. Such a procedure for prime affine algebras was in fact the key in constructing $T(R)$, but now R need not be prime. However $\bar{R}$ is prime, and we want to find some way to adjoin characteristic coefficients in this slightly more general setting. The hurdle is that we do not know that the map from R to its localization need be 1:1, and the kernel may contain a non-nilpotent part of N. The remainder of the proof will be to use several high-powered techniques to "create" enough characteristic coefficients to show {evaluations of the identities of $M_d(C)$ in R} is nilpotent, so that we can then appeal to theorem 6.3.36 by means of the algebra of generic matrices.

Our first task is to carry central localization as far as we can. Take $x_1, \ldots, x_m$ in R such that $z = g_n(x_1, \ldots, x_m) \notin N$. By lemma 6.1.32 we have

$$\bar{z}\bar{r} = \sum_{u=1}^{n^2} (-1)^{u+1} g_n(\bar{r}, \bar{x}_1, \ldots, \bar{x}_{u-1}, \bar{x}_{u+1}, \ldots, \bar{x}_m)\bar{x}_u \qquad (10)$$

for any r in R. Writing $z_{iju} = (-1)^{u+1} g_n(x_i x_j, x_1, \ldots, x_{u-1}, x_{u+1}, \ldots, x_m)$ we see $\bar{z}_{iju} \in Z(\bar{R})$ so by reduction 3 we may assume $z \in Z$ and each $z_{iju} \in Z$. Furthermore, $zx_i x_j - \sum z_{iju} x_u \in N$ so modding out by $\sum R(zx_i x_j - \sum z_{iju} x_u)R$ we may assume by Latyshev's lemma that

$$zx_i x_j = \sum_u z_{iju} x_u$$

for each $1 \le i, j \le n^2$. Likewise, we may assume $zr_i = \sum_u z_{iu} x_u$ for z_{iu} in Z.

Let $B = \{x_1, \ldots, x_n\}$ and let $\tilde{g}$ be as in the proof of lemma 6.1.32. By Latyshev's lemma we also assume $\tilde{g}(B) = 0$ and $[g_n(B), x_u] = 0$ for each u. But $R[z^{-1}] = \sum_{u=1}^{n^2} Z[z^{-1}]x_u$, so $\tilde{g}$ is an identity of $R[z^{-1}]$ and g_n is $R[z^{-1}]$-central, yielding

$$1^{-1}zr = 1^{-1} \sum_{u=1}^{n^2} (-1)^{u+1} g_n(r, x_1, \ldots, x_{u-1}, x_{u+1}, \ldots, x_m)x_u. \qquad (11)$$

We aim for a similar equation in R itself. To this end write $g_n = \sum h_{1i} X_1 h_{2i}$ where the indeterminate X_1 appears in neither h_{1i} nor h_{2i}. Letting $a_{iu} = h_{1i}(x_1, \ldots, x_{u-1}, x_{u+1}, \ldots, x_m)$ and $b_{iu} = h_{2i}(x_1, \ldots, x_{u-1}, x_{u+1}, \ldots, x_m)$ we see $g_n(r, x_1, \ldots, x_{u-1}, x_{u+1}, \ldots, x_m) = \sum_i a_{iu} r b_{iu} = \psi(\sum_i a_{iu} \otimes b_{iu})r$ where $\psi: R \otimes R^{op} \to \mathrm{End}\, R$ is the canonical map. Thus (11) becomes

$$1^{-1}z \otimes 1 = \sum_{u,i} 1^{-1}(-1)^{u+1} a_{iu} \otimes 1^{-1} b_{iu}$$

in $R[z^{-1}]^e \approx R^e[z^{-1}]$ (cf., remark 5.3.28), so there is some d such that

$$z^{d+1} \otimes 1 = \sum_{i,u} (-1)^{u+1} z^d a_{iu} \otimes b_{iu}$$

in R^e, and thus, for all r in R,

$$z^{d+1}r = \sum_{i,u} (-1)^{u+1} z^d a_{iu} r b_{iu} = z^d \sum_{u=1}^{n^2} (-1)^{u+1} g_n(r, x_1, \ldots, x_{u-1}, x_{u+1}, \ldots, x_m)x_u.$$

$$(12)$$

Increasing d if necessary, we may also assume by corollary 5.3.30 for all r in R that

$$z^d g_n(r, x_1, \ldots, x_{u-1}, x_{u+1}, \ldots, x_m) \in Z. \tag{13}$$

Just to clean up notation, we may assume $d = 0$ by passing to $R/\text{Ann } z^d$ (by reduction 4, noting by (13) that $[g_n(r, x_1, \ldots, x_{u-1}, x_{u+1}, \ldots, x_m), R] \subseteq \text{Ann } z^d$.) Thus (12) becomes $zr = \sum_{u=1}^{n^2} (-1)^{u+1} g_n(r, x_1, \ldots, x_{u-1}, x_{u+1}, \ldots, x_m) x_u$, and (13) becomes $g_n(r, x_1, \ldots, x_{u-1}, x_{u+1}, \ldots, x_m) \in Z$.

Reduction 5. Any *finite* set of homogeneous identities of $\bar{R}$ may be assumed to be identities for R. (An identity is *homogeneous* if the total degree of each monomial is the same.) Indeed suppose $f(X_1, \ldots, X_v)$ has degree j in each monomial. Let $B = \{X_i : 1 \le i \le n^2\}$; passing to $R/Rf(B^{(v)})R$ via Latyshev's lemma we may assume $f(B^{(v)}) = 0$. But for any $r'_1, \ldots, r'_v$ in R we have each $zr'_i \in ZB$, so

$$z^{jf}(r'_1, \ldots, r'_v) = f(zr'_1, \ldots, zr'_v) \in Zf(B^{(v)}) = 0,$$

i.e., $f(R) \in \text{Ann } z^j$, which by reduction 4 we may assume is 0.

One particular identity of $\bar{R}$ actually contains enough information to encode the Hamilton-Cayley theorem. Let $n' = n^2$ and, taking the Capelli polynomial $C_{2n'}$ define

$$f(X_1, \ldots, X_{2n'+1}) = X_{n'+1} C_{2n'}(X_1, \ldots, X_{n'}, X_{n'+2}, \ldots, X_{2n'+1})$$

$$= \sum_{\pi \in \text{Sym}(n')} (\text{sg } \pi) X_{n'+1} X_{\pi 1} X_{n'+2} \cdots X_{\pi n'} X_{2n'+1}.$$

Consider n'-tuples $\mathbf{j} = (j(1), \ldots, j(n'))$ where each $j(v) \in \{0, 1\}$ for $1 \le v \le n'$, and write $|\mathbf{j}|$ for the number of $j(v)$ which are 1, i.e., $|\mathbf{j}| = j(1) + \cdots + j(n')$. Picking y_j in R and letting $y = f(y_1, \ldots, y_{2n'+1})$ we have

$$c_u \bar{y} = \sum_{|\mathbf{j}| = u} \overline{f(r^{j(1)} y_1, \ldots, r^{j(n')} y_{n'}, y_{n'+1}, \ldots, y_{2n'+1})}$$

for any r in R by proposition 6.1.19, where $(-1)^u c_u$ denotes the appropriate "characteristic coefficient" of the matrix T corresponding to left multiplication by $\bar{r}$ in $\bar{R}$. But $\sum (-1)^u c_u T^{n'-u} = 0$ so we have

$$0 = \sum (-1)^u c_u \overline{r^{n'-u} y} = \sum_u \overline{(-1)^u r^{n'-u}} \sum_{|\mathbf{j}| = u} \overline{f(r^{j(1)} y_1, \ldots, r^{j(n')} y_{n'}, y_{n'+1} \cdots)}. \tag{14}$$

Thus $\bar{R}$ satisfies the identity

$$\tilde{f} = \sum_u \sum_{|\mathbf{j}| = u} (-1)^u X_0^{n'-u} f(X_0^{j(1)} X_1, \ldots, X_0^{j(n')} X_{n'}, X_{n'+1}, \ldots, X_{2n'+1})$$

which is homogeneous of degree $3n' + 1$.

Reduction 6. We may assume $\tilde{f}$ is an identity of R, by reduction 5.

The last ingredient in the proof is based on an ideal of Razmyslov. Since we have trouble working in R, we switch to the free algebra $C\{X, Y\} = C\{X_1, \ldots, X_t, Y_1, \ldots, Y_{n'}\}$ in noncommuting indeterminates X_i, Y_i over C, and let $W = C\{X_1, \ldots, X_t\} \subset C\{X, Y\}$. We want to "create" characteristic coefficients of the elements of W. Namely, let $\mu = \{\mu_{u,h}: 1 \leq u \leq n', h \in W\}$ be a set of indeterminates which commute with W but not with each other. (The reason the $\mu_{u,h}$ initially are not to commute with each other is to ensure that (16) below is well-defined.) Let A be the ideal in $W\{\mu\}$ generated by $\{\sum_u(-1)^u \mu_{u,h} h^{n'-u}: h \in W\} \cup \{[\mu_{u,h}, \mu_{u',h'}]: 1 \leq u, u' \leq n' \text{ and } h, h' \in W\}$. Note the μ are commutative modulo A. Thus $(W + A)/A$ is integral of bounded degree $\leq n'$ over the integral domain which is the image of $C\{\mu\}$, and thus by example 6.1.9 satisfies some PI of some degree m', which we multilinearize. Let $\tilde{\mu} = \{\mu_{u,h}: 1 \leq u \leq n', h \text{ is a word in the } X_i \text{ of degree } \leq m'\}$, and let $T = W\{\tilde{\mu}\}/(A \cap W\{\tilde{\mu}\})$. Then T satisfies the same PI, so by Shirshov's theorem is f.g. over the Noetherian domain $C\{\tilde{\mu}\}/A \cap C\{\tilde{\mu}\}$. Hence T is Noetherian by remark 6.3.38, implying $\text{Nil}(T)^k = 0$ for some k.

$T/\text{Nil}(T)$ has some PI-degree d. Let $\mathscr{L}$ be the set of identities of $M_d(C)$. $\mathscr{L}$ are all identities of $T/\text{Nil}(T)$ by exercise 6.1.13', letting $\mathscr{L}(W)$ be the set of evaluations of $\mathscr{L}$ in W we see $\mathscr{L}(W)^k \subseteq A$. To utilize this result we let L be the C-submodule of $C\{X, Y\}$ spanned by all $\{f(Y_1, \ldots, Y_{n'}, h_1, \ldots, h_{n'+1}): h_j \in W\}$. L is naturally a W-module since

$$X_u f(Y_1, \ldots, Y_{n'}, h_1, \ldots, h_{n'+1}) = f(Y_1, \ldots, Y_{n'}, X_u h_1, \ldots, h_{n'+1}). \tag{15}$$

On the other hand, each $\mu_{u,h}$ acts on L via

$$\mu_{u,h} f(Y_1, \ldots, Y_{n'}, h_1, \ldots, h_{n'+1}) = \sum_{|j|=u} f(h^{j(1)} Y_1, \ldots, h^{j(n')} Y_{n'}, h_1, \ldots, h_{n'+1}). \tag{16}$$

To see the right-hand side is indeed in L we must observe

$$\sum_{|j|=u} f(h^{j(1)} Y_1, \ldots, h^{j(n')} Y_{n'}, h_1, \ldots, h_{n'+1})$$

$$= \sum_{\pi \in \text{Sym}(n')} \sum_{|j|=u} (\text{sg } \pi) h_1 h^{j(\pi 1)} Y_{\pi 1} \cdots h_{n'} h^{j(\pi n')} Y_{\pi n'} h_{n'+1}$$

$$= \sum_{\pi \in \text{Sym}(n')} \sum_{|j|=u} (\text{sg } \pi) h_1 h^{j(1)} Y_{\pi 1} \cdots h_{n'} h^{j(n')} Y_{\pi n'} h_{n'+1}$$

$$= \sum_{|j|=u} f(Y_1, \ldots, Y_{n'}, h_1 h^{j(1)}, \ldots, h_{n'} h^{j(n')}, h_{n'+1}) \in L.$$

(The middle equality is obtained by noting that $(j(\pi 1), \ldots, j(\pi n'))$ and $(j(1), \ldots, j(n'))$ each have precisely u entries which are 1.)

By iterating (16) we extend the action to monomials in μ and thus get a $W\{\mu\}$-module action on L. Also note $(-1)^u \mu_{u,h} f(Y_1, \ldots, Y_{n'}, h_1, \ldots, h_{n'+1})$ is the coefficient of $\lambda^{n'-u}$ in $f((\lambda - h)Y_1, \ldots, (\lambda - h)Y_{n'}, h_1, \ldots, h_{n'+1})$.

There are a finite number of polynomials $[\mu_{u,h}, \mu_{u',h'}]f$ for $\mu_{u,h}, \mu_{u'h'} \in \tilde{\mu}$; by reduction 5 we may assume they all are identities of R. Now let I be the ideal of $C\{X, Y\}$ generated by all evaluations of $\tilde{f}$ and all evaluations of these $[\mu_{u,h}, \mu_{u',h'}]f$. Then (15) shows $I \supseteq AL$; we saw above $AL \supseteq \mathscr{Z}(W)^k L$. Hence $I \supseteq \mathscr{Z}(W)^k L$. Specializing down to R then yields $0 \supseteq \mathscr{Z}(R)^k f(R) \supseteq I(R)^k g_n(R)$, since g_n is clearly a formal consequence of the Capelli polynomial, cf., exercise 6.1.4. But $g_n(R) \not\subseteq N$, so by reduction 4 we may assume $\mathscr{Z}(R)^k = 0$. On the other hand, $R/\mathscr{Z}(R)$ satisfies the identities of $d \times d$ matrices, so $N/\mathscr{Z}(R)$ is nilpotent by theorem 6.3.36. Hence N is nilpotent, as desired.   Q.E.D.

Crucial ingredients in this proof were Shirshov's theorem, the Artin-Procesi theorem, the properties of the Capelli polynomial which lead to the construction of the central polynomial, the Razmyslov-Schelter theorem (6.3.36), the construction of $\tilde{f}$ to incorporate a Hamilton-Cayley theorem into a polynomial identity, and the generic matrix algebras. It is safe to say this proof requires all the machinery of PI-theory developed here.

Dimension Theory of Affine PI-Algebras

Although we have so far obtained a considerable amount of information about affine PI-rings, we have not yet built up a dimension theory. One obvious candidate is the GK dimension, which we have seen exists, by theorem 6.3.25. Thus we can draw on the theory of GK dimension, especially in view of the following result:

Proposition 6.3.40: If R is a prime PI-algebra (not necessarily affine) over a field F then GK dim R = tr deg $Z(R)/F$ (the transcendence degree).

Proof: By proposition 6.2.23 we can pass to the ring of central quotients of R, and assume R is central simple over the field $Z = Z(R)$. But then GK dim R = GK dim Z = tr deg Z/F by remark 6.2.20.   Q.E.D.

When R is affine as well this is an integer by theorem 6.3.25; however, GK dim R need not be an integer for R non-semiprime affine PI. Let us bring in now the classical Krull dimension (the maximal length of a chain of primes), which *is* always an integer.

Theorem 6.3.41:   *If R is prime affine PI over a field F then* GK dim $R =$
cl. K-dim $R =$ tr deg $Z(R)/F$. *Furthermore, a chain of prime ideals of maximal
length can be obtained from* $\operatorname{Spec}_n(R)$ *where* $n =$ PI deg R.

Proof: Induction on $d =$ GK dim R. We just saw $d =$ tr deg $Z(R)/F$, and
$d \geq$ cl. K-dim R by theorem 6.2.25. It remains to show that there is a chain
$0 \subset P_1 \subset \cdots \subset P_d$ in $\operatorname{Spec}_n(R)$. If $d = 0$ then there is nothing to prove, so
assume $d \geq 1$. We claim $g_n(R)$ contains an element z_1 transcendental over
F. Indeed, this is clear unless each r in $g_n(R)$ is algebraic over F; but then
$F \cap \langle r \rangle \neq 0$ by remark 6.3.32 so r invertible and thus $1 \in g_n(R)$, so $g_n(R)^+ =$
$Z(R)$ and the claim follows at once.

Expand z_1 to a transcendence base $z_1, \ldots, z_d$ of $Z(R)$ over F, and let
$S = \{p(z_1, \ldots, z_{d-1}): 0 \neq p \in F[\lambda_1, \ldots, \lambda_{d-1}]\}$. Then $S^{-1}R$ is affine over the
ring $S^{-1}F$. If $S^{-1}R$ were simple then by corollary 6.3.2 $S^{-1}R$ would be alge-
braic over $S^{-1}F$, contrary to $d \geq 1$. Thus $S^{-1}R$ has a nonzero maximal ideal
$S^{-1}P$, where $P \in \operatorname{Spec}(R)$ with $P \cap S = \varnothing$. In particular, $P \in \operatorname{Spec}_n(R)$. Let
$\bar{R} = R/P$, which contains the algebraically independent elements $\bar{z}_1, \ldots, \bar{z}_{d-1}$.
Thus

$$d - 1 \leq \frac{\text{tr deg } Z(\bar{R})}{F} = \text{GK dim } \bar{R} \leq d - 1$$

by induction and prosition 6.2.24. Hence equality holds, and by induction
there is a chain $0 \subset \bar{P}_1 \subset \cdots \subset \bar{P}_{d-1}$ in $\operatorname{Spec}_n(\bar{R})$. This lifts to a chain $0 \subset P \subset$
$P_1 \subset \cdots \subset P_{d-1}$ in $\operatorname{Spec}_n(R)$, of length d as desired.   Q.E.D.

Corollary 6.3.42:   *(R prime affine PI)* cl K-dim $R =$ cl K-dim $T(R)$ *where*
$T(R)$ *is as in 6.3.28(iv).*

Proof: Chains from Spec_n lift, by theorem 6.3.31(i), and these determine
cl K-dim, by the theorem.   Q.E.D.

Let us apply these results to obtain a lovely theorem of Schelter.

Theorem 6.3.43:   *(Schelter) Catenarity of affine PI-algebras. If R is affine
PI over a field F, and $P \in \operatorname{Spec}(R)$ then*

$$\text{cl. K-dim } R = \text{cl. K-dim}\left(\frac{R}{P}\right) + \text{height } P,$$

i.e., P can be put into a chain of primes of maximal length.

Proof: Note $\geq$ is clear, so we only need to prove $\leq$. Let $d = $ cl. K-dim R, and $h = $ height(P). If P contains a nonzero prime ideal P_1 then by induction on d applied to $\bar{R} = R/P_1$ we would have

$$\text{cl. K-dim } \bar{R} - \text{height } \bar{P} = \text{cl. K-dim}\left(\frac{\bar{R}}{\bar{P}}\right) = \text{cl. K-dim }\frac{R}{P},$$

and by induction on h we would have

$$\text{cl. K-dim } R = \text{cl. K-dim } \bar{R} + \text{height } P_1$$
$$= \left(\text{cl. K-dim }\frac{\bar{R}}{\bar{P}} + \text{height } \bar{P}\right) + \text{height } P_1$$
$$\leq \text{cl. K-dim }\frac{R}{P} + \text{height } P$$

as desired.

Thus we may assume P is minimal prime, i.e., $h = 1$, and we want to prove $d \leq$ cl. K-dim $R/P + 1$. Now either P is in $\text{Spec}_n(R)$ where $n = $ PI degree(R), or P is minimal over $g_n(R)$; thus there is P' in $T(R)$ lying over P, by theorem 6.3.31(i). Take P' maximal such. Then every nonzero ideal of $T(R)/P'$ intersects R/P nontrivially.

We claim cl. K-dim $T(R)/P' = $ cl. K-dim R/P. Indeed letting $R' = T(R)/P'$ and $S = Z(R/P) - \{0\}$, we see any nonzero ideal of R' intersects S nontrivially, so $S^{-1}R'$ is simple and by construction is affine over the field $F' = S^{-1}Z(R/P)$. Hence $S^{-1}R'$ is algebraic over F' by corollary 6.3.2, so tr deg $Z(R')/F = $ tr deg $Z(R/P)/F$ and hence cl. K-dim $R' = $ cl. K-dim R/P by theorem 6.3.41, as desired.

In view of the claim we need to show $d \leq$ cl. K-dim $T(R)/P' + 1$. Note $T(R)$ is integral over C' which is affine over F. Let $A_2 = P' \cap C'$ and $A_1 = A_2 \cap g_n(T(R))^+$, nonzero ideals of C' since $Z(T(R))$ is integral over C'. By the Noether-Bourbaki normalization theorem quoted at the beginning of §6.2, we have a transcendence base $c_1, \ldots, c_d$ of C' over F and $n(1), n(2)$ in $\mathbb{N}$, such that taking $C_0 = F[c_1, \ldots, c_d]$ we have $A_i \cap C_0 = \sum_{u=1}^{n(i)} C_0 c_u$ for $i = 1, 2$, and C' is integral over C_0.

$T(R)$ is integral over C_0, which is "normal" in the terminology of commutative algebra, so one can modify the standard commutative proof to obtain a version of "going down," cf., exercise 18. Hence taking $A_0 = \sum_{u=2}^{n(2)} C_0 c_u$ we have some $P_0' \neq 0$ in $\text{Spec}(T(R))$ lying over A_0 with $P_0' \subset P'$. Moreover,

$P_0' \in \operatorname{Spec}_n(T(R))$ since $c_1 \notin P_0'$, and using theorem 6.3.30 we have

$$\text{cl. K-dim} \frac{T(R)}{P_0'} = \text{cl. K-dim} \left(\frac{C_0}{A_0} \right) = d + 1 - n(2)$$

$$= \text{cl. K-dim} \frac{C_0}{C_0 \cap P} + 1 = \text{cl. K-dim} \frac{T(R)}{P'} + 1.$$

Thus it remains to show $d \leq \text{cl. K-dim } T(R)/P_0'$. But $R \cap P_0'$ is a prime ideal of $\operatorname{Spec}_n(R)$ contained in P; applying theorem 6.3.31(ii) (strong INC) to $R \cap P_0'$ we see $R \cap P_0' \subset P$, so $R \cap P_0' = 0$ implying $P_0' = 0$. Hence cl. K-dim $T(R)/P_0' = \text{cl. K-dim } T(R) = d$ by theorem 6.3.40, as desired.   Q.E.D.

Having seen that the affine PI-theory is so close to the commutative theory, one might wonder whether a meaningful algebraic geometry can be developed for affine PI-algebras. This has been accomplished to some extent in a major series of papers of Artin-Schelter.

Digression 6.3.44:   Gabriel dimension. Suppose R is prime PI of Krull dimension α. Then K-dim $T(R) \leq \alpha$ by remark 6.3.29, and for R affine we thus have   K-dim $T(R) = \text{K-dim } C' = \text{cl. K-dim } C' = \operatorname{tr deg} C'/F = \operatorname{tr deg} Z(R)/F = \operatorname{GK dim} R$. Thus one might expect the noncommutative Krull dimension to be a useful tool for studying affine PI-algebras. Unfortunately it may fail to exist! Indeed we saw in exercise 3.5.29 that an affine PI-algebra satisfying the identities of 2×2 matrices may lack K-dim, and thus the affine algebra of generic 2×2 matrices is prime PI but lacks K-dim.

When R is affine PI and has K-dim then applying exercise 3.5.33 to the proof of proposition 3.5.51 one has K-dim $R = \text{cl. K-dim } R = \text{cl. K-dim } T(R) = \text{cl. K-dim } C' = \operatorname{GK dim} R$ as above, so we have the peculiar situation that K-dim may fail to exist, but equals GK dim when it does exist. Of course this motivates us to consider the Gabriel dimension, which we recall is very close to K-dim except that it exists more often. Gordon-Small [84] proved every affine PI-algebra over a field has Gabriel dimension, by reducing to the prime case and using the theory of left-bounded Goldie rings, cf., exercise 19.

Exercises

§6.1

1. Every identity is a sum of blended identities. (Hint: Induction on the number of indeterminates in which f is *not* blended, applied to the identities $f(X_1 \mapsto 0)$ and $f - f(X_1 \mapsto 0)$.)

2. If h is a monomial of degree $d > 1$ in X_i then $(2^d - 2)h = \Delta_i h(X_1, \ldots, X_n, X_i)$. Hence some multiple of h can be recovered from a multilinearization of h. Conclude from this that in characteristic 0 any central polynomial can be multilinearized to a multilinear central polynomial, and any identity can be recovered from multilinear identities.

3. Write $f_{(ij)}$ for the polynomial in which X_i and X_j are interchanged. Show that f is t-alternating iff $f_{(ij)} = -f$ for all $1 \le i < j \le t$, iff $f_{(i, i+1)} = -f$ for all $1 \le i < t$.

4. Write $\bar{f}$ for the sum of those polynomials of f for which $X_i, \ldots, X_t$ occur in ascending order. f is t-alternating iff $f = \sum_{\pi \in \mathrm{Sym}(t)} (\mathrm{sg}\,\pi) \bar{f}(X_{\pi 1}, \ldots, X_{\pi t})$, by exercise 3. Conclude that every t-normal polynomial can be written in the form $\sum_i h_{i0} C_{2t}(X_1, \ldots, X_t, h_{i1}, \ldots, h_{it})$ for suitable monomials h_{ij}. Thus Capelli identities are "initial" in the sense that if C_{2t} is an identity of R then every t-normal polynomial is an identity of R.

6. Arguing as in remark 6.1.18 show the only possible multilinear identity of $M_n(C)$ having degree n is S_{2n}.

7. The polynomial $\sum_{\pi, \sigma \in \mathrm{Sym}(t)} (\mathrm{sg}\,\pi)(\mathrm{sg}\,\sigma)[X_{\pi 1}, X_{t+\sigma 1}] \cdots [X_{\pi t}, X_{t+\sigma t}]$ is a nonzero multiple of the standard polynomial S_{2t}. (Hint: It is t-normal, with no cancellation of terms.)

8. $S_t = \sum_{i=1}^t (-1)^{i-1} X_i S_{t-1}(X_1, \ldots, X_{i-1}, X_{i+1}, \ldots, X_t) = \sum_{i=1}^t (-1)^{t-i} S_{t-1}(X_1, \ldots, X_{i-1}, X_{i+1}, \ldots, X_t) X_i$. Consequently if S_{t-1} is an identity then so is S_t.

9. $2\,\mathrm{tr}\, S_{2t}(x_1, \ldots, x_{2t}) = 0$ for all x_i in $M_n(C)$. (Hint: Exercise 8 shows $2\,\mathrm{tr}\, S_{2t}(x_1, \ldots, x_{2t}) = \sum (-1)^{i-1} \mathrm{tr}[x_i, S_{2t-1}(x_1, \ldots, x_{i-1}, x_{i+1}, \ldots, x_{2t})] = 0$.)

10. (Amitsur-Levitzki theorem.) S_{2n} is an identity of $M_n(C)$ for every commutative ring C. (Hint (Razmyslov): By exercise 5 take $C = \mathbb{Z}$. Any matrix x satisfies the Hamilton-Cayley equation $x^n + \sum_{i=0}^{n-1} \alpha_i x^i = 0$ where α_i can be written in terms of traces by Newton's formulae (cf., Rowen [80B, p. 18]). The multilinearization technique can be applied to this equation to yield $0 = \sum_{\pi \in \mathrm{Sym}(n)} x_{\pi 1} \cdots x_{\pi n} + \sum$ terms involving traces, for all x_i in $M_n(\mathbb{Z})$. Now substituting commutators $[x_i, x_{n+\pi i}]$ in place of the x_i and summing over suitable permutations σ, one obtains standard polynomials by exercise 7, and the terms involving traces drop out by exercise 9. Thus $0 = \sum_{\pi, \sigma} [x_{\pi 1}, x_{n+\sigma 1}] \cdots [x_{\pi n}, x_{n+\sigma n}] = m S_{2n}(x_1, \ldots, x_{2n})$ for some m, so S_{2n} is an identity.)

11. $S_{n-1}([X_1^{n-1}, X_2], \ldots, [X_1, X_2])$ is not an identity of $M_n(C)$ for any commutative ring C. (Hint: Passing to homomorphic images one may assume C is a field. If $|C| \ge n$ take a diagonal matrix x_1 having distinct eigenvalues and $x_2 = \sum_{i=1}^{n-1} e_{i, i+1}$. Even when $|C| < n$ one can maneuver into a position to use this substitution by means of "companion matrices.")

12. (Amitsur) A PI-ring R satisfying the identities of $n \times n$ matrices with $\mathrm{Nil}(R)^{n-1} \not\subseteq N(R)$. Let C be a commutative ring containing a nil ideal I which is not of bounded index, e.g., $C = H/\sum_{i \in \mathbb{N}} H\lambda_i^i$ where $H = F[\lambda_1, \lambda_2, \ldots]$. Let $R = \{(c_{ij}) \in M_n(C): c_{ij} \in I$ for all $i > j\}$, a subring of $M_n(C)$ containing the nil ideal $A = \{(c_{ij}) \in M_n(C): c_{ij} \in I$ for all $i \ge j\}$. Then $e_{1n} = e_{12} e_{23} \cdots e_{n-1, n} \in A^{n-1}$ but $e_{1n} \notin N(R)$ since $I \approx I e_{nn} = (I e_{n1}) e_{1n} \subseteq R e_{1n}$. Thus $A^{n-1} \not\subseteq N(R)$.)

13 (Remark 6.1.47' in the unabridged text). Suppose R is a semiprime C-algebra of PI-degree n, and every prime image of C is infinite. Then R satisfies all identities of $M_n(C)$. (Hint: One may assume R is prime and then simple, taking central quotients; apply theorem 6.1.44.)

PI-Rings Satisfying ACC

23. If R is semiprime PI with the ACC on annihilators of ideals then R is semiprime Goldie, and its semisimple Artinian ring of fractions can be had by localizing at the regular elements of $Z(R)$. (Hint: a finite intersection of primes is 0, so use the results of §3.2.)

24. (Cauchon) If R is a semiprime PI-ring satisfying ACC(ideals) then R is Noetherian. (Hint: Reduce to the prime case. Pick $z = g_n(r_1, \dots, r_m) \neq 0$ and let $t = n^2$. Given $L < R$ and a in L let $z(a)$ denote the vector $(z_1, \dots, z_t) \in Z^{(t)}$ where $za = \sum_{i=1}^t z_i a_i$. Then $RM \leq R^{(t)}$ is generated by $\{z(a_u): 1 \leq u \leq k\}$ for suitable $a_u \in L$ and suitable k; show $L = \sum_{u=1}^k Ra_u$. Here M is as in proposition 6.1.51.)

24'. (Cauchon) If R is a PI-ring satisfying ACC(ideals) then $\bigcap_{i \in \mathbb{N}} \mathrm{Jac}(R)^n$ is nilpotent. (Hint: One may assume R is semiprime by lemma 2.6.22; then apply exercise 24.) Of course, this is sharp in view of Herstein's counterexample.

25. If R is prime of PI-degree n satisfying ACC(ideals) then $\bigcap_{i \in \mathbb{N}} P^i = 0$ for all P in $\mathrm{Spec}_n(R)$. (Hint: Pass to the Azumaya algebra $R[s^{-1}]$.) Using induction, conclude that there are only a finite number of idempotent prime ideals in any PI-ring satisfying ACC(ideals), where we say $A \lhd R$ is *idempotent* if $A^2 = A$.

26. (Robson) Let C be a local Noetherian integral domain with $J = \mathrm{Jac}(C) \neq 0$. Then $\begin{pmatrix} C & C \\ J & C \end{pmatrix}$ is prime Noetherian but $\begin{pmatrix} J & C \\ J & C \end{pmatrix}$ is an idempotent maximal ideal. Thus exercise 25 is sharp.

27. (Schelter) Working in $C = \mathbb{Q}(\sqrt{2}, \sqrt{3})[\lambda_1, \lambda_2]$ let $C_1 = \mathbb{Q}[\sqrt{6}, \sqrt{2} + \lambda_1, \lambda_2, \lambda_2\sqrt{2}]$ and $C_2 = \mathbb{Q}[\sqrt{6}, \sqrt{3} + \lambda_1, \lambda_2, \lambda_2\sqrt{2}]$, and $A = C_1\lambda_2 + C_1\lambda_2\sqrt{2}$. Then $A = \lambda_2 C = C_2\lambda_2 + C_2\lambda_2\sqrt{3}$, and $C_1 \cap C_2 = \mathbb{Q}(\sqrt{6}) + A$, which is not Noetherian. Let $R = \begin{pmatrix} C_1 & A \\ A & C_2 \end{pmatrix} \subset M_2(C)$. R is f.g. as $C_1 \times C_2$-module so is Noetherian, and every prime ideal of R has height ≤ 2. But the "principal ideal theorem" fails for $Z(R)$ since A is λ_2-minimal but not minimal. In particular $Z(R)$ is not Noetherian.

§6.2

0. R has subexpoential growth iff $\lim_{n \to \infty} (\log G_s(n))/n = 0$.

Some results of Montgomery-Small [81] when a ring is affine.

1. Suppose R is affine over a commutative Noetherian ring C. Then eRe is affine over eCe under either of the following conditions: (i) eR is f.g. as eRe-module; (ii) ReR is f.g. as R-module. (Hint: (i) Write $R = C\{r_1, \dots, r_t\}$ let $eR = \sum_{i=1}^m eRer'_i$ with $r'_1 = 1$, and $er'_i er_j = \sum ea_{ijk}er'_k$. Then the $\{er'_k e, ea_{ijk}e\}$ generate eRe. (ii) Write $ReR = \sum Rx_i$ and $x_i = \sum_{ij} r''_{ij} er'_{ij}$ for r'_{ij}, r''_{ij} in R. Then the er'_{ij} generate eR as ReR-module, so concude by (i)).

2. Suppose R is left Noetherian and affine over a commutative Noetherian ring C. If G is a finite group of automorphisms on R with $|G|^{-1} \in R$ then the fixed ring R^G is affine. (Hint: The skew group ring $R*G$ is affine Noetherian, so R^G is affine by exercise 1 and exercise 2.6.14.)

3. If T is an affine algebra over a commutative Noetherian ring C and $L < T$ satisfies $LT = T$ then $C + L$ is affine over C. (Hint: Apply example 6.2.3, noting $Re_{11}R = R$ so that exercise 1(ii) applies.) In particular, this provides a host of affine algebras when T is simple, such as the Weyl algebra $\mathbb{A}_1$. It is an open question if every affine subalgebra of $\mathbb{A}_1$ is Noetherian.

4. (Resco) An affine domain which is left and right Ore, and left but not right Noetherian. Let $\mathbb{A}_1$ be the Weyl algebra $F\{\lambda, \mu\}$ satisfying $\mu\lambda - \lambda\mu = 1$; then $R_0 = F + \mathbb{A}_1\mu$ is left and right Noetherian and affine by exercise 3. Let t denote a commuting indeterminate over $\mathbb{A}_1$, and let $R = \{p \in \mathbb{A}_1[t]$: the constant coefficient of p is in $R_0\}$. Then R is affine and left Noetherian since R is a submodule of $\mathbb{A}_1$ which is f.g. as R_0-module; however, R is not right Noetherian since $\mathbb{A}_1$ is not f.g. as right R_0-module.)

5. (Small) Suppose R is affine over C but not f.g. over C. Then there is P in $\mathrm{Spec}(R)$ for which R/P is not f.g. over $C/(P \cap C)$. (Hint: Write $R = C\{r_1, \ldots, r_n\}$. $\{A \lhd R$: R/A is not f.g. over $C/(A \cap P)\}$ is Zorn, arguing as in Artin-Tate. Thus there is $P \lhd R$ maximal with respect to R/P not f.g. over $C/(P \cap C)$. To show P is prime suppose $P \subset A \lhd R$ and $Ar \subseteq P$. Then RrR is cyclic over $(R/A) \otimes (R/A)^{\mathrm{op}}$ which is f.g. over C; hence RrR is f.g. over C implying $R/(P + RrR)$ is not f.g. over the image of C, and therefore $r \in P$.)

6. (Martindale) Applying example 2.1.30 to Golod-Shafarevich yields a primitive algebraic algebra which is not locally finite.

7. (Irving-Small [83]) A primitive, affine algebra R with GK-dim 2, having socle $\neq 0$ and failing ACC(Ann). Let $R = F\{r_1, r_2\}$ where r_1, r_2 satisfy the relations $r_1^2 = 0$, $r_1 r_2^m r_1 = 0$ for m not a power of 2, and $r_1 r_2^m r_1 = r_1$ for m a power of 2. Let M be the F-vector space with base $\{x_0, x_1, \ldots\}$, made into R-module by the actions $r_2 x_i = x_{i+1}, r_1 x_i = 0$ for i not a power of 2, and $r_1 x_i = x_0$ for i a power of 2. Then M is faithful and simple since any submodule contains x_0 and thus M. Hence R is a primitive ring. Also $\{\mathrm{Ann}\, r_2^m r_1 : m$ is a power of $2\}$ is an infinite ascending chain of annihilators. Furthermore Rr_1 is a minimal left ideal since $Rr_1 \approx M$ under the correspondence $r_2^j r_1 \mapsto x_i$.

8. Prove theorem 6.2.17 without the hypothesis ACC(Ann). (Hint: Continue the given proof. By exercise 3.3.14 it is enough to show $\mathrm{Sing}(R) = 0$. But if $z \in \mathrm{Sing}(R)$ then $0 \neq L \cap \mathrm{Ann}\, z \subseteq L$ so $L \cap \mathrm{Ann}\, z = L$ and thus $L\,\mathrm{Sing}(R) = 0$, implying $\mathrm{Sing}(R) = 0$.)

9. An affine F-algebra of GK dim 2 and classical Krull dimension 1, but not having K-dim. Let $R = F\{X_1, X_2\}/\langle X_2\rangle^2$. Then taking $S = \{\bar{X}_1, \bar{X}_2\}$ we have $G_S(n) = (n + 1)(n + 2)/2$ so GK dim$(R) = 2$, but R has infinite uniform dimension and thus lacks K-dim. Also note GK dim $R/N = 1$ where $N = \langle X_2\rangle$ is nilpotent.

10. If T is the ring of differential polynomials $R[\lambda; \delta]$ with R affine then GK dim$(T) = 1 + $ GK dim(R). (Hint: ($\geq$) as in remark 6.2.22'. ($\leq$) Note that if $\delta S \subseteq V_t(S)$ then $\delta V_n(S) \subseteq V_{n+t}(S)$.)

13. (Bergman [81]) An affine PI-algebra R with an f.g. module M of GK-dim 2, having submodule N for which GK dim$(N) = $ GK dim$(M/N) = 1$. Let $R = F\{r_1, r_2\}$ where $r_2 r_1 = 0$ and let $M = Rx + Ry$ satisfy the relations $r_1^{n+1} r_2^n x = 0$ and $r_1 r_2^n y = 0$ unless n is a square m^2 in which case $r_1 r_2^n y = r_1 r_2^m x$. Let $N = Ry$. Taking $S = \{1, r_1, r_2\}$ note $S^n x + S^n y$ has base $\{r_1^i r_2^j x$ and $r_2^j y: i \leq j$ and $i + j \leq n\}$ so GK dim$(M) = 2$. But every $r_1 r_2^j x \in N$ so GK dim$(M/N) = 1$. On the other

hand, GK dim$(N) = 1$ since to get $r_1^i r_2^j x$ in $S^n y$ one needs only $i + j^2 \leq n$. R is PI since $[R, R]^2 = 0$. Bergman also shows how to modify this example to get M arbitrary.

14. (Lenagan) If R is a fully bounded Noetherian ring then GK dim$(M) =$ GK dim$(R/\text{Ann}_R M)$. (Hint: By exercise 3.5.13 R is an H-ring. Let $A = \text{Ann}_R M = \text{Ann}_R\{x_1, \ldots, x_t\}$. Then $R/A \hookrightarrow \bigoplus Rx_i \subseteq M^{(t)}$ so GK dim$(R/A) \leq$ GK dim M.)

17. If R is prime Noetherian then GK dim $R = $ GK dim L for each $0 \neq L \lhd R$. (Hint: Analogous to the proof of proposition 3.5.46.)

§6.3

The Bergman-Sarraille Counterexample

1. (Bergman) Let $R = \mathbb{Q}\{x, y\}/\langle[x, z], [y, z], z^2\rangle$ where $z = [x, y]$. R is an affine Noetherian PI-ring not f.g. over any commutative subring. (Hint: $z \in Z(R)$, $(Rz)^2 = 0$, and R/Rz is commutative Noetherian. Hence $[X_1, X_2]^2$ is a PI of R, and R is Noetherian since Rz is an f.g. module over R/Rz.

   To get the negative properties let T be the subring of $M_3(\mathbb{Q}[\lambda_1, \lambda_2])$ generated by $x = e_{12} + \lambda_1 \cdot 1$ and $y = e_{23} + \lambda_2 \cdot 1$. Then $[x, y] = e_{13} \in Z(T)$ and $[x, y]^2 = 0$, so there is an obvious surjection $\varphi: R \to T$ which is shown to be an isomorphism by describing T explicitly as $\{p \cdot 1 + p_1 e_{12} + p_2 e_{23} + se_{13} : p, s \in \mathbb{Q}[\lambda_1, \lambda_2]\}$ where p_i denotes $\partial p/\partial\lambda_i$.

   One can now show that R cannot be f.g. over a commutative subring H. For $\bar{R} = R/Rz$ would be f.g. over $\bar{H} = (H + Rz)/Rz$, so $\bar{H}$ is affine by Artin-Tate; then $\bar{H}$ is f.g. over a subalgebra $\mathbb{Q}[\bar{f}, \bar{g}]$ where $\bar{f}, \bar{g}$ are algebraically independent over $\mathbb{Q}$. But taking $f = p_1 \cdot 1 + \cdots$ and $g = q \cdot 1 + \cdots$ in $H \subset T$ one sees the derivation $\delta = p_1\partial/\partial_2 - p_2\partial/\partial_1$ is 0 on $\mathbb{Q}[\bar{f}, \bar{g}]$ and thus on $\bar{R}$, since $\bar{R}$ is integral over $\mathbb{Q}[\bar{f}, \bar{g}]$. Then $0 = \delta\lambda_1 = \delta\lambda_2$ implying $p_1 = p_2 = 0$ so $\bar{f} \in \mathbb{Q}$, contradiction.) Actually this argument shows R is not integral over any commutative subring.

2. (Sarraille) A prime affine Noetherian PI-ring not f.g. (or even integral) over any commutative subring. Let $R' = R[\lambda_3]$ where λ_3 is a new commuting indeterminate, and R is as in exercise 1, and let $R'' = R' + \lambda_3 M_3(\mathbb{Q}[\lambda_1, \lambda_2, \lambda_3])$ using the description of R' as a subring of $M_3(\mathbb{Q}[\lambda_1, \lambda_2, \lambda_3])$ of example 1. Then R'' is a prime PI-ring since localization gives $M_3(\mathbb{Q}(\lambda_1, \lambda_2, \lambda_3))$; however R is a homomorphic image of R'', so R'' cannot be integral over a commutative subring. To prove R'' is affine and Noetherian it suffices to show $\lambda_3 M_3(\mathbb{Q}[\lambda_1, \lambda_2, \lambda_3])$ is an f.g. R'-module, as in example 6.2.3. This is done by showing the $\lambda_3 e_{ij}$ span, by a straightforward computation.) Sarraille's example has K-dim 3. This is the lowest dimension for a counterexample, in light of Braun [81].

Kurosch's Problem and the Trace Ring

12. Fast PI-solution to Kurosch's problem (Small). By exercise 6.2.5 one may assume R is prime. Then $Z(R)$ is affine algebraic so is a field, and thus R is simple. Conclude with corollary 6.3.2.

13. (Schelter) Suppose W is a given subring of R. We say $r \in R$ is *integral over W of degree u* if there are generalized W-monomials $f_i(X_1)$ of degree i for $0 \le i < u$ such that $r^u = \sum_{i=0}^{u-1} f_i(r)$. R is *W-integral* if each element of R is integral over W. Prove the more general version of theorem 6.3.23 where R is a centralizing W-integral extension of W.

14. (Schelter) If $R = W\{r_1, \ldots, r_t\}$ is a prime centralizing extension of W then modifying construction 6.3.28 build a "trace ring" $T(R)$ which is f.g. over WC'. (Hint: Use exercise 13.) Thus R and $T(R)$ have the common ideal $Rg_n(R)$, and $T(R)$ inherits many of the properties of W via WC', including the analogue of theorem 6.3.31.

15. Notation as in construction 6.3.28, suppose $\lambda_1, \ldots, \lambda_t$ are commuting indeterminates over R and $a_1, \ldots, a_t \in R$ are arbitrary. Then each characteristic coefficient of $\sum_{i=1}^t a_i \lambda_i$ (as an element of $R[\lambda_1, \ldots, \lambda_t]$) can be expressed as a polynomial in $\lambda_1, \ldots, \lambda_t$ and the characteristic coefficients of the a_i. (Hint: This can be read off as a special case of Amitsur [80], but seems to have an easier proof in this special case by multiplying by any z in $g_n(R)$ and applying proposition 6.1.19 and the theory of elementary symmetric functions.)

16. If R is a domain and $W \subseteq R$ then PI-deg(W) divides PI-deg(R). (Hint: Passing to ring of fractions of R one may assume R is a division ring; replacing W by $Z(R)W$ one may assume W is a division ring. Take a maximal subfield E of W and a maximal subfield F of R containing E. Then PI-deg(R)/PI-deg(W) = $[F:E][Z(W):Z(R)]$.)

17. The reduction of the Bergman-Small theorem to $T(R)$.

Step I. It suffices to prove the weaker assertion

$$(*)\quad n - m = \sum k_i n_i \qquad \text{where } k_i \in \mathbb{N} \text{ and } n_i = \text{PI-deg}(R/P_i) \text{ for } P_i \in \text{Spec}(R).$$

(Hint: Take this such that the subsum over the maximal ideals P_i is as large as possible. If there is some nonmaximal P_i then taking a maximal ideal $Q_i \supset P_i$ and letting $m' = \text{PI-deg}(R/Q) = \text{PI-deg}(\bar{R}/\bar{Q})$ where $\bar{R} = R/P_i$, one has $n_i - m' = \sum k_i' n_i'$ so substitute back $n_i = m' + \sum k_i' n_i'$ to get a contradiction.)

Step II. Take P minimal with respect to PI-deg(R/P) = m. Arguing inductively on $n - m$ one may assume PI-deg(R/Q) = n for all prime ideals $Q \subset P$. But then LO(P) holds from R to $T(R)$ by theorem 6.3.31(i), so the passage to $T(R)$ can be made.

18. (Going Down) Suppose R is a prime PI-ring and is integral over an integral domain $C \subseteq Z(R)$, with C "normal" in the sense that C is integrally closed in its field of fractions. If $P_0 \subset P_1$ in Spec(C) and $P_1' \in \text{Spec}(R)$ lies over P_1 then there is some $P_0' \in \text{Spec}(R)$ lying over P_0, with $P_0' \subseteq P_1'$. (Hint, cf., Rowen [80B, theorem 4.4.24]: Let $R_1 = \{r \in R : 0 \ne r + P_1' \in Z(R/P_1')\}$ and $S = C - P_0$, and let $S_1 = \{rs : r \in R_1, s \in S\}$, a submonoid of R. It is enough to show $S_1 \cap P_0 R = \emptyset$. Otherwise, $rs \in P_0 R$ for some r in R_1, s in S. The minimal monic $p(\lambda)$ for rs over C is irreducible by Gauss' lemma. On the other hand, write $rs = \sum a_i r_i$ for a_i in P_0 and r_i in R. $C\{r_1, \ldots, r_t\}$ is f.g., so using the regular representation and taking determinants show rs satisfies a monic polynomial $q(\lambda)$ all of whose non-leading coefficients are in P_0. Hence $p | q$ by Gauss' lemma. Write $p = \lambda^d + \sum_{i=0}^{d-1} c_i \lambda^i$. Passing to the domain $\bar{C} = C/P_0$ note $\bar{q}$ is a power of λ, so $\bar{p} = \lambda^d$ and hence each $c_i \in P_0$.

Now pass to $S^{-1}C$, which is also normal. $\lambda^d + \sum c_i s^{i-d} \lambda^i$ is the minimal polynomial of r; since r is integral over C each $c_i s^{i-d} \in C$, and thus $\in P_0$. Thus $r^d \in P_0 R \subseteq P_1'$ so $r + P_1'$ is nilpotent, contradiction.)

19. (Gordon-Small) Every affine PI-algebra over a field has Gabriel dimension. (Hint: By 6.3.36' and 6.3.39 there are prime ideals $P_1, \ldots, P_t$ with $P_1 \ldots P_t = 0$. Considering each $P_{i+1} \cdots P_t / P_i \cdots P_t$ one may pass to R/P_i and assume R is prime. By induction on cl. K-dim one may assume G-dim R/L exists for every large left ideal L of R. Thus it suffices to show G-dim U exists for each cyclic uniform left ideal U. For any $0 \neq U' < U$ note U/U' is torsion so has the form R/L' for L' large, implying G-dim U/U' exists; conclude G-dim U exists.)

7 Central Simple Algebras

This chapter deals with the theory of (finite dimensional) central simple algebras and the Brauer group, in particular, with division algebras. There is some question among experts as to whether this theory belongs more properly to ring theory, field theory, cohomology theory, or algebraic K-theory. Accordingly we give an abbreviated account of classical parts of the subject readily found elsewhere; a very thorough treatment of the subject is to be found in the forthcoming book by Jacobson and Saltman, referred to in the sequel as JAC-SAL; certain key parts of this chapter (such as Brauer factor sets) draw on Jacobson's notes for that book. There also has been growing interest in infinite dimensional division algebras, which we shall discuss briefly in 7.1.46ff.

§7.1 Structure of Central Simple Algebras

In this section we develop the tools for studying the basic structure theory of central simple algebras.

Let us recall some of the properties already proved about central simple algebras. Any simple ring R can be viewed as an algebra over the field $F = Z(R)$, and we consider the case when $[R:F] < \infty$, in which case we say R is a *central simple F-algebra*. Then $[R:F] = n^2$ for some n which we call the *degree* of R (cf., corollary 2.3.25), written $\deg(R)$. By Wedderburn's theorem we can write $R = M_t(D)$ for a division algebra D and suitable t. Note that $D \approx \operatorname{End}_R D^{(t)}$ from which we see by means of proposition 2.1.15 that D is unique up isomorphism, and t is uniquely determined since D has IBN. In particular, $\deg D$ is a uniquely determined integer ($= n/t$) called the *index* of R and is a very important invariant of R. We call D the *underlying division algebra* of R. A central simple division algebra is called a *central division algebra*, for short.

Centralizers and Splitting Fields

We start off with some fundamental structural theorems describing R in terms of its subalgebras. If R is a central simple F-algebra and R' is a finite dimensional simple F-algebra with center F' then by corollary 1.7.24 and theorem 1.7.27 $R \otimes_F R'$ is a central simple F'-algebra, whose dimension over F' is clearly $[R:F][R':F']$. In particular, theorem 2.3.27 says $R \otimes_F R^{\text{op}} \approx M_{n^2}(F) \approx \operatorname{End}_F R$.

This result has an immediate consequence. Recall $C_R(A)$ denotes the centralizer of A in R.

Proposition 7.1.1: *Suppose A is a central simple F-subalgebra of R.*

(i) $R \approx A \otimes_F C_R(A)$; *in particular*, $\deg(A) \mid \deg(R)$.

(ii) *If A' centralizes A with $[A':F][A:F] \geq [R:F]$ then $R \approx A \otimes A'$ and $A' = C_R(A)$.*

Proof:

(i) Let $m = [A:F]$. Then $A^{\text{op}} \otimes_F R$ is central simple and contains $A^{\text{op}} \otimes_F A \approx M_m(F)$, and thus has a set of $m \times m$ matric units, implying by proposition 1.1.3 there is an F-algebra R_1 for which $A^{\text{op}} \otimes R \approx M_m(R_1) \approx M_m(F) \otimes_F R_1 \approx A^{\text{op}} \otimes A \otimes R_1$. By corollary 1.7.29 we see R and $A \otimes R_1$ are each the centralizer of $A^{\text{op}} \otimes 1$ in $A^{\text{op}} \otimes R$; hence $R \approx A \otimes R_1$; implying $R_1 \approx C_R(A)$ by corollary 1.7.29 again. Thus $R \approx A \otimes_F C_R(A)$ and so $\deg R = (\deg A)(\deg C_R(A))$.

(ii) $R \approx A \otimes C_R(A)$ and $A' \subseteq C_R(A)$; counting dimensions shows $A' = C_R(A)$.
                                                                                Q.E.D.

Of course the isomorphism $R \otimes R^{\text{op}} \approx \operatorname{End}_F R$ pertains to the theory of separable subalgebras, which would permit an even faster proof of proposition 7.1.1. We turn now to the key fact that every maximal commutative separable subalgebra K of R splits R and $[K:F] = \deg(R)$. There are several ways of approaching this result: (i) the generalized density theorem (exercise 2.4.1), as utilized in Jacobson [80B, theorem 4.8]; (ii) basic properties of tensor products, cf., exercise 4; (iii) properties of separable algebras, which we present here, following Knus-Ojanguren [74B].

Remark 7.1.2: Suppose T is any ring containing R. Viewing T as $R - R$ bimodule we have $T \approx R \otimes_F C_T(R)$, by theorem 5.3.24(iii) and remark 5.3.24'. If $T \subseteq R^e = R \otimes R^{op}$ then identifying R with $R \otimes 1$ yields $T = R \otimes C_T(R)$.

Proposition 7.1.3: $C_R(K) \otimes_F R^{op} \approx \text{End}_K R$ for any subalgebra K of R.

Proof: Let T be the centralizer of $K \otimes 1$ in $R^e \approx \text{End}_F R$; thus $T \approx \text{End}_K R$. Let $R' = C_T(1 \otimes R^{op})$. Then $T \approx R' \otimes R^{op}$ by the opposite of remark 7.1.2. It remains to show $R' = C_R(K) \otimes 1$. ($\supseteq$) holds by inspection. On the other hand, corollary 1.7.29 yields $R \otimes 1 = C_{R^e}(1 \otimes R^{op}) \supseteq R'$; since $R' \subseteq T$ we have $R' \subseteq C_R(K) \otimes 1$.   Q.E.D.

Corollary 7.1.4: If K is a subfield of R then $C_R(K)$ is a central simple K-algebra of degree $n/[K:F]$, where $n = \deg R$.

Proof: $C_R(K)$ is simple since $\text{End}_K(R)$ is simple, and likewise $Z(C_R(K)) = K$. Finally if $m = \deg C_R(K)$ then $m^2 n^2 [K:F] = [C_R(K) \otimes_F R^{op}:F] = [\text{End}_K R:F] = [R:K]^2[K:F] = n^4/[K:F]$, implying $m = n/[K:F]$.   Q.E.D.

Corollary 7.1.5: If a given maximal commutative subalgebra K of R is a field then K splits R and $[K:F] = \deg R = [R:K]$.

Proof: $C_R(K) = K$ by remark 0.0.6, so K splits R^{op}, and thus $K^{op} = K$ splits R. Furthermore $1 = \deg C_R(K) = n/[K:F]$, so $[K:F] = n$ and $[R:K] = [R:F]/[K:F] = n^2/n = n$.   Q.E.D.

What if K is not a field? If K is separable over F we have essentially the same outcome, cf., exercise 1. However if K is not separable one may have $[K:F] > \deg R$.

Example 7.1.6: Let $R = \text{End}_F V \approx M_n(F)$ where $[V:F] = n$, and let V' be any subspace of V having dimension $[n/2]$ over F. Taking a complement V'' of V' we define

$$N = \{f \in R: fV' \subseteq V'' \text{ and } fV'' = 0\}$$

Then $N^2 = 0$ so $F \cdot 1 + N$ is a commutative subalgebra of R whose dimension is $1 + [n/2](n - [n/2]) = 1 + [n^2/4]$. Schur proved this is indeed the maximal possible dimension, and Gustafson [76] found an elegant structure-theoretic proof which we give in exercise 3.

Of course it may be impossible to find a maximal commutative subring which is a field; if $R = M_2(\mathbb{C})$ then R has no subfields properly containing $\mathbb{C}$. However, we can bypass this difficulty by passing to the underlying division algebra D, for we have

Remark 7.1.7: Any commutative subalgebra of a central division algebra D is a field, by remark 2.3.23(ii); consequently, if K is a maximal subfield of D then $C_D(K) = K$ and K splits D.

Remark 7.1.8: Any splitting field of the underlying division algebra of R splits R (for if $D \otimes_F K \approx M_n(K)$ then $M_t(D) \otimes_F K \approx M_t(F) \otimes D \otimes K \approx M_{tn}(K)$).

Our final result about centralizers is

Theorem 7.1.9: *(Double centralizer theorem) If A is any simple subalgebra of R then*

  (i) $C_R(A)$ *is simple and* $Z(A)$*-central;*
  (ii) $C_R(C_R(A)) = A$;
  (iii) $[A:F][C_R(A):F] = [R:F]$.

Proof:

  (i) Let $n = \deg(R)$, $K = Z(A)$, and $t = [K:F]$. Let $S = C_R(K)$, which is simple and K-central of degree n/t by corollary 7.1.4. Clearly $A \subseteq S$ so $S \approx A \otimes_K C_S(A)$ by proposition 7.1.1, implying $C_S(A)$ is simple and $Z(C_S(A)) = K$. But $C_R(A) \subseteq C_R(K) = S$ so $C_S(A) = C_R(A)$ proving (i).
  (ii) First assume $K = F$. Write $A' = C_R(A)$. Then $R \approx A \otimes A'$ so $A = C_R(A')$ by proposition 7.1.1(ii). In general note $C_R(A') \subseteq C_R(K) = S$ and $A' = C_R(A) \subseteq S$ so $C_R(A') = C_S(A') = C_S(C_S(A)) = A$ as just shown.
  (iii) $(n/t)^2 = [S:K] = [A:K][A':K] = [A:F][A':F]/t^2$.   Q.E.D.

Another key tool was presaged in corollary 2.9.2:

Theorem 7.1.10: *(Skolem-Noether theorem) Any isomorphism of simple F-algebras of R can be extended to an inner automorphism of R.*

The proof, which can be found in any standard reference such as Jacobson [80B, p. 222], is sketched in exercise 5.

One useful consequence of the Skolem-Noether theorem is

Theorem 7.1.11: (*Wedderburn's theorem*) *Every finite simple ring R has the form $M_n(F)$ for a suitable field F.*

Proof: It suffices to show every finite division ring D is a field, since we could take D to be the underlying division ring of R. Let K be a maximal subfield of D and assume $|K| = m < |D|$. The number of subgroups of $D - \{0\}$ (as multiplicative group) conjugate to $K - \{0\}$ is at most $|D - \{0\}|/|K - \{0\}| = (|D| - 1)/(m - 1)$. But each subgroup contains the element 1 so the number of conjugates of elements of $K - \{0\}$ is at most $(m - 2)(|D| - 1)/(m - 1) + 1 < |D| - 1$, so some element d of D is not conjugate to any element of K.

Let $F = Z(D)$, and let K' be a maximal subfield of D containing $F[d]$. Then $|K'| = |F|^{[K':F]} = |F|^{\deg D} = |F|^{[K:F]} = |K|$ so $K' \approx K$ since finite fields of the same order are isomorphic. Hence $K' - \{0\}$ and $K - \{0\}$ are conjugate by the Skolem-Noether theorem, contrary to choice of d.   Q.E.D.

This theorem reduces the structure theory of central simple F-algebras to the case F is infinite. This enables us to bring in the Zariski topology and to prove the following important result which ties up some loose ends in our previous discussion.

Theorem 7.1.12: *Suppose $\deg(R) = n$ and F is infinite then $\{a \in R: F[a]$ is separable of dimension n over $F\}$ is dense in the Zariski topology.*

Proof: Take a splitting field K of R. The generic reduced characteristic polynomials of R and $R \otimes_F K \approx M_n(K)$ are formally the same (since we use the same base); let us call them both m_x. But $M_n(K)$ has elements of degree n, so n is the degree of m_x. It follows from proposition 2.3.35 that $\{$elements of degree $n\}$ is a Zariski open subset of R. For these elements the minimal and reduced characteristic polynomials are the same.

On the other hand, $F(r)$ is a separable extension of R iff the roots $r_1, \ldots, r_t$ of the minimal polynomial are distinct, iff the discriminant

$$\prod_{1 \le i < j \le t} (r_i - r_j)^2 \ne 0.$$

There is a way of calculating the discriminant from the minimal polynomial (cf., Jacobson [85B, p. 258]. Since $M_n(K)$ has elements of degree n of discriminant $\neq 0$, we see the discriminant of the generic element is nonzero. Hence $\{r \in R: \text{discriminant}(r) \neq 0\}$ is Zariski open and nonempty and thus dense. Thus

$$\{r \in R: F(r) \text{ is separable of degree } n\}$$

$$= \{r \in R: \text{discriminant}(r) \neq 0\} \cap \{r \in R: r \text{ has degree } n\}$$

is Zariski dense, as desired.   Q.E.D.

Corollary 7.1.12' (Koethe's theorem): *Suppose K is a subfield of a central division algebra D and is separable over F. Then K is a subfield of a maximal subfield of D separable over F.*

Proof: K is infinite by Wedderburn's theorem, so the division algebra $C_D(K)$ has a maximal subfield L separable over F, by theorem 7.1.11. Clearly L is separable over F and $[L:F] = [L:K][K:F] = (\deg C_D(K))[K:F] = \deg D$ by corollary 7.1.4.   Q.E.D.

Note: The proof relied on the fact every central division algebra D contains a separable field extension of F. Another proof was already given in exercise 2.5.10. A very quick proof due to Serre runs as follows: Otherwise $F(d)$ is purely inseparable over F for all d in D so $d^n \in F$ for $n = \deg D$. But then X^n is a central polynomial of D, and thus of $M_n(F)$ (seen by splitting D), which is absurd.

We shall also need criteria for R to be split.

Lemma 7.1.13: *Suppose $\deg(R) = n$.*

(i) *R is split if R has n orthogonal idempotents.*

(ii) *If there is some $r \in R$ whose minimal polynomial over F splits into n distinct linear factors, then $F[r] \approx F^{(n)}$ and R is split.*

Proof:

(i) R is a direct sum of n left ideals; writing $R = M_t(D)$ we see $t \geq n$, so $t = n$ and thus $D = F$

(ii) $F[r] \approx F[\lambda]/\langle f(\lambda) \rangle \approx \prod_{i=1}^{n} F$ by the Chinese Remainder Theorem; hence we can apply (i).   Q.E.D.

Examples of Central Simple Algebras

The first known noncommutative division algebra was Hamilton's *quaternions* $\mathbb{H} = \mathbb{R} \oplus \mathbb{R}i \oplus \mathbb{R}j \oplus \mathbb{R}k$ where multiplication is given by $i^2 = j^2 = k^2 = -1$ and $ij = -ji = k$. $\mathbb{H}$ is 4-dimensional over $\mathbb{R}$ and is a ring because we can identify $\mathbb{H}$ as the subalgebra of $M_2(\mathbb{C})$ spanned by $\begin{pmatrix} i & 0 \\ 0 & -i \end{pmatrix}$, $\begin{pmatrix} 0 & 1 \\ -1 & 0 \end{pmatrix}$, and $\begin{pmatrix} 0 & i \\ i & 0 \end{pmatrix}$ (where i is $\sqrt{-1}$ in $\mathbb{C}$), which take on the respective roles of i, j, and k; $\mathbb{H}$ is a division ring since every nonzero element is invertible by means of the "norm" formula

$$(a_1 + a_2 i + a_3 j + a_4 k)(a_1 - a_2 i - a_3 j - a_4 k) = a_1^2 + a_2^2 + a_3^2 + a_4^2 \in \mathbb{R}^+$$

for $a_1, \ldots, a_4$ in $\mathbb{R}$. Note $\deg \mathbb{H} = 2$, and $\mathbb{C} = \mathbb{R} \oplus \mathbb{R}i$ is a maximal subfield of $\mathbb{H}$.

Inspired by this example we say R is a *quaternion algebra* if R is central simple of degree 2. Quaternion algebras are often studied in the context of the theory of quadratic forms, cf., exercises 1.9.9ff.

Quaternions were later generalized to the *cyclic algebras* (K, σ, α) of example 1.6.28, for any field K with automorphism σ of order n, and any α in $F = K^\sigma$. We recall (K, σ, α) is formally $\sum_{i=0}^{n-1} Kz^i$, satisfying the multiplication rules

$$a_1 z^i a_2 z^j = \begin{cases} (a_1 \sigma^i a_2) z^{i+j} & \text{if } i + j < n \\ (\alpha a_1 \sigma^i a_2) z^{i+j-n} & \text{if } i + j \geq n \end{cases}$$

for a_1, a_2 in K. We shall see that this very simple construction can be obtained under quite general circumstances.

Note: Under this definition $M_2(\mathbb{C})$ would not be cyclic because it has no maximal subfields! To resolve this difficulty we could study maximal commutative separable subalgebras in place of maximal subfields, and we shall carry out this idea below. However, it is easier (and customary) to pass instead to the underlying division algebra, since each of its commutative subalgebras is a subfield.

Remark 7.1.14: By definition the cyclic algebra $R = (K, \sigma, \alpha)$ contains

(i) a cyclic field extension K of F of degree n, and
(ii) an element $z \notin F$ for which $z^n \in F$.

Perhaps surprisingly, it turns out that

(i) already ensures R is cyclic, and

(ii) by itself ensures R is cyclic when n is prime. The first of these results is easy and will be given now; the second is a theorem of Albert, cf., [JAC-SAL, theorem 2.19]. (Note (ii) implies (i) if F contains a primitive n-th root ζ of 1; the proof of Albert's theorem in characteristic $\neq n$ entails adjoining ζ and then applying field theory to pass back. When $\text{char}(F) = n$ the proof is easy, cf., exercise 8.)

Proposition 7.1.15: *If R is central simple and has a maximal subfield cyclic over F then R is cyclic.*

Proof: Let $\text{Gal}(K/F) = \langle \sigma \rangle$. By the Skolem-Noether theorem there is z in R for which $zaz^{-1} = \sigma a$ for all a in K. Let $\alpha = z^n$ where $n = [K:F]$. Then $\alpha a \alpha^{-1} = \sigma^n a = a$ so $\alpha \in C_R(K) = K$; in fact, $\alpha \in F$ since $\sigma \alpha = zz^n z^{-1} = r^n = \alpha$. We claim $R \approx (K, \sigma, \alpha)$. Indeed, there is a homomorphism from the skew polynomial ring $K[\lambda; \sigma] \to R$ given by $\lambda \mapsto z$; the kernel contains $\lambda^n - \alpha$ so we get a homomorphism $(K, \sigma, \alpha) \to R$ which is an injection since (K, σ, α) is simple. But $[R:F] = [K:F]^2 = n^2 = [(K, \sigma, \alpha): F]$ so $(K, \sigma, \alpha) \approx R$.   Q.E.D.

Corollary 7.1.15′:   *Every quaternion algebra is cyclic (by Koethe's theorem, since every separable quadratic field extension is cyclic).*

Symbols

The cyclic algebra $R = (K, \sigma, \alpha)$ has a better form when F has a primitive n-th root $\zeta = \zeta_n$ of 1. Indeed, then $K = F(y)$ for some y such that $\beta = y^n \in F$, and taking z for which $zyz^{-1} = \sigma y = \zeta y$ we see multiplication in R is now given by the exceedingly simple rules

$$y^n = \alpha, \qquad z^n = \beta, \qquad \text{and} \qquad zy = \zeta yz \qquad (0)$$

Definition 7.1.16: The *symbol* $(\alpha, \beta)_n$ denotes the cyclic algebra described by the equations (0), where $\alpha, \beta \in F - \{0\}$. Write $(\alpha, \beta)_n \sim 1$ if $(\alpha, \beta)_n$ is split. (Strictly speaking, the definition depends on the specific choice of ζ, but we shall not worry about this matter.)

Every quaternion algebra of characteristic $\neq 2$ is a symbol by corollary 7.1.15′, since $-1 \in F$. When $\text{char}(F) = n$ clearly F cannot have a primitive n-th root of 1, but in this case there is also a version of the symbol, cf., exercise 8. However, we shall assume $\text{char}(F)$ is prime to n.

Jumping the gun a bit, let us say two central simple algebras R_1 and R_2

are *similar*, written $R_1 \sim R_2$, if their underlying division algebras are isomorphic. If F has "enough" roots of 1 then the very deep Merkurjev-Suslin
theorem says any central simple F-algebra is similar to a tensor product of
symbols; thus symbols are the building blocks of the theory, and we would
like to see just how easily we can compute with them.

To start off, given y, z as in (0) let us compute $(y + z)^n$. Certainly we can
write $(y + z)^n = \sum_{i=0}^{n} f_{i,n}(y, z)$ where the $f_{i,n}$ are polynomials in y and z,
homogeneous of degree i in y and degree $(n - i)$ in z. We shall show $f_{i,n} = 0$
whenever $1 \leq i < n$. Indeed let us write a typical monomial of $f_{i,n}$ as $v_1 \cdots v_n$,
where i of the v_i are y and $n - i$ of the v_i are z. Then $v_n v_1 \cdots v_{n-1}$ is also a
monomial of $f_{i,n}$, and $v_n v_1 \cdots v_{n-1} = \zeta^i v_1 \cdots v_n$. (Indeed if $v_n = z$ then in
moving z to the right we must pass over y i times, each time replacing zy by
ζyz and thereby multiplying altogether by ζ^i; if instead $v_n = y$ then we replace
yz by $\zeta^{-1}zy$ a total of $(n - i)$ times, thereby multiplying altogether by
$(\zeta^{-1})^{n-i} = \zeta^i$.) Continuing the process shows

$$v_1 \cdots v_n + v_n v_1 \cdots v_{n-1} + \cdots + v_2 \cdots v_n v_1 = (1 + \zeta^i + \cdots + \zeta^{(n-1)i})v_1 \cdots v_n = 0$$

so partitioning $f_{i,n}$ in this manner shows $f_{i,n}$ is a sum of zeroes and is thus 0!
(This lovely argument of G. Bergman was communicated by P. M. Cohn.)
Consequently we have

$$(y + z)^n = y^n + z^n = \alpha + \beta.$$

Proposition 7.1.17:   (*Basic properties of symbols*):

(i) $(\alpha, \beta)_n \sim 1$ *if there is r in $(\alpha, \beta)_n$ such that r^n is an n-th power in F and*
$1, r, \ldots, r^{n-1}$ *are independent.*

(ii) *If α is a norm of the extension $F(\beta^{1/n})$ over F then $(\alpha, \beta)_n \sim 1$. In particular $(1, \beta)_n \sim 1$*

(iii) *If $\alpha + \beta$ is an n-th power in F then $(\alpha, \beta)_n \sim 1$*

(iv) $(\alpha, 1 - \alpha)_n \sim 1$

(v) $(\alpha, \beta)_n \otimes (\alpha, \gamma)_n \sim (\alpha, \beta\gamma)_n$

(v') $(\beta, \alpha)_n \otimes (\gamma, \alpha)_n \sim (\beta\gamma, \alpha)_n$

(vi) $(\alpha, -\alpha)_n \sim 1$

(vii) $(\alpha, \beta)_n \sim (\beta^{-1}, \alpha)_n$

(viii) $(\alpha, \beta)_n \otimes (\beta, \alpha^t)_{nt} \sim 1$ *for any t; thus $(\alpha, \beta)_n \sim (\alpha, \beta)_{nt}^{\otimes t}$*

Proof:

(i) The minimal polynomial of r is $\lambda^n - \gamma^n$ where $\gamma^n = r^n$, so we are done
by lemma 7.1.13(ii).

(ii) Special case of (i) for if $\alpha = N(a)$ then $(a^{-1}z)^n = N(a)^{-1}z^n = \alpha^{-1}\alpha = 1 = 1^n$.

(iii) Take y, z as in (0). As computed above $(y + z)^n = \alpha + \beta$ is an n-th power in F so we are done by (i).

(iv) $\alpha + (1 - \alpha) = 1$, so done by (iii).

(v) Take y, z in $(\alpha, \beta)_n$ satisfying (0), and take y', z' in $(\alpha, \gamma)_n$ satisfying $z'y' = \zeta y'z'$ and $(y')^n = \alpha$ and $(z')^n = \gamma$. Working in $R = (\alpha, \beta)_n \otimes (\alpha, \gamma)_n$ let R_1 be the subalgebra generated by $y \otimes 1$ and $z \otimes z'$, and R_2 be the subalgebra generated by $y^{-1} \otimes y'$ and $1 \otimes z'$. Visibly R_1 and R_2 centralize each other, and $R_1 \approx (\alpha, \beta\gamma)$ and $R_2 \approx (1, \gamma)_n \sim 1$, so $R \approx R_1 \otimes R_2$ by proposition 7.1.1.

(v') As in (v).

(vi) Taking y, z as in (0) where $\beta = -\alpha$ we have $(yz^{-1})^n = \zeta^{-n(n+1)}y^n z^{-n} = -\zeta^{-n(n+1)/2} = (-1)^n$ as seen by considering n even and n odd separately. Hence $(\alpha, -\alpha)$ is split by (i).

(vii) If y, z are as in (0) then $yz^{-1} = \zeta z^{-1}y$ so we could replace y, z, respectively, by z^{-1}, y which yield $(\beta^{-1}, \alpha)_n$.

(viii) Take y, z in $(\alpha, \beta)_n$ as in (0), and y', z' such that $(y')^{nt} = \beta$, $(z')^{nt} = \alpha^t$, and $z'y' = \zeta_{nt}y'z'$. Then $R = (\alpha, \beta)_n \otimes (\beta, \alpha^t)_{nt}$ has the centralizing subalgebras R_1 generated by $y^{-1} \otimes z'$ and $1 \otimes y'$, and R_2 generated by $y \otimes 1$ and $z^{-1} \otimes (z')^t$. Then $R_1 \sim (1, \beta)_{nt}$ and $R_2 \sim (\alpha^{-1}, 1)_n$ so $R \approx R_1 \otimes R_2$ is split. The second assertion follows from (vii).   Q.E.D.

Certain of these properties follow formally from others, e.g., the reader might try deriving (vii) from (v), (v'), and (vi). This becomes significant in the K_2-theory in §7.2.

In fact the converse of (ii) also holds. For convenience we assume $F(\beta^{1/n})$ is a field, although this assumption could be removed if we apply digression 7.1.10' in the proof below instead of the Skolem-Noether theorem.

Proposition 7.1.18: (*Suppose $F(\beta^{1/n})$ is a field.*) $(\alpha, \beta)_n$ *is split iff α is a norm from $F(\beta^{1/n})$ to F.* (*This is called Wedderburn's criterion.*)

Proof: ($\Leftarrow$) is (ii) above. ($\Rightarrow$) Let $R = (\alpha, \beta)_n$. By assumption $R \sim 1 \sim (1, \beta)_n$, so in addition to elements y, z satisfying (0) we also have y', z' in R satisfying $(y')^n = 1$, $(z')^n = \beta$, and $z'y' = \zeta y'z'$. But $F(z')$ and $F(z)$ are each isomorphic to $F(\beta^{1/n})$ so by the Skolem-Noether theorem we may apply a suitable inner automorphism of R and assume $z' = z$. Now let $x = y'y^{-1} \in C_R(z) = F(z)$. Then $1 = (y')^n = (xy)^n = N(x)y^n = N(x)\alpha$ so $\alpha = N(x^{-1})$ as desired.   Q.E.D.

Overview of the Theory of Central Simple Algebras

In view of major recent advances by Merkurjev and Suslin the theory of central simple algebras can now be presented in a remarkably clear picture. Let us sketch the picture here and spend the rest of the chapter in elaborating the sketch. The underlying idea is to describe central simple algebras in terms of cyclic algebras.

Using "generic" methods one can readily construct a central division algebra which is not cyclic (cf., after remark 7.1.20), leading to the question of whether central division algebras need have maximal subfields Galois over the center. This also has a counterexample, which we shall present shortly.

These examples lead us back to matrices over a division algebra, for perhaps some $M_t(D)$ will be cyclic even when D is not. Accordingly we define $\text{Br}(F)$ as {similarity classes of central simple F-algebras}, where we recall $R_1 \sim R_2$ if they have isomorphic underlying division algebras. $\text{Br}(F)$ has multiplication given by $[R_1][R_2] = [R_1 \otimes R_2]$, and is a group since $[R^{\text{op}}] = [R]^{-1}$ by theorem 2.3.27. Thus $[M_n(F)] = [F] = 1$ in $\text{Br}(F)$. $\text{Br}(F)$ is called the *Brauer group*.

The Brauer group turns out to be torsion, i.e., for every $[R]$ in $\text{Br}(F)$ there is m such that $[R]^m = 1$ or, equivalently, $R^{\otimes m} \sim 1$; m is called the *exponent* of R. If D is a division algebra of degree n then $m \mid n$ and, furthermore, every prime factor of n divides m. Conversely, if m and n satisfy these two requirements then Brauer constructed a cyclic division algebra D of degree n and exponent m, cf. §7.3 of the unabridged edition.

A fundamental property asked of division algebras is *decomposability*. Namely, if $\deg(D) = n = n_1 n_2$, are there division subalgebras D_i of respective degree $n_i \neq 1$ for which $D \approx D_1 \otimes D_2$? If n_1 and n_2 are relatively prime the answer is, "Yes," and we thereby can reduce the theory to the case where n is a power of a prime number p. But now if $D \approx D_1 \otimes D_2$ and each $n_i \neq 1$ then n_i divides n/p and thus m divides n/p. Hence any decomposable division algebra of degree n must have exponent $< n$. In fact there are indecomposable division algebras of exponent p and degree p^2 for every odd prime p; for exponent 2 there are indecomposable division algebras of degree 8, but every division algebra of degree 4 and exponent 2 is decomposable.

Instead of asking for a division algebra itself to be indecomposable we might instead ask whether it can be decomposed *in the Brauer group*. This leads us to

Merkurjev-Suslin Theorem: *If F has a primitive m-th root ζ of 1 then every*

division algebra of exponent m is similar to a tensor product of cyclic algebras of degree m. (It follows that $\text{Br}(F)$ *is a divisible torsion abelian group when F has "enough" roots of* 1.)

Thus $M_t(D)$ is decomposable as a tensor product of cyclics for suitable t. When $\deg(D) = p$ for $p = 3$ then D is already cyclic; for larger primes p this is unknown, but $t \leq (p - 1)!/2$.

Although we shall not go into detail here, there is a fundamental identification of the Brauer group with a second cohomology group, by means of Noether's description of crossed products in terms of factor sets. This is discussed in the unabridged edition on pages 208 and 213.

The First Generic Constructions

There are two main ways of building and studying central simple algebras. The more classical method is the "arithmetic method" using tools of algebraic number theory such as local field theory. The second method is the "generic method," which is to attach indeterminates to key quantities and then study leading coefficients of polynomials. Both methods are of the utmost importance. In the text we shall focus on the generic method, which gives quick results. The arithmetic method will be discussed all too briefly in Appendix A.

The usual way of employing the generic method is to build first a domain of PI-degree n resembling a skew polynomial ring and the passing to the ring of fractions, as illustrated in the following basic example.

Example 7.1.19: Let λ_1, λ_2 be commuting indeterminates over a field F_0 which contains a primitive n-th root ζ of 1. Then the symbol $D = (\lambda_1, \lambda_2)_n$ is a division algebra, where $F = F_0(\lambda_1, \lambda_2)$. Indeed, let $D = \sum_{i,j=0}^{n-1} F y^i z^j$ where $y^n = \lambda_1, z^n = \lambda_2$, and $zyz^{-1} = \zeta y$. Letting $C = F_0[y]$ a polynomial ring in y one sees easily that D is the ring of central fractions of the skew polynomial ring $C[z; \sigma]$ where σ is given on C by $\sigma(\sum \alpha_i y^i) = \sum \zeta^i \alpha_i y^i$. But $C[z; \sigma]$ is a domain by proposition 1.6.15 so D is a division algebra by the following observation:

Remark 7.1.20: If a central simple algebra D is the ring of central fractions of a domain then D is a division algebra. (Indeed, obviously D is a domain, but then D cannot have any nontrivial idempotent so the Wedderburn-Artin theorem says D must be its own underlying division ring.)

Generic methods enable one to readily produce the noncyclic algebra $(\lambda_1, \lambda_2)_m \otimes (\lambda_3, \lambda_4)_n$ of degree mn over the rational fraction field $F = F_0(\lambda_1, \lambda_2, \lambda_3, \lambda_4)$; this is a special case of theorem 7.1.29(ii).

Digression 7.1.21: (Tignol) Let us modify this example by taking F to be the field of Laurent series in $\lambda_1, \ldots, \lambda_4$ over F_0. Amitsur showed $D = (\lambda_1, \lambda_2)_m \otimes (\lambda_3, \lambda_4)_n$ is a noncyclic division ring, cf., Jacobson [75B, p. 102]. But Tignol-Amitsur [84] implies D cannot have a cyclic splitting field, for this would have a subfield isomorphic to a subfield of D (since F is Henselian), contrary to D not cyclic. Thus D is *not* similar to a cyclic algebra.

Historically (in view of Koethe's theorem), the next question was whether any central simple algebra has a maximal subfield Galois over the center; such an algebra is called a *crossed product*. This has a negative answer, but first we look for a sufficiently general counterexample. We can find a clue by tidying up a loose end from the Artin-Procesi theorem.

Proposition 7.1.22: *Suppose* $\deg R = n$. *Then there are elements* $a, b_1, \ldots, b_n$ *in R for which* $\{a^{i-1}b_j : 1 \le i, j \le n\}$ *are a base for R over F.*

Proof: *Case I.* R is split, i.e., $R = M_n(F)$. Taking $b_i = e_{ii}$ and $a = e_{n1} + \sum_{i=1}^{n-1} e_{i,i+1}$ we see $a^n = 1$ and each $e_{ij} = a^{j-i}b_j$.

Case II. R is *not* split. In particular, F is infinite by theorem 7.1.11. We shall translate the assertion to one on PI's. We want to show the $a^{i-1}b_j$ are independent, so, in other words, the Capelli polynomial evaluation

$$C_{2t}(b_1, ab_1, \ldots, a^{n-1}b_1, b_2, \ldots, a^{n-1}b_n, r_1, \ldots, r_t) \ne 0$$

for suitable $r_1, \ldots, r_t$ in R where $t = n^2$. This means we want to show

$$C_{2t}(X_2, X_1 X_2, \ldots, X_1^{n-1}X_2, X_3, \ldots, X_1^{n-1}X_{n+1}, X_{t+1}, \ldots, X_{2t})$$

is a nonidentity of R, or, equivalent, by theorem 6.1.44 that this is a nonidentity of $M_n(F)$. But this is clear by case I and theorem 6.1.44. Q.E.D.

We shall now introduce the "most general" division algebra of degree n. Let $F_n\{Y\}$ denote the F-algebra of generic $n \times n$ matrices $Y_1, Y_2, \ldots$. By proposition 6.1.41 $F_n\{Y\}$ is prime of PI-degree n, so its algebra of central fractions is a simple algebra which we denote as $UD(F, n)$. $UD(F, n)$ is central simple of degree n over its center, whatever that may be, but even without knowing the center we can study $UD(F,n)$ effectively by means of PI-theory. This is done in the unabridged edition; the most famous application, which we state here without proof, is

Theorem 7.1.30: *(Amitsur) Suppose n is divisible by p^3 and $p \nmid \mathrm{char}(F)$. Then* $\mathrm{UD}(F, n)$ *does not have a subfield Galois over* $Z(\mathrm{UD}(F, n))$ *of dimension* p^3; *in particular,* $\mathrm{UD}(F, n)$ *does not have a maximal subfield Galois over the center.*

Proof: If $\mathrm{UD}(F, n)$ had a subfield Galois over the center with Galois group G of order p^3 then G is abelian of exponent p by theorem 7.1.29(ii) applied to lemma 7.1.27 (taking $n_1 = n_2 = \cdots = p$); but then $G \approx \mathbb{Z}_p \times \mathbb{Z}_p \times \mathbb{Z}_p$, which is impossible by theorem 7.1.29(iii) (taking $n_1 = n$) applied to lemma 7.1.27.
                                                                                   Q.E.D.

A direct non-crossed product construction has been found recently by Jacob-Wadsworth [86].

Note that these counterexamples all have composite degree, leaving open

Question 7.1.31: (The major open question in division algebras) (i) Is every division algebra of prime degree cyclic? (ii) Failing this, is it similar to a cyclic algebra?

Algebras of Degree 2, 3, 4

Due to lack of space in this abridged edition, we shall close the section here by giving a very brief description of the known cyclicity results, modifying an idea of D. Haile. Assume throughout that D is a division algebra over F. We make use of the reduced trace of an element d of D, written $\mathrm{tr}(d)$, which is the usual trace of d in $D \otimes \bar{F} = M_n(\bar{F})$, where $\bar{F}$ is the algebraic closure of F.

Remark 7.1.32. Clearly $\mathrm{tr}(ab) = \mathrm{tr}(ba)$ for a, b in D, and thus $\mathrm{tr}[a,b] = 0$.

Remark 7.1.33: If a, b commute in D, then $[a, d]b = [a, db]$ and $b[a, d] = [a, bd]$, as seen by direct computation.

Theorem 7.1.34. *(Compare with Haile [88]). Suppose K is a subfield of D. Then for any a in $K - F$ there is an element d in $D - K$ such that* $\mathrm{tr}[a, d] = \mathrm{tr}[a, d]^{-1} = 0.$

Proof. Since $a \notin F$, there is d_0 in D for which $[a, d_0] \neq 0$. Define the map $f: K \to F$ by $fk = \mathrm{tr}([a, d_0]^{-1}k)$. Counting dimensions, we see $\ker f$ is $([K:F] - 1)$-dimensional over F. Take b in $\ker f$, and let $d = b^{-1}d_0$. Then

$$[a, d]^{-1} = [a, b^{-1}d_0]^{-1} = (b^{-1}[a, d_0])^{-1} = [a, d_0]^{-1}b,$$

so $\mathrm{tr}([a, d]^{-1}) = 0$. On the other hand $\mathrm{tr}[a, d] = 0$ for any a and any d.

Corollary 7.1.35. *Any division algebra D has an element d such that* $tr(d) = tr(d^{-1}) = 0$.

Corollary 7.1.36. (i) *Any division algebra of degree 2 is cyclic.*
(ii) *Any division algebra of degree 3 is cyclic.*
(iii) *Any division algebra of degree 4 is a crossed product with respect to* $\mathbb{Z}_2 \times \mathbb{Z}_2$.

Proof. Take an element d of D with $tr(d) = tr(d^{-1}) = 0$.

(i) by corollary 7.1.15′.

(ii) For convenience, assume F has a primitive cube root of 1; then clearly $F(d)$ is cyclic over F. (In general one has $d \in F$ and appeals to remark 7.1.14.)

(iii) For convenience assume $\mathrm{char}(F) \neq 2$. The coefficients of d and d^3 in the minimal polynomial of d are 0. But this means d^2 has degree 2. Adding a suitable scalar to d^2 gives us an element a such that $a^2 \in F$. By theorem 7.1.10, take b in D such that $bab^{-1} = a$. Then b^2 commutes with a, and $F(a,b^2)$ is the desired field unless $b^2 \in F(a)$. But then a and b generate an F-central subalgebra Q of D of degree 2, and by proposition 7.1.1, $Q_1 = C_D(Q)$ has degree 2, with $D \approx Q \otimes Q_1$. We are done by (i). Q.E.D

There are two other methods for obtaining these results, one via Wedderburn's method (exercises 15ff) and one using Brauer factor sets, which are treated in depth in the unabridged volume. However, the cyclicity question for division algebras of prime degree ≥ 5 is still wide open.

§7.2 The Brauer Group

In the overview in §7.1 we defined the Brauer group of a field, denoted $\mathrm{Br}(F)$, as the set of similarity classes of central simple algebras (two algebras are *similar* if they have the same underlying division algebra); the group operation is tensor product. Our goal in this section is to obtain some of the properties of the Brauer group.

One obvious question to ask is, "What is the group structure of $\mathrm{Br}(F)$, for a given field F?" If F is algebraically closed then $\mathrm{Br}(F) = (1)$ since there are no nontrivial F-central division algebras. The Tsen-Lang theorem (Jacobson [80B, p. 649]) implies more generally that $\mathrm{Br}(F) = (1)$ whenever F is an algebraic extension of $F_0(\lambda)$ where F_0 is algebraically closed. Another instance of $\mathrm{Br}(F) = (1)$ is when F is a finite field, by theorem 7.1.11. It is also easy to check $\mathrm{Br}(\mathbb{R}) \approx \mathbb{Z}/2\mathbb{Z}$, cf., exercise 1. Usually $\mathrm{Br}(F)$ is infinite and difficult to ascertain, as one sees when investigating $\mathrm{Br}(\mathbb{Q})$.

General Properties of the Brauer Group

There are three major approaches to the Brauer group. The first is by looking internally at each algebra; the second is by means of crossed products; and the third is via cyclic algebras. This last approach is both the oldest (used in the classic books by Dickson and by Albert) and the freshest, bringing in algebraic K-theory. We shall pursue it here, but only so far as we can with elementary methods.

Remark 7.2.1: For any fields $F \subseteq K$ there is a group homomorphism $\mathrm{res}_{K/F} \, \mathrm{Br}(F) \to \mathrm{Br}(K)$ called the *restriction*, given by $[R] \to [R \otimes_F K]$. Thus Br is a functor from {fields} to {abelian groups}. $\mathrm{Ker}(\mathrm{res}_{K/F}) = \{[R] \in \mathrm{Br}(F) :$ R is split by $K\}$. Thus the functor Br contains the theory of splitting fields.

One fundamental result needed later to work with restriction is

Proposition 7.2.2:   *Suppose $K \supseteq F$ is a subfield of R. Then $R \otimes_F K \sim C_R(K)$.*

Proof: By proposition 7.1.3 $C_R(K) \otimes_K K \otimes_F R^{\mathrm{op}} \approx \mathrm{End}_K R$ is split as K-algebra, so $C_R(K) \sim R \otimes K$.   Q.E.D.

First Invariant of $\mathrm{Br}(F)$: The Index

Recall index(R) is defined as the degree of the underlying division algebra D of R. Note for any field $K \supset F$ that index$(R \otimes K) = $ index$(D \otimes K)$ divides degree$(D \otimes K) = $ degree$(D) = $ index(R).

Theorem 7.2.3:   *Let $t = $ index$(R)/$index$(R \otimes_F K)$. Then $t \, | \, [K:F]$ and K is isomorphic to a subfield of $M_{[K:F]/t}(D)$, where D is the underlying division algebra of R.*

Proof: Let $k = [K:F]$. By the regular representation $K \subseteq M_k(F)$, so $M_k(D) \supseteq D \otimes K = M_t(D_1) \supseteq M_t(F)$ for some division algebra D_1. By proposition 7.1.1 there is a division algebra D' for which $M_k(D) \approx M_t(F) \otimes M_u(D') \approx M_{tu}(D')$; hence $u = k/t$ and $D' \approx D$. But $K = Z(D \otimes K)$ centralizes $M_t(F)$ so $K \subseteq M_u(D') \approx M_u(D)$.   Q.E.D.

Corollary 7.2.4:   *If $[K:F]$ and index(R) are relatively prime then* index$(R \otimes K) = $ index(R).

Second Invariant of $\mathrm{Br}(F)$: The Exponent

One of the major features of the Brauer group is that it is torsion. This can be seen at once by combining exercise 5.2.2 with theorem 7.2.6, but better results can be obtained directly. To wit, we define the *exponent* of a central simple algebra R, written $\exp(R)$, to be the order of $[R]$ in the Brauer group; i.e., $\exp(R)$ is the smallest number m for which $R^{\otimes m} \sim 1$, where we recall $R^{\otimes m}$ denotes $R \otimes_F \cdots \otimes_F R$ taken over m copies of R.

Example 7.2.9: If $R = (\alpha, \beta)_n$ and $F(\beta^{1/n})$ is a field then $\exp(R)$ is the smallest m for which α^m is a norm from $F(\beta^{1/n})$ to F, by Wedderburn's criterion (proposition 7.1.18).

On the other hand, $m|n$ since (by proposition 7.1.17) $R^{\otimes n} \approx (\alpha^n, \beta) \sim 1$.

Lemma 7.2.10. *If* $index(R) = n$, *and* L *is a field extension of* F *with* $[L:F]$ *prime to* n, *then* $\exp(R \otimes_F L) = \exp(R)$.

Proof. Let $m = \exp(R)$, and $m' = \exp(R \otimes_F L)$. Then $(R \otimes_F L)^{\otimes m} \approx R^{\otimes m} \otimes_F L \sim 1$, so m' divides m. On the other hand, $R^{\otimes m'} \otimes_F L \approx (R \otimes_F L)^{\otimes m'} \sim 1$, so $R^{\otimes m'} \sim 1$ by corollary 7.2.4. Q.E.D.

Corollary 7.2.10'. *If* $index(R)$ *is prime, then* $\exp(R) = index(R)$.

Proof. If F has a primitive pth root of 1, then this is example 7.2.9; if char$(F) \neq p$ then one can adjoin a pth root of 1 by a field extension of dimension p-1, so use lemma 7.2.10. Finally, if char$(F) = p$, then write $R = (K, \sigma, \alpha)$; in analogy to proposition 7.1.17(v) we see $R^{\otimes p} \approx (K, \sigma, \alpha^p)$ which by lemma 7.1.13(ii) is split since $\lambda^p - \alpha^p = (\lambda - \alpha)^p$.

Proposition 7.2.11: *Suppose* $index(R) = p^j t$ *where* p *is a prime number not dividing* t. *Then there is a field extension* L *over* F *whose dimension is relatively prime to* p, *for which* $index(R \otimes_F L) = p^j$.

Furthermore, for any splitting subfield K *separable over* F *(which exists by Koethe's theorem),* KL *is a splitting subfield of* $R \otimes_F L$, *and in particular there is a descending chain of subfields* $K_0 = KL > K_1 > \ldots > K_j = L$, *with each* $[K_i : K_{i+1}] = p$.

Proof: Let E be the normal closure of a separable splitting subfield K of R; $[E:F] = [E:K][K:F] = p^{j'} t'$ for suitable $j' \geq j$ and t' prime to p. Let H be a Sylow p-subgroup of $\mathrm{Gal}(E/F)$ and let L be the fixed subfield of E under H. By Galois theory $[L:F] = t'$ is prime to p, so by theorem 7.2.3 index$(R \otimes_F L) = p^j t''$ for some t'' dividing t. On the other hand, $(R \otimes_F L) \otimes_L E \approx R \otimes_F E$ is split so $p^j t''$ divides $[E:L] = p^j$, implying $t'' = 1$, proving index$(R \otimes_F L) = p^j$.

To prove the additional assertion, note KL contains both K and L, implying $[KL:F]$ is divisible by both p^j and t', so $[KL:F] = p^j t'$, and thus $[KL:L] = p^{j}$. This proves KL is a maximal subfield of $R \otimes_F L$. But KL is contained in E, which is Galois over L with Galois group H. Since H (being a p-group) is solvable, there is a chain of subgroups of H each having index p in the previous subgroup. By Galois theory, the corresponding fixed subfields of E constitute a chain of subfields $E = E_0 > E_1 > \ldots > E_j = L$, with each $[E_i:E_{i+1}] = p$. Let $K_i = K \cap E_i$. By the Jordan–Holder theorem $[K_i:K_{i+1}] = [E_{i+1} + K_i : E_{i+1}]$, which divides p, so is 1 or p. If $[K_i:K_{i+1}] = 1$ then $K_i = K_{i+1}$, in which case we discard K_{i+1}. The result follows.   Q.E.D.

Corollary 7.2.11'   *Suppose the prime number p divides index(R). Then p divides exp(R), and index$(R^{\otimes p})$ divides index$(R)/p$.*

Proof.   We may assume R is a division ring. In the above notation let $R_1 = C_{R \otimes L}(K_1)$, which has degree p (since it has center K_1 and maximal subfield $K_0 = KL$, and $[K_0:K_1] = p$). Hence exp$(R_1) = p$. But $R_1 \sim R \otimes K_1$ by proposition 7.2.2, so p divides exp(R); moreover,

$$1 = \text{index}(R_1^{\otimes p}) = \text{index}((R \otimes K_1)^{\otimes p}) = \text{index}((R^{\otimes p} \otimes_F L) \otimes_L K_1),$$

implying index$(R^{\otimes p})$ divides $[K_1:L] = [KL:L]/p = [K:F]/p = \text{index}(R)/p$.
                                                                                      Q.E.D.

Theorem 7.2.12   *If n, m, are the respective index and exponent of a central simple algebra R, then the following two conditions hold:*

(i) $m \mid n$
(ii) Every prime divisor of n also divides m.

Proof.   Let $R_1 = R^{\otimes p}$, where p is a prime dividing n. Then p divides m by corollary 7.2.11', proving (ii). We prove (i) by induction on m. Clearly exp$(R_1) = m/p$, which by induction divides index(R_1) which in turn divides n/p, by corollary 7.2.11'. Hence, $m \mid n$.   Q.E.D.

We close this brief treatment with a useful decomposition result.

Theorem 7.2.13: *Suppose D is a central division algebra of degree n =*
$p_1^{\alpha(1)} \cdots p_k^{\alpha(k)}$ *for suitable distinct* $p_1, \ldots, p_k$ *primes and* $\alpha(1), \ldots, \alpha(k)$ *in* $\mathbb{N} - \{0\}$.
Then $D \approx D_1 \otimes \cdots \otimes D_k$ *for central division subalgebras* D_i *of degree* $p_i^{\alpha(i)}$,
$1 \le i \le k$.

Proof: It suffices to prove that if $\deg(D) = n = n'n''$ where n' and n'' are
relatively prime then $D \approx D_1 \otimes D_2$ where $\deg(D_1) = n'$ and $\deg(D_2) = n''$,
for then we can continue the decomposition by induction. Take u, v in $\mathbb{N}$ for
which $un' + vn'' \equiv 1 \pmod{n}$ and take $D_1 \sim D^{\otimes vn''}$ and $D_2 \sim D^{\otimes un'}$. Then
$D_1 \otimes D_2 \sim D^{\otimes (un' + vn'')} \sim D$. On the other hand, $D_1^{\otimes n'} \sim D^{\otimes vn} \sim 1$ so $\exp(D_1) | n'$,
implying by theorem 7.2.12 $\deg(D_1)$ is prime to n''; but $\deg(D_1)$ divides
$\deg(D) = n'n''$ so $\deg(D_1)$ divides n'. Likewise, $\deg(D_2)$ divides n''. Thus

$$\deg(D_1 \otimes D_2) \le n'n'' = n = \text{index}(D) = \text{index}(D_1 \otimes D_2) \le \deg(D_1 \otimes D_2)$$

so equality holds; in particular, $D_1 \otimes D_2$ is a division algebra so $D_1 \otimes D_2 \approx D$.
$$\text{Q.E.D.}$$

In view of theorem 7.2.13 the structure theory of division algebras reduces
to the case where the degree (and exponent) is a prime power.

Exercises

§7.1

Maximal Commutative Subalgebras

1. Suppose K is a maximal commutative subalgebra of a central simple F-algebra
   R. If K is separable over F then R is projective as K-module, and $[K:F] = \deg R$.
   (Hint: K is projective as K^e-module so $R \approx R \otimes_K K$ is projective over $R \otimes_K K^e \approx$
   $R \otimes_K (K \otimes_F K) \approx R \otimes_F K$, which is free over K. Now using exercise 2.11.15
   note $[R:F] = [R \otimes_F K:K] = [\text{End}_K R:K] = [\text{End}_{K_P} R_P:K_P] = [R_P:K_P]^2$ for any P
   in $\text{Spec}(K)$; but $[R \otimes_F K:K] = [R \otimes_K K^e:K] = [R_P \otimes (K^e)_P:K_P] = [R_P:K_P][K:F]$.)
2. Any central simple algebra has a maximal commutative subalgebra which is
   separable. (Hint: Write $R = M_t(D)$ and take $K^{(t)}$ where K is a maximal separable subfield of D. Appeal to the counting argument of exercise 1 to show
   $C_R(K^{(t)}) = K^{(t)}$.
3. Prove *Schur's inequality* $[C:F] \le [n^2/4] + 1$ for any commutative subalgebra
   C of $M_n(F)$. (Hint: (Gustafson) Writing $M_n(F) \approx \text{End}_F V$ where $[V:F] = n$ note
   V is a faithful C-module. Switching points of view, one need only show if C is a
   commutative F-algebra and V is a faithful C-module with $[V:F] \le n$ then
   $[C:F] \le [n^2/4] + 1$. C is Artinian and thus semiperfect commutative, implying
   C is a direct product of local rings. It suffices to assume C is local. Let $J = \text{Jac}(C)$.
   Let $\bar{b}_1, \ldots, \bar{b}_m$ be a base for V/JV as C/J-module; choosing b_i such that $\bar{b}_i =$
   $b_i + JV$ define $f: J \to \text{Hom}(V/JV, JV)$ sending a to the map f_a given by $f_a \bar{b}_i = ab_i$.

Then f is monic by Nakayama's lemma so $[J:F] \le m(n-m) \le [(n/2)^2]$. But $[C:F] = 1 + [J:F]$.)

4. (Sweedler) An alternate approach to showing $D \otimes_F K \approx \operatorname{End}_K D$ for a maximal subfield K of a central division algebra D. First show if L is a subfield with $C_D(C_D(L)) = L$ and $L \subseteq T \subseteq D$ then there is a monic $\psi: T \otimes_L D^{op} \to \operatorname{Hom}_{C_D(T)}(C_D(L), D)$ given by $\psi(a \otimes d)x = axd$. (To prove ψ is monic take $0 \ne r = \sum_{i=1}^n a_i \otimes d_i \in \ker \psi$ with n minimal. One may assume $d_n = 1$, so $\sum_{i=1}^{n-1} a_i \otimes [x, d_i] \in \ker \psi$ for any x in $C_D(L)$, implying each $d_i \in L$; hence $r = (\sum a_i d_i) \otimes 1 = 0$, contradiction.

Apply this fact twice. First $T = K$ and $L = F$ yields $\psi_1: K \otimes_F D \hookrightarrow \operatorname{End}_K D$. But taking $T \ne D$ and $L = K$ yields $[D:K]^2[K:F] \le [K:F][D:F]$ so $[D:K] \le [K:F]$ and ψ_1 is thus onto.)

Skolem-Noether Theorem

5. Prove the Skolem Noether theorem. (Hint: Suppose $\psi: A \to A'$ is an isomorphism of simple F-subalgebras and form $T = A \otimes R^{op}$. R is an $A - R$ bimodule and thus T-module in two ways: the obvious way and by the action $(a \otimes r)x = (\psi a)rx$. These module actions are isomorphic since T has a unique simple module whose multiplicity in R is determined by $[R:F]$. Hence there is $f: R \to R$ satisfying $f(axr) = (\psi a)(fx)r$. Take $a = 1$ to show f is given by left multiplication by an invertible element u of R; then take $r = 1$.)

6. Weaken the hypothesis in the Skolem-Noether theorem to "finite dimensional simple subalgebras of an arbitrary simple Artinian aglebra R." (Hint: Use the underlying division ring of R in place of F.)

7. Counterexample to generalizing Skolem-Noether to semisimple subalgebras.

   Let $A = F[a]$ and $A' = F[a']$ where $R = M_3(F)$, $a = \begin{bmatrix} 1 & 0 & 0 \\ 0 & 2 & 0 \\ 0 & 0 & 2 \end{bmatrix}$ and $a' = \begin{bmatrix} 1 & 0 & 0 \\ 0 & 1 & 0 \\ 0 & 0 & 2 \end{bmatrix}$. Then a and a' have the same minimal polynomial so $A \approx A'$ but $ar \ne ra'$ for all invertible r in R.

8. Suppose $\operatorname{char}(F) \ne p$. Define a p-symbol $[\alpha, \beta)_n$ (where n is a power of p) to be the cyclic algebra (K, σ, β) where $K = F(y)$, $y^p = y + \alpha$, $zyz^{-1} = y + 1$, and $z^p = \beta$. If $\deg(R) = p$ and R has an element $r \notin F$ with $r^p \in F$ then R is a p-symbol. (Hint: exercise 2.5.10.) Also check $[\alpha_1, \beta) \otimes [\alpha_2, \beta) \sim [\alpha_1 + \alpha_2, \beta)$ and $[\alpha, \beta_1) \otimes [\alpha, \beta_2) \sim [\alpha, \beta_1 \beta_2)$.

Wedderburn's Division Algorithm (see Rowen [80B, pp. 178–179]

15. We say f divides g in $D[\lambda]$ if $g = qf$ for some q in $D[\lambda]$. Using the Euclidean algorithm (remark 1.6.20) show $\lambda - d$ divides $g(\lambda) - g(d)$, and thus $(\lambda - d)$ divides g iff $g(d) = 0$.

16. If $\lambda - d$ divides gh and not h then $\lambda - h(d)dh(d)^{-1}$ divides g. (Hint: $\lambda - d$ divides $g(\lambda)h(d)$ by exercise 15.)

17. If $h(d) = 0$ for all conjugates d of d_1 then the minimal polynomial p of d_1 over $Z(D)$ divides h. (Hint: Take a counterexample h of minimal degree; writing $h = qp + r$ for $\deg r < \deg p$ show the coefficients of r lie in $Z(D)$.)

18. (Wedderburn [21]) If $g(\lambda)$ is irreducible in $F[\lambda]$ and g has a root d_1 in some F-central division algebra D then g can be written as a product $(\lambda - d_n)\cdots(\lambda - d_1)$ where the d_i are each conjugates of d_1. (Hint: Write $g = q(\lambda)(\lambda - d_k)\cdots(\lambda - d_1)$ for k as large as possible, and let $h = (\lambda - d_k)\cdots(\lambda - d_1)$. Then h satisfies the criterion of exercise 17, so the minimal polynomial p of d_1 over $Z(D)$ divides h and thus g. Hence $p = g$ implying $h = g$.) Actually, the proof does not require $[D:F] < \infty$.

19. Give Wedderburn's proof that every central division algebra of degree 3 is cyclic, using exercise 18. (Hint: The minimal polynomial of d_1 can be written $(\lambda - d_3)(\lambda - d_2)(\lambda - d_1)$ where d_2 can be taken to be $[d, d_1]d_1[d, d_1]^{-1}$ for arbitrary d, cf., exercise 16. Let $x = [d_1, d_2]$. $d_1 + d_2 + d_3 \in F$, so commuting with d_i shows $[d_{\pi 1}, d_{\pi 2}] = \pm x$ for every π in Sym(3). But then $d_3 = xd_2x^{-1}$ by exercise 16, and, likewise, $d_{i+1} = xd_ix^{-1}$ for each i. Hence $x^3 \in F$. By Zariski topology there is d_1 such that $x \notin F$.)

20. Using Wedderburn's method reprove the fact that every D of degree 4 is a crossed product. (Hint: Take $d \in D$ of degree 4. The minimal polynomial is of the form $(\lambda^2 + a'\lambda + b')(\lambda^2 + a\lambda + b)$, and matching parts one can show a^2 is quadratic over F, cf., Rowen [80B, p. 181]. One concludes easily D is a crossed product.)

Division Rings The next exercises contain a collection of results about arbitrary division rings. We start with a very easy proof of the theorem of Cartan-Brauer-Hua.

21. Suppose V is a vector space over a division ring D, and A is an additive subgroup of V containing two D-independent elements. If $f: A \to V$ is a group homomorphism satisfying $fa \in Da$ for every a in A then there is d in D such that $fa = da$ for all a in A. (Hint: If a_1, a_2 are D-independent and $fa_i = d_ia_i$ then $f(a_1 + a_2) = d_1a_1 + d_2a_2$ implies $d_1 = d_2$.)

22. (Treur [77]) Suppose $D \subseteq E$ are division rings, such that D is invariant with respect to all inner automorphisms determined by an additive subgroup A of E. Assume $1 \in A$. Then either $A \subseteq D$ or $D \subseteq C_E(A)$. (Hint: Fixing d in D, let $f_d: A \to D$ be right multiplication by d. Exercise 21 implies either $ad = da$ for all a in A, or else every element of A is D-dependent on 1 so that $A \subseteq D$.)

23. (Cartan-Brauer-Hua) Suppose D is a division subring of E, invariant with respect to all inner automorphisms of E. Then $D = E$ or $D \subseteq Z(E)$.

24. If $\frac{1}{2} \in E$ prove the analogue of exercise 23, using derivations instead of automorphisms.

25. (Faith [58]) If E is a division ring which is not a finite field and if D is a proper division subring then the multiplicative group $D - \{0\}$ is of infinite index in $E - \{0\}$. (Hint: E is infinite by Wedderburn's theorem, so one may assume D is infinite. If $[E - \{0\}: D - \{0\}] < \infty$ then for any x in E one can find $d_1 \neq d_2$ in D with $x + d_1 \in D(x + d_2)$; solve to get $x \in D$.)

26. (Herstein) Any noncentral element of a division ring has infinitely many conjugates. (Hint (Faith): In exercise 25 take D to be the centralizer.)

§7.2

1. (Frobenius) The only central division algebras over $\mathbb{R}$ are $\mathbb{R}$ and the quaternion algebra $\mathbb{H}$. (Hint: If $D \neq \mathbb{R}$ then $\mathbb{C}$ is a maximal subfield of D, so $\deg D = 2$ and thus D is isomorphic to the symbol $(\alpha, -1)_2 \approx (\pm 1, -1)_2$.) Thus $|\text{Br}(\mathbb{R})| = 2$.
2. $eRe \sim R$ for any idempotent e of R. (Hint: Let $L = Re \approx L_0^{(t)}$ where L_0 is a minimal left ideal of R; $eRe \approx \text{End}_R Re \approx \text{End}_R L_0^{(t)} \approx M_t(\text{End}_R L_0)$.)
3. (Cancellation) If $[K:F]$ is prime to $\deg(R)$ and $R \otimes K \approx R' \otimes K$ then $R \approx R'$. (Hint: $R^{\text{op}} \otimes R' \otimes K$ is split.)
4. Suppose $\deg(R) = n = m_1 \cdots m_t$ where $m_1, \ldots, m_t$ are distinct prime powers. If there are fields K_i with $[K_i : F] = n/m_i$ such that each $C_R(K_i)$ is a division ring, then R is also a division ring. (Hint: Otherwise, let $m = n/\text{index}(R)$; take some m_i not prime to m and reach a contradiction.)
5. Suppose R has a maximal subfield K Galois over F with Galois group G. Letting K_p be the fixed subfield of K under the Sylow p-subgroup of G, show that if each $C_R(K_p)$ is a division ring then R is also a division ring. (Hint: exercise 4.)

Divisibility in the Brauer Group

35. Suppose ζ is a primitive m-th root of 1. Then $F(\zeta)$ is cyclic over F provided m is odd or $\sqrt{-1} \in F$. (Hint: Use Jacobson [85B, theorem 7.19 and 4.20].)
36. Suppose K_0 is a cyclic field extension of F with $[K_0 : F] = p^t$, and F has a primitive p-th root ζ of 1. Then K_0 is contained in a cyclic field extension K of F with $[K:F] = p^{t+1}$, iff $\zeta = N_0(a_0)$ for some a_0 in K_0, where N_0 is the norm from K_0 to F. (Hint: ($\Rightarrow$) by lemma 7.3.1. ($\Leftarrow$) Let σ be the generating automorphism of $\text{Gal}(K_0/F)$. By Hilbert's theorem 90 there is b_0 in K_0 for which $b_0^{-1}\sigma b_0 = a_0^p$. Let $K = K_0(b)$ where b is a root of the polynomial $f(\lambda) = \lambda^p - b_0$. Then $\sigma f = \lambda^p - \sigma b_0$ so σ extends to an automorphism of K given by $\sigma b = \zeta^j a_0 b$. Applying N_0 shows $b \notin K_0$ so $[K:K_0] = p$.)
37. Suppose K is a cyclic field extension of F having a subfield K_0 with $[K:F] = n$ and $[K_0 : F] = m$. Then $[(K, \sigma, \alpha)]^{n/m} = [(K_0, \sigma, \alpha)]$ in $\text{Br}(F)$ where σ generates $\text{Gal}(K/F)$. (Hint: Let $k = n/m$. It suffices to show $R = (K_0, \sigma, \alpha) \otimes (K, \sigma, \alpha^{-k})$ is split. Take z_0 in (K_0, σ, α) with $z_0^m = \alpha$ and $z_0 a z_0^{-1} = \sigma a$ for all a in K_0; take z in (K, σ, α^{-k}) with $z^n = \alpha^{-k}$ and $zaz^{-1} = \sigma a$ for all a in K. Let R_1 be the sub-algebra of R generated by $z_0 \otimes z$ and $1 \otimes K$; R_1 is split since $(z_0 \otimes z)^n = 1$. Let $K' = (K_0 \otimes K_0)^{\sigma \otimes \sigma}$ and let R_2 be the subalgebra of R generated by K' and $z_0 \otimes 1$. It suffices to prove K' has m orthogonal idempotents, for then $[R_2 : F] \geq m^2$ and thus $R_2 = C_R(R_1)$ and is also split, proving R is split. Take the primitive idempotent $e \in K'$ as in exercise 6; $\{(\sigma^i \otimes 1)e : 1 \leq i \leq n\}$ are distinct primitive idempotents, as desired.)
38. Suppose F has a primitive p-th root ζ of 1. $[R]$ has a cyclic p-th root in $\text{Br}(F)$ if R has a cyclic splitting field from which ζ is a norm. (Hint: One may assume $\deg(R) = q^t$ for q prime, and the result is trivial unless $q = p$. If $R \sim (K_0, \sigma, \alpha)$ take K as in exercise 36 and apply exercise 37.)
39. If ζ is a reduced norm from a splitting field of R then $[R]$ has a p-th root in $\text{Br}(F)$. In particular, if F has a primitive p^{t+1}-root of 1 then the p^t-torsion subgroup of $\text{Br}(F)$ is p-divisible. (Hint: Apply Merkurjev-Suslin to exercise 36.)

8 Rings from Representation Theory

§8.1 General Structure Theory of Group Algebras

In Chapter 1 we defined monoid algebras as a tool for studying polynomial rings. In this section we turn to the case where the monoid is a group $G \neq (1)$; $C[G]$ is then called a *group algebra* and is a C-algebra with base consisting of the elements of G. We shall concentrate on the situation for which the base ring C is a field F and shall study group algebras in terms of their ring-theoretic structure.

$F[G]$ is never simple since there is a surjection $F[G] \to F$ sending each g in G to 1. On the other hand, when G is finite $F[G]$ is finite dimensional and thus Artinian and, in fact, is semisimple Artinian when char$(F) = 0$ by Maschke's theorem (8.1.8′ below).

Thus many interesting structural questions pertain only to infinite groups. On the other hand, the study of group algebras of finite groups is enhanced by the rich theory of characters of group representations, and we would like that theory to be available. We shall try to circumvent this paradox by outlining some of the basic character theory, emphasizing that part which will yield intuition for some of the later results. However, we shall focus on general structure theory, which does not rely much on the character theory. Structural results on Noetherian group algebras are to be had in §8.2. Most of this material follows Passman [77B], which is the standard text (as well as reference for results published through 1976). A few newer results are given in the exercises.

Throughout this section we assume that C is a commutative ring, F is a field, and G is a group, written multiplicatively. We identify G canonically as a subgroup of $C[G]$ or $F[G]$, by the injection $g \mapsto 1g$. In this sense the group algebra contains all of the information about the group. On the other hand, sometimes $\mathbb{C}[G] \approx \mathbb{C}[H]$ even when $G \not\approx H$.

Remark 8.1.1: If R is a C-algebra and H is a multiplicative subgroup of R then any group homomorphism $f: G \to H$ extends uniquely to an algebra homomorphism $C[G] \to R$ given by $\sum c_g g \mapsto \sum c_g fg$. (This is a special case of proposition 1.3.16'.)

In particular, any group homomorphism $G \to H$ extends to an algebra homomorphism $C[G] \to C[H]$, thereby yielding a functor $\mathcal{Grp} \to C\text{-}\mathcal{Alg}$ given by $G \to C[G]$. On the other hand, the functor $Unit: C\text{-}\mathcal{Alg} \to \mathcal{Grp}$, sending R to its multiplicative group of invertible elements (and restricting homomorphisms), also enters into the picture.

Remark 8.1.2: If G is a group then $C[G]$, together with the canonical injection $G \to C[G]$, is a universal from G to the functor $Unit$. (Indeed, any group homomorphism $f: G \to Unit(R)$ extends uniquely to an algebra homomorphism $\hat{f}: C[G] \to R$, which restricts, in turn, to a group homomorphism $Unit(C[G]) \to Unit(R)$; this completes the appropriate diagram and is unique.)

In view of these observations it makes sense to transfer some terminology from algebras to groups, and in definition 5.2.12 we introduced G-modules; we see at once that any G-module becomes a $\mathbb{Z}[G]$-module under the action $(\sum_g n_g g)x = \sum n_g(gx)$, and thus $G\text{-}\mathcal{Mod}$ and $\mathbb{Z}[G]\text{-}\mathcal{Mod}$ are isomorphic categories.

To proceed further we shall consider some of the basic structural properties of groups. Some minor ambiguities arise since the same terminology has a different connotation for groups than for rings and will depend on the context. In particular,

$N \lhd G$ denotes N is a normal subgroup of G.

$[g_1, g_2]$ will denote the *group* commutator $g_1 g_2 g_1^{-1} g_2^{-1}$.

The *center* $Z(G) = \{z \in G : [z, G] = (1)\}$.

The subgroup of G generated by a subset S is denoted $\langle S \rangle$ or as $\langle g_1, \ldots, g_m \rangle$ if $S = \{g_1, \ldots, g_m\}$.

$C_G(A)$ denotes the *centralizer* of a subset A of G, which is $\{g \in G: [g, A] = (1)\}$; recall for any g in G that $[G:C_G(g)]$ is the number of conjugates of g.

Recall $\operatorname{supp}(\sum c_g g) = \{g : c_g \neq 0\}$.

The order of an element g in G is also called the *period* of g; G is called a *p'-group* if no element has period p. It turns out that many of the structural results hold for $F[G]$ when $\operatorname{char}(F) = 0$ or when $p = \operatorname{char}(F) > 0$ and G is a *p'-group*. We shall unify these two situations by means of the terminology, "G is a char$(F)'$-group".

An Introduction to the Zero-Divisor Question

One of the most elementary questions one can ask about $F[G]$ is whether or not it has zero-divisors. If $g \in G$ has period n then $0 = 1 - g^n = (1 - g) \cdot (1 + g + \cdots + g^{n-1})$, so $C[G]$ certainly is *not* a domain. What about the torsion-free case?

Questions for Torsion-Free Groups:

(i) *Zero-divisor question.* Is $F[G]$ necessarily a domain?
(ii) Failing (i), can $F[G]$ have nontrivial idempotents?

Amazingly, these questions are still open, although $F[G]$ is now known to be a domain in many cases, to be discussed in §8.2. When G is abelian the answer is easy.

Proposition 8.1.3: *If G is torsion-free abelian and C is an integral domain then $C[G]$ is an integral domain.*

Proof: Suppose $r_1 r_2 = 0$ for r_1, r_2 in $C[G]$. Replacing G by the subgroup generated by $\operatorname{supp}(r_1) \cup \operatorname{supp}(r_2)$, we may assume G is finitely generated abelian, so $G \approx \langle g_1 \rangle \times \cdots \times \langle g_t \rangle$ for suitable g_i in G. But now $C[G]$ can be identified with a subring of the Laurent series ring $C((G))$, which is a domain. Q.E.D.

An alternate conclusion to the previous proof is by the following convenient observation.

Remark 8.1.4: $C[G_1 \times G_2] \approx C[G_1] \otimes C[G_2]$. (Indeed, define $\varphi : C[G_1] \times C[G_2] \to C[G_1 \times G_2]$ by $\varphi(\sum c_g g, \sum c'_h h) = \sum c_g c'_h (g, h)$. Clearly φ is a balanced map which induces a surjection $C[G_1] \otimes C[G_2] \to C[G_1 \times G_2]$, whose inverse is given by $\sum c_{(g,h)} (g, h) \mapsto \sum c_{(g,h)} g \otimes h$.

Maschke's Theorem and the Regular Representation

Group algebras have an extra piece of structure. An *involution* of a ring is an anti-automorphism (*) of degree 2, i.e. $(ab)^* = b^*a^*$ and $a^{**} = a$.

Remark 8.1.5: Suppose C has a given automorphism σ of degree 2. (In particular, one could take $\sigma = 1$). Then $C[G]$ has an involution given by $\left(\sum_{g \in G} c_g g\right)^* = \sum (\sigma c_g) g^{-1}$.

Proposition 8.1.6: *Suppose C is a subring of $\mathbb{C}$ closed under complex conjugation, and take (*) as in remark 8.1.5. If $a \in C[G]$ and $a^*a = 0$ then $a = 0$. In particular, $C[G]$ is semiprime.*

Proof: Write $a = \sum a_g g$. The coefficient of 1 in a^*a is $0 = \sum_g \bar{a}_g a_g = \sum |a_q|^2$ so each $a_g = 0$. Hence $a = 0$. Hence $C[G]$ is semiprime: If $rC[G]r = 0$ then $(r^*r)^*(r^*r) = r^*rr^*r = 0$; taking $a = r^*r$ shows $r^*r = 0$, and thus $r = 0$.
                                                                                                Q.E.D.

Corollary 8.1.7: *(Maschke's theorem for $\mathbb{C}$) $\mathbb{C}[G]$ is semisimple Artinian for any finite group G.*

Proof: $\mathbb{C}[G]$ is semiprime and Artinian.   Q.E.D.

Maschke's theorem is the fundamental theorem in the subject of group algebras. Let us try to generalize it in various directions. Using field theory we shall first show that the main hypothesis in proposition 8.1.6 is superfluous. For that proof to work, the only property required of $\mathbb{C}$ was that $\mathbb{C} = K[\sqrt{-1}]$ for a suitable field K in which 0 cannot be written as a sum of nonzero squares; in the proof we took $K = \mathbb{R}$. We shall now see that *all* algebraically closed fields of characteristic 0 have this property.

Theorem from the Artin-Schreier Theory of Real-Closed Fields: *Any algebraically closed field E of characteristic 0 has a subfield K satisfying the following properties:*

(1) *Any sum of nonzero squares is a nonzero square.*
(2) *$E = K[i]$ where $i^2 = -1$.*

Proof: Obviously -1 is not a sum of squares in $\mathbb{Q}$; so by Zorn's lemma, E has a subfield K maximal with respect to -1 is not a sum of squares. By Lang [65B, theorem 11.2.1] K is real closed; in particular, every sum of nonzero squares in K is a nonzero square in K, and $K[i]$ is algebraically

closed. It remains to show E is an algebraic extension of K, for then we shall
have $K[i] = E$.

Suppose $a \in E - K$. Then $K(a) \supset K$ so

$$-1 = \sum (p_i(a)/q(a))^2$$

for p_i, q in $K[\lambda]$ taken with $\deg q$ minimal. Thus $-q(a)^2 = \sum p_i(a)^2$, so $q(a)^2 + \sum p_i(a)^2 = 0$.

We claim the polynomial $q(\lambda)^2 + \sum p_i(\lambda)^2$ is nonzero. Otherwise, taking
the constant terms p_{i0}, q_0 of p_i, q we get $0 = q_0^2 + \sum p_{i0}^2$ in K, implying each
$p_{i0} = q_0 = 0$. But then we can cancel λ from q and from each p_i, contrary
to supposition, so the claim is established. It follows at once that a is algebraic over K, as desired.   Q.E.D.

Lemma 8.1.8: *If F is algebraically closed of characteristic 0 then $F[G]$ has
no nil left ideal $L \neq 0$.*

Proof: Otherwise take $0 \neq a \in L$. As seen above we can write $F = F_0[i]$
where F_0 is real closed. Then the proof of proposition 8.1.6 is available (for
$C = F_0$), implying $b = a^*a \neq 0$, so $b^2 = b^*b \neq 0$, and by iteration each
power of $b \neq 0$, contrary to L being nil.   Q.E.D.

Theorem 8.1.8': (*Maschke's theorem in characteristic 0 for arbitrary groups*).
*If $\mathrm{char}(F) = 0$ then $F[G]$ is semiprime. In particular, if G is finite then F is
semisimple Artinian.*

Proof: Let $\bar{F}$ be the algebraic closure of F. Then $\bar{F}[G] \approx F[G] \otimes \bar{F}$ has
no nonzero nilpotent ideals, so $F[G]$ has no nonzero nilpotent ideals.
                                                                   Q.E.D.

Next we aim for an alternate proof of Maschke's theorem, which also
deals with characteristic p, by looking at conjugacy classes.

Proposition 8.1.9: *If $h \in G$ has only a finite number of conjugates $h_1, \ldots, h_t$
then $\sum_{i=1}^{t} h_i \in Z(C[G])$, and conjugation by any g in G permutes the
conjugates.*

Proof: Let $h_{\pi i} = gh_i g^{-1}$, a conjugate of h. Clearly $\pi i \neq \pi j$ for $i \neq j$, so
π permutes the subscripts. Write $a = \sum_{i=1}^{n} h_i$. Then $gag^{-1} = \sum_{i=1}^{n} h_{\pi i} = \sum h_i = a$, showing a commutes with each g in G; thus $a \in Z(C[G])$.
                                                                   Q.E.D.

This leads us to the following important definition.

Definition 8.1.10: $\Delta(G) = \{g \in G : g \text{ has only a finite number of conjugates}\} = \{g \in G : C_G(g) \text{ has finite index in } G\}$.

Remark 8.1.11: $\Delta(G)$ is a characteristic subgroup of G, and $\Delta(\Delta(G)) = \Delta(G)$. More generally, if g_i has n_i conjugates for $i = 1, 2$ then g_1^{-1} has n_1 conjugates and $g_1 g_2$ has at most $n_1 n_2$ conjugates (seen by observing $g(g_1 g_2)g^{-1} = (gg_1^{-1}g)(gg_2^{-1}g)$).

Proposition 8.1.12: (*Converse to proposition 8.1.9*) *Suppose* $z \in Z(C[G])$. *If* $h \in \text{supp}(z)$ *then every conjugate of* h *is in* $\text{supp}(z)$, *so* $h \in \Delta(G)$. *Consequently,* $Z(C[G])$ *is a free C-module whose base consists of all elements of the form* $\sum \{\text{all conjugates of } h\}$ *where* $h \in \Delta(G)$; *in particular,* $Z(C[G]) \subseteq C[\Delta(G)]$.

Proof: Write $z = \sum c_h h$. For any g in G we have $\sum c_h h = z = gzg^{-1} = \sum c_h ghg^{-1}$, implying every conjugate of h appears in $\text{supp}(z)$ with the same coefficient. Since $\text{supp}(z)$ is finite this proves the first assertion. Moreover, writing z_h for $\sum \{\text{conjugates of } h\}$ we can rewrite $z = \sum c_h z_h$, summed over representatives of the conjugacy classes in $\text{supp}(z)$. Thus all the z_h span $Z(C[G])$ and are independent because distinct conjugacy classes are disjoint.   Q.E.D.

Corollary 8.1.13: *Suppose H is a finite subgroup of G, and $r = \sum_{h \in H} h$. Then $r^2 = |H|r$; moreover, $r \in Z(C[G])$ iff $H \triangleleft G$.*

Proof: For each h in H we have $rh = \sum_{h' \in H} h'h = \sum_{h' \in H} h' = r$, so $r^2 = \sum_{h \in H} rh = \sum_{h \in H} r = |H|r$. The other assertion follows at once from propositions 8.1.9 and 8.1.12.   Q.E.D.

The element $\sum_{h \in H} h$ used in this proof has fundamental significance in the theory of group algebras over finite groups. To apply this result, suppose $|G| = n$. Recall that there is a group injection $\text{Sym}(n) \to GL(n, F)$ sending a permutation π to the *permutation matrix* $\sum_{i=1}^{n} e_{\pi i, i}$; composing this with the group injection $G \to \text{Sym}(n)$ given by Cayley's theorem yields a (faithful) representation $G \to GL(n, F)$ called the *regular representation*, which is the restriction to G of the regular representation of $F[G]$ as F-algebra. Writing $G = \{g_1, \ldots, g_n\}$ we see that the matrix $\hat{g}$ corresponding to g is $\sum e_{\pi i, i}$ where $gg_i = g_{\pi i}$. Note $\text{tr}(\hat{g}) = 0$ if $g \neq 1$, since then $gg_i \neq g_i$ for all i; $\text{tr}(1) = n$.

Theorem 8.1.14:   (*Maschke's theorem for arbitrary fields*) *Suppose* $|G| = n$. $\text{Nil}(F[G]) = 0$ *iff* $\text{char}(F) \nmid n$, *in which case* $F[G]$ *is semisimple Artinian.*

Proof: ($\Rightarrow$) Let $z = \sum_{g \in G} g$. Then $0 \neq z \in Z(F[G])$, so $0 \neq z^2 = |G|z = nz$, implying $\text{char}(F) \nmid n$.

($\Leftarrow$) Take $r = \sum_{g \in G} \alpha_g g$ in $\text{Nil}(F[G])$. For any h in $\text{supp}(r)$ we have $h^{-1}r = \alpha_h 1 + \sum_{g \neq h} \alpha_g h^{-1} g$, so (taking traces via the regular representation) we get

$$\text{tr}(h^{-1}r) = \text{tr}(\alpha_h 1) + \sum_{g \neq h} \alpha_g \, \text{tr}(h^{-1}g) = n\alpha_h + 0 = n\alpha_h.$$

But $h^{-1}r$ is nilpotent so $\text{tr}(h^{-1}r) = 0$. Thus $n\alpha_h = 0$ implying $\alpha_h = 0$. This proves $r = 0$, as desired.

The last assertion is clear since $F[G]$ is Artinian.   Q.E.D.

Group Algebras of Subgroups

Later on we shall consider the question of when a group algebra satisfies a polynomial identity. For the present we record one easy result which places much of the PI-theory at our disposal. Recall a *transversal* of a subgroup H in G is a set of representatives of the left cosets of H in G.

Remark 8.1.15: Any transversal B of H is a $C[H]$-base for $C[G]$; if $H \triangleleft G$ then B is $C[H]$-normalizing. (Indeed $G = \bigcup_{b \in B} Hb$ implying $C[G] = C[H]B$, proving B spans. To prove B is a base suppose $\sum r_i b_i = 0$ where $r_i = \sum c_{ih} h \in C[H]$ and $b_i \in B$; then $0 = \sum c_{ih} h b_i$ so each $c_{ih} = 0$ implying $r_i = 0$. The last assertion is clear.)

This has an immediate application.

Proposition 8.1.16: *If* $[G:H] = n$ *then there is a canonical injection* $C[G] \rightarrow M_n(C[H])$ *since* $C[G]$ *is a free* $C[H]$-*module of dimension* n; *in particular, if* G *has an abelian subgroup* H *of index* n *then* $C[G]$ *satisfies all polynomial identities of* $n \times n$ *matrices.*

Proof: The first assertion follows from remark 8.1.15 by means of corollary 1.5.14; the second assertion is then immediate since $C[H]$ is abelian.
                                                                    Q.E.D.

Isaacs–Passman (cf. Passman [77B, theorem 3.8]) proved the striking theorem that if $F[G]$ is a PI-ring then G has an abelian subgroup of finite index bounded by a function of the PI-degree. In the unabridged edition we consider certain key aspects of this result, including the reduction to the case where G is finite.

Augmentation Ideals and the Characteristic p Case

It remains to consider the case where $\mathrm{char}(F) = p$ and G has p-torsion. The situation becomes much more complicated, although the answer is very neat in one particular case. If $H \lhd G$ the group homomorphism $G \to G/H$ extends to a unique C-algebra surjection $\rho_H \colon C[G] \to C[G/H]$. In particular, the surjection $\rho_G \colon C[G] \to C$ is given by $\rho(\sum c_g g) = \sum c_g$.

Definition 8.1.17: The *augmentation ideal* $\omega(F[G])$ is

$$\ker \rho_G = \{\textstyle\sum c_g g \colon \sum c_g = 0\}.$$

Remark 8.1.18: If F is a field then $\omega(F[G])$ is a maximal ideal since $F[G]/\omega(F[G]) \approx F$. More generally, if $\mathrm{char}(F) = p$ and G/H is a finite p'-group then $\ker \rho_H$ is an intersection of a finite number of maximal ideals, by Maschke's theorem.

Remark 8.1.19: $\omega(C[G]) = \sum_{g \in G} C(g - 1)$. (Indeed $(\supseteq)$ is clear; conversely, if $r = \sum c_g g \in \omega(C[G])$ then $r = \sum c_g g - \sum c_g = \sum c_g(g - 1)$.)

Theorem 8.1.20: *Suppose G is a finite p-group and $\mathrm{char}(F) = p$. Then $F[G]$ is a local ring whose Jacobson radical is nilpotent.*

Proof: For any g in G we have $g^q = 1$ for a suitable power q of p, so $(g - 1)^q = g^q - 1 = 0$. Thus $\omega(F[G]) = \sum F(g - 1)$ is spanned by nilpotent elements and thus is nilpotent by proposition 2.6.32. We are done since $\omega(F[G])$ is a maximal ideal.   Q.E.D.

One can actually use the augmentation ideal to compute $\ker \rho_H$ for $H \lhd G$.

Proposition 8.1.21: $\ker \rho_H = \omega(C[H])C[G] = C[G]\omega(C[H])$.

Proof: By symmetry we need only show the first equality. Let $\{g_i \colon i \in I\}$ be a transversal and write $\bar{G} = G/H$. By remark 8.1.15 any r of $C[G]$ can then be written as $\sum r_i g_i$ where $r_i \in C[H]$, and we have $\rho_H r = \sum (\rho_H r_i)\bar{g}_i$. Thus $r \in \ker \rho_H$ iff each $\rho_H r_i = 0$ in C, i.e., iff each $r_i \in \omega(C[H])$. This proves $\ker \rho_H \subseteq \omega(C[H])C[G]$, and the reverse inclusion is clear.   Q.E.D.

The Kaplansky Trace Map and Nil Left Ideals

To generalize theorem 8.1.14 to infinite groups we need a trace map which does not depend on matrices.

Definition 8.1.22:   Define the (Kaplansky) *trace map* tr: $C[G] \to C$ for any group G and any commutative ring C, by tr $\sum c_g g = c_1$. (For finite groups this differs from the trace of the regular representation by a factor of $|G|$.) For the remainder of this section tr *always* denotes the Kaplansky trace map.

Remark 8.1.23:   tr is a trace map according to definition 1.3.28. (Indeed

$$\text{tr}\left(\sum c_g g\right)\left(\sum c'_g g\right) = \sum_{gh=1} c_g c'_h = \sum_{hg=1} c'_h c_g = \text{tr}\left(\sum c'_g g\right)\left(\sum c_g g\right).\right)$$

Consequently, every group algebra has IBN by proposition 1.3.29.

The following computation helps in characteristic p.

Lemma 8.1.24:   *Suppose* $r = \sum c_g g \in C[G]$. *If* char$(C) = p$ *is prime and* q *is a power of* p *then* tr$(r^q) = \sum\{c_g^q : g^q = 1\}$.

Proof: It is enough to show tr$(r^q) = \text{tr}\left(\sum c_g^q g^q\right)$. Expanding $\left(\sum c_g g\right)^q$ yields terms of the form $c_1 \cdots c_q g_1 \cdots g_q$, where c_i is written in place of c_{g_i}. When two g_i are distinct we also have $c_1 \cdots c_q g_j \cdots g_q g_1 \cdots g_{j-1}$ for each $j \le q$. Since tr$(g_j \cdots g_q g_1 \cdots g_{j-1}) = \text{tr}(g_1 \cdots g_q)$ and char $C = p$ we see that the terms in tr(r^q) drop out except those from tr$\left(\sum c_q^q g^q\right)$, as desired.

Proposition 8.1.25:   *Suppose* C *is a domain. For any nilpotent* r *in* $C[G]$ *we have* tr$(r) = 0$ *provided the following condition is satisfied:*

char$(C) = p \ne 0$ *and* supp(r) *does not have a nonidentity element whose period is a power of* p.

Proof: Take a large enough power q of p such that $r^q = 0$. Writing $r = \sum c_g g$ we have (by hypothesis) $g^q \ne 1$ for every $g \ne 1$ in supp(r), so $0 = \text{tr}(0) = \text{tr}(r^q) = \sum\{c_g^q : g^q = 1\} = c_1^q$, implying $0 = c_1 = \text{tr}(r)$.   Q.E.D.

Remark 8.1.25':   The conclusion of proposition 8.1.25 also holds in characteristic 0. Indeed, suppose, on the contrary, $r \ne 0$ is nilpotent with tr$(r) \ne 0$. Then $r^m = 0$ for some m. Write $r = \sum_{i=1}^t c_i g_i$. The fact $r^m = 0$ can be expressed as a polynomial equality $f(c_1, \ldots, c_t) = 0$. Taking a language containing a constant symbol for each element of G, we can appeal to remark 2.12.37' as follows:

Let φ denote the sentence $f(c_1,\ldots,c_t) = 0 \Rightarrow c_1 = 0$. Any maximal ideal P of C yields an ideal $C[G]P$ of $C[G]$; picking P such that C/P has characteristic $> \max\{\text{periods of } g_1,\ldots,g_t\}$ we see by proposition 8.1.25 that φ holds in $(C/P)[G]$. But remark 2.12.37' shows there are "enough" such P to conclude φ holds in characteristic 0, i.e., $\text{tr}(r) = 0$, contradiction.)

This argument of "passing to characteristic p" is extremely useful in group algebras.

Armed with this "nilpotent implies trace 0" result, we are finally ready for a result which combines and improves both theorem 8.1.8' and 8.1.14.

Theorem 8.1.26: $F[G]$ *has no nonzero nil left ideals, if G is a* char$(F)'$-*group.*

Proof: Suppose L is a nil left ideal and $a = \sum \alpha_g g \in L$. For each g in supp(r) we have $g^{-1}r \in L$ so $\alpha_g = \text{tr}(g^{-1}r) = 0$ by proposition 8.1.25, implying $a = 0$. Thus $L = 0$.   Q.E.D.

(Of course, the characteristic 0 part of this result relies on remark 8.1.25' which is quite deep; in most of our applications we shall fall back on theorem 8.1.8'.)

Having considered finite groups, we turn to infinite groups for the remainder of this section.

Subgroups of Finite Index

One of the techniques of studying infinite groups is in terms of "nice" subgroups of finite index, especially in terms of transversals; see remark 8.1.15 and proposition 8.1.16, for example. We take the opportunity of collecting several results along this vein, in preparation for the major structure theorems. We are interested in relating the structure of $C[G]$ to $C[H]$ where $[G:H] < \infty$, and start with some easy general observations about cosets of groups.

Remark 8.1.27:

(i) If $[G:H] = n$ then for any subgroup G_1 of G we have $[G_1:H \cap G_1] \leq n$ (since any transversal of $H \cap G_1$ in G_1 lifts to distinct coset representatives of H in G.)

(ii) If $[G:H_i] = n_i$ for $1 \leq i \leq t$ then $[G: \bigcap H_i] \leq n_1 \cdots n_t$. (Indeed, any coset $(\bigcap H_i)g = \bigcap H_i g$ is determined by the cosets $H_1 g, \ldots, H_t g$, so there are $\leq n_1 \cdots n_t$ possible choices.)

Proposition 8.1.28:   *Any subgroup of finite index contains a normal subgroup of finite index. In fact, if $[G:H] = n$ and $H_1 = \bigcap\{g^{-1}Hg : g \in G\}$ then $H_1 \triangleleft G$ and $[G:H_1] \leq n!$*

Proof: Right multiplication by any element of G permutes the cosets of H, yielding a group homomorphism $\varphi: G \to \text{Sym}(n)$; since H_1 is the largest normal subgroup of G contained in H we have $[G:H_1] \leq [G:\ker \varphi] \leq n!$
Q.E.D.

To continue the investigation we recall from proposition 8.1.16 that there is an injection $\psi: C[G] \to M_n(C[H])$ when $[G:H] = n$. Let us describe the action of ψ on G. Fixing a transversal $g_1, \ldots, g_n$ we note $g_i g = h_i g_{\sigma i}$ for suitable h_i in H and $1 \leq \sigma i \leq n$. Moreover, $\sigma: \{1, \ldots, n\} \to \{1, \ldots, n\}$ is 1:1 for if $\sigma i = \sigma j$ then $h_i^{-1} g_i g = h_j^{-1} g_j g$ so $h_i^{-1} g_i \in Hg_i \cap Hg_j$, implying $i = j$. Thus σ is a permutation and $\psi g = \sum_{i=1}^{n} h_i e_{\sigma i, i}$. In case H is abelian we can take determinants to get $|\psi g| = (sg\,\sigma) h_1 \cdots h_n$; forgetting about $sg\,\sigma$ gives us a group homomorphism $\tilde{\psi}: G \to H$ defined by $\tilde{\psi} g = h_1 \cdots h_n$. Note $G' \subseteq \ker \tilde{\psi}$.

Proposition 8.1.29: If $[G:Z(G)] = n$ then $G' \cap Z(G)$ has exponent n.

Proof: Taking $H = Z(G)$ in the preceding paragraph we see $\sigma = (1)$ and each $h_i = g$ whenever $g \in G' \cap H$, so $1 = \tilde{\psi} g = g^n$. Q.E.D.

Digression 8.1.30: We have made a superficial application of a useful tool. Even if H is not abelian we could pass to $C[H/H']$ before taking the determinant, thereby yielding a map $\tilde{\psi}: G \to H/H'$; since $G' \subseteq \ker \tilde{\psi}$ we have a map $\text{ver}: G/G' \to H/H'$ which is called the *transfer map* in Passman [77B].

Before utilizing proposition 8.1.29 we need another group-theoretic result.

Lemma 8.1.31: If G is a finitely generated group then every subgroup of finite index is finitely generated. More precisely, if $G = \langle a_1, \ldots, a_t \rangle$ and $[G:H] = n$ then H is generated by tn elements.

Proof: Take a transversal $\{1 = g_1, g_2, \ldots, g_n\}$. Then $(Hg_i)a_j = Hg_{\sigma i}$ for a suitable permutation σ, implying $g_i a_j = h_{ij} g_{\sigma i}$ for suitable h_{ij} in H. Let

$$H_0 = \langle h_{ij} : 1 \leq i \leq t, 1 \leq j \leq n \rangle \quad \text{and} \quad G_0 = \bigcup_{i=1}^{n} H_0 g_i.$$

To show $H_0 = H$ it suffices to show $G_0 = G$. Then $H = H \cap (\bigcup H_0 g_i) = H_0$. But

$$G_0 a_j = \bigcup H_0 g_i a_j = \bigcup H_0 h_{ij} g_{\sigma i} = \bigcup H_0 g_{\sigma i} = G_0,$$

implying $G_0 = G_0 G \supseteq G$. Q.E.D.

The commutator subgroup G' now enters in an interesting fashion.

Theorem 8.1.32: If $[G:Z(G)] = n$ then $|G'| \leq n^{n^3+1}$.

Proof: Take a transversal $\{g_1, \ldots, g_n\}$ of G over $Z = Z(G)$. Any group commutator has the form $[z_1 g_i, z_2 g_j] = z_1 g_i z_2 g_j (z_1 g_i)^{-1}(z_2 g_j)^{-1} = [g_i, g_j]$ where $z_1, z_2 \in Z$; hence there are at most n^2 distinct commutators, and these generate G'. Let $t = n^3$. The abelian group $G' \cap Z$ has index $\leq n$ in G' and thus is generated by $\leq t$ elements by lemma 8.1.31, implying $G' \cap Z$ is a direct product of t cyclic groups each of exponent (and thus order) $\leq n$, by proposition 8.1.29. Thus $|G' \cap Z| \leq n^t$ so $|G'| \leq n^{t+1}$ as desired.   Q.E.D.

Corollary 8.1.33: *Suppose* $H = \langle h_1, \ldots, h_t \rangle$ *is a subgroup of* $\Delta(G)$.

(i) $[H:Z(H)] \leq [G:C_G(H)] \leq \prod [G:C_G(h_i)]$.

(ii) H' *is a finite group.*

(iii) $\operatorname{tor} H = \{$*elements of* H *of finite order*$\}$ *is a finite normal subgroup of* H, *and* $H/\operatorname{tor} H$ *is torsion-free abelian.*

Proof:

(i) Clear from remark 8.1.27 since $Z(H) = H \cap C_G(H)$ and $C_G(H) = \bigcap_{i=1}^{t} C_G(h_i)$.

(ii) Apply (i) to theorem 8.1.32, since each $C_G(h_i)$ has finite index.

(iii) Since H' is finite we have $(\operatorname{tor} H)/H' \approx \operatorname{tor}(H/H')$, which is finite and normal in the finitely generated abelian group H/H'. Thus $\operatorname{tor}(H)$ is finite and normal in H, and $H/\operatorname{tor} H \approx (H/H')/\operatorname{tor}(H/H')$ is torsion-free abelian.

Q.E.D.

Lemma 8.1.34: *Suppose* G *is a finite union of cosets of subgroups* $H_1, \ldots, H_m$. *Then some* H_i *has finite index in* G.

Proof: Induction on m. Since the assertion is obvious for $m = 1$, we may assume $m > 1$ and $[G:H_m]$ is infinite. Write $G = \bigcup_{i=1}^{m} \bigcup_{j=1}^{t_i} H_i g_{ij}$. Some coset $H_m g$ does not appear among the $H_m g_{mj}$; since the cosets of H_m are disjoint we have

$$H_m g \subseteq \bigcup_{i=1}^{m-1} \bigcup_{j=1}^{t_i} H_i g_{ij}.$$

But each $H_m g_{mj} = H_m g g^{-1} g_{mj} \subseteq \bigcup_{i=1}^{m-1} \bigcup_{j=1}^{t_i} H_i g_{ij} g^{-1} g_{mj}$ so G is a finite union of cosets of $H_1, \ldots, H_{m-1}$, and we are done by induction.   Q.E.D.

Prime Group Rings

Let us turn to the question of when $C[G]$ is prime, which turns out to be much easier than the zero-divisor question. Of course, C must be a domain. Otherwise, the answer depends only on G and requires passing to $\Delta(G)$ by means of the following projection map.

Definition 8.1.35: Suppose H is a subgroup of G. Define $\pi_H: C[G] \to C[H]$ by $\pi_H \sum c_g g = \sum_{g \in H} c_g g$, i.e., we restrict the support to H. Write π_Δ for $\pi_{\Delta(G)}$.

Remark 8.1.36: π_H is a map both in $C[H]$-$\mathcal{M}od$ and $\mathcal{M}od$-$C[H]$. (Indeed given h in H we have $g \in H$ iff $hg \in H$ so $\pi_H(h \sum_{g \in G} c_g g) = \pi_H(\sum c_g hg) = \sum_{g \in H} c_g hg = h\pi_H(\sum c_g g)$. The other side is symmetric.)

Lemma 8.1.37: Suppose r_{i1}, r_{i2} in $C[G]$ satisfy $\sum_{i=1}^t r_{i1} g r_{i2} = 0$ for all g in G. Then $\sum_{i=1}^t r_{i1} \pi_\Delta r_{i2} = 0$ and $\sum_{i=1}^t (\pi_\Delta r_{i1})(\pi_\Delta r_{i2}) = 0$.

Proof: It is enough to prove the first assertion (for then

$$0 = \pi_\Delta\left(\sum_{i=1}^t r_{i1} \pi_\Delta r_{i2}\right) = \sum \pi_\Delta(r_{i1} \pi_\Delta r_{i2}) = \sum(\pi_\Delta r_{i1})(\pi_\Delta r_{i2}).$$

Let $A = (\bigcup_{i=1}^t \operatorname{supp}(r_{i2})) - \Delta(G)$, and $B = \bigcup_{i=1}^t \operatorname{supp}(r_{i1})$, both finite subsets of G. Thus there are c_{ai} in C such that $r_{i2} = \pi_\Delta r_{i2} + \sum_{a \in A} c_{ai} a$.

For any a in A and any b in B such that $b^{-1} g_0$ is a conjugate of a we pick a specific g_{ab} in G such that $b^{-1} g_0 = g_{ab}^{-1} a g_{ab}$. Since A and B are finite we have a finite number of the g_{ab}.

Let $H = \bigcap \{C_G(g): g \in \operatorname{supp} \pi_\Delta r_{i2} \text{ for some } i\}$. Then H centralizes all $\pi_\Delta r_{i2}$; by hypothesis each h in H satisfies

$$0 = \left(\sum_{i=1}^t r_{i1} h^{-1} r_{i2}\right)h = \sum_{i=1}^t r_{i1}(h^{-1} r_{i2} h)$$

$$= \sum_{i=1}^t r_{i1} \pi_\Delta r_{i2} + \sum_{i=1}^t \sum_{a \in A} r_{i1} c_{ai} h^{-1} a h.$$

Suppose $\sum r_{i1} \pi_\Delta r_{i2} \neq 0$; we aim for a contradiction. There is some $g_0 \neq 0$ in its support, so $g_0 \in \operatorname{supp}(r_{i1} c_{ai} h^{-1} a h)$ for some i and some a. Then $g_0 = bh^{-1}ah$ for some b in B, implying $g_{ab}^{-1} a g_{ab} = b^{-1} g_0 = h^{-1}ah$. Hence $hg_{ab}^{-1} \in C_G(a)$, i.e., $h \in C_G(a)g_{ab}$. Thus H is a finite union of cosets of the $C_G(a)$. But H has finite index in G by remark 8.1.27(ii), so by lemma 8.1.34 some $C_G(a)$ has finite index, i.e., $a \in \Delta(G)$ contrary to $a \in A$.   Q.E.D.

Corollary 8.1.38: If $A_1, A_2 \lhd C[G]$ with $A_1 A_2 = 0$ then $\pi_\Delta A_1 \pi_\Delta A_2 = 0$.

Theorem 8.1.39: (Connell's theorem) Suppose C is an integral domain. The following assertions are equivalent:

(i) $C[G]$ is prime.
(ii) $Z(C[G])$ is an integral domain.
(iii) G has no finite normal subgroup $\neq (1)$.

(iv) $\Delta(G)$ *is torsion-free abelian* (*or* (1)).

(v) $C[\Delta(G)]$ *is an integral domain.*

Proof: (i) $\Rightarrow$ (ii) is clear; (ii) $\Rightarrow$ (iii) follows from corollary 8.1.13 since in its notation if $H \lhd G$ is finite then $r(r - |H|1) = 0$ implying $r = |H|1$ and thus $r = 1 = |H|$.

(iii) $\Rightarrow$ (iv) Suppose $g_1, g_2 \in \Delta(G)$. The conjugates of g_1 and g_2 generate a normal subgroup H of G. But corollary 8.1.33(iii) shows tor H is a finite normal subgroup of G, so tor $H = (1)$ and $H \approx H/$tor H is torsion-free abelian. This shows $g_1 g_2 = g_2 g_1$ and g_1 is torsion-free; since g_i were arbitrary in $\Delta(G)$ we conclude $\Delta(G)$ is torsion-free abelian.

(iv) $\Rightarrow$ (v) by proposition 8.1.3.

(v) $\Rightarrow$ (i) by corollary 8.1.38.   Q.E.D.

We saw in theorem 8.1.8′ that $F[G]$ is semiprime if char$(F) = 0$; an analysis of the characteristic p case is given in exercise 14.

F Supplement: The Jacobson Radical of Group Algebras

Other than the "zero divisor problem," the leading question in group algebras is probably whether the Jacobson radical is necessarily nil. A positive answer would show that all the group algebras of theorem 8.1.26 were semiprimitive. Although there is no known counterexample, a positive solution seems to be out of reach. We shall present here some of the basic positive results. Some of these rely on the structure theory of rings, using only the most superficial properties of groups.

Theorem 8.1.43: *If $H < G$ then $C[H] \cap \text{Jac}(C[G]) \subseteq \text{Jac}(C[H])$ (by lemma 2.5.32 applied to lemma 2.5.17(i), for $C[H]$ is a summand of $C[G]$.)*

Remark 8.1.44: If $\text{Jac}(C[H])$ is nil for every finitely generated subgroup H of G then $\text{Jac}(C[G])$ is nil. (Indeed, take any $r \in \text{Jac}(C[G])$ and let $H = \langle \text{supp}(r) \rangle$. Then $r \in C[H] \cap \text{Jac}(C[G]) \subseteq \text{Jac}(C[H])$ is nilpotent.)

Theorem 8.1.45: *(Amitsur, Herstein) If F is uncountable then $\text{Jac}(F[G])$ is nil.*

Proof: By remark 8.1.44 we may assume G is finitely generated and thus countable, so $\text{Jac}(F[G])$ is nil by theorem 2.5.22.   Q.E.D.

Corollary 8.1.46: *If F is uncountable and G is a char$(F)'$-group then $F[G]$ is semiprimitive.*

Proof: Apply theorem 8.1.45 to theorem 8.1.26.   Q.E.D.

The characteristic 0 case can be improved to

Theorem 8.1.47: (*Amitsur*) *Suppose* $\text{char}(F) = 0$ *and* F *is not algebraic over its characteristic subfield* $F_0 \approx \mathbb{Q}$. *Then* $F[G]$ *is semiprimitive.*

Proof: Let $\Lambda = \{\lambda : i \in I\}$ be a transcendence base of F over F_0 and let $F_1 = F_0[\Lambda]$. Then $F_1[G] \approx (F_0[G])(\Lambda)$ is semiprimitive by corollary 2.5.42 applied to theorem 8.1.10'; thus $F[G]$ is semiprimitive by theorem 2.5.36.
Q.E.D.

§8.2 Noetherian Group Rings

This section is concerned with group algebras of polycyclic-by-finite groups, to be defined presently. These are the only known Noetherian group rings and as such have been the subject of intensive research. The main result presented here is the Farkas-Snider solution of the zero-divisor question for torsion-free polycyclic-by-finite groups.

Remark 8.2.0: Since $F[G]$ has an involution, $F[G]$ is left Noetherian iff it is right Noetherian. We shall use this observation implicitly.

Polycyclic-by-Finite Groups

Definition 8.2.1: A *subnormal series* of subgroups of G is a chain $G = G_m \triangleright G_{m-1} \triangleright \cdots \triangleright G_0 = (1)$, i.e., each $G_{i-1} \triangleleft G_i$. G is *polycyclic* if G has a subnormal series with each factor G_i/G_{i-1} cyclic; G is *poly-{infinite cyclic}* if each factor in the series is infinite cyclic. G is *polycyclic-by-finite*, or *virtually polycyclic*, if G has a polycyclic group of finite index.

G is *solvable* if there is a subnormal series with each factor abelian. G is *nilpotent* if the lower central series terminates at G. Thus every nilpotent group is polycyclic, and every polycyclic group is solvable. On the other hand, any finitely generated abelian group is a finite direct product of cyclic groups and thus nilpotent. An easy but enlightening example of a non-nilpotent polycyclic group is the infinite dihedral group $G = \langle ab : b^{-1}ab = a^{-1} \text{ and } b^2 = 1 \rangle$.

Theorem 8.2.2: *If* G *is polycyclic-by-finite then* $F[G]$ *is Noetherian.*

Proof: Let H be polycyclic of finite index. Then $F[G]$ is an f.g. left and right $F[H]$-module, so it suffices to prove $F[H]$ is Noetherian. Taking a subnormal series $H = H_m \triangleright \cdots \triangleright H_0 = (1)$ with each H_i/H_{i-1} cyclic, we shall prove by induction that each $F[H_i]$ is Noetherian. This is obvious for $i=0$, so suppose inductively that $R = F[H_{i-1}]$ is Noetherian. If H_i/H_{i-1} is finite we are done as before, so assume H_i/H_{i-1} is infinite cyclic, generated by some element g. Since $H_{i-1} \triangleleft H_i$ we see conjugation by g produces an automorphism σ of R. Let $T = R[\lambda; \sigma]$. Then T is left Noetherian by proposition 3.5.2; but $S = [\lambda^i : i \in \mathbb{N}\}$ is clearly a left denominator set so $S^{-1}T$ also is left Noetherian by proposition 3.1.13.

There is a ring homomorphism $\varphi : T \to F[H_i]$ given by $\varphi(\sum r_i \lambda^i) = \sum r_i g^i$ for r_i in R; extending this naturally to a surjection $S^{-1}T \to F[H_i]$ shows $F[H_i]$ is left (and thus right) Noetherian.   Q.E.D.

The idea behind this proof was to obtain a polycyclic-by-finite group ring from an iterated skew Laurent extension; we return to this idea in §8.4. Conversely, let us see what can be said of G when $F[G]$ is Noetherian.

Lemma 8.2.3: *If $R = F[G]$ is Noetherian then G has no infinite chain $H_1 < H_2 < \cdots$ of subgroups. In particular G is finitely generated.*

Proof: $R\omega F[H_1] \subseteq R\omega F[H_2] \subseteq \cdots$ must terminate so suppose

$$R\omega(F[H_i]) = R\omega(F[H_{i+1}]).$$

Write $H = H_i$. For any h in H_{i+1} we have $\pi_H(h-1) \in \pi_H(\omega(F[H_{i+1}])) \subseteq \pi_H(R\omega(F[H])) = \omega(F[H])$ so $\pi_H(h-1) \neq -1$, thereby implying $h \in H$. Thus $H_i = H_{i+1}$.   Q.E.D.

Proposition 8.2.4: *If $F[G]$ is Noetherian and G is solvable then G is polycyclic.*

Proof: Take a subnormal chain with each factor G_i/G_{i+1} abelian. $F[G_i]$ is Noetherian by sublemma 2.5.32' since $F[G]$ is free over $F[G_i]$. Hence $F[G_i/G_{i+1}]$ is Noetherian. By lemma 8.2.3 each G_i/G_{i+1} is finitely generated abelian and thus polycyclic, implying G is polycyclic.   Q.E.D.

Remark 8.2.5: The familiar proofs about finite solvable groups show that any subgroup or homomorphic image of a polycyclic group is polycyclic. Explicitly, if G has a subnormal series $G = G_m \triangleright \cdots \triangleright G_0 = (1)$ and $H < G$ then taking $H_i = H \cap G_i$ we see $H = H_m \triangleright \cdots \triangleright H_0 = (1)$ is a subnormal series of H, and the factor H_i/H_{i+1} is isomorphic to H_iG_{i+1}/G_{i+1}, a subgroup of G_i/G_{i+1}.

Proposition 8.2.6: *If G has a subnormal series each of whose factors is finite or cyclic then G is polycyclic-by-finite. Moreover, any polycyclic-by-finite group has a characteristic poly-{infinite cyclic} subgroup H of finite index.*

Proof: Let $G = G_m \triangleright \cdots \triangleright G_0 = (1)$. We proceed by induction on m; assuming G_{i-1} has a characteristic poly-{infinite cyclic} subgroup H_{i-1} of finite index, we want to prove this for G_i. If G_i/G_{i-1} is finite we merely take $H_i = H_{i-1}$, so assume G_i/G_{i-1} is cyclic. Pick $g \in G_i$ such that the coset $G_{i-1}g$ generates G_i/G_{i-1}. Note $H_{i-1} \triangleleft G_i$.

Let $\tilde{H} = H_{i-1}\langle g \rangle$, a subgroup of G_i of finite index $\leq [G_{i-1}:H_{i-1}]$. By induction $\tilde{H}_i$ has a subnormal series each of whose factors is infinite cyclic, but $\tilde{H}_i$ need not be characteristic in G. We finish at once by means of the following two lemmas, the first of which improves on proposition 8.1.28.

Lemma 8.2.7: *Suppose G is finitely generated. For any n there are only finitely many subgroups of index n in G. In particular, any subgroup H of finite index contains a characteristic subgroup of finite index.*

Proof: Write $G = \langle g_1, \ldots, g_t \rangle$, and suppose $[G:H] = n$. As observed in proposition 8.1.28 right multiplication by elements of G permutes the cosets of H, thereby yielding a group homomorphism $\varphi: G \to \text{Sym}(n)$ with $\ker \varphi \subseteq H$. But φ is determined by its action on $g_1, \ldots, g_t$ so there are $\leq t^{n!}$ possible homomorphisms. Moreover, H must be the preimage of a subset of $\text{Sym}(n)$, of which there are $2^{n!}$, so there are $\leq (2t)^{n!}$ possibilities for H.

Thus $\bigcap\{$subgroups of G having index $n\}$ is a characteristic subgroup (contained in H) of finite index in G.   Q.E.D.

Lemma 8.2.8: *Suppose H is poly-{infinite cyclic}. Then the same holds for every subgroup of H.*

Proof: By remark 8.2.5, noting every subgroup of an infinite cyclic group is infinite cyclic.   Q.E.D.

The infinite factors actually give us an important invariant.

Definition 8.2.9: The *Hirsch number* $h(G)$ of a polycyclic-by-finite group G is the number of infinite cyclic factors in a subnormal series each of whose factors is infinite cyclic or finite.

Proposition 8.2.10:   *Suppose G is polycyclic-by-finite. Then h(G) is well-defined, and h(G) = 0 iff G is finite. If N ◁ G then h(G) = h(N) + h(G/N).*

Proof: We borrow the idea of proof of the Jordan-Holder theorem, dealing now with subnormal series whose factors are either infinite cyclic or finite. First suppose we have

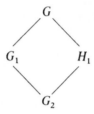

i.e., $G_2 = G_1 \cap H_1$. Recall from remark 8.1.27 that $[G:G_1] < \infty$ iff $[H_1:G_2] < \infty$; likewise, $[G:H_1] < \infty$ iff $[G_1:G_2] < \infty$. Thus we see that the number of infinite cyclic factors is the same for the left side as for the right side.

In general, suppose we have two subnormal series $G \triangleright G_1 \triangleright \cdots \triangleright G_m = (1)$ and $G \triangleright H_1 \triangleright \cdots > H_n = (1)$. Then $G_1 \cap H_1$ is polycyclic-by-finite by proposition 8.2.6 and has a subnormal series whose factors are all infinite cyclic except the first, which is finite. Using the above observation plus induction applied to G_1 and H_1 show $h(G)$ is well-defined; the finishing touches are left to the reader.

Clearly $h(G) = 0$ iff there are no infinite factors, iff G is finite. To prove the last assertion note N and G/N each are polycyclic-by-finite; putting together their subnormal chains shows by inspection $h(N) + h(G/N) = h(G)$.

Q.E.D.

To illustrate the use of the Hirsch number we present the next proposition.

Remark 8.2.11:   Any infinite polycyclic-by-finite group G has a characteristic torsion-free abelian subgroup $\neq (1)$. (Indeed, take H as in proposition 8.2.6. The next-to-last subgroup in the derived series of H is clearly characteristic abelian and is torsion-free since H clearly is torsion-free.)

Proposition 8.2.12:   *Any polycyclic-by-finite group G is residually finite.*

Proof: Remark 8.2.11 shows G has a torsion-free characteristic abelian subgroup $A \neq (1)$. Let $A_i = \{a^i : a \in A\}$. Then $h(G/A_i) = h(G) - h(A_i) \leq h(G) - 1$

so by induction each G/A_i is residually finite. Since $\bigcap A_i = (1)$ we conclude G is residually finite.   Q.E.D.

Digression: Growth of Groups

An obvious question to ask at this juncture is whether polycyclic-by-finite groups have finite GK-dimension, for then we could apply the theory of §6.2 (which, in fact, will be very useful for enveloping algebras). Let us back-track a bit.

Definition 8.2.13: Suppose G is a finitely generated group generated by $g_1, \ldots, g_t$. The *growth function* φ_G is defined by taking $\varphi_G(n)$ to be the number of products of length n of the g_i and the g_i^{-1} for $1 \le i \le t$.

Remark 8.2.14: We can define polynomial, subexponential, and exponential growth just as in definition 6.2.10. In fact, the growth of G precisely matches the growth of $F[G]$ (as an affine algebra) since we can use the generating set of $g_1, \ldots, g_n$ and their inverses.

Although growth of groups is an older subject than growth of algebras, we shall frame our results for group algebras since then we can draw on the results from §6.2. In particular we have

Remark 8.2.15: If $[G:H] < \infty$ then $F[G]$ and $F[H]$ have the same growth, by remark 6.2.14(iv).

One might have expected polycyclic-by-finite group algebras to have polynomial growth, by a skew version of proposition 6.2.22. However, a completely different story emerges.

Theorem 8.2.16: *(Milnor-Wolf-Gromov) A finitely generated group has polynomial growth iff it is nilpotent-by-finite* (Proof omitted.)

Homological Dimension of Group Algebras

Proposition 8.2.18: *If G possesses a subnormal series each of whose factors is infinite cyclic then* gl. dim $F[G] \le h(G)$.

Proof: Let $G = G_m \rhd \cdots \rhd G_0 = (1)$ be the given series; clearly $m = h(G)$. By induction on $h(G)$ we have gl. dim $F[G_{m-1}] \le m - 1$. But there is g in G

such that $G/G_{m-1} \approx \langle \bar{g} \rangle$. Conjugating by g induces an automorphism σ on $F[G_{m-1}]$. The subring R of $F[G]$ generated by $F[G_{m-1}]$ and g is isomorphic to $(F[G_{m-1}])[\lambda; \sigma]$ and thus has gl. dim $\leq m$ by corollary 5.1.25. But $F[G] = S^{-1}R$ where $S = \{g^i : i \in \mathbb{N}\}$, so gl. dim $F[G] \leq m$ by proposition 5.1.26.

Q.E.D.

We want to improve this to polycyclic-by-finite groups. First let us show it is enough to prove $pd_{F[G]} F < \infty$. To see this we shall make some general observations about modules over arbitrary group rings.

Proposition 8.2.19:   gl. dim $F[G] = pd_{F[G]} F$ for any group G, where F has the trivial G-action $g\alpha = \alpha$.

Proof: $\geq$ is true by definition. On the other hand, let

$$0 \to P_n \to \cdots P_0 \to F \to 0$$

be a projective resolution of F as $F[G]$-module. Consider the functor $\underline{\hspace{1em}} \otimes_F M$ from $F[G]$-$\mathcal{M}od$ to $F[G]$-$\mathcal{M}od$ where any $N \otimes M$ is viewed as $F[G]$-module via the diagonal action $g(x \otimes y) = gx \otimes gy$. This functor is exact since it is obviously exact as a functor from F-$\mathcal{M}od$ to F-$\mathcal{M}od$, since F is a field. In particular, each $P_i \otimes_F M$ is a projective $F[G]$-module, and $M \approx F \otimes_F M$ has the projective resolution

$$0 \to P_n \otimes_F M \to \cdots \to P_0 \otimes_F M \to M \to 0.$$

Thus $pd\, M \leq n$, proving gl. dim $F[G] \leq n = pd_{F[G]} F$.   Q.E.D.

Theorem 8.2.20:   *If G is polycyclic-by-finite and is a* char$(F)'$*-group then* gl. dim $F[G] < \infty$.

Proof: Letting H be as in proposition 8.2.6 we have $[G:H] < \infty$ and gl. dim $F[H] < \infty$ by proposition 8.2.18. Thus it suffices to prove

Theorem 8.2.21:   *(Serre) Suppose G has a subgroup H of finite index such that* gl. dim $F[H] < \infty$. *If G is a* char$(F)'$*-group then* gl. dim $F[G] < \infty$.

Proof: By proposition 8.2.19 it suffices to find a projective resolution of finite length of F as $F[G]$-module. We are given a projective resolution $0 \to P_n \xrightarrow{d_n} \cdots \to P_0 \xrightarrow{d_0} F \to 0$ as $F[H]$ module. View this is an acyclic complex (P, d) where $P = (\bigoplus_{u=0}^{n} P_u) \oplus F$, and $d = \bigoplus_{i=1}^{n} d_u$. Let $m = [G:H]$. Then the acyclic complex $\tilde{P} = (\bigotimes P, \tilde{d})$ can be constructed as in corollary 5.2.57. Write $\tilde{P} = \tilde{P}' \oplus F$ (identifying $F \otimes_F \cdots \otimes_F F$ with F). We shall give $\tilde{P}$ a natural $F[G]$-module structure with respect to which $\tilde{P}'$ is projective, and this provides a projective resolution of length $\leq mn$ over F, by remark 5.2.55.

Write $G = \bigcup_{i=1}^{m} g_i H$; then for any g in G we have $g^{-1}g_i = g_{\sigma i}h_{\sigma i}^{-1}$ where $\sigma \in \text{Sym}(m)$ depends on g. G acts on $\tilde{P}$ by

$$g(a_1 \otimes \cdots \otimes a_m) = \pm h_{\sigma 1}a_{\sigma 1} \otimes \cdots \otimes h_{\sigma m}a_{\sigma m} \tag{1}$$

for a_i homogeneous in P_i where the sign arises because of the sign switch in graded tensor products (see after 5.2.57) and can be computed to be (-1) to the power $\sum \deg a_i \deg a_j$ summed over all $i < j$ for which $\sigma i > \sigma j$. Then $\tilde{P}$ is a G-module and thus an $F[G]$-module, and $\tilde{d}$ is an $F[G]$-module map. (We skip the straightforward but tedious computation that the sign works out; complete details are given in Passman [77B, pp. 443–448].)

Write $P = P' \oplus F$ where $P' = \bigoplus_{i=1}^{n} P_i$ is projective as $F[H]$-module, and take P'' such that $M_0 = P' \oplus P''$ is a free $F[H]$-module. Let $M = M_0 \oplus F$, and form $\tilde{M} = \bigotimes_{i=1}^{m} M$ which can be written as $\tilde{M}' \oplus F$. $\tilde{M}'$ is an $F[G]$-module by (1), and $\tilde{P}'$ is a summand of $\tilde{M}'$ since the canonical epic $\tilde{M}' \to \tilde{P}'$ splits. Hence it suffices to show $\tilde{M}'$ is projective.

Explicitly $\tilde{M}'$ is the direct sum of all tensor products involving j copies of M_0 and $(m - j)$ copies of F, for each $j \geq 1$. Thus taking a base B for M_0 over $F[H]$ we see $\tilde{M}'$ has a base over F consisting of all $w = w_1 \otimes \cdots \otimes w_m$ where w_i has the form hb for $h \in H$ and $b \in B \cup \{1\}$, with not all $w_i = 1$. The given action of G on $\tilde{P}'$ extends naturally to $\tilde{M}'$. Let $K_w = \{g \in G : gw \in Fw\}$. The G-orbit of w spans an $F[G]$-module isomorphic to $F[G] \otimes_{F[K_w]} Fw$; since $\tilde{M}'$ decomposes as the direct sum of these it suffices to prove each $F[G] \otimes_{F[K_w]} Fw$ is projective. Hence it suffices to show each Fw is projective as $F[K_w]$-module.

We claim K_w is finite. Indeed H contains a normal subgroup N of G of finite index by proposition 8.1.28 and it suffices to prove $N \cap K_w = (1)$. But if $g \in N \cap K_w$ then $g^{-1}g_i = g_i h_i'$ where $h_i' = g_i^{-1}g^{-1}g_i \in N$, so the corresponding permutation σ is (1); writing $w = w_1 \otimes \cdots \otimes w_m$ with $w_i = h_i b_i$ we get $\alpha w = gw = h_1'h_1 b_1 \otimes \cdots \otimes h_m'h_m b_m$ for some α in F. Matching components for $b_i \neq 1$ shows some $h_i' = 1$; hence $g^{-1} = 1$, proving the claim.

By hypothesis $|K_w| \neq 0$ in F. By Maschke's theorem $F[K_w]$ is semisimple Artinian, so each of its modules (including Fw) is projective.   Q.E.D.

The fact that polycyclic-by-finite group algebras have finite global dimension is one of the keys to their structure. The proof given above, due to Serre, was taken from Swan [69].

Traces and Projectives

We shall now begin to focus on the zero-divisor question. Formanek [73] recognized that it might first be wise to consider idempotents, since a domain cannot have nontrivial idempotents; he proved in characteristic 0 that every Noetherian group algebra over a torsion-free group has no nontrivial idempotents. The proof (exercise 2) is an ingenious application of the (Kaplansky) trace map. Actually a slight generalization of Formanek's theorem is needed. Let tr: $M_n(F[G]) \to F[G]$ be the usual matrix trace, i.e.,

$$\text{tr}(r_{ij}) = \sum_{i=1}^{n} r_{ii} \in F[G].$$

Remark 8.2.22:   Any trace map $t: F[G] \to F$ yields a trace map

$$\text{Tr}: M_n(F[G]) \to F$$

given by $\text{Tr}(a) = t(\text{tr}\, a)$ for $a \in M_n(F)$.

Our first use of remark 8.2.27 is somewhat technical.

Lemma 8.2.23:   (*Formanek* [73]). *Define* $t_h: F[G] \to F$ *by* $t_h(\sum \alpha_g g) = \sum\{\alpha_g : g$ *is conjugate to* $h\}$, *where* $h \in G$. *If for all* $n \neq \pm 1$ h *is not conjugate to* h^n *then* $\text{Tr}_h e = 0$ *for every idempotent* e *of* $M_n(F[G])$.

Proof: First assume char$(F) = p$. Write $e = (r_{ij})$ and $\text{tr}(e) = \sum \alpha_g g = \sum_{i=1}^{n} r_{ii}$. We claim there is some $q = p^m$ such that no g^q is conjugate to h, for all g in supp(tr e). This surely follows if we can show for any g in G there is at most one positive power q' of p such that $g^{q'}$ is conjugate to h (for then take q greater than all of the q' pertaining to supp(tr e).) So assume $g^{q'}$ and $g^{q''}$ each are conjugate to h, with $q' > 0$ minimal possible. Then $q' | q''$; letting $n = q''/q'$ we see h^n is conjugate to $(g^{q'})^n = g^{q''}$ so by hypothesis $n = \pm 1$. But then $q'' = q'$ as desired, proving the claim.

   Note $e_{ij} = [e_{ii}, e_{ij}]$ for any $i \neq j$. Thus 8.1.5(i) yields

$$e = e^q \in \sum r_{ii}^q e_{ii} + [M_n(F[G]), M_n(F[G])],$$

so $\text{Tr}_h e = \sum t_h(r_{ii}^q) = 0$, proved as in lemma 8.1.24.

   Having proved the result for characteristic p, we get the result for characteristic 0 by a "Nullstellensatz" argument, as in remark 8.1.25'.   Q.E.D.

Proposition 8.2.24:   *If* G *is torsion-free and* $F[G]$ *is Noetherian then* $\text{Tr}_h e = 0$ *for every idempotent* e *of* $M_n(F[G])$, *and all* $h \neq 1$ *in* G.

Proof: We need to verify h is not conjugate to h^n. Indeed, if $ghg^{-1} = h^n$ let $h_i = g^{-i}hg^i$ and $H_i = \langle h_i \rangle$. Then

$$h_i^n = g^{-i}h^n g^i = g^{-i}ghg^{-1}g^i = g^{-(i-1)}hg^{i-1} = h_{i-1}$$

implying $H_{i-1} \le H_i$, and $H_{i-1} < H_i$ since G is torsion-free. But this gives an infinite ascending chain of subgroups of G contrary to lemma 8.2.3.

Q.E.D.

Now let t_G denote the Kaplansky trace map, so that if $r = \sum \alpha_g g$ then $\alpha_g = t_G(rg^{-1})$. (Note in §8.1 we used the notation "tr" instead.)

Corollary 8.2.25: *If $F[G]$ is Noetherian and G is torsion-free then $\sum_{g \ne 1} t_G((\operatorname{tr} e)g^{-1}) = 0$ for any idempotent e of $M_n(F[G])$.*

Suppose P is a projective $F[G]$-module. Then $P = eM_n(F[G])$ for a suitable idempotent e of $M_n(F[G])$. We are now ready for an important invariant for projective modules.

Definition 8.2.26: If P is $F[G]$-projective define $\operatorname{Tr}_G P = \operatorname{Tr}_G e = t_G(\operatorname{tr} e)$.

Theorem 8.2.27: *Suppose G is torsion-free and $F[G]$ is Noetherian. Then $\operatorname{Tr}_G P = [F \otimes_{F[G]} P : F]$ and thus is a well-defined integer (independent of n or the choice of e).*

Proof: Let $\pi: F[G] \to F$ be the augmentation homomorphism given by $\pi g = 1$. Using corollary 8.2.25 we have

$$\operatorname{Tr}_G P = t_G(\operatorname{tr} e) = \sum_{g \in G} t_G((\operatorname{tr} e)g^{-1}) = \operatorname{tr}(\pi e) = [F \otimes_{F[G]} P : F],$$

the last equality holding since the trace of an idempotent in $M_n(F)$ is its rank, and $\pi e \in M_n(F)$ is idempotent.   Q.E.D.

Corollary 8.2.28: $\operatorname{Tr}_G(P_1 \oplus P_2) = \operatorname{Tr}_G P_1 + \operatorname{Tr}_G P_2$ *for any two projective modules P_1 and P_2, provided G is torsion free.*

Theorem 8.2.35: *(Farkas-Snider) If G is torsion-free polycyclic-by-finite and $\operatorname{char}(F) = 0$ then $F[G]$ is a domain.*

Proof: The idea behind the Farkas-Snider proof is quite intuitive. $R = F[G]$ is a prime Noetherian ring by Connell's theorem and theorem 8.2.2 and thus has a classical ring of fractions $S^{-1}R = M_n(D)$. Suppose R is not a domain. There is a left ideal L of R with $0 < [S^{-1}L:D] \leq n$. We would like to conclude as in the proof of theorem 5.1.45 with the contradiction that n^2 divides $[S^{-1}L:D]$, but we do not know if every projective $F[G]$-module is stably free.

Nevertheless, by proposition 8.2.6 there is a normal polycyclic subgroup H of finite index t, each of whose factors is infinite cyclic. Thus $F[H]$ is a Noetherian domain and so $S_1^{-1}F[H]$ is a division ring D_1, where $S_1 = \{$regular elements of $F[H]\}$. But $S_1 \subset S$ in view of remark 8.1.36. Hence $0 < [S_1^{-1}L:D_1] < [S_1^{-1}R:D_1] = t$ (for if $S_1^{-1}L = S_1^{-1}R$ then localizing further at S would yield $S^{-1}L = S^{-1}R$, contrary to assumption). Thus to arrive at the desired contradiction it suffices to prove $t \mid [S_1^{-1}L:D_1]$.

By theorem 8.2.20 there is an f.g. projective resolution $0 \to P_m \to \cdots \to P_0 \to L \to 0$ of L in $R\text{-}\mathcal{M}od$. Each P_i is f.g. projective as $F[H]$-module, so $0 \to S_1^{-1}P_m \to \cdots \to S_1^{-1}P_0 \to S_1^{-1}L \to 0$ is a projective resolution in $D_1\text{-}\mathcal{M}od$. By remark 5.1.43 we need prove the following.

Claim 1. If P is a projective R-module then t divides $[S_1^{-1}P:D_1]$.

To prove claim 1 we use Tr_G of definition 8.2.26. Note P is also $F[H]$-projective so we can also define $\text{Tr}_H P$. Coupling theorem 5.1.34 with proposition 5.1.32 one sees every projective $F[H]$-module is stably free. Writing $P \oplus F[H]^{(u)} \approx F[H]^{(v)}$ we tensor on the left by $D_1 = S_1^{-1}F[H]$ to get

$$S_1^{-1}P \oplus D_1^{(u)} \approx D_1^{(v)}.$$

Thus $[S_1^{-1}P:D_1] = v - u$. On the other hand,

$$v = \text{Tr}_H(F[H]^{(v)}) = \text{Tr}_H(P \oplus F[H]^{(u)}) = \text{Tr}_H P + u,$$

so $\text{Tr}_H P = [S_1^{-1}P:D_1]$. Thus to prove claim 1 we need to show t divides $\text{Tr}_H P$. Since $\text{Tr}_G P$ is an integer (theorem 8.2.27) it suffices to prove

Claim 2. $\text{Tr}_H P = (\text{Tr}_G P)t$.

Proof of Claim 2. Let $e_1, \ldots, e_n$ be a canonical right base for $R^{(n)}$ over R. Writing $P = eM_n(R)$ and $ee_i = \sum_{j=1}^n e_j r_{ji}$ we can identify e with the matrix (r_{ji}), and thus $\text{Tr}_G P = \sum t_G r_{ii}$. Now let $g_1, \ldots, g_t$ be a left transversal of G over H. Then $\{e_i g_k : 1 \leq k \leq t, 1 \leq i \leq n\}$ is a right base of $R^{(n)}$ over $F[H]$ and

$$e(e_i g_k) = (ee_i)g_k = \left(\sum e_j r_{ji}\right)g_k = \sum_j e_j g_k(g_k^{-1} r_{ji} g_k),$$

so we can identify e with the $nt \times nt$ matrix whose diagonal has the entries $g_k^{-1} r_{ii} g_k$ for $1 \leq k \leq t$ and $1 \leq i \leq n$. Thus

$$\text{Tr}_H P = \sum_{k=1}^{t} \sum_{i=1}^{n} t_G(g_k^{-1} r_{ii} g_k) = \sum_{k=1}^{t} \sum_{i=1}^{n} t_G r_{ii} = \sum_{k=1}^{t} \text{Tr}_G P = (\text{Tr}_G P)t,$$

as desired. This proves claim 2 and thus concludes the proof of the Farkas-Snider theorem.   Q.E.D.

Cliff [80] has verified the zero-divisor conjecture in characteristic $\neq 0$ for torsion-free polycyclic-by-finite groups.

These results greatly enhanced the status of group algebras of polycyclic-by-finite groups, especially since they are the only known examples of Noetherian group algebras, and below we shall describe the flurry of activity in this area. Special attention should be paid to the following situation. Suppose $\text{char}(F) = 0$ and G is polycyclic-by-finite. $F[G]$ is a semiprime Goldie ring and thus has a semisimple Artinian ring of fractions Q. If G has no finite normal subgroups $\neq (1)$ then $F[G]$ is prime so $Q \approx M_n(D)$ for some division ring D and suitable n. J. Moody [86] has proved the following result (c.f. Farkas [80, #18] and Cliff-Weiss [88]):

Goldie Rank Theorem: *n is the least common multiple of the orders of the finite subgroups of G. (Equivalently, $\text{Tr}_G P \in \mathbb{Z}$ for every projective module P over $F[G]$.)*

Futher progress also has been made on the zero-divisor question for solvable groups of finite rank, cf., Crawley-Boerey-Kropholler-Linnell [88].

Strojnowski [86] found a proof of claim 1 which appeals more directly to dependence of elements and thus applies in more general settings, cf., exercises 12–15; he also obtains results concerning when idempotents are trivial, cf., exercise 16.

§8.3 Enveloping Algebras

The object of this section is to describe some of the theory of enveloping algebras of Lie algebras using methods of ring theory; the next section continues this project from an even more ring-theoretic point of view. Unfortunately, we do not have the space to deal adequately with semisimple Lie algebras; a brief overview of the subject is given at the end of the section.

We shall prove only the basic general facts about enveloping algebras and then focus on nilpotent and solvable Lie algebras, which are more amenable to ring theoretic techniques. We draw heavily from the classic Dixmier [74B] and the excellent reference Borho-Gabriel-Rentschler [73B], henceforth referred to respectively as DIX and BGR. These books have a large overlap, and we usually refer to BGR. The classic text on Lie algebras is still Jacobson [62B]. Recall the basic definitions from 1.6.10ff, and suppose L is a Lie algebra over a commutative ring C. The following concept also will be useful.

Definition 8.3.0:   A *Lie L-module* is a C-module M endowed with scalar multiplication $L \times M \to M$ satisfying $[ab]x = a(bx) - b(ax)$ for all a, b in L and all x in M.

Remark 8.3.1:   There is a functor $F: C\text{-}\mathscr{Alg} \to C\text{-}\mathscr{Lie}$ given by $FR = R^-$, where given a morphism $f: R_1 \to R_2$ we define $Ff: R_1^- \to R_2^-$ to be the same map viewed as a Lie homomorphism.

We are led to consider the universal from a given Lie algebra L to the functor of remark 8.3.1; in other words, we want an algebra $U(L)$ together with a morphism $u: L \to U(L)^-$, such that for any associative algebra R and any Lie homomorphism $f: L \to R^-$ there is a unique homomorphism $g: U(L) \to R$ such that $f = gu$.

Proposition 8.3.2: $U(L)$ *exists and can be taken to be* $T(L)/A$ *where* $T(L) = T_C(L)$ *is the tensor algebra for L (as C-module) and A is the ideal generated by all* $a \otimes b - b \otimes a - [ab]$ *for all a, b in L; $u: L \to T(L)/A$ is the map* $a \mapsto a + A$.

Proof: Suppose $f: L \to R^-$ is given. By proposition 1.9.14 there is an algebra homomorphism $\tilde{f}: T(L) \to R$ given by $\tilde{f}(a_1 \otimes \cdots \otimes a_t) = fa_1 \cdots fa_t$. In particular, $\tilde{f}(a \otimes b - b \otimes a - [ab]) = [fa, fb] - f[ab] = 0$ for all a, b in L, so $A \subseteq \ker \tilde{f}$. This gives us the desired homomorphism $g: T(L)/A \to R$, and uniqueness of g follows from the fact that uL generates $T(L)/A$ as an algebra.   Q.E.D.

By abstract nonsense $U(L)$ is unique up to isomorphism; the key additional fact (cf. [DIX] or Jacobson [62B] for the proof) is

Theorem 8.3.3.:   (*Poincaré-Birkhoff-Witt = PBW theorem*) *Suppose L is free in $C\text{-}\mathscr{Mod}$ (e.g., if C is a field). Then $u: L \to U(L)$ is monic. Explicitly if $X = \{x_i : i \in I\}$ is a base for L over C then well-ordering I and writing y_i for ux_i in $U(L)$ we have the base $B = \{y_{i_1} \cdots y_{i_t} : t \in \mathbb{N}$ and $i_1 \leq \cdots \leq i_t\} \cup \{1\}$ of $U(L)$.*

We say a *Lie representation* is a Lie map $L \to \text{End}_C V$ where V is an arbitrary C-module.

Corollary 8.3.4: *Any Lie representation* $L \to \text{End}_C V$ *extends uniquely to an algebra homomorphism* $U(L) \to \text{End}_C V$.

Corollary 8.3.5: *Any Lie map* $f: L_1 \to L_2$ *extends uniquely to an algebra homomorphism* $Uf: U(L_1) \to U(L_2)$, *thereby yielding a functor* $U: C\text{-}\mathscr{L}\!\mathit{ie} \to C\text{-}\mathscr{A}\!\mathit{lg}$. *If* L_1, L_2 *are free C-modules and* f *is monic (resp. epic) then* Uf *is an injection (resp. surjection).*

Proof: The composition $L_1 \to L_2 \hookrightarrow U(L_2)$ yields the desired homomorphism $U(L_1) \to U(L_2)$. The other assertions are straightforward.   Q.E.D.

Basic Properties of Enveloping Algebras

The sequel will be limited to a Lie algebra L over a field F, and we shall view $L \subset U(L)$ by means of the PBW theorem. As in group algebras, we strike the dilemma that the structure of $U(L)$ must be intricate enough so as to include the representation theory of L; thus the study of $U(L)$ is both desirable and forbidding. Certain properties however are easy consequences of the PBW theorem.

Proposition 8.3.6: $U(L_1 \times \cdots \times L_t) \approx U(L_1) \otimes \cdots \otimes U(L_t)$.

Proof: It suffices to prove this for $t = 2$. Then the map $f: L_1 \times L_2 \to U(L_1) \otimes U(L_2)$ given by $f(a_1, a_2) = ua_1 \otimes ua_2$ extends to a homomorphism $U(L_1 \times L_2) \to U(L_1) \otimes U(L_2)$. Conversely $U(L_i) \subseteq U(L_1 \times L_2)$ for $i = 1, 2$ by corollary 8.3.5, and the balanced map $g: U(L_1) \times U(L_2) \to U(L_1 \times L_2)$ given by $g(r_1, r_2) = r_1 r_2$ yields a homomorphism $g: U(L_1) \otimes U(L_2) \to U(L_1 \times L_2)$ which is the inverse of f on $L_1 \times L_2$ and thus on $U(L_1 \times L_2)$.   Q.E.D.

Remark 8.3.7: If K is a field extension of F then $L \otimes_F K$ is a Lie algebra under "extension of scalars," and $U(L) \otimes_F K$ is isomorphic to the enveloping algebra $U(L \otimes_F K)$ of $L \otimes_F K$ over K. Indeed, there is a monic $L \otimes K \to U(L) \otimes K$; by corollary 8.3.5 this extends to an injection $U(L \otimes K) \to U(L) \otimes K$ which is onto by the PBW.)

Remark 8.3.8: If L_1 is a Lie subalgebra of L then expanding a base B_1 of L_1 to a base B of L we see by the PBW theorem that $U(L)$ is free as $U(L_1)$-module. (Indeed, write $B' = B - B_1 = \{x_i : i \in I\}$, and well-order I. Then $\{x_{i_1} \cdots x_{i_m} : i_1 \leq \cdots \leq i_m\}$ is a base, since elements of B_1 can be pushed to the left by means of the Lie multiplication.)

Proposition 8.3.9: *If A is a Lie ideal of L then $U(L)A = AU(L) \lhd U(L)$ and $U(L/A) \approx U(L)/U(L)A$.*

Proof: The Lie map $L \to L/A$ gives a surjection $\varphi : U(L) \to U(L/A)$ as in corollary 8.3.5; viewing $U(L)$ as a free $U(A)$-module we see the base of remark 8.3.8 is sent to a base of $U(L/A)$, so $U(L)A = \ker \varphi \lhd U(L)$; symmetrically $AU(L) = \ker \varphi$.   Q.E.D.

Definition 8.3.10: Let $U_m(L)$ be the subspace of $U(L)$ generated by all products $\{a_1 \cdots a_n : n \leq m$ and $a_i \in L\}$. $\{U_m : m \in \mathbb{N}\}$ will be called the *standard filtration* of $U(L)$. We say the *order* of $r \in U(L)$ is the smallest m for which $r \in U_m$.

Note: This is a filtration according to definition 1.2.13, if we change the indices to negative number as in remark 1.2.13'; thus $U(L)$ is filtered according to definition 5.1.36.

Remark 8.3.11: Each $U_m(L)$ is an L-module under Lie multiplication, since $[a, a_1 \cdots a_m] = \sum a_1 \cdots a_{i-1}[aa_i]a_{i+1} \cdots a_m$. Of course, if L is finite dimensional as F-algebra then $U_m(L)$ also is f.d.

We shall now use the associated graded algebra of $U(L)$, denoted as $G(L)$ (cf., definition 3.5.30) to study $U(L)$. Recall $G(L) = \bigoplus G_i$ where $G_i = U_i(L)/U_{i-1}(L)$, in particular, $G_0 = F$ and $G_1 = L$; multiplication is transferred from $U(L)$.

Remark 8.3.12: $G(L)$ is commutative, since $[r_i, r_j] \in U_{i+j-1}(L)$ for all r_i in $U_{i-1}(L)$ and r_j in $U_{j-1}(L)$ (as in remark 8.3.11). Moreover, $G_n = (G_1)^n$.

Proposition 8.3.13: *$G(L)$ is canonically isomorphic (as graded algebra) to the symmetric algebra $S(L)$ defined in example 1.9.17.*

Proof: Let $\{x_i : i \in I\}$ be a base of L. Define $\varphi : T_F(L) \to G(L)$ by sending a monomial $x_{i_1} \otimes \cdots \otimes x_{i_m}$ to its image in G_m. Note each $x_i \otimes x_j - x_j \otimes x_i \in$

ker φ since, viewed in $U(L)$, $x_i x_j + U_1(L) = x_j x_i + U_1(L)$ (because $[x_i, x_j] \in L$). Thus we have a homomorphism $\bar{\varphi}: S(L) \to G(L)$ which sends base to base (since $\{x_{i_1} \cdots x_{i_m} : i_i \leq \cdots \leq i_m\}$ can be viewed as a base for the m-th component in each). Thus $\bar{\varphi}$ is an isomorphism.   Q.E.D.

(We shall show below this implies $U(L)$ has GK-dim when L is finite dimensional.) Since $S(L)$ is a domain we see $G(L)$ is a domain, providing the following consequence.

Theorem 8.3.14: $U(L)$ *is a domain.* $\text{Jac}(R) = 0$ *for every* subring R *of* $U(L)$ *not contained in* F. *In fact, if* $r_i \in U(L)$ *has order* m_i *for* $i = 1, 2$ *then* $r_1 r_2$ *has order* $m_1 + m_2$.

Proof: The last assertion is clear by passing to "leading terms" in the domain $G(L)$. Hence $U(L)$ is a domain. Furthermore, if $0 \neq r \in \text{Jac}(R)$ then taking $r_1 \in R - F$ we see $r_1 r$ has order > 0 so $1 - r_1 r$ also has order > 0 and cannot be invertible, contradiction; thus $\text{Jac}(R) = 0$.   Q.E.D.

Corollary 8.3.14' (Irving): *Suppose* L *is a Lie algebra over an integral domain* C *and is faithful (as* C-*module). Then* $\text{Jac}(U(L)) = 0$.

This result generalizes the commutative version of Amitsur's theorem (by taking L to be 1-dimensional).

Proof: Let $S = C - \{0\}$ and $F = S^{-1}C$. Then $S^{-1}U(L)$ satisfies the universal property of the universal enveloping algebra of the Lie algebra $L \otimes_C F$ over F; making this identification we see $\text{Jac}(U(L)) = 0$ by theorem 8.3.14. Furthermore, $L \to L \otimes F \to U(L \otimes F)$ is monic so $L \to U(L)$ is monic.   Q.E.D.

Proposition 8.3.15: *If* $[L:F] < \infty$ *then* $U(L)$ *is left and right Noetherian.*

Proof: $G(L)$ is Noetherian by the Hilbert basis theorem so apply corollary 3.5.32(ii).   Q.E.D.

Remark 8.3.16: There is an anti-automorphism of L given by $a \to -a$, which can be viewed as an isomorphism of L to the opposite algebra L^{op}. This extends to an automorphism $U(L) \to U(L)^{\text{op}}$, i.e., an involution of $U(L)$, called the *principal involution*.

One more general result of interest is

Theorem 8.3.23: *(Quillen) If L is a finite dimensional algebra over a field then U(L) is Jacobson.*

The proof involves certain ideas of Duflo [73] which have then been extended by McConnell to much more general classes of rings and will be discussed in §8.4, cf., corollary 8.4.14. This result enhances interest in the primitive ideals (since they are now seen to be dense in the prime spectrum).

Nilpotent and Solvable Lie Algebras

Before continuing, we should quote some of the standard Lie algebra theory. First some definitions in analogy to group theory. *Assume that L is finite dimensional over* a field *F* of characteristic 0.

Definition 8.3.24: Writing L^2 for $[LL]$ and inductively L^n for $[L^{n-1}L]$, we say *L* is *nilpotent of index n* if $L^n = 0$ for *n* minimal such. Likewise, writing $L^{(1)} = L$ and inductively $L^{(n)}$ for $[L^{(n-1)}L^{(n-1)}]$, we say *L* is *solvable of length n* if $L^{(n)} = 0$ for some *n*. In particular, every nilpotent algebra is solvable.

In analogy to the associative radical theory, every finite dimensional Lie algebra has a unique largest solvable ideal, called the *radical* or rad(*L*); rad(*L*/rad *L*) = 0.

Consider a Lie algebra *L* of endomorphisms of a finite dimensional vector space *V* over *F*. We say *L* is *triangularizable* (resp. *strictly triangularizable*) if there is a change of base of *V* with respect to which each element of *L* is upper triangular (resp. strictly upper triangular). If every element of *L* is nilpotent as a matrix then $L^n = 0$ for some *n*, by proposition 2.6.30. It follows easily that *L* is strictly triangularizable. Indeed, take *k* maximal for which $L^k \neq 0$, and take $0 \neq v_1 \in L^k V$; noting $Lv_1 = 0$ we see $[LV:F] < [V:F]$, so proceeding by induction on $[V:F]$ we can expand v_1 to a base with respect to which *L* is strictly upper triangular. Since the hypothesis is satisfied by the adjoint representation of a nilpotent Lie algebra, we have proved the following:

L is nilpotent iff Ad *L* is nilpotent as a set of matrices, in which case Ad *L* is strictly triangularizable.

For solvable Lie algebras the situation is almost as nice. Suppose $\rho: L \to$ End$_F V$ is a representation. Given a subalgebra L_1 of *L* we write L_1^* for the dual space $\mathrm{Hom}(L_1, F)$ of linear functionals. If $\alpha \in L_1^*$ we define the *eigenspace* $V_\alpha = \{v \in V : (\rho a)v = \alpha(a)v \text{ for all } a \text{ in } L_1\}$, clearly a subspace invariant under the action of L_1.

Lie's Lemma: [*DIX, lemma* 1.3.11] *If* $[V : F] < \infty$ *and* $L_1 \lhd L$ *then in the above notation* V_α *is invariant under the action of* L.

Lie's Theorem: [*DIX, theorem* 1.3.12] *Any solvable Lie algebra of matrices over an algebraically closed field is triangularizable.*

The proof is a straightforward induction based on Lie's lemma. Let us now call a Lie algebra L *completely solvable* if $\mathrm{Ad}\, L$ is triangularizable. Every completely solvable Lie algebra is solvable; conversely, every solvable Lie algebra over an algebraically closed field is completely solvable. For this reason the study of enveloping algebras over solvable Lie algebras usually is based on the hypothesis F is algebraically closed. Lie's theorem provides $a \in L$ for which $Fa \lhd L$.

Remark 8.3.25: If $Fa \lhd L$ then $U(L)a \lhd U(L)$ and $U(L/Fa) \approx U(L)/U(L)a$. (By proposition 8.3.9.)

This remark is exceedingly useful in the study of enveloping algebras of solvable Lie algebras. One application ties in with McConnell's theory of completions discussed in §3.5. We say an ideal A of R is *polynormal* if $A = \sum_{i=1}^{t} Ra_i$ where a_1 is R-normalizing and inductively the image of a_k in $\bar{R} = R/\sum_{i=1}^{k-1} Ra_i$ is $\bar{R}$-normalizing for each $1 < k \le t$. (Compare to "polycentral," definition 3.5.34.) The a_i are called the *polynormal generating set* of A.

Theorem 8.3.26: (*McConnell* [68]) (i) *If* L *is a nilpotent Lie algebra then every ideal of* $U(L)$ *is polycentral.* (ii) *If* L *is solvable and* F *is algebraically closed then every ideal of* $U(L)$ *is polynormal.*

Proof: Suppose $A \lhd U(L)$. L acts on A by the adjoint action. Let us prove (ii), for (i) is analogous. As noted above there is some a_1 in A for which $Fa_1 \lhd L$ and $U(L/Fa_1) \approx U(L)/U(L)a_1$. Proceeding by induction on $[L : F]$ we see A/Fa_1 is polynormal with a suitable polynormal base $a_2 + Fa_1, \ldots,$ $a_t + Fa_1$, so $a_1, \ldots, a_t$ is a polynormal generating set of A.   Q.E.D.

Corollary 8.3.27: *The completion of* $U(L)$ *with respect to any ideal is Noetherian if* L *is nilpotent.*

Proof: Apply theorem 3.5.36 to theorem 8.3.26.   Q.E.D.

We shall continue this train of thought in §8.4.

Nilpotent Lie algebras are "better behaved" in Dixmier's theory than solvable Lie algebras. For example, ψ of corollary 8.3.22 *is* an associative homomorphism for L nilpotent, cf., [DIX, proposition 4.8.12].

Solvable Lie Algebras as Iterated Differential Polynomial Rings

Differential polynomial rings were introduced in §1.6, under the notation $R[\lambda; 1, \delta]$ where δ is a derivation of R. Since "1" is always understood here we shall now use the abbreviated notation $R[\lambda; \delta]$. In studying enveloping algebras of solvable Lie algebras the following remark is important. We say a ring R is an *iterated differential polynomial ring* if there is a chain of subrings R_α of R such that $R_{\alpha+1} \approx R_\alpha[\lambda_\alpha; \delta_\alpha]$ is a differential polynomial ring for every α.

Proposition 8.3.28:   *If L is a solvable Lie algebra over a field F then $U(L)$ is an iterated differential polynomial ring.*

Proof: $L^{(t)} = 0$ for some t. By induction on t we have $U(L')$ is an iterated differential polynomial ring. Take any a in $L - L'$. Then $L_1 = L' + Fa$ is a Lie subalgebra of L, and ad a induces a derivation of L_1 which thus extends to a derivation δ of $U(L_1)$. Let $R = U(L')[\lambda; \delta]$. There is an epimorphism $\varphi: R \to U(L_1)$ given by $\lambda \mapsto a$; but $\ker \varphi = 0$ for if $0 = \varphi(\sum r_i \lambda^i) = \sum r_i a^i$ for r_i in $U(L')$ then each $r_i = 0$ by remark 8.3.8. Continuing inductively, given $L_i \supset L'$ we can take a_{i+1} in $L - L_i$ and let $L_{i+1} = L_i + Fa_i$; then $U(L_{i+1}) \approx U(L_i)[\lambda_i; \text{ad } a_{i+1}]$ and eventually we reach L.   Q.E.D.

This result is the cornerstone of our study of enveloping algebras of solvable Lie algebras and leads us to look a bit closer at properties of a derivation δ on a ring R.

Remark 8.3.29:   If $A \lhd R$ then $\delta(A^2) \subseteq A$. (Indeed

$$\delta(a_1 a_2) = a_1 \delta a_2 + (\delta a_1) a_2 \in A.)$$

Lemma 8.3.30:   *If S is a denominator set on R then δ extends uniquely to a derivation on $S^{-1}R$.*

Proof: We start with uniqueness. Clearly $0 = \delta(s^{-1}s) = s^{-1}\delta s + (\delta s^{-1})s$ so we must have

$$\delta s^{-1} = -s^{-1}(\delta s)s^{-1} \text{ and thus}$$

$$\delta(s^{-1}r) = s^{-1}\delta r - s^{-1}(\delta s)s^{-1}r. \tag{1}$$

To verify that (1) indeed defines a derivation, first we show it is well-defined. In view of theorem 3.1.4(iii) we need show that $\delta((r's)^{-1}r'r) = \delta(s^{-1}r)$ for every r' in R such that $r's \in S$. Indeed

$$(r's)^{-1}\delta(r'r) - (r's)^{-1}\delta(r's)(r's)^{-1}r'r$$

$$= (r's)^{-1}(r'\delta r + (\delta r')r) - (r's)^{-1}(r'\delta s + (\delta r')s)(r's)^{-1}r'r$$

$$= s^{-1}\delta r + (r's)^{-1}(\delta r')r - s^{-1}(\delta s)s^{-1}r - (r's)^{-1}(\delta r')r$$

$$= s^{-1}\delta r - s^{-1}(\delta s)s^{-1}r$$

as desired.   Q.E.D.

This lemma also holds for more general quotient ring constructions, as one may expect.

Proposition 8.3.31: *Suppose R is semiprime Goldie and δ is a derivation on R. Then $\delta P \subseteq P$ for every minimal prime ideal P of R.*

Proof: Let Q be the semisimple Artinian ring of fractions of R. By theorem 3.2.27 we know some simple component $Q \approx Q/A$ is the ring of fractions of R/P, where A is a maximal ideal of Q containing P. But δ extends to Q by lemma 8.3.30, and $A = Qe$ for some central idempotent e of Q. Hence $A^2 = A$ so $\delta A \subseteq A$ by remark 8.3.29; hence $\delta P \subseteq R \cap \delta A \subseteq R \cap A = P$.   Q.E.D.

We shall say $P \in \operatorname{Spec}(R)$ is δ-*invariant* if $\delta P \subseteq P$ and write δ-$\operatorname{Spec}(R)$ for $\{\delta\text{-invariant primes}\}$. We just proved δ-$\operatorname{Spec}(R)$ contains the minimal primes for R semiprime Goldie.

Proposition 8.3.32: *Suppose $\operatorname{char}(F) = 0$ and R is a Noetherian F-algebra with a derivation δ. Let $R' = R[\lambda; \delta]$.*

  (i) *If $A \lhd R$ and $\delta A \subseteq A$ then $R'A = AR' \lhd R'$.*
  (ii) *If $P \in \delta$-$\operatorname{Spec}(R)$ then $PR' \in \operatorname{Spec}(R')$.*
  (iii) *There is a surjection $\operatorname{Spec}(R') \to \delta$-$\operatorname{Spec}(R)$ given by $P' \mapsto P' \cap R$.*

Proof:

  (i) $\lambda a = a\lambda + \delta a$ for all a in A, implying $\lambda(AR') \subseteq AR'$ and thus $AR' \lhd R'$. Hence $R'A \subseteq AR'$, and by symmetry $AR' \subseteq R'A$.

(ii) $PR' \lhd R'$ by (i). It remains to show the ideal PR' is prime. Passing to R/P we may assume R is prime and need to show R' is prime. But this follows from a leading term argument analogous to that preceding definition 1.6.14; if $\left(\sum_{i=0}^{m} r_i \lambda^i\right) R \left(\sum_{j=0}^{n} r'_j \lambda^j\right) = 0$ with $r_m, r'_n \neq 0$ then $0 = r_m R r'_n \lambda^{m+n} +$ terms of lower degree, so $r_m R r'_n = 0$, contrary to R prime.

(iii) $P = PR' \cap R$ is seen by checking the constant term of PR', so LO holds from δ-Spec(R) to Spec(R'). It remains to show for each P' in Spec(R') that $P' \cap R \in \delta$-Spec(R). Let $A = P' \cap R$. Clearly $\delta A = [\lambda, A] \subseteq P' \cap R = A$, so we need to show A is prime. Passing to R/A and R'/AR' we may assume $A = 0$. Let $N = \text{Nil}(R)$. $N^t = 0$ for some t, and N is δ-invariant by proposition 2.6.28. By (i) we have $(NR')^t \subseteq N^t R' = 0$ so $NR' \subseteq P'$ and thus $N \subseteq P' \cap R = 0$. Let $P_1, \ldots, P_m$ be the minimal primes of R, which are δ-invariant by proposition 8.3.31. The same argument as above shows $(P_1 R') \cdots (P_m R') \subseteq P_1 \cdots P_m R' = 0$, so some $P_i R' = P'$ and thus $P_i = 0$, proving R is indeed prime.   Q.E.D.

We say a prime ideal P of R is *principal* if $P = Ra = aR$ for some a in P.

Corollary 8.3.33: *Suppose R is prime Noetherian. If $P' \in \text{Spec}(R[\lambda; \delta])$ has height 1 and $P' \cap R$ is nonzero principal then P' is principal.*

Proof: Let $P = P' \cap R$, which by hypothesis has the form $aR = Ra$. Let $R' = R[\lambda; \delta]$. Then $aR' = PR' \in \text{Spec}(R')$ so $aR' = P'$; analogously $P' = R'a$.
                                                                            Q.E.D.

One can frame these results in a setting more appropriate to arbitrary enveloping algebras, cf., exercise 9. Now we shall couple these results with proposition 1.6.36.

Completely Prime Ideals (Following BGR)

Definition 8.3.34:   An ideal P of R is *completely prime* if R/P is a domain.

Although we studied this idea in §2.6, we did not have many examples of completely prime ideals of domains. We shall see now that *every* prime ideal of an enveloping algebra of a solvable Lie algebra is completely prime, thereby showing how the solvable theory begins to diverge from the semisimple theory, in view of the next example.

Example 8.3.35: Suppose $F = \mathbb{C}$ and $R = U(L)$ where L is a Lie algebra which is *not* solvable. Then L has a simple module M with $n = [M:\mathbb{C}] > 1$, and we view M as R-module. Let $P = \text{Ann } M$, a primitive ideal. By the density theorem R/P is dense in $\text{End}_\mathbb{C} M \approx M_n(\mathbb{C})$, and so $R/P \approx M_n(\mathbb{C})$ which surely is not a domain.

Theorem 8.3.36: *If $R = U(L)$ with L solvable then every prime ideal of R is completely prime.*

Proof: Induction on $n = [L:F]$; the assertion is trivial for $n = 1$. Let $F = L_0 \lhd L_1 \lhd \cdots \lhd L_n = L$, where each $[L_i:F] = i$, and let $R_i = U(L_i)$. We shall show by induction on i that each prime ideal of R_i is completely prime. This assertion is trivial for $i = 0$; assuming it holds for i we shall prove it for $i + 1$. By proposition 8.3.28 we have $R_{i+1} \approx R_i[\lambda:\delta]$ for a suitable derivation δ of R_i. Suppose $P \in \text{Spec}(R_{i+1})$, and let $P_0 = P \cap R_i \in \delta\text{-Spec}(R_i)$ by proposition 8.3.32(iii).

Let $\bar{R}_i = R_i/P_0$ and $\bar{R}_{i+1} = R_{i+1}/P_0 R_{i+1} \approx \bar{R}_i[\lambda;\delta]$. By induction $\bar{R}_i$ is a domain, so $\bar{R}_{i+1}$ is a domain, and we are done unless $\bar{P} \neq 0$. Let $S = \bar{R}_i - \{0\}$ and let $D = S^{-1}\bar{R}_i$, the division ring of fractions of $\bar{R}_i$. By lemma 8.3.30 δ extends to D, so we can view $\bar{R}_{i+1} \subseteq D[\lambda;\delta]$. Furthermore, $0 \neq S^{-1}\bar{P} \lhd D[\lambda;\delta]$. Hence proposition 1.6.36 shows δ is inner on D, and $D[\lambda;\delta]$ is a polynomial ring $D[\mu]$ where $\mu = \lambda - d$, and also $S^{-1}\bar{P}$ generated by some central element $f \neq 0$. Let $Z = Z(D)$ and $K = Z[\mu]/Z[\mu]f$, a field.

$D[\lambda;\delta]/S^{-1}\bar{P} \approx D \otimes_Z K$ is simple by theorem 1.7.27. On the other hand, there is a composition

$$\psi: U(L_i \otimes_F K) \approx U(L_i) \otimes_F K \subseteq D \otimes_F K \to D \otimes_Z K.$$

$\psi U(L_i \otimes_F K)$ is prime since $S^{-1}\psi U(L_i \otimes_F K) \approx D \otimes_Z K$ is simple, and therefore $\ker \psi \in \text{Spec}(U(L_i \otimes_F K))$. But $[L_i \otimes_F K:K] \leq i$, implying $\ker \psi$ is completely prime by the induction hypothesis. Hence $D \otimes_Z K$ is a domain, and consequently $S^{-1}\bar{P}$ (and thus $\bar{P}$) is completely prime.   Q.E.D.

One strategy of BGR is beginning to shape up. Given a finite dimensional solvable Lie algebra L we can write $U(L)$ in the form $U(L_1)[\lambda;\delta]$ where L_1 is any subspace of codimension 1 containing L'. But localizing at suitable elements inside $U(L_1)$ will give this differential polynomial ring better form, thereby enabling us to break $\text{Spec}(U(L))$ into simpler and better-known spectra.

Dimension Theory on U(L)

The two dimensions used for enveloping algebras are Krull dimension and Gelfand-Kirillov dimension. Some of the basic results can be obtained using K-dim, cf., exercise 10, but we shall focus on GK dim. *We assume throughout that L is finite dimensional over F.*

Remark 8.3.37: GK dim $U(L) = [L:F]$. (Indeed, by proposition 8.3.13 the associated graded algebra $\approx F[\lambda_1,\ldots,\lambda_n]$ where $n = [L:F]$, and GK dim $F[\lambda_1,\ldots,\lambda_n] = n$ by proposition 6.2.22. Hence GK dim $U(L) = n$ by remark 6.2.14(vi).)

When $[L:F]$ is infinite dimensional $U(L)$ does not have GK dim, cf., exercise 13. Nonetheless, there are instances when $U(L)$ has subexponential growth, cf., exercises 12, 14, 15.

Lemma 8.3.38: *Suppose L is solvable and let $0 = L_0 \lhd L_1 \lhd \cdots \lhd L_n = L$ where each $[L_i:F] = i$. Suppose $P \in \mathrm{Spec}(U(L))$ and let $P_i = (P \cap U(L_i))U(L)$. Then the chain $P_0 \subseteq P_1 \subseteq \cdots \subseteq P_n = P$ in $U(L)$ has length $n - \mathrm{GK}\,\dim(U(L)/P)$ and is in $\mathrm{Spec}(U(L))$.*

Proof: Induction on n. Let $R = U(L)$, $R_i = U(L_i)$, and $Q_i = (P \cap U(L_i))R_{n-1}$. Note $P_{n-1} = Q_{n-1}R$. By induction $Q_0 \subseteq Q_1 \subseteq \cdots \subseteq Q_{n-1}$ has length $m = (n-1) - \mathrm{GK}\,\dim(R_{n-1}/Q_{n-1})$ in $\mathrm{Spec}(R_{n-1})$. But $R \approx R_{n-1}[\lambda;\delta]$ and each Q_i is δ-invariant; hence $P_i = Q_i R \in \mathrm{Spec}(R)$. Clearly $P_0 \subseteq \cdots \subseteq P_{n-1}$ also has length m. Let $d = \mathrm{GK}\,\dim(R_{n-1}/Q_{n-1})$; thus $m = n - 1 - d$.

$R/P_{n-1} = R/Q_{n-1}R \approx (R_{n-1}/Q_{n-1})[\lambda;\delta]$. Hence GK dim $R/P_{n-1} = d + 1$ by exercise 6.2.10 (a minor modification of proposition 6.2.22). Therefore $m = n - \mathrm{GK}\,\dim(R/P_{n-1})$, and we are done if $P = P_{n-1}$. Thus we may assume $P_n \supset P_{n-1}$. Now the chain $P_0 \subseteq \cdots \subseteq P_n$ has length $m + 1$, so it remains to prove GK dim $R/P = d$. But "$\geq$" is clear since $Q_{n-1} = P \cap R_{n-1}$ (so $R_{n-1}/Q_{n-1} \hookrightarrow R/P$), and "$\leq$" follows from theorem 6.2.25 which shows

$$\mathrm{GK}\,\dim R/P \leq \mathrm{GK}\,\dim R/P_{n-1} - 1 = d. \qquad \text{Q.E.D.}$$

Theorem 8.3.39: (*Tauvel's height formula*) *If L is a solvable Lie algebra over a field F of characteristic 0 then*

$$[L:F] = \mathrm{height}(P) + \mathrm{GK}\,\dim U(L)/P$$

for every P in $\mathrm{Spec}(R)$.

Proof: First assume F is algebraically closed. $[L:F] = \mathrm{GK}\,\dim U(L)$ by remark 8.3.37. Thus we have ($\geq$) by theorem 6.2.25, and ($\leq$) follows from the chain constructed in lemma 8.3.38 (in view of Lie's theorem).

In general we tensor by the algebraic closure of F, noting this affects neither GK dim (by remark 6.2.26') nor height (by theorem 3.4.13').   Q.E.D.

This result leads us to wonder next whether or not $U(L)$ is catenary. The catenarity of $U(L)$ for L nilpotent was proved by Lorenz-Rentschler using the properties of the Dixmier Map (to be described below) and by Malliavin using Smith's theory of localization at local homologically regular rings. Recently Gabber has proved catenarity for enveloping algebras of solvable Lie algebras over $\mathbb{C}$. The reader can consult Krause-Lenagan [85B, pp. 129–135] for a proof of Gabber's result and its application to computing GK-dimensions of modules.

§8.4 General Ring Theoretic Methods

In this section we touch on one of the more promising directions of current endeavor in ring theory. Throughout the history of the subject, ring theorists have tried to prove deep theorems in other subjects with economy of effort. As the allied subjects have progressed, this task has become more and more difficult, with the gaping gulf of triviality always dangerously nearby. Nevertheless, several elegant theories have been constructed recently in an effort to unify the results of polycyclic group algebras and enveloping algebras. Since these often involve Noetherian rings, parts of this section could be viewed as an extension of §3.5, although the outlook is completely different. Most of the results also apply to other interesting classes of rings such as affine PI-algebras.

Vector Generic Flatness and the Nullstellensatz

Following McConnell [82a] we shall begin by unifying a large number of verifications of the Nullstellensatz. Note that these results are characteristic-free.

Definition 8.4.1: A ring T is called an *almost normalizing extension* of R if T is generated over R *as a ring* by elements $a_1, \ldots, a_t$ such that

$$\text{(i)} \quad R + Ra_i = R + a_i R \quad\quad \text{for } 1 \leq i \leq t$$

$$\text{(ii)} \quad [a_i, a_j] \in R + \sum Ra_i \quad\quad \text{for } 1 \leq i, j \leq t.$$

Example 8.4.2:

(i) Any Ore extension $R[\lambda; \sigma, \delta]$ is an almost normalizing extension of R.

(ii) Any enveloping algebra $U(L)$ of a finite dimensional Lie algebra L is an almost normalizing extension, where we take the a_i to be a base of L.

(iii) If $S = \{\lambda^i : i \in \mathbb{N}\}$ then $S^{-1}R[\lambda; \sigma] = R[\lambda, \lambda^{-1}; \sigma]$ is an almost normalizing extension of R.

(iv) Polycyclic group algebras are built as a series of almost normalizing extensions in the form of (iii).

Remark 8.4.3: Any almost normalizing extension T of R is spanned by the monomials $a_1^{i(1)} \cdots a_t^{i(t)}$ as left (or right) R-module, for $i(1), \ldots, i(t)$ in $\mathbb{N}$, notation as in definition 8.4.1. Write $\mathbf{i}$ for $(i(1), \ldots, i(t))$ and $a^{\mathbf{i}}$ for $a_1^{i(1)} \cdots a_t^{i(t)}$, and write $|\mathbf{i}|$ for $i(1) + \cdots + i(t)$. Then T is a filtered ring, seen by taking $T_n = \{\sum Ra^{\mathbf{i}} : |\mathbf{i}| \leq n\}$, cf., definition 5.1.36 and remark 1.2.13′.

Theorem 8.4.4: *Any almost normalizing extension T of a left Noetherian ring R is left Noetherian.*

Proof: Using the above filtration we see the associated graded ring Gr T is generated over R by elements $\bar{a}_1, \ldots, \bar{a}_t$ such that $R\bar{a}_u = \bar{a}_u R$ and $[\bar{a}_u, \bar{a}_v] = 0$ for $1 \leq u, v \leq t$; thus Gr T is left Noetherian by proposition 3.5.2, implying T is left Noetherian by corollary 3.5.32(i).   Q.E.D.

This gives us a very broad class of Noetherian rings, and we should check them for Nullstellensatz properties. We would like to use generic flatness and theorem 2.12.36, but it is not obvious how to lift generic flatness directly. McConnell's solution was to require a stronger condition, which does lift easily. To this end we examine the filtration on T. Using the total order of example 1.2.18, we can speak of the *leading coefficient* of $\sum r_i a^{\mathbf{i}}$ to be that nonzero $r_{\mathbf{i}}$ for $\mathbf{i}$ maximal. Also we define $T_{\mathbf{i}} = \sum_{\mathbf{j} \leq \mathbf{i}} Ra^{\mathbf{j}}$ and $T_{\mathbf{i}}^- = \sum_{\mathbf{j} < \mathbf{i}} Ra^{\mathbf{j}}$. Given $L < T$ let $\text{lead}_{\mathbf{i}}(L) = \{r \in R : r$ is the leading coefficient of some element $\sum r_{\mathbf{i}} a^{\mathbf{i}}$ of $L \cap T^{\mathbf{i}}\}$. Note the "leading coefficient" is not well-defined on $T_{\mathbf{i}}$, but may vary according to the way that we write an element. However, it *is* well-defined modulo $\text{lead}_{\mathbf{i}}(0)$, and thus modulo $\text{lead}_{\mathbf{i}}(L)$ for any left ideal L.

Let us say a set $\{L(\mathbf{i}) : \mathbf{i} \in \mathbb{N}^{(t)}\}$ of left ideals is *weakly ascending* if $L_{(\mathbf{i})} \subseteq L_{(\mathbf{j})}$ whenever $i(u) \leq j(u)$ for all $1 \leq u \leq t$.

Remark 8.4.5: $\{\text{lead}_i(L): i \in \mathbb{N}^{(t)}\}$ is a weakly ascending set of left ideals of R. If R is left Noetherian then $\{\text{lead}_i(L): i \in \mathbb{N}^{(t)}\}$ is finite. (Indeed, the first assertion is immediate, and the second assertion follows at once from the following combinatorial fact:

Claim 8.4.6: An infinite sequence of $\mathbb{N}^{(t)}$ has an infinite subsequence which is increasing (or stationary) in each component. (It suffices to prove the claim for $t = 1$, for then we could refine the sequence to an infinite subsequence increasing in the last component and are done by induction. For $t = 1$ suppose $\{n_1, n_2, \ldots\}$ is a sequence in $\mathbb{N}$. Some n_u is smallest in the sequence; cutting off the first $(u - 1)$ terms we may assume n_1 is minimal. Now we choose n_v minimal among $n_2, n_3, \ldots$ and throwing away $n_2, \ldots, n_{v-1}$ we may assume $n_v = n_2$. Continuing this selection procedure indefinitely yields $n_1 \leq n_2 \leq n_3 \leq \cdots$.)

Lemma 8.4.7: *Suppose T is an almost normalizing extension of R. For $L < T$ let $\bar{}$ denote the canonical image in the R-module $M = T/L$. Then $\bar{T_i}/\bar{T_i^-} \approx R/\text{lead}_i(L)$.*

Proof: Define $T_i \to R/\text{lead}_i(L)$ by taking the leading coefficient of any element of T_i modulo $\text{lead}_i(L)$. Clearly $L \to 0$ so we have a map $\varphi: \bar{T_i} \to R/\text{lead}_i(L)$. Suppose $x = \sum_{j \leq i} r_j a^j \in T_i$. Then $\bar{x} \in \ker \varphi$ iff $\bar{r}_i = 0$; so that $\bar{x} \in \bar{T_i^-}$.

Definition 8.4.8: An algebra R over an integral domain C satisfies *vector generic flatness* if for any $k \in \mathbb{N}$ and for any weakly ascending set $\{L(i): i \in \mathbb{N}^{(k)}\}$ of left ideals of R one can find s in $C - \{0\}$ such that $(R/L(i))[s^{-1}]$ is a free $C[s^{-1}]$-module for all $\mathbf{i}$.

It turns out that vector generic flatness is satisfied by the rings in which we are interested but is a stronger condition than generic flatness. First note any integral domain C satisfies vector generic flatness over itself. Indeed, by claim 8.4.6 any weakly ascending set of ideals $\{L(\mathbf{i}): i \in \mathbb{N}^{(k)}\}$ has only finitely many minimal nonzero members, which we call $L_1, \ldots, L_m$; taking $0 \neq s \in \bigcap_{i=1}^m L_u$ yields $(C/L(\mathbf{i}))[s^{-1}] = 0$ whenever $L(\mathbf{i}) \neq 0$, and clearly $C[s^{-1}]$ is free as $C[s^{-1}]$-module. To see vector generic flatness implies generic flatness we need the following observation:

Remark 8.4.9: If $M = M_0 > \cdots > M_t = N$ is a chain of modules for which every factor module M_{i-1}/M_i is free then M/N is free. (Indeed, we may assume $N = 0$. Suppose B_i is a set of elements of M_{i-1} which modulo M_i is a base in M_{i-1}/M_i. Then $\bigcup B_i$ surely spans M, and we claim it is a base of M. For, otherwise, $\sum_{i,j} r_{ij} b_{ij} = 0$ for suitable b_{ij} in B_i; then $\sum_j r_{1j} b_{1j} \in M_1$ implying each $r_{1j} = 0$; continuing inductively, once we have proved $r_{uj} = 0$ for all $u < i$ we can conclude the $r_{ij} = 0$.)

Remark 8.4.10: Vector generic flatness implies generic flatness. (Indeed, suppose M is an f.g. R-module. Then there is a chain $M = M_0 > M_1 > \cdots > M_t = 0$ with each factor cyclic. But any cyclic module has the form R/L, so we can write each $M_{i-1}/M_i = R/L_i$ for suitable $L_i < R$. Picking s such that each $(R/L_i)[s^{-1}]$ is free over $C[s^{-1}]$, we see $M[s^{-1}] = M_0[s^{-1}] > M_1[s^{-1}] > \cdots > M_t[s^{-1}] = 0$ satisfies the hypothesis of remark 8.4.9, so $M[s^{-1}]$ is free, as desired.)

Having seen that vector generic flatness is merely an infinitistic version of generic flatness, we now feel confident in improving the results of §2.12.

Theorem 8.4.11: *Suppose $R \subseteq T$ are algebras over an integral domain C, and R satisfies vector generic flatness over C. Then T also satisfies vector generic flatness over C, given either of the additional hypothesis:*

   (i) *T is f.g. as R-module.*
   (ii) *T is an almost normalizing extension of R.*

Proof: Let $\{L(\mathbf{i}) : \mathbf{i} \in \mathbb{N}^{(k)}\}$ be a weakly ascending set of left ideals of T. Our strategy in either case is to build for each $T/L(\mathbf{i})$ a chain of cyclic modules in R, apply vector generic flatness to the corresponding factors, and conclude by means of remark 8.4.9.

   (i) Write $T = \sum_{i=1}^k Rx_i$ and for each $j \le k$ let $L_j(\mathbf{i})$ denote $\{r \in R : rx_j \in L(\mathbf{i}) + \sum_{u=1}^{j-1} Rx_u\}$. For each j the set $\{L_j(\mathbf{i}) : \mathbf{i} \in \mathbb{N}^{(k)}\}$ is weakly ascending, so there is $s(j)$ in C such that $(R/L_j(\mathbf{i}))[s(j)^{-1}]$ is free over $C[s(j)^{-1}]$. Taking $s = s(1) \cdots s(t)$ we see each $(R/L_j(\mathbf{i}))[s^{-1}]$ is free over $C[s^{-1}]$. But by definition right multiplication by x_j gives an isomorphism $R/L_j(\mathbf{i}) \approx (L(\mathbf{i}) + \sum_{u=1}^j Rx_u)/(L(\mathbf{i}) + \sum_{u=1}^{j-1} Rx_u)$, so we have a chain from T to $L(\mathbf{i})$ whose factors are $R/L_j(\mathbf{i})$, implying by remark 8.4.9 that each $(T/L(\mathbf{i}))[s^{-1}]$ is free.

   (ii) Let $a_1, \ldots, a_t$ be as in definition 8.4.1, and let $L(\mathbf{i}, \mathbf{j}) = \text{lead}_j(L(\mathbf{i})) \le R$. Then $\{L(\mathbf{i}, \mathbf{j}) : \mathbf{i} \in \mathbb{N}^{(k)}, \mathbf{j} \in \mathbb{N}^{(t)}\}$ can be viewed as a weakly ascending set of left ideals of R, indexed by $\mathbb{N}^{(k+t)}$, so there is $0 \ne s \in C$ such that each $(R/L(\mathbf{i}, \mathbf{j}))[s^{-1}]$ is free over $C[s^{-1}]$. But iterating lemma 8.4.7 shows there is a chain from T to $L(\mathbf{i})$ whose factors are isomorphic to the $R/L(\mathbf{i}, \mathbf{j})$; consequently $(T/L(\mathbf{i}))[s^{-1}]$ is free by remark 8.4.9.   **Q.E.D.**

Theorem 8.4.12: *Suppose $R \subseteq T$ are C-algebras, and T can be obtained from R by a finite series of extensions as in (i) or (ii) of theorem 8.4.11. Furthermore,* *assume* $R[\lambda_1, \lambda_2]/PR[\lambda_1, \lambda_2]$ *satisfies vector generic flatness over* $C[\lambda_1]/P$ *for all* P *in* $\mathrm{Spec}(C[\lambda_1])$. *If* C *is a Jacobson ring then* T *satisfies the weak Nullstellensatz and MN (the maximal Nullstellensatz, cf., definition 2.12.26).*

Proof: It is enough to show these vector generic flatness hypotheses lift to T, since then we are done by remark 8.4.10 and theorem 2.12.36. By iteration we may assume T satisfies hypothesis (i) or (ii) of theorem 8.4.11, which enables us to obtain the vector generic flatness. Q.E.D.

Corollary 8.4.13: *If* C *is a Jacobson ring and* T *is built from* C *by a finite series of extensions as in* (i) *or* (ii) *of theorem* 8.4.11, *then* T *is Jacobson and satisfies MN over* C.

Proof: Passing to prime homomorphic images of C, we may assume C is an integral domain. Taking $R = C$ we see T satisfies the weak Nullstellensatz and MN over C. It remains to show for every prime ideal P of T that $\mathrm{Nil}(T/P) = 0$ (for then $\mathrm{Jac}(T/P) = 0$ by the weak Nullstellensatz). Let $P_0 = C \cap P \in \mathrm{Spec}(C)$. Passing to T/P and C/P_0 we may assume $P = P_0 = 0$, so C is an integral domain. Let $S = C - \{0\}$. Then $S^{-1}C$ is a field so $S^{-1}T$ is Noetherian by theorem 8.4.4. But $S^{-1}\mathrm{Nil}(T)$ is a nil ideal and is thus 0, so $\mathrm{Nil}(T/P) = 0$ as desired. Q.E.D.

Corollary 8.4.14:

(i) *(Quillen-Duflo) Every enveloping algebra of a finite dimensional Lie algebra is Jacobson and satisfies the Nullstellensatz.*

(ii) *(Roseblade) Every group algebra of a polycyclic-by-finite group is Jacobson and satisfies the Nullstellensatz. (Note the base ring can be taken to be arbitrary Jacobson.)*

Proof:

(i) By example 8.4.2(ii).

(ii) Each cyclic group corresponds to an almost normalizing extension; each finite group factor corresponds to an f.g. extension. Q.E.D.

The generic flatness techniques have been seen to apply to almost normalizing extensions of commutative Jacobson rings and to affine PI-algebras (theorem 6.3.3). There is a nice theorem which unifies these results.

Theorem 8.4.15: *(McConnell [84a]) Suppose* R *is an affine PI-algebra over a commutative Jacobson ring* C. *Suppose* T *is built from* R *by a finite series of* *extensions as in (i) or (ii) of theorem 8.4.11, such that at each stage the new generators centralize R. Then T is Jacobson and satisfies MN.*

Proof: Any primitive ideal of T intersects R at a prime ideal by proposition 2.12.39 (since the generators centralize R), so we may assume R is prime PI of some PI-degree n. By theorem 6.1.33 if we take s in $g_n(R)$ then $R[s^{-1}]$ is f.g. free over $Z(R)[s^{-1}]$. $Z(R)[s^{-1}]$ is affine over C by proposition 6.3.1 and is certainly an almost normalizing extension of C; hence $R[s^{-1}]$ and thus $T[s^{-1}]$ satisfies the hypotheses of corollary 8.4.13. Therefore $T[s^{-1}]$ is Jacobson and satisfies MN.

To prove MN for T, suppose M is a simple T-module. We want to show $\text{End}_T M$ is algebraic. As observed in the first sentence we may assume M is R-faithful, and thus M is naturally a $T[s^{-1}]$-module by lemma 2.12.31. Obviously M is simple over $T[s^{-1}]$, so MN for $T[s^{-1}]$ implies $0 = \text{Ann}_C M$ is maximal in C. Therefore C is a field, and $\text{End}_T M \approx \text{End}_{T[s^{-1}]} M$ is algebraic over C, proving MN for T.

$T[\lambda]$ satisfies MN (using $T[\lambda]$ instead of T above), so the weak Nullstellensatz holds by lemma 2.12.35. It remains to prove $\text{Nil}(T/P) = 0$ for every P in $\text{Spec}(T)$. Passing to T/P we may assume T is prime. But then taking $S = Z(R) - \{0\}$ we see from the first paragraph that $S^{-1}T$ is obtained by a series of almost normalizing and f.g. extensions of the field $S^{-1}C$, so is Noetherian by theorem 8.4.4; hence $\text{Nil}(S^{-1}T) = 0$ implying $\text{Nil}(T) = 0$.
                                                                                                   Q.E.D.

The Primitive Spectrum

Having gotten past the Nullstellensatz, we can look for even finer information about the primitive spectrum; for example, is it locally closed in $\text{Spec}(R)$? This question can be formulated algebraically. Since P in $\text{Spec}(R)$ is the intersection of an open set and a closed set iff P is a G-ideal, we are in effect asking if every primitive ideal is a G-ideal. This is patently true if every primitive ideal is maximal but is much more difficult for enveloping algebras of non-nilpotent Lie algebras and can be false for polycyclic group algebras. Thus a very interesting challenge is to find exactly how ring theory can be brought in to bear on the matter. We shall discuss this in some detail, getting at once to the heart of the matter.

Definition 8.4.16: The *heart* of a prime ideal P, written $\text{Heart}(P)$, is the extended centroid of R/P.

Usually the heart is defined only for Noetherian rings, as the center of the classical ring of fractions Q of R/P. Definition 8.4.16 is a straightforward generalization and also has the following immediate application.

Remark 8.4.17: Suppose P is a primitive ideal. Writing $P = \text{Ann}_R M$ for a simple module M and $D = \text{End}_R M$, we have a canonical injection $\varphi: \text{Heart}(P) \to Z(D)$ given in example 3.4.16.

We are now ready to examine the following possible properties for a prime ideal P in an F-algebra R;

(1) P is primitive.
(2) Heart(P) is algebraic over F.
(3) P is a G-ideal.

These properties are called *Dixmier's conditions.* Dixmier's condition (2) is less intuitive than (1) and (3), but could be thought of as follows: If F is algebraically closed and (2) holds then Heart(P) = F; thus R/P is centrally closed with center F. Thus in utilizing (2) we should expect to rely on the theory of centrally closed algebras developed in §3.4, and this will indeed be the case.

Proposition 8.4.18: *If P is semiprimitive then* (3) $\Rightarrow$ (1). *If R satisfies the Nullstellensatz then* (1) $\Rightarrow$ (2).

Proof: (3) $\Rightarrow$ (1). P is an intersection of primitive ideals, one of which must be P, by (3). (1) $\Rightarrow$ (2) by remark 8.4.17.   Q.E.D.

Thus (3)$\Rightarrow$(1)$\Rightarrow$(2) whenever R is Jacobson and satisfies the Nullstellensatz. To complete the chain of implications we would like to show (2) $\Rightarrow$ (3). This is a rather touchy issue; fortunately, there is a weakening of (3) which is more amenable.

Definition 8.4.19: We say a set $S \subset R - \{0\}$ *separates* Spec(R) if every non-zero prime ideal contains an element of S. We say $P \in R$ *satisfies property* (3′) if there is a countable set separating Spec(R/P).

Obviously, any G-ideal satisfies property (3′) so we have (3) $\Rightarrow$ (3′); the reason we weaken (3) is the following result of Irving. (The proof is modified according to an idea of M. Smith.)

Proposition 8.4.20: *If R is an F-algebra with a countable base B then* (2) $\Rightarrow$ (3′).

Proof: Writing $B = \{b_i : i \in \mathbb{N}\}$ and $b_i b_j = \sum_k \alpha_{ijk} b_k$ with α_{ijk} in F, let F_0 be the subfield of F generated by the α_{ijk}, and let $R_0 = F_0 B$. Then R_0 is a countable subalgebra of R. We shall prove R_0 itself separates Spec($\hat{R}$) where $\hat{R}_0, \hat{R}$ are the respective central closures of R_0, R. Note every ideal of $\hat{R}$ intersects $\hat{R}_0$ nontrivially by theorem 3.4.11, and obviously every ideal of $\hat{R}_0$ intersects

R_0 nontrivially since $\hat{R}_0$ is an essential extension of R_0. Hence R_0 separates $\text{Spec}(\hat{R})$.

To prove the proposition we may assume $P = 0$. Then (2) shows $Z(\hat{R})$ is algebraic over F. Hence LO holds from R to $RZ(\hat{R}) = \hat{R}$, by theorem 2.12.48 and proposition 2.12.49. We conclude R_0 separates $\text{Spec}(R)$.   Q.E.D.

Thus far we have proved $(3) \Rightarrow (1) \Rightarrow (2) \Rightarrow (3')$ under quite general circumstances and should note that $(3') \Rightarrow (3)$ can fail, also, cf., Lorenz [77].) Thus we led first to examine $(3') \Rightarrow (1)$, to try to complete a general chain of equivalence. Note that we should require F uncountable, since for F countable $F[\lambda]$ is countable (so that $(3')$ is trivial for $R = F[\lambda]$), whereas $F[\lambda]$ is not primitive. When F is uncountable there is a host of examples of $(3') \Rightarrow (1)$, including enveloping algebras (Dixmier [77]), polycyclic-by-finite group algebras (Farkas [79]), and related constructions (Irving [79]). Unfortunately, there is not yet a sweeping result stating that $(3') \Rightarrow (1)$ for Noetherian algebras over an uncountable field. To stimulate such a result let us pose

Conjecture 8.4.21:   *Suppose R is an affine algebra over an uncountable field F. Then $(3') \Rightarrow (1)$ for any prime, semiprimitive ideal P.*

In the absence of an answer to this conjecture, let us consider a general strategy mapped out by Farkas, which in fact lies behind all the known positive results.

Call a ring R *Kaplansky* if the primitive spectrum of each prime homomorphic image of R is a Baire space. (Recall a topological space is Baire if any countable insersection of dense open sets is dense.)

Proposition 8.2.22:   *If R is Jacobson and Kaplansky over an uncountable field then $(3') \Rightarrow (1)$.*

Proof: Let $P \in \text{Spec}(R)$ such that $\text{Spec}(R/P)$ has a countable separating set S disjoint from P. Passing to R/P we may assume $P = 0$, and $0 \notin S$. Given any s in S let $U_s = \{$primitive ideals not containing $s\}$ a dense subset of $\text{Spec}(R)$ by corollary 2.12.5'. By hypothesis there is some $P' \in \bigcap_{s \in S} U_s$, and since each $s \notin P'$ by construction we see $0 = P'$ is primitive.   Q.E.D.

Thus we are reduced to showing that the rings of interest to us are Kaplansky. Extending Farkas [79], Lorenz-Passman [79] proves that polycyclic-by-finite group algebras are Kaplansky. Irving [79] gives many other instances. Also, cf., Dixmier [77]. In summary we have

Fact 8.4.23: Over an algebraically closed field of characteristic 0, (2) ⇒ (1) for enveloping algebras and polycyclic-by-finite group algebras.

Let us return now to the question of when (1), (2), and (3) are equivalent. In view of proposition 8.4.18 the "missing link" is (2) ⇒ (3), which by Lorenz [77] fails for various polycyclic group algebras. (2) ⇒ (3) does hold for any group algebra of finitely generated nilpotent by-finite groups, for which, in fact, primitive ideals are maximal.

We shall focus here on the story for enveloping algebras, which involves an interesting blend of Lie techniques with pure ring theory. Assume char(F) = 0. We just observed (2) ⇒ (1) when F is algebraically closed; more recently Moeglin [80a] proved (1) ⇒ (3) when F is uncountable and algebraically closed, thereby yielding (2) ⇒ (3) in this case. Using these results we can apply some purely ring-theoretic methods of Irving-Small [80] to remove these extra restrictions from F.

Reduction Techniques

Proposition 8.4.24: *Suppose K is a separable field extension of F, R is an F-algebra, and $R' = R \otimes_F K$ satisfies* ACC(ideals).

(i) *If B is a semiprime ideal of R then BR' is a semiprime ideal of R' which can be written as $P'_1 \cap \cdots \cap P'_t$ for suitable t, where each P'_j is a minimal prime over BR'.*

(ii) *If, furthermore, $B \in \operatorname{Spec}(R)$ then $B = P'_j \cap R$ for some j; if B is primitive and $[K:F] < \infty$ then P'_j is primitive.*

Proof:

(i) Passing to R'/BR' we may assume $B = 0$, i.e., R is semiprime. Then R' is semiprime by theorem 2.12.52 and has a finite number of minimal prime ideals $P'_1, \ldots, P'_t$ by corollary 3.2.26.

(ii) Continuing the proof of (i) we may assume R is prime and

$$(P'_1 \cap R) \cdots (P'_t \cap R) = 0,$$

implying some $P'_j \cap R = 0 = B$. The last assertion is corollary 3.4.14.
Q.E.D.

Note: The separability assumption was needed only in showing R' is semiprime, but, anyway, this is satisfied when $\text{char}(F) = 0$.

We can improve proposition 8.4.24(ii) if we bring in the Nullstellensatz (cf., definition 2.12.26).

Proposition 8.4.24′: *Let $\bar{F}$ be the algebraic closure of a field F of characteristic 0. Suppose R is an F-algebra, $R \otimes_F \bar{F}$ satisfies the Nullstellensatz and is left Noetherian, and K is a field extension of F such that $R' = R \otimes_F K$ is left Noetherian. Let M be any simple R-module. Then*

(i) $[\text{End}_R M : F] < \infty$,
(ii) $M' = M \otimes_F K$ *is cyclic and completely reducible as R'-module,*
(iii) *If R' is prime and R is primitive then R' is primitive.*

Proof: First we prove (ii) for $K = \bar{F}$. View $M = M \otimes 1 \subset M'$. If $M = Rx$ then $M' = R'x$, so M' is cyclic and thus Noetherian. Suppose N' is a submodule of M'. Then N' is f.g. and thus is contained in $M \otimes F'$ for some finitely generated (and thus finite dimensional) field extension F' of F. But $M \otimes F'$ is completely reducible (proved in analogy to corollary 3.4.14), and thus contains a complement N'' of N', and N'' tensors up to a complement of N' in M', thereby proving M' is completely reducible.

To prove (i), now take a simple submodule N' of M' (still taking $K = \bar{F}$). As above we view N' in $M \otimes F'$; passing to the normal closure of F' we may assume F' is Galois over F. Let $G = \text{Gal}(F'/F)$. The group $\tilde{G} = \{1 \otimes \sigma : \sigma \in G\}$ acts naturally on $M \otimes F'$; $\sum (1 \otimes \sigma)N'$ is a submodule of $M \otimes F'$ fixed under $\tilde{G}$ and thus equal to $M \otimes F'$ (since M is simple). Hence M' is a finite sum of copies of $N' \otimes \bar{F}$, implying $\text{End}_{R'} M'$ is finite dimensional over $\bar{F}$. But M' is cyclic over R' so $\text{End}_{R'} M' \approx (\text{End}_R M) \otimes \bar{F}$, proving (i).

To prove (ii) for arbitrary K, let $D = \text{End}_R M$ and note

$$M' \approx (M \otimes_D D) \otimes_F K \approx M \otimes_D (D \otimes K).$$

One proves in analogy to theorem 3.4.11 that each submodule of M' has the form $M \otimes L$ for some left ideal L of $D \otimes K$. But $D \otimes K$ is semisimple Artinian by (i), so M' is a finite direct sum of simple submodules $M'_1, \ldots, M'_t$ proving (ii). Now let $P'_i = \text{Ann } M'_i$. $P'_1 \cdots P'_t = 0$, so some $P'_i = 0$ if R' is prime, proving (iii).   Q.E.D.

Proposition 8.4.25: *If $R' = R \otimes_F K$ is primitive and left Noetherian then $R \otimes_F K_1$ is primitive for some finitely generated field extension K_1 of F.*

Proof: Take a maximal left ideal L of R' having core 0. If $L = \sum_{i=1}^{t} R' a_i$ then writing $a_i = \sum_{j=1}^{t(i)} r_{ij} \otimes a_{ij}$ for a_{ij} in K let K_1 be the subfield of K generated by the a_{ij}. R' is a central extension of $R \otimes K_1$. Let $L_0 = L \cap (R \otimes K_1)$, clearly a proper left ideal of $R \otimes K_1$ with core 0 (for if $A = \text{core}(L_0)$ then $L \supseteq AK \lhd R'$). L_0 is also maximal (for if $L_1 \supset L_0$ then $R'L_1 = R'$ implying $L_1 = R \otimes K_1$). Hence $R \otimes K_1$ is primitive.   Q.E.D.

Lemma 8.4.26: *Suppose R is an algebra over a field F, such that $R \otimes_F K$ is semiprime, left Noetherian for every algebraic extension K of F. Let Q be the classical ring of fractions of R. If $Z(Q)$ is algebraic over F then there is a finite dimensional field extension K_1 of F such that $Q \otimes_F K_1$ is split, i.e., a direct product of matrix rings over K_1.*

Proof: We claim the ring of fractions of the semiprime Goldie ring $R \otimes K$ is $Q \otimes K$. This is clear for K finite over F for then $Q \otimes K$ is semisimple Artinian; the claim is established by passing to direct limits.

Now take K to be the algebraic closure of F. By corollary 1.7.24 $Z(Q) \otimes K = Z(Q \otimes K)$, a direct product of fields since $Q \otimes K$ is semisimple Artinian. But by assumption $Z(Q)$ is algebraic over F, implying $Z(Q) \otimes K$ is a finite direct product of algebraic extensions of K and thus a direct product of copies of K itself. Writing out the idempotents explicitly one obtains these in a suitable finite dimensional extension of F, and this is the desired K_1. Counting dimensions shows that if $Z(Q) \otimes K \approx K^{(n)}$ then $Z(Q) \otimes K_1 \approx K_1^{(n)}$. Thus $Q \otimes K_1$ is split.   Q.E.D.

We also need some results concerning G-ideals. We say a prime ring R is *G-prime* if 0 is a G-ideal.

Lemma 8.4.26′:

(i) *If a G-prime ring T is a direct limit of finite centralizing extensions of R then R is also G-prime.*

(ii) *If $R \otimes_F K$ is G-prime for a field extension K of F then R is also G-prime.*

Proof:

(i) Let $A = \bigcap \{\text{nonzero primes of } T\}$. Then $R \cap A \neq 0$ by the theorem 3.4.13 and is contained in every nonzero prime P of R by LO. (There is some P' lying over P by theorem 2.12.48 and proposition 2.12.49, and $R \cap A \subset R \cap P' = P$.)

(ii) K is algebraic over a purely transcendental extension K_0 of R, and $R \otimes K_0$ is G-prime by (i). Write $K_0 = F(\Lambda)$ for a suitable set of commuting indeterminates Λ over F; then $R \otimes K_0 \approx R(\Lambda)$, and some $\sum r_i \lambda_1^{i(1)} \cdots \lambda_t^{i(t)}$ is in every nonzero prime ideal of $R(\Lambda)$. But if $0 \neq P \in \operatorname{Spec}(R)$ then $P(\Lambda) \in \operatorname{Spec} R(\Lambda)$ since $R(\Lambda)/P(\Lambda) \approx (R/P)(\Lambda)$. Hence $\sum r_i \lambda_1^{i(1)} \cdots \lambda_t^{i(t)} \in P(\Lambda)$ implying each $r_i \in P$, i.e., each $r_i \in \bigcap \{\text{nonzero prime ideals of } R\}$, as desired.

$$\text{Q.E.D.}$$

Theorem 8.4.27: *(Irving-Small [80]) Assume R is a countable dimensional F-algebra and $R \otimes_F K$ is left Noetherian and Jacobson satisfying the Nullstellensatz for every field extension K of F. Furthermore, assume $\operatorname{char}(F) = 0$, and $(2) \Rightarrow (1)$ in $R \otimes \bar{F}$ where $\bar{F}$ is the algebraic closure of F, and $(2) \Rightarrow (3)$ in $R \otimes \bar{K}$ for some uncountable algebraically closed field $\bar{K} \supseteq F$. Then Dixmier's conditions (1), (2), and (3) are equivalent in R.*

Proof: We shall make use of two lemmas, assuming the hypothesis of the first sentence.

Lemma 8.4.28: *If $(1) \Rightarrow (3)$ in $R \otimes_F K$ for some separable (not necessarily algebraic) field extension K of F then $(1) \Rightarrow (3)$ in R.*

Proof of Lemma 8.4.28: Let P be a primitive ideal of R, and let $\bar{R} = R/P$ and Q be the ring of fractions of $\bar{R}$. Then $Z(Q)$ is algebraic over F by proposition 8.4.18, so by lemma 8.4.26 we can split Q by a finite dimensional field extension K_1 of F. By proposition 8.4.24 $R \otimes K_1$ contains a primitive ideal P' lying over P. Cleary $\operatorname{heart}(P') = K_1$; in view of lemma 8.4.26'(ii) we may replace R by $R \otimes K_1$, and thereby assume $Z(Q) = F$.

Now $\bar{R} \otimes_F K$ is prime and thus primitive by proposition 8.4.24', so by hypothesis is G-prime; hence $\bar{R}$ is G-prime by lemma 8.4.26'(ii).   Q.E.D.

Lemma 8.4.29: *If $(2) \Rightarrow (1)$ in $R \otimes_F K$ where K is the separable algebraic closure of F then $(2) \Rightarrow (1)$ in R.*

Proof of Lemma 8.4.29: Suppose $P \in \operatorname{Spec}(R)$ and $\bar{R} = R/P$. By proposition 8.4.24 we see $\bar{R} \otimes_F K$ is semiprime and satisfies ACC(ideals) (since it is an image of $R \otimes_F K$). Let Q be the ring of fractions of $\bar{R}$. By lemma 8.4.26 we see $Q \otimes K_1$ is split for some field K_1 finite over F. By proposition 8.4.24 there is a minimal prime ideal P' of $\bar{R} \otimes K_1$ with $P' \cap \bar{R} = (0)$, so the ring of fractions of $(\bar{R} \otimes K_1)/P'$ is one of the (split) components of Q. This implies $K_1 = \operatorname{heart}(P')$ so, by hypothesis, P' is primitive. Hence $\bar{R}$ is primitive by corollary 2.5.30.   Q.E.D.

Proof of Theorem 8.4.27: By hypothesis we have $(2) \Rightarrow (3)$ in $R \otimes \bar{K}$, and $(1) \Rightarrow (2)$ by proposition 8.4.18, so $(1) \Rightarrow (3)$ holds in R by lemma 8.4.28. $(3) \Rightarrow (2)$ by proposition 8.4.18, and $(2) \Rightarrow (1)$ by lemma 8.4.29.   Q.E.D.

Corollary 8.4.30: (*modulo Moeglin's theorem*). *Dixmier's conditions are equivalent for any enveloping algebra of a finite dimensional Lie algebra over a field of characteristic* 0.

Proof: The hypotheses of theorem 8.4.27 are satisfied by fact 8.4.23 and Moeglin's theorem.   Q.E.D.

An Additivity Principle for Goldie Rank

In §6.3 we discussed the Bergman-Small "additivity principle" which expresses the PI-degree of R/P as a linear combination of PI-degrees of prime homomorphic images of R. Bergman-Small [75] also discuss the analogous result concerning the PI-degree of a prime subring W of R, and the question arose in the study of Jantzen's conjecture for enveloping algebras, whether a suitable non-PI version of this easier Bergman-Small theorem is available. Although we do not have the PI-degree to work with, we could content ourselves with the *Goldie rank* of a prime ring R, which we recall is that n for which the classical ring of fractions $Q(R) \approx M_n(D)$ for a suitable division ring D. (This is also the uniform dimension of R.) Joseph-Small [78] indeed proved such an additivity principle for enveloping algebras by means of GK dimension, and Borho [82] found a general axiomatic approach. We reproduce an elegant treatment of Warfield [83] which penetrates quickly to the main result, with minimal requirements on the rings.

Warfield's approach uses bimodules, a subject that we have skipped so far in this abridged edition. In particular we need the following result:

Proposition 3.5.77. *Suppose M is an $R - T$ bimodule that is f.g. as right module over T, and R is prime Goldie. M is torsion iff $\operatorname{Ann}_R M \neq 0$.*

Proof: $(\Rightarrow)$ $\operatorname{Ann}_R M$ is an ideal so has a regular element.
  $(\Leftarrow)$ Write $M = \Sigma_{\text{finite}} x_i T$. Each $\operatorname{Ann} x_i$ contains a regular element and thus is large in R, so $\operatorname{Ann} M = \cap \operatorname{Ann} x_i$ is large.   Q.E.D.

Let us carry the following notation: R is prime left Goldie, and $Q(\ )$ denotes the classical ring of fractions. Write *rank* for the Goldie rank. Thus $Q(R) \approx M_n(D)$ for a suitable division ring D and for $n = \operatorname{rank}(R)$.

Remark 8.4.31: If R contains a simple Artinian ring of rank t then $t \mid \operatorname{rank}(R)$. (Indeed, we have a set of $t \times t$ matric units so proposition 1.1.3

shows $R \approx M_t(R')$ for a suitable subring R' of R; then $M_n(D) \approx Q(R) \approx Q(M_t(R')) \approx M_t(Q(R')$ so rank$(R) = t \cdot$ rank(R').)

Theorem 8.4.32: (*Warfield*) *Suppose* $W \subseteq R$ *and all prime images of* W *are left Goldie. Let* $Q = Q(R)$, *and let* $Q = M_0 > M_1 > \cdots > M_t = 0$ *be a composition series of* Q *as* $W - Q$ *bimodule, and let* $P_i = \text{Ann}_w M_i/M_{i+1}$ *and* $\mathscr{P} = \{P_i : 0 \le i < t\}$. *Then* $\mathscr{P} \subseteq \text{Spec}(W)$, *every minimal prime of* W *belongs to* $\mathscr{P}$, *and there are positive integers* m_P *such that*

$$\text{rank}(R) = \sum_{P \in \mathscr{P}} m_P \text{rank}(W/P)$$

Moreover, each M_i/M_{i+1} *is torsion-free as* W/P_i-*module.*

Proof: First note each P_i is prime, for if $AB \subseteq P_i$ with $A, B \supset P_i$ then $M_i \supset M_{i+1} + BM_i \supset M_{i+1}$, contrary to hypothesis. Also $P_t \cdots P_1 \subseteq \text{Ann } Q = 0$ so each minimal prime ideal of W belongs to $\mathscr{P}$. It remains to prove the last assertion. Let $W_i = R/P_i$ and $\bar{M}_i = M_i/M_{i+1}$ a faithful W_i-module. $\bar{M}_i$ is simple as a bimodule and is torsion-free as W_i-module by proposition 3.5.77. Left multiplication by any regular element of W_i thus yields a bijection from $\bar{M}_i$ to itself, and thus $Q(W_i)$ is contained in the endomorphism ring E_i of $\bar{M}_i$ over Q. Hence remark 8.4.31 yields m_i such that m_i rank$(Q(W_i)) = $ rank(E_i); since rank $W_i = $ rank $Q(W_i)$ we conclude

$$\sum m_i \text{rank}(W_i) = \sum \text{rank}(E_i) = \text{rank}(Q) = \text{rank}(R)$$

as desired.   Q.E.D.

The natural question arises to when $\mathscr{P}$ consists *only* of the minimal primes of W. Equivalently, we want to show the primes in $\mathscr{P}$ are incomparable. To answer this we require a new notion.

Definition 8.4.32': R is a *restricted extension* of a subring W if WrW is f.g. both as left W-module and right W-module.

Example 8.4.33:   Let $L_1 \subseteq L$ be finite dimensional Lie algebras. Then $U(L)$ is a restricted extension of $U(L_1)$. Indeed, clearly $L_1 U_n(L) = U_n(L)L_1$ where $U_n(L)$ is the component under the standard filtration. Hence $U(L_1)U_n(L) = U_n(L)U(L_1)$ is f.g. and thus Noetherian as left and right $U(L_1)$-module. But any r in $U(L)$ lies in some $U_n(L)$ and thus $U(L_1)rU(L_1) \subseteq U(L_1)U_n(L)$ is f.g., as desired.

The same argument shows that any Ore extension of a Noetherian ring is restricted. On the other hand, Warfield [83] also gives some easy examples of extensions of Noetherian rings which are not restricted. In the next result "Goldie" means left and right Goldie.

Proposition 8.4.34: *Suppose R is prime Goldie and is a restricted extension of W, and all prime images of W are Goldie. Then each P_i and P_j of theorem 8.4.32 are bonded primes, i.e., there is a $W/P_i - W/P_j$ bimodule that is f.g. torsion-free on both sides.*

Proof: Since $Q = Q(R)$ is simple Artinian each M_{i+1} has a complement I_i in M_i as right Q-module, i.e., $I_i \oplus M_{i+1} = M_i$. Since $I_i \leq Q$ we have $I_i = e_i Q$ for suitable idempotents e_i, and $\{e_0, \ldots, e_{t-1}\}$ are orthogonal idempotents of Q with $\sum e_i = 1$. $R \cap Qe_i$ and $R \cap e_i Q$ are, respectively, nonzero left and right ideals of R, since Q is left and right essential over R; since R is prime we have $0 \neq (R \cap e_i Q)(R \cap Qe_j) \subseteq R \cap e_i Qe_j$ for all $1 \leq i, j \leq t$; select $0 \neq r_{ij}$ in $R \cap e_i Qe_j$. Also note for $j < i$ that $e_j M_i \subseteq e_j M_{j+1} = 0$.

We now use the fact M_i is a W-module to establish the key fact:

$$e_j W e_i = 0 \qquad \text{for any } j < i \qquad \text{since } e_j W e_i Q \subseteq e_j W M_i \subseteq e_j M_i = 0.$$

Letting $E = \bigoplus_{j \geq i} e_j Qe_i$ we have $W \subseteq E$ by applying the Pierce decomposition $Q = \bigoplus_{i,j=1}^{t} e_i Qe_j$ to (1).

Note each $e_i Qe_i$ is simple Artinian, by lemma 2.7.12(iv). Let

$$J = \sum_{j > i} e_j Qe_i \lhd E.$$

Then $J^t = 0$ and $E/J \approx \sum_i e_i Qe_i$ is semisimple Artinian. Fixing i let $P_i' = J + \sum_{u \neq i} e_u Qe_u$. Clearly $E/P_i' \simeq e_i Qe_i$ so P_i' is a maximal ideal of E.

$M_i = \sum_{u=i}^{t-1} e_u Q$ and is thus naturally an $E - E$ bimodule, and $P_i' M_i \subseteq \sum_{u=i+1}^{t-1} e_u Q = M_{i+1}$. Hence $\text{Ann}_E(M_i/M_{i+1}) = P_i'$ since P_i' is maximal so

$$W \cap P_i' = W \cap \text{Ann}_E(M_i/M_{i+1}) = \text{Ann}_W(M_i/M_{i+1}) = P_i,$$

and we have $W/P_i \hookrightarrow E/P_i'$. But M/M_{i+1} is torsion-free over W/P_i by theorem 8.4.32, so each regular element of W/P_i remains regular in the simple Artinian ring E/P_i'. Hence we can form the simple Artinian ring of fractions Q_i of W/P_i inside E/P_i'.

Now define the $E - E$ bimodule $L_i = \sum_{u=1}^{i} Qe_u$; let $N_{ij} = M_{i+1} + L_{j-1}$, $N_{ij}' = N_{ij} + W r_{ij} W$, and $\bar{N}_{ij} = N_{ij}'/N_{ij}$, which are $W - W$ bimodules since $W \subseteq E$. Furthermore, $\bar{N}_{ij}$ is an image of $W r_{ij} W$ so (by restrictedness) is f.g. both as left and right module. To conclude we shall display $\bar{N}_{ij}$ as a submodule of an $E - E$ bimodule annihilated, respectively, on the left and right by P_i' and P_j'; then $\bar{N}_{ij}$ is a $Q_i - Q_j$ bimodule and thus a torsion-free $W/P_i - W/P_j$ bimodule.

Let $N_{ij}'' = N_{ij} + e_i Qe_j$, clearly an $E - E$ bimodule containing N_{ij}'; we have

$$P'_i e_i Q e_j \subseteq P'_i e_i Q \subseteq \sum_{u=i+1}^{t-1} e_u Q = M_{i+1}$$

$$e_i Q e_j P'_j \subseteq Q e_j P'_j \subseteq \sum_{u=1}^{j-1} Q e_u = L_{j-1}.$$

Hence N''_{ij}/N_{ij} is annihilated respectively on the left and right by P'_i and P'_j, as desired.   Q.E.D.

The conclusion of proposition 8.4.34 raises the following

Question 8.4.35:   Are bonded prime ideals of a given Noetherian ring W necessarily incomparable?

A positive answer would show that $\mathscr{P}$ (of theorem 8.4.32) is precisely the set of minimal primes of W whenever R is a restricted extension of W. In particular, this is the case when GK dim$(W) < \infty$, as we shall see presently.

Proposition 8.4.36:   *(Borho, Lenagan) Suppose M is an $R - T$ bimodule, f.g. as R-module, where R, T are affine algebras over a field F. Then GK-dim$_R M \geq$ GK-dim$_T M$, the right-handed GK-dimension of M as right T-module.*

Proof: Choose finite dimensional F-subspaces M_0, T_0 of M, T, respectively. Since M is f.g. it has an f.d. subspace $M'_0 \supseteq M_0$ such that $RM'_0 = M \supseteq M'_0 T_0$. Thus R has an f.d. subspace R_0 such that $R_0 M'_0 \supseteq M'_0 T_0$ (since $M'_0 T_0$ is f.d.), implying $R_0^n M'_0 \supseteq M'_0 T_0^n$ for all n. Taking limits yields the desired result.   Q.E.D.

Corollary 8.4.36' (GK Symmetry):   *If M is an $R - R$ bimodule f.g. on each side and GK-dim $R < \infty$ then the left and right GK module dimensions of M are equal.*

Corollary 8.4.37:   *Question 8.4.35 has an affirmative answer for all W of finite GK dim. Hence, for any prime restricted extension R of such W, $\mathscr{P}$ is the set of minimal primes of W.*

Proof: Write ℓ-GK and r-GK for the left and right GK module dimensions, respectively. If M is an f.g. torsion-free $W/P_1 - W/P_2$ bimodule then GK dim $R/P_2 = r$-GK $M = \ell$-GK $M =$ GK dim R/P_1, so P_1 and P_2 are incomparable by theorem 6.2.25.   Q.E.D.

When GK dim $R = \infty$ we would hope to make do with K-dim. Question 8.4.35 has a positive answer for almost fully bounded rings, seen via Jategaonkar's theory (*J* supplement in §3.5), and this yields an additivity principle for group algebras of polycyclic-by-finite groups. However, a general attempt to answer question 8.4.35 using K-dim runs up against the hoary Krull-symmetry problem. (For solvable Lie algebras P. Polo [86] has proved Krull symmetry for f.g. $U(L)$-bimodules.)

R Supplement: Hopf Algebras

The material we shall now present is of a completely different flavor than the rest of this section and does not have any obvious connection to Noetherian ring theory. Nevertheless, it is of considerable interest, providing a unified view of group algebras and enveloping algebras and has many important applications. The standard reference is Sweedler [69B], which we shall to refer to as SW.

The underlying idea is to dualize the operations of an algebra. For convenience we work in the category of algebras over a field F. Multiplication in an associative algebra A defines the map $\mu: A \otimes A \to A$ in which the following diagram is commutative (all tensors over F):

$$\begin{array}{ccc} A \otimes A \otimes A & \xrightarrow{\mu \otimes 1} & A \otimes A \\ \downarrow{\scriptstyle 1 \otimes \mu} & & \downarrow{\scriptstyle \mu} \\ A \otimes A & \xrightarrow{\mu} & A \end{array}$$

Furthermore, the canonical injection $u: F \to A$ (given by $\alpha \mapsto \alpha \cdot 1$) satisfies the commutative diagram

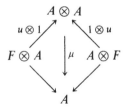

where the bottom arrows are the natural isomorphisms. Actually, we saw in exercise 1.7.4 that these properties characterized an F-vector space A being an associative algebra. Dualizing yields the following definition:

Definition 8.4.51: A *coalgebra* is a triple (A, Δ, ε) where A is a vector space (over F), $\Delta: A \to A \otimes A$ is called the *diagonalization* (or *comultiplication*) and $\varepsilon: A \to F$ is called the *augmentation* (or *counit*), satisfying the following commutative diagrams:

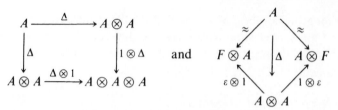

Let as adopt Sweedler's shorthand $\Delta a = \sum a_{(1)} \otimes a_{(2)}$ for a in A. We thus have $\sum \Delta a_{(1)} \otimes a_{(2)} = \sum a_{(1)} \otimes \Delta a_{(2)}$ and $a = \sum (\varepsilon a_{(1)}) a_{(2)} = \sum (\varepsilon a_{(2)}) a_1$. An algebra A is a *bialgebra* if A is also a coalgebra such that Δ and ε are algebra homomorphisms. (This is equivalent to saying μ and u are coalgebra morphisms, as is seen by writing these conditions diagrammatically.)

Example 8.4.52: If S is a monoid then $F[S]$ is a coalgebra under the maps given by $\Delta s = s \otimes s$ and $\varepsilon s = 1$ for all s in S. (Note that ε corresponds to the usual augmentation map.)

In what follows A^* *always* denotes $\text{Hom}(A, F)$; given $\varphi: A \to B$ we let $\varphi^{\#}$ denote the canonical map $B^* \to A^*$ given by $\varphi^{\#} f = f\varphi$, cf., example 0.1.8. We shall often rely on the natural identification $(A^*)^* \approx A$ when $[A:F] < \infty$.

Remark 8.4.53: If A is a coalgebra then A^* is an algebra where multiplication is the composite $A^* \otimes A^* \to (A \otimes A)^* \overset{\Delta^{\#}}{\to} A^*$ and $u = \varepsilon^{\#}$. This relies on the fact there is a monic $\rho: A^* \otimes A^* \to (A \otimes A)^*$ sending $f \otimes g$ to the map $h: A \otimes A \to F$ defined by $h(a_1 \otimes a_2) = fa_1 ga_2$. In case A is finite dimensional over F, we see by counting dimensions that ρ is bijective; reversing the arrows we see that if A is an algebra then A^* is a coalgebra, and if A is a bialgebra then A^* is a bialgebra.

Definition 8.4.54: An *antipode* (if it exists) of a bialgebra A is a map $S: A \to A$ such that $\mu(S \otimes 1_A)\Delta = u\varepsilon = \mu(1_A \otimes S)\Delta$, i.e., $\sum (Sa_{(1)})a_{(2)} = u\varepsilon a = \sum a_{(1)} Sa_{(2)}$ for all a in A. A *Hopf algebra* is a bialgebra with antipode.

Example 8.4.55: If G is a group the involution of $F[G]$ given by $g \to g^{-1}$ is an antipode, so $F[G]$ is a Hopf algebra.

If L is a Lie algebra then $U(L)$ is a Hopf algebra, where $\Delta a = a \otimes 1 + 1 \otimes a$ for a in L; ε is given by $\varepsilon L = 0$, and S is the principal involution.

Example 8.4.55': In §8.3 we briefly described the theory of algebraic Lie algebras of linear algebraic groups. The general theory can be described concisely in terms of Hopf algebras, as was shown by Hochschild [71B]; more full details are to be found in Abe [77B]. Let G be any group and write $M_F(G)$ for the F-algebra of functions from G to F. G acts on $M_F(G)$ by translations, i.e., for x in G and f in $M_F(G)$ we define xf by $(xf)g = f(gx^{-1})$ for g in G. Define $\mathscr{R}_F(G) = \{f \in M_F(G): Gf$ spans a finite dimensional vector space over $F\}$.

We have a homomorphism $\sigma: M_F(G) \to M_F(G \times G)$ given by $\sigma f(g_1, g_2) = f(g_1 g_2)$, and we have a monic $\pi: M_F(G) \otimes M_F(G) \to M_F(G \times G)$ given by $\pi(f_1 \otimes f_2)(g_1, g_2) = f_1 g_1 f_2 g_2$. Then $f \in \mathscr{R}_F(G)$ iff $\sigma f \in \pi(M_F(G) \otimes (M_F(G))$. Now we define the following Hopf algebra structure on $\mathscr{R}_F(G)$:

comultiplication $\Delta = \pi^{-1}\sigma: \mathscr{R}_F(G) \to \mathscr{R}_F(G) \otimes \mathscr{R}_F(G)$;
counit $\varepsilon: \mathscr{R}_F \to F$ given by $\varepsilon f = fe$ where e is the neutral element of G;
antipode S given by $(Sf)g = fg^{-1}$.

Given any Hopf subalgebra A of $\mathscr{R}_F(G)$ we let A^0 denote the dual space $A^* = \operatorname{Hom}(A, F)$ endowed with the convolution product (exercise 29). Call $h: A \to A$ *proper* if h commutes with each translation by G or equivalently the diagram

$$
\begin{array}{ccc}
A & \xrightarrow{h} & A \\
\downarrow{\scriptstyle\Delta} & & \downarrow{\scriptstyle\Delta} \\
A \otimes A & \xrightarrow{1 \otimes h} & A \otimes A
\end{array}
$$

is commutative. Then {proper endomorphisms of A} is a subalgebra of $\operatorname{End}_F A$, and there is an algebra isomorphism $\Psi:$ {proper endomorphisms of $A\} \to A^0$ given by $h \mapsto \varepsilon \circ h$ (Ψ^{-1} is given by $\varphi \mapsto (1_A \otimes \varphi) \circ \Delta$.) Let $\mathscr{G}(A) = $ {algebra homomorphisms from A to F}, a subgroup of A^0 since $\varphi^{-1} = \varphi \circ S$ for φ in $\mathscr{G}(A)$; Ψ^{-1} maps $\mathscr{G}(A)$ onto {proper automorphisms of A}. A *differentiation* of A is a map $\partial: A \to F$ satisfying $\partial(a_1 a_2) = \varepsilon a_1 \partial a_2 + (\partial a_1)\varepsilon a_2$. $\delta = \Psi^{-1}\partial$ then satisfies $\delta(a_1 a_2) = a_1 \delta a_2 + (\delta a_1)a_2$, i.e., $\delta \in \operatorname{Deriv}(A)$. $\operatorname{Deriv}(A)$ is called the *Lie algebra* of $\mathscr{G}(A)$ and is isomorphic to the Lie algebra of differentiations of A under the Lie product

$$[\partial_1 \partial_2] = (\partial_1 \otimes \partial_2 - \partial_2 \otimes \partial_1) \circ \Delta.$$

Now we can define "algebraic Lie algebra" in this general setting, and Hochschild [71B, Chapter 13] has rather quick proofs of the basic facts about algebraic Lie algebras.

Let us return to the Hopf algebras of example 8.4.55.

Definition 8.4.56:   A coalgebra A is *cocommutative* if the following diagram (dualizing commutativity of multiplication) is commutative:

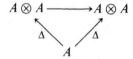

where the top arrow is the isomorphism given by $a_1 \otimes a_2 \to a_2 \otimes a_1$; in other words, $\sum a_{(1)} \otimes a_{(2)} = \sum a_{(2)} \otimes a_{(1)}$.

Remark 8.4.57:   It is enough to check cocommutativity on a base of A; thus group algebras are cocommutative (checked on G), and enveloping algebras are cocommutative.

So far we have generalized the augmentation of a group algebra and its canonical involution. To generalize the trace map we need another definition. Given any Hopf algebra A define the homomorphism $\pi: A^* \to F$ by $\pi f = f1$. View A^* as algebra via remark 8.4.53.

Definition 8.4.58:   A *(left) integral of* A^* is an element f_0 in A^* satisfying $ff_0 = (\pi f)f_0$ for every f in A^*. {left integrals of $A^\#$} is denoted as $\int$, and is an F-vector space called the *space of* integrals.

For *any* algebra homomorphism $\varphi: A \to F$ one has $\ker \varphi = \{a - \varphi a : a \in A\}$. Thus $\int = \text{Ann}' \ker \pi$.

Example 8.4.59:   The trace map tr of $F[G]^*$ is a left integral by direct verification.

There is a monic $\int \otimes A \to A^*$ sending $f_0 \otimes a$ to the map $f: A \to F$ given by $fa = f_0 Sa$. In case A is finite dimensional this is an isomorphism (cf., [SW, corollary 5.1.6], from which one concludes S is bijective and $\dim \int = 1$. In this case, however, we have a canonical isomorphism $A \approx (A^*)^*$ so A possesses a left integral (arising from A^*); noting ε takes the place of π above we have {left integrals of A} $\approx \text{Ann}' \ker \varepsilon$, a 1-dimensional vector space. For $|G|$ finite one sees $\sum_{g \in G} g$ is a left (and right) integral of $F[G]$. Since $\varepsilon(\sum g) = |G|$ exercise 32 generalizes Maschke's theorem.

To give the reader a better picture of Hopf algebras, we sketch some of the other important definitions and results, leading to a statement of Kostant's theorem.

Definition 8.4.60:   A *subcoalgebra* of a coalgebra is a subspace V satisfying $\Delta V \subseteq V \otimes V$.

If B is a subcoalgebra of a coalgebra A then the inclusion map $i: B \to A$ is a morphism yielding an algebra homomorphism $i^{\#}: A^* \to B^*$ whose kernel is $B^{\perp} = \{ f \in A^*: fB = 0 \}$. In particular $B^{\perp} \triangleleft A^*$. Conversely, if B is a subspace of A such that $B^{\perp} \triangleleft A^*$ then B is a subcoalgebra of A. (This verification is longer but straightforward, cf., SW, proposition 1.4.3). Thus the ideal structure of A^* corresponds to the subcoalgebra structure of A.

One is thus led to dualize the structure theory of algebras; note that the subcoalgebra theory is "weaker" than the ideal theory, since any theorem about ideals (stated in A^*) yields a theorem about subcoalgebras, but not vice versa. For *finite dimensional* bialgebras A we have $A \approx (A^*)^*$, so the subcoalgebra theory and the ideal theory are equivalent in this case. This is significant because subcoalgebras turn out to be quite tractable, as we shall see now.

Definition 8.4.61:   A coalgebra A is *irreducible* if any two nonzero subcoalgebras intersect nontrivially. A is *simple* if A has no proper subcoalgebras $\neq 0$; A is *pointed* if all simple subcoalgebras are 1-dimensional.

The *fundamental theory of coalgebras* (cf., exercise 34 or [SW, theorem 2.2.1]) states that any element of a coalgebra is contained in a finite dimensional subcoalgebra. Consequently, any simple coalgebra A is finite dimensional.

Corollary 8.4.62:   *Any cocommutative coalgebra A over an algebraically closed field F is pointed.*

Proof: Let B be a simple subcoalgebra. Then $[B:F] < \infty$. Hence B^* is a commutative, simple finite dimensional F-algebra and thus by the Wedderburn-Artin theorem is a field extension of F; since F is algebraically closed $B^* = F$ so $[B:F] = 1$.   Q.E.D.

The last ingredient needed is the smash product, a generalization of the group algebra crossed product (exercise 8.1.35).

Definition 8.4.63:   Suppose H is a Hopf algebra. An F-algebra R is an H-*module algebra* if R is an H-module such that for all h in H and r, s in R we have

$$h \cdot (rs) = \sum (h_{(1)} \cdot r)(h_{(2)} \cdot s) \qquad \text{and} \qquad h \cdot 1 = \varepsilon(h)1$$

where $\cdot$ denotes the scalar multiplication. In this case define the *smash product* $R \# H$ to be the vector space $R \otimes_F H$ provided with multiplication

$$(r \# g)(s \# h) = \sum r(g_{(1)} \cdot s) \# g_{(2)}h$$

for r, s in R and g, h in H. It is easy to check this really is an algebra.

Kostant's Theorem: ([SW, *theorems 8.1.5 and 13.0.1*] *Any cocommutative pointed Hopf algebra H is isomorphic (as an algebra) to a smash product of an enveloping algebra (of a suitable Lie algebra L) and a group algebra $F[G]$. (One obtains G and L via exercise 31.)*

In an effort to study enveloping algebras an $\mathbb{N}$-graded version of Hopf algebras has been introduced, where the graded tensor product is used (i.e., there is a sign switch for the odd components).

In view of Kostant's theorem we should expect the theory of left Noetherian, cocommutative pointed Hopf algebras to be rich enough to yield many of the results of this chapter. I do not know of results in this direction.

Recently Hopf algebras have been seen to be the "correct" setting for many results about group actions on rings, see especially Bergman [85] and work by M. Cohen and Montgomery.

Exercises

§8.1

1. (Snider [81]) If char$(F)=0$ then $F[G]$ is a nonsingular ring. (Hint: Passing to the algebraic closure one may assume F is algebraically closed; Thus $F[G]$ has an involution (*) for which $r*r=0$ implies $r=0$. If $a \in \text{Sing}(R)$ then $Ra* \cap \text{Ann } a \neq 0$; but if $ra*a = 0$ one has $(ra*)(ra*)* = 0$ so $ra* = 0$, contradiction.) His paper also deals with characteristic p.
2. If C is a semiprime $\mathbb{Q}$-algebra then $\text{Nil}(C[G]) = 0$. (Hint: Pass to prime images of C and then to their fields of fractions.)

The Kaplansky Trace Map and Idempotents

8. (Kaplansky) Suppose char$(F) = 0$ and $0 \neq e \in F[G]$ is idempotent. Then tr(e) is real and algebraic over $\mathbb{Q}$, and $0 < \text{tr } e < 1$. (Hint: It suffices to show tr e is real and positive; if so then $1 - \text{tr } e > 0$ since $1 - e$ is also idempotent, and tr e is algebraic since if transcendental its image under a suitable automorphism would be nonreal. Replacing F by the field extension generated by supp(e), one may assume F is a subfield of $\mathbb{C}$. Define the inner product $\langle, \rangle$ on $\mathbb{C}[G]$ by $\langle r_1, r_2 \rangle = \text{tr}(r_1^* r_2)$, cf., remark 8.1.5. If $L = \mathbb{C}[G]e$ then $L^{\perp}$ is also a left ideal of $\mathbb{C}[G]$, so there is a decomposition $1 = f + f'$ where $f \in L$ and $f' \in L^{\perp}$ are idempotents. $f^*(1 - f) = 0$ so $f^* = f^*f$ implying $f^* = f^{**} = f$ and tr $f = \text{tr}(f^*f) > 0$. But tr$(e - f) = \text{tr}[f, e] = 0$, so tr $e = \text{tr } f$.)
9. (Kaplansky) If char$(F) = 0$ then $F[G]$ is weakly 1-finite. (Hint: If $ab = 1$ then tr$(ba) = 1$.)

10. (Zaleskii [72]) If $e \in F[G]$ is idempotent then $\mathrm{tr}\, e$ is in the characteristic subfield of F. (Hint: Reduce to the case $\mathrm{char}(F) > 0$ using remark 8.1.25′. Write $e = \sum \alpha_g g$ and let $S = \{1\} \cup p\text{-supp}(e)$. Taking q large enough one has $\mathrm{tr}\, e = \mathrm{tr}\, e^q = \sum \{\alpha_g^q : g \in S\}$. Then $(\mathrm{tr}\, e)^p = \mathrm{tr}\, e$, as desired.)

The Center of the Group Algebra

11. $Z(K[G]) \approx K \otimes_C Z(C[G])$ for any commutative C-algebra K. If $A \lhd C$ then $Z((C/A)[G])$ is the image of $Z(C[G])$ under the canonical map $C[G] \to (C/A)[G]$. (Hint: Compare bases, using IBN.)

12. Suppose C is Noetherian and H is a finitely generated subgroup of $\Delta(G)$ which is normal in G. Then $Z(C[G]) \cap C[H]$ is an affine C-algebra. (Hint: Let $Z = Z(H)$. Then $[H:Z] < \infty$ so Z is finitely generated. Hence $C[Z]$ is commutative Noetherian, over which $C[H]$ is f.g. and thus Noetherian. Thus $T = Z(C[H])$ is affine by Artin-Tate, and the fixed ring T^G is affine by exercise 6.2.2, where G acts by conjugation on $C[H]$. But $C[H]^G = Z(C[G]) \cap C[H]$ is a T^G-submodule of T.)

13. $\mathrm{Jac}(Z(C[G]))$ is nil whenever C is Noetherian. (Hint: Write $Z = Z(C[G])$. If $r \in \mathrm{Jac}(Z)$ then taking $H = \langle \mathrm{supp}(r) \rangle \subseteq \Delta(G)$ note $Z \cap C[H]$ is affine over C by exercise 12. Hence $\mathrm{Jac}(Z \cap C[H])$ is nil, implying r is nilpotent.)

14. Suppose $\mathrm{char}(F) = p > 0$. The following are equivalent: (i) $F[G]$ is semiprime. (ii) $Z(F[G])$ is semiprime. (iii) $Z(F[G])$ is semiprimitive. (iv) The order of every normal finite subgroup of G is prime to p. (v) $\Delta(G)$ is a p'-group. (vi) $Z(F[\Delta(G)])$ is semiprime. (Hint: (ii) $\Rightarrow$ (iii) by exercise 13. (iii) $\Rightarrow$ (iv) $\Rightarrow$ (v) and (vi) $\Rightarrow$ (i) as in theorem 8.1.39.)

The Augmentation Ideal

15. If $H = \langle h_i : i \in I \rangle \subset G$ then $\omega(C[H])C[G] = \sum_{i \in I}(h_i - 1)C[G]$. (Hint: Let $H_0 = \{h \in H : h - 1 \in \sum (h_i - 1)C[G]\}$. Show $h_i h \in H_0$ and $h_i^{-1} h \in H_0$ for any h in H_0; thus $HH_0 = H_0$. Since $1 \in H_0$ conclude $H = H_0$.)

16. A subgroup H of G is finite iff $\mathrm{Ann}\, \omega(C[H]) \neq 0$, in which case $\mathrm{Ann}\, \omega(C[H]) = C[G] \sum_{h \in H} h$. (Hint: Suppose $0 \neq r = \sum_{i=1}^{n} c_i g_i \in \mathrm{Ann}\, \omega(C[H])$. Then $r = rh$ for all h in H, implying $g_1 h \in \{g_1, \ldots, g_n\}$ and thus H is finite. Continue along these lines.)

17. If $\omega(F[G])$ is nilpotent then $\mathrm{char}(F) = p > 0$ and G is a finite group.

18. (Auslander-Villamayor) $F[G]$ is (von Neumann) regular iff G is locally finite and is a $\mathrm{char}(F)'$-group. (Hint: ($\Rightarrow$) If $H = \langle h_1, \ldots, h_n \rangle$ then $\omega(F[H])F[G]$ is an f.g. right ideal and thus has nonzero annihilator, implying H is finite. To check $|H| \neq 0$ in F let $a = \sum_{h \in H} h$ and take b such that $1 - ab \in \mathrm{Ann}'\, a = C[G]\omega(F[H])$; apply the augmentation.)

Injective Modules

19. $F[G]$ is a symmetric algebra and thus Frobenius for any finite group G. (Hint: Apply the trace map to exercise 3.3.30.) Renault showed $F[G]$ is self-injective iff G is finite (cf., Passman [77B, p. 79] for Farkas' short proof).

20. If E is an injective $F[G]$-module and $\omega(F[G])E \neq E$ then G is locally finite.

(Hint: Suppose $H = \langle h_1, \ldots, h_n \rangle$. Then $E \neq \sum (h_i - 1)E$, implying $\text{Ann}\{h_1 - 1, \ldots, h_n - 1\} \neq 0$. Thus $\text{Ann}\,\omega(F[H]) \neq 0$ so H is finite by exercise 16.)

21. (Farkas-Snider [74]) View F as an $F[G]$-module by means of the augmentation map. F is injective iff $F[G]$ is a regular ring. (Hint: ($\Leftarrow$) By exercise 2.11.21. ($\Rightarrow$) To verify the criteria of exercise 18, first note G is locally finite by exercise 20. Furthermore, if H is a finite subgroup and $a = \sum_{h \in H} h$ then there is a map $f: F[G]a \to F$ sending a to 1 (in view of exercise 16); thus f extends to $\hat{f}: F[G] \to E$ and $|H|\hat{f}1 = a\hat{f}1 = \hat{f}a \neq 0$.) Conclude that if $F[G]$ is a V-ring then it is a regular ring.

Primitive and Semiprimitive Group Rings

22. Suppose G is the group free product $G_1 \coprod G_2$. Then $F[G]$ is the free product (as F-algebras) of $F[G_1]$ and $F[G_2]$.
23. (Formanek) If $|G_1| > 2$ and $G = G_1 \coprod G_2$ then $F[G]$ is primitive (cf., exercise 2.1.15).
24. (Formanek) Suppose $G = \mathbb{Z} \times (A \coprod B)$ where $|A| > 2$ and $|B| > 1$. Then $F[G] \approx F[\mathbb{Z}] \otimes F[A \coprod B]$ is primitive.
25. (Farkas-Snider [81]) If $\text{char}(F) = 0$ then $F[G]$ cannot be primitive with non-zero socle. (Hint: It suffices to show there are only a finite number of independent minimal left ideals, for then $F[G]$ would be simple Artinian, impossible. So assume there are an infinite number. This yields an infinite number of orthogonal primitive idempotents e_1, e_2, all with the same Kaplansky trace α, since minimal left ideals are isomorphic. But α is real > 0; taking $n > 1/\alpha$ yields $1 < n\alpha = \sum \text{tr}\,e_i \leq 1$, contradiction.) Their paper also deals with characteristic p.

§8.2

Noetherian Group Rings

1. If G is polycyclic-by-finite with Hirsch number t then K-dim $F[G] \leq t$. (Hint: As in theorem 8.2.2.)
2. (Formanek [73]) If G is torsion-free and $F[G]$ is left Noetherian then $F[G]$ has no nontrivial idempotents. (Hint: By proposition 8.2.24 the Kaplansky trace of an idempotent of $F[G]$ is an integer; apply exercise 8.1.8.) Compare to exercise 16.

Polycyclic-by-Finite Group Algebras

12. (Snider) Suppose R is a prime ring of finite gl. dim, which is free as a module over a Noetherian domain $W \subseteq R$, with normalizing base $a_1, \ldots, a_n$. Then R is a domain if $n \mid u\text{-dim}\,P$ for every f.g. projective R-module P. Here u-dim denotes uniform dimension. (Hint: As in the beginning of the proof of theorem 8.2.35.)
13. Exercises are from Strojnowski [86]: A group H is called an $E(F)$-group if every free module $Q = F[H]^{(n)}$ satisfies the property that dependent elements remain dependent modulo the augmentation ideal, i.e., $\{x_i + \omega(F[H])Q : 1 \leq i \leq n\}$ are F-dependent in $Q/\omega(F[H])Q$ if $\{x_1, \ldots, x_n\}$ are F-dependent in Q. Suppose $S \subset F[H] - \omega(F[H])$ is a denominator set of regular elements. Then for

any f.g. projective $S^{-1}F[H]$-module P one has u-dim $P = [P/\omega(F[H])P:F]$. (Hint: Take $P \oplus P'$ free.)

14. Suppose the $E(F)$-group H is a subgroup of index n in G, and suppose $R = F[G]$ and $W = F[H]$ satisfy the hypothesis of exercise 12. Then R is a domain if $n \mid [P/\omega(F[H])P:F]$ for every f.g. projective $F[G]$-module P.

15. A group G is an $E(F)$-group under any of the following conditions: (i) G is infinite cyclic. (ii) G is a subgroup of an $E(F)$-group. (iii) Every nontrivial f.g. subgroup has a homomorphic image which is an $E(F)$-group $\neq (1)$. (iv) G has a normal subgroup N such that N and G/N are $E(F)$-groups. (v) G is solvable with torsion free abelian factors. (Hint: (i) If $G = \langle g \rangle$ then cancel a suitable power of $(1 - g)$ from any dependence relation. (ii) Write $F[G]$ as a summand of $F[H]$ where H is an $E(F)$-group, and match components. (iii) Induct on the support of a dependence relation. (iv) by (ii) and (iii).)

Idempotents

16. If G is an $E(F)$ group then $F[G]$ has no nontrivial idempotents. (Hint: Clearly $\omega(F[G])$ contains no projective left ideal $\neq 0$ and thus no nontrivial idempotent. Now let $\pi: F[G] \to F$ be the augmentation map. If $e \neq 0$ is idempotent then πe is idempotent implying $\pi e = 1$; hence $\pi(1 - e) = 0$ implying $1 - e = 0$.)

§8.3

1. $U(\varinjlim L_i) \approx \varinjlim U(L_i)$ canonically.

2. If L_i are Lie subalgebras of L such that $L_1 + L_2 = L$ then taking $H = L_1 \cap L_2$ one has a unique bijective map $f: U(L_1) \otimes_{U(H)} U(L_2) \to U(L)$ such that $f(r_1 \otimes r_2) = r_1 r_2$ for r_i in $U(L_i)$. (Hint: To prove f exists, view $U_i = U(L_i)$ as $U(H)$-modules and define the balanced map $U_1 \times U_2 \to U(L)$ by $(r_1, r_2) \mapsto r_1 r_2$. Show f is an isomorphism by matching bases.)

8. (BGR, example 5.9) Let L be as in exercise 3. Localizing at z show Spec$(U(L))$ is isomorphic to Spec $F[z, z^{-1}, \lambda]$ where $\lambda = w + z^{-1} yx$.

9. Suppose L acts as a Lie algebra of derivations on a ring R. Then we can define the *Lie algebra of differential polynomials over R* as the F-vector space $R \otimes_F V(L)$ with multiplication agreeing with that in R and in $U(L)$, and satisfying the rule $ar = ra + \delta_a r$ for every a in L and r in R. Show this defines an algebra structure and generalize proposition 8.3.32 to this setting.

10. If $[L:F] = n$ then K-dim $U(L) \geq n$. (Hint: Define a lattice injection from $\mathscr{L}(U(L))$ to $\mathscr{L}(G(L))$ by $I \mapsto \bigoplus_{n \geq 1}(I \cap U_n(L))/(I \cap U_{n-1}(L))$. Since $G(L) \approx F[\lambda_1, \ldots, \lambda_n]$, conclude K-dim $U(L) \leq$ K-dim $F[\lambda_1, \ldots, \lambda_n] = n$.)

11. Show every chain in Spec$(U(L))$ has length $\leq [L:F]$ by using exercise 10.

Growth of Infinite Dimensional Lie Algebras

12. Define "growth" of a Lie algebra as in §6.2. If L has subexponential growth then $U(L)$ also has subexponential growth (by PBW).

13. (M. Smith [76]) If L is infinite dimensional but finitely generated as Lie algebra then $U(L)$ does not have polynomial bounded growth. (Hint: The growth function

in L is strictly increasing, so the growth function in $U(L)$ increases at least as fast as $p(n)$, the number of ways of writing n as a sum of nondecreasing positive integers. But $p(n)$ is not polynomially bounded.)

14. (M. Smith [76]) Let L be a Lie algebra with base $a_0, a_1, \ldots$ such that $[a_i a_j] = 0$ and $[a_0 a_j] = a_{j+1}$ for $i, j \geq 1$. Then $U(L)$ is subexponential by exercise 12 but not polynomial bounded, by exercise 13. As observed by Bergen-Montgomery-Passman [86], this example can be altered to give an example in which the differential polynomial ring $E[\lambda; \delta]$ does not have polynomial bounded growth, where E is the infinite dimensional exterior algebra and thus has GK dim 0. (λ replaces a_0, and x_i replace a_i.)

The Ore Condition for Infinite Dimensional Lie Algebras

15. If L has subexponential growth then $U(L)$ is Ore, by exercise 12 and proposition 6.2.15.

16. Suppose L_1 is a Lie ideal of a Lie algebra L and L/L_1 is solvable. If $U(L_1)$ is Ore then $U(L)$ is Ore. (Hint: One may assume L/L_1 is abelian; furthermore, since the Ore condition involves a finite number of elements one may assume L/L_1 is finitely generated. But $U(L)$ is a free $U(L_1)$-module and matching coefficients in the base one sees $U(L_1)$ is a left denominator set for $U(L)$; localizing puts one in a position to use the Hilbert basis theorem.)

§8.4

1. (Amitsur-Small [78]) Suppose D is a division ring. The polynomial ring $R = D[\lambda_1, \ldots, \lambda_t]$ is primitive iff $M_n(D)$ contains a subfield of transcendence degree t over $F = Z(D)$ for some n. (Hint: ($\Leftarrow$) is exercise 2.3.5. ($\Rightarrow$) Suppose R has maximal left ideal L with core 0. Then $F[\lambda_1, \ldots, \lambda_t]$ can be embedded into $\mathrm{End}_R(R/L)$, so conclude by means of the Nullstellensatz.) This result is used to prove that there are division rings D for which $D[\lambda_1, \ldots, \lambda_t]$ is primitive but $D[\lambda_1, \ldots, \lambda_{t+1}]$ is not primitive.

F.g. Overrings of Left Noetherian Rings

12. (Small) Suppose R is left Noetherian and $T \supseteq R$ is a prime ring which is f.g. (as R-module). Then $A \cap R \neq 0$ for every nonzero $A \lhd T$. (Hint: T is a Noetherian R-module so any a in T satisfies $\sum_{i \leq n} Ra^i = \sum_{i \leq n+1} Ra^i$ for suitable n; thus $a^{n+1} \in \sum_{i \leq n} Ra^i$, so take a regular in A.) Using a similar argument show every element of R invertible in T is invertible in R, and thus $R \cap \mathrm{Jac}(T) \subseteq \mathrm{Jac}(R)$ by proposition 2.5.17.

12'. (Resco) If R is left Noetherian and $T \supseteq R$ is f.g. then INC holds from R to T. (Hint: By exercise 12.)

13. (Small) If R is left Noetherian, Jacobson and $T \supseteq R$ is f.g. then T is left Noetherian, Jacobson. (Hint: One may assume T is prime. Let $A = R \cap \mathrm{Jac}(T)$. Exercise 12 implies $0 \neq A \subseteq \mathrm{Jac}(R) = \mathrm{Nil}(R)$. Then

$$\text{K-dim } T = \text{K-dim } R = \text{K-dim } R/A = \text{K-dim } T/\mathrm{Jac}(T)$$

so $\mathrm{Jac}(T)$ is nilpotent.)

Hopf Algebras (See SW)

27. Any Hopf algebra has IBN.

28. If A, B are coalgebras then so is $A \otimes B$. (Hint: Take Δ to be the composite $A \otimes B \to (A \otimes A) \otimes (B \otimes B) \to (A \otimes B) \otimes (A \otimes B)$.)

29. If $C = (C, \Delta, \varepsilon)$ is a coalgebra and $A = (A, \mu, u)$ is an algebra then $\mathrm{Hom}(C, A)$ is an algebra under the *convolution product*, where the convolution product h of f and g is given by $hc = \mu \sum f c_{(1)} \otimes g c_{(2)}$, i.e., $h = \mu(f \otimes g)\Delta$. The unit element of this algebra is $u\varepsilon$. Taking $C = A$ in case A is a bialgebra, show an antipode S is an inverse of 1_A under the convolution, and thus is unique (if it exists).

30. The antipode S of a Hopf algebra A is an anti-homomorphism and is an involution if A is cocommutative. (Hint: define maps $\eta, v: A \otimes A \to A$ by $\eta(a \otimes b) = S(ab)$ and $v(a_1 \otimes a_2) = SbSa$. Taking $A \otimes A$ as a coalgebra as in exercise 28 and letting $\cdot$ denote the convolution product in $\mathrm{Hom}(A \otimes A, A)$ show $\eta \cdot \mu = \mu \cdot v$; in fact, each evaluation on $a \otimes b$ is $\varepsilon a \varepsilon b$ for all a, b in A. This implies $S(ab) = SbSa$. Next note $S1 = 1$, implying $Su = u$. Finally, to prove $S^2 = 1_A$ it is enough to show the convolution product in $\mathrm{Hom}(A, A)$ of S^2 and S is $u\varepsilon$. But this is $\sum S^2 a_{(1)} S a_{(2)} = S(\sum a_{(2)} S a_{(1)}) = Su\varepsilon a = u\varepsilon a$; note cocommutativity was used.)

31. An element a of a Hopf algebra A is *group-like* if $\Delta a = a \otimes a$; a is *primitive* if $\Delta a = a \otimes 1 + 1 \otimes a$. Show $G = \{$group-like elements of $A\}$ is a subgroup of $\mathrm{Unit}(A)$, where $a^{-1} = Sa$, and $F[G]$ is canonically isomorphic to the F-subspace of A spanned by G. Also $L = \{$primitive elements of $A\}$ is a Lie algebra. (Hint: If $g \in G$ then $\varepsilon g = 1$. It remains to show that $\{g \in G\}$ are linearly independent over F. Otherwise, take a dependence relation $\sum_{i=1}^{n} \alpha_i g_i = 0$ with n minimal such that $\alpha_n \neq 0$. Assuming $n \neq 1$ get $\sum_{i=1}^{n-1} \alpha_i g_i \otimes g_i = \sum \alpha_i \Delta g_i = -\Delta g = -g \otimes g = -\sum_{i,j=1}^{n-1} \alpha_i \alpha_j g_i \otimes g_j$, implying $n - 1 = 1$, i.e., $g_2 = -\alpha_1 g_1$. Then $-\alpha_1 = \varepsilon(g_2) = 1$, contradiction.)

32. (Hopf version of Maschke's theorem) A finite-dimensional Hopf algebra A is semisimple Artinian iff $\varepsilon \int \neq 0$. (Hint: ($\Rightarrow$) Write $A = L \oplus \ker \varepsilon$. Then $(\ker \varepsilon)L = 0$ so $L \subseteq \int$, implying $L = \int$. ($\Leftarrow$) Pick an integral a of A with $\varepsilon a = 1$. Given any left ideal L of A take a projection $\pi: A \to L$ as F-vector spaces: then there is an A-module projection $\pi': A \to L$ given by $\pi'ab = \sum a_{(1)} \pi((Sa_{(2)})b)$. This proof uses an important averaging process arising from the integral; however, for R cocommutative and $F = \mathbb{C}$ there is an easier proof.)

33. Any finite dimensional Hopf algebra is a symmetric algebra and thus Frobenius. (Hint: as in exercise 8.1.19.)

34. (The fundamental theorem of coalgebras, generalized) Any f.d. subspace V of a coalgebra is contained in an f.d. subcoalgebra of A. (Hint, following notes of Kaplansky: Given any subspace W of A let LW (resp. RW) denote the subspace of A generated by all $a_{(1)}$ (resp. by all $a_{(2)}$) for each a in W, writing $\Delta a = \sum a_{(1)} \otimes a_{(2)}$. Note $a = \sum (\varepsilon a_{(2)}) a_{(1)} \in LW$, so $W \subseteq LW$ and likewise $W \subseteq RW$. Applying coassociativity and matching components shows $LLV = LV$, and $LRV = RLV$. Hence LRV is the desired subcoalgebra.)

Dimensions for Modules and Rings

(Parenthetical numbers refer to results in text)

Module Dimensions

Note: There are two important possible properties of a module dimension "dim":

Additive: $\dim M = \dim N + \dim M/N$ for $N \leq M$;
Exact: $\dim M = \max(\dim N, \dim M/N)$ for $N \leq M$

Name	Abbr.	Kind of module with minimal dimension	Properties
composition length	$\ell(M)$	simple	additive (2.3.4)
uniform = Goldie dim.	u-dim M	uniform	
reduced rank	$\rho(M)$	torsion	additive (3.5.6)
Krull dimension	K-dim M	Artinian	exact (3.5.41)
basic dimension	B-dim M	unfaithful (for R prime)	exact (3.5.59)
Gabriel dimension	G-dim M	completely reducible	exact (3.5.96f)
projective dimension	pd M	projective	see summary 5.1.13
injective dimension	id M	injective	
Gelfand-Kirillov dim.	GK dim M		exact over Noetherian PI-ring (6.3.46)

Ring Dimensions

Name	Abbr.	Ring of minimal dim.	Remarks
Gelfand-Kirillov dim.	GK dim	finite dim. algebra	integer valued for certain classes of rings
Krull dimension	K-dim	Artinian	$\leq$ gl. dim for Noetherian PI-rings (6.3.46)
classical Krull	cl. K-dim	every prime is maximal	$\leq k$ dim
little Krull dimension	k dim	every prime is maximal	$\leq$ K-dim
Gabriel dimension	G-dim	direct prod. of Artinian	$\leq K$-dim $+$ 1
global dimension	gl. dim.	semisimple Artinian	$=\sup\{\text{pd } M\}$ $=\sup\{\text{id } M\}$
weak dimension	w. dim.	von Neumann regular	$\leq$ gl. dim.
PI degree	PI deg	commutative	$=$ degree, for f.d. division algebra
Goldie rank		Ore domain	$=$ length, for simple Artinian ring

Sketch of theory for an affine prime PI-algebra over a field F (cf., 6.3.40 ff).
GK dim R = cl. K-dim R = tr. deg $Z(R)/F$, and equals K-dim R if the latter exists!; although K-dim R may fail to exist, G-dim R always exists. Also cl. K-dim R = cl. K-dim R/P + height P for any prime ideal P of R.

Major Ring- and Module-Theoretic Results Proved in Volume I (Theorems and Counterexamples; also cf. "Characterizations")

E means "exercise". Result preceded by (*) appears in stronger form elsewhere in the text.

Theorems

Chapter 0

0.3.2 Existence and "uniqueness" of a base.

Chapter 1

1.1.17 $R\text{-}\mathcal{M}od$ and $M_n(R)\text{-}\mathcal{M}od$ are equivalent categories.
1.1.25   Matrix units can be lifted via idempotent-lifting ideals.
1.2.23 $R((S))$ is a domain, for any domain R and ordered monoid S.
1.2.24' $D((G))$ is a division ring, for any ordered group G.

Chapter 2

*2.9.15   ("Exchange property") The summands in two LE-decompositions can be interchanged.

2.9.17   (Wedderburn-Krull-Schmidt-Remak-Azumaya = Krull-Schmidt) All decompositions of a module having an LE-decomposition are equivalent.

2.9.18   A complete set of idempotents of a semiperfect ring is unique, up to inner automorphism.

2.9.29   (Harada–Sai) Any sequence of nonisomorphisms of indecomposables having bounded composition length is finite.

2.9.40   First Brauer Thrall conjecture verified (bounded representation type implies finite representation type).

(Also sesqui-BT is proved modulo 2.9.42; other major theorems from the representation theory of Artin algebras are described but not proved)

2.10.3′   (Baer's criterion) Injectivity can be checked on left ideals.

2.10.10   There are "enough" injectives.

2.10.20   Injective hull = maximal essential extension.

2.10.33   "Uniqueness" of decomposition of injective.

2.11.13 F is a flat module iff $F^{\#}$ is injective.

2.11.14 F flat iff $I \otimes F \approx IF$ for any f.g. right ideal I (also cf. E2.11.5′).

2.12.22   The rank of f.g. projective modules is locally constant.

2.12.28   Nullstellensatz for algebras over "large" fields.

2.12.36   "Generic flatness" implies the Nullstellensatz.

2.12.48   LO and GU hold for finite centralizing extensions.

2.12.50   LO and GU hold when tensoring up by a field.

2.12.52   Analogue of 2.5.35, 2.5.36 for prime ideals.

2.13.21   Structure of semiprime (*)-rings with nonzero socle, in terms of a sesquilinear form.

2.13.29   Characterizations of involutions on matrix rings.

E2.1.15   A free product of algebras over a field is usually primitive.

E2.4.1   Density theorem for completely reducible modules.

E2.5.10   (Koethe–Noether–Jacobson) Every algebraic division algebra contains a separable element.

E2.9.4   Exchange property for a module having an LE-decomposition into countable submodules. (This implies every projective over a local ring is free, generalizing example 2.8.9.)

Chapter 3

*3.1.4 $S^{-1}R$ is a ring of fractions for R with respect to S, for any denominator set S.

3.5.82   Any Noetherian bimodule has a large bimodule which is a direct sum of cells, and this decomposition is "unique".

3.5.89   Jacobson's conjecture holds for almost fully bounded Noetherian rings.

E3.2.31   (Faith-Utumi) Any order in $M_n(D)$ contains a subring (without 1) of the form $M_n(T)$ where T is an order in D.

E3.2.38   Any nilpotent algebra over a field is embeddible into an algebra of upper triangular matrices.

E3.3.14   Goldie = nonsingular, for any semiprime ring satisfying $ACC(\bigoplus)$.

E3.3.40   (Goodearl) The ideals of Q/P are totally ordered, for any prime ideal P of a regular self-injective ring Q.

E3.3.53   (Goodearl) The lattice of ideals of a prime regular self-injective ring is isomorphic to an interval of ordinals.

Chapter 4

4.1.4   (Morita's theorem) R' is Morita equivalent to R if $R' \approx \text{End } P$ for a progenerator P.

4.1.17   More explicit description of Morita equivalence.

4.2.9   A left adjoint preserves direct limits.

4.2.11   A right adjoint preserves inverse limits.

E4.2.8   (Watt) Any right exact functor preserving direct sums is a tensor functor.

E4.2.11   (Watt) Any left exact functor preserving direct products is a Hom functor.

E4.2.17   (Popescu-Gabriel) Any Grothendieck category with generator C is a category of modules over the ring $\text{Hom}(C, C)$.

Counterexamples

Chapter 0

0.1.4   A non-onto epic in $\mathscr{Ring}$.

Chapter 1

1.1.29   (Zelinski) An uncountable set of idempotents cannot be lifted.

1.3.33   A ring lacking IBN.

Chapter 3

3.2.48	A finite ring which cannot be embedded into a matrix ring.
3.2.49	A domain not embeddable into a division ring.
3.2.54	An irreducible Noetherian ring which is not an order in an Artinian ring.
3.3.11	A nonsingular, but not semiprime, Artinian ring.
3.5.8	A noncommutative Noetherian ring failing the principal ideal theorem.
3.5.23	A left Noetherian ring failing Jacobson's conjecture.
3.5.29	A Noetherian ring having a non-Artinian module which is an essential extension of a simple module.
3.5.38′	A Noetherian ring whose radical fails the AR-property.
3.5.53	cl K-dim $< K$ dim for the Weyl algebra.
3.5.92	Nonlocalizable prime ideals of an Artinian ring.
E3.2.8	A commutative ring with ACC (Ann) but having arbitrarily long chains of annihilators.
E3.2.12	A simple Noetherian PLID which is a V-ring with a unique simple module, but which is not Artinian.
E3.2.40	A left Noetherian ring which cannot be embedded in a left Artinian ring. (A more powerful example has recently been discovered by Dean–Stafford.)
E3.3.9	A left but not right nonsingular ring.
E3.3.10	A ring whose injective hull does not have a ring structure.
E3.4.1	A large submodule which is not dense.
E3.5.5	A Noetherian ring whose nilradical has unequal left and right reduced ranks.
E3.5.9, 10	A local PLID for which Jacobson's conjecture fails, and in which primary decomposition fails.
E3.5.17′	A Noetherian ring whose P-adic completion is non-Noetherian.
E3.5.18	A prime ideal failing the AR-property.
E3.5.27	A PLID with arbitrary large K-dim.
E3.5.29	A *right* Noetherian ring lacking (left) K-dim.

Chapter 4

4.1.22	A simple Noetherian hereditary ring not Morita equivalent to a domain (given without proof).
E4.1.2	A faithful cyclic projective module which is not a progenerator.

Major Theorems and Counterexamples for Volume II

E before number denotes "exercise"; otherwise the result is from the main text.

Theorems

The vast majority of these results are proved in the text. In this list, precision may be sacrificed for conciseness; please check the text for the full hypotheses.

§5.1

5.1.13. The relationships of pd among modules of an exact sequence.

5.1.25, E5.1.4. (See also E5.1.8, E5.1.16) $\text{gl.} \dim R[\lambda; \sigma] = \text{gl.} \dim R + 1$.

5.1.34, 5.1.41. (Serre-Quillen, see also E5.1.13 and the appendix to Ch. 5). For R homologically regular and filtered, if $K_0(R_0)$ is trivial then so is $K_0(R)$. (In fact $K_0(R) \approx K_0(R_0)$ in general.)

5.1.45 (Walker). A prime, homologically regular ring with trivial K_0 is a domain.

5.1.53 (Quillen-Suslin). f.g. projectives over $F[\lambda_1, \ldots, \lambda_n]$ are free.

5.1.59, 5.1.60, 5.1.61 (Stafford-Coutinho, generalizing results of Bass and Swan). The main tool for proving a f.g. R-module has a summand isomorphic to R, and for proving cancellation.

5.1.67 (Bass, see also E5.1.28). "Uniformly big" projectives are free.

§5.2

5.2.10, 5.2.11. Existence of the "long exact" homology and cohomology sequence.

5.2.22, 5.2.25.   Existence of the "long exact" sequence for the derived func-
                  tors of a left or right exact functor. (This is applied later to Tor
                  and Ext.)

5.2.34, 5.2.37.   The definitions of $\text{Ext}(M, N)$ and $\text{Tor}(M, N)$ are well-defined.

5.2.40.   The following are equivalent for a module M:

    (i) pd $M \leq n$
    (ii) $\text{Ext}^k(M, \_) = 0$ for all $k > n$
    (iii) $\text{Ext}^{n+1}(M, \_) = 0$
    (iv) The $(n-1)$ syzygy of any projective resolution of M is
    projective.

5.2.40'.   The following are equivalent for a module N:

    (i) id $N \leq n$
    (ii) $\text{Ext}^k(\_, N) = 0$ for all $k > n$
    (iii) $\text{Ext}^{n+1}(\_, N) = 0$
    (iv) The $(n-1)$ cosyzygy of every injective resolution is injective.

5.2.42   (Auslander). gl. dim $R = \sup\{\text{pd } R/L : L < R\}$.

5.2.48.   Any pair of right exact functors that commutes with Hom also
                  commutes with Ext^n for all $n \in \mathbb{N}$.

5.2.49.   Any pair of left exact functors that commutes with $\otimes$ also com-
                  mutes with Tor_n for all $n \in \mathbb{N}$.

E5.2.21.   Structure of homomorphic images of hereditary semiprimary
                  rings.

§5.3

5.3.7.   Equivalent conditions for an algebra to be separable.

5.3.18.   The "modern" definition of "separable" coincides with the classical
          definition for algebras over a field.

5.3.20.   Cohomological setting of Wedderburn's principal theorem. ($R = S \oplus N$ where S is semisimple Artinian and $N = \text{Jac}(R)$.)

5.3.21.   Uniqueness of S in 5.3.20.

5.3.24   (See also E5.3.9, E5.3.11). Equivalent conditions for an algebra to be
          Azumaya.

§6.1

6.1.20   (See also 6.1.23, E6.1.22). A multilinear central polynomial for
          matrices.

6.1.25 (Kaplansky's Theorem, see also E6.1.17′). Primitive PI-rings are central simple.

6.1.26. $R/N(R)$ is embeddible in $n \times n$ matrices for R PI. (cf., definition 2.6.25.)

6.1.27, 6.1.28. If R is semiprime PI then $\text{Nil}(R) = 0$ and every ideal of R intersects the center nontrivially.

6.1.30. The ring of central fractions of a prime PI-ring is central simple.

6.1.33. Suppose R has PI-degree n. If an n^2-normal central polynomial takes on the value 1 then R is a free $Z(R)$-module.

6.1.35, 6.1.35′ ("Artin-Procesi"). Various PI-conditions equivalent to being Azumaya of unique rank.

6.1.38. $\text{Spec}_n(R)$ is intimately related to the center.

6.1.44. R is PI-equivalent to $R \otimes_C H$ for H commutative.

6.1.49 (see also E6.1.24′, E6.1.25). Jacobson's conjecture is true for Noetherian PI-rings.

6.1.50. In a left Noetherian PI-ring every prime ideal has finite height.

6.1.53. Technical criterion used to prove a ring is PI.

6.1.59, 6.1.61. (*)-version of 6.1.27 and 6.1.28.

E6.1.10 (Amitsur-Levitzki). The standard polynomial S_{2n} is an identity of $n \times n$ matrices.

E6.1.31 (Amitsur). A primitive ring is strongly primitive iff it satisfies a proper generalized identity.

§6.2

6.2.5 (Artin-Tate, see also 6.3.1). If R is affine and f.g. over Z then Z is affine.

6.2.17 (Irving-Small, see also E6.2.8). Any affine, semiprime F-algebra of subexponential growth and satisfying ACC on left annihilators, is Goldie.

6.2.25. $\text{GK dim}(R) \geq \text{GK dim}(R/P) + \text{height}(P)$ for any P in $\text{Spec}(R)$, if prime images of R are Goldie.

6.2.27, 6.2.33 (Bergman). If $\text{GK dim}(R) \leq 2$ then $\text{GK dim}(R)$ is an integer.

6.2.38. GK module dimension is exact over any filtered affine F-algebra whose associated graded algebra is affine and left Noetherian.

§6.3

6.3.3 (Amitsur-Procesi). Any affine PI-algebra is Jacobson and satisfies the Nullstellensatzes.

6.3.11.   Embedding a ring into a generalized triangular matrix ring, by means of "universal derivations."

6.3.14   (Lewin). If $F\{X\}/A$ and $F\{X\}/B$ are embeddible, respectively, in $m \times m$ and $n \times n$ matrices then $F\{X\}/AB$ is embeddible in $(m + n) \times (m + n)$ matrices.

6.3.16   (Lewin). If R is PI and Nil(R) is nilpotent then R satisfies the identities of $k \times k$ matrices for some k.

6.3.23.   (Levitzki-Kaplansky-Shirshov, see also E6.3.12, E6.3.13). Suppose $R = C\{r_1,\ldots,r_t\}$ satisfies a multilinear polynomial identity, and each monomial in the r_i of length $<m$ is integral over C. Then R is f.g. as C-module, spanned by the monomials in the x_i of a certain bounded length.

6.3.24   (Shirshov). Technical but useful related result, which holds for arbitrary affine PI-algebras.

6.3.25.   Any affine PI-algebra has finite Gelfand-Kirillov dimension.

6.3.30.   If a PI-ring R is integral over C then GU, LO and INC hold from C to R.

6.3.31   (See also E6.3.14). Versions of LO, GU, and INC holding from a prime PI-ring to its trace ring.

6.3.34   (Schelter, see also 6.3.36′). Any affine PI-algebra over a Noetherian ring satisfies ACC (prime ideals).

6.3.35, 6.3.36, 6.3.39   (Razmyslov-Braun). The Jacobson radical of an affine PI-algebra is nilpotent.

6.3.41   (See also 6.3.42, 6.3.44). If R is prime affine PI over a field F then GK dim R = cl. K-dim R = tr deg $Z(R)/F$.

6.3.43   (Schelter). Catenarity of affine PI-algebras.

E6.3.17.   Reduction of Bergman-Small theorem (describing the PI-degree in terms of its prime images) to trace ring.

E6.3.18.   "Going down" from a prime PI-ring to its center.

E6.3.19.   Affine PI-algebras have Gabriel dimension.

§6.4

6.4.5. A is a T-ideal of $C\{X\}$ iff $C\{X\}/A$ is relatively free.

6.4.7.   (*)-version of 6.4.5.

6.4.8.   (*)-version of 6.1.44.

6.4.19   (Regev, see also 6.4.18, E6.4.4-E6.4.7). The tensor product of PI-algebras is PI.

E6.4.2   (Nagata-Higman). In characteristic 0 nil of bounded degree implies nilpotent.

Note: We write "csa" below for "central simple algebra."

§7.1

7.1.9   (Double centralizer theorem). $C_R(C_R(A)) = A$ for any simple subalgebra A of the csa R.

7.1.10   (Skolem-Noether, see also 7.1.10′, E7.1.5). Any isomorphism of simple F-subalgebras of R can be extended to an inner automorphism of R.

7.1.11   (Wedderburn's theorem). Every finite division ring is commutative.

7.1.12.   {separable elements of maximal degree} is dense in the Zariski topology.

7.1.12′   (Koethe). Any separable subfield of a central division algebra is contained in a separable maximal subfield.

7.1.18   (Wedderburn's criterion, see also E7.1.12). The symbol $(\alpha, \beta)_n$ is split iff α is a norm from $F(\beta^{1/n})$ to F.

7.1.24. $UD(F, n)$ is a division algebra of degree n.

7.1.37   (See also E7.1.10). The correspondence between central simple algebras and Brauer factor sets.

7.1.40   (See also E7.1.11). Central simple algebras are isomorphic iff their Brauer factor sets are associates.

7.1.44.   (Wedderburn). Every division algebra of degree 2 or 3 is cyclic.

7.1.45.   (Albert). Every csa of index 4 is split by the Galois group $(\mathbb{Z}/2\mathbb{Z})^{(2)}$.

E7.1.18   (Wedderburn's factorization theorem). The minimal polynomial of d over $Z(D)$ splits into linear factors over D.

E7.1.23   (Cartan-Brauer-Hua). Any proper division subring, invariant with respect to all inner automorphisms, is central.

§7.2

7.2.3.   The effect on the index of tensoring by a field extension.

7.2.6.   Identification of the Brauer group with the second cohomology group.

7.2.10. $\exp(R)$ divides $\operatorname{index}(R)$.

7.2.12.   If n, m are the respective index and exponent then

     (i) $m \mid n$ and
     (ii) every prime divisor of n also divides m.

7.2.13.   Any csa decomposes as a tensor product of algebras of prime power index.

7.2.20.   The explicit algebraic description of $\mathrm{cor}_{L/F}R$ is well defined.

7.2.26. $\mathrm{cor} \circ \mathrm{res} = [L:F]$.

7.2.32   ("Projection formula"). $\mathrm{cor}_{L/F}(\alpha, \beta; L) \sim (\alpha, N_{L/F}\beta; F)$.

7.2.37   (Rosset-Tate). $\mathrm{cor}_{L/F}(a, b)$ is a tensor product of $\leq [L:F]$ symbols.

7.2.43   (mostly Rosset). Any csa of prime degree p is similar to a tensor product of $(p - 1)!/2$ cyclic algebras.

7.2.44.   A csa R has exponent 2 iff R has an involution of the first kind.

7.2.48   (Albert-Riehm-Scharlau). Description of csa with anti-automorphism, in terms of the corestriction.

7.2.49'. Suppose $[L:F] = 2$. A csa R (over L) has an involution (*) of second kind over F iff $\mathrm{cor}_{L/F}R \sim 1$.

7.2.50.   A csa R containing a quadratic field extension K of F has exponent 2, iff $C_R(K)$ has an involution of second kind over F.

7.2.57   (Albert). Every division algebra of degree 4 and exponent 2 is a tensor product of quaternion subalgebras.

7.2.60   (Tignol). If R has degree 8 and exponent 2 then $M_2(R)$ is a tensor product of quaternions.

7.2.61.   If $\sqrt{-1} \in F$ then any csa of index 4 is similar to a tensor product of a cyclic of degree 4 and four quaternion algebras.

7.2.62   (Merkurjev-Suslin, see also Appendix B, E7.2.18, and E7.2.33.). Every csa is similar to a tensor product of cyclic algebras. (Proved modulo "Hilbert's theorem in K_2".)

E7.2.32   (Rowen, see also E7.2.43). Every csa of index 8 and exponent 2 is split by $(\mathbb{Z}/2\mathbb{Z})^{(3)}$.

Appendices to Chapter 7

B2 (The real Merkurjev-Suslin Theorem). The m-torsion part of the Brauer group is isomorphic to a K_2-group.

C3. Properties of generics.

§8.1

8.1.8'   (Maschke's theorem). $F[G]$ is semiprime if $\mathrm{char}(F) = 0$.

8.1.14   (Maschke's theorem for arbitrary fields, G finite, see also E8.4.32). $F[G]$ is semisimple Artinian iff $|G| \neq 0$ in F.

8.1.26. $F[G]$ has no nonzero left ideals if G is a $\mathrm{char}(F)'$-group.

8.1.32.   If $Z(G)$ is of finite index in G then G' is finite.

8.1.39   (Connell, see also E8.1.14). Characterization of prime group rings.

8.1.41   (Passman, see also E8.1.25). Criteria for $K[G]$ to be primitive if $K \supseteq F$ and $F[G]$ is primitive.

8.1.45   (Amitsur, Herstein). $\mathrm{Jac}(F[G])$ is nil for F uncountable.

8.1.47. $F[G]$ is semiprimitive when F is a $\mathbb{Q}$-algebra not algebraic over $\mathbb{Q}$.

8.1.52   (See also 8.1.56). If $C[G]$ is semiprime PI then $\Delta(G)$ is of finite index in G. (This is part of the Isaacs-Passman Theorem in char 0, that G has an Abelian subgroup of finite index).

§8.2

8.2.2.   If G is polycyclic-by-finite then $F[G]$ is Noetherian.

8.2.16   (Milnor-Wolf-Gromov). A finitely generated group has polynomial growth iff it is nilpotent-by-finite (not proved in full here).

8.2.20.   If G is polycyclic-by-finite and is a $\mathrm{char}(F)'$-group then gl. dim. $F[G] < \infty$.

8.2.21   (Serre). If G is a $\mathrm{char}(F)'$-group with a subgroup H of finite index such that gl. dim. $F[H] < \infty$, then gl. dim. $F[G] < \infty$.

8.2.27.   If G is torsion-free and $F[G]$ is Noetherian then $\mathrm{Tr}_G P$ is a well defined integer for each f.g. projective P.

8.2.35.   (Farkas-Snider, see also appendix). If G is torsion-free polycyclic-by-finite and $\mathrm{char}(F) = 0$ then $F[G]$ is a domain.

§8.3

8.3.3   (Poincare-Birkhoff-Witt = PBW theorem, see also 8.3.51). $u: L \to U(L)$ is monic.

8.3.14, 8.3.15. $U(L)$ is a Noetherian domain, and its subrings are semiprimitive.

8.3.23   (Quillen). $U(L)$ is Jacobson.

8.3.27   (McConnell). If L is a nilpotent Lie algebra then every ideal of $U(L)$ is polycentral; hence any A-adic completion of $U(L)$ is Noetherian.

8.3.36.   If L is solvable then every prime ideal of $U(L)$ is completely prime.

8.3.39.   If L is solvable then $[L:F] = \mathrm{height}(P) + \mathrm{GK}$ dim $U(L)/P$ for any prime ideal P of $U(L)$. (In particular, GK dim $U(L) = [L:F]$.)

§8.4

8.4.4.   Any almost normalizing extension of a left Noetherian ring is left Noetherian.

8.4.11   (McConnell). Conditions for vector generic flatness to pass up from a subring.

8.4.12, 8.4.15   (See also E8.4.3–E8.4.10, E8.4.13, E8.4.14). Theorems concluding that a ring is Jacobson and/or satisfies the Nullstellensatz.

8.4.18, 8.4.20, 8.4.22.   Various ring-theoretic implications of Dixmier's conditions.

8.4.31   (Irving-Small). Reduction of the proof of the equivalence of Dixmier's conditions to the algebraically closed case.

8.4.32ff   (Warfield). The Goldie rank of R in terms of its prime images.

8.4.38.   If R is the enveloping algebra of a finite dimensional Lie algebra or the group algebra of a poly-{infinite cyclic} group then every f.g. projective R-module is stably free.

E8.4.34   (Fundamental theorem of coalgebras). Any element of a coalgebra is contained in a finite dimensional subcoalgebra. Consequently, any simple coalgebra A is finite dimensional.

Appendix to Chapter 8

Moody's theorem: (Verification of the Goldie rank conjecture): *For G polycyclic-by-finite, the Goldie rank of the ring of fractions of $F[G]$ is the least common multiple of the orders of the finite subgroups of G. (Highlights of the proof are given.)*

Counterexamples

Chapter 5

E5.1.1.   Schanuel's lemma fails for flat modules.

E5.1.9.   See also 8.4.43. A f.g. projective $D[\lambda_1, \lambda_2]$ that is not free.

Chapter 6

E6.1.12.   A PI-ring failing $\mathrm{Nil}(R)^{n-1} \subseteq \mathrm{Nil}(R)$.

E6.1.26.   A PI-ring with an idempotent maximal ideal.

E6.1.27.   A Noetherian PI-ring whose center fails the "principal ideal theorem."

6.2.4.   An affine algebra failing the Nullstellensatz.

6.2.9   (Golod). A nil algebra R_0 which is not locally nilpotent, but satisfying $\bigcap_{n \in \mathbb{N}} R_0^n = 0$. (Also $M_2(R_0)$ is nil, see 6.2.9'.)

E6.2.4.   An affine domain that is left and right Ore, and left but not right Noetherian.

Chapter 7

Chapter 8

The Basic Ring-Theoretic Notions and Their Characterizations

A ring R is *left Artinian* if R satisfies the DCC (or, equivalently, the minimum condition) on left ideals. Other characterizations:

1. Every f.g. R-module is Artinian (Corollary 0.2.21);
2. R has a composition series as left R-module (Remark 2.3.2 and Characterization 4.);
3. Every f.g. R-module has a composition series (Remark 2.3.2);
4. Left Noetherian and semiprimary (Theorem 2.7.2);
5. Left Noetherian and perfect (Exercise 2.7.7);
6. R is perfect and $\text{Jac}(R)/\text{Jac}(R)^2$ is f.g. as R-module (Exercise 2.7.8);
7. K-dim $R = 0$ (Example 3.5.40).

A ring R is *simple Artinian* if R is a simple ring which is left Artinian. Other characterizations:

1. Matrix ring over a division ring (Remark 2.3.8)
   (Note: This criterion is left-right symmetric, so the right-handed version of all subsequent characterizations are applicable;
2. Prime ring and left Artinian (Theorem 2.3.9.);
3. Morita equivalent to division ring (Morita's Theorem + Proposition 4.1.18);
4. Simple ring with minimal left ideal (Corollary 2.1.25′);
5. Ring of fractions of prime (left) Goldie ring (Goldie's Theorem);

6. Primitive with ACC (Ann) and socle $\neq 0$ (Digression 3.2.9);
7. Prime regular with all primitive images Artinian (Exercise 2.11.22).

A ring R is *semisimple* (*Artinian*) if it satisfies properties (1), (2), or (3) below, all equivalent by Theorem 2.3.10. (The name is justified by property (4)). This is the most important kind of ring. Other characterizations:

1. Finite direct product of simple Artinian rings
   Note: This criterion is left-right symmetric, so the right-handed versions of all subsequent characterizations are applicable);
2. Semiprime and left Artinian;
3. Completely reducible as R-module, i.e. $R = \text{soc}(R)$ (Theorem 2.3.10);
4. Left Artinian, and the intersection of the maximal ideals is 0 (by theorem 2.3.9 + Chinese Remainder Theorem);
5. Every R-module is completely reducible (Theorem 2.4.9);
6. Complemented as R-module (Theorem 2.4.8);
7. Every R-module is complemented (Theorem 2.4.9);
8. R has no proper large left ideal (Theorem 2.4.8);
9. Ring of fractions of semiprime (left) Goldie ring (Goldie's Theorem);
10. R is semiprime, each nonzero left ideal contains a minimal left ideal, and $\text{soc}(R)$ is a sum of a finite number of minimal left ideals (Corollary 2.3.11);
11. Semiprime with a complete set of orthogonal primitive idempotents (Corollary 2.3.12);
12. Every short exact sequence of R-modules splits (Theorem 2.4.9; Remark 2.11.7);
13. Every R-module is injective (Remark 2.11.7);
14. Every cyclic R-module is injective (Exercise 2.11.2);
15. Every left ideal is injective as R-module (by Theorem 2.4.9 + Proposition 2.10.14);
16. Every R-module is projective (Remark 2.11.7);
17. All cyclic modules are projective (since every left ideal is then a summand);
18. Left Noetherian and (von Neumann) regular ring (Corollary 2.11.21);
19. Regular ring with a complete set of orthogonal primitive idempotents (Proposition 2.11.20 + Theorem 2.4.8);
20. Regular ring without an infinite set of orthogonal idempotents (Exercise 2.11.3);
21. Every additive functor from R-$\mathcal{M}od$ to W-$\mathcal{M}od$ is exact (Proposition 2.11.8);
22. Every left exact functor from R-$\mathcal{M}od$ to $\mathbb{Z}$-$\mathcal{M}od$ is exact (proof of Proposition 2.11.8);
23. Global dimension is 0 (Example 5.1.17).

A ring R is *left Noetherian* if R satisfies the ACC (or equivalently the maximum condition) on left ideals:

1. Every f.g. R-module is Noetherian (Proposition 0.2.21);
2. ACC on f.g. left ideals (Exercise 3.3.21);
3. The direct sum of injective modules must be injective (Exercise 2.10.10);
4. Every injective is a direct sum of indecomposable injectives (Exercises 2.10.11, 2.10.13);
5. There is a cardinal c such that every injective R-module is a direct sum of modules each spanned by at most c elements. (Exercise 2.10.12);
6. The polynomial ring $R[\lambda]$ has K-dim (Exercise 3.5.30).

A ring R is *primitive* if R has a faithful simple module. Other characterizations:

1. R has a maximal left ideal whose core is 0 (proposition 2.1.9);
2. R has a left ideal with core 0, which is comaximal with all ideals (Proposition 2.1.11);
3. R is prime and has a faithful module of finite length (Exercise 2.3.3).

The *Jacobson radical* Jac(R) is the intersection of the primitive ideals of R. Other characterizations:

1. Intersection of the right primitive ideals, by note below.
2. The quasi-invertible ideal which contains all quasi-invertible left ideals (Proposition 2.5.4);
3. The right-handed version of property 2 (by discussion after Proposition 2.5.4)
   Note: This criterion is left-right symmetric, so the right-handed versions of all subsequent characterizations are applicable;
4. The intersection of the maximal left ideals (Proposition 2.5.2);
5. The intersection of the left annihilators of all simple modules (Proposition 0.2.14 + Proposition 2.5.2);
6. The intersection of the cores of maximal left ideals (Exercise 2.5.3);
7. The intersection of the cores of maximal inner ideals (Exercise 2.5.4);
8. The sum of the small left ideals of R (Exercise 2.8.11). Note the Jacobson radical itself is a small left ideal, by property 4;
9. $\{r \in R : r + a$ is invertible for all invertible a in $R\}$ (Exercise 2.5.2).

References

Bibliography of Books

This list contains most of the advanced ring theory books, plus books from other subjects which bear heavily on the material.

ABE, E.
  [77] *Hopf Algebras*. Cambridge University Press, Cambridge
ALBERT, A. A.
  [61] *Structure of Algebras*. AMS Colloq. Pub. 24. American Mathematical Society, Providence
ANDERSON, F., and FULLER, K.
  [74] *Rings and Categories of Modules*. Springer-Verlag, Berlin
ARTIN, E., NESBIT, C. J., and THRALL, R.
  [44] *Rings with Minimum Condition*. University of Michigan Press, Ann Arbor
BARBILIAN, D.
  [46] *Teoria Aritmetica a Idealilor* (in inele necomutative). Ed. Acad. Rep. Pop. Romine, Bucaresti
BARWISE, J. (Ed.)
  [78] *Handbook of Logic*. North-Holland, Amsterdam
BASS, H.
  [68] *Algebraic K-Theory*. Benjamin, New York
BJORK, J.
  [79] *Rings of Differential Operators*. North-Holland, Amsterdam
BOREL, A.
  [69] *Linear Algebraic Groups*. Benjamin, New York
BORHO, W., GABRIEL, P., and RENTSCHLER, R.
  [73] *Primideale in einhüllenden auflösbarer Lie-Algebren*. Lecture Notes in Mathematics 357, Springer-Verlag, Berlin
BOURBAKI, N.
  [72] *Commutative Algebra* (transl. from French). Elements de Mathematique. Hermann, Paris
CARTAN, H., and EILENBERG, S.
  [56] *Homological Algebra*. Princeton University Press, Princeton
CHATTERS, A. W., and HAJARNAVIS, C.
  [80] *Rings with Chain Conditions*. Pitman, London
CHEVALLEY, C.
  [68] *Théorie des groupes de Lie*. Hermann, Paris

COHN, P. M.

[74] *Algebra I*. Wiley, London

[77] *Skew Field Construction*, Cambridge University Press, Cambridge

[81] *Universal Algebra* (second edition). Reidel, Dordrecht, Holland

[85] *Free Rings and Their Relations* (second edition), Academic Press, New York

[88] *Algebra II, III* (second edition, in press). Wiley, London

COZZENS, J., AND FAITH, C.

[75] *Simple Noetherian Rings*. Cambridge University Press, Cambridge University Press, Cambridge

CURTIS, C., and REINER, I.

[62] *Representation Theory of Finite Groups and Associative Algebras*, Interscience, New York

[81] *Methods of Representation Theory*, Vol. 1. Wiley, New York (Vol. 2 in press)

DEMEYER, F., and INGRAHAM, E.

[71] *Separable Algebras over Commutative Rings*. Lecture Notes in Mathematics 181. Springer-Verlag, Berlin

DEURING, M.

[66] *Algebren*, Springer-Verlag, Berlin

DICKSON, L. E.

[23] *Algebras and Their Arithmetics*, University of Chicago, Chicago

DIVINSKY, N. J.

[65] *Rings and Radicals*. Toronto University, Toronto

DIXMIER, J.

[77] *Enveloping Algebras*. North Holland, Amsterdam (Transl. of *Algebres Enveloppantes*, Gauthier-Villars, Paris)

DRAXL, P. K.

[83] *Skew Fields*. London Mathematical Society Lecture Notes 81. Cambridge University Press, Cambridge

FAITH, C.

[67] *Lectures on Injective Modules and Quotient Rings*. Lecture Notes in Mathematics 49, Springer-Verlag, Berlin

[73] *Algebra: Rings, Modules, and Categories*. Springer-Verlag, Berlin

[76] *Algebra II: Ring Theory*. Springer-Verlag, Berlin

[82] *Injective Modules and Injective Quotient Rings*. Marcel Dekker, New York

FAITH, C., and PAGE, S.

[84] *FPF Ring Theory*. London Mathematical Society Lecture Notes 88. Cambridge University Press, Cambridge

FREYD, P.

[64] *Abelian Categories: An Introduction to the Theory of Functors*. Academic Press, New York

GILLMAN, L., and JERISON, M.

[60] *Rings of Continuous Functions*. Academic Press, New York

GOLAN, J.

[75] *Localization of Noncommutative Rings*. Marcel Dekker, New York

[86] *Torsion Theories*. Longman Scientific & Technical, Harlow, England

GOODEARL, K.

[76] *Ring Theory: Nonsingular Rings and Modules*. Marcel Dekker, New York

[79] *von Neumann Regular Rings*. Pitman, New York

HALL, M.

[59] *The Theory of Groups*. MacMillan, New York

HARTSHORNE, R.

[75] *Algebraic Geometry*. Springer-Verlag, New York

HERSTEIN, I. N.

  [64] *Topics in Algebra* (second edition). Xerox, Lexington, Massachussetts

  [68] *Noncommutative Rings.* Carus Mathematical Monographs 15. Mathematical Association of America, Wiley, New York

  [69] *Topics in Ring Theory.* University of Chicago Press, Chicago

  [76] *Rings with Involutions.* University of Chicago Press, Chicago

HOCHSCHILD, G.

  [71] *Introduction to Affine Algebraic Groups.* Holden-Day, San Francisco

ISAACS, M.

  [76] *Character Theory of Finite Groups.* Academic Press, New York

JACOBSON, N.

  [43] *Theory of Rings.* AMS Surveys 1. American Mathematical Society, Providence

  [62] *Lie Algebras.* Wiley, New York

  [64] *Structure of Rings* (second edition). AMS Colloq. Pub. 37. American Mathematical Society, Providence

  [75] *PI-Algebras: An Introduction.* Lecture Notes in Mathematics 441. Springer-Verlag, Berlin

  [80] *Basic Algebra II.* Freeman, San Francisco

  [85] *Basic Algebra I* (second edition). Freeman, San Francisco

JACOBSON, N., and SALTMAN, D.

  [88] *Division algebras* (forthcoming)

JANTZEN, J. C.

  [83] *Einhüllende Algebren halbeinfacher Lie-Algebren.* Springer Verlag, Berlin

JATEGAONKAR, A. V.

  [70] *Left Principal Ideal Rings.* Lecture Notes in Mathematics 123. Springer-Verlag, Berlin

  [86] *Localization in Noetherian Rings.* London Mathematical Society Lecture Notes 98. Cambridge University Press, Cambridge

KAPLANSKY, I.

  [68] *Rings of Operators.* Benjamin, New York

  [70] *Commutative Rings.* Allyn and Bacon, Boston

  [72] *Fields and Rings.* University of Chicago, Chicago

KASCH, F.

  [82] *Modules and Rings* (transl. from German). London Mathematical Society Monographs 17. Academic Press, New York

KELLEY, J.

  [55] *General Topology.* van Nostrand, New York

KNUS, M.-A., and OJANGUREN, M.

  [74] *Théorie de la descente et algèbres d'Azumaya.* Lecture Notes in Mathematics 389, Springer-Verlag, Berlin

KRAUSE, G. R., and LENAGAN, T. H.

  [85] *Growth of Algebras and Gelfand-Kirillov Dimesion.* Research Notes in Mathematics 116. Pitnam, London

KRUSE, R. L., and PRICE, D.

  [69] *Nilpotent Rings.* Gordon & Breach, New York

LAM, T. Y.

  [78] *Serre's Conjecture.* Lecture Notes in Mathematics 635. Springer-Verlag, Berlin

LAMBEK, J.

  [66] *Lectures on Rings and Modules.* Blaisdell, Waltham, Massachussetts

LANG, S.

  [84] *Algebra* (second edition). Addison-Wesley, Reading, Massachussetts

MACLANE, S.

  [75] *Homology,* Springer-Verlag, Berlin

MATLIS, E.

[72] *Torsion-Free Modules.* University of Chicago Press, Chicago

MCCONNELL, J. and ROBSON, J. C.

[88] *Noetherian Rings.* Wiley, London

MCDONALD, B.

[74] *Finite Rings with Identity.* Marcel Dekker, New York

MILNOR, J.

[71] *Introduction to Algebraic K-Theory.* Annals of Mathematical Studies. Princeton University Press, Princeton

MITCHELL, B.

[65] *Theory of Categories.* Pure and Applied Mathematics 17. Academic Press, New York.

MONTGOMERY, S.

[80] *Fixed Rings of Finite Automorphism Groups of Associative Rings.* Lecture Notes in Mathematics 818, Springer-Verlag, Berlin

NASTASESCU, C., and VAN OYSTAEYEN, F

[82] *Graded Ring Theory.* Mathematical Library 28. North-Holland, Amsterdam

OSOFSKY, B.

[72] *Homological Dimension of Modules.* AMS Regional Conference Series in Mathematics 12, American Mathematical Society, Providence

PASSMAN, D.

[77] *The Algebraic Structure of Group Rings.* Wiley, New York

PIERCE, R. S.

[82] *Associative Algebras.* Springer-Verlag, Berlin

PROCESI, C.

[73] *Rings with Polynomial Identities.* Marcel Dekker, New York

REINER, I.

[75] *Maximal Orders.* Academic Press, London

RENAULT, J.

[75] *Algèbre Noncommutative.* Gauthier-Villars, Paris

ROTMAN, J.

[79] *An Introduction to Homological Algebra.* Academic Press, New York

ROWEN, L. H.

[80] *Polynomial Identities in Ring Theory.* Academic Press, New York

RUDIN, W.

[66] *Real and Complex Analysis.* McGraw Hill, New York

SCHOFIELD, A. H.

[85] *Representations of Rings over Skew Fields.* London Mathematical Society Lecture Notes 92. Cambridge University Press, Cambridge

SERRE, J. P.

[79] *Local Fields.* Springer-Verlag, Berlin (transl. of *Corps Locaux*)

SHARPE, D. W., and VAMOS, P.

[72] *Injective Modules.* Cambridge University Press, Cambridge

SMALL, L. (Ed.)

[81] *Reviews in Ring Theory.* American Mathematical Society, Providence

[86] *Reviews in Ring Theory II.* American Mathematical Society, Providence

STENSTRÖM, B.

[75] *Rings of Quotients: An Introduction to Methods of Rings Theory.* Springer-Verlag, Berlin

SWEEDLER, M.

[69] *Hopf Algebras.* Benjamin, New York

VAN DER WAERDEN, B. L.

[49] *Modern Algebra.* Ungar, New York

VAN OYSTAEYEN, F. M. J., and VERSCHOREN, A. H.

[81] *Noncommutative Algebraic Geometry: An Introduction.* Springer-Verlag, Berlin

VASCONCELOS, W.

[76] *The Rings of Dimension Two.* Marcel Dekker, New York

WEHRFRITZ, B. A. F.

[73] *Infinite Linear Groups.* Ergebnisse der Mathematik 76. Springer-Verlag, Berlin

ZARISKI, O., and SAMUEL, P.

[58] *Commutative Algebra.* Van Nostrand-Reinhold, Princeton (reprinted Springer-Verlag, Berlin)

ZHEVLAKOV, K. A., SLINKO, A. M., SHESTAKOV, I. P., and SHIRSHOV A. I.

[82] *Rings That Are Nearly Associative.* Academic Press, New York

Collections of Papers

Collections which are in print, by year of publication.

1972

Proceedings of Ring Theory Conference, Tulane University: Lectures on the Applications of Sheaves to Ring Theory. Lecture Notes in Mathematics 248. Springer-Verlag, Berlin

Proceedings of Ring Theory Conference, Tulane University: Lectures on Rings and Modules. Lecture Notes in Mathematics 246. Springer-Verlag, Berlin

Conference on Orders, Group Rings, and Related Topics. Lecture Notes in Mathematics 353. Springer-Verlag, Berlin

R. Gordon (Ed.). *Ring Theory.* Academic Press, New York

1973

A. Kertesz (Ed.). *Rings Modules and Radicals.* Colloquium Mathematica Societatis Janos Bolyai 6. North-Holland, Amsterdam

1974

B. R. McDonald, A. Magid, and K. Smith (Eds.). *Ring Theory: Oklahoma University Conference.* Lecture Notes in Mathematics 7. Marcel Dekker, New York

V. Dlab and P. Gabriel (Eds.). *Representations of Algebras: University of Ottawa.* Lecture Notes in Mathematics 488. Springer-Verlag, Berlin

1975

B. R. McDonald and R. Morris (Eds.). *Ring Theory II: Second Oklahoma University Conference.* Lecture Notes in Mathematics 26. Marcel Dekker, New York

D. Zelinsky (Ed.). *Brauer Group Conference.* Lecture Notes in Mathematics 549. Springer-Verlag, Berlin

J. H. Cozzens and F. L. Sandomierski (Eds.). *Noncommutative Ring Theory: Kent State.* Lecture Notes in Mathematics 545. Springer-Verlag, Berlin

1977

S. K. Jain (Ed.). *Ring Theory: Ohio State University Conference, 1976.* Lecture Notes in Mathematics 25. Marcel Dekker, New York

M. P. Malliavin (Ed.). *Séminaire d'Algèbre Paul Dubreil, Paris (1975–76),* Lecture Notes in Mathematics 586. Springer-Verlag, Berlin

1978

F. Van Oystaeyen (Ed.). *Ring Theory: Antwerp Conference, 1977.* Lecture Notes in Mathematics 40. Marcel Dekker, New York

M. P. Malliavin (Ed.). *Séminaire d'Algèbre Paul Dubreil, Paris (1976–77).* Lecture Notes in Mathematics 641. Springer-Verlag, Berlin

R. Gordon (Ed.). *Representation Theory of Algebras: Temple University Conference, 1976.* Lecture Notes in Mathematics 37. Marcel Dekker, New York

1979

D. Handelman and J. Lawrence (eds.). *Ring Theory: Waterloo Conference, 1978.* Lecture Notes in Mathematics 734. Springer-Verlag, Berlin

F. Van Oystaeyen (Ed.). *Ring Theory: Antwerp Conference, 1978.* Lecture Notes in Mathematics 51. Marcel Dekker, New York

M. P. Malliavin (Ed.). *Séminaire d'Algèbre Paul Dubreil, Paris (1977–78).* Notes in Mathematics 740. Springer-Verlag, Berlin

1980

V. Dlab (Ed.). *Representation Theory I: Carleton University Conference, 1979.* Lecture Notes in Mathematics 831. Springer-Verlag, Berlin.

V. Dlab (Ed.). *Representation Theory II: Carleton University Conference, 1979.* Lecture Notes in Mathematics 832. Springer-Verlag, Berlin

B. R. McDonald (Ed.). *Ring Theory and Algebra III: Third Oklahoma University Conference.* Lecture Notes in Mathematics 55. Marcel Dekker, New York

M. P. Malliavin (Ed.). *Séminaire d'Algèbre Paul Dubreil et Marie-Paule Malliavin, Paris (1978–79).* Lecture Notes in Mathematics 795. Springer-Verlag, Berlin.

F. Van Oystaeyen (Ed.). *Ring Theoy: Antwerp Conference, 1980.* Lecture Notes in Mathematics 825. Springer-Verlag, Berlin

1981

M. Kervaire and M. Ojanguren (Eds.). *Groupe de Brauer: Séminaire, Les Plans-Sur-Bex.* Lecture Notes in Mathematics 844. Springer-Verlag, Berlin

M. P. Malliavin (Ed.). *Séminaire d'Algèbre Paul Dubreil et Marie-Paule Malliavin, Paris (1979–80).* Lecture Notes in Mathematics 867. Springer-Verlag, Berlin

1982

S. A. Amitsur, G. Seligman, and D. Saltman (Eds.). *N. Jacobson: An Algebraists's Homage: Yale University 1980.* Contemporary Mathematics 13. American Mathematical Society, Providence

P. J. Fleury (Ed.). *Advances in Noncommutative Ring Theory: Plattsburgh Conference, 1981.* Lecture Notes in Mathematics 951. Springer-Verlag, Berlin

M. P. Malliavin (Ed.). *Séminaire d'Algèbre Paul Dubreil et Marie-Paule Malliavin, Paris (1981–82).* Lecture Notes in Mathematics 924. Springer-Verlag, Berlin

Representations of Algebras (Pueblo). Lecture Notes in Mathematics 944. Springer-Verlag, Berlin

F. Van Oystaeyen and A. Verschoren (Eds.). *Brauer Groups in Ring Theory and Algebraic Geometry.* Lecture Notes in Mathematics 917. Springer-Verlag, Berlin

1983

M. P. Malliavin (Ed.). *Séminaire d'Algèbre Paul Dubreil et Marie-Paule Malliavin, Paris (1982–83).* Lecture Notes in Mathematics 1029, Springer-Verlag, Berlin

1984

F. Van Oystaeyen (Ed.). *Methods in Ring Theory.* NATO Advanced Science Institutes, Reidel, Dordrecht

B. Srinivasan and J. Sally (Eds.). *Emmy Noether in Bryn Mawr.* Springer-Verlag, Berlin

1985

M. P. Malliavin (Ed.). *Séminaire d'Algèbre Paul Dubreil et Marie-Paule Malliavin, Paris (1983–84).* Lecture Notes in Mathematics 1146. Springer-Verlag, Berlin

S. Montgomery (Ed.). *Group Actions on Rings.* Contemporary Mathematics 43. American Mathematical Society, Providence

L. Marki and R. Wiegandt (Eds.). *Radical Theory, Colloquium Mathematica Societatis Janos Bolyai 38.* North-Holland, Amsterdam

1986

V. Dlab, P. Gabriel, and G. Michler (Eds.) *Representation Theory I: Carleton University Conference, 1984.* Lecture Notes in Mathematics 1177. Springer-Verlag, Berlin.

F. Van Oystaeyen (Ed.). *Ring Theory: Antwerp Conference, 1985.* Lecture Notes in Mathematics 1197. Springer-Verlag, Berlin

1987

L. W. Small (Ed.). *Noetherian Rings and Their Applications.* Surveys and Monographs 24. American Mathematical Society, Providence

Bibliography of Articles

In order to keep down the size (and price) of this student edition, we have omitted the bibliography of articles. However, three recent articles not in the original edition are particularly relevant to this text:

CLIFF, G. AND WEISS, A.

[88] Moody's induction theorem, *Illinois J. Math.* **32**, 489–500 MR 89h:16009

This is a very readable account of Moody's theorem, verifying the Goldie rank conjecture, cf. §8.2, p. 543.

HAILE, D.

[89] A useful proposition for division algebras of small degree, *Proc. Amer. Math. Soc.* **106**, 317–319. MR 89K:16037

This note provided the proof used in this edition for the cyclicity of division algebras of degree 3.

PASSMAN, D.,

[89] Crossed products and enveloping algebras satisfying a polynomial identity, in Ring Theory 1989, *Israel Mathematical Conference Proceedings* Vol. 1, 61–73.

This engaging survey article also contains a 2-page proof of the Isaacs–Passman Theorem (modulo theorem 8.1.52) that for a field F of characteristic 0, a group algebra F[G] is a PI-ring iff G has an Abelian subgroup of finite index.

Index

Printed and bound by CPI Group (UK) Ltd, Croydon, CR0 4YY

17/10/2024

01775564-0001